# CELL DIFFERENTIATION

## Molecular Basis and Problems

Edited by

**Lutz Nover, Martin Luckner, Benno Parthier**

With Contributions by

**W. Berg, G. Butschak, W. Heese, R. Krauspe, W. Lerbs, M. Luckner, K. Müntz, L. Nover, B. Parthier, R. Piechocki, H. Reinbothe, R. Richter, K.-P. Rindt, E. Serfling, R. Wollgiehn, H. Zühlke**

With 228 Figures and 65 Tables

SPRINGER-VERLAG BERLIN HEIDELBERG GMBH 1982

Dr. sc. Lutz Nover
Prof. Dr. sc. Benno Parthier

Akademie der Wissenschaften der DDR
Forschungszentrum für Molekularbiologie und Medizin
Institut für Biochemie der Pflanzen
Weinberg
DDR - 4010 Halle

Prof. Dr. sc. Martin Luckner

Sektion Pharmazie der Martin-Luther-Universität
Weinbergweg
DDR - 4010 Halle

Title of the German original edition: Zelldifferenzierung


**Additional material to this book can be downloaded from http://extras.springer.com**

ISBN 978-3-642-68022-9 ISBN 978-3-642-68020-5 (eBook)
DOI 10.1007/978-3-642-68020-5

Library of Congress Cataloging in Publication Data. Main entry under title: Cell differentiation. Translation of Zelldifferenzierung. Bibliography: p. Includes index. 1. Cell differentiation. 2. Molecular biology. I. Nover, Lutz 1935 –. II. Luckner, Martin. III. Parthier, Benno. IV. Title. QH607.Z4413. 574.87'612. 81-8862. ISBN-13:978-3-642-68022-9 (U.S.). AACR2



Chapter 3, 8, 9, 11, 12, 16, 18 and 19 translated from German by Joy Wieser.


Softcover reprint of the hardcover 1st edition 1982



2131/3140-543210

# List of Contributors


Dr. Wieland Berg, Archiv für Geschichte der Naturforschung und Medizin, Deutsche Akademie der Naturforscher Leopoldina, DDR - 4010 Halle, August-Bebel-Straße 50a

Dr. Günter Butschak, Zentralinstitut für Krebsforschung der AdW der DDR, DDR - 1115 Berlin Buch, Lindenberger Weg

Dipl. Biol. Wolfgang Heese, Archiv für Geschichte der Naturforschung und Medizin, Deutsche Akademie der Naturforscher Leopoldina, DDR - 4010 Halle, August-Bebel-Straße 50a

Dr. Reinald Krauspe, Institut für Biochemie der Pflanzen der AdW der DDR, DDR - 4010 Halle, Weinberg

Dr. Werner Lerbs, Sektion Pharmazie der Martin-Luther-Universität, DDR - 4010 Halle, Weinbergweg

Prof. Dr. Martin Luckner, Sektion Pharmazie der Martin-Luther-Universität, DDR - 4010 Halle, Weinbergweg

Prof. Dr. Klaus Müntz, Zentralinstitut für Genetik und Kulturpflanzenforschung der AdW der DDR - 4325 Gatersleben

Dr. Lutz Nover, Institut für Biochemie der Pflanzen der AdW der DDR, DDR - 4010 Halle, Weinberg

Prof. Dr. Benno Parthier, Institut für Biochemie der Pflanzen der AdW der DDR, DDR - 4010 Halle, Weinberg

Dr. Reinhard Piechocki, Lehrbereich Genetik, Sektion Biowissenschaften der Martin-Luther-Universität, DDR - 4020 Halle, Domplatz

Prof. Dr. Horst Reinbothe, Sektion Biowissenschaften der Martin-Luther-Universität, DDR - 4010 Halle, Neuwerk 1

Dr. Roland Richter, Laboratorium für Immunbiologie, Sektion Biowissenschaften der Karl-Marx-Universität, DDR - 7010 Leipzig, Talstraße 33

Dr. Klaus-Peter Rindt, Sektion Biowissenschaften der Martin-Luther-Universität, DDR - 4010 Halle, Am Kirchtor 1

Dr. Edgar Serfling, Zentralinstitut für Genetik und Kulturpflanzenforschung der AdW der DDR, DDR - 4325 Gatersleben

Dr. Reinhold Wollgiehn, Institut für Biochemie der Pflanzen der AdW der DDR, DDR - 4010 Halle, Weinberg

Dr. Harald Zühlke, Zentralinstitut für Diabetes, DDR - 2201 Karlsburg

# Preface

Three years have elapsed since the publication of the first edition of this book (in German). The continued interest of our readers and the rapid progress of our knowledge in many fields necessitates a thoroughly revised and somewhat enlarged new edition.

Cell differentiation is a prerequisite of life. It is defined as the *process* leading to the qualitatively and quantitatively selective realization of distinct parts of a given genetic material. Cell differentiation comprises five main aspects: (1) signal reception and transformation, (2) selective rearrangements of the genetic material, (3) differential gene expression, (4) organization of gene expression programs and (5) intercellular coordination of cell differentiation within the developmental programs of multicellular organisms. Despite the bewildering multiplicity of its results, i.e., the differentiated phenotypes of cells, there are apparently fundamental similarities with respect to the molecular mechanisms of the process itself. These mechanisms constitute the central subject of this book.

After two introductory chapters (Part 1) and a short historical reflection on the term the editors decided to treat the complex problem of cell differentiation by presenting experimental results in two different ways: (1) The central chapter in Part 2 is aimed at outlining a unifying picture of the molecular basis of cell differentiation using experimental data obtained with various kinds of prokaryotic and eukaryotic organisms. Hence, in its entirety this chapter does not reflect a concrete experimental system. (2) Part 3 contains 15 selected experimental systems. Each of them is distinguished by specified regulatory details and derives its identity from its typical combination of regulatory elements. These special chapters include a multiplicity of diverse aspects characteristic of cell differentiation, e.g., the cell division cycle, metabolic and developmental adaptation of the protein synthesis pattern to nutrient or hormonal signals, numerous examples of cell specialization, intercellular signal exchange, programmed gene expression in virus-infected cells as well as tumorigenesis in plants and animals.

The book is written as an advanced textbook for students and young scientists interested in biology and medicine. Therefore numerous experimental details, methodological remarks and selected references to reviews, books and some, mostly recent, original papers, as well as an extensive subject index, have been included to facilitate a deeper penetration into special problems treated in this book.

The editors gratefully acknowledge critical remarks and valuable support of many colleagues during the preparation of this second edition. To name only a few: N. Amrhein, H. Bielka, M. Delbrück, H. G. Floss, R. Hagemann, J. Heslop-Harrison,

J. Hofemeister, R. Lindigkeit, H. Lodish, K. Mothes and U. Wobus. Our gratitude also extends to Mrs. G. Luckner for preparing most of the line drawings, to Mrs. J. Wieser for translating part of the manuscripts as well as to Mrs. J. von dem Bussche and Dr. D. Czeschlik of the Springer Verlag and Mrs. J. Schlüter and Mr. H. Kramer of the VEB G. Fischer Verlag for their important contributions to the preparation of this English edition.

Halle, January 1981

L. Nover,
M. Luckner,
B. Parthier

**Note added in proof (to Max Delbrück)**

It was not unexpected but none the less very hard to learn about Max Delbrück's death on March 3, 1981. His outstanding contribution to the development of modern biology will be remembered for a long time. With deep gratitude we experienced his encouragement and constant direct support during the preparation of the revised manuscripts for this book. At the same time his own scientific activities were already seriously influenced by his advancing disease.

... I had a series of medical problems last fall and have been homebound for several months but now gradually I am improving a little. I still have a Phycomyces research group in the lab and there things are going along quite well ... As to your book, I am happy that you have finished the chore of getting the English edition done ... (letter to L. N. of February 18, 1981).

Halle, August 1981

L. Nover
M. Luckner
B. Parthier

## Prefatory Note to the First German Edition

We are all aware that individual cells or groups of cells of a wide variety of plant tissues can be isolated and grown under suitable physical and chemical conditions. As a rule the morphological and chemical specificity of the original material is quickly lost, and a mass of "dedifferentiated" callus cells develops. In principle, these cells can be kept alive and actively dividing indefinitely. We also know that, under appropriate chemical and physical conditions, such plant cell cultures can be caused to undergo further orderly cell divisions. The degree of organisation can be such that an entire organism may develop, corresponding in every way to the plant it originated from.

Thus the development of a plant or, in other words, the specialization of its organs, tissues and cells does not involve a loss of genes. The information content of the genetic material remains unaltered throughout the undoubtedly complicated process of cell specialization and despecialization associated with formation of callus cells, which contain the entire developmental potential of the fertilized egg cell. Thus, cell specialization in the course of regeneration is the result of a highly ordered realization of different parts of the information coded in the genetic material.

Cells of animals and of prokaryotic and eukaryotic microorganisms differ from those of plants only in the degree of complexity of their genetic material and the interactions between the individual cells. The basic mechanisms of storage and selective realization of genetic information, on the other hand, are very similar in all living organisms. Cell differentiation is therefore one of the central problems of general biology, and one whose further understanding is of decisive importance for medicine, agriculture and industrial microbiology.

Editors and authors of this book have made the intrepid attempt to show, with the help of relatively simple but also highly complex experimental systems, how, in the most varied groups of organisms, the ordered realization of genetic information leads to the development of new biochemical and cytological structures, i.e., to cell differentiation. In this process, forming the basis for the development of organismic structures and functions, a predominant role is played by proteins. But in addition, innumerable other factors exerting inductive or repressive effects form vital elements in coordinating the processes of single cells into the specialized functions of tissues and organs.

The young biologists to-day are in the lucky position of living in an era in which it is possible to do more than merely describe these fundamental events. May this book provide a stimulus to pursue their experimental approach or to enter into this exciting field of modern biology.

Halle, January 1977 Kurth Mothes

# Contents

## General Part

## Special Part

# INTRODUCTORY PART

# 1. Biochemistry of Gene Expression

L. Nover and H. Reinbothe

## 1.1 Introduction: Biological Information

During the evolution of living systems nucleic acids have become the universal macromolecules for storage of genetic information (DNA and in rare cases RNA) as well as for information amplification and transfer within cells (mostly RNA). Nucleic acid information, stored in the specific sequence of four code letters (cf. Fig. 1.2) is principally exchangeable, provided the corresponding types of information-copying enzymes are present:

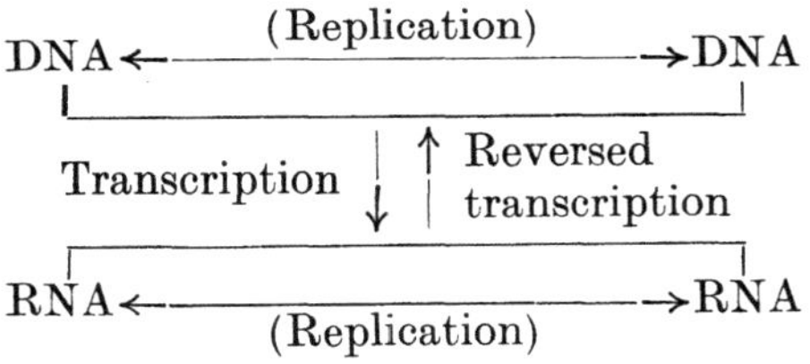


There is increasing evidence that, in addition to the linear information in the code sequence, the spatial organization of nucleic acids and its reversible changes contribute decisively to the overall information contained in these macromolecules. This is particularly evident for the different types of RNA (e.g., rRNA, tRNA, mRNA) involved in the complex process of a regulated and fidel translation of the nucleic acid into the protein code consisting of twenty amino acids (gene expression).

Proteins are the universal carriers and generators of three-dimensional biological information. They act as structural, regulatory and, above all, catalytic components of all living systems. Due to their particular primary sequence of amino acid residues, they form highly elaborate and specific three dimensional structures. They are capable of associating with each other to multiprotein complexes, undergoing multifold conformational changes, and of reversibly binding and eventually transforming all other types of molecules.

In cells of all living systems storage of the genetic information and its realization, i.e., its translation into proteins, proceed by a basically similar mechanism (Fig. 1.1). In the overwhelming number of cases, this means (1) transcription of the codogene strand of DNA into a complementary copy (mRNA) and (2) translation of this copy into the corresponding protein.

---

Abbreviations: aa, amino acid or aminoacyl, e.g., in aatRNA; bp, base pairs; kb, kilobase pairs; DNP, RNP, DNA-protein and RNA-protein complexes; H1, H2A, H2B, H3, H4, histones (cf. Table 1.2); HMG 1, 2, 14, 17, high-mobility group proteins (cf. Table 1.3); d, kd, Md, dalton, kilodalton, megadalton; MW, molecular weight; $M_r$, relative molecular weight (determined, e.g., by electrophoretic methods).

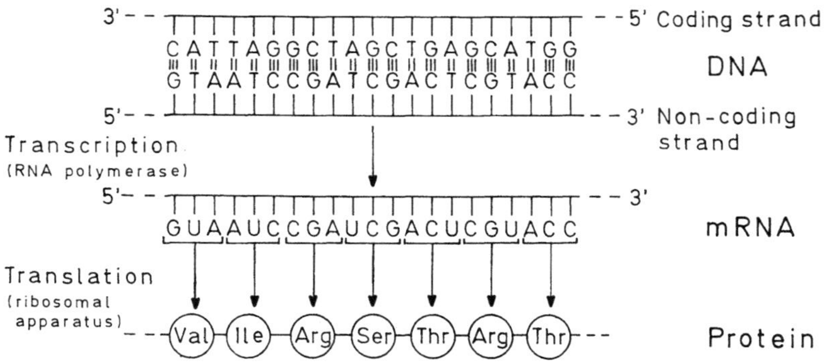


**Fig. 1.1:** Realization of genetic information coded by DNA.
Transcription of a complementary mRNA copy of one DNA strand (codogen strand) is catalyzed by RNA polymerase. The transcript, which serves as means of information transport and amplification, is translated by the ribosomal apparatus into the amino acid sequence of the corresponding protein.

As will be outlined very briefly in the following sections, the practically error-free transfer of biological information (cf. Nature *276* (1978): 442 and Sprinzl & Cramer [76]) between different types of nucleic acids but also from nucleic acids to proteins depends on the ordered interaction of nucleic acids and catalytic proteins involved in the transfer process. Moreover, fidelity of all steps, i.e., of replication, transcription and translation, is not only dependent on the Watson-crick type of base pairing (cf. Fig. **1**.2) but also on proteins that prevent errors or eliminate them (proof reading), a process consuming a considerable part of the overall energy required for gene expression.

## 1.2 The Genetic Material

### 1.2.1 Structure of DNA

With the exception of certain viruses (cf. e.g., Table **14**.1 with examples for different types of bacteriophages) the genetic material consists of double-stranded DNA, forming a regular right-handed helix of two continuous polymeric chains with a 2-deoxyribose-phosphate backbone and the four aromatic base residues (adenine (A), guanine (G), thymine (T) and cytosine (C)) attached to it (Fig. **1**.2). The diameter of the double-helix in the B-form is about 2.2 nm with 10.5 nucleotides per helical turn and a distance of 0.34 nm between two adjacent nucleotide bases. The remarkable regularity and stability of double-stranded DNA stems from two types of strong interactions between its aromatic bases:

1. The stacking of the flat, hydrophobic aromatic residues of each chain leads to an overlap of their $\pi$-electron orbitals which form a continuum.
2. The two strands are connected by hydrogen bonds between each two complementary bases. Adenine is always paired with thymine and guanine with cytosine respectively (Fig. **1**.2).

The steric constraints of the DNA double helix allows only these pairings and results in a complementary relation of the two intertwined chains. Furthermore, it requires that the two chains run in opposite directions (cf. Fig. **1**.2). The principle of complementarity governs the whole process of information transfer between nucleic acids but also from nucleic acids to proteins (cf. **1**.3.1).

Regulatory or enzymatic proteins interacting with DNA mostly bind to the major groove (Fig. **1**.2) which allows a certain access to the base residues buried inside the double helix. They may lead to locally restricted changes of the DNA conformation with intrastrand hairpin structures formed in regions with palindrome symmetry (cf. e.g., Fig. **1**.7). Further, the local unwinding of the two DNA chains is a necessary step during replication and transcription. It is promoted by unwinding proteins.

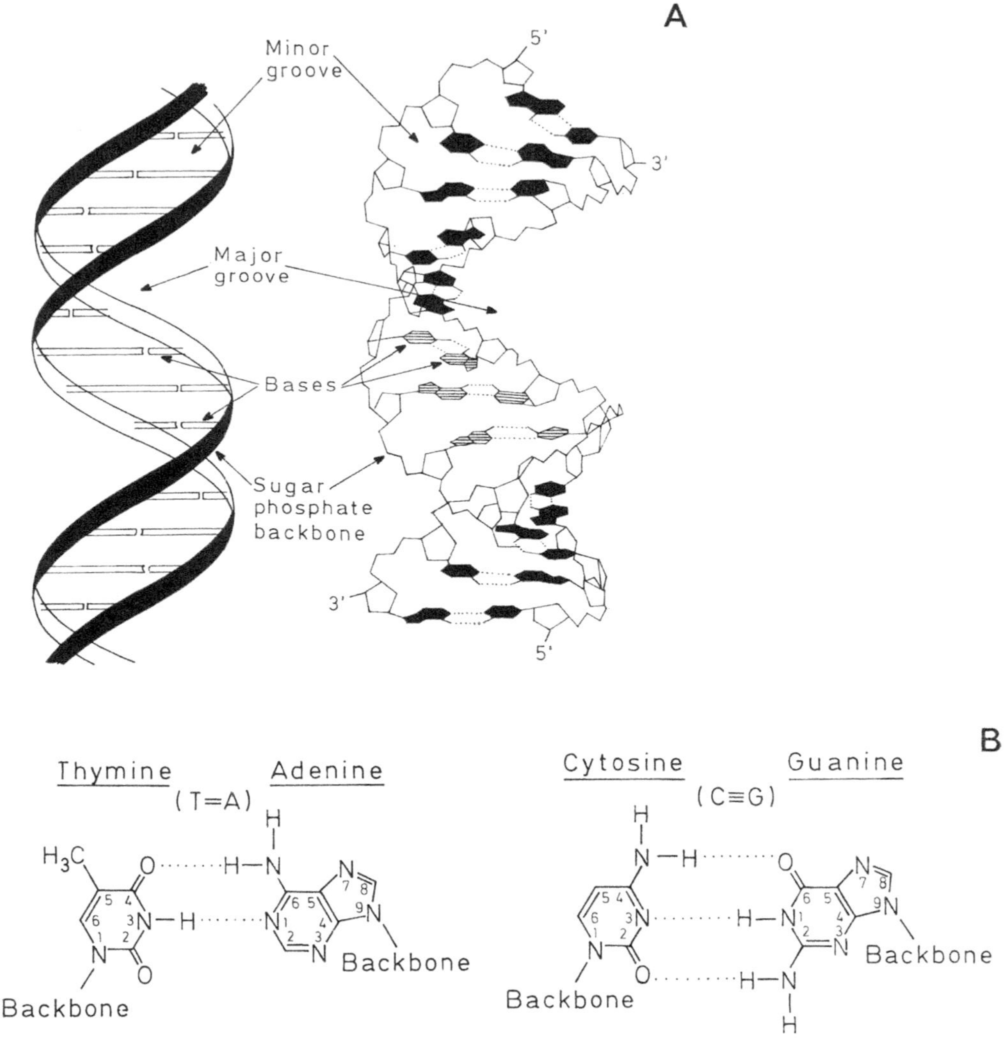


**Fig. 1.2**: Scheme of DNA double helix and of its complementary base pairs (from Watson [1], modified).

The two DNA strands in the so-called B conformation are twisted to a right handed helix with a minor and a major groove (**A**). The base residues filling the inner part of the helix form complementary pairs (A=T and G≡C) connected by hydrogen bonds (**B**).

It is of considerable interest that, besides some minor variations in the classical B-conformation, there is experimental evidence for a strongly divergent DNA form (Z-DNA) obtained by crystallization of the deoxy hexamer $d(GC)_3$. It was shown to exist as left-handed helix[1]. As the G-C base pairs form part of the outer surface of Z-DNA they are much more accessible, e.g., to protein binding and/or to modification, than the base residues in B-DNA. It is tempting to speculate that the frequently encountered G-C clusters in DNA exist permanently or transiently in the alternative Z-conformation and thus form distinct landmarks for site-specific DNA-protein interactions or that alternating segments of right-handed and left-handed DNA contribute to solve the topological problems of the DNA structure for replication and transcription.

Practically all DNA in prokaryotic and eukaryotic cells exists in a supercoiled state (cf. the review by Cozzarelli [2]). Remarkably, the superhelical twists are in the opposite direction to that of the double helix, with about one negative supercoil per 15 double helical twists. Special enzymes (DNA topoisomerases) are involved in the introduction and relaxation of supercoils by transient double-strand breaks. In bacteria, negative supercoiling of DNA by DNA gyrase was shown to influence all processes involving DNA (replication, recombination, transcription)[2]. As will be discussed in the following sections the DNA supercoils of the prokaryotic and eukaryotic chromosomes are stabilized by association with basic proteins.

### 1.2.2 Organisation of Bacterial Chromosomes

The DNA amount in bacterial cells varies between 0.002 and 0.06 pg. In rapidly dividing cells of *Escherichia coli* 0.004 pg DNA are distributed between two highly condensed, membrane bound DNP-complexes, the nucleoids (cf. Fig. **4.11**A). Each nucleoid of about 1 $\mu$m diameter contains a circular DNA molecule of 1.4 mm length corresponding to $4.3 \times 10^6$ base pairs. The DNA condensation in the nucleoid is stabilized by association with RNA and above all with two histone-like basic proteins (HLPI with $M_r$ 17 kd and HLPII with $M_r$ 9 kd, [3, 4]).

Such bacterial "histones" are structurally highly conserved between different prokaryotes so far analysed. Comparable to their eukaryotic counterparts they are able to introduce negative superhelical turns into naked DNA and to form nucleosome-like structures (cf. **1.2.3.4**) which stabilize DNA against nuclease attack. Surprisingly the HLP content of stationary phase *Escherichia coli* with $\sim$ 20,000 molecules HLPI and 80–120,00 molecules HLPII predicts a ratio of appoximately one copy of HLPI and four copies of HLPII per 200 base pairs of DNA. This is a remarkable agreement with the stoichiometry of eukaryotic nucleosomes (cf. Fig. **1.4**).

In addition to the nucleoid, bacterial cells frequently contain several extrachromosomal small circular replicons, e.g., F-plasmids which code for proteins involved in conjugation or plasmids conferring resistance to various antibiotics (r-plasmids) or to colicins (Col-plasmids). Several of these plasmids have been adapted to serve as vectors for gene cloning (cf. **2**.4).

### 1.2.3 Organisation of the Genetic Material in Eukaryotic Cells

Disregarding some special cases, the genetic material of eukaryotic cells is distributed unevenly, with a small part in the double-membraned organelles (mitochondria,

[1] Wang, A. H.-J. et al.: Science *211* (1981): 171; Arnott, S. et al.: Nature *283* (1980): 743; Crawford, J. L. et al.: Proc. Natl. Acad. Sci. USA *77* (1980): 4016.

[2] Kubo, M. et al.: Gene *7* (1979): 153; Jolly, S. O. & L. Bogorad: Proc. Natl. Acad. Sci. USA *77* (1980): 822; Baldi, M. I.: Cell *20* (1980): 461; Smith, G. R., Cell *24* (1981): 599.

plastids) and the overwhelming part in the nucleus. The extranuclear genomes of mitochondria and plastids are structurally and, to a certain extent, also functionally related to the bacterial nucleoids. They are not further discussed in this section (for some details cf. Chap. **6**).

### 1.2.3.1 DNA Sequence Classes

By denaturation-renaturation experiments with nuclear DNA it was shown many years ago that DNA sequences of the eukaryotic genome can be classified in several classes (cf. Table **1**.1):

1. *Unique sequences* with 1–2 copies per haploid genome that contain most structural genes coding for proteins. They make about 70 % of the total genome in man but only 12 % in parsley.
2. *Repetitive sequences* that fall into several subclasses (cf. Table **1**.1) depending on their length and degree of repetition [20–$10^5$ ($10^6$) copies per haploid genome].

   Frequently middle repetitive ($10^1$–$10^3$ copies per haploid genome) and highly repetitive sequences ($10^4$–$10^6$ copies) are discerned. Importantly two types of structural genes are found in this class: (1) unique genes with internal repetitive structure, e.g., the silk fibroin gene (cf. **9**.4.2) and (2) 10- to many 1,000-fold repeated genes or gene segments, e.g., genes coding for rRNA, 5S RNA, tRNA, histones, actins or immunoglobulin V-gene segments (cf. Fig. **14**.4). In some cases the contents of rRNA genes is additionally increased by selectively replicating the nucleolar rDNP complex (gene amplification, cf. Chaps. **15** and **16** and Table **4**.6).
3. *Foldback sequences* that represent only a few percent of the total genome and contain inverse repeated sequences (palindromes) capable of forming intrastrand hairpins.

   Because palindromes are typical constituents of prokaryotic DNA sequences interacting with regulatory proteins or indicating points of origin for DNA replication (cf. **4**.4.2.2), similar functions are also suggested for at least some of the foldback sequences.

The survey of sequence classes in some selected organisms (Table **1**.1) shows the great variability of the eukaryotic genome organisation. This is especially evident from the remarks on the interspersion patterns of unique and repetitive sequences. Short (200–500 bp) and longer (1–5 kb) middle repetitive sequences are scattered throughout the genome. They are typical constituents of functional chromatin domains (cf. **4**.4.2.5.2), in some cases behave like transposable elements (cf. **4**.3.4) and may be transcribed as separate transcription units (cf. **4**.3.4). There is increasing evidence that they have an important role for chromatin structure and function and possibly also for precursor RNA processing (cf. Fig. **4**.21).

### 1.2.3.2 Basic Proteins Associated with DNA

The main structural proteins of the eukaryotic DNP-complexes (chromatin) are the five types of histones (H1, H2A, H2B, H3 and H4, Table **1**.2). They are low molecular weight proteins with high amounts of basic amino acid residues. All are subject to extensive post-synthetic modifications, i.e., to methylation, acetylation, phosphorylation and poly-ADP-ribosylation (cf. Table **1**.2 and the discussion **4**.4.2.5.3). The scheme of the histone amino acid sequences from calf thymus (Fig. **1**.3) and the comparison of histones from different sources, e.g., from pea, sea urchin and calf, lead to the following conclusions:

1. Histone Hl molecules are composed of three distinct structural domains: (1) a random-coil, basic N-terminal segment that encompasses amino acid residues 1–39, (2) a central globular region of about 2.8 nm diameter that extends from

**Table 1.1:** DNA sequence classes and interspersion patterns in eukaryotes (compiled by E. Serfling)

| Species (genome size) | Sequence classes[a] | % of total DNA | Kinetic complexity[b] (bp) | Repetition per haploid genome | Interspersion pattern[c] |
|---|---|---|---|---|---|
| *Mammals:* | Foldback | 2 | 530 | $>10^5$ | 70% of the genome consist of repetitive sequences of 200–400 bp (–1,500 bp) interspersed with unique sequences of 2,500 bp; cf. Ref. [1] and **4.4.2.5.2** |
| Rat | Repeated | 21 | $1.85\times10^5$ | $3\times10^3$ | |
| (very similar | | 10 | $2.12\times10^7$ | 125 | |
| to man) | Unique | 67 | $1.79\times10^9$ | 1 | |
| (2.9 pg) | | | | | |
| *Amphibia:* | Foldback | 3 | | | 50% of the genome consist of 300 bp repeated sequences interspersed with 700 bp unique sequences; 50% consist of unique DNA interspersed at a space of 400 bp by repetitive elements; cf. Ref. [2] |
| *Xenopus laevis* | Repeated | 6 | $5.25\times10^3$ | $3.2\times10^4$ | |
| (3.1 pg) | | 31 | $5.4\times10^5$ | $1.6\times10^3$ | |
| | | 6 | $1.4\times10^7$ | 20 | |
| | Unique | 54 | $1.54\times10^9$ | 1 | |
| *Insects:* | | | | | About one third of middle repetitive sequences (average length 5.6 kb) are interspersed with unique sequences; the average distance is greater than 13 kb; cf. Ref. [3, 4] and **4.3.4** |
| *Drosophila* | Repeated | 7 | $3.1\times10^3$ | $2.6\times10^3$ | |
| *melanogaster* | | 15 | $5.1\times10^5$ | 35 | |
| (0.12 pg) | Unique | 78 | $9.2\times10^7$ | 1 | |
| *Sarcophaga* | Foldback | 4 | | | Moderately repetitive sequences of 5 kb are interspersed with unique DNA of more than 10 kb; cf. Ref. [5] |
| *bullata* | Repeated | 8.5 | $1.92\times10^4$ | $2.4\times10^3$ | |
| (0.6 pg) | | 6 | $5.4\times10^5$ | 60 | |
| | Unique | 81 | $4.4\times10^8$ | 1 | |
| *Plants:* | Foldback | 5 | | | 47% of the genome consist of 300 bp repeated sequences interspersed with unique sequences of about 1,000 bp; long repetitive sequences constitute 33% of the genome, 13% are organized in a satellite-like manner; cf. Ref. [6] |
| *Petroselinum* | Repeated | 13 | $1.7\times10^3$ | $1.3\times10^5$ | |
| *sativum* | | 48 | $3\times10^5$ | $3\times10^3$ | |
| (2.0 pg) | | 16 | $7.4\times10^6$ | 42 | |
| | Unique | 12 | $2.2\times10^8$ | 1 | |

| | | | | | |
|---|---|---|---|---|---|
| *Nicotiana tabacum* (1.65 pg) | Foldback | 21 | | | 55% of the genome consist of repetitive sequences of 300 bp interspersed with unique sequences of 1,400 bp; another 25% contain long repetitive sequences with a minimal length of 1,500 bp; cf. Ref. [7] |
| | Repeated | ~ 5 | $4.6\times10^{3}$ | $1.24\times10^{4}$ | |
| | | ~52 | $2.5\times10^{6}$ | 252 | |
| | Unique | 22 | $2.8\times10^{8}$ | 1 | |
| *Water mold: Achlya bisexualis* (0.046 pg) | Foldback | 2 | | | The vast majority of single copy sequences are not adjacent to repetitive DNA at a fragment length up to 7,500 bp; interspersion of repetitive and single copy sequences at longer intervals (>10 kb); cf. Ref. [8] |
| | Repeated | 2 | 380 | $2.2\times10^{3}$ | |
| | | 14 | $7.6\times10^{4}$ | 80 | |
| | Unique | 82 | $3.4\times10^{7}$ | 1 | |

[a] Foldback sequences represent short inverse repetitive sequences. The repeated sequences are frequently classified into middle repetitive ($10$–$10^{4}$-fold repeated) and highly repetitive ($10^{4}$–$10^{6}$-fold repeated) sequences.

[b] Figures for the kinetic complexity derive from the DNA reassociation kinetics and are proportional to the Cot $^{1}/_{2}$ values ($C_0t$), i.e., the product of the initial DNA concentration ($C_0$, mol/l) multiplied with the time (s) at which half of the DNA has renaturated. They reflect the single copy complexity, i.e., repeated sequences are calculated only once. Example: 21% of the rat genome, corresponding to totally $5.65\times10^{8}$ bp, consist of $3\times10^{3}$-fold repeated sequences. Hence, the kinetic complexity of this sequence class is $5.65\times10^{8}/3\times10^{3} = 1.8\times10^{5}$ bp.

[c] The data were taken from the following papers:


1. Pearson, W, R. et al.: Biochemistry *17* (1978): 51–59
2. Davidson, E. H. et al.: J. Mol. Biol. *77* (1973): 1–23
3. Wu, J.-R. et al.: J. Mol. Biol. *64* (1972): 211–219
4. Manning, J. E. et al.: Cell *4* (1975): 141–155
5. Samols, D. & H. Swift: Chromosoma *75* (1979): 129–143
6. Kiper, M. & F. Herzfeld: Chromosoma *65* (1978): 335–351
7. Zimmermann, J. L. & R. B. Goldberg: Chromosoma *59* (1977): 227–252
8. Hudspeth, M. E. S. et al.: Proc. Natl. Acad. Sci. USA *74* (1977): 4332–4336

**Table 1.2:** Properties and classification of histones from calf thymus (cf. CRC Handbook of Biochemistry and Mol. Biol., 3rd. edn. 1976, Proteins Vol. II, pp. 293–300 and Isenberg [7])

| Histone | MW (kd) | Amino acid residues | | | | Co- and post-translational modifications[b] | | | | Evolutionary conservation[c]) |
|---|---|---|---|---|---|---|---|---|---|---|
| | | total | basic[a] | Lys | Arg | Me | Ac | P | ADP-R | |
| H1 (very Lys-rich) | 22 | 220 | 65 | 59 | 4 | – | – | +++ | ++ | variable |
| H2A (Lys-rich) | 14.0 | 129 | 30 | 14 | 12 | – | + | + | + | N-terminus variable, C-terminus conserved |
| H2B (Lys-rich) | 13.8 | 125 | 30 | 20 | 7 | – | +++ | + | + | |
| H3 (Arg-rich) | 15.3 | 135 | 33 | 13 | 18 | + | +++ | + | + | Highly conserved |
| H4 (Arg-rich) | 11.3 | 102 | 26 | 10 | 14 | + | +++ | + | + | |

[a] Including Lys, Arg, His
[b] Me, N-methylation of ε-amino groups of Lys; Ac, N-acetylation of ε-amino groups of Lys; P, O-phosphorylation of hydroxy groups of Ser and Thr, N-phosphorylation of His questionable; ADP-R, Poly(ADP)ribosylation of glutamic acid residues (ester linkage). –, no modification; +, ++, +++, indicates increasing frequency of modifications (cf. Fig. 1.3).
[c]) Evaluated by comparison with histone sequences from pea, trout and echinoderms (cf. details in Fig. 1.3). Homology between calf thymus and pea is 97% for H3 and 98% for H4.

residues 39 to 116 is rich in apolar residues and contains most of the acidic amino acid residues and (3) a highly basic, random-coil C-terminal domain that contains about 40 mol% of lysine. The latter segment binds strongly to purified DNA.
2. Histones H2A and H2B appear as evolutionary hybrids. The basic N-terminal regions are variable, whereas the amino acid sequences in the much less basic but rather hydrophobic C-terminal regions are more conserved among different species.
3. Histones H3 and H4 are extremely conservative proteins with only 4 and 3 variable positions respectively. The structural organisation into a basic domain at the N-terminus and a hydrophobic domain at the C-terminus resembles that of H2A and H2B.
4. The amino acid residues subject to postsynthetic modification are mainly localized at the N-termini of H2A, H2B, H3 and H4 (cf. Fig. 1.3).

Besides the five major basic proteins of eukaryotic DNP, which occur with $10^7$–$10^8$ copies per nucleus, there are at least four minor proteins associated with chromatin

---

▶

**Fig. 1.3:** Amino acid distribution in histones (from Isenberg [7], modified).
The sequence of H1 represents rabbit thymus H1, all others represent histones from calf thymus. Different types of amino acid residues are indicated with symbols (cf. *bottom* of the figure) to illustrate the uneven distribution of hydrophobic, basic and acidic residues which contribute to the functional domains of histones. The residues so far known to be subject to postsynthetic modification (cf. *bottom* of the figure) are mostly concentrated at the N-termini [7, 13].

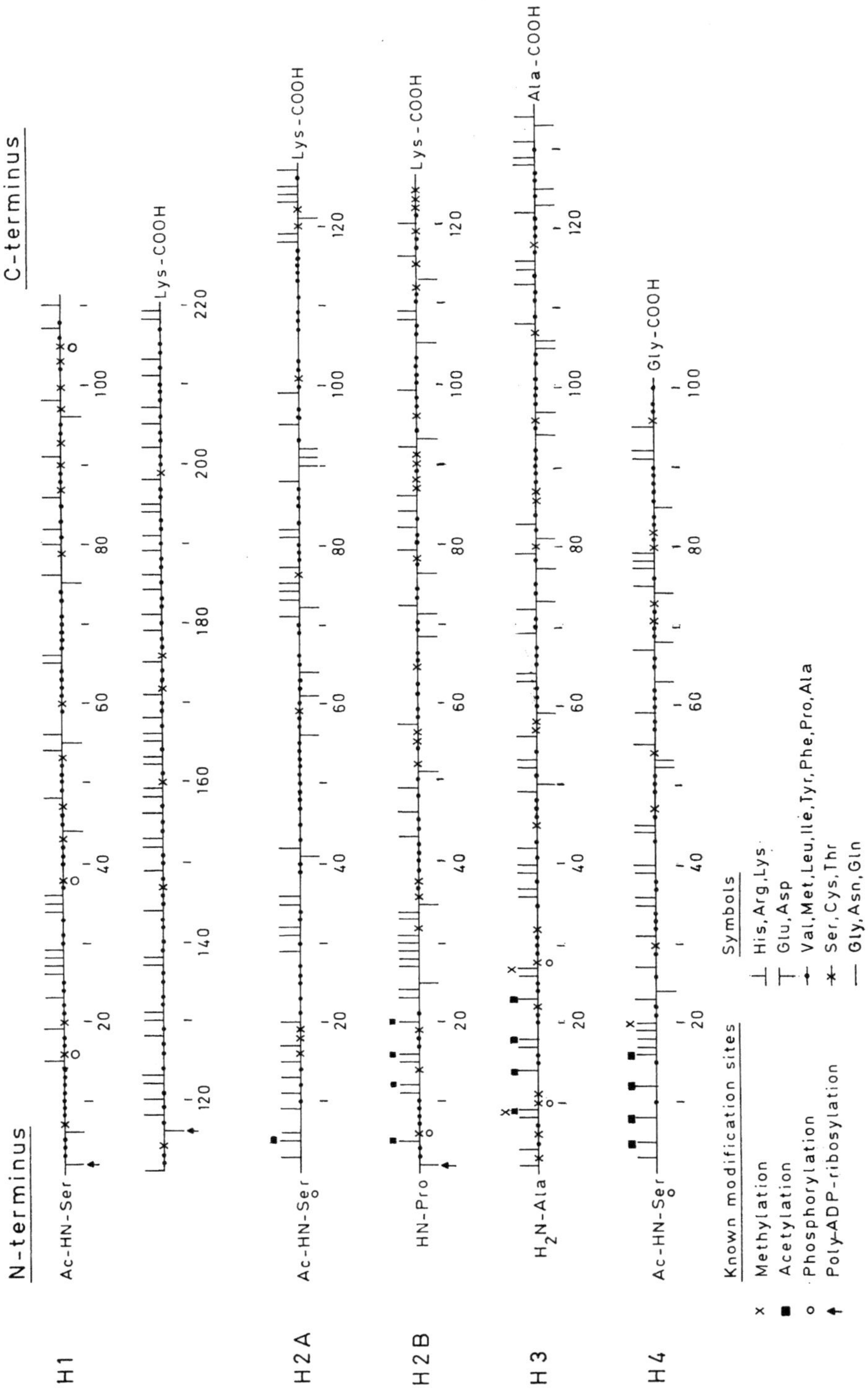
N-terminus
C-terminus
H1
H2A
H2B
H3
H4
Ac-HN-Ser
Lys-COOH
Ac-HN-Ser
Lys-COOH
HN-Pro
Lys-COOH
$H_2$N-Ala
Ala-COOH
Ac-HN-Ser
Gly-COOH
Known modification sites
Methylation
Acetylation
Phosphorylation
Poly-ADP-ribosylation
Symbols
His, Arg, Lys
Glu, Asp
Val, Met, Leu, Ile, Tyr, Phe, Pro, Ala
Ser, Cys, Thr
Gly, Asn, Gln

**Table 1.3**: Properties of minor basic proteins of chromatin (HMG-proteins)

| Protein | $M_r$ (kd) | Amino acid residues[2] | | | | | | Selected references[3] |
|---|---|---|---|---|---|---|---|---|
| | | total | Pro (% of total residues) | Asp | Glu | Lys | Arg | |
| HMG1[1] | 26 | 220 | 5–7 | 11–12 | 17–18 | 17–21 | 4–6 | a, b, c, e, h, i |
| HMGT | 29 | 263 | (isoform of HMG1 in trout) | | | | | d |
| HMG2[1] | 26 | | 6–8 | 10–11 | 16–19 | 17–22 | 4–6 | b, c, h, i, k |
| HMG14 | 10.5 | 100 | 8–10 | 8–11 | 13–17 | 18–21 | 4–6 | f |
| HMG17 | 9.25 | 89 | 10–16 | 10–12 | 10–17 | 18–24 | 4–5 | g, h |
| H6 | 7.2 | 69 | (isoform of HMG17 in trout testis) | | | | | d |

[1] HMG1 and 2 are closely related with respect to their sequence data and function. Both contain a highly acidic C-terminal part with 41 and 35 continuous aspartic plus glutamic acid residues.

[2] The average composition is given to indicate the preponderance of Pro, Asp, Glu, Lys and Arg residues. HMG proteins contain very low amounts of Cys, His, Phe, Tyr, Ile and Met some of which may even be totally lacking in HMG14 and 17.

[3] Selected references: a) Walker, J. M. et al.: Nature *271* (1978): 281; b) Romani, M. et al.: J. Biol. Chem. *254* (1979): 2918; c) Isackson, P. J. et al.: J. Biol. Chem. *254* (1979): 5569; d) Peters, E. H. et al.: J. Biol. Chem. *254* (1979): 3358; e) Rechsteiner, M. & Le Roy Kuehl; Cell *16* (1979): 901; f) Walker, J. M. et al.: FEBS Letters *100* (1979): 394; g) Walker, J. M. et al.: Europ. J. Biochem. *76* (1977): 461; h) Rabbani, A. et al.: Biochem. J. *173* (1978): 497; i) Javaherian, K. et al.: Nucleic Acids Res. *6* (1979): 3569; k) Walker, J. M. et al.: Biochem. J. *181* (1979): 659.

in all eukaryotic cells. These are the so called high-mobility-group proteins (HMG cf. Table **1**.3), which are preferentially associated with active chromatin regions (cf. **4**.4.2.5.2).

Disregarding some special forms (HMGT, HMGE and H6) the four major representatives HMG1, 2, 14 and 17 are found in all eukaryotic cells so far investigated. They constitute about 1%–5% of total chromatin proteins corresponding to $10^5$–$10^6$ molecules per cell nucleus. HMG proteins are extractable from chromatin with 0.35 M NaCl, are soluble in 2% trichloroacetic acid and, because of their low molecular weights, have a high mobility upon SDS gel electrophoresis. Sequencing of HMG proteins revealed them to be rather conservative with some extraordinary structural features: Cys, His, Met, Phe and Tyr residues are scarce or even lacking. The amino acid sequence is dominated by Pro and above all by Glu, Asp and Lys residues. The former two prevail at the C-terminus, whereas basic residues (Lys, Arg) prevail at the N-terminus of the very polar molecules. Hence two functional domains can be assumed. HMG1, 2 and E show remarkable affinity for single-stranded DNA, help to unwind double-stranded DNA and may have a crucial role for the reversible dissociation-reassociation of nucleosomes (cf. Fig. **4**.17).

### 1.2.3.3 Non-Histone Chromosomal Proteins

Besides the major structural components of chromatin (histones and HMG-proteins) there are numerous minor protein species more or less tightly bound to the chromatin network, which is attached to the nuclear membrane and spans the whole interphase nucleus. These proteins are frequently referred to as non-histone proteins (NHP). They are mostly non-basic and highly heterogeneous in size and function (Table **1**.4). Most NHP-species are not cell-specific, with less than 10% of them being somehow involved in regulation of gene expression. Of special interest are regulatory proteins (about $10^3$ molecules/nucleus or less) which contribute to the temporal and spatial control of chromosomal protein modification (cf. **4**.4.2.5.3), influence the transitions between distinct structural levels and hence activity states of chromatin (cf. **4**.4.2.5.2.),

**Table 1.4**: Major components of the non-histone chromatin fraction (NHP fraction). Compiled from Elgin & Weintraub [9] and Grunicke [8]

| Functional group | Examples | Ref. to in this book |
|---|---|---|
| Enzymes of DNA metabolism | DNA polymerases,<br>DNA ligase,<br>DNA topoisomerases,<br>Endo- and exonucleases,<br>Terminal DNA-nucleotidyl transferase | <br><br>1.2.1 |
| Enzymes of RNA metabolism | RNA polymerases,<br>Poly(A)-polymerase,<br>Enzymes of 5′-cap synthesis,<br>Splicing and processing enzymes,<br>RNases | Tables **1**.6 and **1**.7<br>**1**.3.3.1<br>**1**.3.3.1<br>**1**.3.3.1 and **4**.4.4.4 |
| Histone and NHP modifying and demodifying enzymes | Protein kinases and phosphoprotein phosphatases,<br>Protein-acetyl transferases and esterases,<br>Poly (ADP ribose) synthetase and corresponding hydrolase,<br>Proteases | Figs. **4**.2 and **10**.6<br>**4**.4.2.5.3<br>**4**.4.2.5.3 |
| Structural proteins | Chromosomal spindel proteins (actin, myosin, tropomyosin, α- and β-tubulin),<br>Nuclear matrix proteins<br>Chromosomal scaffold proteins,<br>Proteins of RNP particles | <br><br>**1**.3.3.1<br>Figs. **1**.5 and **1**.6<br>**1**.3.3.1 |

act as intracellular hormone receptors or acceptors (cf. Chap. **7**) or which might provide specificity to mRNP processing and export into the cytoplasm (cf. **4**.4.3 and **4**.4.4.4).

### 1.2.3.4 Organizational Levels of Chromatin

Basic structural element of chromatin is the nucleosome (Fig. **1**.4). It contains an octamer of the so-called core histones $(H2A, H2B, H3, H4)_2$, one molecule of histone H1 and ~200 bp ob DNA. 145 bp of DNA form a negative supercoil of 1.75 turns around the core histone complex whose central part consists of the hydrophobic C-terminal domains of the eight histone molecules, whereas the highly basic N-terminal parts on the surface interact with DNA in a non-sequence-specific manner. Between two adjacent nucleosomes there is a DNA linker of variable length (average 60 bp). Hence chromatin in this extended form appears as beads-on-a-string with 5.5–11 nm diameter (cf. state 2 in Fig. **1**.5). By interaction with histone H1 the inter-nucleosomal linker DNA is also coiled thus generating the 10–11 nm unit fiber of chromatin (state 3 in Fig. **1**.5). One molecule H1 per nucleosome is assumed to interact by its C-terminal segment with about 20 bp of linker DNA on the exit side of the nucleosome and by its N-terminal segment with about 10 bp of linker DNA near the entry side of the nucleosome. The central globular region of H1 intimately contacts the histone octamer mainly by interactions with H2A [10].

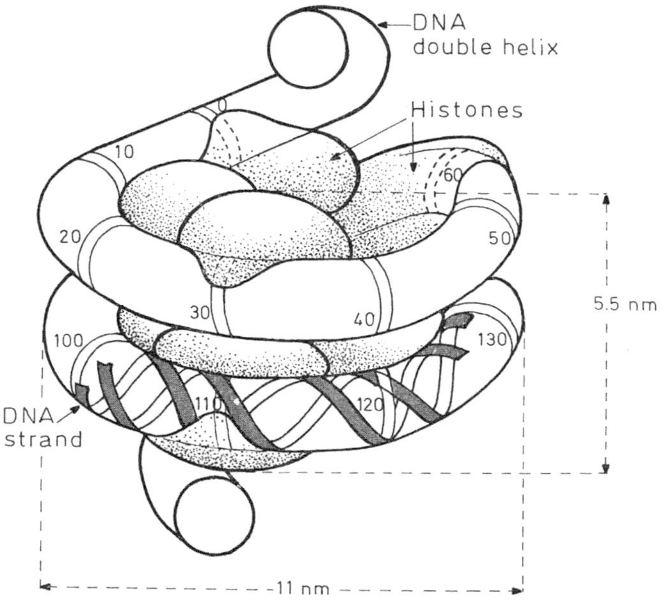


**Fig. 1.4:** Model of the nucleosomal organisation of DNA in chromatin (from Rindt & Nover [11]). Sequential degradation of chromatin with micrococcal nuclease first generates a particulate fraction *(nucleosome)* containing 200 bp of DNA and, besides the octamer of the core histones, one molecule of H1. Further degradation results in the release of H1 and cleavage of the linker DNA leaving the *nucleosomal core particle* with 145 bp of DNA. The DNA double-helix is supercoiled around the core histone octamer formed of two tetramers of each one molecule of histones H2A, H2B, H3 and H4. The $11 \times 5.5$ nm nucleosomal core particle contains 1.75 superhelical turns of DNA with a pitch of 2.8 nm and an average of 145 base pairs. The basic N-terminal domains of the histones at the surface interact with the DNA thus providing a local protection against DNase I attack (sites 30, 60, 80 and 110). Other, non-protected, sites are easily cleaved (sites 10, 20, 40, 50 and 130). The preferential cleavage at base numbers which are multiples of 10 results from the fact that the DNA helix contains 10 base pairs per helical turn [14]. Thus, each strand rises to the nucleosome surface every 10 nucleotides. The bands on the DNA denote this distance of ten base pairs for one of the two DNA strands. A recent, more detailed analysis of the nuclease cleavage patterns showed that, in fact, the helical periodicity of DNA changes from 10.5 bp per helical turn to 10 bp when the DNA is supercoiled in the nucleosome (cf. Rhodes, D. & S. Klug: Nature *286* (1980): 573).

By self-assembly of nucleosomes from naked DNA and an excess of the four core histones it can be shown that the steric information for this process resides solely in these components. However, there is experimental evidence that under natural conditions assembly factors are involved [15]. Such are DNA topoisomerases, aiding in DNA supercoiling, and an abundant phosphoprotein in the nucleoplasm of a wide range of vertebrate cell types. It was called nucleoplasmin ($M_r$ 30 kd) and probably promotes ordered histone – histone interactions by competing against the strong electrostatic attraction between histones and DNA. A similar function was suggested for HMG1 and HMG2, because of their long stretches of acidic amino acid residues in the C-terminal domains (cf. Table 1.3). The rapid and ordered assembly and disassembly of nucleosomes is essential for chromatin replication and transcription (cf. 4.4.2.5.3).

Histone H1 is evidently also involved in the further coiling of the unit fiber to the regular solenoids or to the irregularly formed "superbeads" of 25–30 nm with 8–10

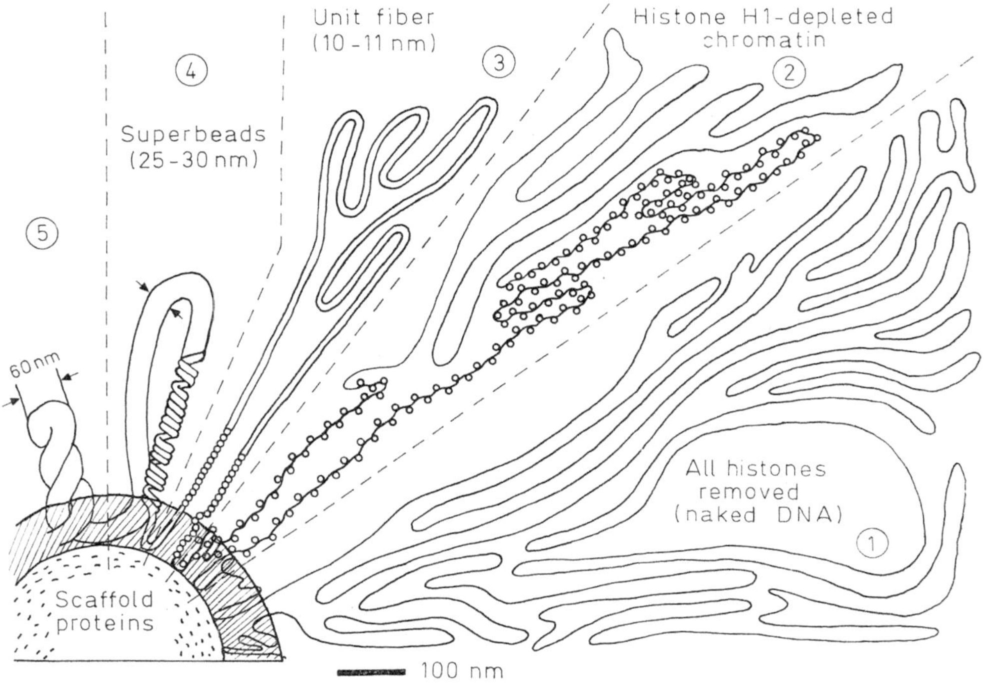


**Fig. 1.5**: Schematic cross section through a eukaryotic chromosome showing successive steps of DNA compaction (from Rindt & Nover [11]).

The scheme is based mainly on data and electron microscopic pictures obtained by Laemmli and coworkers (cf. Fig. **1**.6). DNA of interphase and metaphase chromosomes is attached to a protein scaffold. It is condensed manyfold starting from naked DNA *(state 1)* to the H1-free nucleosomal chain (*state 2*, beads-on-a-string conformation), to the 10–11 nm unit fiber of nucleosomes tightly packed by association with H1 *(state 3)*, to the 25–30 nm thick fiber of superbeads *(state 4)* and, finally, to 50–60 nm projections *(state 5)* extending from the chromosomal scaffold. The overall condensation in length from double-helical DNA to the chromosome multiplies to 5000 to 8000 fold:

| | | | | | | |
|---|---|---|---|---|---|---|
| States: | 1 ⟶ | 2 ⟶ | 3 ⟶ | 4 ⟶ | 5 → → | Chromosome |
| Compaction: | 2× | 2.5–3.5× | 3–10× | 2× | 50–80× | |

State 5 is the only state in metaphase chromosomes, states 4, 3 and 2 prevail in interphase chromosomes, whereas state 1 (in this extended form) is an experimental artefact.

nucleosomes per turn (state 4, Fig. **1**.5). The H1 molecules are located in the inside of solenoids or superbeads and it may well be that the aggregation of H1 into a helical skeleton accompanies and, actually, may control the formation of these chromatin superstructures [16]. Beads-on-a-string (state 2), unit fiber (state 3) and superbeads (state 4) are the states of functional chromatin in interphase nuclei (cf. **4**.4.2.5.3). Their formation and stability reflects the tendency of nucleosomes to become stacked to higher order structures by protein-protein interactions between their top and bottom planes.

Even if all histones and soluble proteins are removed from chromatin the naked DNA (state 1) remains attached to a protein scaffold that forms the core of the interphase and metaphase eukaryotic chromosomes (Fig. **1**.5 and **1**.6A). In a current

**Fig. 1.6**: Electron micrographs visualizing different structural levels of chromatin (Courtesy of U. K. Laemmli, with permission, © M. I. T. Press).

HeLa cell chromosomes were isolated and pretreated in the indicated manner before preparing for electron microscopy.

**A** Spread, histone-depleted metaphase chromosome after treatment with dextran sulfate and heparin. The central protein scaffold and part of the surrounding naked DNA loops (state 1 of **Fig. 1.5**) are clearly visible. *Bar* represents 1 $\mu$m (from Paulson, J. R. & U. K. Laemmli: Cell *12* (1977): 817).

**B** Cross section through EDTA-swollen chromatid showing the radial arrangement of the 10–11 nm unit fibers (state **3** of Fig. **1.5**). *Bar* represents 200 nm.

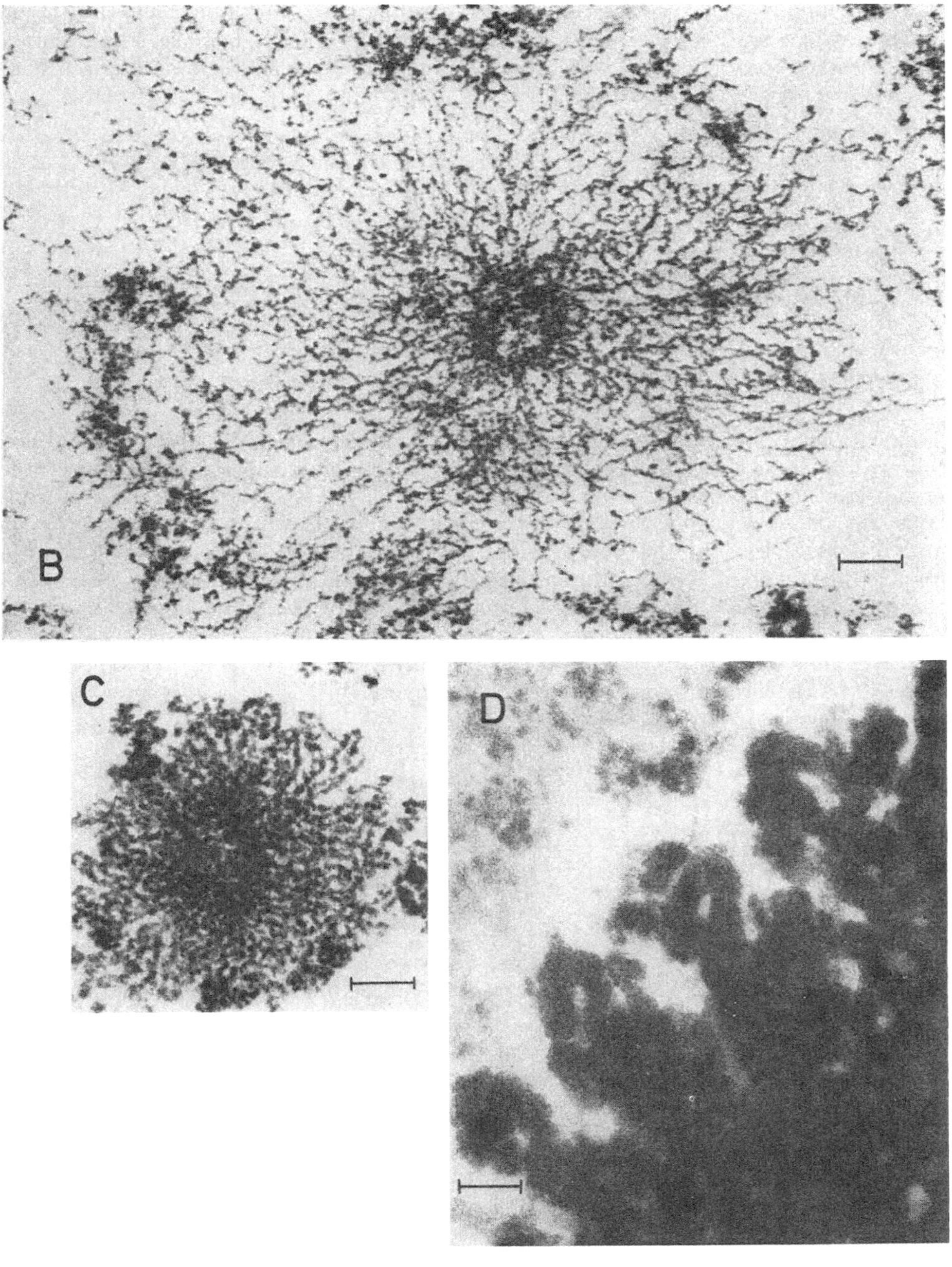


**C** Cross section through a chromatid fixed in a buffer containing 1 mM $MgCl_2$. Chromatin is shown as 20–30 nm fibers (state **4** of Fig. **1**.5). *Bar* represents 200 nm.

**D** Cross section through peripheral projections emerging from a chromosome isolated with hexylene glycol (state 5 of Fig. **1**.5). *Bar* represents 100 nm. (Figs. **B–D** are from Marsden & Laemmli [18]).

model of the metaphase chromosome structure DNA loops of 15–30 μm length (45,000–90,000 base pairs, cf. Fig. 1.6A) are condensed over several steps (Fig. 1.6B, C) finally to knob-like projections of 50–60 nm width and 300–800 nm length attached to the protein scaffold (cf. Fig. 1.5, state 5 and Fig. 1.6D). For the largest human chromosomes (e.g., chromosomes 1 or 2) the $8 \times 10^4$ μm DNA per chromatid are assumed to form about 4000 knobs each containing an average of 20 μm DNA.

The scaffold proteins make about 1% of the total chromosomal proteins. About 20–30 proteins were detected in scaffolds purified from Hela cells. The molecular weights range from 35–150 kd with three major components of 35, 51 and 53 kd [18].

## 1.3 Realization of the Genetic Information

### 1.3.1 The Genetic Code

Translation of the four-lettered nucleic acid code into the 20-lettered protein code follows the same rules of base complementarity in antiparallel nucleic acid strands as outlined for the DNA double helix (cf. Fig. 1.2). The nucleic acid code is a triplet code, i.e., three adjacent nucleotides comprise one code word corresponding to one amino acid residue in the polypeptide chain (Table 1.5). If four letters are used for a triplet code, $4^3 = 64$ different code words can be formed which is more than three times the amount required for incorporation of 20 amino acids. Due to the particularities of genetic information processing and the rules of base pairing (cf. Fig. 1.2) gene expression involves a two-fold inversion of the genetic code (exemplified by the triplet coding for methionine incorporation):

3′ · · TAC · · 5′ → 5′ · · AUG · · 3′ → 3′ · · UAC · · 5′ → · · Met · ·

| Codogen (DNA) | Codon (mRNA) | Anticodon ($tRNA^{Met}$) |
|---|---|---|
| Transcription | | Translation |

The genetic code (mRNA-code, Table 1.5) has the following characteristics:

- It is universal for all living systems, non overlapping and comma-free, i.e., the first triplet (initiator triplet) determines the reading frame for the translation machinery for a given cistron.

  With the advancing techniques of gene analyses it became apparent that the non-overlapping is true only for the triplets within one cistron of the mRNA. In contrast, overlapping genes with identical but also with divergent reading frames are well known. Examples are given elsewhere in this book, e.g., phage MS2 (Fig. 4.24), phage λ (14.2.3.10) and simian virus 40 (Fig. 10.8). The most sophisticated use of codons in all three possible reading frames was found for three overlapping genes of bacteriophages G4 and ∅X174 (Shaw, D. C. et al.: Nature *272* (1978): 510; Borrias, W. E. et al.: J. Virol. *31* (1979): 288) and for some animal viruses, e.g., polyoma virus (Soeda, E. et al.: Nature *283* (1980): 445). Thus, the three early proteins of polyoma virus (large, middle and small T antigens) are coded by overlapping genes with all three possible reading frames used. The final reading frames for the large and middle T antigens, generated only by mRNA splicing, are composed of each two reading frames of the coding sequence.

- The code is un-ambiguous, though a certain flexibility of base pairing between mRNA and tRNA was found for the third base (Wobble-base, cf. Fig. 1.14).

**Table 1.5:** Genetic code and codon usage of selected genes

| First position (5′-end) | Second position | | | | | | | | | | | | | | | | | | | | | | | | Third position (3′-end) |
|---|---|---|---|---|---|---|---|---|---|---|---|---|---|---|---|---|---|---|---|---|---|---|---|---|---|
| | **U** | | | | | | **C** | | | | | | **A** | | | | | | **G** | | | | | | |
| | | MS | LR | M | Gb | SV | | MS | LR | M | Gb | SV | | MS | LR | M | Gb | SV | | MS | LR | M | Gb | SV | |
| **U** | Phe | 12 | 3 | 8 | 5 | 22 | Ser | 7 | 7 | 4 | 0 | 16 | Tyr | 5* | 5 | 14 | 2 | 16 | Cys | 5 | 2 | 4 | 2 | 3 | **U** |
| | Phe | 16 | 1 | 4 | 3 | 1 | Ser | 12 | 6 | 0 | 2 | 4 | Tyr | 16 | 3 | 0 | 1 | 9 | Cys | 2 | 1 | 0 | 0 | 4 | **C** |
| | Leu | 8 | 5 | 27 | 0 | 16 | Ser | 6 | 3 | 7 | 0 | 6 | ochre | 0 | 0 | 1 | 1 | 1 | opal | 0 | 1 | 5 | 0 | 1 | **A** |
| | Leu | 5* | 5 | 2 | 0 | 16 | Ser | 10 | 5 | 0 | 0 | 0 | amber | 1 | 0 | 0 | 0 | 0 | Trp | 9 | 2 | 0 | 2 | 10 | **G** |
| **C** | Leu | 7 | 2 | 0 | 0 | 10 | Pro | 10 | 0 | 7 | 5 | 21 | His | 4* | 3 | 5 | 2 | 6 | Arg | 11 | 2 | 0 | 0 | 2 | **U** |
| | Leu | 15 | 3 | 0 | 3 | 3 | Pro | 4 | 6 | 0 | 0 | 8 | His | 6* | 4 | 0 | 7 | 5 | Arg | 13 | 10 | 0 | 0 | 0 | **C** |
| | Leu | 8 | 1 | 0 | 0 | 6 | Pro | 3 | 2 | 6 | 2 | 15 | Gln | 7 | 14 | 8 | 0 | 26 | Arg | 4* | 4 | 0 | 0 | 0 | **A** |
| | Leu | 7* | 25 | 0 | 14 | 6 | Pro | 9 | 6 | 0 | 0 | 0 | Gln | 8 | 14 | 1 | 3 | 16 | Arg | 8* | 2 | 0 | 0 | 1 | **G** |
| **A** | Ile | 7 | 10 | 24 | 0 | 23 | Thr | 4 | 3 | 6 | 3 | 25 | Asn | 11 | 7 | 8 | 1 | 19 | Ser | 4* | 5 | 2 | 2 | 16 | **U** |
| | Ile | 13 | 7 | 2 | 0 | 0 | Thr | 12 | 11 | 0 | 3 | 15 | Asn | 7 | 5 | 0 | 5 | 11 | Ser | 9 | 6 | 0 | 0 | 6 | **C** |
| | Ile | 12* | 1 | 3 | 0 | 8 | Thr | 8* | 1 | 10 | 1 | 16 | Lys | 9 | 10 | 7 | 3 | 28 | Arg | 4* | 1 | 5 | 0 | 15 | **A** |
| | Met | 9 | 9 | 5 | 1 | 13 | Thr | 7 | 4 | 0 | 0 | 0 | Lys | 16 | 1 | 1 | 8 | 8 | Arg | 2* | 0 | 0 | 3 | 17 | **G** |
| **G** | Val | 9 | 8 | 11 | 3 | 24 | Ala | 15 | 3 | 6 | 4 | 45 | Asp | 19 | 10 | 14 | 5 | 18 | Gly | 19 | 6 | 7 | 4 | 12 | **U** |
| | Val | 10 | 9 | 1 | 2 | 1 | Ala | 7 | 15 | 0 | 7 | 6 | Asp | 14 | 7 | 0 | 2 | 17 | Gly | 7 | 11 | 0 | 8 | 7 | **C** |
| | Val | 6 | 3 | 7 | 0 | 11 | Ala | 8 | 4 | 9 | 1 | 12 | Glu | 9 | 10 | 15 | 2 | 22 | Gly | 8 | 1 | 3 | 0 | 19 | **A** |
| | Val | 7 | 15 | 0 | 13 | 18 | Ala | 12 | 22 | 0 | 0 | 0 | Glu | 13 | 5 | 1 | 6 | 19 | Gly | 10 | 4 | 2 | 1 | 15 | **G** |

The data for the codon usage were derived from the following papers:

*MS: E. coli* phage MS2, genes coding for replicase β-subunit and coat proteins;
* codons not used in the coat protein gene; cf. Fiers, W. et al.: Nature *260*(1976): 500;
*LR: E. coli lac i* gene coding for *lac* repressor, codon AUA codes for Met instead of Ile; cf. Farabaugh, Th. J.: Nature *274* (1978): 765;
*M:* Yeast mitochondrial gene for cytochrome c oxidase subunit II. The codon UGA ist used as Trp codon instead of termination codon; cf. Fox, T.D.: Proc. Natl. Acad. Sci. USA *76* (1979): 6534;
*Gb:* Human β-globin gene, cf. Kafatos, F. C. et al.: Proc. Natl. Acad. Sci. USA *74* (1977): 5618;
*SV:* Viral coat protein genes $VP_1$ and $VP_2$ of simian virus 40, cf. Fiers, W. et al.: Nature *273* (1978): 113.

- There are codons signalling initiation (usually AUG, coding for methionine) and termination of the polypeptide chain (codons UAA, UGA and UAG).
- The code is degenerated, i.e., with the exception of tryptophane and methionine, there is more than one triplet for all amino acid residues.

Comparison of the actual use of the different codons in two prokaryotic, one mitochondrial and two eukaryotic systems shows striking differences (Table **1.5**). The very large $\beta$-subunit (544 aa residues) of *E. coli* phage MS2 replicase (MS, cf. **4.4.4.3**) lacks striking preferences for any one of the codons. This contrasts to the much smaller, very rapidly translated coat protein message of the same phage avoiding certain triplets recognized by minor tRNA species of the host cell (MS, triplets marked by an asterisk). But this contrasts also to the message of the *E. coli* lac repressor with 360 aa residues (LR, cf. **4.4.2.2**).

Codon usage in the yeast mitochondrial gene for cytochrome c oxidase subunit II with 251 aa residues (M) is quite extraordinary in that many codons are not used at all and that the termination codon UGA (opal) serves for trp incorporation. Similar results were reported for mitochondria from human cells indicating that the modern organelles have inherited and conserved a somewhat altered genetic code from their common ancestors.

The two eukaryotic systems included as examples in Table **1.5** are the human $\beta$-globin chain (Gb, 146 aa residues, cf. **4.4.2.5.2**) and the two coat proteins of simian virus 40 (SV, Protein VP2 with 352 and VP1 with 361 aa residues, cf. Fig. **10.8**). Both sequences are characterized by a remarkable deficit of CG-containing codons, i.e., of XCG or CGX. Codon preference is most pronounced for the $\beta$-globin, whose efficient synthesis depends on an appropriate adaptation of the tRNA isoacceptor pattern (cf. **4.5.4.3**).

## 1.3.2 Transcription

### 1.3.2.1 RNA Polymerases in Prokaryotic and Eukaryotic Cells

Compared to their eukaryotic counterparts RNA polymerases (RNAP) in prokaryotes look relatively simple. In all bacteria investigated so far enzymes with two larger ($\beta$, $\beta'$) and two smaller subunits ($\alpha$ and $\sigma$) and $M_r$ of 400–500 kd were reported (details cf. Table **1.6**). Generally two different forms function sequentially in initiation (holoenzyme, $\alpha_2\beta\beta'\sigma$) and elongation (core enzyme, $\alpha_2\beta\beta'$) of transcription respectively. Modification of subunits ($\alpha$, $\beta$ and $\beta'$), exchange of $\sigma$ or addition of subunits are means for selective transcriptional control (cf. **4.4.2.6**). In fact, the particular subunit composition of bacterial RNAP seems to be primarily related to the multiplicity of transcriptional control mechanisms, since the same catalytic reaction can be achieved also with much simpler enzymes containing only one subunit, e.g., the phage T7-specific RNAP (cf. **14.2.2.2**).

Analyses of numerous mutants and experiments with individual subunits and complexes of them resulted in a rather detailed knowledge about the differential function of RNAP subunits [20, 25]:

The *$\alpha$ subunit* is the major structural part of the enzyme essential for subunit assembly upon biogenesis and for keeping $\beta$ and $\beta'$ in proper spatial positions. Thus it influences indirectly the fidelity of transcription.

The *$\beta$ and $\beta'$ subunits* contain the catalytic sites of the enzyme. Both influence initiation and termination. The $\beta$ subunit represents the site of sensitivity against the antibiotic rifampicin. It interacts with termination factor rho, the phage $\lambda$ antitermination protein gpN (cf. **14.2.3.4**) and the $\sigma$ subunit. The $\beta'$ subunit is essential for the stability of DNA-RNAP binding (core enzyme).

The *$\sigma$ subunit* determines promoter selection and the DNA binding affinity of the $\beta$, $\beta'$ subunits. It is required for the assembly of the core enzyme during RNAP biosynthesis. The $\sigma$ subunit nteracts with regulatory proteins modifying the rate of transcription initiation or termination, e.g., with the CAP · cAMP complex (cf. **4.4.2.2**) or with the antitermination protein gpN (cf. **14.2.3.4**).

**Table 1.6**: Prokaryotic types of RNA polymerases (from Polya, G. M. in Ref. [24], pp. 7–41)

| | Subunit composition [a] ($M_r$ in kd) | | | |
|---|---|---|---|---|
| Bacteria | $2\alpha$ | $\beta$ | $\beta'$ | $\sigma$ |
| *Escherichia coli* | (39) | (155) | (165) | (95) |
| *Pseudomonas putida* | (44) | (155) | (165) | (98) |
| *Bacillus subtilis*[b] | (45) | (155) | (155) | (57) |
| *Anacystis nidulans* | (39) | (147) | (125) | (86) |
| Bacteriophage T7 (cf. Table **14.2**) | 1 polypeptide (107) | | | |
| Plant chloroplasts (maize)[c] | 4 polypeptides (180, 140, 42, 27) | | | |
| Mitochondria[d] | $M_r \sim 500$ kd with two large subunits (in yeast $M_r$ 135, 190 kd) and several smaller subunits | | | |

[a] The sigma subunit is only associated with the holoenzyme (cf. **1.3.2.2**) and may be replaced by $\sigma$-like factors (cf. **4.4.2.6**).

[b] *Bacillus subtilis* RNAP contains additional subunits ($\omega_1$, $\omega_2$ and $\delta$, cf. Fig. **4.18**). The $\delta$ subunit cooperates with the $\sigma$ subunit in transcription initiation. In fact, $\delta$ and $\sigma$ of *B. subtilis* and related bacteria may represent two functional domains of the larger type of $\sigma$ subunit encountered, e.g., in *E. coli*.

[c] Kidd, G. H. & L. Bogorad: Proc. Natl. Acad. Sci USA *76* (1979): 4890.

[d] Mitochondrial RNAP are not well characterized. The data stem from Stewart, P. R. in ref. [24], pp. 271–295.

Recently an additional subunit was detected, which combines with the core enzyme to form the termination form of *E. coli* RNAP [31]. This highly acidic subunit ($M_r$ 69 kd) is coded by the *nusA* gene. It binds, presumably site-specifically, after RNAP has started transcription and afterwards is displaced by the $\sigma$ subunit to regenerate the initiation form:

$$\underset{\text{Initiation}}{\alpha_2\beta\beta'\sigma} \xrightarrow{-\sigma} \underset{\text{Elongation}}{\alpha_2\beta\beta'} \xrightarrow[+\text{nusA}]{} \underset{\text{Termination}}{\alpha_2\beta\beta'\ \text{nusA}} \xrightarrow[+\sigma]{-\text{nusA}} \underset{\text{Initiation}}{\alpha_2\beta\beta'\sigma}$$

Enzymes structurally and functionally related to the bacterial RNAP are found in mitochondria and chloroplasts of eukaryotic cells, too (cf. Table **1.6**). In addition, there are three classes of nuclear RNA polymerases, whose major characteristics are summarized in Table **1.7**. Their complex subunit composition is strikingly similar in many eukaryotic systems analysed so far, though the molecular weights of the subunits and the stoichiometry of subunit composition evidently vary. In most cells isoforms of the three classes were found, but their biological role is unclear. Immunological investigations point out that some of the small subunits of different classes are closely related or even identical. The detailed functional analysis of the eukaryotic transcription machinery has just started with the establishment of faithful in vitro transcription systems with purified RNAP II and RNAP III from different sources (for a summary cf. [32, 33]).

### 1.3.2.2 The Transcription Cycle

Detailed investigations on the individual steps of the transcription cycle are so far limited to some few bacterial systems, mostly to *Escherichia coli*. Four subsequent phases can be discerned: (1) promoter recognition and stable RNAP binding, (2) initiation, (3) elongation and (4) termination and release [20].

**Table 1.7**: Properties of nuclear RNA polymerases (from Roeder [21])

| Characteristics | *RNAP classes* | | |
|---|---|---|---|
| | I | II | III |
| Transcripts | Pre-rRNA. (cf. Fig. **1**.12) | Pre-mRNA (cf. Fig. **1**.8) | Pre-5S-rRNA, pre-tRNA |
| Localization | Nucleolus | Nucleoplasm | Nucleoplasm |
| Subunit composition ($M_r$ in kd) | | | |
| Mouse plasmacytoma cells | 195, 117, 61[a] 49, 29, 19 | 205[b], 140, 41, 29, 27, 22, 19.5, 19, 16.5 | 155, 138, 89, 70, 53, 49, 41 32[c], 29, 19 |
| Yeast[d] | 185, 137, 48 44, 41, 36, 28 24, 20, 14.5, 12.3, 9 | 205, 145, 46, 33.5, 28, 24 18, 14.5, 12.5, 9 | 160, 128, 82 41, 37, 34, 28 24, 20, 14.5, 11, 9 |
| Catalytic properties | | | |
| $(NH_4)_2SO_4$ optima (M) | ~0.05 | ~0.10 | 0.05–0.20 |
| $Mn^{2+}/Mg^{2+}$ activity ratio | 1–2 | 5–10 | 2 |
| α-Amanitin sensitivities (μg/ml for 50% inhibition) | | | |
| Animals | Insensitive | 0.01–0.05 | 10–25 |
| Insects | Insensitive | 0.03–0.06 | Insensitive |
| Yeast | 300–600 | ~1 | Insensitive |

[a] Two types of RNAP I were found. RNAP $I_B$ lacks the subunit with 61 kd.
[b] Three types of RNAP II were found, differing in the $M_r$ of the largest subunit: 240 for RNAP $II_0$, 205 for $II_A$ and 170 for $II_B$.
[c] In type B of RNAP III $M_r$ of this subunit was 33 kd.
[d] Data compiled by Thorner, J., In: T. Leighton & W. F. Loomis (eds.). The Molecular Genetics of Development, Acad. Press, New York 1980, pp. 119–178.

1. *Promoter recognition* and stable RNAP binding: By virtue of its general binding affinity for DNA there is a perpetuated binding and release of RNAP holoenzyme. This "random hunt" for promoters leads to a "closed promoter complex" if RNAP catches up with a promoter recognition site located 30–35 bp upstream from the transcription initiation site (cf. Table **1**.8). From there binding extends to a DNA region (Pribnow box) about 10 bp upstream of the initiator base (+1). "*Open promoter complex*" formation involves the unwinding of about one helical turn of DNA (about bp −7 to +3). DNA-protein cross-linking in this complex by U.V. irradiation demonstrated the σ-subunit as binding to base residue −3 and the β-subunit as binding to residue + 3 of the non-coding DNA strand [30].
2. *Initiation:* The first nucleotide, which is almost exclusively ATP or GTP, is connected preferably with a pyrimidine nucleotide to give the start dinucleotide (5′pppPupPy-OH 3′). The initiation steps are blocked by rifampicin.
3. *Elongation:* During or shortly after chain initiation the RNAP σ-subunit is released. The core enzyme ($\alpha_2\beta\beta'$), which catalyzes chain elongation, has a much higher binding affinity for DNA than the holoenzyme ($K_D$: $2\times10^{-11}$ M for the former and $\sim10^{-7}$ M for the latter). These differences are highly relevant when considering the peculiar functions of the two enzyme forms: stable elongation complex vs. reversible promoter selection.
4. *Termination:* The RNA chain grows from 5′→3′ end untill a termination signal is reached (a cluster of T residues on the non-coding DNA strand) preceded by sequence elements that allow the non-coding DNA strand or the nascent RNA to form stem-loop structures by intra-strand base-pairing (cf. Fig. **1**.7). Termination signals cause a slow-down in the propagation

rate of RNAP (cf. Table **14**.4) after the core enzyme was modified by association with the termination subunit (product of *nusA* gene). Pausing after the base-paired region is assumed to precede a conformational transition of the RNAP and, finally, to result in the release of the enzyme and RNA from the DNA template [31].

In bacteria a proteinaceous termination factor (rho, $M_r$ 50 kd) is required for termination at weak termination sites. It forms a disc-shaped hexameric complex ($M_r$ 300 kd) binding to the nascent RNA chain and thus acquiring ATPase activity [29a]. ATP binding and hydrolysis by the rho factor are necessary for termination to occur. There is evidence that, after a site-specific primary contact of rho with the nascent RNA chain, ATP hydrolysis accompanies a progressive wrapping of the RNA around the hexameric protein. If the rho factor catches up with RNAP, slowed down at the termination signal, it could either trigger a conformational change with subsequent release of RNAP and RNA or it could pull the RNA out of the transcription complex. It is a remarkable fact that both rho-dependent terminators shown in Fig. **1**.7 (examples 2 and 3) are sites of specific termination control mechanisms (cf. **4**.4.2.7 for the *trp* operon of *E. coli* and **14**.2.3.4 for the coliphage $\sigma$ terminator $t_{R1}$). The requirement for a primary binding site for rho makes this type of termination operon-specific. By genetic evidence such sites for "rho-utilization" *(rut)* were mapped, e.g., in the middle of the coliphage $\lambda$ *cro* gene. It exhibits the following sequence (non-coding DNA strand): promoter · 115 nucl. · CCATTCATGCAGGCC · 180 nucl. · $t_{R1}$ (cf. Bektesh, S. L. & J. P. Richardson, Nature *283* (1980): 102 and Sect. **14**.2.3.4 for the role of $t_{R1}$ in the transcription program of phage $\lambda$).

### 1.3.2.3 Initiation and Termination Sequences for Transcription

A selection of prokaryotic promoters out of the great number of known sequences is presented in Table **1**.8. It is evident that there are only three regions with a certain sequence homology. These are the "-35 region" (recognition site), the "-10 region" (Pribnow box) and the transcription initiation point. In addition, a close inspection reveals a pattern of GC pairs (indicated by asterisks) separated from each other by about 10 bp. Hence, they are localized at the same face of the DNA double helix, which evidently represents the contact face for RNAP (cf. Fig. **4**.10).

By analyzing with chemical probes the base residues and phosphate groups, which are protected by RNAP binding, it was indeed shown that all DNA-RNAP contacts appear on one face of the DNA molecule (cf. the details for the promoter of the *lac* operon given in Fig. **4**.10). This face is given by the recognition site (around bp −35) and extends therefrom to the adjacent turns of the double helix, i.e., to the −45, −25 (?) and −15 regions. All known promoter mutations map in the −45, −35 and −15 to −10 regions. It is interesting to notice that only in the −10 region RNAP seemingly binds to the opposite face of the DNA strand. However, these backside contacts might reflect the DNA unwinding [30].

Our knowledge on eukaryotic promoter sequences and their possible function for RNAP binding and transcription initiation is much more restricted. In Table **1**.8 5′-flanking sequences of genes (promoters) recognized by all three types of nuclear RNAP (cf. Table **1**.7) are compared. The most striking similarities with the prokaryotic promoters are the same type of GC pair pattern at one face of the DNA helix and the initiation site with a pyrimidine preceding the purine start nucleotide. In addition, all protein-coding genes contain in their 5′-flanking regions an A-T-rich sequence (5′-TATAAA-3′, Hogness box) with a function similar to the -35 region in prokaryotic promoters. For a complete survey of homologous sequence elements so far detected in the 5′-flanking region of eukaryotic protein-coding genes see the legend to Fig. **4**.14.

Special notice deserve genes transcribed by RNAP III. This group includes genes coding for cytoplasmic tRNA's, 5S RNA's, small virus-associated RNA's as well as some short repetitive DNA sequences actively transcribed in eukaryotic cells (cf. Table **1**.1 and Sect. **4**.3.4). Comparison of the four 5′-flanking sequences given in Table **1**.8 reveals essentially no sequence homology.

**Table 1.8**: Promoter sequences in prokaryotes and eukaryotes

Sequences present the forty nucleotides of the non-coding DNA strand preceding the initiation points (+ 1). G or C residues spaced at intervals of 10 ± 1 bp are marked by asterisks. This was only done to indicate a regular pattern of base pairs at one face of the DNA double helix, which for the prokaryotic promoters was shown to be the site for interaction with RNAP (cf. Fig. **4.10**B).

The prokaryotic sequences stem from Siebenlist et al. [30]. Lipoprotein: promoter of the *E. coli* major outer membrane lipoprotein gene (> 75% A–T bp); $\lambda$ $p_R$ and $\lambda$ $p_{rm}$: $\lambda$ phage promoters (cf. Fig. **14.8**); *trp* operon: promoter of *E. coli* tryptophan operon (cf. Fig. **4.19**); $tRNA^{Tyr}$, promoter of *E. coli* suppressor $tRNA^{Tyr}$. For further promoter sequences cf. Figs. **4.10**, **4.18**, **5.9**, Table **14.3** and Sect. **14.2.3.9**. The prokaryotic consensus sequence for the RNAP recognition (RS) and binding sites (PB = Pribnow box) as well as transcription initiation sites (I) were derived from comparison of more than 50 promoters (cf. Siebenlist et al [30]).

The eukaryotic promoters are grouped into those used by *RNAP type I: Xenopus laevis*, rRNA gene, clone pXIr 14 (Sollner-Webb, B. & R. H. Reeder: Cell *18* (1979): 485); *RNAP type II:* chicken, ovalbumin gene; rabbit, β-globin gene; adenovirus II, major late promoter (all sequences taken from Lai, E. C. et al.: Cell *18* (1979): 829); rat, insulin gene II (Lomedico, P. et al.: Cell *18* (1979): 545) and *Strongylocentrotus purpuratus*, histone H1 gene (Sures, I. et al.: Proc. Natl. Acad. Sci. USA *77* (1980): 1265); *RNAP type III: Xenopus laevis*, 5S rRNA gene, clone Xlo; adenovirus II, gene coding for the VA RNAI(G); *Saccharomyces cerevisiae*, 5S rRNA gene (all three sequences taken from Korn, H. J. & D. D. Brown: Cell *15* (1978): 1145) and *Drosophila* $tRNA^{Lys}$ 4 gene (De Franco et al.: Proc. Natl. Acad. Sci. USA *77* (1980): 3365). Part of the sequences around −30 which might serve as RNAP recognition sites are printed bold-faced (group II). For genes transcribed by RNAP III accumulating evidence points at an intragenic control site for transcription initiation, whereas the 5′-flanking sequences are dispensible in this respect (cf. details given in text).

| | |
|---|---|
| *Prokaryotes* | |
| (*E. coli* RNAP) | 40 · 30 · 20 · 10 · 1 +1 |
| Lipoprotein | 5′ - AATATTCTCAACATAAAAAACTTTGTGTAATACTTGTAAC GC - 3′ |
| Phage λ, $p_R$ | GCGTGTTGACTATTTTACCTCTGGCGGTGATAATGGTTGC AT |
| Phage λ, $p_{rm}$ | GGTGTTAGATATTTATCCCTTGCGGTGATAGATTTAACGT AT |
| trp operon | GCTGTTGACAATTAATCATCGAACTAGTTAACTAGTACGC AA |
| $tRNA^{Tyr}$ | TAACACTTTACAGCGGCGCGTCATTTGATATGATGCGCCC GC |
| Consensus sequence | ---TGTTGACAATT---------------TATAAT----- C/T AC/GT |
| | RS · PB · I |
| *Eukaryotes* | |
| I *Xenopus l.*, rDNA | CTCCATGCTACGCTTTTTTGGCATGTGCGGGCAGGAAGGT AG |
| II Chicken, Ov-gene | GCTATATATTCCCCAGGGCTCAGCCAGTGTCTGT AC |
| Rabbit, β-Gl-gene | CTTGGGGGGCATAAAAGGCAGAGCAGGGCAGCTGCTGCTT AC |
| Adv II, ML-prom. | GGCTATAAAAGGGGGTGGGGGCGCGTTCGTCCTC AC |
| Rat, Insulin II-gene | TGCTTTGGACTATAAAGCTAGTGGGGATTCAGTAACTCCC AG |
| *Str. purp.*, H1-gene | CTTATATTGAGCGTTGCCGAGCCGATGGTT AT |
| III *Xen, l.*, 5S DNA | CGCTGACAAGTCAAGAAGCCGAAAAGTGCCGCTGTTCATC |
| Adv II, VA RNAI (G) | GTTGACGCTTTAGACCGTGCAAAAGGAGAGCCTGTAAGCG G |
| *Sacch. cer.*, 5S DNA | GACCCTGCCCTCATATCACCTGCGTTTCCGTTAAACTATC |
| *Dros.*, $tRNA^{Lys}$ 4 gene | AATCAGCAAATGTGAGCTTTTATCGGCGATCTTCACAAGT GC |
| | 40 · 30 · 20 · 10 · 1 +1 |

In fact, in vitro transcription experiments with rearranged DNA templates, containing partial or total deletions of the 5′-flanking sequences, showed the transcription activity of RNAP III to be barely influenced. Instead, intragenic sequences important for RNAP binding and transcription initiation have been mapped by point mutations and deletions. This is true for the 5S RNA genes of *Xenopus laevis* [32], for tRNA genes of yeast (Koski, R.A. et al.: Cell **22** (1980): **415**) and *Droso-*

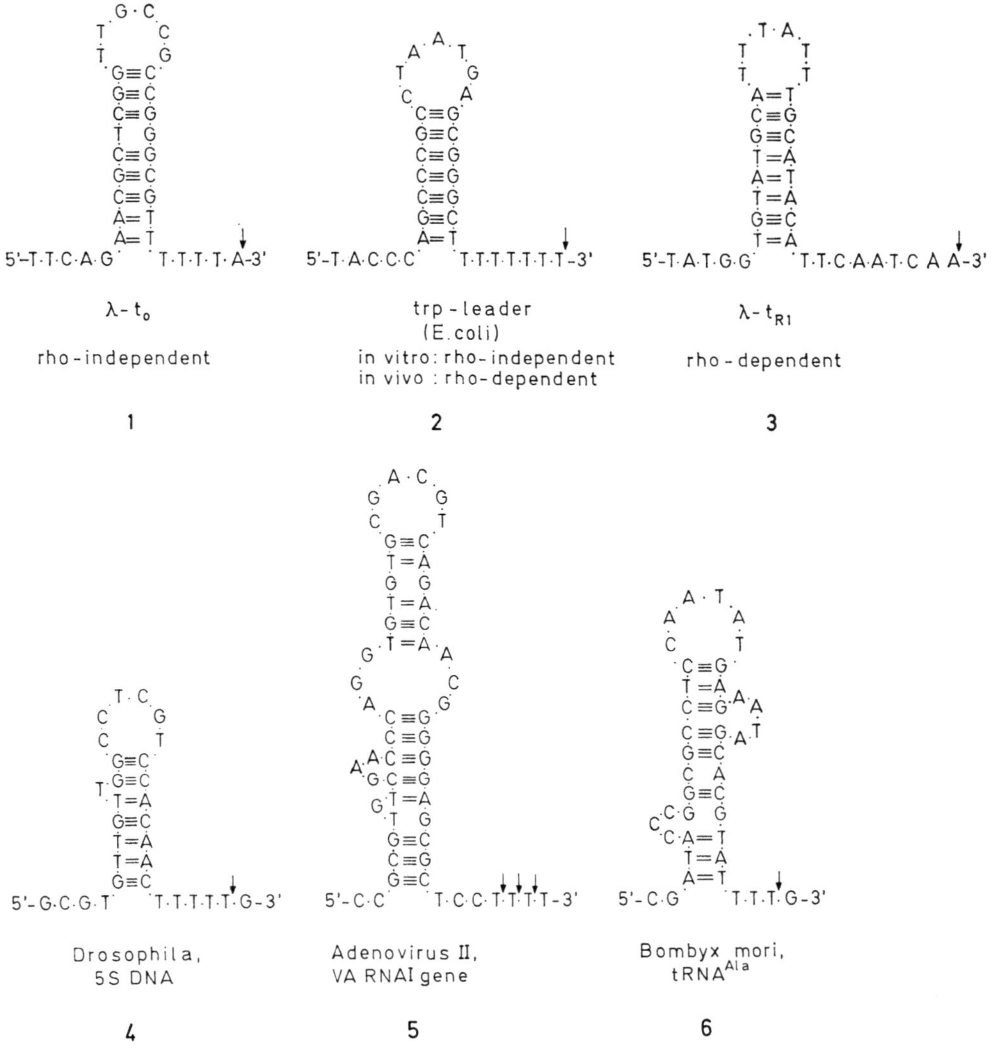


**Fig. 1.7**: Examples of stem-loop structures of the non-coding DNA strand preceding transcription termination points in prokaryotes (terminators 1–3) and eukaryotes (terminators 4–6).

Termination signals are GC-rich sequences which by virtue of rotational symmetry elements can form stem-loop structures. They are followed by a cluster of 6–8 T-residues (U-residues in the mRNA) after which transcription stops (marked by the *arrow*). For sequences 1–3 cf. Adhya & Gottesman [27] and Rosenberg & Court [28]. Genetics and function of the two terminators $t_0$ and $t_{R1}$ of phage $\lambda$ are discussed in Chap. **14**. The role of the terminator of the *trp* leader sequence is outlined in Fig. **4.19**.

For the eukaryotic terminator sequences, all three recognized by RNAP III, cf. Korn, L. J. & D. D. Brown: Cell *15* (1978): 1145 (sequences 4, 5) and Hagenbüchle, O. et al.: Cell *18* (1979): 1217 (sequence 6).

*phila* (De Franco. D. et al.: Proc. Natl. Acad. Sci. USA *77* (1980): 3365) and for the adenovirus VA RNA I gene (Fowlkes, D. M. & T. Shenk: Cell *22* (1980): 405). Sequence comparison within this class of genes led to the definition of two homology boxes of about 12 nucleotides, one 6–15 nucleotides upstream of the 5′-end and one in the middle of the gene. In addition, from *Xenopus laevis* ovaries a 37 kd transcription control protein was isolated and shown to associate with the intragenic binding region 50–80 bp upstream of the transcription initiation site (cf. Engelke, D. R. et al.: Cell *19* (1980): 717).

The termination sites of RNAP in *E. coli* (cf. Fig. **1**.7) comprise an oligo (dT) stretch preceded by a frequently GC-rich region with palindrome symmetry. Formation of a stem-loop structure is assumed either by the non-coding DNA strand (as shown in Fig. **1**.7) or by the nascent RNA strand (e.g., Figs. **4**.19 and **14**.7). For the factor rho-dependent termination sites (e.g., example 3 in Fig. **1**.7) the stem-loop structure appears less stable and the oligo(dT) stretch is reduced in length (for details of rho-dependent termination cf. **14**.2.3.4).

It is a remarkable result that the two types of eukaryotic termination signals (for RNAP I and III, respectively) analysed so far have structures very similar to the prokaryotic terminators: an oligo(dT) stretch with a GC-rich region upstream (Fig. **1**.7, examples 4 and 5). Even the formulation of putative stem-loop structures, though somewhat distorted in example 5, is possible.

Evidently the termination signals (Fig. **1**.7) reflect only one part of the information supplied by the DNA sequence in the vicinity of termination points. It was an unexpected finding that the DNA region immediately following the termination point in the tryptophane operon of *E. coli* is somehow required for proper termination [29]. Several deletions with start points 12–73 base pairs *beyond* the termination point and extending from there downstream cause 10%–95% read-through. Thus, the list of features used to specify termination signals must be extended to include DNA regions many base pairs downstream from the actual termination site. Possible explanations for these results may be looked for (1) in the size of RNAP which is able to cover 60–75 base pairs of DNA, (2) in the participation of an additional termination protein that binds downstream of the termination point or (3) in effects that were described as telestability, i.e., DNA conformation at a given site is dependent on the base composition of remote portions of the double helix.

### 1.3.3 Transcription Products and RNA Processing

Only in rare cases can RNA exert its biological functions in the form generated by the transcription apparatus (primary transcript). It is rather the rule that transcription is followed by a more or less elaborate and highly specific processing of the primary transcript which involves endonucleolytic removal of excess sequences, terminal or interspersed (intervening sequences), endonucleolytic cutting of multicistronic RNA species, additions to the 5′- and/or 3′-ends and multifold modifications of nucleotides. The specificity of the overall process results only partly from the enzymes involved. In addition, RNA shows distinct elements of a three-dimensional structure and is always associated with structural and regulatory proteins forming RNP-particles.

#### 1.3.3.1 Pre-mRNA Processing

A typical feature of prokaryotic gene expression is the coupling of transcription and translation in a giant complex of DNA, the nascent mRNA chain and numerous proteins (cf. Figs. **4**.11 and **4**.19). In this case mRNA serves as template for translation without further processing. However, there are experimental findings that an endonucleolytic cleavage of polycistronic mRNA by RNase III contributes to or may

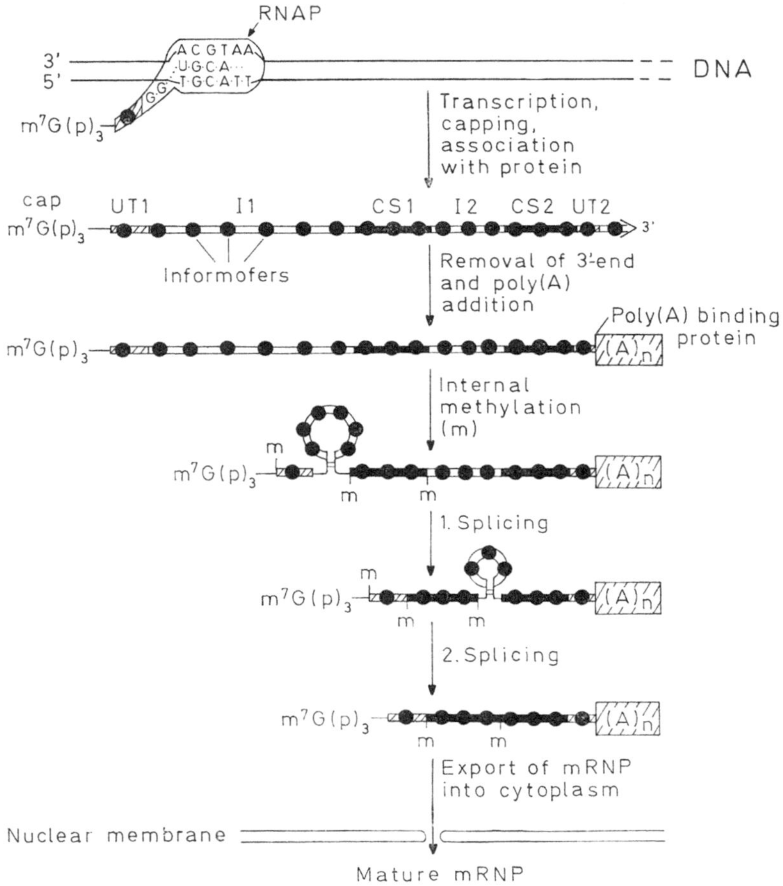


**Fig. 1.8:** Scheme of nuclear pre-mRNP processing.

Addition of the cap at the 5'-end of the nascent RNA and association with the major structural proteins ($M_r$ 30–50 kd) of hnRNP particles (informofers) proceed already co-transcriptionally. The hypothetic pre-mRNP released from the transcription complex contains (from 5' → 3'-end) the cap ($m^7G(p)_3$, cf. Fig. 1.9), an untranslated sequence (UT1), two intervening sequences (introns I1 and I2), two coding sequences (CS1 and CS2) and an untranslated sequence UT2 plus an excess RNA piece at the 3'-end. All parts of the sequence retained in the mature mRNA are called exons, i.e., the two untranslated sequences (UT1 and UT2) plus the two coding sequences (CS1 and CS2). Further processing is assumed to involve (1) removal of excess RNA at the 3'-end, addition of 50–200 AMP residues and binding of the poly(A)-specific protein ($M_r$ 75 kd), (2) methylation of internal adenosine residues in 6-position as well as eventual methylation of the 5' penultimate nucleotide in the 2'-position and (3) two splicing events to remove the intervening sequences.

even be a prerequisite for the efficient translation of some prokaryotic messengers, too (cf. Table **4**.9 and Sect. **14**.2.2.4).

In contrast to prokaryotic systems, the multistep processing of eukaryotic mRNA is the rule. The principles are shown in Fig. **1**.8 and discussed in more detail in the following (cf. the summaries in Ref. [35–37] and [39]).

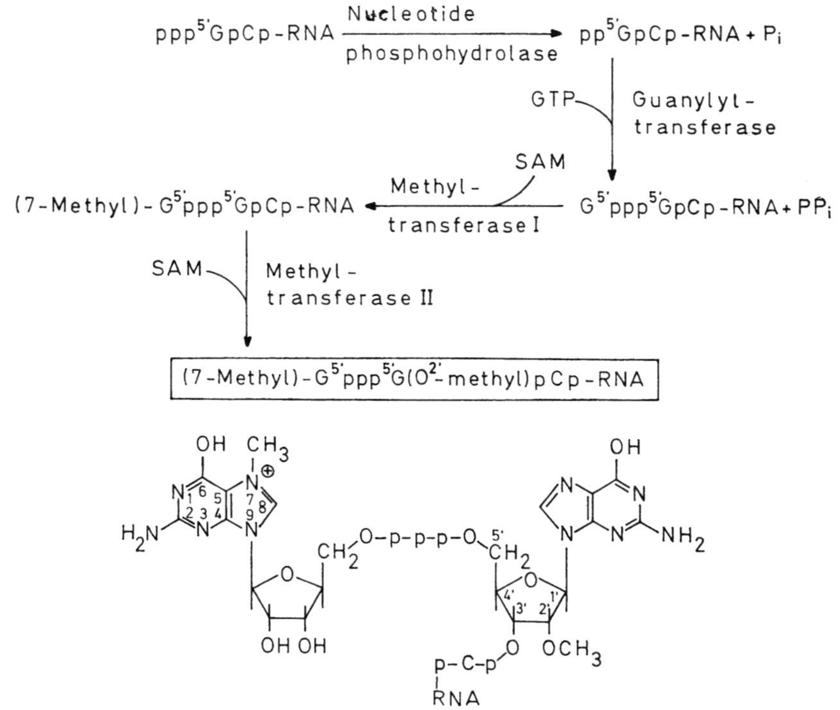


**Fig. 1.9:** Major pathway of cap-formation in animal cells (from Furuichi & Shatkin [42], modified). The eukaryotic mRNP cap is a 7-methyl-guanosine residue bound 5′ → 5′ by a triphosphate bridge to the 5′-end of the RNA. The second methylation of the penultimate nucleotide in 2′-position is not obligatory. SAM, S-adenosyl methionine. The two nucleotides at the 5′-end are not necessarily G and C as found in this example of reovirus mRNA.

*Pre-mRNA Associated Proteins:* During transcription the nascent pre-mRNA chains are complexed with proteins to generate the so-called heterogeneous nuclear ribonucleic acid-protein complexes (hnRNP). It seems that in this way the cell can handle (by condensing) and protect (by covering with protein) long, unstable RNA while still leaving it available for further enzymatic modification. Additional stabilization of the pre-mRNP is presumably obtained by binding to nuclear skeletal structures thought to consist of a sponge-like protein matrix connected with the nuclear membrane[3]. The nuclear skeleton may be also a major factor for the ordered processing and transport of the mRNP complexes through the nuclear pores to the cytoplasm.

Depending on the size of the primary transcript intact hnRNP complexes have sedimentation coefficients up to 400S. They exhibit a polysome-like appearance in the electron microscope with subunits of $20 \times 8$ nm. The buoyant density of these 30S–40S subparticles (1.4 g/cm$^2$ in CsCl gradients) indicate a protein to RNA ratio of 3 : 1.

A few major proteins with $M_r$ 30–50 kd form the core of the hnRNP subparticles (informofers) that contain about 400–900 nucleotides of RNA [36, 37, 40]. Six proteins in a simple stoichiometric ratio constitute the protein cores of hnRNP: two highly conserved basic proteins with $M_r$ 32

[3] Long, B. H. et al.: Cell *18* (1979): 1079; Clark, T. G. & J. L. Rosenbaum: Cell *18* (1979): 1101.

and 34 kd making about 60 % of total proteins, two methylated basic proteins with $M_r$ 36 and 37 kd and two acidic proteins with $M_r$ 42 and 44 kd. In addition, the intact pre-mRNP complex contains numerous minor proteins including poly(A) polymerase, enzymes of the splicing complex, methylases. A special protein ($M_r$ 75 kd) is found associated with the poly(A)-stretch added post-transcriptionally to the 3′-end. Except this poly(A)-binding protein, most or all other protein components are exchanged upon transport of the completed mRNP from the nucleus into the cytoplasm. Immunological evidence point out that poly(A) polymerase and the 75 kd poly(A)-associated protein may in fact be identical.

*Cap Formation:* Analysis of the 5′-terminus of eukaryotic mRNPs has revealed in almost all cases the presence of a blocked structure, $m^7G5'ppp5'N$, referred to as "cap" (cf. Fig. 1.9). Cap formation seems to be an early event, occurring on the RNA while transcription proceeds (cf. Fig. 1.8). It is possible that early capping protects the nascent pre-mRNA chains from 5′-exonucleolytic degradation.

In many instances caps are conserved in the corresponding cytoplasmic mRNP molecules. The preservation of the 5′-terminal segments is an indication that the 5′-end is also conserved and that the cap is important for mRNA transport, translation and/or stability [36]. On the other hand it is known that some animal and plant viral RNAs are not capped, e.g., RNA of polyoma, picorna and cowpea mosaic viruses. In the picorna viruses (cf. Fig. 4.32) the RNA present in the virion carries a protein covalently bound to the 5′-terminus, while picorna virus-specific mRNA found in polyribosomes has neither protein nor cap at its 5′-end.

*Poly(A) Addition:* To the 3′-ends of most, but not all (e.g., histone mRNA) pre-mRNA molecules AMP residues are added by nuclear poly(A) polymerases. Twenty to 200 adenyl residues were reported (cf. Fig. 1.8). In most cases polyadenylation

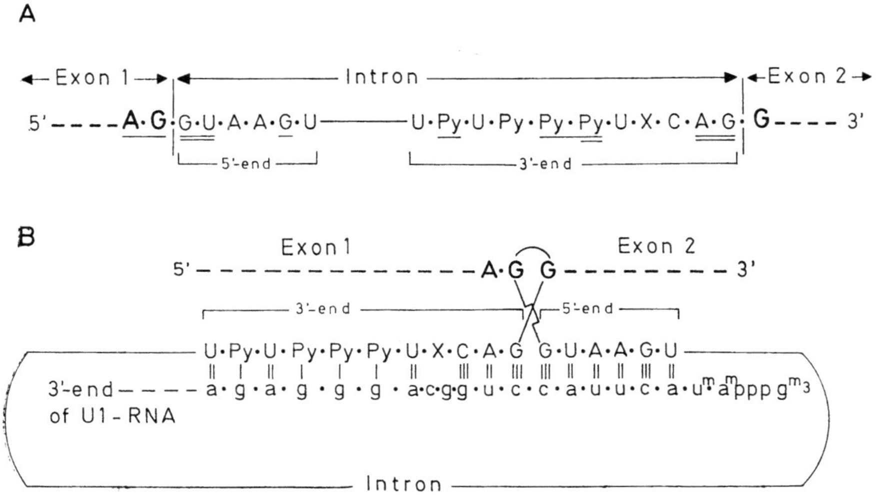


**Fig. 1.10**: Possible role of snRNP containing U1-RNA for RNA splicing (from Lerner et al. [38], modified).

**A** Generalized sequence at exon-intron borders as deduced from comparison of about 40 pre-mRNA sequences. The indicated bases are those most common in the particular position occurring with a frequency of at least 45 %, underlined bases with 75 % and twice underlined with 95 % or more. Py, pyrimidines (U or C); X, marks the single non-conserved base in the sequence.

**B** Model of interaction by base-pairing between the 5′-end of U1-RNA and the bases of the exon-intron border. The U1-RNA sequence is indicated with low case letters. It contains a highly methylated 5′-end with a cap-like structure formed by 2,2,7-trimethyl guanosine. U1-RNA is hydrogen-bonded to the nucleotides situated at the 3′- and 5′-borders of the intron (nucleotides denoted with capital letters). The AG 3′-end of exon 1 and the 5′-end (G) of exon 2 to be joined by the splicing event are indicated by *bold-faced letters.*

occurs after transcription was terminated and excess RNA at the 3'-terminus, if present, was removed. Poly(A) polymerase catalyzes the following reaction:

$$\text{Primer-RNA-3'-OH} + \text{nATP} \xrightarrow[\text{Mg}^{2+}]{\text{Poly(A)-P}} \text{Primer-RNA-(AMP)}_n + \text{nPP}_i$$

Experimental results indicate that the poly(A)-protein complex at the 3'-end may function in the attachment of mRNP to nuclear and cytoplasmic elements of the cytoskeleton and in the control of mRNP half-lives (cf. **4**.4.3.2). A characteristic hexanucleotide (5'-AAUAAA-3') is localized about 10 nucleotides upstream from the poly(A) addition site and might serve as recognition signal for poly(A) polymerase.

*Methylation:* Besides the methylation of the cap-structure there are few internal adenosine residues methylated in 6-position (Fig. **1**.8). In HeLa cells the methylated sequence was found to be pPupApC. In analogy to the situation for rRNA a role of methylation for site-specific endonucleolytic processing is assumed.

*Splicing:* One of the most startling peculiarities of many eukaryotic genes is the discontinuous arrangement of the nucleotide sequences coding for mRNA [35, 38]. Hence, mRNA processing involves removal of the so-called intervening sequences (or introns) and covalent jouning of the coding sequences (or exons). When comparing the partial sequences around the splicing points of different pre-mRNP (cf. Fig. **1**.10) a striking homology with the 5'-end of certain low molecular weight RNAs found as snRNPs in eukaryotic nuclei was detected. It appears reasonable to assume a guide role of these small nuclear RNPs in directing site-specific splicing (Fig. **1**.10). snRNPs may even contain the necessary cleaving and joining enzymes (cf. **4**.4.3.1).

### 1.3.3.2 Processing of rRNA and tRNA Precursors

Ribosomes contain three and four types of RNAs respectively (cf. Table **1**.9) which serve structural and catalytic functions in the translation cycle. In bacteria the three types of rRNAs (5S, 18S and 23S rRNA respectively) are transcribed from a single transcription unit that, in addition, contains varying numbers and species of tRNA genes (Fig. **1**.11 A). The *rrnX* operon of *E. coli* is one out of seven unlinked rRNA

---

▶

**Fig. 1.11:** Nucleolytic processing of rRNA precursors in prokaryotic and eukaryotic cells.

**A** *Escherichia coli* rRNA genes (cf. Nomura et al. [43] and Abelson [35]): The *E. coli rrnX* operon contains from left to right: two tandem promoters ($P_1$, $P_2$), the 16S rRNA gene, a spacer with two tRNA genes, the 23S rRNA gene, the 5S rRNA gene and two further tRNA genes at the 5'-end. The putative polycistronic precursor RNA (30S) can be isolated from mutants defective in RNase III and RNase E functions, only. Processing normally starts already in statu nascendi. Cleavage by RNase III requires formation of a dsRNA stem with a giant loop containing the entire 16S and 23S rRNA respectively. Intermediates of the primary cleavage events are further acted on by several RNases as indicated at the *arrows* (M16, M23, D, P, E, X). RNases M16 and M23 evidently generate the mature 5'-ends of the 16S and 23S rRNA respectively, whereas a special, unknown enzyme "X" is needed for the 3'-end. More or less extensive base modifications accompany the synthesis and nucleolytic processing of all RNA species. They are not included into the scheme.

**B** rRNA processing in eukaryotes (HeLa cells, from Perry [34]): In contrast to the prokaryotes the rRNA transcription unit codes only for the three rRNA genes (5.8S, 18S and 28S rRNA). The polycistronic primary transcript is a regular intermediate. *1*, *2*, *3* and *4* represent endonudeolytic cleavage sites of the precursor molecule. After the first step two alternative processing routes are possible. In addition to the 5.8S RNA, which is equivalent to the prokaryotic 5S rRNA, eukaryotic ribosomes contain another small RNA (5S). It is coded by an unlinked non--nucleosomal locus that usually contains many tandemly repeated 5S RNA genes. o, RNA polymerase; p denotes precursors of the mature RNA characterized by their sedimentation coefficients (S).

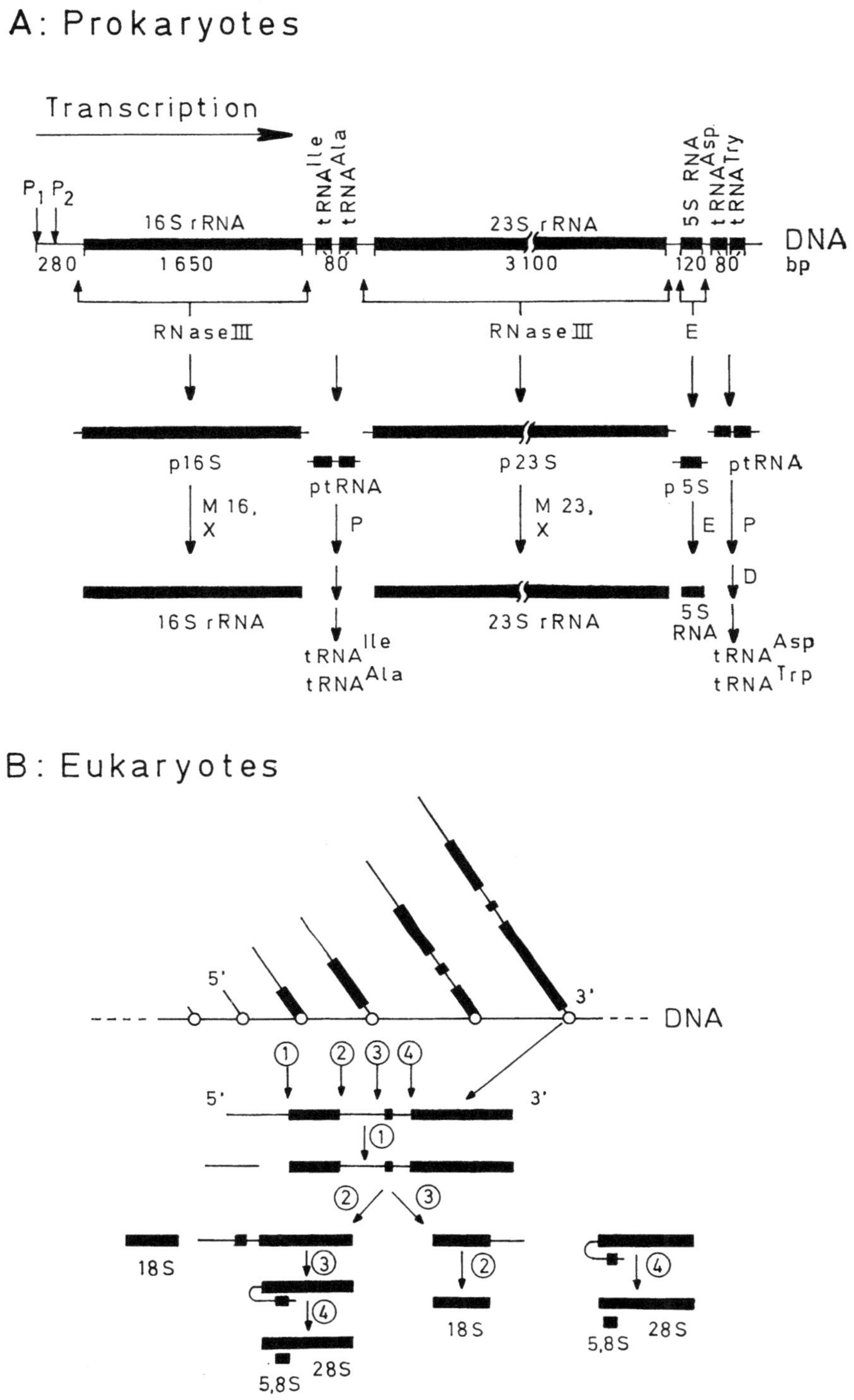
A: Prokaryotes
Transcription
P1 P2
16S rRNA
tRNA Ile
tRNA Ala
23S rRNA
5S RNA
tRNA Asp
tRNA Try
DNA
280 1650 80 3100 120 80
bp
RNase III
RNase III
E
p16S
ptRNA
p23S
p5S
ptRNA
M 16, X
P
M 23, X
E
P
D
16S rRNA
tRNA Ile
tRNA Ala
23S rRNA
5S RNA
tRNA Asp
tRNA Trp
B: Eukaryotes
5'
3'
DNA
① ② ③ ④
5'
3'
①
② ③
18S
③
②
④
④
18S
28S
5,8S
28S
5,8S

Fig. 1.11

operons. Processing of the primary transcript by endonucleolytic cleavage and enzymatic base modifications starts already during transcription so that normally no high molecular weight precursor can be detected.

An important step in the co-translational processing is the RNase III catalyzed site-specific excision of the two pre-rRNAs. It appears that, to accomplish this, dsRNA hairpins with giant loops containing the whole rRNA sequences are formed. Thus, for base-pairing in the hairpins base residues separated by 1,600 and 2,800 nucleotides come together [35, 44]. RNase III-like enzymes have also been characterized from animal cells [46].

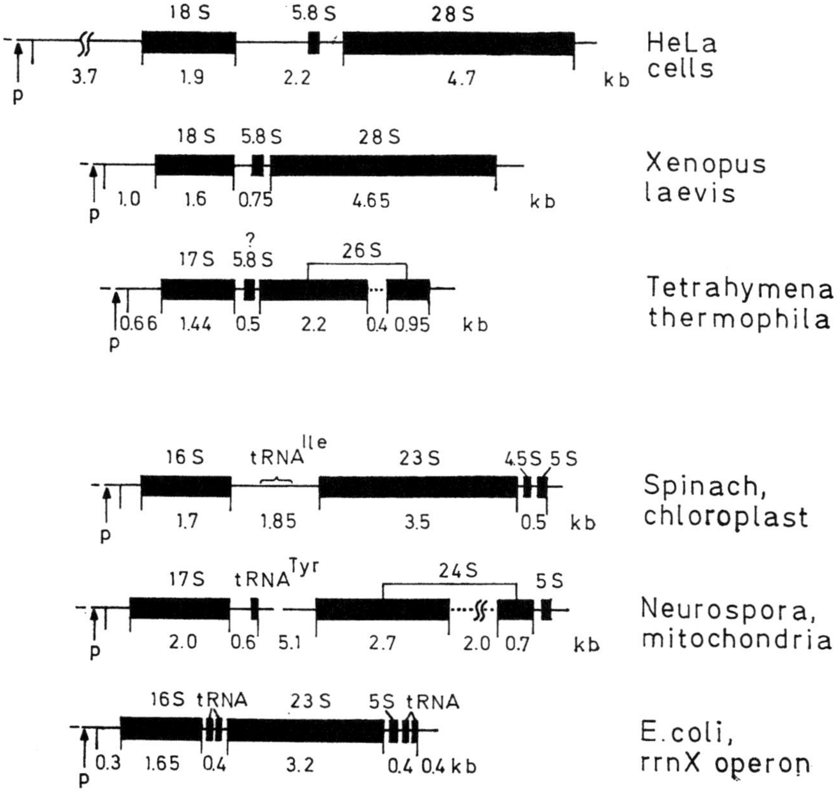


**Fig. 1.12:** Physical maps of transcription units coding for rRNA.

*Solid boxes* denote the coding sequences for the large rRNA (23–28S), for the small rRNA (16–18S), for 4.5S, 5S and 5.8S RNA as well as for tRNA's. *Lines* represent transcribed spacer RNA that is removed during processing of the RNA precursor molecules. *Dotted lines* indicate intervening sequences in some cases interrupting the large rRNA coding sequence. Sizes of genes and spacers are given in kb. Promoters (p) are on the *left site.*

Data were obtained from the following sources: HeLa cells, nucleolus: from Perry [34]; *Xenopus laevis,* nucleolus: from Wellauer, P. K. & I. B. Dawid, J. Mol. Biol. *89* (1974): 379 and Boseley, P. et al., Cell *17* (1979): 19; *Tetrahymena thermophila,* nucleolus: from Dinn, N. et al., Cell *18* (1979): 525 and Cech, T. R. & D. C. Rio, Proc. Natl. Acad. Sci. USA *76* (1979): 5051; spinach, chloroplast: from Driesel, A. J., Gene *6* (1979): 285; *Neurospora crassa,* mitochondria: from Heckman, J. E. & U. L. Raj Bhandary, Cell 17 (1979): 583. Besides the $tRNA^{Tyr}$ 9 further tRNA genes have been mapped in the 5.7 kb spacer between the small and large rRNA. There is no indication so far that the total gene complex is read as a single trancsription unit. *E. coli, rrnX* operon is one of the 9 rRNA coding operons. Its fine structure (cf. Fig. 1.11A) can be given as following: promoter – 320 bp – 16S rRNA – 70 bp – $tRNA^{Ile}$ – 30 bp – $tRNA^{Ala}$ – 130 bp – 23S rRNA – 100 bp – 5S RNA – 30 bp – $tRNA^{Asp}$ – 10 bp – $tRNA^{Trp}$ – ~400 bp.

In contrast to prokaryotes, eukaryotic rRNA synthesis always involves a high molecular weight precursor. The example given in Fig. **1.11** B reflects the pre-rRNA processing in HeLa cell nuclei, but basically similar schemes were reported for many other systems including fungi and plants. The major differences concern the sizes of the primary transcript and hence of the processing intermediates. Fig. **1.12** summarizes schematically the structural data of some well known rRNA transcription units of prokaryotic and eukaryotic cells.

In all cells the final steps of rRNA maturation proceed in rRNP particles, so-called pre-ribosomes. Selectivity of the limited endonucleolytic cleavages is essentially influenced by the rRNA secondary structure, the methylation pattern and the association with ribosomal proteins. Details of this coordinate ribosome assembly are just emerging.

As shown in Figs. **1.11** and **1.12**, tRNA genes in bacteria as well as in chloroplasts and mitochondria may be part of the rRNA transcription units. But most of them, especially those coded by eukaryotic nuclear genes, form separate transcription units. In bacteria multicistronic tRNA precursors with two (cf. Fig. **1.11** A) to seven (*E. coli* phage T4 [35]) different tRNA species have been found. A prominent enzyme of bacterial tRNA processing is RNase P which requires an internal RNA as necessary component for its activity [45]. It is tempting to infer a role as guide RNA (cf. **1.3.3.1**)

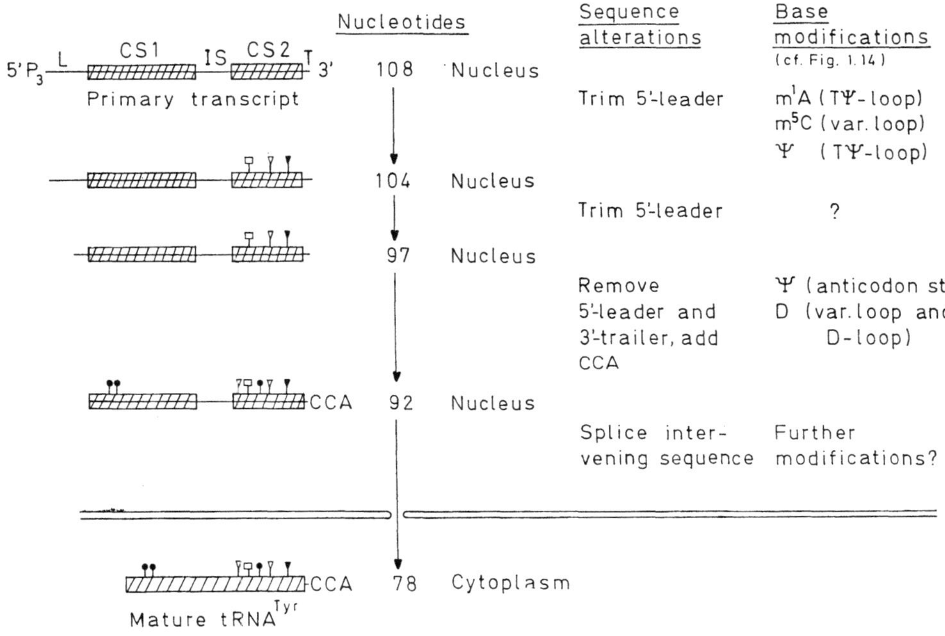


**Fig. 1.13**: Order and intracellular sites of processing of yeast pre-$tRNA^{Tyr}$ in *Xenopus* oocytes (from Melton et al. [49], modified).

Cloned $tDNA^{Tyr}$ of *Saccharomyces cerevisiae* and [$^{32}$P]-GTP were injected into nuclei of *Xenopus* oocytes (about 5 ng per oocyte, cf. **2.8**). After 6 h labelling RNA was extracted and purified by gel electrophoresis. The primary transcript contains besides the two coding sequences (CS1, CS2) a 5′-leader (L, about 19 nucleotides), an intervening sequence adjacent to the anticodon (IS, 14 nucleotides) and a 3′-trailer (T, 2–3 uridins). The processing steps follow a strict program detailed on the *right site* of the figure. Only some of the base modifications have been analysed by RNA fingerprints of the precursor and processing intermediates: ▼, 1-methyladenosine; □, 5-methylcytosine; ▽, pseudouridine; ●, dihydrouridine.

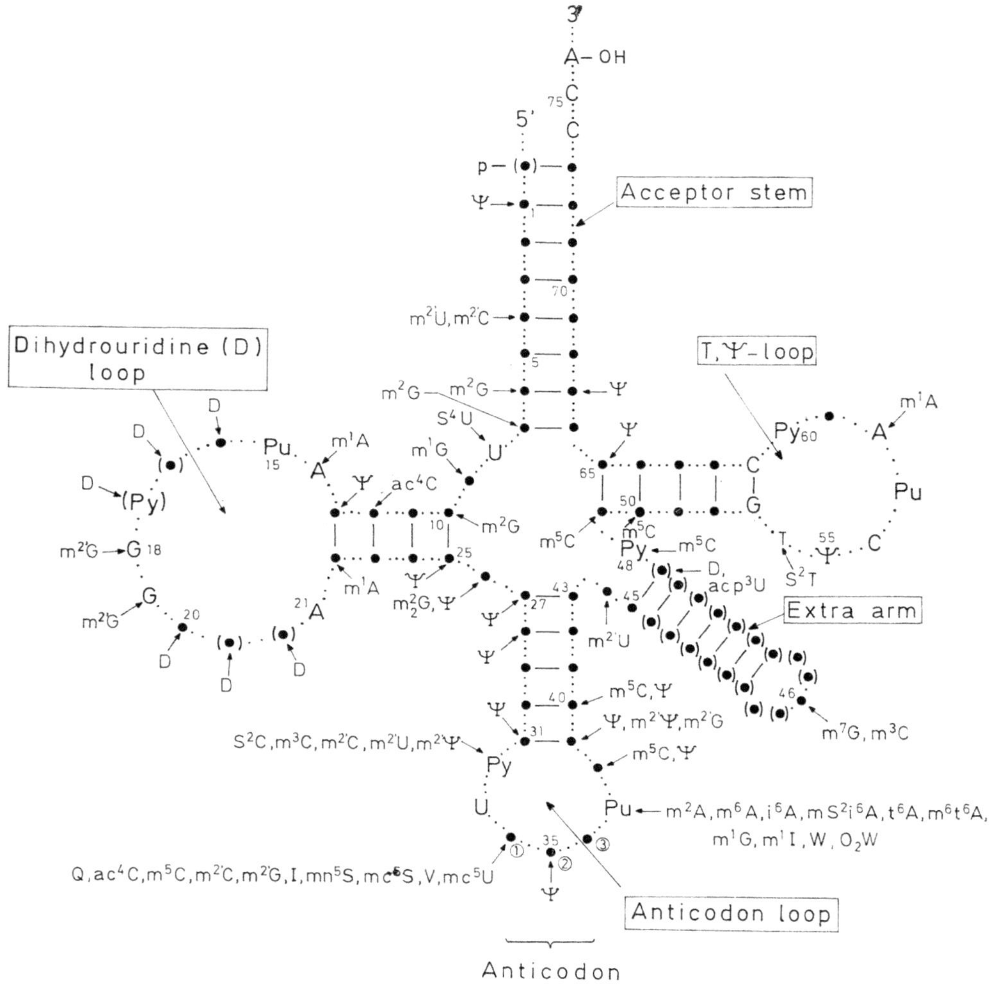


**Fig. 1.14**: Basic two-dimensional formula of tRNA (clover-leaf model) (from Sodd, M. A.: CRC Handbook of Biochem. Mol. Biol., Nucleic Acids II, CRC Press, Cleveland, Ohio 1976, pp. 423–456, and Gauss, D. H. et al.: Nucleic Acids Res. *6* (1979) r1).

The model shows the base-paired stem and the loop regions of the clover leaf. Numerous bases undergo postsynthetic modifications, most extensively in the anticodon loop (positions 32, 34 and 37). Nucleotide 34 represents the first base of the anticodon which allows a certain flexibility in the base-pairing with the third position of the codon (Wobble base).

Nucleotides in brackets are not obligatory in all tRNA (cf. especially the extra arm which may be totally or partially lacking). Symbols in the nucleotide chain denote: ●, adenosine, guanosine, cytidine, uridine; *Pu*, always purin riboside; *Py*, always pyrimidine riboside; *A*, *G*, *C*, *T*, *U*, *ψ*, always adenosine, guanosine, cytidine, thymine, uridine and pseudouridine resp. The numbering system follows that used by Gauss et al. (cf. also Fig. 1.15). Most of the non-obligatory nucleotides are not counted.

for ensuring site-specific cleavage. Another enzyme, a 3′-exonuclease (RNase D) is involved to generate the mature 3′-end.

Two types of tRNA precursor genes are found: type I, which prevails in *E. coli*, contains the 3′-CCA end in the coding sequence, whereas type II is encountered in most eukaryotic cells but also in *E. coli* phage T4. It does not encode the 3′-CCA end, which is added postsynthetically by terminal nucleotidyl transferase (cf. Fig. 1.13).

In addition to the nucleolytic reactions, pre-tRNA processing is dominated by elaborate base modifications (cf. legend to Fig. 1.14), whose enzymatic bases and structural requirements for sequence specificity are largely unknown. Practically all modified bases are thought to arise by transformation of parental nucleosides originally present in tRNA precursors. A clear exception is the hypermodified base Q found in the first position of the anticodons of $tRNA^{Tyr}$, $tRNA^{His}$, $tRNA^{Asn}$ and $tRNA^{Asp}$ of most, if not all, organisms. It is unique in that its purine skeleton itself is modified to a 7-deaza structure. The Q base (cf. formula in the legend to Fig. 1.14) can be further modified by mannose or galactose residues attached to the 4- or 5-position of the cyclopentenediol ring. Such glycosylations are frequently observed in animal tumor cells, but not in normal cells. The generation of Q does not proceed by in situ modification, but rather by insertion of queuine

---

◀

The following symbols are used for the modified bases in Fig. 1.14:

$m^1A$, 1-methyladenosine
$m^2A$, 2-methyladenosine
$t^6A$, $N^6$-threoninocarbonyladenosine
$m^6t^6A$, $N^6$-methyl-$N^6$-threoninocarbonyladenosine
$i^6A$, $N^6$-(Δ2-isopentenyl) adenosine
$mS^2i^6A$, $N^6$-(Δ2-isopentenyl)-2-methylthioadenosine
$m^6A$, $N^6$-methyladenosine
$S^2C$, 2-thiocytidine
$m^3C$, 3-methylcytidine
$ac^4C$, $N^4$-acetylcytidine
$m^5C$, 5-methylcytidine
$m^{2'}C$, 2′-methylcytidine
$m^1G$, 1-methylguanosine
$m^2G$, $N^2$-methylguanosine
$m_2{}^2G$, $N^2$,$N^2$-dimethylguanosine
$m^7G$, 7-methylguanosine
$m^{2'}G$, 2′-methylguanosine
I, inosine
$m^1I$, 1-methylinosine
Q, queuosine, 7-(3,4-trans-4,5-cis-dihydroxy-1-cyclopenten-3-ylaminomethyl)-7-deazaguanosine (cf. formula)
$mc^5S$, 5-(methoxycarbonylmethyl)-2-thiouridine
$mn^5S$, 5-(methylaminomethyl)-)2-thiouridine
T, thymidine
$S^2T$, 2-thiothymidine
$acp^3U$, 3-(3-amino-3-carboxypropyl)-uridine
$S^4U$, 4-thiouridine
$mc^5U$, 5-(methoxycarbonylmethyl)-uridine
V, 4-carboxymethoxy-uridine
$m^5U$, 5-methyluridine
D, dihydrouridine
$m^{2'}U$, 2′-methyluridine
ψ, pseudouridine
$m^{2'}ψ$, 2′-methylpseudouridine
W, wybutosine (cf. formula)

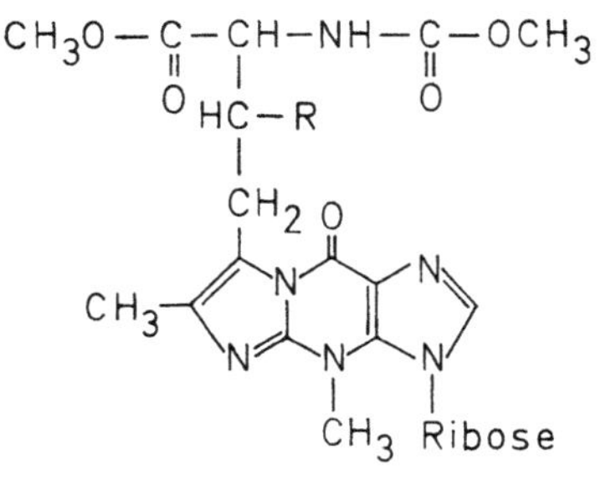


R:H=Wybutosine
R:OOH=Wybutoxosine ($O_2W$)

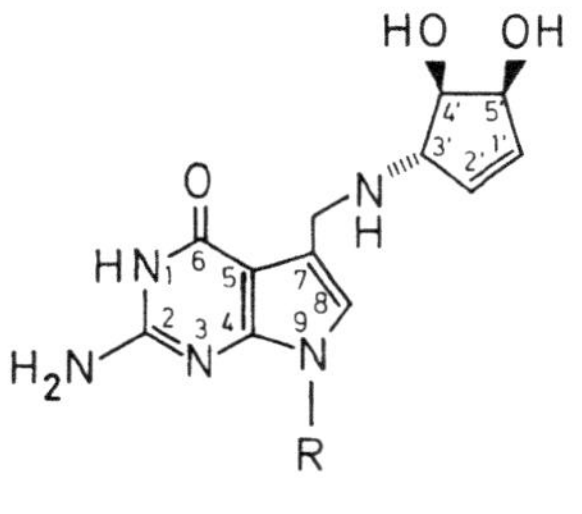


R: H = Queuine
R: Ribose = Queuosine (Q)

or, in *E. coli*, of 7-(aminomethyl)-7-deazaguanosine into pre-tRNA in place of guanine originally present in this position [49a]. The transglycosylating enzyme (tRNA-guanine transferase) cleaves the N-glycosyl bond and replaces guanine by preformed queuine and its precursor respectively. This modification of the pre-tRNA polynucleotide chain by transglycosylation without breakage of the phosphodiester backbone is a novel type of processing reaction. It will be interesting to reinvestigate whether other strongly modified bases are also generated by similar base exchange.

Much less is known about pre-tRNA processing in eukaryotes. RNase P-like enzymes have been characterized from different animal sources [46, 48]. Considerable interest have evoked some yeast tRNA genes, which contain shorter or longer intervening sequences. The subtle interplay of nucleolytic cleavages, splicing and base modifications in the processing program is evident from the scheme in Fig. **1**.13 showing the maturation of such a yeast pre-$tRNA^{Tyr}$ in the heterologous system of *Xenopus* oocytes (cf. **2**.8). Its intervening sequence adjacent to the anticodon is removed as a very late step of processing only, immediately before the mature tRNA enters the cytoplasm. Evidently the *Xenopus* system is capable of accurately transcribing the yeast gene for $tRNA^{Tyr}$ and processing its precursor on a pathway very similar to that observed in the yeast cell itself.

### 1.3.4 Translation

Biochemistry of translation and the basic structure of the translation machinery are highly conserved between prokaryotic and eukaryotic cells. Non-ribosomal steps (amino acid activation and selective transfer to the correct tRNA) and ribosomal steps (initiation with mRNA binding, elongation and termination with release of the completed polypeptide chain) can be distinguished.

#### 1.3.4.1 tRNA Structure and Function

Mature tRNAs consist of 73 to 88 nucleotides, about 20% of which are modified (Fig. **1**.14). The particular base sequences of all known tRNAs allows the molecule to fold back in itself forming a clover leaf-like structure with a number of basepaired stems and loops (Fig. **1**.14). As evident from X-ray analysis, this clover leaf is organized in a three-dimensional structure (L-form, Fig. **1**.15) with the anticodon loop at the end of the long and the CCA acceptor stem at the end of the short arms. The correct three-dimensional structure as determined by the base sequence and modification pattern is crucial for the multifold functions of the tRNAs in the translation cycle. These are:

- the highly specific recognition by the corresponding tRNA synthetase for charging with the correct amino acid,
- interaction with the ribosome acceptor site (A-site) in dependence of the codon presented by mRNA (Fig. **1**.19),
- participation in the general reactions of the ribosomal elongation cycle, i.e., in peptidyl transfer, translocation from the A-site to the P-site and release of the uncharged tRNA (Fig. **1**.19).

It is remarkable that all organisms use a specialized tRNA for initiation of protein synthesis: $tRNA_f^{Met}$. Despite significant differences in their primary structures, prokaryotic and eukaryotic initiator tRNA are readily interchangeable in in vitro systems. Hence they constitute a distinct functional class of tRNA distinguished from the chain-elongating tRNA in several respects, perhaps the most striking of which is the direct interaction with the ribosomal P-site while all chain elongating tRNA bind initially to the A-site (cf. Sect. **1.3.4.3**). Analysis of the three-dimensional structure of E. coli $tRNA_f^{Met}$ [54] revealed substantial differences with respect to the L-form shown

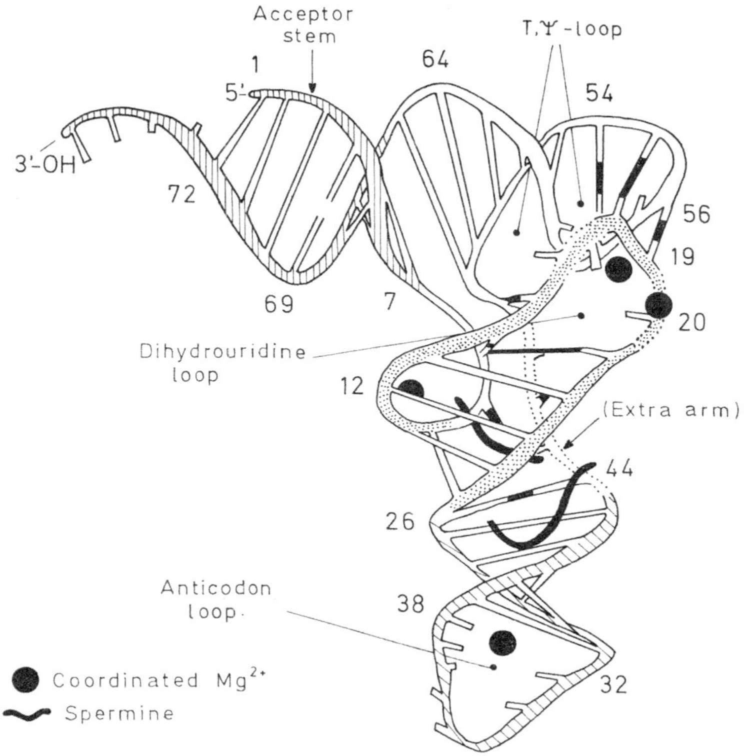


**Fig. 1.15**: Three-dimensional structure of tRNA (yeast $tRNA^{Phe}$) (from Kim [52] and Rich et al. [53]).

The spatial model was derived from X-ray difraction patterns. The L-form results from the clover leaf (Fig. **1.14**) by folding back the T,ψ- and dihydro-uridine loops. It is stabilized by "secondary" hydrogen bonds *(black cross rungs)* and by the coordinate binding of $Mg^{2+}$ and **fspermine.**

or the elongating $tRNA^{Phe}$ of yeast (Fig. **1.15**): (1) In the initiator tRNA, the last five nucleotides at the 3′-end curl back toward the acceptor end, rather than continuing the helical part of the acceptor stem, as in yeast $tRNA^{Phe}$. This shows the conformational flexibility of this part which may be of crucial importance for the biological function in peptidyl transfer. (2) A second deviation concerns the conformation of the anticodon loop, in particular the position of the constant uridine residue adjacent to the anticodon (cf. Fig. **1.14**). In the initiator tRNA the base projects away from the loop while in the elongating yeast $tRNA^{Phe}$ it points into the loop (Fig. **1.15**). It is hypothesized that a swivel of uridine 33 might characteristically accompany the translocation of the peptidyl tRNA from the A- to the P-site (cf. Sect. **1.3.4.3**). Thus, both differences detected between the two types of tRNA, reveal points of conformational flexibility which may be essential for the correct steric accomodation of charged tRNAs that participate in the ribosomal elongation cycle (Fig. **1.19**).

For incorporation of the 20 proteinogenous amino acids bacteria contain about 60 different tRNA species. Due to the organelle-specific protein synthesis apparatus in eukaryotic cells even 80–100 different species are found. Different tRNA species transferrring the same amino acid within one cell are called isoacceptor tRNA. tRNA makes about 10 % of total cellular RNA with about $10^5$ and $7.5 \times 10^7$ molecules in *E. coli* and rat liver cells respectively.

The charging of tRNA by the aminoacyl-tRNA synthetases (L-amino acid: tRNA ligases (AMP)) proceeds in two steps with enzyme-bound aminoacyl AMP as inter-

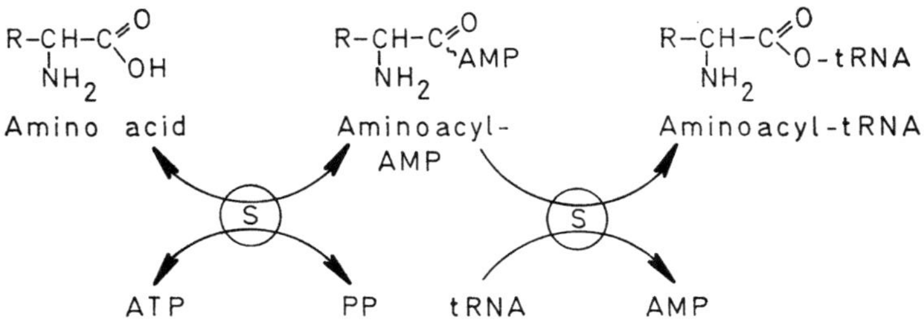


**Fig. 1.16**: Two steps of amino acid activation and charging of tRNA.
Both steps are catalyzed by the same aminoacyl-tRNA synthetase (*S*) that is specific for the particular amino acid and its cognate tRNA. The aminoacyl residue is bound either to the 2'- or 3'-hydroxy group of the 3'-terminal adenosine residue of the tRNA [76].

mediate (Fig. **1**.16). The high specificity of the reaction results from the interaction of aminoacyl AMP, tRNA and synthetase in a ternary complex.

The aminoacyl-tRNA synthetases, although catalyzing the same type of reaction, are structurally unrelated. They are composed of one or two types of subunits ($\alpha$ and $\beta$) and exist as monomers ($\alpha$), dimers ($\alpha_2$ or $\alpha\beta$) and tetramers ($\alpha_2\beta_2$ or $\alpha_4$) with $M_r$ ranging from 60 to 280 kd. Only one synthetase type for each amino acid is found in bacteria, whereas eukaryotic cells contain different sets of synthetases [57] in cytoplasm, mitochondria and, in plants, also in chloroplasts (cf. Chap. **6**). Furthermore, eukaryotic cytoplasmic synthetases from diverse sources (e.g., rat liver, sheep liver, rabbit reticulocytes, Hela cells, chicken embryos) have been isolated as multienzyme complexes, which might represent the actual tRNA charging factories associated with polysomes [55, 56].

### 1.3.4.2 Structure and Function of Ribosomes

Ribosomes are highly ordered ribonucleoprotein particles built of two subunits, a smaller (s) and a larger (l) subunit. In addition to 3–4 RNA molecules they contain 50–80 proteins (cf. details in Table **1**.9 and Figs. **1**.17 and **1**.18). The mostly basic proteins of the two subunits can be separated by two-dimensional gel electrophoresis and are numbered consecutively according to their relative positions in the electropherograms (Fig. **1**.18). Their molecular weights range from 6–60 kd.

Ribosomes generally contain one molecule of each ribosomal protein or in some cases less. A prominent exception is the acidic protein L7/L12 in the bacterial ribosome that makes about 4 molecules per large subunit and exists in two forms with free N-terminus (L12) or with acetylated N-terminus (L7).

Two principle types of ribosomes are distinguished with respect to their size and sedimentation coefficients (70S- and 80S-types respectively). They also differ in the size of their RNA components (cf. Table **1**.9), the differential effects of antibiotics on translation (Table **1**.9 and Fig. **1**.19), the number of structural proteins (Fig. **1**.18) and their natural occurrence: the 70S-type is found in all prokaryotes and in mitochondria as well as chloroplasts of eukaryotic cells, whereas the 80S-type characterizes the cytosol of eukaryotic cells.

The ribosomal RNAs represent 80% of total cell RNA and make about 65% *(E. coli)* and 50% (rat liver) of total ribosomal dry mass. They form the structural backbone of ribosomes and their dynamic secondary structure and interaction with ribosomal proteins contribute decisively to the versatile functions of these multiprotein complexes. However, most details of these structural and functional interactions are unclear.

**Table 1.9:** Size and composition of 70S- and 80S-type ribosomes[a]

| | 70S-type ribosomes | | | | | | | |
|---|---|---|---|---|---|---|---|---|
| | *Escherichia coli* | | Rat liver mitochondria | | *Euglena gracilis* chloroplasts | | Wheat mitochondria | |
| Sed. coefficient | 70S | | 55S | | 68S | | 78S | |
| Size (nm)[b] | $21 \times 18$ | | $26 \times 20$ | | $21 \times 17$ | | | |
| Subunits: | *l* | *s* | *l* | *s* | *l* | *s* | *l* | *s* |
| Sed. coefficient | 50S | 30S | 40S | 28S | 49S | 32S | | |
| Number of proteins | 34 | 21 | 36–40 | 30 | 32–34 | 22 | | |
| RNA | | | | | | | | |
| Sed. coefficient | 23S | 16S | 16S | 13S | 23S | 16S | 26S | 18S |
| (MW in Md) | (1.1) | (0.56) | (0.65) | (0.36) | (1.07) | (0.56) | (1.3) | (0.75) |
| | 5S | | 5S (?) | | 5S | | 5S | |
| | (0.04) | | | | (0.04) | | (0.04) | |
| | | | | | no 4.5S[c] | | | |
| Translation inhibitor[d] | CLP | | CLP (?) | | CLP | | CLP (?) | |
| | 80S-type ribosomes (cytosol) | | | | | | | |
| | Yeast | | Rat liver | | *Euglena gracilis* | | Pea | |
| Sed. coefficient | 80.4S | | 79S | | 88S | | 78S | |
| Size (nm)[b] | | | $31 \times 26$ | | $27 \times 22$ | | | |
| Subunits: | *l* | *s* | *l* | *s* | *l* | *s* | *l* | *s* |
| Sed. coefficient | 61S | 38S | 59S | 41S | 62S | 41S | 56S | 36S |
| Number of proteins | 37–45 | 28–34 | 39–46 | 28–42 | 42–45 | 30–34 | 44–55 | 32–40 |
| RNA | | | | | | | | |
| Sed. coefficient | 28S | 18S | 28S | 18S | 25S | 20S | 28S | 18S |
| (MW in Md) | (1.21) | (0.72) | (1.44) | (0.66) | (1.2) | (0.8) | (1.32) | (0.70) |
| | 5.8S | | 5.8S | | 5.8S | | 5.8S | |
| | (0.05) | | (0.05) | | (0.05) | | (0.05) | |
| | 5S | | 5S | | 5S | | 5S | |
| | (0.04) | | (0.04) | | (0.04) | | (0.04) | |
| Translation inhibitor[d] | CH | | CH | | CH | | CH | |

[a] Data compiled from: Boschetti, A.: Biogenese der Chloroplasten und Mitochondrien, G. Fischer, Stuttgart 1978; Kurland [58], Wittmann [60] and Bielka [63].
[b] The approximate size of the highly asymmetric ribosomes was derived from electron microscopic measurements of negatively stained preparations.
[c] Most higher plant chloroplast ribosomes contain an additional 4.5S RNA (~ 100 nucleotides) in their large ribosomal subunits (cf. [68] for references).
[d] CLP, chloramphenicol; CH, cycloheximide.

Besides the large rRNA in the large ribosomal subunit there are one or, in many cases, two small RNAs. The 5S RNA (120 nucleotides) in 70S-type ribosomes and the 5.8S RNA (140 nucleotides) in 80S-type ribosomes respectively are integral parts of the entrance site (A-site) for aminoacyl tRNA. They exhibit similar features of secondary structure with a central region capable of base-pairing with the T$\psi$-loop of the tRNA [67]. An additional small RNA (5S, 120 nucleotides) was reported for 80S-type ribosomes. It participates together with protein L5 in the binding of initiator Met $tRNA_f^{Met}$ to the large ribosomal subunit. The L5–5S RNA complex has ATPase activity, a fact that might point out that eukaryotic initiation needs an additional ATP for association of the two ribosomal subunits. Recently a second small RNA (4.5S RNA) was also reported for chloroplast ribosomes (e.g., in maize, spinach and the alga *Chlamydomonas*). Its function is unknown [66, 68].

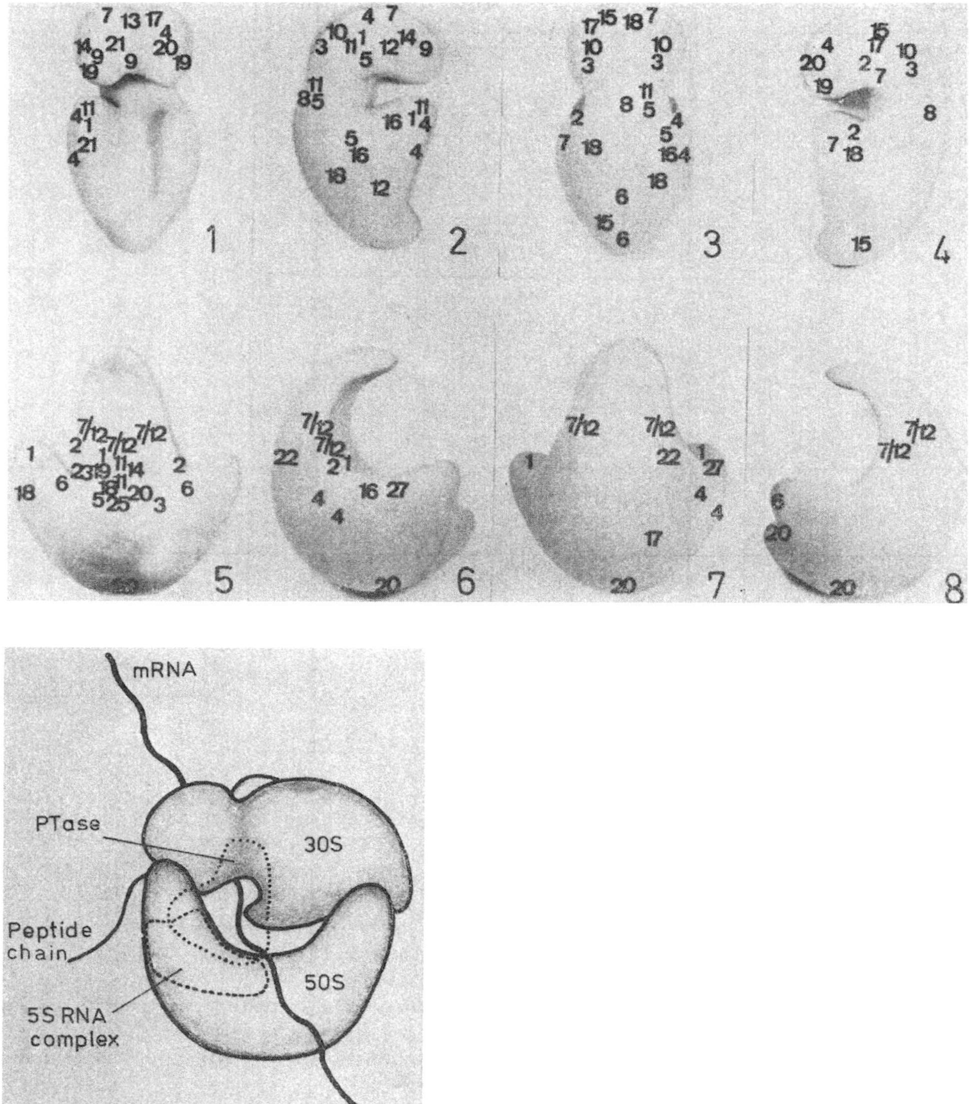


**Fig. 1.17**: Three dimensional models of the *E. coli* ribosome (from Wittmann [60] and Tischendorf et al. [65]).

*Numbers* denote the position of ribosomal proteins as determined by binding specific, non cross--reacting antibodies to distinct proteins of the ribosomal subunits, purifying the antibody--ribosome complexes and preparing them for electron microscopic visualization [65].

*Above* (1–4): Four views of the 30S subunit obtained by successive rotation of the model by 90°.
*Middle* (5–8): Four views of the 50S subunit.
*Below left* (9): Teddy bear – arm chair model of the translating ribosome as proposed by Tischendorf et al. [65]. The 30S subunit is arranged horizontally transverse with respect to the 50S subunit, thus forming a channel between the two subunits where the mRNA is bound.

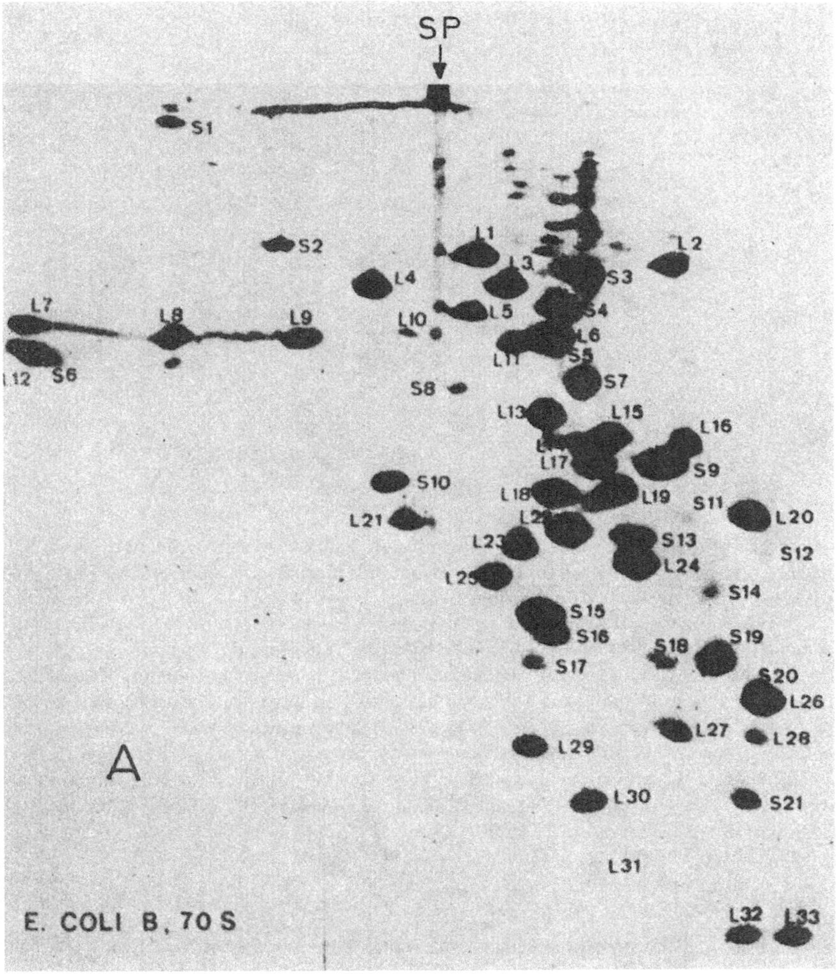


**Fig. 1.18:** Two-dimensional electropherograms of prokaryotic (**A**) and eukaryotic (**B, C**) ribosomal proteins.

**A** Total proteins of *Escherichia coli* 70S ribosomes (from Wittmann [60], with permission). Ribosomal proteins were prepared and separated as described by Kaltschmidt, E. & H. G. Wittmann: Anal. Biochem. *36* (1970): 401. 1. dimension: 4% polyacrylamide gel, urea-boric acid-Tris buffer, *p*H 8.6. The starting point is indicated by *SP*. Acidic proteins migrate to the cathode *(left)*, basic proteins migrate to the anode *(right)*. 2. dimension (from *top* (+) to *bottom* (−)): 18% polyacrylamide gel with urea-acetic acid-KOH buffer, *p*H 4.6 in the gel and urea-glycine-acetic acid, *p*H 4.6 as electrode buffer. The proteins are numbered from the upper left corner to the lower right corner with *L* and *S* denoting proteins of the large and small ribosomal subunits respectively.

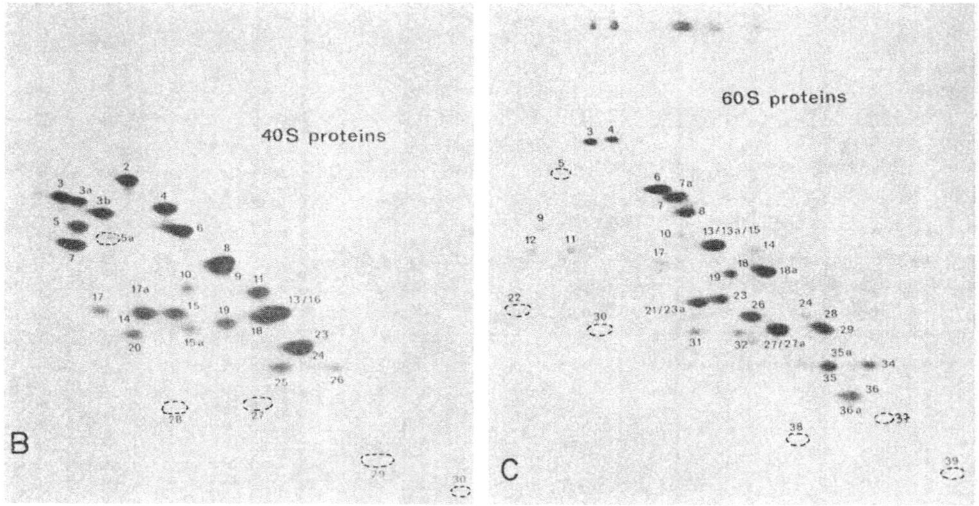


**(Fig. 1.18)**
**B, C** Proteins of the large (60S) and small (40S) subunits of rat liver cytoplasmic ribosomes. They were prepared and separated according to Welfle, H. et al.: Mol. Gen. Genetics *163* (1978): 101 by a modified Kaltschmidt-Wittmann technique. Starting point is in the upper left corner. Only the basic proteins are shown. 1. dimension (*left* to *right*): 5% polyacrylamide, urea-boric acid--Tris buffer, *p*H 8.3; 2. dimension (*top* to *bottom*): 15% polyacrylamide with urea-acetic acid--KOH buffer, *p*H 4.5 in the gel and 0.1M glycin-HCl, *p*H 2.0 as electrode buffer. Though it is generally assumed that ribosomes contain 1 : 1 stoichiometric amounts of the ribosomal proteins some spots are barely visible *(encircled spots)* in the electropherogram. Besides a factural micro-heterogeneity of the ribosome population with respect to the protein composition, many short--comings of the experimental procedure may contribute to this picture. Protein nomenclature follows the proposal by McConkey, E. H. et al.: Mol. Gen. Genetics *169* (1979): 1. B and C by courtesy of H. Bielka and H. Welfle.

Elucidation of the highly complex ribosomal structure is far advanced especially for the *E. coli* ribosome [58–61]. All proteins have been purified to homogeneity and sequenced. With specific antibodies raised against individual proteins it was possible to localize distinct ribosomal proteins by electron microscopy of the antibody-linked complexes (cf. Fig. **1**.17). Further RNA-RNA, RNA-protein and protein-protein cross-linking with bifunctional chemical reagents as well as neutron scattering analysis of ribosomes marked with deuterated proteins led to data about the internal structural relations of the ribosomal components. These methods were complemented by studies on the binding of distinct ribosomal proteins to rRNA and the ordered in vitro assembly from their components of both ribosomal subunits in an active state [60].

### 1.3.4.3 The Translation Cycle

The fidel translation of the nucleic acid information of the mRNA into proteins requires the concerted interaction of the ribosomal machinery with numerous cytoplasmic factors (Table **1**.10). This is detailed in Fig. **1**.19 for prokaryotic systems. But the scheme is also basically valid for translation by 80S-type ribosomes with the coresponding factors exchanged (cf. Table **1**.10).

**Table 1.10**: Properties of prokaryotic and eukaryotic translation factors (compiled from [63, 64, 68a])

| | Prokaryotes (*E. coli*) (70S-type ribosomes) | | Eukaryotes (80S-type ribosomes) | |
|---|---|---|---|---|
| | Factor ($M_r$, kd) | Functions (cf. Fig. **1.19**) | Factor ($M_r$, kd) | Functions, remarks |
| INITIATION | IF-1 (8.1) | Dissociation of 70S ribosome | eIF-1 (15) | mRNA binding, stabilization of initiation complex |
| | IF-2 (90) | fMet-tRNA$_f^{Met}$ binding to 30S subunits, GTPase activity | eIF-2 (86) (SU-$\alpha$: 32–38, SU-$\beta$: 47–52) | GTP-dependent binding of Met-tRNA$_f^{Met}$ to 40S subunit, mRNA recognition |
| | IF-3 (20.7) | mRNA binding to 30S subunits, antiassociation | eIF-3 (500–750 kd, 7–10 different SU) | Antiassociation, stabilization of initiation complex, associates with cap-binding factor[a] |
| | | | eIF-4A (48–53), eIF-4B (80–82), eIF-4C (19), eIF-4D (17) | mRNA binding to 40S subunit, stabilization of initiation complex, eIF-4B associates with cap-binding factor[a] |
| | | | eIF-5 (125–160) | Release of eIF-2 and eIF-3 from initiation complex, binding of 60S subunit |
| ELONGATION | EF-Tu (47) | aatRNA binding to translation complex, GTPase activity | eEF-Tu (47–60) | Formerly EF-1 or EF-1$\alpha$, GTP-dependent binding of aatRNA to translation complex |
| | EF-Ts (34) | Regeneration of EF-Tu · GTP complex | eEF-Ts (two SU with 53 and 30 kd) | Formerly EF-1$\beta$, regeneration of eEF-Tu · GTP complex |
| | EF-G (83) | Translocation and GTPase activity | eEF-G (65–112) | Formerly EF-2, translocation and GTPase activity, ADP-ribosylated by Diphtheria toxin |
| TERMINATION | RF-1 (44) RF-2 (47) RF-3 | Termination of elongation, release of deacylated tRNA, GTPase activity | eRF | Eukaryotic release factors have been described but not further characterized |

Abbreviations: aa, aminoacyl; SU, subunit; e, eukaryotic; IF, EF, RF, factors involved in initiation, elongation, termination.

[a] The cap-binding factor ($M_r$ 24 kd) is a further essential component of the eukaryotic initiation process [73] that controls the interaction of capped RNA with the 40S ribosomal subunit (cf. text).

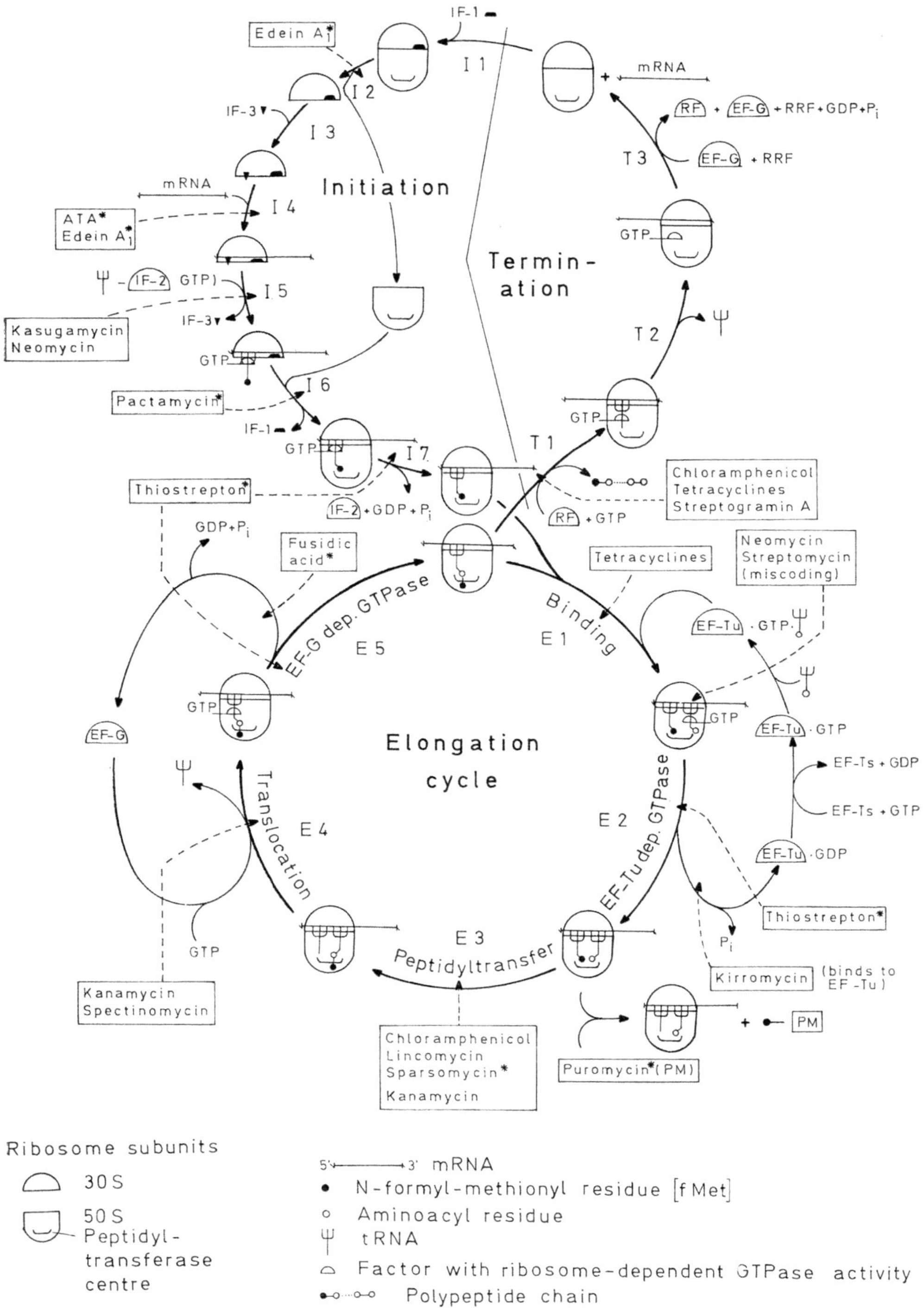

Initiation
Termin-ation
Elongation cycle
I 1
I 2
I 3
I 4
I 5
I 6
I 7
T 1
T 2
T 3
E 1
E 2
E 3
E 4
E 5
Binding
EF-Tu dep. GTPase
Peptidyltransfer
Translocation
EF-G dep. GTPase
IF-1
IF-3
IF-2
mRNA
GTP
GDP+P_i
RF
EF-G
RRF
EF-Tu
EF-Ts + GDP
EF-Ts + GTP
Edein A_1*
ATA*
Kasugamycin
Neomycin
Pactamycin*
Thiostrepton*
Fusidic acid*
Chloramphenicol
Tetracyclines
Streptogramin A
Neomycin
Streptomycin
(miscoding)
Kirromycin
(binds to EF-Tu)
Puromycin*(PM)
PM
Chloramphenicol
Lincomycin
Sparsomycin*
Kanamycin
Spectinomycin
Ribosome subunits
30S
50S
Peptidyl-transferase centre
5'—3' mRNA
N-formyl-methionyl residue [fMet]
Aminoacyl residue
tRNA
Factor with ribosome-dependent GTPase activity
Polypeptide chain

*Initiation:* The sequence of initiation reactions leads to the association of the ribosomal subunits with mRNA and the charged initiator tRNA. With rare exceptions this is N-formylmethionyl $tRNA_f^{Met}$ for the 70S-type and methionyl $tRNA_f^{Met}$ for the 80S-type ribosomes. A characteristic step of initiation is the interaction of mRNA with the 16S rRNA of the small ribosomal subunit catalyzed by protein Sl (Fig. **1**.20) that also assists in the subsequent binding of the initiator tRNA [69–71]. The affinity of mRNA for interaction with the various components of the initiation process can be an essential means for gene expression control (cf. **4**.4.4).

In contrast to their prokaryotic counterparts (cf. Fig. **1**.20) eukaryotic mRNAs contain no recognizable sequence element that might base-pair with the 18S rRNA [72]. Instead, with the exception of few viral systems, mRNAs are always modified at the 5′-end by a 7-methylguanyl residue (cap, cf. Fig. **1**.9). This cap protects the

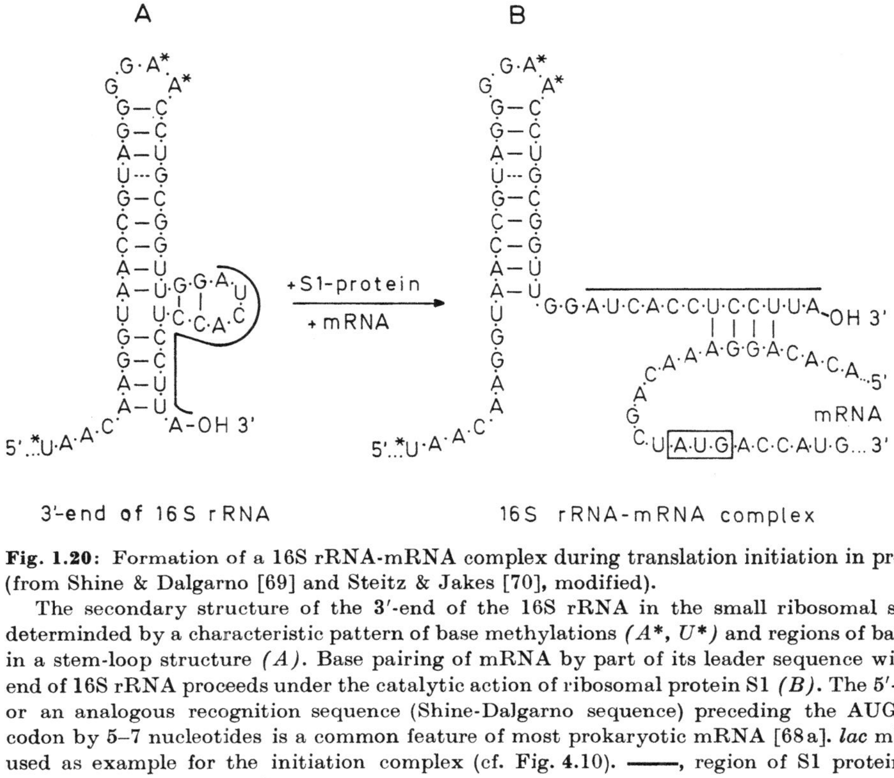


**Fig. 1.20:** Formation of a 16S rRNA-mRNA complex during translation initiation in prokaryotes (from Shine & Dalgarno [69] and Steitz & Jakes [70], modified).

The secondary structure of the 3′-end of the 16S rRNA in the small ribosomal subunit is determinded by a characteristic pattern of base methylations *(A*, U*)* and regions of base pairing in a stem-loop structure *(A)*. Base pairing of mRNA by part of its leader sequence with the 3′-end of 16S rRNA proceeds under the catalytic action of ribosomal protein S1 *(B)*. The 5′-AGGA-3′ or an analogous recognition sequence (Shine-Dalgarno sequence) preceding the AUG initiator codon by 5–7 nucleotides is a common feature of most prokaryotic mRNA [68a]. *lac* mRNA was used as example for the initiation complex (cf. Fig. **4**.10). ——, region of S1 protein binding to the 3′-end of 16S rRNA.

---

◀

**Fig. 1.19:** Functional scheme of the *E. coli* ribosome and antibiotics interfering with individual steps of the translation cycle (from Nierhaus & Wittmann [59], modified).

The steps at each functional domain are numbered consecutively (*I1–I7* for initiation, *E1–E5* for the elongation cycle and *T1–T3* for termination). The scheme applies also for eukaryotic ribosomes, except that during initiation initiator tRNA binding precedes mRNA binding. The termination steps are not well characterized. Antibiotics that act as inhibitors of translation are boxed and indicated with their sites of action. Antibiotics interfering with translation at 70S- and 80S-type ribosomes as well (cf. Table **1**.9) are marked with a *dot*, the others are specific for 70S-type ribosomes.

5′-end from exonucleolytic degradation and serves as an essential signal for translation initiation. A special cap-binding protein ($M_r$ 24 kd) is involved. It associates with eukaryotic initiation factors eIF-3 and eIF-4B. The experimentally well founded "scanning hypothesis" of eukaryotic initiation [72a] assumes that the 40S ribosomal subunit, after binding to mRNA at the cap site complexed to the cap-binding protein, travels to the next AUG codon where the initiation complex is completed by addition of the 60S ribosomal subunit. Both, binding of the 40S ribosomal subunit with concomitant changes of the mRNA secondary structure and its subsequent migration to the first AUG codon are ATP-dependent processes.

The scanning mechanism implies that all eukaryotic mRNA's behave as monocistronic mRNA's [73, 74]. If they are transcribed as functionally polycistronic (pre)-mRNA's they are either translated into a single polyprotein that is processed to the individual proteins during or after translation (cf. **4.4.5.1** and **4.4.5.3**) or they are spliced before translation into monocistronic mRNA's while preserving their 5′-capped ends (cf. Fig. **10.8** and Sect. **4.4.3.1**).

*Elongation:* The role of IF-2 for GTP-dependent binding of initiator tRNA is to a certain extent analogous to the function of EF-Tu in the elongation cycle. Both factors have GTPase activity. After a "check in" of the in-coming aatRNA an important step (E 2) is a "fitting" of the aatRNA bound in the A-site to the steric requirements for peptidyl transfer (E 3). E 2 may eventually involve transacylation of the aminoacyl residue from the 2′- to the 3′-position of the terminal adenosine residue [76]. Fitting of aatRNA (E 2) and translocation of peptidyl tRNA from the A- into the P-site (E 4, E 5), but not the peptide bond formation itself (E 3), require an energy input by GTP. The peptidyl transferase is the ribosomal protein L11 neighbored in the A-site by proteins L6 and L16 and in the P-site by proteins L2 and L27 (cf. Fig. **1.17**).

A special role of the four copies of the acidic protein L7/L12 with respect to the GTP cleavage in steps E2 and E4/E5 was suggested [77]. The pentameric complex $(L7/L12)_2 \cdot L10 \cdot (L7/L12)_2$ at the interphase between large and small ribosomal subunits (cf. Fig. **1.17**) might alternatively interact by one L7/L12 dimer with the EF-Tu · GTP · aatRNA complex in the A-site or by the other dimer with EF-G · GTP in the P-site thus providing a "mechanical switch" for ribosome movement along the mRNA.

Important components of the mRNA decoding site at the small ribosomal subunits are proteins S1, 3, 4, 5, 11, 12, 18 and 21. Especially the concerted interactions of proteins S3, 4, 5, 11 and S12 are crucial for translation fidelity. S12 has a dual role in both reactions E 1 and E 2 that are essential for tRNA discrimination [59, 79].

*Termination* of translation is indicated by appearance of one of the termination codons in the ribosomal A-site. By interaction of release factors (RF-1 in the presence of UAG or UAA and RF-2 in the presence of UAA or UGA) the peptidyl residue is cleaved of the tRNA followed by release of the deacylated tRNA and dissociation of the ribosome-mRNA complex. Finally the ribosomes dissociate into subunits with the help of IF-3. The three so-called nonsense codons differ in their stringency to cause termination. Usually low levels of suppressor tRNAs for UGA and for UAG are found in prokaryotic and eukaryotic cells so that only tandem terminator codons are effective. In contrast, no suppression of UAA has so far been reported [80]. Terminator read-through may be an efficient means of translation control (cf. **4.4.4.3**).

### 1.3.5 Proteinogen Processing

The advancing analyses of the biochemistry and spatial organization of the gene expression chain revealed many proteins to be synthesized by the polysomal complex as precursors that must be processed during translation (co-translationally) and/or after release from polysomes (post-translationally). "End point of protein synthesis is the biologically active protein in its proper compartment" [86]. There is increasing evidence that both, activity and cellular localization of proteins, are dependent on their reversible or irreversible modifications by three types of reactions (Table **1**.11) I: limited proteolysis, II: protein-protein cross-links (mainly formation of disulphide and $\gamma$-glutamyl isopeptide bonds) and III: numerous types of modifications of amino acid residues.

**Table 1.11**: Modification of proteinogenous amino acids prior, during or after their incorporation into proteins (from Rosa & Wold [86])

| Types of modification | Stage of modification: pre- translational | co- | post- | Amino acid residues | Reference to sections of this book |
|---|---|---|---|---|---|
| I Limited proteolysis[a] | | + | + | depends on enzyme involved | **4.4.5**, **6.4.2**, **8.4.4**, **13.3.3**, Table **14.5** |
| II Protein cross-links[b] | | + | + | mainly Cys (formation of –S–S-bonds), but also Asp, Lys, Glu | |
| III Modification of amino acid residues | | | | | |
| Acetylation (N)[c] | + | + | + | Ala, Asp, Cys, Gly, Lys, Met, Ser, Thr, Val | Table **1**.2, **4.4.2.5.2**. |
| Formylation (N) | + | | | Gly, Met | **1.3.4.3**. |
| Phosphorylation | | + | + | | Table **1**.2, Figs. |
| (N) | | | | Lys, His | **4**.2, **4**.4, **4**.26, |
| (O) | | | | Asp, Ser, Thr, Tyr | **10**.6 and **10**.9, Chap. **10** |
| Methylation | | + ? | + | | |
| (N) | | | | Ala, Arg, Lys, His | Table **1**.2 |
| (O) | | | | Glu, Ser | |
| Glycosylation | | + | + | | Fig. **13**.10 |
| (N) | | | | Asn, Gly | and Sect. |
| (O) | | | | HO-Lys, HO-Pro, $\beta$ HO- -Phe, Ser, Thr, $\beta$ HO-Tyr | **12.2.4**. |
| (S) | | | | Cys | |
| ADP-ribosylation | | | + | | Table **1**.2, |
| (N) | | | | Arg | Table **1**.10 |
| (O) | | | | Glu, Ser | |
| Adenylylation (O) | | | + | Tyr | Fig. **4**.3 |
| Uridylylation (O) | | | + | Tyr | Fig. **4**.3 |
| Hydroxylation (C) | | + | + ? | Lys, Phe, Pro, Tyr | |
| Halogenation (C) | | | + | Tyr | |
| Carboxylation (C) | | ? | | Glu | |

[a] Limited proteolysis is a general phenomenon in all living cells (cf. the numerous examples in this book). In prokaryotic systems a special type of "limited proteolysis" is the removal of the N-formyl methionine residue at the N-terminus of practically all proteins.

[b] Besides ubiquitous formation of disulphide bridges between protein subunits, covalent protein cross-linking is mainly found in extracellular protein systems serving mechanical functions (e.g., collagen). Some special cases are cited in this book: cf. the fusion of two proteins involved into phage $\lambda$ morphogenesis (**14.2.3.10**), the role of histone A24 for chromatin inactivation (**4.4.2.5.3**) and the protein cross-linking during receptor internalization by coated vesicles (Fig. **8**.12).

[c] Letters in brackets indicate the atom modified in the particular amino acid.

**Table 1.12:** Intracellular proteolytic enzymes of yeast and their natural inhibitors (compiled from Ref. [85]).

| Enzyme (E.C. number) | Characteristics | | | | Natural polypeptide inhibitors | | | | |
|---|---|---|---|---|---|---|---|---|---|
| | Active centre[a] | $M_r$ (kd) | Inhibitors[b] | Compartment | Name[c] | IP[c] ($p$H) | $M_r$ (kd) | Inactivated by | Compartment |
| Proteinase A (3.4.23.6) | Asp | 60 | Pepstatin | Vacuole | I $A_2$<br>I $A_3$ | 5.7<br>6.3 | 7.7 | Proteinase B | Cytosol |
| Proteinase B (3.4.22.9) | Ser or Cys (?) | 32–44 | IPF, PMSF, CMB | Vacuole | I $B_1$<br>I $B_2$ | 8.0<br>7.0 | 8.5 | Proteinase A | Cytosol |
| Carboxypeptidase Y (3.4.16.1) | Ser | 61 | IPF, PMSF, CMB, TPCK | Vacuole | I C | 6.6 | 23.8 | Proteinases A and B | Cytosol |
| Carboxypeptidase S ? | $Zn^{2+}$ | n.d. | EDTA | ? | ? | | | | |
| Aminopeptidase I[d] | $Zn^{2+}$ | 200 | EDTA, CMB | Vacuole | ? | | | | |

[a] The biological activity depends on the presence of distinct amino acid residues or metal ions in the active centers, i.e., of Asp (acid proteases), Cys (thiolproteases), Ser (serine proteases) or $Zn^{2+}$ (metal proteases).

[b] Pepstatin is a peptide aldehyde antibiotic produced by *Actinomyces* spec. Abbreviations of other inhibitors: IPF, diisoproyl phosphofluoridate; PMSF, phenylmethylsulfonylfluoride; CMB, p-chloromercuribenzoate; TPCK, N-tosyl-L-phenylalanine chloromethylketone.

[c] The natural isoinhibitors differ in their amide contents and hence isoelectric points (IP), but not in their amino acid composition and end groups.

[d] Two more aminopeptidases were detected, but less well characterized.

The extent of the latter type of proteinogen processing is impressively demonstrated by the fact that about 140 amino acid derivatives have been found so far in different proteins, but all originate from precursors containing only the common 20 amino acid residues. Very few modifications occur pre-translationally, in some instances the growing peptide chain is modified (co-translationally), but in most cases the processing steps proceed post-translationally. Table **1**.11 contains also numerous references, where concrete examples of protein modifications and the enzymes involved are discussed elsewhere in this book.

### 1.3.6 Protein Degradation

Protein degradation in all living cells is a stepwise process requiring the activity of many proteinases (cf. Tables **1**.12 and **1**.13). It is experimentally well established that more or less pronounced protein break-down always proceeds but that it can drastically increase under certain physiological conditions, e.g., sporulation of bacteria (cf. **4**.5.3) or amino acid or carbohydrate starvation of animal cells (cf. **4**.4.6) etc. In many cells proteinases are found together with specific polypeptide inhibitors (cf. Table **1**.12), though both are restricted to different compartments. This surely reflects the tight regulation of the proteolytic system.

**Table 1.13:** Mammalian intracellular proteinases (cathepsins) (compiled by P. Bohley)

| Enzyme (E.C. number) | Active center | $M_r$ (kd) | Inhibitors[a] | Compartment |
|---|---|---|---|---|
| Cathepsin B (3.4.22.1) | Cys | 25 | $TBR^+$, leupeptin, antipain | Lysosomes |
| Cathepsin D (3.4.23.5) | Asp | 42 | Pepstatin, diazoacetylnor-leucinmethylester + $Cu^{2+}$ | Lysosomes |
| Cathepsin E (3.4.23.-) | Asp | 100 | Pepstatin, diazoacetylnor-leucinmethylester + $Cu^{2+}$ | Lysosomes (?) |
| Cathepsin F (3.4.99.-) | ? | 50–70 | $\alpha_2$-Macroglobulin | ? |
| Cathepsin G (3.4.21.20) | Ser | 30 | IPF, CPBK | ? |
| Cathepsin H (3.4.22.-) | Cys | 28 | TBR, LCK | Lysosomes |
| Cathepsin L (3.4.22.15) | Cys | 24 | TBR, leupeptin, CPNK | Lysosomes |
| Cathepsin N (3.4.22.-) | Cys | 18–20 | TBR, TLCK | Lysosomes (?) |
| Cathepsin R (3.4.21.-) | Ser | 12 | IPF | Ribosomes |
| Cathepsin S (3.4.22.) | Cys | 14–17 | TBR | ? |

[a]) Natural inhibitors were isolated but not well characterized, except two inhibitors for cathepsins B and H respectively (Lenney, J. F. in [83], pp. 73–86).

Abbreviations: TBR, thiol-blocking agents, e.g., p-chloromercuribenzoate, iodoacetamide, $Hg^{2+}$; IPF, diisopropyl phosphofluoridate; CPBK, N-carbobenzoxy-L-phenylalanine bromomethylketone; LCK, L-leucyl chloromethylketone; CPNK, N-carbobenzoxy-L-phenylalanyl-L-phenylalanine diazomethylketone; TLCK, N-tosyl-L-lysyl chloromethylketone.

Several minor proteinases are found in different compartments of the cell. Many of them, e.g., those associated with membranes of mitochondria, with the endoplasmic reticulum or with ribosomes (cf. **4**.4.5., and the references given in Table **1**.10, part I), are agents of the limited proteolytic processing of proteinogens. However, it becomes more and more evident that the initial step of the biological inactivation of proteins is also a limited endoproteolytic cleavage yielding few high molecular weight products. These are taken up and further degraded by endoproteinases, exoproteinases and peptidases in the lytic compartments of the cell (vacuoles). In bacteria the enzyme for the primary endoproteolytic attack is membrane-bound, whereas the final breakdown to amino acids proceeds in the cytoplasm. A remarkable feature of the initial reaction is its ATP-dependence reported for the membrane-bound proteinase complex of *E. coli* as well as for the cytosolic complex of animal cells [84, 89, 90].

The reticulocyte system was shown to be composed of three indispensible proteins [89], one of them interacting with ATP ($M_r$ 450 kd) and one being a heat-stabil polypeptide (APF-1) of $M_r$ 8.5 kd. The latter is evidently identical with ubiquitin [91], a highly conserved polypeptide found universally in nature. In an ATP-dependent reaction APF-1 is covalently linked by an isopeptide bond to the protein to be degraded. The following hypothetical sequence of reactions can be given [90]:

$$\begin{matrix} \text{Protein} \\ + \\ n\text{APF-1} \end{matrix} \xrightarrow[+\,n\text{ATP}]{1} (\text{APF-1})_n \cdot \text{Protein} \xrightarrow{2} \begin{matrix} \text{Peptides or amino acids} \\ + \\ n\text{APF-1} \cdot \text{x} \end{matrix}$$

$$n\text{APF-1} \cdot \text{x} \xrightarrow{3} n\text{APF-1}$$

The "labelling" of the protein by several molecules of APF-1 is catalyzed by APF-1-protein amide synthetase (1). The proteinase or peptidase (2) acts strongly only on protein derivates when $n > 1$ or 2. Amidase (3) regenerates the APF-1 factor for a new degradation cycle [90].

## 1.4 Selected References

### 1.4.1 DNA and Chromatin


1. Watson, J. D.: Molecular Biology of the Gene. W. A. Benjamin, Inc., New York-Amsterdam 1965.
2. Cozzarelli, N. R.: DNA gyrase and the supercoiling of DNA. Science *207* (1980): 953–960.
3. Rouvière-Yaniv, J., M. Yaniv & J. E. Germond: E. coli DNA binding protein HU forms nucleosome-like structure with circular double-stranded DNA. Cell *17* (1979): 265–274.
4. Lathe, R., H. Buc, J.-P. Lecocq & E. K. F. Bautz: Prokaryotic histone-like protein interacting with RNA polymerase. Proc. Natl. Acad. Sci. USA *77* (1980): 3548–3552.
5. Britten, R. J. & E. H. Davidson: Gene regulation for higher cells: A theory. Science *165* (1969): 349–357.
6. Nagel, W.: Zellkern und Zellzyklen. E. Ulmer, Stuttgart 1976.
7. Isenberg, I.: Histones. Ann. Rev. Biochem. *48* (1979): 159–191.
8. Grunicke, H.: Regulatorische Proteine des Chromatins höherer Zellen. Naturwiss. *66* (1979): 347–353.
9. Elgin, S. C. R. & H. Weintraub: Chromosomal proteins and chromatin structure. Ann. Rev. Biochem. *44* (1975): 725–774.
10. Boulikas, T., J. M. Wiseman & W. T. Garrard: Points of contact between histone H1 and the histone octamer. Proc. Natl. Acad. Sci. USA *77* (1980): 127–131.
11. Rindt, K. P. & L. Nover: Chromatin structure and function. Biol. Zentralbl. *99* (1980): 641–673.
12. Chromatin. Cold Spring Harbor Symp. Quant. Biol. *42* (1978).

13. McGhee, J. D. & G. Felsenfeld: Nucleosome structure. Ann. Rev. Biochem. *49* (1980): 1115–1156.
14. Lutter, L. C.: Kinetic analysis of dexoxyribonuclease I cleavages in the nucleosome core: evidence for a DNA superhelix. J. Mol. Biol. *124* (1978): 391–420.
15. Laskey, R. A. & W. C. Earnshaw: Nucleosome assembly. Nature *286* (1980): 763–767.
16. Thoma, F., Th. Koller & A. Klug: Involvement of histone H1 in the organization of the nucleosome and of the salt-dependent superstructures of chromatin. J. Cell Biol. *83* (1979): 403–427.
17. Laemmli, U. K., S. M. Cheng, K. W. Adolph, J. R. Paulson, J. A. Brown & W. R. Baumbach: Metaphase chromosome structure: The role of nonhistone proteins. Cold Spring Harbor Symp. Quant. Biol. *42* (1978): 351–360.
18. Marsden, M. P. F. & U. K. Laemmli: Metaphase chromosome structure: Evidence for a radial loop model. Cell *17* (1979): 849–858.
19. Bak, A. L., P. Bak & J. Zeuthen: Higher levels of organization in chromosomes. J. Theor. Biol. *76* (1979): 205–217.

### 1.4.2 RNA Polymerases and Transcription

20. Chamberlin, M. J.: RNA polymerase – An overview. In: Losick, R. & M. Chamberlin (eds.): RNA Polymerase. Cold Spring Harbor Laboratory, Cold Spring Harbor, New York 1976, pp. 17–67.
21. Roeder, R. G.: Eukaryotic nuclear RNA polymerases: In: Losick, R. & M. Chamberlin (eds.): RNA Polymerase. Cold Spring Harbor Laboratory, Cold Spring Harbor, New York 1976, pp. 285–329.
22. Chambon, P.: Eukaryotic nuclear RNA polymerases. Ann. Rev. Biochem. *44* (1975): 613–638.
23. Becker, W. M.: RNA polymerases in plants. In: Hall, T. C. & J. Davis (eds.): Nucleic Acids in Plants. CRC Press Inc. Boca Raton, Florida 1979, pp. 111–141.
24. Stewart, P. R. & D. S. Letham (eds.): The Ribonucleic Acids, 2nd ed., Springer, New York 1978, pp. 7–41.
25. Yura, T. & A. Ishihama: Genetics of bacterial RNA polymerases. Ann. Rev. Genetics *13* (1979): 59–97.
26. Wiggs, J. L., J. W. Bush & M. J. Chamberlin: Utilization of promoter and terminator sites on bacteriophage T7 DNA by RNA polymerases from a variety of bacterial orders. Cell *16* (1979): 97–109.
27. Adhya, S. & M. Gottesman: Control of transcription termination. Ann. Rev. Biochem. *47* (1978): 967–996.
28. Rosenberg, M. & D. Court: Regulatory sequences involved in the promotion and termination of RNA transcription. Ann. Rev. Genetics *13* (1979): 319–353.
29. Wu, A. M., A. B. Chapman, T. Platt, L. P. Guarente & J. Beckwith: Deletions of distal sequence affect termination of transcription at the end of the tryptophan operon in *E. coli*. Cell *19* (1980): 829–836.
29a. Gallupi, G. & J. P. Richardson: ATP-induced changes in the binding of RNA synthesis termination protein rho to RNA. J. Mol. Biol. *138* (1980): 513–539.
30. Siebenlist, U., R. B. Simpson & W. Gilbert: *E. coli* RNA polymerase interacts homologously with two different promoters. Cell *20* (1980): 269–281.
31. Platt, T.: Termination of transcription and its regulation in the tryptophan operon of *E. coli*. Cell *24* (1981): 10–23.
32. Sakonju, Sh., D. F. Bogenhagen & D. D. Brown: A control region in the center of the 5S RNA gene directs specific initiation of transcription: I. the 5′ border of the region. Cell *19* (1980): 13–25; Bogenhagen, D. F., Sh. Sakonju & D. D. Brown: A control region in the center of the 5S RNA gene directs specific initiation of transcription: II. The 3′ border of the region. Cell *19* (1980): 27–45.
33. Luse, D. S. & G. Roeder: Accurate transcription initiation on a purified mouse $\beta$-globin DNA fragment in a cell-free system. Cell *20* (1980): 691–700.

### 1.4.3 RNA Processing

34. Perry, R. P.: Processing of RNA. Ann. Rev. Biochem. *45* (1976): 605–629.
35. Abelson, J.: RNA processing and the intervening sequence problem. Ann. Rev. Biochem. *48* (1979): 1035–1069.

36. Van Venroij, W. J. & E. C. Mariman: Some aspects of hnRNA processing. In: Nover, L., F. Lynen & K. Mothes (eds.): Cell Compartmentation and Metabolic Channeling. G. Fischer Verlag, Jena and Elsevier/North-Holland Publ. Comp. Amsterdam 1980, pp. 427–438.
37. Preobrazhensky, A. A. & A. S. Spirin: Informosomes and their protein components: The present state of knowledge. Progr. Nucleic Acid Res. Mol. Biol. *21* (1978): 1–38.
38. Lerner, M. P., J. A. Boyle, S. M. Mount, S. L. Wolin & J. A. Steitz: Are snRNPs involved in splicing? Nature *283* (1980): 220–224.
39. Scherrer, K., M. T. Imaizumi-Scherer, C. A. Reynaud & A. Therwath: On pre-messenger RNA and transcription. A review. Mol. Biol. Rep. *5* (1979): 5–28.
40. LeStourgeon, W. M., A. L. Beyer, M. E. Christensen, B. W. Walker, S. M. Pourpore & L. P. Daniels: The packaging proteins of core hnRNP particles and the maintenance of proliferative cell states. Cold Spring Harbor Symp. Quant. Biol. *42* (1978): 885–898.
41. Brawermann, G.: Characteristics and significance of the polyadenylate sequence in mammalian mRNA. Progr. Nucleic Acid Res. Mol. Biol. *17* (1976): 118–148.
42. Furuichi, Y. & A. J. Shatkin: Differential synthesis of blocked and unblocked 5′-termini in reovirus mRNA: Effect of pyrophosphate and pyrophosphatase. Proc. Natl. Acad. Sci. USA *73* (1976): 3448–3452.
43. Nomura, M., E. A. Morgan & S. R. Jaskunas: Genetics of bacterial ribosomes. Ann. Rev. Genetics *11* (1977): 297–347.
44. Braun, R. J., R. A. Young & J. A. Steitz: The ribonuclease III site flanking 23S sequences in the 30S ribosomal precursor RNA of *E. coli*. Cell *19* (1980): 393–401.
45. Stark, B. C., R. Kole, E. J. Bowman & S. Altman: Ribonuclease P: An enzyme with an essential RNA component. Proc. Natl. Acad. Sci. USA *75* (1978): 3717–3721.
46. Sierakowska, H. & D. Shugar: Mammalian nucleolytic enzymes. Progr. Nucleic Acid Res. Mol. Biol. *20* (1977): 59–130.
47. Cudny, H. & M. P. Deutscher: Apparent involvement of ribonuclease D in the 3′ processing of tRNA precursors. Proc. Natl. Acad. Sci. USA *77* (1980): 837–841.
48. Garber, R. & S. Altman: In vitro processing of *B. mori* transfer RNA precursor molecules. Cell *17* (1979): 389–397.
49. Melton, D. A., E. M. De Robertis & R. Cortese: Order and intracellular location of the events involved in the maturation of a spliced tRNA. Nature *284* (1980): 143–148.
49a. Katze, J. R. & W. R. Farkas: A factor in serum and amniotic fluid is a substrate for the tRNA-modifying enzyme tRNA-guanine transferase. Proc. Natl. Acad. Sci. USA *76* (1979): 3271–3275; Okada, N., S. Noguchi, H. Kasai, N. Shindo-Okada, T. Ohgi, T. Goto & S. Nishimura: Novel mechanism of post-transcriptional modification of tRNA. Insertion of bases of Q precursors into tRNA by a specific tRNA-transglycosylase reaction. J. Biol. Chem. *254* (1979): 3061–3073.

**1.4.4 tRNA Structure and Aminoacyl tRNA Synthetases**

50. Moldave, K. & L. Grossman (eds.): Colowick-Kaplan, Methods in Enzymology, Vol. 60, Part G. Academic Press, New York 1979.
51. Schimmel, P. R., D. Söll & J. N. Abelson (eds.): Transfer RNA: Structure, Properties, and Recognition. Cold Spring Harbor Lab. 1979.
52. Kim, S. H.: Crystal structure of yeast $tRNA^{Phe}$ and general structural features of other tRNAs. In [51], pp. 83–100.
53. Rich, A., G. J. Quigley, M. M. Teeter, A. Ducruix & N. Woo: Recent progress in tRNA structural analysis. In [51], pp. 101–113.
54. Woo, N. H., B. A. Roe & A. Rich: Three-dimensional structure of *Escherichia coli* initiator $tRNA_f^{Met}$. Nature *286* (1980): 346–352.
55. Schimmel, P. R., & D. Söll: Aminoacyl-tRNA synthetases: General features and recognition of transfer RNAs. Ann. Rev. Biochem. *48* (1979): 601-648.
56. Kellermann, O., A. Brevet, H. Tonetti & J. P. Waller: Eukaryotic aminoacyl-tRNA synthetase complexes: Limited proteolysis as a probe of the structural organization. In: [83], pp. 235–243.
57. Barnett, W. E., S. D. Schwartzbach & L. I. Hecker: The transfer RNAs of eukaryotic organelles. Progr. Nucleic Acid Res. Mol. Biol. *21* (1978): 143–179.

**1.4.5 Ribosome Structure and Translation**


58. Kurland, C. G.: Structure and function of the bacterial ribosome. Ann. Rev. Biochem. *46* (1977): 173–200.
59. Nierhaus, K. H. & H. G. Wittmann: Ribosomal function and its inhibition by antibiotics in prokaryotes. Naturwissenschaften *67* (1980): 234–250.
60. Wittmann, H. G.: Structure of ribosomes. In: Nover, L., F. Lynen & K. Mothes (eds.): Cell Compartmentation and Metabolic Channeling. G. Fischer Verlag and Elsevier/North-Holland Biomed. Press, Jena and Amsterdam 1980, pp. 185–200.
61. Wool, I. G.: The structure and function of eukaryotic ribosomes. Ann. Rev. Biochem. *48* (1979): 719–754.
62. Moldave, K. & L. Grossman (eds.): Colowick-Kaplan, Methods in Enzymology. Vol. 60, Part H. Academic Press, New York 1979.
63. Bielka, H. (ed.): The Eukaryotic Ribosome. Akademie-Verlag, Berlin 1982.
64. Weissbach, H. & S. Ochoa: Soluble factors required for eukaryotic protein synthesis. Ann. Rev. Biochem. *45* (1976): 191–216.
65. Tischendorf, G. W., H. Zeichardt & G. Stöffler: Architecture of *Escherichia coli* ribosome as determined by immune electron microscopy. Proc. Natl. Acad. Sci. USA *72* (1975): 4820–4824.
66. Dyer, T. A., C. M. Bowman & P. I. Pyne: The low-molecular weight RNAs of plant ribosomes. Their structure, function and evolution. In: Bogorad, L. & J. H. Weil (eds.): Nucleic Acids and Protein Synthesis in Plants. Plenum Press, New York 1977, pp. 121–133.
67. Luoma, G. A. & A. G. Marshall: Laser Raman evidence for new cloverleaf secondary structures for eukaryotic 5.8S RNA and prokaryotic 5S RNA. Proc. Natl. Acad. Sci. USA *75* (1978): 4901–4905.
68. Bedbrook, J. R. & J. Kolodner: The structure of chloroplast DNA. Ann. Rev. Plant Physiol. *30* (1979): 593–620.
68a. Grunberg-Manago, M. & F. Gros: Initiation mechanisms of protein synthesis. Progr. Nucleic Acid Res. Mol. Biol. *20* (1977): 209–284.
69. Shine, J. & L. Dalgarno: Determinant of cistron specificity in bacterial ribosomes. Nature *254* (1975): 34–38.
70. Steitz, J. A. & K. Jakes: How ribosomes select initiator regions in mRNA: Base pair formation between the 3′ terminus of 16S rRNA and the mRNA during initiation of protein synthesis in *Escherichia coli*. Proc. Natl. Acad. Sci. USA *72* (1975): 4734–4738.
71. Li, P. T., T. Shea, St. Ellis & T. W. Conway: Formylmethionyl-tRNA binding properties of *Escherichia coli* ribosomal protein S1. Europ. J. Biochem. *98* (1979): 155–163.
72. Baralle, F. E. & G. G. Brownlee: AUG is the only recognisable signal sequence in the 5′ non-coding regions of eukaryotic mRNA. Nature *279* (1978): 84–87.
72a. Kozak, M.: Role of ATP in binding and migration of 40S ribosomal subunits. Cell *22* (1980): 459–467.
73. Sonenberg, N. & K. M. Rupprecht, S. M. Hecht, A. J. Shatkin: Eukaryotic mRNA cap binding protein: Purification by affinity chromatography on Sepharose-coupled $m^7$GDP. Proc. Natl. Acad. Sci. USA *76* (1979): 4345–4349.
74. Rosenberg, M. & B. M. Paterson: Efficient cap-dependent translation of polycistronic prokaryotic mRNAs is restricted to the first gene in the operon. Nature *279* (1979): 696–701.
75. Lührmann, R., H. Eckhardt & G. Stöffler: Codon-anticodon interaction at the ribosomal peptidyl-site. Nature *280* (1979): 423–425.
76. Sprinzl, M. & F. Cramer: The —C—C—A end of tRNA and its role in protein biosynthesis. Progr. Nucleic Acid Res. Mol. Biol. *22* (1979): 1–69.
77. Donner, D., C. G. Kurland, A. Liljas & I. Petterson: An allosteric ribosome domain. Proc. FEBS Meet. Vol. 51, Pergamon Press, Oxford 1979, pp. 379–385.
78. Bernek, E.: Mechanisms in polypeptide chain elongation on ribosomes. Progr. Nucleic Acid Res. Mol. Biol. *21* (1978): 63–100.
79. Yates, J. L.: Role of ribosomal protein S12 in discrimination of aminoacyl-tRNA. J. Biol. Chem. *254* (1979): 11550–11554.
80. Geller, A. I. & A. Rich: The UGA termination suppression $tRNA^{Trp}$ active in rabbit reticulocytes. Nature *283* (1980): 51–46.

**1.4.6 Proteinogen Processing and Protein Breakdown**

81. Goldberg, A. L. & J. F. Dice: Intracellular protein degradation in mammalian and bacterial cells. Ann. Rev. Biochem. *43* (1974): 835–869.
82. Goldberg, A. L. & A. C. St. John: Intracellular protein degradation in mammalian and bacterial cells: part 2. Ann. Rev. Biochem. *45* (1976): 747–803.
83. Cohen, G. N. & H. Holzer (eds.): Limited Proteolysis in Microorganisms. DHEW Publ. No. (NIH) 79–1591, Bethesda, Md. 1979.
84. Voellmy, R., K. Murakami & A. L. Goldberg: The ATP-dependent degradation of proteins in *Escherichia coli* and mammalian cells. In: [83], pp. 7–16.
85. Holzer, H.: The proteolytic system in yeast. In: [83], pp. 57–60.
86. Rosa, U. & F. Wold: Posttranslational covalent modification of proteins. Science *198* (1977): 890–896.
87. Bohley, P., H. Kirschke, J. Langner, M. Miehe, S. Riemann, Z. Salama, E. Schön, B. Wiederanders & S. Ansorge: Intracellular protein turnover. In: Holzer, H. & H. Tschesche (eds.): Biological Functions of Proteinases. Springer Verlag, Berlin 1979, pp. 17–34.
88. Waterlow, J. C., P. J. Garlick & D. J. Millward: Protein Turnover in Mammalian Tissues and in the Whole Body. North-Holland Publ. Comp., Amsterdam 1978.
89. Hershko, A., A. Ciechanover & I. A. Rose: Resolution of the ATP-dependent proteolytic system from reticulocytes: A component that interacts with ATP. Proc. Natl. Acad. Sci. USA *76* (1979): 3107–3110.
90. Hershko, A., A. Ciechanover, H. Heller, A. L. Haas & I. A. Rose: Proposed role of ATP in protein-breakdown: Conjugation of proteins with multiple chains of the polypeptide of ATP-dependent proteolysis. Proc. Natl. Acad. Sci. USA *77* (1980): 1783–1786.
91. Wilkinson, K. D., M. K. Urban & A. L. Haas: Ubiquitin is the ATP-dependent proteolysis factor I of rabbit reticulocytes. J. Biol. Chem. *255* (1980): 7529–7532.

# 2. Restriction Enzymes and Techniques for Gene Analysis and Synthesis

L. Nover

## 2.1 Introduction

The enormous progress of our knowledge about structure and function of genes in pro- and eukarytotic systems can be traced back – as so often – to a few decisive methodological improvements, which will be briefly outlined in the following [1, 2]. They all start with the discovery and characterization of enzymes selectively cutting DNA molecules into defined pieces (restriction endonucleases). Experimental results obtained by application of these techniques are essential for many parts of this book and will be discussed at appropriate places.

## 2.2 Restriction Endonucleases

Restriction endonucleases were detected in many bacteria (cf. summaries [3, 4]) They fall into two classes: type I and type II. Only the latter type will be discussed here (for some details about type I restriction enzymes cf. **14.2.1**). Most restriction endonucleases are part of a restriction-modification system (R-M-system) designed to protect the host cell against invasion of foreign DNA molecules (cf. **14.2.1**). The endonucleases, cutting DNA at or near distinct recognition sites with inverse rotational symmetry, are combined with specific methylases which protect by methylation the recognition sites in the host DNA. Table **2.1** presents a collection of most frequently used type II restriction endonucleases. These are relatively simple $Mg^{2+}$-dependent enzymes composed of two or four subunits. They cleave the DNA double helix at (all enzymes of Table **2.1**) or near the recognition site so as to produce 3′ hydroxyl or 5′ phosphoryl termini. Cuts are either at two opposite bases or staggered by one to four bases, thus generating even (or blunt) and cohesive (or sticky) ends respectively. By using suitable combinations of such restriction enzymes large DNA molecules can be dissected into pieces of any convenient size. That opens many new experimental possibilities for sequencing of DNA, for genetic fine-structure analysis, for gene transfer and synthesis as well as for studies on differential gene expression.

## 2.3 Cloning of DNA Fragments and Generation of Clone Libraries

To multiply such restriction DNA fragments, bacterial vectors, e.g., plasmids or phages, can be used which, besides qualified recognition sequences for restriction enzymes, contain one or more selection markers, e.g., genes conferring antibiotic resistance to the host cell. Figure **2.1** shows the principle cloning procedure.

**Table 2.1**: Collection of frequently used restriction endonucleases of type II and their mode of cleavage (compiled from Smith [4] and Zabeau & Roberts [3]).

| Organism | Enzyme | Recognition sites and mode of cleavage: Sites[a] | Ends generated: Cohesive 3′ | Cohesive 5′ | Blunt |
|---|---|---|---|---|---|
| *Arthrobacter luteus* | Alu I | 5′–AG↓CT–<br>–TC↑GA–5′ | | | + |
| *Bacillus amyloliquefaciens* H | Bam HI | 5′–G↓GATC C–<br>–C CTAG↑G–5′ | | + | |
| *Bacillus globiggi* | Bgl II | 5′–A↓GATC T–<br>–T CTAG↑A–5′ | | + | |
| *Escherichia coli* RY 13 | Eco RI | 5′–G↓AATT C–<br>–C TTAA↑G–5′ | | + | |
| *Haemophilus aegypticus* | Hae II | 5′–Pu GCGC↓Py–<br>–Py↑CGCG Pu–5′ | + | | |
| *Haemophilus aegypticus* | Hae III | 5′–GG↓CC–<br>–CC↑GG–5′ | | | + |
| *Haemophilus haemolyticus* | Hha I | 5′–G CG↓C–<br>–C↑GC G–5′ | + | | |
| *Haemophilus influenzae* $R_d$ | Hind II | 5′–CTPy↓PuAC–<br>–CAPu↑PyTG–5′ | | | + |
| *Haemophilus influenzae* $R_d$ | Hind III | 5′–A↓AGCT T–<br>–T TCGA↑A–5′ | | + | |
| *Haemophilus parainfluenzae* | Hpa I | 5′–GTT↓AAC–<br>–CAA↑TTG–5′ | | | + |
| *Haemophilus parainfluenzae* | Hpa II[b] | 5′–C↓CG G–<br>–G GC↑C–5′ | | + | |
| *Klebsiella pneumoniae* OK 8 | Kpn I | 5′–G GTAC↓C–<br>–C↑CATG G–5′ | + | | |
| *Proteus vulgaris* | Pvu II | 5′–CAG↓CTG–<br>–GTC↑GAC–5′ | | | + |
| *Providencia stuartii* 164 | Pst I | 5′–C TGCA↓G–<br>–G↑ACGT C–5′ | + | | |
| *Serratia marcescens* $S_b$ | Sma I[c] | 5′–CCC↓GGG–<br>–GGG↑CCC–5′ | | | + |
| *Streptomyces achromogenes* | Sac I | 5′–G AGCT↓C–<br>–C↑TCGA G–5′ | + | | |
| *Xanthomonas badrii* | Xba I | 5′–T↓CTAG A–<br>–A GATC↑T–5′ | | + | |
| *Xanthomonas malvacearum* | Xma I[c] | 5′–C↓CCGG G–<br>–G GGCC↑C–5′ | | + | |

[a] Dots indicate sites of modification by the corresponding methylases.

[b] Msp I (from *Moraxella* spec.) is an isoschizomer of Hpa II, i.e., it recognizes and cleaves the same sequence: 5′–CCGG–3′. But contrasting to the latter Msp I is also capable of cutting the methylated sequence.

[c] Sma I and Xma I recognize and cut the same sequence but with different results.

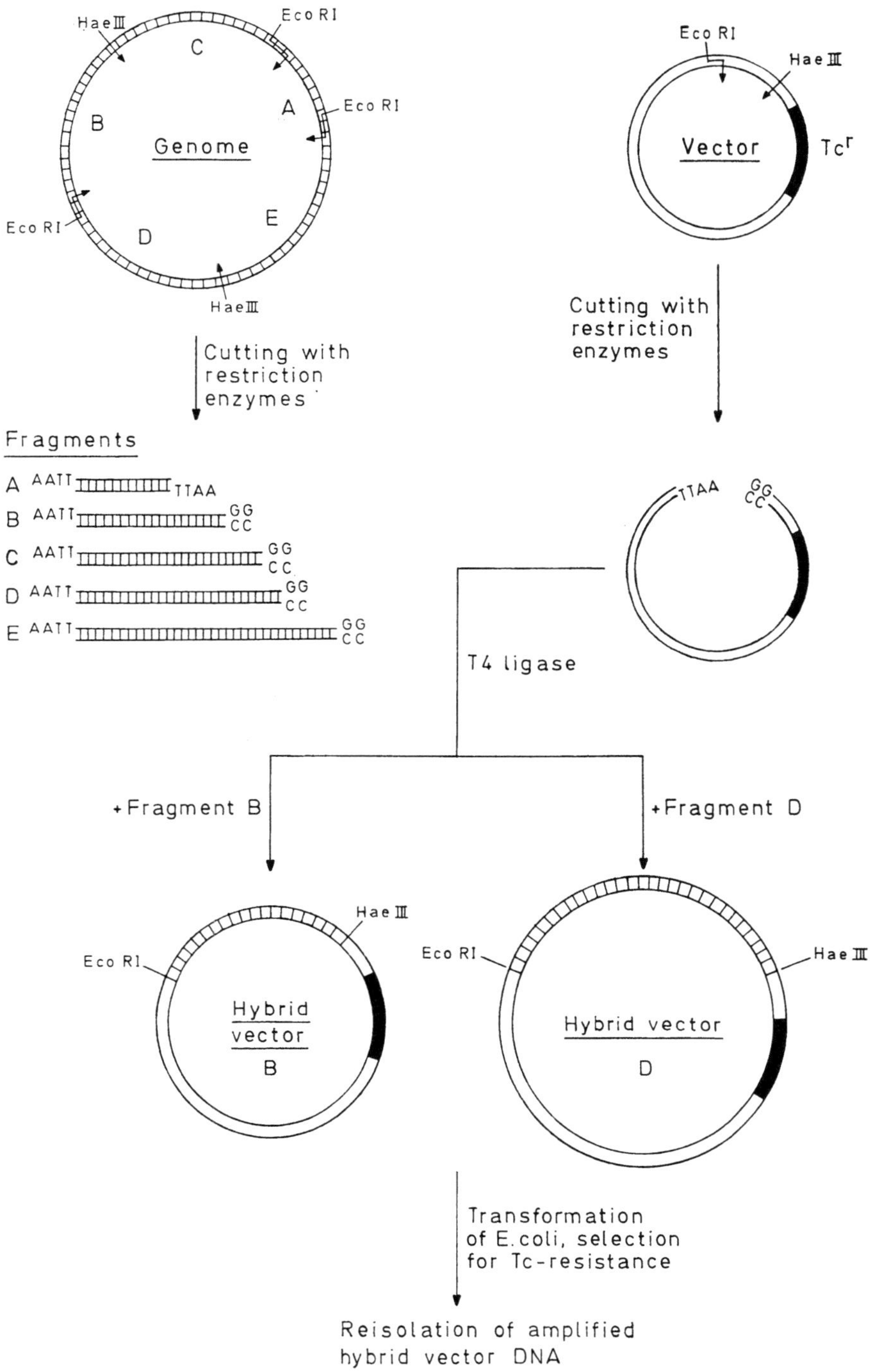


**Fig. 2.1:** Scheme of cloning restriction fragments of a hypothetic circular genome.

The circular genome on the *upper left* is assumed to contain three restriction sites for Eco RI and two for Hae III. Cutting with the two enzymes yields 5 fragments that are ordered according to their size *(A–E)*. The vector used for cloning of these fragments is assumed to contain a gene for tetracycline resistance *(Tc)* and each one site for cleavage by the same two enzymes. Incorporation of the purified fragments, e.g., fragments *B* and *D* into the linearized vector generates the circular hybrid vector DNA of type *B* and *D* respectively. They are amplified by transforming suitable host cells.

Cleavage by restriction endonucleases of a hypothetical circular DNA with three recognition sequences for Eco RI and two for Hae III yields five fragments with 5'-cohesive ends (Eco RI) and/blunt ends respectively (Hae III). Four fragments (B–D) have one Eco RI and one Hae III end each. Therefore these can be selectively incorporated by means of a DNA ligase into the DNA of a plasmid, whose DNA had likewise been treated with Eco RI and Hae III. The resulting hybrid plasmids are multiplied by transformation of a suitable bacterial host, e.g., *Escherichia coli*, and selection of antibiotic resistant cells. After reisolation of the plasmids the incorporated DNA fragments can be selectively removed by treating with the two restriction enzymes Eco RI and Hae III.

In an analogous manner it is possible to obtain complete *clone libraries* even of the highly complicated genomes of eukaryotes. However, contrasting to the very simple example presented in Fig. **2**.1 thousands of different hybrid vectors must be selected and propagated by repeatedly plating single host cells and screening the generated clones for the partial information wanted (colony hybridization technique, cf. **2**.6). The procedure is illustrated by a single example showing essential steps for generation of a mice clone library [6]:

1. Homogenization of mice embryos, extraction of total DNA and purification in a CsCl-gradient.
2. Methylation of all recognition sequences for Eco RI with Eco RI methylase (cf. Table **2**.1).
3. Long-term cleavage under conditions (high *p*H-value, low concentration of $Mg^{2+}$) leading to a degeneration of the recognition specificity of Eco RI from formerly six bases (5'-GAATTC-3') to the relatively often occurring sequence of the central four bases (5'-AATT-3'). The result is a great number of different DNA-fragments of mainly 12–14 kb to be purified by gel electrophoresis and sucrose gradient centrifugation.
4. Incorporation of the DNA fragments into a special λ phage vector (Charon 4A, cf. **2**.4) whose DNA has been cleved with Eco RI.
5. Now the effective transfer of the hybrid vectors into a host cell is of decisive importance in order to conserve the total spectrum of DNA fragments possible. This goal is elegantly achieved by in vitro packaging the Charon 4A hybrid DNA into empty head shells of the λ phage (cf. **2**.5).
6. After infection of *E. coli* the resulting plaques are screened for the desired piece of information, e.g., with cDNA of the immunoglobulin V-region (cf. **2**.6). The phages of positively reacting plaques are purified and multiplied by repeated low density plating. Finally, the mice-DNA fragments incorporated at the beginning can be regained by treatment with Eco RI.

## 2.4 Cloning Vectors

Appropriate vectors for cloning must have properties of a replicon, i.e., they must contain a site for replication origin which interacts with the host DNA replication complex. They must have defined restriction sites for integration of foreign DNA tolerated either because part of the vector can be deleted instead or because an enlargement of the vector DNA does not interfere with its replication functions. Finally, they must contain genetic markers altered by the DNA insertion and/or suitable for selecting transformed host cells. Representative for a great number of different vectors currently used, four types will be described in some detail: plasmid pBR 322, phage λ Charon 4A, cosmids and simian virus 40 (SV 40).

A necessary complement to the vectors are suitable host strains, e.g., of *E. coli*, which are altered by multiple deletions so as to provide a good system for vector propagation under laboratory conditions with a minimum chance for multiplying outside the laboratory. Specialized vectors, together with their incapacitated host strains, form recognized safety systems to be used for gene cloning under strictly defined laboratory conditions.

*Plasmid pBR 322* [7] was constructed from the *E. coli* colicinogenic plasmid Col E1 by incorporating from two different resistance plasmids the genes for ampicillin ($Ap^r$) and tetracycline ($Tc^r$) resistance. pBR 322 is a relatively small circular DNA (4.36 kb) with singular restriction sites for Pst I (in the $Ap^r$ gene), Hind III (in the $Tc^r$ gene promoter region) and Eco RI (cf. Fig. 2.3). As observed for other plasmids replication of pBR 322 in suitable host cells can be "induced" if the host chromosomal DNA replication is stopped by inhibitors of protein synthesis, e.g., chloramphenicol. Up to 800 copies of plasmid DNA per cell can be accumulated. The cloning capacity of pBR 322 is about 0–20 kb.

*Phage λ Charon 4A* [8] is a hybrid phage of λ and the related phage Φ 80 containing two insertions of *E. coli* DNA from the *lac* and *bio* regions (coding for β-galactosidase and enzymes of biotin synthesis). These serve as marker genes. The normal propagation cycle of the λ phage is prohibited by amber mutations in genes *A* and *B* of the morphogenesis region and by deletions, e.g., of the *red*, *xis*, *int* region (for details of the gene functions cf. 14.2.3). Charon 4A contains three Eco RI sites bordering the dispensable DNA region which includes the two insertions of *lac* and *bio* genes. The cloning capacity is 8.2–22.2 kb. Hybrid vectors are selected by using the $lac^-$ and $bio^-$ phenotype of the infected host cells.

*Cosmids* are hybrid plasmids derived from Col E1 by incorporation of the λ cohesive sites (cos, cf. Fig. 14.5). This provides the opportunity to combine the very efficient λ in vitro packaging method (cf. 2.5) with a high cloning capacity. Genetic markers are either those contained in the plasmid DNA, e.g., ampicillin resistance, or an *E. coli* chromosomal gene coding for a rifampicin resistant RNA polymerase (*rpo* B). The cosmid DNA can be as small as 12 kb allowing for cloning of up to 45 kb of DNA.

Though stable transformations of animal and yeast cells have been achieved by means of bacterial vectors as well [11], the use of *simian virus 40* as vector for small DNA fragments in animals cells is much more efficient [10]. SV 40 replicates rapidly, e.g., in monkey kidney cells, reaching levels up to 100,000 copies per cell late in infection. After removal of most of the late genes by cleaving with restriction enzymes DNA can be incorporated. As the strong late viral promoters (cf. Fig. 10.8) are retained, genes on the inserted DNA fragment are expressed quite effectively. For DNA plaque infection the hybrid vector is added together with DNA of a helper virus containing a *ts* mutation in the early genes.

## 2.5 In vitro Packaging of λ DNA

A decisive advantage of λ phage vectors and cosmids is a method for in vitro packaging of the vector DNA. This provides the means for a very efficient and complete transfer of all possible components of a fragment population, e.g., when generating a clone library. The procedure is based on in vitro phage morphogenesis in the presence of the hybrid vector DNA after mixing lysates of two *E. coli* cultures previously infected with two different morphogenesis-defective λ phages. Both lysates complement each other in the defective function and thus give viable phage particles with the vector DNA incorporated. Up to $10^5$ infectious hybrid phages per μg foreign DNA are obtained. In vitro packaging strongly favours incorporation of the hybrid vector DNA containing fairly large DNA fragments. This results from the fact that only DNA which is in the size range of a normal λ genome is efficiently packaged.

## 2.6 Detection of DNA Fragments by Hybridization Techniques

Four basically related hybridization techniques facilitate the screening of *E. coli* colonies or plaques generated by lytic development of phages and of gel electropherograms for the desired DNA sequence or gene. These are the colony hybridization, single plaque hybridization, the Southern blotting techniques, and the two-dimensional gel blotting technique.

For *colony hybridization* [13] the plasmid-containing bacterial colonies are grown on nitrocellulose filters. After a reference set of these colonies has been prepared by replica plating, they are lysed by treatment with 0.5 N NaOH, and their DNA is denatured and fixed to the filter in situ. The resulting DNA prints of the colonies are then hybridized to radioactive RNA or DNA defining the sequence to be screened for. Hybridization is assayed by autoradiography and the corresponding colonies can be isolated from the reference plate.

For *single plaque hybridization* [14] phage infected *E. coli* cells are cultivated on soft agar. Replicas of the plaques generated by lytic phage development are obtained by placing dry nitrocellulose filters on the lawn cells. Phages and their DNA absorbed to the filters are denatured and fixed by treatment with 0.1 N NaOH + 1.5 M NaCl. Hybridization with [$^{32}$P]-labelled probe dissolved in buffer containing 50 % formamide and afterwards removal of unbound radioactivity by washing with buffer is followed by autoradiography. A basically similar technique can also be applied to animal cells infected with SV40 hybrid vectors [15].

For application of the *Southern blotting technique* [12] DNA restriction fragments are separated according to their size by electrophoresis on agarose gels. The DNA is denatured by mild alkali treatment and transferred to nitrocellulose strips. The baked nitrocellulose replica of the agarose gels can be probed by hybridizing with radioactively labelled gene-specific mRNA, DNA or cDNA. After extensive washing, sometimes combined with nuclease digestion, the strip is dried and subjected to autoradiography. For small DNA fragments (~200 bp) interaction with nitrocellulose filters is too weak to withstand the washing procedure. Instead covalent linkage to diazobenzyloxymethylcellulose-paper is preferred.

The *two-dimensional electrophoresis blotting technique* [16, 17] allows for the direct sequence comparison of two DNA pieces digested with restriction enzymes. Restriction fragments of probe 1 are loaded in a well on a 1 % agarose slab gel. After electrophoresis the DNA bands are transferred to nitrocellulose filter paper by the Southern technique. Restriction fragments of probe 2 are treated by the same procedure but labelled by nick translation (cf. **2.7**) before electrophoresis. The DNA fragments of the second gel are denaturated by treatment with 1 M NaCl/0.5 N NaOH. The filter paper of the first gel is laid on the second gel with the two directions of electrophoresis perpendicular to each other. The radioactive DNA of the second gel is blotted through the filter paper under hybridization conditions. After this blotting-hybridization the filter paper is washed several times and exposed to X-ray autoradiography. Fragments with sequence homology are exhibited by spots on the radioautograph.

## 2.7 In vitro Labelling of DNA Fragments

Several types of in vitro labelling methods have been described for obtaining [$^{32}$P]-labelled DNA fragments necessary for hybridization procedures and DNA sequencing. Especially 5′-end labelling by means of polynucleotide kinase and the "nick translation" are frequently used.

For *5′-end labelling* [19] DNA samples are treated with bacterial alkaline phosphatase to remove the 5′-phosphate residues, the incubation mixture is extracted with phenol and precipitated with ethanol. In the next step the dephosphorylated fragments are labelled with [$^{32}$P] by incubating in a Tris-HCl buffer with [$\gamma$-$^{32}$P]-ATP and polynucleotide kinase. Extraction with phenol is followed again by ethanol precipitation.

*Nick translation* [18] is based on the property of *E. coli* DNA polymerase I to repair single strand gaps of DNA. For labelling the DNA fragment is incubated in the presence of DNA polymerase I after adding three unlabelled desoxynucleoside triphosphates (dNTP) plus one [$\alpha$-$^{32}$P]-labelled dNTP. The reaction is stopped by adding phenol for protein denaturation and the labelled DNA is precipitated with ethanol.

## 2.8 Functional Characterization of mRNA and DNA Fragments in *Xenopus* Oocytes and in the HART Assay

Oocytes of the South African clawed frog, *Xenopus laevis*, represent giant cells of 1–2 $mm^3$ volume which synthesize huge amounts of rRNA, 5S RNA, tRNA and histones for the following steps of embryogenesis. These stored and inactive facilities for gene expression can be activated by injection of foreign mRNA's into the cytoplasm or of DNA into the nucleus of oocytes. Thus, *Xenopus* oocytes were introduced by Gurdon [20–22] as highly capable "test tubes" for in vivo studies on transcription and translation including the fidel processing of most eukaryotic RNA and protein precursors.

*Translation of Eukaryotic mRNAs* [20,21]: In contrast to cell-free systems, e.g., the wheat germ or the reticulocyte lysate systems, which translate eukaryotic mRNAs only once (or at most a few times) eukaryotic mRNA's are translated in *Xenopus* oocytes with a significantly higher efficiency, i.e., up to 500 times. Due to this fact 1–10 ng RNA are enough for the demonstration of mRNA activity and the identification of synthesized polypeptides. The mRNA dissolved in 50 $\mu$l water or in Gurdons injection solution is injected into the oocyte cytoplasm by means of a glass capillary of 15–30 $\mu$m diameter held by a micromanipulator. The oocytes are incubated in the presence of labelled amino acids for 20 h at 20 °C. The labelled polypeptides are isolated and analysed by gel electrophoresis. Due to the intensive endogeneous protein synthesis of oocytes enrichment of the desired polypeptides by immunoprecipitation may be necessary.

*Gene Expression with Cloned DNA* [23–26]: By injection into the oocyte nucleus of DNA, e.g., SV40 DNA, 5S DNA or eukaryotic protein-coding genes cloned in an appropriate plasmid, synthesis of complementary RNAs and eventually of their translation products can be investigated. For this end 5–10 $\mu$g of DNA dissolved in 15 $\mu$l buffer are injected into the nucleus together with [$\alpha$-$^{32}$P]-GTP. The oocytes are incubated for 3 to 6 h in a modified Barth medium. Afterwards the labelled RNA is isolated and analysed by gel electrophoresis and hybridization assay. Up to now mostly genes transcribed by RNA polymerases I and III have been tested in this system, e.g., rRNA-, 5S RNA- and tRNA-genes. However, the *Xenopus* oocytes have also been used for transcribing and eventually translating protein-coding genes, e.g., SV40 and polyoma virus DNA or histone and ovalbumin genes [24–26].

The hybrid arrested translation system *(HART assay)* was proposed by Paterson et al. [27] to test the information content of a cloned DNA fragment in an in vitro translation system, e.g., in the wheat germ or reticulocyte lysate systems. For the assay plasmid DNA with the cloned fragment is linearized by treatment with restriction endonuclease. Optimum amounts of the DNA are denatured in the presence of formamide and hybridized with a population of cytosolic mRNA. Those mRNA species that are complementary to the cloned DNA will be hybridized and thus inhibited to serve as messengers in the translation assay. Gel electrophoretic comparison of the translation products plus and minus DNA will exhibit those protein bands that are lacking because they correspond to the genetic information in the DNA fragment. The HART assay is a "negative" assay. A "positive" assay is achieved by binding cloned DNA fragments to nitrocellulose filters and hybridizing the mRNA population to the "impregnated" filters. After washing the complementary mRNA can be eluted and tested in an in vitro translation system.

## 2.9 DNA Restriction Mapping

Cleavage of DNA with restriction endonucleases has also created a new kind of genetic fine-structure analysis [5]. When the fragments, e.g., of the circular chromosome shown in Figl **2**.1, are subjected to gel electrophoresis they are separated according to their size. Thus, with the corresponding standard fragments, it is possible to determine directly the length of the cleavage products from the distance of their migration on electropherograms. By comparison of size and number of the fragments resulting from treatment with a single or a mixture of two endonucleases a restriction map of the genome can be constructed. Since such electropherograms are accessible to the Southern blotting technique (cf. **2**.6) the site of distinct parts of genetic information can be directly localized with suitable samples of radioactively labelled cDNA or mRNA (cf. Chaps. **6** and **18**).

## 2.10 DNA Sequencing

The detailed characterization of DNA sequences which serve as recognition sites for regulatory proteins, for enzymes of gene expression or DNA synthesis and recombination, the development of methods for direct DNA sequencing [28–32] in connection with the application of restriction enzymes has become very essential. Besides the enzymatic plus-minus technique of Sanger & Coulson [31], especially the chemical sequencing according to Maxam & Gilbert [32] (cf. Fig. **2**.2) has contributed decisively to a rapid progress in this field.

The *plus-minus method* is based on in vitro synthesis of a radioactively labelled complementary DNA strand by using DNA polymerase and the template DNA strand to be sequenced. After fixed intervals synthesis is interrupted by removing the four deoxyribonucleotide triphosphates and then continued (a) with addition of one of the four precursors (dATP or dGTP or dTTP or dCTP; plus system) and (b) with addition of three of the precursors (i.e., minus system with respect to the fourth nucleotide). Further synthesis of the complementary strand naturally ends at those sites where the corresponding precursor is lacking. Gel electrophoretic separation and comparison of the DNA fragments directly informs about the nucleotide sequence. Principally similar methods can be used for sequencing of RNA, too.

Faster and less complicated for sequencing of single- and double-stranded DNA, is the specific chemical degradation of DNA fragments according to the *Maxam & Gilbert technique* (Fig. **2**.2). It makes use of the differences in chemical sensitivity of the four DNA bases. After selectively degrading pyrimidine rings (cytosine and thymin) by hydrazine treatment the surgar-phosphate backbone at these sites can be cleaved with piperidine. This process concerns equally cytosine

---

▶

**Fig. 2.2**: Scheme of DNA sequencing according to the Maxam – Gilbert technique (from Scherer [28], modified).

The scheme shows preparation of appropriate 5′-end labelled (p*-5′) DNA fragments and sequencing of the hypothetic subfragment 1 (5′-GCTGTA-3′). Chemical degradation leads to a number of unlabelled products. They are either deleted (fragments of the complementary strand) or put in brackets. The labelled cleavage products are analysed by polyacrylamide gel electrophoresis. The main products are denoted by *heavy lines* and *bold-faced letters*. *Arrows* indicate sites of cleavage.

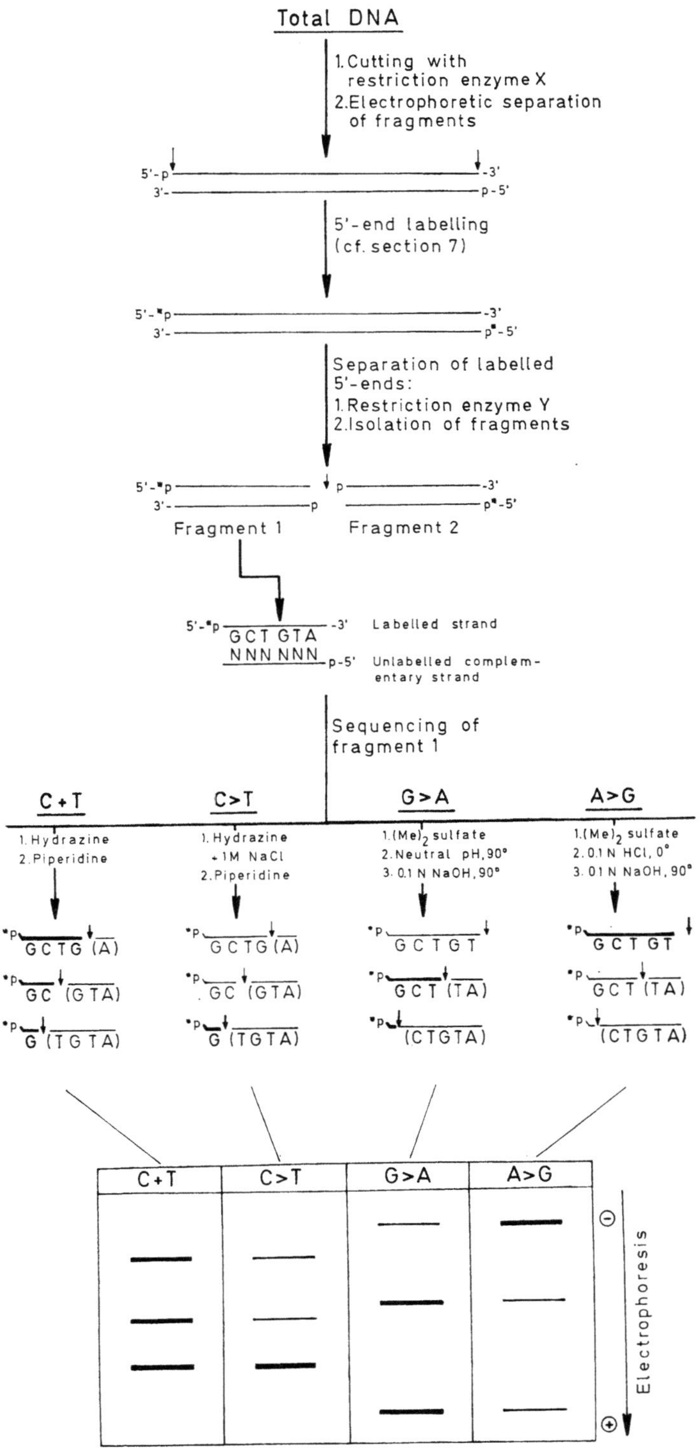
Total DNA
1. Cutting with restriction enzyme X
2. Electrophoretic separation of fragments
5'-end labelling (cf. section 7)
Separation of labelled 5'-ends:
1. Restriction enzyme Y
2. Isolation of fragments
Fragment 1
Fragment 2
GCT GTA
NNN NNN
Labelled strand
Unlabelled complementary strand
Sequencing of fragment 1
C+T
C>T
G>A
A>G
1. Hydrazine
2. Piperidine
1. Hydrazine + 1M NaCl
2. Piperidine
1. (Me)2 sulfate
2. Neutral pH, 90°
3. 0.1 N NaOH, 90°
1. (Me)2 sulfate
2. 0.1 N HCl, 0°
3. 0.1 N NaOH, 90°
GCTG (A)
GC (GTA)
G (TGTA)
GCTGT
GCT (TA)
(CTGTA)
Electrophoresis

Fig. 2.2

and thymin residues (C + T cleavage) or else it prefers cytosine (C > T), if the hydrazine treatment is performed in the presence of 1 M NaCl. In order to remove purin residues selectively, the nitrogen atoms of guanine residues in 7-position and of adenine residues in 3-position are methylated with dimethylsulphate. The following treatment of the modified DNA leads (a) at 90 °C in neutral milieu mostly to the elimination of the N7-methylguanosine residues (G > A) and (b) at 0 °C in 0.1 N HCl preferably to the elimination of the N3-methyladenine residues (A > G). At the purin-free sites the DNA backbone can be cleaved with 0.1 N NaOH at 90 °C. The appropriate dosage of the base specific reagents, i.e., of hydrazine and dimethylsulphate respectively, is decisive for the outcome of the procedure. It must be ensured that statistically only one of the many possible sites is hit in a particular DNA molecule and thus that all possible fragments appear on the following gel electrophoretic analysis of cleavage products (Fig. 2.2). Up to 250 nucleotides can be sequenced in one experiment. Therefore it is necessary to compose longer sequences of nucleotides from the analyses of several small fragments whose relative position on a restriction map is known.

## 2.11 Gene Synthesis

In addition to the methods for site-specific cutting of large DNA molecules into defined fragments and for the cloning and sequencing of these fragments procedures have been designed to synthesize desired genes from such fragments and/or by chemical methods. The most exciting example from the laboratory of Khorana [33] is the total synthesis of a 207 base-pair long bihelical DNA including the 126 nucleotide information of a tyrosine suppressor tRNA precursor plus promoter and control elements at the 5′- and 3′-ends plus two Eco RI integration sites. Cloning of the synthetic gene into a defective λ-vector with amber mutations in two structural genes gave viable phages, i.e., the gene defect was cured by insertion of the synthetic amber suppressor tRNA gene. However, synthesis even of such a small gene needs much time. It requires stepwise condensation of the deoxyribonucleotides supplied with various protecting groups to oligonucleotides with 10–15 bases. These can be phosphorylated at the 5′-ends with polynucleotide kinase and, after reciprocal complementary arrangement, connected with each other in presence of DNA ligase.

The strategy for synthesis and cloning of the much larger gene for the human growth hormone (HGH) involves chemical synthesis of a small part at the 5′-end of the gene and its joining with a large DNA fragment containing the 3′-part of the gene [34]. The mature HGH contains 191 amino acids (Fig. **2.3** A). It is synthesized in the pituitary as a prehormone with a signal sequence of 26 amino acids at the amino end (cf. Table 4.10). The total strategy of synthesis and expression of the HGH gene in *E. coli* is outlined in Fig. **2.3** parts A–D (cf. back cover of book):

A: The DNA sequence for the first 24 amino acids of HGH plus the obligatory initiator amino acid (formyl methionine) plus an Eco RI recognition sequence at one end and a Hind III sequence at the other end are synthesized by chemical methods. Omitting the signal sequence allows to directly synthesize the mature hormone in the bacterial cell.

B: The 2 × 6 oligonucleotides are stepwise connected to the 84 bp DNA sequence (HGH 1–24) and afterwards inserted into the plasmid vector pBR 322 (cf. **2.4**), by means of the cohesive 5′-ends of the Eco RI and Hind III sites. Integration of the DNA fragment at the Hind III site inactivates the gene for tetracycline resistance. The Hind III site separates the promoter and structural gene for tetracycline resistance.

C: The DNA sequence for the large C-terminal part of HGH (HGH 24–191) plus parts of the untranslated sequence (UT) is obtained by synthesis of complementary DNA (cDNA) with purified HGH mRNA as primer for reverse transcriptase. The total sequence is cleaved at Hae III sites in the triplet coding for amino acid residue 24 and in the untranslated sequence, and the DNA fragment of 591 bp is purified by gel electrophoresis. Because of the blunt ends generated by Hae III, terminal deoxynucleotidyl transferase is used to add approximately 20 dC residues per 3′-end of the cDNA fragment. Annealing of the dC-tailed cDNA with the Pst I cleaved and dG-tailed vector gives a hybrid vector (pHGH 31), in which the gene for ampicillin resistance is inactivated.

D: After multiplying both vectors in *E. coli* the two partial fragments are regained by treatment with restriction enzymes and electrophoretically purified. Subsequent connection and integration into the new vector pGH 6 naturally must occur in the correct orientation. For fragment HGH 1–24 this is defined by the Eco RI site at the 5′-end and the Hae III site at the 3′-end. In contrast, polarity of the HGH 24–191 fragment must be restored by treatment with Xma I cleaving a sequence in direct neighbourhood of the stop codon and producing a 5′-cohesive end. Ligation of both fragments results not only in the desired connection of the two Hae III sites but also in all possible combinations of Eco RI and Xma I sites. Hence after treatment with ligase the cleavage with Eco RI and Sma I is repeated. The latter enzyme recognizes the same sequence as Xma I but leaves blunt ends. Electrophoretic purification yields a 591 bp fragment with an Eco RI cohesive and a Sma I even end. The new vector used for integration (pGH 6) originates from pBR 322 by inserting two strong *lac* promoters ($p_{lac}$ UV-5, cf. **4**.4.2.2) in close vicinity of the Eco RI site. For integration the $Tc^r$ gene promoter is removed by cutting with Hind III and degrading the cohesive end with S 1 nuclease. Ligation generates a plasmid vector (pHGH 107) with a new genetic unit containing the two tandem *lac* promoters, the total hybrid gene for the human growth hormone and the gene for tetracycline resistance. Both genes are transcribed as a dicistronic mRNA starting at the *lac* promoter. The HGH concentration in the *E. coli* clone containing vector pHGH 107 was 186,000 molecules/cell or 2.4 $\mu$g/ml. HGH synthesis was inducible in cells containing the *lac* repressor protein.

## 2.12 Conclusions

The experimental techniques related to the application of restriction endonucleases have profoundly influenced the development of most topics to be discussed in the main chapters of this book. The editors therefore considered it necessary to outline the basic principles of these rapidly advancing techniques in a separate chapter of the introductory part. The selection of topics was adapted to the needs of this book. Many more details can be found in two recent books [1, 2]. Furthermore, other techniques frequently complement those described in this chapter. They are treated in connection with experiments discussed in the following chapters, e.g., in vitro translation of mRNA in the wheat germ system (cf. **12**.2.2), synthesis of complementary DNA (cDNA) by reverse transcriptase (cf. **7**.4.5), bacterial in vitro transcription-translation systems (cf. **5**.5) and the electron microscopic visualization of transcriptionally active genes (cf. Figs. **4**.11 and **4**.12) and of DNA-DNA and DNA-RNA hybrids (R-loop technique, cf. Figs. **7**.3 and **9**.10).

## 2.13 Selected References

### 2.13.1 Books and Reviews

1. Grossman, L. & K. Moldave (eds.): Colowick-Kaplan: Methods in Enzymology, Vol. 65, Part I. Nucleic Acids. Academic Press, New York 1980.
2. Wu, R. (ed.): Colowick-Kaplan: Methods in Enzymology, Vol. 68. Recombinant DNA. Academic Press, New York 1980.

3. Zabeau, M. & R. J. Roberts: The role of restriction endonucleases in molecular genetics. In: Gelinas, M. (ed.): Molecular Genetics, Part III. Academic Press, New York 1979, pp. 1–63.
4. Smith, H. O.: Nucleotide sequence specificity of restriction endonucleases. Science *205* (1979): 455–462.
5. Nathans, D.: Restriction endonucleases, simian virus 40, and the new genetics. Science *206* (1979): 903–909.

**2.13.2 Special References**

6. Kemp, D. J., S. Cory & J. M. Adams: Cloned pairs of variable region genes for immunoglobulin heavy chains isolated from a clone library of the entire mouse genome. Proc. Natl. Acad. Sci. USA *76* (1979): 4627–4631.
7. Bolivar, F. et al.: Construction and characterization of new cloning vehicles. II. A multipurpose cloning system. Gene *2* (1977): 95–113.
8. Blattner, F. R. et al.: Charon phages. Safer derivatives of bacteriophage lambda for DNA cloning. Science *196* (1977): 161–169.
9. Collins, J. & B. Hohn: Cosmids: A type of plasmid gene-cloning vector that is packageable in vitro in bacteriophage heads. Proc. Natl. Acad. Sci. USA *75* (1978): 4242–4246.
10. Hamer, D. H. & Ph. Leder: Expression of the chromosomal mouse $\beta^{maj}$-globin gene cloned in SV40. Nature *281* (1979): 35–40.
11. Lai, E. C., S. L. C. Woo, M. E. Bordelon-Riser, T. H. Fraser & B. W. O'Malley: Ovalbumin is synthesized in mouse cells transformed with the natural chicken ovalbumin gene. Proc. Natl. Acad. Sci. USA *77* (1980): 244–248.
12. Southern, E. M.: Detection of specific sequences among DNA fragments separated by gel electrophoresis. J. Mol. Biol. *98* (1975): 503–517.
13. Grundstein, M. & D. S. Hogness: Colony hybridization: A method for the isolation of cloned DNAs that contain a specific gene. Proc. Natl. Acad. Sci. USA *72* (1975): 3961–3965.
14. Benton, W. D. & R. W. Davis: Screening $\lambda$gt recombinant clones by hybridization to single plaques in situ. Science *196* (1977): 180–182.
15. Villarreal, L. P. & P. Berg: Hybridization in situ of SV 40 plaques: Detection of recombinant SV40 virus carrying specific sequences of nonviral DNA. Science *196* (1977): 183–185.
16. Sato, S., C. A. Hutchison & J. I. Harris: A thermostable sequence-specific endonuclease from *Thermus aquaticus*. Proc. Natl. Acad. Sci. USA *74* (1977): 542–546.
17. Shen, C.-K. J. & T. Maniatis: The organization of repetitive sequences in a cluster of rabbit $\beta$-like globin genes. Cell *19* (1980): 379–391.
18. Maniatis, T., A. Jeffrey & D. G. Kleid: Nucleotide sequence of the rightward operator of phage $\lambda$. Proc. Natl. Acad. Sci. USA *72* (1975): 1184–1188.
19. Talkington, C. & J. Pero: Promoter recognition by phage SPO1-modified RNA polymerase. Proc. Natl. Acad. Sci. USA *75* (1978): 1185–1189.
20. Lane, C. D., G. Marbaix & J. B. Gurdon: Rabbit haemoglobin synthesis in frog cells: The translation of reticulocyte 9S RNA in frog oocytes. J. Mol. Biol. *61* (1971): 73–91.
21. Lane, C. D. & J. S. Knowland: The injection of RNA into living cells; the use of frog oocytes for the assay of mRNA and the study of the control of gene expression. In: Weber, R. (ed.): Biochemistry of Development, Vol. 3. Academic Press, New York 1975, pp. 145–181.
22. Gurdon, J. B.: The Control of Gene Expression in Animal Development. Clarendon Press, Oxford 1974.
23. Mertz, J. E. & J. B. Gurdon: Purified DNAs are transcribed after microinjection into *Xenopus* oocytes. Proc. Natl. Acad. Sci. USA *74* (1977): 1502–1506.
24. Rungger, D. & H. Türler: DNAs of simian virus 40 and polyoma direct the synthesis of viral tumour antigens and capsid proteins in *Xenopus* oocytes. Proc. Natl. Acad. Sci. USA *75* (1978): 6073–6077.
25. Kressmann, A., S. G. Clarkson, J. L. Telford & M. L. Birnstiel: Transcription of *Xenopus* tDNA$^{met}$ and sea urchin histone DNA injected into the *Xenopus* oocyte nucleus. Cold Spring Harbor Symp. Quant. Biol. *42* (1978): 1077–1082.
26. Probst, E., A. Kressmann & M. L. Birnstiel: Expression of sea urchin histone genes in the oocyte of *Xenopus laevis*. J. Mol. Biol. *135* (1979): 709–732.

27. Paterson, B. M., B. E. Roberts & E. L. Kuff: Structural gene identification and mapping by DNA-mRNA hybrid-arrested cell-free translation. Proc. Natl. Acad. Sci. USA *74* (1977): 4370–4374.
28. Scherer, G.: Sequenzierung von DNA – Methoden und Ergebnisse. Biologie uns. Zeit *7* (1977): 97–105.
29. Wu, R.: DNA sequence analysis. Ann. Rev. Biochem. *47* (1978): 607–634.
30. Brownlee, G. G.: Sequencing eukaryotic genes or the anatomy of DNA. Biochem. Soc. Trans. *7* (1979): 279–296.
31. Sanger, F. & J. Coulson: A rapid method for determining sequences in DNA by primed synthesis with DNA polymerase. J. Mol. Biol. *94* (1975): 441–448.
32. Maxam, A. M. & W. Gilbert: A new method for sequencing DNA. Proc. Natl. Acad. Sci. USA *74* (1977): 560–564.
33. Khorana, H. G.: Total synthesis of a gene. Science *203* (1979): 614–625.
34. Goeddel, D. V., H. L. Heyneker, T. Hozumi, R. Arentzen, K. Itakura, D. G. Yansura, M. J. Ross, G. Miozzari, R. Crea & P. H. Seeburg: Direct expression in *Escherichia coli* of DNA sequence coding for human growth hormone. Nature *281* (1979): 544–548.

# GENERAL PART

# 3. "Sonderung", Differentiation, Cell Differentiation – a Sketch of the History of Concepts

W. Berg and W. Heese

## 3.1 New Methods – New Conceptions

That it is nowadays possible to do more than merely speculate as to the molecular basis of life is largely due to the development of biochemical and biophysical methods. But the progress of a scientific discipline and the significance that it achieves are determined not only by refinement in its methods, but also by its theoretical definitions. The rapid development of molecular biology over the past three decades made it necessary to redefine the related branches of biology in terms providing an adequate framework for experimental studies.

This is well illustrated in the case of developmental biology, that to-day has some reason to regard the "ultimate goal of physiological embryology" as description of the "exact mechanics of developmental processes (Mechanik der Entwicklungsprozesse), to be explained down to the active molecular forces involved" (Vetter 1885, [28]). This optimism stems from a time when the mere description of isolated factors in animal and plant development had just started.

A fundamental reorientation of any branch of science seldom rests upon a single discovery. It is usually the outcome of experiments and thoughts of an entire generation of scientists. The scene is in fact already set, and when the time is ripe a new theory will be eagerly accepted. However, a single discovery can exert a decisive influence. This is exactly what happened in the case of the theory of descent in the last century. In many respects the way had already been prepared, but the fact that it so rapidly gained a foothold was due to the overwhelming quantity of material evidence presented by Darwin.

A comparable situation led to the establishment of the "developmental mechanics" (Entwicklungsmechanik, Roux 1885, [21]), a branch of science mainly concerned with "differentiation". Again, it was very largely connected with the achievements and convictions of a single scientist – Wilhelm Roux.

The history of concepts and terminology, respectively, inevitably reflects the development of the underlying theories, since a technical term is coined to cover the facts or contemporary state of knowledge. Thus it is not enough to ask when and why a particular term first appeared in the literature, but rather an attempt should be made to trace the changes undergone by the terms used to describe a particular fact. This is the key to understand the development of theories themselves.

## 3.2 Etymology

The word "differentiation" (in German Differenzierung, former Differenzirung) derives from the Latin verb "diferre" = "to spread abroad, to divide" (transitive form) and "to be distinguished from, to be different from" (intransitive form). From these forms are derived the nouns "differens" (-tis, n. = divergence, deviation) and "differentia" (-ae, f. = difference, diversity). From the latter the scholastic philosophers of the 16th century derived the verb " ,ifferentiare" = "to distinguish, to show a distinction, to make a distinction".

From this Neo-Latin verb comes the English verb "to differentiate" and the German "differenzi(e)ren". The corresponding noun "differentiation" covers both the transitive and intransitive meanings of the verb. It is used for the process of differentiation ("to become different") as well as for the state ("to be different"). This is summed up in Webster's Dictionary [31] as: "differentiation . . . the act or process or result of differentiating".

## 3.3 Landmarks in the History of Developmental Biology

"Traditionally, the term 'differentiation' is used to denote the process by which the cells of a multicellular organism become functionally and structurally different during ontogeny; it is usually emphasized that the 'mechanism' of differentiation is one of the major problems of biology" (Mohr and Sitte, 1971). The history of this traditional subject can conveniently be divided into three fundamentally distinct periods, best described as the comparative, the causal and the molecular periods of developmental biology.

### 3.3.1 Comparative Developmental Biology

This period began with the discovery that every organism develops from a primitive germ stage. Although this was implicit in the disputes of the scholastic philosophers, the actual discovery was made by William Harvey (1578–1657). He carried out comparative studies on various species of animals for the first time including also experimental series on mammalian embryos. In the course of this experiments he came to the conclusion that all animals arise from an undifferentiated egg-like anlage. He described both the "worm" (i.e., the larva) of insects and the worm-like embryo of mammals, including man, as a "vegetative primordial form". He regarded what we nowadays term differentiation as the successive formation of parts within a simple and uniform "seed substance" (epigenesis, [4]). Harvey was a typical protagonist of the new empirical approach to science at the beginning of the 17th century combining new experiments with mostly Aristotelian thoughts. Nevertheless he failed to use the microscope newly invented in his time, so that his observations were soon recognized to be superficial and were rejected, together with his conclusions.

Jan Swammerdam (1637–1680) observed insect pupae under the microscope, and like Harvey, assumed them to be eggs. He described how all of the limbs and body segments of the future adult were artistically folded up within them, thus refuting Harvey's postulation of a homogeneous primordial form. Instead, he considered the so-called metamorphosis of insects as the successive growth of organs already preformed in the egg [26]. On the other hand, Antony van Leeuwenhoek (1632–1723), another famous microscopist, thought he saw the future organism already formed in the actively swimming animalcules (spermatozoa) [11]. Later observers even claimed to recognize tiny human forms or homunculi within the spermatozoa.

The theory of preformation began its triumphant progress. "Evolution" or "Differenziation" (Leibniz, [19]) were now used in the sense of the laying bare or becoming obvious of something that had been present all the time but "folded up". The terms had a completely different meaning than in the theory of epigenesis of Harvey. An equivalent term for "differentiation" did not exist, since it was not believed that there was anything homogeneous from which to differentiate.

The fact that the microscope continually delivered new and better arguments in favour of the theory of preformation illustrates very convincingly that the progress of science does not always follow a straight course. The theory of epigenesis, so clearly expounded by Harvey, was now completely forgotten and the great dispute of the time were between the two parties of the preformists, the ovulists and the animalculists. The question was whether the organism is preformed in the egg (ovum) or in the sperms (animalculi). The rivalry provided a stimulus to microscopy and resulted in striking discoveries.

Nevertheless, it was only with the aid of the microscope that the epigenetic theory of development could be reestablished. Caspar Friedrich Wolff (1733–1794) observed the developmental stages in incubated hen's eggs and in 1759, in the "Theoria generationis", he described how during "formation of the parts of the animal body"... "first appears one part which has then to become organized" [33]. In his attempt to explain "how natural causes could account for the formation of complex organic bodies" he used in fact some speculative terms (perhaps one of the reasons why the epigenetic view did not gain a foothold in his lifetime).

His idea about the process of differentiation was that disintegration of the yolk, deposition (depositio) of nutrients, excretion (excretio) and separation (separatio) of substances resulted in the formation (productio, constitutio; organisatio) of distinct parts (distinctae partes) of the animal [33]. He apparently did not realize the fundamental significance of the gradual development of differences between the various parts.

Indeed this remained unrecognized until the beginning of the 19th century. With morphology in a flourishing phase the enormous variety exhibited by the plant and animal kingdoms was interpreted as resulting from a series of metamorphoses of ideal "types" (in philosophical meaning, not historico-genetically). Refinements of the microscope (achromatic lens systems, 1824) and the development of histological staining techniques greatly contributed to the progress of embryological studies. The germ-layer theory (Ch. J. Pander 1817, [18]; K. E. v. Baer 1828, [1]) suggested the common principle for the development of all animals. With the discovery of the mammalian egg (v. Baer, 1827) the very starting point of this development had been found, although the egg, at the time, was not recognized to be a cell. K. E. v. Baer showed convincingly that in all animals the characteristics basic to its type are first developed and then gradually differentiated. This is why the embryos of different species resemble each other more closely than the adult individuals.

The term differentiation (Differenzi(e)rung) was in the German literature hardly used at all during this important era of developmental biology. The vocabulary in use was descriptive and variable. Leading German text books of the time used such words as "Genese" (genesis) or "Bildung" (formation) of the organs, "Spaltung der Blätter" (cleavage of the germ layers), "Scheidung der Urflüssigkeit" (partitioning of the primitive fluid) or "Entwicklung", sometimes even "histologische Entwicklung" (development or histological development). The nearest approach to a technical term was the use of "Sonderung" (separation) and later of "histologische Sonderung". On the other hand, the words "Differenz" and "differiren" were often used for differences and for the state to be different, respectively. Sometimes an author describing "differente Bildungen" (different formations) calls the result a "Differenzirung" (Valentin 1835 [27]), but not yet in a terminological meaning.

In fact the word first advanced to terminological rank in a completely different context. H. Milne-Edwards postulated the "localization of functions" and the "division of physiological labour" as important principles in the "perfection of the organism" [15]. In his opinion this principle was "one that has exerted a most happy influence on progress and human industry". In the middle of the last century this idea had already been enthusiastically adopted by the progressive, popular press of

Germany. Phylogenetically, morphological differentiation was equated to the division of physiological labour. Thus the first use of the term differentiation in the realm of biology was primarily to denote states (of being different). The first mention in a German conversation lexicon of "differentiation in the biological sense" merely refers to the article on "division of labour" (Meyer 1888, [14]).

### 3.3.2 The Causal Phase of Developmental Biology

In the meantime the science of embryology had been undergoing decisive changes. The cell, defined by Max Schultze in 1861 as a lump of protoplasm with a nucleus [23], had become the main interest of biologists as an "elementary organism". Physiologists were now employing physical and chemical methods and chemists were analysing the basic elements of living organisms. They had even synthesized the first organic substances (Wöhler 1828, [32]). The epoche-making achievements of Darwin led to the acceptance of the theory of descent as the phylogenetic explanation of the hierarchy and diversity of organisms. More than once, embryologists had pointed out the need for a causal approach to developmental biology.

That a fundamentally new line of research urgently requires its own terminology is demonstrated by the fact that two attempts were made in close succession. In both cases "differentiation" occupied a central position. The failure of the first attempt clearly shows that a terminology can only survive if it is confirmed by successful experimentation.

Gustav Jäger (1832–1916), a firm adherent of Darwin's theory of descent, developed a peculiar hypothesis concerning the origin of life on the earth [9]: From an aqueous solution of ammonium carbonate (the contemporary idea of the "primeval broth") and with the participation of sulphur and phosphorous compounds a protein-like substance appeared and formed "a droplet that floated in the aqueous medium where it was subjected to concentric differentiation (concentrische Differenzirung) which conferred upon it the familiar cell form". What he meant by "concentric differentiation" was the process by which the outer and inner layers of a body take on a different form under the action of the surrounding medium. In this way the differences between the cell nucleus and the plasma covering should have emerged. The development of different layers of tissue ("Strata") was termed "sociological concentric differentiation". Although he performed experiments concerned with the process of growth, none of them in any way affected his terminology. It is not surprising to find that the "Handwörterbuch der Zoologie, Anthropologie und Ethnologie", Vol. 2, 1883 [10] contains his definitions under the heading of "concentrische Differenzirung" because he was at that time the editor. However, in subsequent volumes, edited by A. Reichenow, the cross-references to "Stratum", "Stratigraphie", "geocentrische" and "polare Differenzirung" were omitted. A purely hypothetical mechanism as an explanation of the differentiation of nucleus and cytoplasm was no longer acceptable after the final rejection of the idea of free cell formation. Instead, in the third volume of the same encyclopedia (1885), under the heading of "Embryologie", the aims of developmental mechanics are formulated in accordance with the views of Wilhelm Roux. But it was pointed out that any attempt even to outline the basic aspects of the subject was premature (Vetter 1885, [28]).

Unlike Jäger, Wilhelm Roux (1850–1924) performed brilliant experiments. He showed that a frog's egg can be manipulated if stabbed with a needle in the early phase of cleavage. His most enduring achievement was the foundation of the "Entwickelungsmechanik" (1885, [21]), for which he outlined a program even before any causal mechanism was really established. Like Jäger, Roux possessed a strong deductive talent which, combined with his consistent experimentation, brought him continual success.

Roux was convinced that the primary concern of causal analytical research should be to localize the specific causes underlying alterations. This was clearly reflected in his definition of the two alternatives "autodifferentiation" ("Selbstdifferenzierung" = determining factors contained within the changing part) and "dependent differentiation" ("abhängige Differenzierung" = specific nature of alteration determined from without). His terminology provided the necessary framework upon which the experimental search for the developmental factors could proceed without an anticipation of the mechanisms itself.

Roux' conception was adopted by many biologists over the ensuing decades under the heading of "developmental mechanics", and Roux himself continued to publish prolifically. Brillant experimenters like Hans Driesch (1867–1941) and Hans Spemann (1869–1941) enlarged the standard repertory of methods by designing isolation experiments (Driesch [3]) and transplantation experiments (Spemann [24]), although their results led to widely differing philosophical interpretations causing an extensive dispute between "mechanists" and "vitalists". Thus the causal approach now dominated embryological research. From 1904 onwards the two better-known German conversation lexica (Meyer, Brockhaus) contained exhaustive entries on the subjects reflecting Roux' ideas, although the biological dictionaries lagged behind (e.g., Linsbauer 1917, [12]).

Of course, Roux' terminology was not adopted by all biologists. Spemann used the alternatives of "ortsgemäße Entwicklung" and "herkunftsgemäße Entwicklung" (development according to localization and according to origin, respectively). Driesch formulated the contrasting terms "prospektive Bedeutung" (prospective significance) and "prospektive Potenz" (prospective potency). Both Spemann and Driesch used other terms for in fact the same conception as Roux.

Similarly, in the ten editions of his "Lehrbuch der Entwicklungsgeschichte" (1888–1915, [5]) Oscar Hertwig (1849–1922) avoided the use of the words "autodifferentiation" and "dependent differentiation" altogether. Nevertheless, almost every author concerned with developmental biology used the word "differentiation". The word became so popular that at the beginning of the 20th century in Germany the words that had previously often been used instead of "living", i.e., "organic, organized" were frequently replaced by "differentiated". With regard to the use of the word the situation was essentially the same as today: "differentiation" being employed as a technical term, but also, in a broader sense, as synonymous for specialization, for tissue- and organ-development (histological and morphological differentiation replaced "Sonderung", e.g., in Hertwig 1913 [6]), and for the development of races in the process of evolution. This is how the word is defined in Roux' own dictionary [22]. Additional entries under the heading of "qualitative Differenzierung" (qualitative differentiation), were "gewebliche Differenzierung" (tissue differentiation) and "Differenzierung der Organe" (differentiation of the organs). Beyond the confines of developmental biology it was now customary to speak, for example, of differentiation of the cell into nucleus and cell body, of the differentiation of individuals into distinct species and of the differentiation of the individuals of a single stock of animals to fulfil different functions.

In the English literature the use of the word differentiation for tissue and organ development was established at an early date due to Thomas H. Huxley's use of the "very suggestive term differentiation" for the German "Sonderung" in his translation of the works of Karl Ernst von Baer (1853, [13]).

It was certainly due to the common use of the term in its broader sense that, although there was an enormous increase in the amount of relevant factual material the definition of the phenomena involved was inadequate. This was clearly reflected in Waddington's (1962) attempts to banish the term altogether from the vocabulary of developmental biology [29]. He suggested replacing it by the terms "histogenesis" and "regionalization" to distinguish between the temporal and spatial aspects of differentiation in cells. In fact this was a more concrete form of Roux' use of "autodifferentiation" and "dependent differentiation" to distinguish between intracellular changes and organizer effects.

Following the discovery of such organizer effects (Spemann 1921, [25]), causal developmental biology went through an enormously successful phase, and with the introduction of important new methods (e.g., in vitro tissue culture, ultracentrifugation, electron microscopy, chromatography, electrophoresis, and immunological techniques) the way was open in the 1930's for the beginning of molecular developmental biology.

### 3.3.3 Molecular Developmental Biology

Molecular genetics provides us with yet another example of the far-reaching consequences of a single discovery. Despite years of experimentation it was still not known whether proteins or nucleic acids represent the genetic material. The two requirements of coding and identical reduplication seemed to be incompatible. The problem was solved in one step by the brilliant feat of Watson and Crick (1953, [30]) in combining the chemical and physicochemical properties of DNA to produce the model of the double helix.

The decisive consequence for developmental biology was a model for a regulatory mechanism (Jacob & Monod 1961, [7]), that settled the preformation or epigenesis dispute at the molecular level. According to this model by means of the regulated mobilization of information that is already present, new enzyme patterns are continously formed to serve as the basis of the functional structures of the cell. The study of regulatory mechanisms and their interactions now occupied a central position in molecular developmental biology. Embryologists, who had so far only been interested in multicellular organisms, now turned their attention to bacteria and phage-infected bacteria which provided them with useful models. It became clear that the key to differentiation had to be looked for at the cellular level, i.e., in the exchange of information between nucleus and cytoplasm. Logically, the term "cell differentiation" or "cytodifferentiation" was now coined to ensure distinction from the development of organs or tissues.

The first demands for this new conception – just as in the case of causal developmental biology – had been made much earlier. In 1904, at the Versammlung Deutscher Naturforscher und Ärzte, Ludwig Rhumbler claimed that the success of physiologists in the realm of "organ mechanics" should be encouragement enough to warrant the establishment of "cellular mechanics" [20]. In fact what he had in mind was the "fluid mechanics" of streaming protoplasm. However, it was not until the 1940's that it was possible to formulate the problem of enzyme induction and enzyme suppression (Monod 1947, [17]). Little by little "cell differentiation" came to be used to cover the genesis of different cell phenotypes in the course of individual development, and to describe the induction processes in eukaryotes in biochemical terms.

The tendency to use "cell differentiation" in biological texts first became noticeable at the beginning of the 1950's, and the first "International Symposium on Cytodifferentiation" took place in 1956.

Since the discovery of differential gene expression in prokaryotes and its manifold possibilities "cell differentiation" has been increasingly employed to describe the intracellular mechanisms of nucleoplasmatic interactions. Thus Jacob and Monod (1963) wrote "that two cells are differentiated with respect to each other if, while they harbour the same genome, the protein pattern which they synthesize is different" [8]. However, this does not prevent "differentiation" from being used to describe histogenesis and organogenesis as well. The situation is very similar to that of a century ago: although the term "cell differentiation" refers more and more specifically to the primary mechanisms involved in gene expression regulation in prokaryotes and eukaryotes, "differentiation" is used in a broader sense for various aspects of animal and plant development. This results from the fact that, while the beginning is made in the analysis of molecular mechanisms, there is still a wide scope for phenomenological studies of differentiating organisms.

In cases where "cell differentiation" is still used to describe the phenomenon itself (e.g., certain text books of the 70's) it is used as a synonym of "histological differentiation". The large number of terms such as "terminal differentiation", "redifferentiation", "dedifferentiation", "transdifferentiation", "chemodifferentiation", "primary and secondary differentiation" might give rise to the impression that "cell differentiation" is only one of a number of related variations. But this view would ignore the fact that modern developmental biology has principally complete conceptions only in the field of molecular biology, in particular with respect to the genetic material and its expression. These concepts are described as cell differentiation in this book.

## 3.4 Selected References


1. Baer, K. E. v.: De ovi mammalium et hominis genesi. Epistolam ad Academiam Imperialem Scientiarum Petropolitanam. Leipzig 1827; –, Ueber Entwickelungsgeschichte der Thiere: Beobachtung und Reflexion. Königsberg 1828.
2. Brockhaus Konversations-Lexikon. 14th ed., vol. 17, Suppl., Leipzig 1904.
3. Driesch, H.: Entwicklungsmechanische Studien I. Der Wert der ersten beiden Furchungszellen in der Echinodermen-Entwicklung. Zschr. wiss. Zool. *53* (1891): 160.
4. Harvey, W.: Exercitationes de generatione animalium. London 1651; cf. Pagel, W.: William Harvey's Biological Ideas. Basel 1967.
5. Hertwig, O.: Lehrbuch der Entwicklungsgeschichte. 1st–10th eds., Jena 1888–1915.
6. Hertwig, O.: Allgemeine und experimentelle Morphologie und Entwicklungslehre der Tiere. In: Hertwig, O. (ed.): Kultur der Gegenwart. Teil 3, Abt. 4, vol. 2, II. Leipzig 1913, pp. 94–175.
7. Jacob, F. & J. Monod: Genetic regulatory mechanism in the synthesis of proteins. J. Mol. Biol. *3* (1961): 318–356.
8. Jacob, F. & J. Monod: Genetic repression, allosteric inhibition, and cellular differentiation. In: Locke, M. (ed.): Cytodifferentiation and Macromolecular Synthesis. New York 1963, pp. 30–64.
9. Jäger, G.: Zoologische Briefe. Wien 1864.
10. Jäger, G. (ed.): Handwörterbuch der Zoologie, Anthropologie und Ethnologie. Vol. 1–2, 1880, 1883, continued by A. Reichenow, vol. 3–6, 1885–1892, J. Frenzel, vol. 7, 1897, and P. Matschie, vol. 8, 1900 (= Encyklopädie der Naturwissenschaften, Abt. I., Theil III, Breslau 1880–1900).
11. Leeuwenhoek, A. v.: Opera omnia, seu Arcana Naturae ... Ed. novissima ... Leyden, 1715–1722; cf. Dobell, C.: Antony van Leeuwenhoek and his "little animals". London, New York 1932; repr. New York 1960.
12. Linsbauer, K. (ed.): C. K. Schneider's illustriertes Handwörterbuch der Botanik. 2nd ed., Leipzig 1917.
13. Merz, J. Th.: A History of European Thought in the Nineteenth Century. Vol. 2, 2nd ed., Edinburgh and London 1912, on p. 305.

14. Meyer's Konversations-Lexikon. 4th ed., 19 vols., Leipzig 1885–1892. – Meyer's großes Konversations-Lexikon. 6th ed., 20 vols., Leipzig 1902–1908.
15. Milne-Edwards, H.: Die Zoologie. In: Beudant, F. S., H. Milne-Edwards, A. v. Jussieu (eds.): Populäre Naturgeschichte der drei Reiche. 4 vols., Stuttgart 1844, esp. vol. 3, p. 6.
16. Mohr, H. & P. Sitte: Molekulare Grundlagen der Entwicklung. Berlin 1971, on p. 62.
17. Monod, J.: The phenomenon of enzymatic adaption and its bearing on problems of cell physiology, genetics and differentiation. Growth *11* (1947): 223–289.
18. Pander, Ch. H. v.: Dissertatio sistens historiam metamorphoseos, quam ovum incubatum prioribusque quinque diebus subit. Würzburg 1817.
19. Radl, E.: Geschichte der biologischen Theorien in der Neuzeit. Teil 1, 2nd ed., Leipzig, Berlin 1913, on p. 221.
20. Rhumbler, L.: Zellenmechanik und Zellenlehre. Verh. Ges. Dtsch. Naturf. Ärzte, 76. Vers., Breslau 1904, T. 1, 88–103 (1904); cf. Querner, H. & H. Schipperges (eds.): Wege der Naturforschung 1822–1972. Berlin, Heidelberg, New York 1972, on p. 195.
21. Roux, W.: Beiträge zur Entwicklungsmechanik des Embryo. Einleitung. Beitrag 1. Zur Orientierung über einige Probleme der embryonalen Entwicklung. Zschr. f. Biol. *21* (1885): 411–524. In: Roux, W.: Gesammelte Abhandlungen über Entwicklungsmechanik der Organismen. 2 vols., Leipzig 1895.
22. Roux, W.: Terminologie der Entwicklungsmechanik der Tiere und Pflanzen. Leipzig 1912.
23. Schultze, M.: Über Muskelkörperchen und das, was man eine Zelle zu nennen habe. Arch. f. Anat. Physiol. u. wiss. Med. (1861): 1–27.
24. Spemann, H.: Über die Determination der ersten Organanlagen des Amphibienembryos. Arch. Entwicklungsmechanik *43* (1918): 448–555.
25. Spemann, H.: Die Erzeugung tierischer Chimären durch heteroplastische embryonale Transplantation zwischen *Triton cristatus* und *taeniatus*. Arch. Entwicklungsmechanik *48*(1921): 533–570.
26. Swammerdam, J.: Biblia naturae, sive historia insectorum. Leyden 1737/38; cf. Schierbeek, A.: Jan Swammerdam, his life and works. Amsterdam 1967.
27. Valentin, G.: Handbuch der Entwicklungsgeschichte des Menschen mit vergleichender Rücksicht der Entwicklung der Säugetiere und Vögel. Berlin 1835, esp. p. 619.
28. Vetter, A.: Embryologie. In: [10], vol. 3, 1885.
29. Waddington, C. H.: New Patterns in Genetics and Development. New York 1962; – (ed.), Towards a Theoretical Biology. Edinburgh 1970, on pp. 184–186.
30. Watson, J. D. & F. Crick: A structure for deoxyribose nucleic acid. Nature *171* (1953): 737–738.
31. Webster's Third International Dictionary. London 1961.
32. Wöhler, F.: Über künstliche Bildung des Harnstoffs. Ann. Physik *12* (1828): 253–256; cf. Neufeldt, S.: Chronologie der Chemie 1800–1970. Weinheim, New York 1977, p. 257.
33. Wolff, C. F.: Theoria generationis (1759). In: Ostwalds Klassiker der exakten Wissenschaften, Nr. 84. Leipzig 1896.

# 4. Molecular Basis of Cell Differentiation

L. Nover

## 4.1 Cell Differentiation as a Prerequisite of Life

### 4.1.1 Definitions

There is no doubt that the existence, as well as the efficient function and propagation of multicellular organisms, depends on the coordinate interaction of their cells specialized for different tasks. The biochemical and morphological peculiarities characterizing the divergent cell types of an organism are the result of cell differentiation, i.e., of the qualitatively and quantitatively selective realization of distinct parts of a given genetic material (differential gene expression). Cell differentiation ultimately proceeds as response on signals from the extracellular milieu which are recognized and transformed if cells are competent, i.e., if they contain the corresponding receptor and signal transformation systems. But in many cases the response includes complex gene expression programs whose progress and completion is relatively autonomous and whose outcome is determinated by inherent regulatory factors of the cell type affected.

However, cell differentiation is by no means restricted to multicellular organisms. Unicellular eukaryotic and prokaryotic organisms respond as well to signals from their surroundings with short-lived adaptive changes in their gene expression patterns but also with profound changes of their developmental states. Moreover, there are several examples of both pro- and eukaryotic unicellular organisms forming, during distinct phases of their life cycle, highly elaborate multicellular aggregates with labour sharing between specialized cells.

Despite many differences in detail there are basic similarities between all living cells with respect to the mechanisms of gene expression, its spatial organization and regulatory means for its selectivity including signal transformation systems. Therefore, when discussing about cell differentiation it appears reasonable to consider prokaryotic as well as eukaryotic systems and to compare, on a molecular basis, aspects as diverse as the cell division cycle (Chap. **15**), metabolic and developmental adaptation of protein synthesis to nutrient or hormonal signals (Sects. **4**.4.2.2, **4**.4.2.7, **4**.4.4.5, **4**.5.3, **4**.6.1, **4**.6.4 and Chaps. **5**–**8**), cell specialization (Chaps. **9** and **11**–**14**), long- and short-range cell–cell interactions (Sects. **4**.2.4.2, **4**.4.4.7, **4**.5.3, **4**.5.4.1, **4**.6 and Chaps. **13**, **17**), programmed gene expression in virus-infected cells (Sects.

---

Abbreviations: AC, adenylate cyclase; bp, kb, base pairs, kilobase pairs (of DNA); cAMP, cyclic adenosine-3′,5′-monophosphate; CAP, catabolite activator protein (interacting with cAMP); ppGpp, guanosine-3′-diphosphate-5′-diphosphate; H1, H2A, H2B, H3, H4, histones (cf. Table **1**.2); HMG, high mobility group proteins (cf. Table **1**.3); NHP, nuclear non-histone proteins (cf. Table **1**.4); hnRNA, heterogeneous nuclear RNA (mRNA precursors); $M_r$, molecular weight, determined by chromatographic methods; d, kd, dalton, kilodalton; FSH, folicle stimulating hormone; LH luteotropic hormone; O,p, operator, promoter (mostly referring to bacterial transcription units); RNAP, RNA polymerase.

**4**.4.2.6, **4**.4.4.3, **4**.4.5.3 and Chaps. **14, 19**) as well as tumorigenesis in plants and animals (Chaps. **18** and **19**). In all cases we are primarily interested in the molecular mechanisms of how cells become different from each other in time and/or in space. We are not so much interested in the results of cell differentiation, the multifold short-lived or long-lived changes unless they contribute themselves, e.g., as regulatory proteins or components of signal transformation chains, to the selectivity of subsequent steps of gene expression programs.

The editors and authors of this book have deliberately decided to tackle the complex problem of cell differentiation by presenting it in two different ways which are mutually dependent: (1) This chapter is aimed at outlining a unifying picture of the molecular basis of cell differentiation by means of experimental results from all living systems. Though it is intended as a summary of all possible aspects of cell differentiation, in its entirety it does not reflect a concrete experimental system. Such systems are discussed in the following special part. Each of these systems reveals certain regulatory details more clearly than others and derives its identity to a considerable extent from a system-specific combination of regulatory elements involved in the developmental control of differential gene expression.

Frequently, deep insights into details of cell differentiation mechanisms result from the analysis of abnormalities, in many cases generated by mutations or virus infection but sometimes also by alterations of the extracellular conditions for gene expression. Many such mutants have been described for systems with an advanced genetics of differentiation and development (e.g., bacteria, yeasts or the fruit fly *Drosophila*). Besides this, we are especially interested in diseased states of man. Thus, in several sections of this chapter we will use the opportunity to discuss data about the molecular pathology of such differentiation abnormalities. In addition two chapters of the special part will be devoted to diabetes (Chap. **8**) and tumorigenesis (Chap. **19**) respectively.

### 4.1.2 Determination and Irreversibility of Cell Differentiation

The term "cell differentiation" is sometimes restricted to examples of irreversible specialization of cells which in extreme cases may include loss of the genetic material (e.g., in mature mammalian erythrocytes, cf. **4**.5.4.4) or even cell death (e.g., host cell lysis as part of phage development, cf. Chap. **14**; stalk cells in slime bacteria and slime moulds, cf. **4**.6.1 and Chap. **17**; xylem cells in vascular plants). This unjustified restriction implies fundamental differences in the regulatory mechanisms of differential gene expression leading to short- and long-lived, seemingly irreversible changes of the cellular protein pattern and function.

It will become apparent that such differences do not exist or, at least, could not be found so far. Determinated gene expression programs and their intercellular coordination in the network of physical and chemical cell-cell interactions within a multicellular organism imply multifold regulatory mechanisms that are outlined in the following sections. Determination appears to be simply a special aspect of cell differentiation, so-called cryptic differentiation, which may include gene translocations (cf. **4**.3 and Chap. **13**) or the formation of regulatory proteins or RNA (cf. **4**.6.3 and **4**.6.4.1) and which precedes "ouvert differentiation" leading to the expression of all characteristics typical for the particular cell type. It is a matter of the analytical procedure that the determinated cell type can be recognized and described in terms of cell differentiation, i.e., by its pattern of differential gene expression especially with regard to those molecular changes which are crucial to subsequent steps of cell specialization.

**Table 4.1**: Proteins as primary products of cell differentiation

| A. Proteins of primary metabolism | B. Proteins of cell specialization |
|---|---|
| (Proteins which themselves or whose products are far spread and play a direct role for the viability of the cell. Regulation of their synthesis and function is mainly an intracellular "problem") | (Proteins, which are restricted to certain cell types or to a few species only. Though they themselves or their products have no direct importance for the cell, they may be essential for the existence and survival of the whole organism. Hence, synthesis and function of these proteins are frequently controlled by extracellular signals). |
| 1. Enzymes<br>– Enzymes of the gene expression apparatus (cf. Chap. **1**)<br>– Enzymes of nucleic acid and protein precursor metabolism<br>– Enzymes of energy metablism (e.g., Chap. **6**)<br>– Enzymes participating in the synthesis and degradation of fatty acids and lipids | 1. Enzymes<br>– Enzymes of secondary metabolism (cf. Chap. **11**) including enzymes involved in cell wall formation (cf. Sect. **4.4.5.2** and Chap. **17**)<br>– Enzymes with morphopoietic functions (e.g., Sect. **14.2.4**)<br>– Synthetic or degradative enzymes or proenzymes secreted to the extracellular milieu by microorganisms, plants and animals<br>– Enzymes involved in virus multiplication (e.g., **4.4.4.3**, **4.4.5.3**, and Chap. **14**)<br>– Adaptive enzymes, which catalyze the conversion or detoxification of nutrients or xenobiotics from the extracellular milieu (cf. Fig. **4.36** and Chaps. **5** and **19**)<br>– Enzymes of signal transformation (cf. **4.2** and Chaps. **10** and **17**)<br>– Enzymes of the photosynthetic apparatus (cf. Chap. **6**) |
| 2. Non-enzymic proteins<br>– Structural proteins of cell membranes<br>– Structural proteins of ribosomes (cf. **1.3.4.2**)<br>– Non-enzymic protein factors of the gene expression apparatus (cf. Chap. **1**)<br>– Structural proteins of the DNP and RNP complexes (cf. **1.2** and **1.3.3**)<br>– Proteins of cytoskeletal systems (e.g., actin, tubulin, cf. **4.1.4**) | 2. Non-enzymic proteins<br>– Proteins which endow cells with a special function within a multicellular organism (e.g., hemoglobin, cf. **4.5.4.3**; muscle proteins; storage proteins in plant seeds, cf. Chap. **12**)<br>– Special proteins in the outer membrane (e.g., antigen receptors, cf. Chap. **13**, or hormone receptors, cf. **4.2.4** and Chap. **8**)<br>– Structural proteins of the photosynthetic apparatus (cf. Chap. **6**)<br>– Regulatory proteins of gene expression (e.g., hormon receptor and acceptor proteins, cf. Chap. **7**; minor proteins of the NHP fraction cf. **4.4.2.5.3**; bacterial repressor or activator proteins, cf. **4.4.2.2** and Chaps **5**, and **14**)<br>– Proteins secreted to the extracellular milieu (e.g., immunoglobulins, cf. Chap. **13**; serum proteins; proteins of the mammalian milk; ovalbumin, cf. Chap. **7**; proteohormones and peptid hormones and their precursors, cf. **4.6.4** and Chap. **8**)<br>– Structural proteins of viruses (cf. **4.4.5.3**) and phages (Chap. **14**) |

### 4.1.3 Proteins and RNA's as Primary Products of Cell Differentiation

End products of gene expression are biologically active proteins and RNAs respectively in their proper compartments. Whereas the latter aspect (compartmentation) is introduced in the following section and further detailed by numerous examples elsewhere in this book, it seems appropriate to deal briefly with a classification of proteins and RNAs whose regulated synthesis, function and degradation is the central aspect of cell differentiation. Table **4.1** considers proteins only. It appears suitable to classify into enzymic and non-enzymic proteins, as well as into proteins of primary metabolism (group A) as opposed to proteins of cell specialization (group B).

One must keep in mind that, besides the gross differences included in the decription of groups and subgroups of Table **4.1**, this classification does not imply basic differences in the regulatory mechanisms involved in the differential gene expression leading to group A and group B proteins. Furthermore, the classification is not absolute. Enzymes or their subunits may serve as structural proteins and, vice versa, structural or regulatory proteins may function as subunits of enzymes. It may be a matter of discussion and of the concrete situation whether to put a particular protein in group A or B, e.g., enzymes of photosynthesis or adaptive enzymes of carbohydrate degradation (group B) or proteins of the cytoskeleton (group A). This ambiguity exists especially for those organisms where the organismic and cellular level are identical, i.e., unicellular organisms or unicellular states of multicellular organisms.

Though proteins clearly dominate as primary products of gene expression, there are important examples of gene expression chains ending at the RNA level. Such are the formation of ribosomal RNAs, of tRNAs (cf. **1**.3.3.2), of primer RNA for DNA synthesis and of regulatory RNA (cf. **1**.3.3.1 and **4**.4.3.1). As, with the exception of the primer RNA, all are directly involved into gene expression and its control, their formation and function will be dealt with in several parts of this book (cf., e.g., Chaps **6**, **12**, **15** and **16**).

### 4.1.4 Compartmentation of Gene Expression

Gene expression probably represents the most complex metabolic chain encountered in nature. Because of the special matter to be processed (genetic information and its derivatives) gene expression must combine highly efficient basal reactions for sequence-specific polymerization of nucleotides (transcription) and amino acids (translation) respectively with the individual handling of each type of information. This includes not only control of gene expression with respect to its qualitative and quantitative parameters (information synthesis, transport and degradation) but as well the proper positioning of intermediates and endproducts in the compartments they are destined for. As we will notice from numerous examples in this book, compartmentation is an essential means to achieve the high selectivity, adaptive flexibility and accurate regulation of gene expression.

Three types of structures are involved:

1. Highly ordered superstructures containing nucleic acids and numerous protein components participate in the two most prominent steps, i.e., in transcription and translation (cf. **1**.3.2, **1**.3.4 and Figs. **4**.11, **4**.12, **4**.17 and **4**.19).
2. More or less extended membrane systems with a remarkable specificity in their structure and functions play decisive roles for signal reception and transformation (cf. **4**.2) as well as for synthesis, postsynthetic modification and transport of proteins (cf. **4**.4.5.1).

3. Most eukaryptoc cells contain cytoplasmic structural proteins which may be organized in a three-dimensional network of filaments and cross-linked micro-trabeculae (cytoskeleton or cellular skeletal framework)[1]. The outer boundary of the cytoskeleton is a lamina of membrane attached proteins. This continuous skeletal system stabilizes the outer membrane as well as the cytoplasmic and nuclear compartments. It is involved in the organization of interphase chromatin (**4**.4.2.5.3), nuclear RNP processing (**1**.3.3.1), in the directed transport of RNP and other particles to the cytoplasm, serves the binding of cytosolic polysomes and presumably participates in signal transformation processes (**4**.2.4.3).

Besides the major proteins of well known structural filaments, i.e., microfilaments (actin, myosin), intermediate filaments and microtubules (tubulin), many further proteins are connected with such skeletal systems. Moreover, cell type-specific proteins are not only typical for the intermediate filaments but they have been described also for the submembranous lamina (cf., e.g., the detailed example given for the mammalian erythrocyte in Sect. **4**.5.4.4). An important property of cytoskeletal systems is their structural flexibility. Rapid changes with dissociation and reassociation of the protein components have been observed. Calcium, protein kinases and special cytoplasmic proteins participate in the control of these gelation processes.

### 4.1.5 Two Types of Regulatory Information

The question about selectivity and control of gene expression is in the center of all experimental and theoretical efforts to describe the molecular mechanisms of cell differentiation. Regulation always requires an input of regulatory information that, with the exception of some extracellular signals, must be encoded in the genetic material. In fact, a considerable part of the genome is devoted to this aim. Basically two types of regulatory information can be discriminated:

1. In most cases it is part of the structural genes themselves or is contained in directly adjacent or interspersed sequences. It may be transferred to the intermediates of the gene expression chain (pre-mRNA, mRNA, proteins) and sequentially interpreted and "consumed" by interaction with components of the gene expression apparatus.
2. Additional regulatory information may be coded by separate genes. In these cases expression of the structural genes to be controlled depends on special regulatory proteins, RNA's and/or effectors and it is frequently influenced or triggered by extracellular signals. The biologically meaningful interpretation of the second type of regulatory information is always bound to the first type, i.e., to the existence of specific receptor structures in the gene system itself.

In the biological reality, however, both types are frequently intermixed and it is not very hepful to stress the differences too much, e.g., by discriminating between control and regulation. Both terms are used as synonyma in this book. The multiplicity of regulatory mechanisms encountered in cell differentiation withstand such a simple classification.

With few, but prominent exceptions of regulatory RNA's (cf. **4**.4.3.2 and **4**.4.4.4), proteins play a dominant role not only in the complex processes of signal transformation (cf. Table **4**.2) but also in transmitting the regulatory information to the proper compartment and target reaction of the gene expression chain. As regulatory proteins

[1] Edelman, G. M.: Science *192* (1976): 218; Webster, R. E., et al.: Proc. Natl. Acad. Sci. USA *75* (1978): 5511; Lazarides, E.: Nature *283* (1980): 249; Fulton, A. B., et al.: Cell *20* (1980): 849; Wolosewick, J. J. & K. Porter: J. Cell. Biol. *82* (1979): 114; Bretscher, A. & K. Weber: Cell *20* (1980): 839.

serve: (a) products of special regulatory genes, (b) structural or enzymatic proteins with a "side role" in regulation and (c) permanent or temporary components of the gene expression machinery itself, e.g., subunits of RNAP, termination factors, ribosomal proteins or translation factors, which contribute to the selectivity of information transfer.

Important for an efficient function of the whole system are mechanisms for control of the cellular level and distribution of the active regulatory proteins. Four types of reactions are involved:

1. interaction with mostly low molecular weight effectors, whose formation and/or removal proceed as parts of signal transformation systems (cf. numerous examples cited in Table **4**.2),
2. reversible enzymatic modification (cf. Figs. **4**.3, **4**.26, **10**.6, **10**.9 and Sect. **14**.2.2.2),
3. control of synthesis and degradation (cf. Figs. **5**.8, **14**.8 and **14**.10) and
4. recompartmentation, e.g., evoked by effector binding and/or modification (cf. Figs. **8**.12, **10**.7 and Sect. **7**.4).

### 4.1.6 Five Main Aspects of Cell Differentiation

For sake of clarity it is useful to treat the following five main aspects of cell differentiation in a consecutive manner:

1. Signal reception and transformation,
2. selective alterations in the genetic material,
3. differential gene expression,
4. organization of gene expression programs and
5. intercellular coordination of cell differentiation within developmental programs of tissues, organs and organisms.

This classification emphasizes one essential feature of cell differentiation: all changes of gene expression patterns proceed in the metabolic and regulatory environment provided by the cytoplasm, which reflects the overall "differentiation history" of the cell and its predecessors. In addition to the structural and catalytic components of the gene expression machinery themselves, cytoplasmic factors play a decisive role for triggering differential gene expression and for controlling its quantitative and qualitative aspects. There is a constant flow of information from the extracellular milieu filtered through signal transformation systems of cell membrane and cytoplasm to the sites where gene expression proceeds. In reverse, the gene expression products influence or, in fact, constitute this cytoplasmic environment and may lead to quantitative or qualitative changes in the signal pattern transmitted by the particular cell to its surroundings. This kind of mutual information exchange and processing is the basis for gene expression programs and their intercellular coordination.

## 4.2 Signal Reception and Transformation

### 4.2.1 Classification of Signal Systems

Differential gene expression, as well as the metabolic activity of practically all cells, is under the perpetual influence of *signals* from the extracellular surroundings (nutrients, hormones, neurotransmitters, light etc.). In most cases these signals do not act directly on their intracellular target reaction. Their effects are rather mediated by more or less elaborate *signal systems* comprising specific receptor proteins, proteins involved into signal transformation, in many instances secondary signals ("second messengers") formed within the cell and, finally, the regulatory protein(s) interacting with parts of the gene expression chain (Table **4.2**). The whole sequence of signal processing provides the means for amplification and for transport of the information to the sites of intracellular action as well as for degradation of the primary or secondary signals. Further, such systems allow to discriminate between "sense" and "nonsense" signals and to integrate the response into the metabolic and developmental state of the target cells. It is not surprising that, with the exception of some simple examples found in bacteria, signal systems generally are rather complex multicomponent systems.

Signal systems in Table **4.2** are classified into three groups of increasing complexity which possibly reflect the evolutionary line of signal systems:

*Group I* comprises substrates or endproducts influencing the levels of proteins that directly participate in the metabolism of these signals. They frequently evoke similar effects in widely different species, though their mode of action may be totally unrelated.

*Group II* contains metabolic integrator systems centered around nutrients which, because of their central role in metabolism, are used to coordinate gene expression as well as enzyme activity in distinct parts of intermediary metabolism, e.g., carbon metabolism or nitrogen supply. Details of the function of such integrator systems are known only for bacteria and some eukaryotic microorganisms. They are frequently involved in the triggering of developmental processes, e.g., bacterial sporulation (cf. **4**.5.3), fruiting body formation in myxobacteria (cf. **4**.5.1), or aggregation and fruiting body formation in cellular slime moulds (cf. Chap. **17**).

*Group III* comprises the most sophisticated hormonal and hormone-like signal systems of higher plants and animals, including hormones, neurotransmitters, interferon, immunogenic signals and light. In all cases the response is determined by the type of target cell. The same group III signal may evoke different types of response by interacting with different target cells of the same organism (for details cf. examples given below). This group also includes all morphogenetic factors discussed in Sect. **4**.6. They were not listed in Table **4.2** because frequently neither their chemical nature nor their mode of action are sufficiently clarified.

### 4.2.2 Group I: Nutrient Signal Systems

Extracellular nutrients are well known for their influences on protein synthesis patterns in cells of all organisms, though their role is most prominent in microorganisms [12, 13]. Nutrient signal systems may be very simple but most of them, especially in eukaryotic cells, involve complex multicomponent chains for signal transformation and their correct function depends on the proper compartmentalization of the participating low and high molecular weight compounds.

In the most simple case (cf. Chap. **5** and Table **4.2**) a particular nutrient (e.g., arabinose) enters the cell *(Escherichia coli)* and interacts directly with a specific regulatory protein (the *ara* C protein) thus promoting synthesis of arabinose degrading enzymes, provided the cell is in a metabolic state (lack of glucose, cf. the description of C-catabolite repression in Sect. **4.2**.3.1) to respond to this signal. The whole system is self-controlling. The signal induces its own degradation, and synthesis of the regulatory protein is submitted to autorepression (cf. details in Chap. **5**).

**Table 4.2:** Survey of signal systems.

| Extracellular signal[a]; Organism | Receptor[b] (compartment) | Factors and processes involved in signal transformation | Regulatory proteins (compartment) | Response | Referred to in this book |
|---|---|---|---|---|---|
| *Group 1* | | | | | |
| Arabinose ↑ (Ars); Bacteria | *ara* C Protein (nucleoid) | No | *ara* C Protein (nucleoid) | Induction of Ars-degrading enzymes | Chap. **5** |
| Lactose ↑ (Lac); Bacteria | β-Galactosidase (cytosol) | β-Galactosidase, allolactose | *lac* repressor (nucleoid) | Induction of lac-degrading enzymes | **4.4.2.2** and Fig. **4.36** |
| Tryptophan ↑ (Trp); Bacteria | *trp* Repressor (nucleoid) | No | *trp* Repressor (nucleoid) | Repression of *trp* operon (operator) | **4.4.2.7** |
| | Trp tRNA synthetase (cytosol) | Synthesis of Trp $tRNA^{Trp}$, translation of leader peptide, RNAP interaction with attenuator | Components of transcription-translation complex | Repression of *trp* operon (attenuator) | Fig. **4.19** |
| Iron ↓ (Fe-transferrin complex); Mammals | Transferrin receptor (membrane) | Coated pit-transglutaminase system (cf. Fig. **8.12**), cytosolic iron chelates | Unknown | Regulation of proteins controlling iron equilibrium | **4.4.4.5** |
| | | Heme synthesis, hemin-dependent repressor | Transl. initiation factor $eIF_2$ | Inhibition of transl. of mRNA e.g., of globin mRNA | Fig. **4.26** |
| Cholesterol (cholesterol-LDL complex); Mammals | LDL receptor (membrane) | Coated pits, coated vesicles, lysosomes, free cytosolic cholesterol | Unknown | Regulation of microsomal enzymes of cholesterol metabolism | Fig. **4.1** |
| Aromatic hydrocarbons ↑ (AH); Mammals | *Ah*-receptor (cytosol) | Unknown | *Ah*-receptor (nucleus) | Induction of enzymes of AH metabolism | Fig. **19.4** |

| | | | | | |
|---|---|---|---|---|---|
| *Group 2* | | | | | |
| Glucose ↑ (Glc); Bacteria | Glc-specific transport protein (cell membrane) | Cytosolic proteins of the PTS-system, adenylate cyclase, cAMP | Catabolite activator protein (nucleoid) | Inducibility of operons subject to catabolite repression | Fig. **4.2** |
| Amino acid ↓ (aa); Bacteria | aa tRNA synthetases | Uncharged tRNA; ribosomal proteins, "stringent" protein, ppGpp | RNA polymerase (nucleoid) | Induction of enzymes of aa biosynthesis, repression of rRNA synthesis | **4.4.2.6** |
| Ammonia ↑ as nitrogen source; Bacteria | Glutamin (Gln) synthetase | Components of the Gln synthetase system (cf. Fig. **4.3**), 2-ketoglutarate | Gln synthetase, interconversion factor $P_{IIA}$ | Repression of enzymes of amino acid degradation, induction of Gln synthetase and Glu dehydrogenase | Fig. **4.3** |
| *Group 3* | | | | | |
| Steroid hormones (estradiol, testosterone); Mammals | Hormone receptor (cytosol) | No | Hormone receptor, chromatin acceptor (nucleus) | Dependent on the target cells | Chap. **7** |
| Insulin; Vertebrates | Insulin receptor (cell membrane) | cf. Chap. **8**, Fig. **8.12** | Largely unknown | Regulation of glucose homeostasis | Chap. **8** |
| Non-steroid hormones (e.g., glucagon, gonadotropins, neurotransmitters); Mammals | Hormone receptor (cell membrane) | Membrane-bound hormone receptor – adenylate cyclase complex (Fig. **4.4**), cAMP, protein kinases | Substrates of protein kinases | Dependent on the target cells | Figs. **4.4** and **10.6** |
| Light; Plants | Phytochrome and other light receptor chromoproteins | Unknown | Unknown | Dependent on the target cells | Chaps. **6** and **11** |
| Antigen; Mammals | Antigen receptor (cell membrane) | cf. Fig. **13.7** | Unknown | Cell proliferation and specialization to antibody secretion | Chap. **13** |
| Interferon; Mammals | Gangliosides (outer cell membrane) | Largely unknown, adenylate cyclase(?), specific protein kinase and oligoisoadenylate synthetase | Tranlation initiation factor $eIF_2$, RNase F | Inhibition of virus multiplication | **4.4.4.7** |

[a] ↑ denotes increase, ↓ denotes decrease of signal concentration.

[b] All proteins that serve as signal-specific primary binding sites are denoted as receptor within a signal system.

Abbreviations: cAMP, cyclic adenosine-3′,5′-monophosphate; LDL, low-density lipoprotein; $eIF_2$, eukaryotic initiation factor 2; PTS, phosphoenolpyruvate: sugar phospho-transferase system; RNAP, RNA polymerase.

In many eukaryotic cells similar reactions of substrate induction are governed by a particular phenomenon of intracellular compartmentation of the inducing agent as illustrated by the induction of nitrate assimilation in plant cell cultures or of enzymes involved in arginine degradation in *Neurospora crassa*. In both systems the inducing signals (nitrate and arginine resp.) are accumulated in considerable amounts in storage compartments (vacuoles) of the cells, but only a very small cytoplasmic pool, directly supplied from the extracellular milieu, has regulatory and metabolic significance. Removing the inducing substrates from the external solution almost immediately stops induction, though the cells are still overcrowded with the substrates (cf. the summaries by Oaks & Gadal [15] and Davis [14]). The mechanisms of this direct measuring of signal concentration in the extracellular milieu and of transferring the information from the cell membrane to the sites where gene expression is controlled are still unknown. The same holds true for the regulatory proteins involved and putative components of signal transformation chains.

The overall complexity of such nutrient signal systems is impressively demonstrated by the regulation of cholesterol and iron equilibrium in animal cells, which may serve as further examples for the decisive role of individual cell compartments for uptake, intracellular release, incorporation and regulatory function of a basic nutrient (cf. below and Sect. **4.4.4.5** for details about the iron uptake system).

### 4.2.2.1 Receptor-Mediated Endocytosis and Regulation of Cholesterol Homeostasis in Animal Cells

Animal cells acquire a large part of cholesterol necessary for cell membrane synthesis by receptor-mediated endocytosis of a cholesterol-lipoprotein complex (low-density lipoprotein, LDL). The mechanisms underlying LDL uptake are illustrated in Fig. **4.1**. They are characteristic of other protein transport systems as well, e.g., for the uptake of the iron-transferrin complex (cf. **4.4.4.5**), of some proteohormones like nerve growth factor and insulin (cf. Chap. **8**) and for the uptake of lysosomal enzymes by certain cells.

LDL is a large spherical particle of 22 nm diameter ($M_r \sim 3 \times 10^3$ kd). It contains a hydrophobic core of approximately 1,500 cholesterol molecules esterified to long-chain fatty acids, mainly linoleate. The core is surrounded by a hydrophobic coat of phospholipid, free cholesterol and a glycoprotein (apoprotein B). The protein makes about 25 % of the particle. The lipoprotein

---

▶

**Fig. 4.1**: Model of cholesterol-LDL uptake by receptor-mediated endocytosis (after Basu et al. [16], Goldstein & Brown [17] and Goldstein et al. [18]).

Cholesterol linoleate circulates in the blood as an integral part of low-density lipoprotein particles *(LDL)*. The rapid internalization of LDL proceeds by receptor-mediated endocytosis *(steps 1–5)*: **(1)** Binding of LDL and **(2)** clustering of LDL receptors over special regions of the cell membrane marked by clathrin (coated pits) leads to **(3)** endocytosis. Fusion of intracellular coated vesicles with primary lysosomes **(4)** results in formation of secondary lysosomes, where the protein coat of LDL is removed, the cholesterol ester is cleaved and part of the LDL receptors are degraded. Recycling **(5)** of LDL receptors and clathrin coated membrane vesicles is very rapid. The free cholesterol of the cell enters three pathways **(7)**: cholesterol storage by reesterification **(7a)**, incorporation into membranes **(7b)** and, in steroidogenic cells, conversion to steroid hormones **(7c)** (cf. Fig. **4.5**).

The intracellular cholesterol level is controlled by the rate of LDL uptake, cholesterol de novo synthesis **(8)** and the reversible esterification **(7a)** by acyl-coenzyme A: cholesterol acyltransferase (ACAT). High levels of intracellular free cholesterol increase activity of ACAT and inhibit formation of the key enzyme of cholesterol biosynthesis **(8)**, i.e., of hydroxy-methylglutaryl coenzyme A reductase (HMG-CoA reductase). A receptor down-regulation is typical of situations with high extracellular LDL concentrations, because degradation of LDL receptors in secondary lysosomes cannot be belanced by their de novo synthesis on membrane-bound polysomes **(9)**.

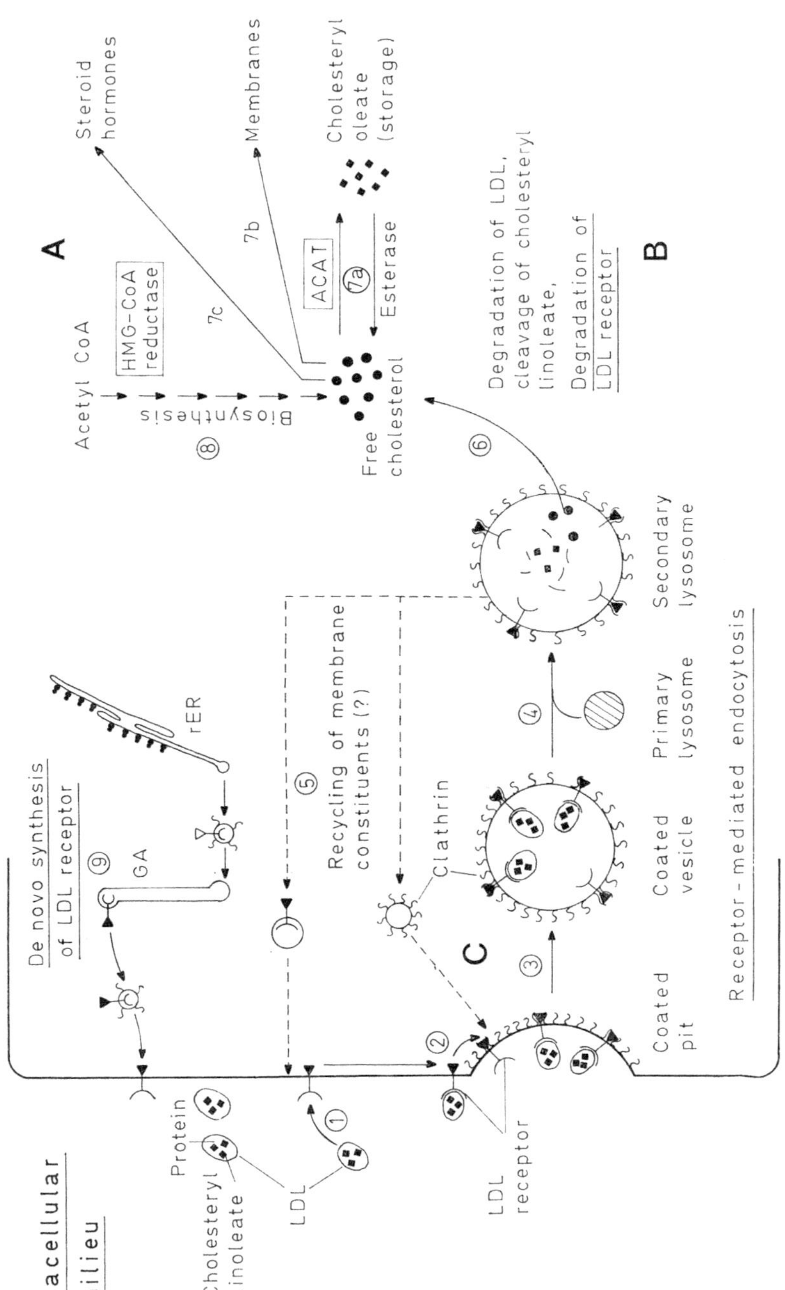
Extracellular milieu
Cell membrane
De novo synthesis of LDL receptor
rER
GA
Protein
Cholesteryl linoleate
LDL
LDL receptor
Coated pit
Coated vesicle
Clathrin
Primary lysosome
Secondary lysosome
Receptor-mediated endocytosis
Recycling of membrane constituents (?)
Degradation of LDL, cleavage of cholesteryl linoleate,
Degradation of LDL receptor
Acetyl CoA
HMG-CoA reductase
Biosynthesis
Free cholesterol
ACAT
Esterase
7a
7b
7c
Steroid hormones
Membranes
Cholesteryl oleate (storage)
A
B
C
1
2
3
4
5
6
8
9

Fig. 4.1

coat of LDL is synthesized and secreted by liver cells as a precursor particle which is transported to adipose tissue to be processed and charged with cholesteryl esters [17].

The LDL uptake depends on the presence of specific high-affinity receptors that are transmembrane glykoproteins of animal cells. Up to 70,000 receptors per human fibroblast have been detected. Uptake proceeds very fast. Half-time of the external LDL-receptor complexes is 3 min at 37 °C. The turnover of the receptors themselves exhibits a half-life of 20 hours. A prominent intermediate of the receptor-mediated LDL uptake is the clustering of the occupied receptors over special regions of the cell membrane (coated pits) that afterwards invaginate to form the coated endocytotic vesicles.

Coated pits and coated vesicles have been found in virtually all animal cells [18–20]. They derive their name from a coat of periodically spaced fibres and bristles radiating from the cytoplasmic face. Coated vesicles are evidently composed of a lipid vesicle and a protein coat that contains one predominant protein with $M_r$ 180 kd (clathrin). Clathrin can be removed from coated vesicles and can be induced to reaggregate, under in vitro conditions, to the same type of regular protein basket normally surrounding coated vesicles. Coated vesicles evidently also function as intracellular carriers of protein precursors on their way from the rER to the compartments they are destined for (cf. 4.4.5.1).

Receptor-mediated endocytosis of LDL evokes a number of regulatory effects (Fig. 4.1) essential for the maintenance of cholesterol homeostasis. After degradation of the lipoprotein coat in secondary lysosomes the free cholesterol serves as an indispensible component for membrane synthesis at the endoplasmic reticulum. In steroidogenic cells it serves as precursor for the biogenesis of hormones (cf. Fig. 4.5) e.g., of cortisol (cells of the adrenal cortex), of testosterone and estradiol (testicular Leydig cells) or of estradiol (luteal cells of the ovary). Upon excess supply of cholesterol from the extracellular milieu the rate of LDL uptake rapidly decreases because of receptor degradation. Further, an increased reesterification of cholesterol in the cells and a stop of the cell's internal synthesis of cholesterol by blocking the de novo formation of hydroxymethylglutaryl-CoA reductase are observed. This key enzyme of cholesterol biosynthesis has a half-life of 2–3 h (cf. Table 4.12) and therefore is well suited for such a rapid regulatory response.

Important contributions to the delineation of the mechanism of receptor-mediated endocytosis of LDL and its intracellular degradation in lysosomes stem from the analysis of human mutant fibroblast cultures obtained from patients with genetic defects in distinct steps of the LDL pathway [17]. Especially revealing were three types of LDL-receptor mutants resulting in familial hypercholesterolaemia[2]. There are (1) mutants with non-functional receptor (no receptor detectable by LDL binding), (2) mutants with receptors showing reduced affinity and (3) mutants whose receptors bind LDL normally but fail to aggregate over coated pits (no internalisation of LDL). Two further mutants with defective LDL pathway were shown to result from a block of either the synthesis or secretion of the apoprotein B of LDL and from the lack or inactivity of the lysosomal lipase that cleaves cholesteryl linoleate (cholesteryl ester storage disease).

The receptor aggregation over coated pits presumably involves direct interactions of a hypothetic cytoplasmic domain of the transmembrane protein with clathrin. Even unoccupied LDL receptors tend to form clusters and are internalized. Evidently the recycling of receptor and other membrane components (Fig. 4.1) functions quite effectively under normal conditions [16]. In con-

---

[2] These patients exhibit extremely high plasma LDL levels with accumulation of free and esterified cholesterol in interstitial spaces and within phagocytic cells of skin and tendons as well as a high incidence of atherosclerosis with myocardial infarction.

trast, there is good evidence that in other systems, e.g., receptor-mediated endocytosis of insulin (cf. Chap. 8), clustering of the occupied receptors and formation of coated vesicles require covalent cross-linking of the receptor molecules with distinct membrane proteins [19]. The linkage is catalysed by locally activated transglutaminase and involves $\varepsilon$-($\gamma$-glutaminyl)-lysine bridges (cf. Fig. 8.12).

This short overview on the mechanisms of LDL uptake demonstrates the function of coated pits and coated vesicles as transport organelles that carry selected receptor-bound proteins and nutrients bound to protein carriers into cells. As will be discussed in Chap. 8, the reason for proteohormones to use the same system are not totally clear. It is evident that the fusion with primary lysosomes contributes decisively to the degradation of both hormones and receptors and thus constitutes an essential element of control, e.g., by receptor down-regulation under conditions of signal excess. But to date no regulatory action of a proteohormone can be clearly attributed to the internalized hormone. Long-term biological effects must be studied in this respect (cf. the details in Chap. 8).

### 4.2.3 Group II: Metabolic Integrator Systems of Bacteria

Besides numerous rather specific nutrient signal systems summarized in group I of Table 4.2 there is a limited number of metabolic integrator systems with pleiotropic regulatory effects. Well known examples of bacteria are (1) N-catabolite repression mediated by a highly complex signal transformation system centered around glutamine synthetase, (2) control of amino acid supply by the ribosome-ppGpp system (cf. 4.4.2.6 and 4.6.1) and (3) C-catabolite repression mediated by cAMP. Though experimental data are very limited, there is good evidence that similar integrator systems play important roles also in many eukaryotic cells (cf., e.g., the maintenance of glucose homeostasis in animals, Chap. 8). We will restrict ourselves to the two bacterial systems responsible for C- and N-catabolite repression.

#### 4.2.3.1 C-Catabolite Repression

It was estimated that C-catabolite repression in *E. coli* controls expression of about 100 different genetic units (operons, regulons) coding for degradation of a vast number of nutrients which can be used as carbon sources. To describe the basic phenomenon briefly: as long as enough glucose is supplied from outside no other carbon source has the chance of being degraded because substrate induction is blocked. Cells are not "competent" to respond to inducing signals, e.g., arabinose or lactose. This effect is mediated by cyclic adenosine-3′,5′-monophosphate (cAMP) and its receptor protein CAP. The cAMP-CAP complex is required as a positive regulatory protein for transcription of all genetic units mentioned above (cf. 4.4.2.2).

Compared to our advanced knowledge about the molecular biology of catabolite repression, details about the signal transformation chain involved as well as its spatial organisation are sparse. A current model is summarized in Fig. 4.2. Glucose and some other carbohydrates are actively transported into *E. coli* cells by a sugar-specific integral membrane protein (EII) that is associated with two cytosolic helper proteins (EI and HPr) to a transport system called PTS (phosphoenolpyruvate: sugar *p*hospho*t*ransferase *s*ystem). The two soluble proteins serve the energy supply from phosphoenolpyruvate (PEP) to the membrane protein. In addition, one of them is supposed to interact, e.g., as a regulatory subunit, with adenylate cyclase (AC) and thus to control synthesis of cAMP. Two states of the PTS-AC system reflect its dual

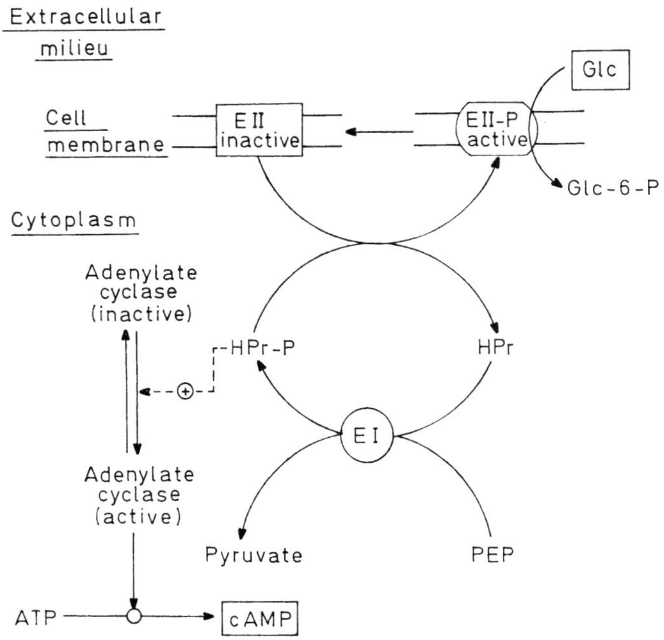


**Fig. 4.2:** Scheme of signal transformation chain controlling cAMP synthesis in bacteria (from Nover [21], after Peterkofsky [24]).

Uptake of glucose *(Glc)* by the energized membrane-bound permease *EII-P* proceeds with concomitant phosphorylation. The inactive permease *EII* is rephosphorylated by phosphoenolpyruvate *(PEP)* in a reaction catalyzed by enzyme I *(EI)* and involving a heat stable, soluble protein cofactor *(HPr)*. The phosphorylated form of HPr (or of *EI*?), which accumulates under conditions of glucose *(Glc)* deficiency, may function as regulatory subunit or activator of adenylate cyclase.

role (1) for sugar transport and (2) for processing the signal about extracellular carbohydrate concentration to the gene expression chain:

1. The PTS is in a highly phosphorylated state, because no glucose transport occurs. Adenylate cyclase is active to synthesize cAMP.
2. The PTS is under-phosphorylated because during transport the phosphate group is rapidly transferred to glucose. Adenylate cyclase is inactive.

Evidence for this model stems from the following observations [24]:

1. Catabolite repression as it is evoked by addition of glucose to *E. coli* cultures is dependent on intact cells.
2. Fructose acts as catabolite repressor only after formation of the fructose-specific isozym of the integral membrane protein (EII) has been induced.
3. Mutants lacking EII always exhibit high activity of adenylate cyclase (high levels of phosphorylated EI and HPr), whereas mutants lacking EI have low adenylate cyclase activity.

This in part rather speculative model of a signal transformation system in *E. coli* is related in its basic features to the well known hormone receptor-adenylate cyclase systems in membranes of animal cells (cf. Fig. 4.4). It explains how the signal about the extracellular glucose concentration is processed and transferred to the sites of gene expression. However, it must be kept in mind that the cAMP level and hence catabolite repression in *E. coli* are influenced not only by the rate of synthesis but also by the rate of degradation, as well as by the rapid excretion of cAMP to the extracellular milieu. The role of glucose for these two processes is still obscure. Moreover, it is well established that the intracellular cAMP concentration is also influenced by the rate of uptake of so-called non-PTS sugars (e.g., lactose, maltose, arabinose) whose transport depends on corresponding permeases and a proton symport. Their transport requires a proton motive force across the membrane [25, 26]. Collapse of the proton gradient either by sugar transport or by addition of protonophores (e.g., $4\,\mu$M carbonyl cyanide m-chlorophenyl hydrazone) evidently also leads to inactivation of adenylate cyclase. The "coupling factor" mediating the energized state of the cell membrane to adenylate cyclase is unknown. Hence, we are left with a second model of signal transformation that couples sugar transport to the synthesis of cAMP.

#### 4.2.3.2 Signal Transformation by the Glutamine Synthetase System

Regulatory and enzymatic center of nitrogen metabolism in *Escherichia coli* and related bacteria (e.g., *Klebsiella aerogenes*, *Salmonella typhimurium*) is the glutamine synthetase system. Signals about the nitrogen supply are transformed by this system (Fig. **4**.3A) and lead to corresponding alterations of the metabolic and gene expression patterns (Fig. **4**.3B).

These bacteria use two basically different reaction sequences to synthesize glutamate (Glu) from ammonia and 2-ketoglutarate (KGA). Both pathways have the same result, but the second needs addition of ATP [13]:

1. Ammonia excess, action of L-glutamate dehydrogenase (GDH):

$$NH_3 + KGA + NAPH + H^+ \xrightarrow{GDH} Glu + NADP^+$$

2. Ammonia deficiency, combined action of glutamine synthetase (GS) and glutamate synthase (GluS):

$$NH_3 + Glu + ATP \xrightarrow{GS} Gln + ADP + Pi$$

$$Gln + KGA + NADPH + H^+ \xrightarrow{GluS} 2\,Glu + NADP^+$$

As is evident from Fig. **4**.3A the activity of the expensive second pathway is controlled by reversible modification of its first enzyme (GS), whose 12 identical subunits serve as substrates of an adenylyl transferase. The fully adenylylated form of GS ($GS \cdot AMP_{12}$) is enzymatically inactive. Important control elements of the enzymatic interconversions are a special protein factor ($P_{IIA}$, $P_{IID}$) and the two key compounds of intermediary metabolism that are indicators of ammonia excess (Gln) and ammonia deficiency (KGA) respectively.

The central role of the GS system for the control of nitrogen metabolism concerns not only the enzymatic level (Fig. **4**.3A) but also the expression of many genes involved in synthesis, uptake and degradation of amino acids [13, 27a]. Ammonia deficiency represses synthesis of GDH and stimulates synthesis of GS itself, but also of many other enzymes and proteins which might improve the ammonia supply for the anabolic needs of the cell, e.g., urease, nitrogenase, asparaginase, enzymes of histidine, prolin and arginine degradation, periplasmic glutamine- and arginine-binding proteins [13]. Under conditions of ammonia excess the reverse holds true. It is worth mentioning that the signal transformation system shown in Fig. **4**.3 does not lead directly to induction of the enzymes mentioned, e.g., for amino acid degradation. It rather makes the cell competent to respond to extracellular signals, e.g., about the presence of arginine or histidine in the medium.

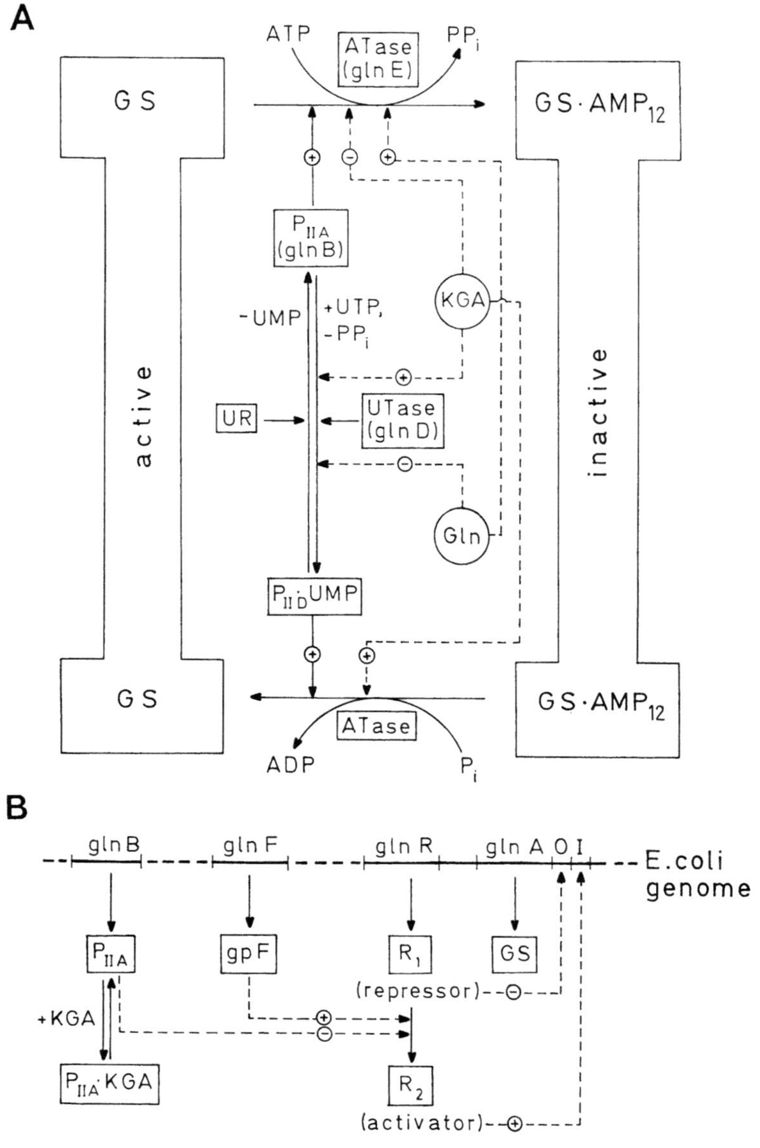


**Fig. 4.3:** Control of enzymes of nitrogen metabolism in bacteria by the glutamin synthetase signal transformation system.

**A** *(above)*: Enzymatic interconversions in the glutamin synthetase *(GS)* system (after Adler et al. [27]).

GS is coded by the *glnA* gene. It exists in an enzymatically active form and an inactive form with all of its 12 subunits ($M_r$ 50 kd) adenylated. The enzymatic interconversion is catalyzed by adenylyl transferase (*ATase*, coded by *glnE* gene), whose activity is adenylating in the presence of $P_{IIA}$ but deadenylating in the presence of $P_{IID}$ (uridylated form of $P_{IIA}$). The interconversion factor $P_{IIA}$ is a protein with $M_r$ 44 kd (product of *glnB* gene). It undergoes reversible interconversions between the two activity states by cycles of uridylation and deuridylation catalysed by uridylyl transferas (*UTase*, product of *glnD* gene) and uridylyl-removing enzyme(s) *(UR)*. The two key metabolites indicating nitrogen deficiency (2-ketoglutarate, *KGA*) and nitrogen excess (glutamine, *Gln*) respectively, act positively (--⊕-->) or negatively (--⊖-->) on distinct reactions of the two interconversion cycles.

The regulatory system acting at the gene expression level is detailed for the *glnA* gene in Fig. **4.3**B. It is centered around a regulatory protein R that acts either as repressor or as activator of *glnA* transcription. The activator conformation arises by interaction with the product of another regulatory gene *(glnF)*. In addition genetic evidence clearly indicates that here again is an important regulatory role of the interconversion factor $P_{IIA}$. It inhibits formation of the *glnR* activator protein and this inhibitory effect is reversed by binding of KGA to $P_{IIA}$. Hence we can discriminate between two alternative regulatory states:

1. Ammonia deficiency: → high level of KGA and low level of Gln, → interconversion factor in $P_{IIA} \cdot$ KGS and $P_{IID}$ form, → activator form of *glnR* product predominates, → transcription of *glnA* and inducibility of other gene systems involved in ammonia supply.
2. Ammonia excess: → low level of KGA and high level of Gln, → interconversion factor in $P_{IIA}$ form, → repressor form of *glnR* product predominates, → no transcription of *glnA* and no inducibility of the other gene systems.

In the original version of this regulatory scheme (cf. the first edition of this book and Ref. [13]) GS itself was thought to play the role not only for signal conversion and ammonia fixation but also as regulatory protein for gene expression. However, the numerous regulatory mutants, originally mapped in the *glnA* gene (coding for GS), by more recent experiments were shown to define a closely neighboured but independently transcribed regulatory gene *(glnR)*. Furthermore, the system was completed by the finding of another regulatory gene *(glnF)* [28, 29] and by the genetic demonstration that the interconversion factor itself serves the transmission of the information about the ammonia situation in the cell to the bacterial genome [30]. But it remains to be shown unequivocally that the regulatory interplay between the products of the three regulatory genes glnB, glnF and glnR as described in Fig. **4.3** really influence the expression of all other gene systems mentioned above. Thus, regulation of the glutamate dehydrogenase level cannot be explained by this scheme. In agreement with the original scheme a possible role of GS as regulatory protein must be discussed. Another major open question concerns the mechanism by which the *glnF* gene product promotes the transformation of the repressor into the activator form of the *glnR* gene product.

### 4.2.4 Group III: Hormonal Signal Systems

Because of the fundamental role of hormones and hormone-like factors for intercellular coordination of cell specialization and maintenance of the specialized state, they are dealt with in many parts of this book (cf. references in Table **4.2**). Thus we will restrict ourselves to the discussion of a single complex, i.e., the mode of action of non-steroid hormones interacting with the adenylate cyclase system in animal cells.

---

◀

**B** *(below)*: Effects of the signal transformation system on gene expression exemplified for the transcription of *glnA* gene (after Kustu et al. [28] and Foor et al. [30]).
Transcription of the gene for glutamin synthetase *(glnA)* is controlled by the product *(R)* of an adjacent regulatory gene *(glnR)*. R functions as repressor *($R_1$)* and as activator *($R_2$)* protein by interacting with the regulatory region O/I of the *glnA* gene. Formation of the activator form of R depends on the activity of a further regulatory protein coded by *glnF* (gp F). It is inhibited by the unmodified form of the interconversion factor ($P_{IIA}$). Binding of KGA destroys this inhibitory effect of $P_{IIA}$. The *glnR* locus is evidently composed of two cistrons coding for two proteins with $M_r$ 36 kd *(ntrB)* and 54 kd *(ntrC)*. Only the latter is sufficient for the activator, whereas both proteins are required for the repressor function [31].

### 4.2.4.1 Model of the Membrane-Bound Hormone Receptor and the Role of Adenylate Cyclase in Animal Cells

For many years a considerable part of experimental efforts with animal cells has concentrated on the mode of action of hormones as intercellular signals. It turned out that all hormones primarily interact with high affinity receptor proteins that are characteristic constituents of their target cells. Other, non-target cells are not competent to respond to the particular hormone because of lacking receptors. These are localized either in the cytoplasm (e.g., for steroid hormones [cf. Chap. **7**] or thyroxine) or, for many other animal hormones, e.g., for peptide hormones, proteo-

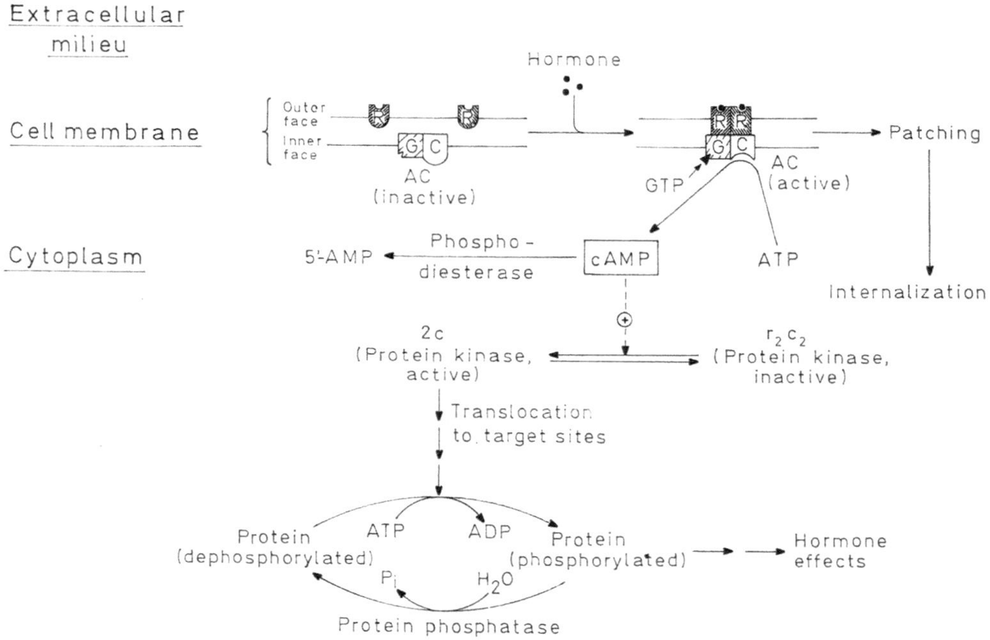


**Fig. 4.4**: Hypothetic scheme of signal transformation triggered by non-steroid hormones acting on animal cells (from Nover [21]).

Unoccupied hormone receptors *(R)* are distributed randomly and diffuse freely in the phospholipid fluid of their target cell membranes. After hormone binding they aggregate to smaller and larger patches which interact with other membrane proteines, e.g., inactive adenylate cyclase *(AC)* or proteins controlling $Ca^{2+}$ ion permeability. Besides rapid changes of ion permeability evoked by some hormones the activation of adenylate cyclase and hence synthesis of the "second messenger" cAMP is a well documented early effect. Activity of adenylate cyclase is evidently not only controlled by integration into a transmembrane complex of the occupied hormone receptor with the catalytic subunit *(C)* of adenylate cyclase but also by a regulatory subunit *(G)* which binds GTP, thus connecting signal transformation with the metabolic state of the cell (cf. details discussed in Chap. **10**). In further progress of the reaction the newly formed cAMP, by binding to the regulatory subunits *(r)* of protein kinases, triggers their activation and migration to cell compartments where their target reaction proceeds (cf. Fig. **10**.6). Phosphorylation of enzymes or regulatory proteins in certain compartments by the catalytic subunits (c) of protein kinases is an important element of the specific hormone response.
A cAMP phosphodiesterase, protein phosphatases and the receptor down-regulation by internalization (cf. text) are essential parts of the signal transformation system that help to destroy the signal itself as well as intracellular products of its transformation.

hormones and growth factors, neurotransmitters, they are integrated into the outer surface of the plasma membrane. Our understanding about the complex sequence of signal transformation triggered after hormone binding to the outer cell membrane has been decisively advanced by a number of theoretical and experimental results (cf. summaries [32–39], [42] and [45]): These are (1) the fluid-mosaic model of cell membranes and the influence of membrane fluidity on signal transformation, (2) the evidence for the existence of cytoskeleton proteins and their possible role for signal transformation (cf. **4**.2.4.3), (3) the discovery that membrane-bound adenylate cyclase (AC) synthesizes cAMP as mediator of hormone action, (4) the demonstration that cAMP-dependent protein kinases regulate, by phosphorylation, the activity of their specific protein substrates (e.g., enzymes or proteins involved into gene expression (cf. Chap. **10**) and, finally, (5) the evidence for a hormone-triggered intracellular recompartmentation of protein kinases (cf. Fig. **10**.7).

These results are combined in a hypothetical model (Fig. **4**.4) that reflects essential aspects of signal transformation processes involving membrane-bound AC. It is based on the coupling between the freely diffusible hormone receptor in the outer layer of the cell membrane and the AC complex in the inner, cytoplasmic layer. The formation of this transmembrane complex resulting in cAMP synthesis is triggered by hormone binding. This mode of action is true for many peptide hormones, e.g., glucagon (cf. Chap. **8**), the gonadotropic hormones (LH, FSH, HCG, cf. below) and ACTH, for prostaglandins, as well as for a number of neurotransmitters, e.g., noradrenaline, histamine, dopamine and acetylcholin.

However, it must be mentioned that this list of signals, whose action is mediated by cAMP, is variable and depends on the cell type investigated. The great differences in the response of different cell types to the same signal result, inter alia, from differences in the signal transformation chains. Thus, catechol amines (dopamine, noradrenaline), by interacting with $\beta$-adrenergic receptors, increase the cAMP level whereas by binding to $\alpha$-adrenergic receptors they evoke effects that are mediated by cGMP or are at least cAMP-independent.

The generation of an intracellular "second messenger" [33], e.g., cAMP, is followed by activation of distinct isoforms of protein kinases, by their eventual intracellular translocation and by phosphorylation of proteins involved in the cell-specific response (cf. Figs. **10**.6 and **10**.7 and Chap. **10** for further details). Control of the biological activity of proteins by phosphorylation seems to be a universal mechanism encountered in all living systems. In many cases enzymes, factors and structural proteins of the gene expression chain are affected.

Essential aspects of signal transformation are its reversibility and control. Degradation of cAMP by corresponding phosphodiesterases, the role of phosphoprotein phosphatases and the participation of GTP and a GTP-dependent regulatory subunit of adenylate cyclase will be discussed in connection with Fig. **10**.6. In addition, two further regulatory phenomena deserve particular attention. They concern the hormone receptor itself: (1) Shortly after signal binding the system is insensitive to respond to the same signal once more or to other signals using the same transformation chain (refractoriness). The mechanism is unclear, but in some cases a temporary inactivation of adenylate cyclase seems to be the cause. Alternatively, changes of the cell membrane, e.g., by phosphorylation of membrane proteins, or massive aggregation of receptors interfering with coupling to adenylate cyclase might be involved. (2) A long-term effect on signal reception observed with peptide hormones is receptor-down regulation by internalization (cf. also the description on receptor-mediated endocytosis via coated pits in Sect. **4**.**2**.**2**.**1**.

### 4.2.4.2 Gonadotropin Control of Steroidogenesis

The complexity of signal transformation processes and our limited knowledge about them will be evident from the following short description of gonadotropin-stimulated testosterone synthesis in the interstitial Leydig cells of mammalian testis (cf. **4.6.4.2**). The example is especially revealing because it concerns the intercellular control of the formation and secretion of another intercellular signal (testosterone) which profoundly influences differentiation and function of distinct target cells within the sexual system of mammals. The regulatory signal acting on Leydig cells is the luteotropic hormone (LH) of the anterior pituitary.

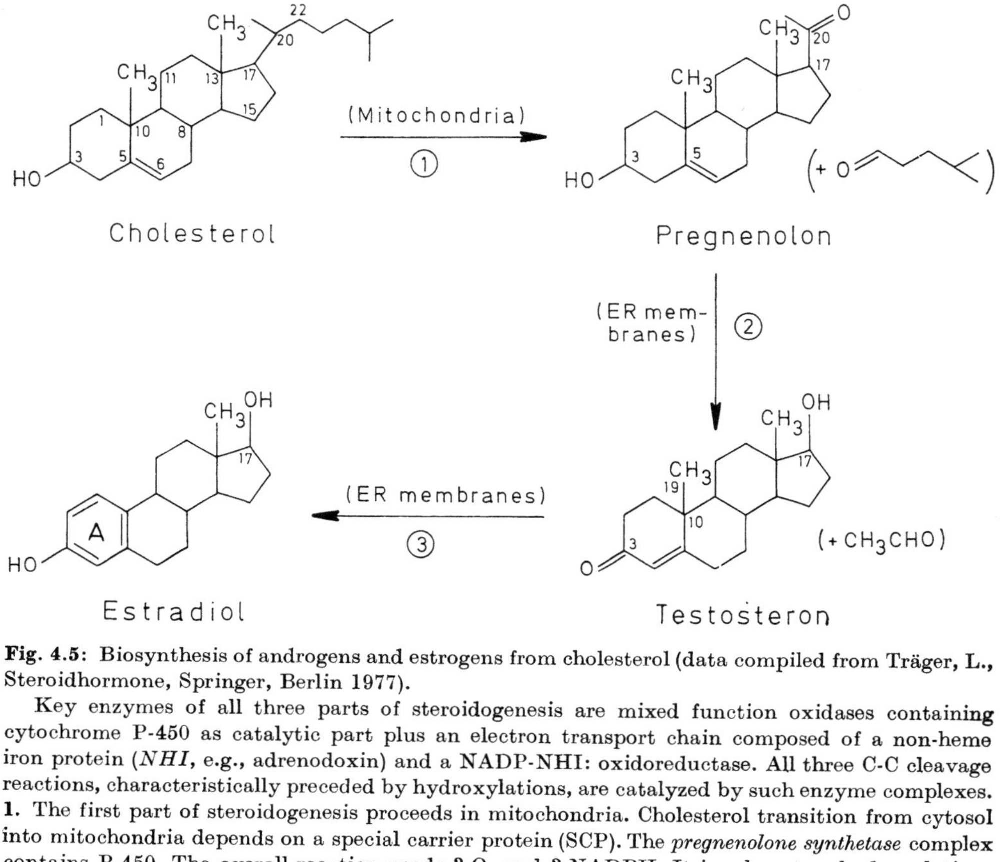


**Fig. 4.5:** Biosynthesis of androgens and estrogens from cholesterol (data compiled from Träger, L., Steroidhormone, Springer, Berlin 1977).

Key enzymes of all three parts of steroidogenesis are mixed function oxidases containing cytochrome P-450 as catalytic part plus an electron transport chain composed of a non-heme iron protein (*NHI*, e.g., adrenodoxin) and a NADP-NHI: oxidoreductase. All three C-C cleavage reactions, characteristically preceded by hydroxylations, are catalyzed by such enzyme complexes.

**1.** The first part of steroidogenesis proceeds in mitochondria. Cholesterol transition from cytosol into mitochondria depends on a special carrier protein (SCP). The *pregnenolone synthetase* complex contains P-450. The overall reaction needs 3 $O_2$ and 3 NADPH. It involves two hydroxylations (in 20$\alpha$- and 22R-positions) and finally side chain cleavage between C-20 and C-22.

**2.** The following two parts of steroidogenesis are catalyzed by enzymes of the endoplasmic reticulum. There is evidence that the *testosterone synthetase* in the ER membranes consists of an ordered multienzyme complex of 3$\beta$-hydroxy-$\Delta^5$-steroid dehydrogenase, steroid-$\Delta^5$-isomerase, the P-450 containing 17$\alpha$-hydroxylase and C-17, 20 lyase and a 17$\beta$-hydroxysteroid dehydrogenase.

**3.** The steroid aromatase is likewise a P-450 containing enzyme complex bound to ER membranes. The reaction needs 3 $O_2$ and 3 NADPH and involves a C-19 hydroxylase, a C-19 oxidase and a C-10, C-19 lyase activity. Final elimination of C-19 (as formic acid) leads to the aromatization of ring A that is characteristic of estrogens.

An initial, rapid phase of reversible LH binding to the specific receptor glykoproteins of the Leydig cells is followed by a prolonged phase of slow and incomplete reversal of receptor-hormone interaction. This suggests that the initial interaction leads to a conformational change in the receptor molecule. Association with the adenylate cyclase complex might be the cause or the result of this altered state of the LH receptor. Hormone binding leads to a rapid increase of the intracellular cAMP level, to an activation of protein kinases, to the stimulation of cholesterol conversion to pregnenolone and, finally, to an enhanced production and secretion of testosterone and its aromatic derivative estradiol (Fig. **4**.5). It is an important finding that Leydig cells, specialized to the production of male sex hormones, are also capable of converting part of the testosterone to the "female" sex hormone oestradiol which exerts distinct regulatory functions within the male sexual system, too (cf. **4**.6.4.2).

The precise mechanisms by which protein phosphorylation regulates steroidogenesis remain obscure. There is some evidence for a concerted internalization of the LH-receptor complex and of cholesterol-lipoprotein bound to corresponding receptors of the target cell membrane. But a major site of control during LH stimulation is an early step in the biosynthetic pathway at the conversion of cholesterol to pregnenolone. At any rate, the LH-stimulated synthesis of androgens in Leydig cells depends on the continuous synthesis of RNA and protein. Further, a labile protein is required for the LH-evoked increase of testosterone secretion.

It is remarkable that many details about the mode of action of LH on Leydig cells can be applied without modification also to other steroidogenic target cells of animals that are stimulated by trophic hormones. Such are cells of the adrenal cortex forming cortisol under the control of adrenocorticotropin (ACTH) or luteal cells of the ovary forming oestradiol under the control of LH. Not only the control points in the steroidogenic pathway but also the mechanisms of signal reception, transformation and internalization are very similar in all these cases.

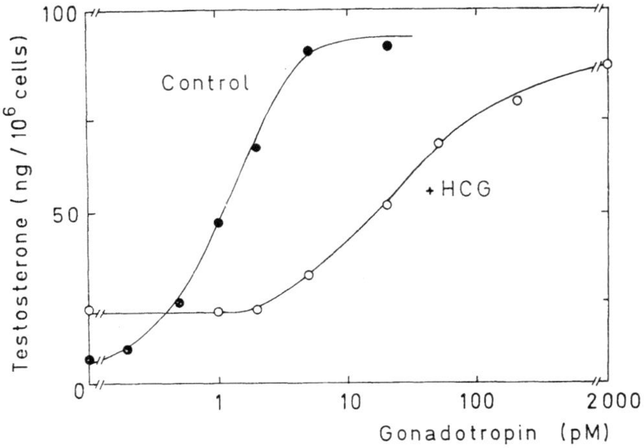


**Fig. 4.6**: Down-regulation of LH receptors in membranes of testicular Leydig cells (after Catt et al. [40]).

Stimulation of testosterone production (100 = 50 ng/$10^6$ Leydig cells) was analysed in isolated Leydig cells from adult male control cats (●) and from cats treated 72 h earlier by subcutaneous injection of 10 $\mu$g HCG (○). Testosterone production was determined by radioimmunoassay. The loss of LH receptors cause a ~20 fold increase of the effective dose of gonadotropin required for 50 % stimulation of the testosterone production. HCG causes a loss of 85 % of the LH receptors after 3 days. Application of the human chorionic gonadotropin (HCG) instead of the luteotropic hormone (LH) to deplete LH receptors of Leydig cells is possible because of the close structural relationship between both polypeptide hormones.

Of particular significance for studies on receptor regulation is the demonstration that the refractory stage includes the occupancy of the LH receptor sites by the initially applied hormone [40, 41]. This stage lasts for about 24 h and precedes a prolonged loss of LH receptors from the cell membrane by internalization (Fig. **4**.6) which proceeds for eight days and is maximal after one to two days. After treatment with low doses of hormone the receptor loss is much greater than calculated from the initial occupancy suggesting an active and cooperative process in the cell membrane. The subsequent return to normal receptor levels, between days 6 and 10 after hormone application, depends on new synthesis of receptor molecules rather than on recycling. It has not been clarified whether LH-receptor internalization by endocytosis proceeds via coated pits (cf. Fig. **4**.1) or requires interaction with elements of the cytoskeleton system. As is true for other hormonal signals, the biological role of hormone-receptor internalization and the following degradation in the lysosomal compartment primarily serves the termination and control of hormone action. But it is very likely that the long-term effects of gonadotropins for cell growth and differentiation as well as maintenance of the differentiated state are not mediated by the adenylate cyclase system but rather depend on intracellular hormone liberated from the lysosomal compartment.

#### 4.2.4.3 Evidence for the Floating Receptor Model

There are two key properties of hormone receptors that are crucial for their function in signal transformation processes as described in the model (Fig. **4**.4): i) They exist as independent mobile entities in the plasma membrane. ii) After hormone binding they tend to interact with the adenylate cyclase (AC) system (cAMP formation) and with other receptor molecules (receptor clustering and internalization). A number of experimental data support this "floating receptor" model [35]:

1. The apparent target sizes of the glucagon receptor and the catalytic unit of adenylate cyclase in liver cell membranes were determined by irradiation inactivation in an electron beam. The size of the functional glucagon-coupled receptor – adenylate cyclase complex was almost precisely the sum of the values obtained for the two uncoupled components determined in the absence of glucagon [42].
2. By cell fusion experiments it could be shown that receptors of signal A of an adenylate cyclase negative cell type can be coupled to the AC system of the fusion partner. Within 3 min, of fusion, cAMP production by the fused cell could be stimulated by signal A. Receptors for catecholamines and prostaglandin $E_1$ and different cell types have been tested in this kind of experiments [42].
3. The catecholamine receptor of turkey erythrocytes was solubilized by treatment with the detergent deoxycholate. By means of phospholipid vesicles it could be implanted into cell membranes of Friend erythroleukemia cells devoid of these $\beta$-adrenergic receptors. A 30-fold stimulation of adenylate cyclase activity was evoked by the implanted receptor after addition of catecholamines [44].
4. Alterations of membrane fluidity, e.g., by manipulating the cholesterol content, have marked effects on the coupling of receptors with the AC system. Moreover, stimulation of $\beta$-adrenergic receptors in rat reticulocyte membranes results in a rapid stimulation of phospholipid methylation and translocation, which, in turn, enhances membrane fluidity and formation of the active receptor-AC complex [42, 43].

The present state of our knowledge allows some general conclusions about signal reception and transformation and its control. However, we are left with the important problem concerning the specificity of the cellular response to distinct signals. Some elements contributing to this specificity emerged from this section, e.g., restriction of receptors to target cells, coupling with different types of signal transformation chains, recompartmentation of protein kinases or hormone internalization. A much broader discussion will follow in Chap. **10**.

Another problem is the possible role of cytoskeletal elements (cf. **4**.1.4) in the clustering of the occupied hormone receptors. We have to discriminate between, at least, two mechanisms:

1. As outlined in Sect. **4**.2.2.1 on receptor-mediated endocytosis of LDL some peptide hormones (e.g., insulin, epidermal growth factor and nerve growth factor), after interaction with their membrane receptors, are clustered and internalized at specific regions of the target cell membrane (coated pits), i.e., clathrin may function as the cluster organizing protein at the cytoplasmic face of the plasma membrane. There is no evidence for the participation of cytoskeletal elements in this process.

2. On the other hand, non-hormonal multivalent signals acting on cell membranes (e.g., antigens, mitogens or lectins) were clearly shown to evoke patching and capping of their occupied receptors with subsequent endocytotic internalization without using coated pits (cf. **13**.3.1). But receptor aggregation in this case is dependent on the interaction with the submembranous microfilaments (actin, myosin and associated proteins). A locally controlled influx of $Ca^{2+}$ ions from the extracellular milieu and its subsequent binding to the calcium receptor protein (calmodulin) have decisive influence on this interaction.

So far, we have no direct experimental evidence for a possible function of cytoskeletal elements either in the aggregation of gonadotropin-occupied hormone receptors with adenylate cyclase or in the clustering and internalization of these types of receptors.

## 4.3 Selective Genome Rearrangements

Metastabile genetic systems have been well known for many years. They oftenly change their functional states (phenotypes) with a frequency several orders of magnitude higher than observed for ordinary mutations. But it was only 20 years ago with the thorough analysis of genetically unstabile control elements in maize by McClintock [67], that such somatic genome rearrangements as means for selective gene expression were more and more recognized as a feature common to all living systems. Many terms have been coined: jumping genes, nomadic or transposable elements, fluid components, unstable genes etc. Without a deeper insight into the molecular mechanisms it was already evident that such controlled rearrangements might play an important role in determinative processes. Hence the definition of cell differentiation (cf. **4**.1.1) as based on the selective realization of the same genetic material must include these types of intracellular genetic rearrangements.

The following examples serve to illustrate the manifold aspects of this exciting field, whose rapid development was decisively enhanced by the methods of restriction analysis and DNA cloning (cf. Chap. **2**). It will be shown that the molecular details of these rearrangements are just beginning to emerge and that in many cases the implications for control of gene expression or determination is far from being clear. Two prominent examples, i.e., the repeated gene rearrangements as prerequisite for selective expression of immunoglobulin genes in animals and the site-specific integration and excision of the genome of phage $\lambda$ as necessary part of its developmental cycle, will be dealth with in special chapters (cf. Chaps. **13** and **14** respectively). Finally, it should be mentioned that the RNA tumor viruses of animals (cf. **19**.2.2.2) exhibit many properties similar to transposable elements [48, 48a].

### 4.3.1 Transposable Elements in Bacteria

The most detailed investigations on transposable elements in bacteria have been accomplished for *Escherichia coli*. Three basic types of such elements were described: (1) insertion sequences (IS elements) of 0.8–1.4 kb, which do not code for proteins (2) transposons (Tn elements), which code for antibiotica resistance and possibly for transposition proteins and (3) transposable phages, e.g., phage Mu. Some details of IS and Tn elements are given in Table 4.3. Their analyses have made important contributions to a deeper understanding of the overall phenomenon, though evidence for a defined role of IS and Tn elements for bacterial gene expression regulation is scarce. One illustrating example is the insertion element controlling flagellar phase variation in *Salmonella* spec. (see below).

**Table 4.3**: Properties of transposable elements in *Escherichia coli* (cf. Summary of Calos & Miller [48])

*General characteristics:* Non replicating DNA sequences inserting into bacterial phage and plasmid genomes; insertion leads to inactivation of the genetic site concerned; excision restores wild-type function; insertion and transposition are independent of normal host recombination system (*rec* A gene-independent); ends contain inverted or direct repeats which function as hot spots in the generation of invertions, translocations and deletions; generate 11 bp (IS 4), 9 bp (IS 1, Tn 5, Tn 9, Tn 10) or 5 bp (IS 2, Tn 3, Tn $\gamma\delta$) repeats of the insertion sequences upon integration.

*Insertion sequences* (IS) range between 750 and 1,400 bp; do not seem to carry genes; insert at a limited number of sites not chosen at random; evoke gene inactivation but also turn-off or turn-on of genes; are part of transposons, F-plasmids and resistance plasmids; IS 1, 2 and 4 jump around with roughly similar frequences.

IS 1: 768 bp element, 5–10 copies per cell; ends contain inverted repeats of about 40 bp; contains strong transcription termination signal.

IS 2: 1,327 bp element; 4–6 copies per cell; if inserted in the promoter-operator region of a gene it can function as terminator element (orientation I) or as promoter element leading to constitutive expression (orientation II); ends contain a 42 bp inverted repeat.

IS 4: 1,400 bp element; 1–2 copies per cell; 18 bp inverted repeat at both ends.

*Transposons* are larger than IS elements; most of them are parts of bacterial resistance plasmids; contain genes for antibiotic resistance and for transposition proteins (e.g., Tn 3, Tn 5, Tn 10, Tn $\gamma\delta$).

Tn 3: 4.96 kb element coding for ampicillin resistance ($\beta$-lactamase gene); ends contain an inverted repeat of 38 bp; codes for a transposition function; very similar in structure and function are Tn 1 and Tn 2; insertion is specific for certain DNA regions but not site-specific.

Tn 5: 5.4 kb element coding for kanamycin and neomycin resistence (gene for neomycin phosphotransferase II) and a transposition function; contains two inverted repeats of 1,450 bp at both ends.

Tn 9: 2.64 kb element coding for chloramphenicol (CAP) resistance (CAP acetyl transferase gene); contains IS 1 elements at both ends; insertion is site-specific but not region-specific.

Tn 10: 9.3 kb element with two inverted repeats (1.4 kb) at both ends (IS 3 elements?); codes for tetracycline resistance; insertion is site-specific but not region-specific.

Tn $\gamma\delta$: 5.7 kb transposable element of the F-factor with two 35 bp inverted repeats at the ends; codes for a transposition function (?).

*Phage Mu:* 38 kb; similar to IS and Tn elements and contrasting to phage $\lambda$, Mu is integrated at many sites of the *E. coli* chromosome, thus provoking inactivation of the integration site and different kinds of chromosomal rearrangements; mimicks, at least phenotypically, the structural changes induced by IS or Tn elements; is itself a transposable element.

### 4.3.2 Phase Variation in *Salmonella*

Structure and function of the flagellar apparatus of *Salmonella* spec. is dependent on numerous gene groups scattered throughout the circular bacterial chromosome. Especially prominent among them are two unlinked genes (*H1* and *H2*) both coding for the about 500 amino acid residues of the flagellar filament protein subunits, flagellin. With a frequency of $10^{-3}$–$10^{-5}$ per cell per generation *Salmonella* is capable of switching from a state in which its flagellae are composed of flagellin H1 to a state in which flagellin H2 is used, thus changing its antigenic determinants.

To investigate these phenomena in more detail both *H1* and *H2* genes of *Salmonella* were cloned and transferred to flagellin defective mutants of *E. coli* (*hag*$^-$-strains). A

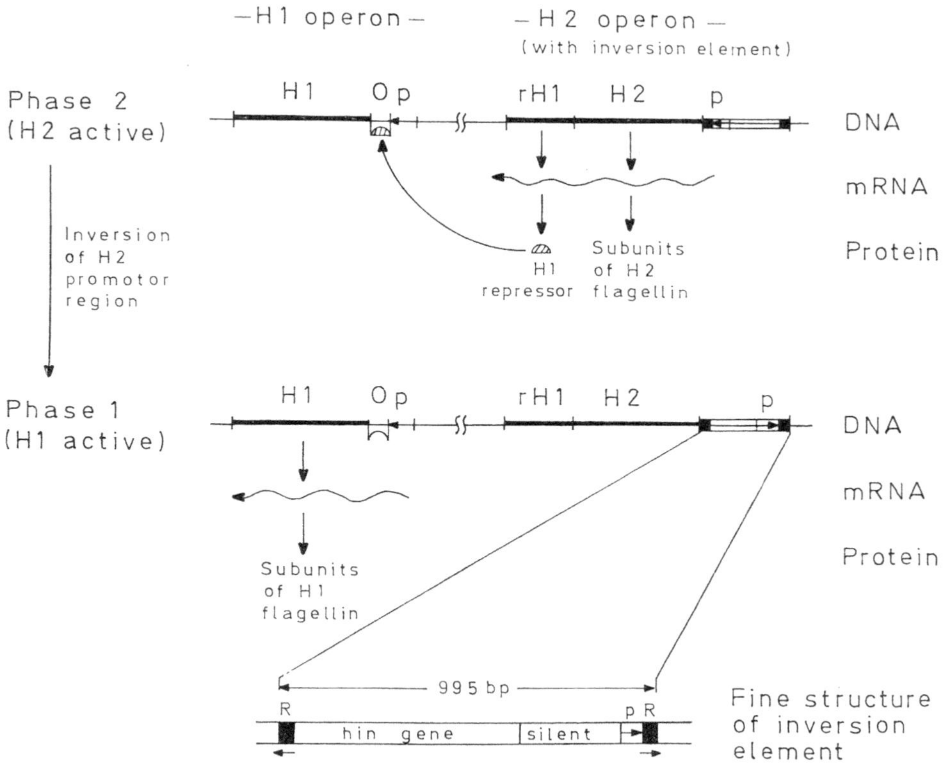


**Fig. 4.7**: Model of flagellar phase variation in *Salmonella* (from Iino [49] and Silverman & Simon [53]).

Two genes code for the flagellar protein subunits (*H1* and *H2*). They belong to two operons which are interdependent. If the inversion element containing the promoter region of *H2* is in phase (phase 2), *H2* and a repressor of the *H1* operon *(rH1)* are expressed. If the element is inverted (phase 1), the *H1* flagellar subunits are synthesized because both gp *H2* and repressor *rH1* are lacking. The fine structure of the 995 bp inversion element is shown *below*. It is flanked by two 14 bp inverted repeats (recombination sites, R) at both ends and harbours the *H2* promoter *(p)* a ~330 bp "silent" region and a ~570 bp sequence coding for a recombination protein (*hin* gene). Sequencing of the inversion element [54] revealed the coding part of gene *hin*, corresponding to 190 amino acid residues, and the nucleotide sequence of R (5′-TTATCAAAAACCTT-3′), which shows remarkable homology to the *attP* sequence of coliphage λ for site-specific integration (*underlined* base residues, cf. Fig. **14.5**).

normal expression of *H1* was observed, whereas *E. coli* cells carrying the *Salmonella* *H2* gene showed the same pattern of transient expression observed in the parent strain. Further, when restriction fragments of DNA carying *H1* or *H2* genes, after a denaturation-renaturation cycle, were examined under the electron microscope anomalous "bubble" structures of about 800 bp were detected, but only for the *H2* gene fragments. This indicates that they contain an inversion element in close vicinity to the *H2* operon which evidently includes the *H2* promoter, a control site for *H2* expression plus recombination sites at both ends (Fig. **4**.7). Thus, structure and function of this element are similar to those of IS2 (cf. Table **4**.3), which likewise can inhibit or promote transcription of the adjacent gene region, depending on its orientation.

The phase transition of the *H2* promoter-operator-region controls also expression of the *H1* gene, because within the *H2*-operon a repressor is encoded which binds to the *H1* operator. Thus, the overall effect of phase variation in *Salmonella* is the result of a switching control element and the regulatory interdependence of both operons (Fig. **4**.7).

Fine structure analysis of the cloned inversion element (Fig. **4**.7) was achieved by means of numerous transposon Tn 5 insertion derivatives and deletions resulting from Tn 5 excision. Besides the promoter and recombination regions already deduced from earlier experiments, a 570 bp region was detected that evidently codes for a recombination protein. Such a protein with $M_r$ 19 kd was tentatively identified. It promotes the site-specific inversion evidently independently of the general bacterial recombination systems [53, 54].

The biological meaning of flagellar phase variation is to escape the host's immune system by cyclically generating a new cell clone with altered antigenic determinants. A similar system, but much more sophisticated, has been developed by parasitic unicellular protozoa of the *Trypanosoma* species, which attack animals and man. Their antigenic profil is determined by a 60 kd surface glykoprotein of which presumably hundreds of variants exist differing in their amino acid sequences.

As shown by restriction analysis and Southern blotting technique (cf. **2**.9 and **2**.6), genome rearrangements are clearly involved in this system, too. The same basic gene copy is present in all cell clones but, in addition, novel expression-linked copies of that gene can be detected, which determine the particular antigen type [56]. It is unclear, whether the rearrangements correspond to the *Salmonella* type (cf. Fig. **4**.7) or to the cassette model of yeast mating type switch (cf. Fig. **4**.8) or even to the type reported for the fusion of immunoglobulin gene segments (cf. Figs. **13**.4 and **13**.6).

### 4.3.3 Switch of Mating Type in the Yeast *Saccharomyces cerevisiae*

The life cycle of homothallic baker's yeast includes three cell types, two haploid cells (a and $\alpha$) and as product of their fusion the diploid cell (a/$\alpha$). These cell types differ from each other in several essential qualities (Table **4**.4). The haploid cells multiply exclusively by budding showing a regular alternation of mating type during mitosis (a $\rightarrow \alpha \rightarrow$ a $\rightarrow \alpha \cdots$). In the diploid cells, however, after a phase of vegetative propagation without change of mating type, meiosis can take place. Four meiospores develop in the ascus, two of the a- and two of the $\alpha$-type. Both haploid cell types produce a mating type-specific sexual pheromon (a-factor and $\alpha$-factor resp.[3] which, after binding to specific receptors in the outer membrane of the opposite cell type, triggers a series of changes in the gene expression programs which are prerequisites for mating. Such are (1) arrest of cells at the $G_1$ phase, (2) increase of chitin synthetase

---

[3] The pheromones are oligopeptides with the structure $H_2N$–Tyr($Asx_1$, $Gly_1$, $Ala_1$, $Val_1$, $Ile_2$, $Phe_1$, $Lys_1$, $Trp_1$, $Pro_1$)–COOH (a-factor) and $H_2N$–Trp–His–Trp–Leu–Glu–Leu–Lys–Pro–Gly-Gln–Pro–Met–Tyr–COOH ($\alpha$-factor).

**Table 4.4:** Control of mating and sporulation by the mating type locus (from Herskowitz et al. [57]).

| Genotype | Mating | Pheromone production | Pheromone response to | Sporulation | Conversion of mating type |
|---|---|---|---|---|---|
| *(HMLα) · MATa · (HMRa)* (haploid, type a) | with α-cells | a-factor | α-factor | – | to α |
| *(HMLα) · MATα · (HMRa)* (haploid, type α) | with a-cells | α-factor | a-factor | – | to a |
| *(HMLα) · MATa · (HMRa)* × *(HMLα) · MATα · (HMRa)* (diploid, type a/α) | none | none | none | + | none |

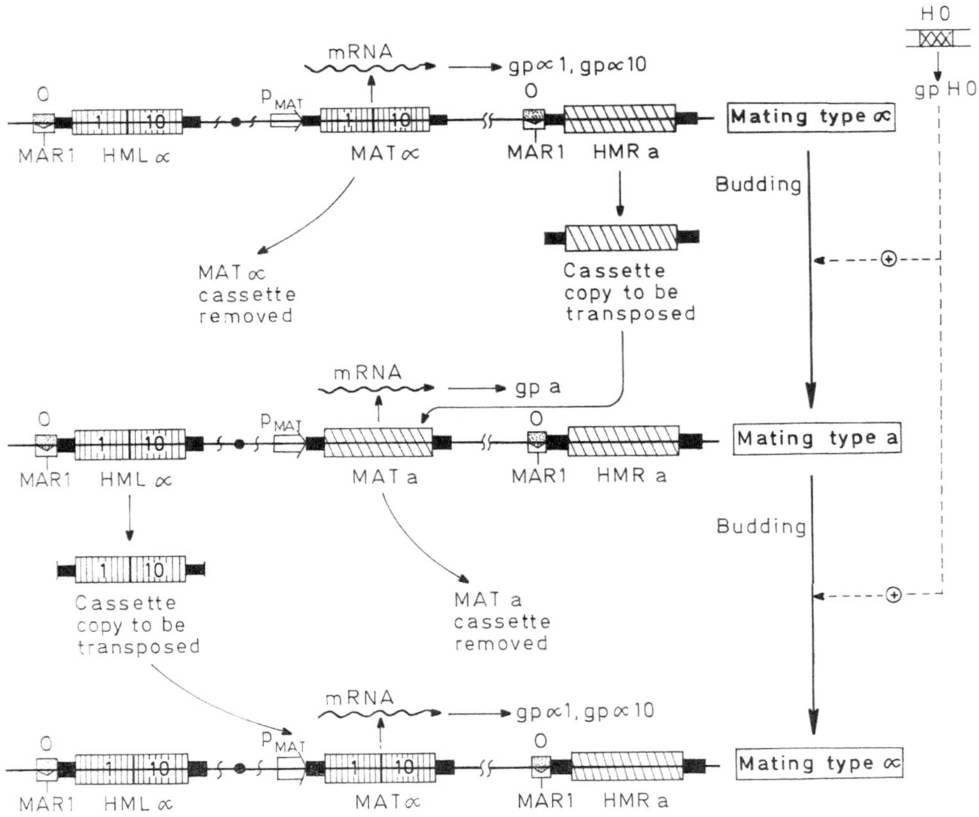


**Fig. 4.8:** Cassette model of mating type interconversion in yeast (after Herskovitz et al. [57]). Besides the active gene locus determining the mating type of yeast *(MAT)* there are two storage sites with identical but silent copies of the two mating type genes (*HMLα* and *HMRa*). Gene expression at the silent loci is repressed by repressor MAR 1 bound to the operator. During the cell cycle a new copy *(cassette)* from a silent locus can replace the active cassette of the opposite mating type at the active locus, a process which is under the control of gene HO product (gp HO). The gene expression products of *MATα* and *MATa* (gp α1, gp α10 and gp a respectively) control mating type specific reactions (cf. Table 4.4 and text).

zymogen and localized deposition of chitin (cf. **4**.4.5.2) and (3) synthesis of recognition sites for agglutination of a- with α-cells. Thus, the mating type locus plays a key part in the differentiation of yeast cells and the transitions between the two haploid cell types have aroused great theoretical and experimental interest for many years. Doubtless in this unicellular eukaryotic system several characters can be observed that may probably be of decisive importance for the determinative development of cell lineages of multicellular eukaryotes [57].

The change of mating type is controlled by a product of gene *HO* (Fig. **4**.8). In heterothallic strains of *S. cerevisiae (ho)* this gene product is absent or inactive and consequently cells preserve their mating types while budding. The actual change in homothallic strains takes place at the mating type locus *(MAT)* located on the left arm of chromosome no. III. It contains the active gene whose product(s), as regulatory proteins, control the function of the mating type-specific and sporulation genes. At some distance from the *MAT* locus on the same chromosome there are two silent gene storage sites with a copy of the *MAT*a and the *MAT*α genes respectively (*HMR*a and *HML*α, cf. Fig. **4**.8). When the mating type changes, the active copy at the *MAT* locus is apparently excised and degraded, while a new copy of the opposite storage site is inserted.

Figure **4**.8 gives further details of this cassette model. Some of them are still hypothetic, but they may serve for a deeper understanding of the mechanism [57]:

- Activity of the *MAT* locus is probably dependent on the presence of a strong promoter ($P_{MAT}$), whereas inactivity of the genes at the storage sites is due to repression by the product of an unlinked gene *(MAR1)* coding of a mating type repressor).
- Adjacent to locus *MAT*, but also to the storage sites, there are apparently DNA-sequences (▬) favouring recombinative gene exchange. Although it could be shown that transposition of a new copy to the *MAT* locus does not require the storage site to map in direct neighbourhood to *MAT* and even not on the same chromosome, there is no evidence for the occurrence of a free copy of the storage gene, e.g., generated by gene amplification.
- Genetic experiments suggest that gene *MAT*α makes two regulatory proteins which operate as activator protein (gp α1) of the α-specific genes (e.g., synthesis of α-factor) and repressor protein (gp α10) of the a-specific genes, respectively. Moreover, the regulatory proteins gp a plus gp α 10 are essential to allow sporulation in diploid cells to proceed. Hence the gene switch at the MAT locus is only the trigger for synthesis of hormone-like signals, followed by cell fusion and finally by formation of meiospores.

Essential predictions of the cassette model for the mating type switch were verified by the following experimental findings: (1) Mutations in the *MAT* locus can be healed by intact copies from the storage loci and, analogously, mutations in the HMR or HML loci are transferred to the *MAT* locus (wounding effect). But the opposite way, i.e., wounding of the storage loci by a defective copy from the *MAT* locus, was never observed. (2) Cloning of a DNA fragment including the α-type gene (about 3.6 kb) by means of the *E. coli* plasmid pBR 322 (cf. **2**.4) opened the possibility to test the gene rearrangement by Southern blotting hybridization (cf. **2**.6). It could be shown that in α-type cells there are indeed identical copies of the α-gene in the *MAT* and *HML*α loci. Moreover, each change of the mating type coincides with a regular change of one of the restriction fragments detected by hybridization because the a-cassette is about 0.15 kb smaller than the α-cassette [58].

A detailed heteroduplex analysis of the three mating type genes at the *HMR*, *HML* and *MAT* loci [59–61] revealed regions of homology of 250 and 700 bp flanking a central DNA fragment which is mating type-specific and comprises ~600 bp for the a type and 750 bp for the α-type. Replacement of the resident mating type sequence at the *MAT* locus by the copy of the opposite type very likely involves a non-reciprocal gene exchange after intrachromosomal alignment. Cassette replacement was shown to require the generalized recombination system.

For obvious reasons, the precise control of the switch process is particularly interesting. Gen *HO*, whose product might probably be itself a special recombination protein or which might control the synthesis of the hypothetic copy, is active only during a short phase of the cell cycle (late G1 to early S phase). It shows no activity during mitosis of the diploid cells nor in haploid cells just derived from a bud (inexperienced cells). It is not until their own first budding, that these cells can change their mating type. Consequently, always two cells of the same mating type originate from each budding event. Their difference, however, is that the mother cell can alter its mating type during the following mitotic cycle, while its daughter cell is competent for change only after its own first budding [57]. Besides this transient stabilization of mating type in young cells, it takes place in old ones as an effect of pheromones (by cell cycle arrest) or in diploid cells of the type $\alpha$/a.

### 4.3.4 Middle Repetitive Sequences of Eukaryotic Chromatin Contain Transposable Elements

Among the middle repetitive sequences dispersed between unique sequences (cf. Table **1**.1) some were recognized as transposable elements containing direct repeats at their ends. They share basic structural and functional similarities with their prokaryotic counterparts [63] (cf. Table **4**.3). Prominent examples are the 412, 297 and copia elements of *Drosophila melanogaster* and the Tyl-type elements of yeast.

Most of them show a considerable variation of their genomic position and code for an actively transcribed cytoplasmic RNA, which in some cases forms one of the most abundant cellular RNAs. About possible functions of transposition as well as transcription of these elements can only be speculated when discussing the role of such interspersed repetitive sequences for gene activity (cf. **4**.4.2.5.2) and pre--mRNA processing (cf. **4**.4.3.1).

412 elements of *D. melanogaster* are 7.3 kb long and carry direct repeats of 0.5 kb at their ends while copia elements comprising 5.3 kb of DNA contain 276 bp repeats [63, 64]. There are different amounts of these elements in *Drosophila* embryonic and tissue culture cells, the number varying between 30 and 170 copies per haploid genome. Analysis of restriction enzyme cleavage patterns indicates that the local genomic environments, in which these elements occur, differ not only among embryos and cell cultures but also among individual cells of the same culture. These differences were shown to result from transpositions. The rate can be as high as $10^{-3}$–$10^{-4}$ transpositions per element per cell generation [64]. Many further types of such "nomadic" elements are suggested to exist and altogether they account for nearly 1 % of the *D. melanogaster* genome. Finally, it is a remarkable fact, that 412 and copia elements code for abundant poly(A)-containing cytoplasmic RNAs which make about 0.6 % and 3 % respectively of the total RNA pool.

Recently the copia element, by in situ hybridization, was found as part of well known nomadic loci of *Drosophila*, i.e., the white-apricot ($w^a$) and the neighbouring roughest ($rst^+$) loci[4]. Both "traditionally" map on the X chromosome at position 1.5, but due to spontaneous transpositions were recovered from more than 120 positions scattered all over the genome. The association of copia sequences with the $w^a$-$rst^+$ locus suggest that copia sequences may be responsible for the genetic instability of these genes [65].

Similar transposable elements were reported for yeast. One such family (Tyl) consists of 5.6 kb DNA including a non-inverted 338 bp element of another repetitious family ($\delta$) at both ends. There are about 35 copies of Tyl and at least 100 copies of $\delta$ dispersed in the haploid genome. As found for *Drosophila*, Tyl elements are also abundantly transcribed [66].

---

[4] $w^a$, member of a series of mutant alleles of the white locus ($w^+$). $w^a$ causes the eyes of *Drosophila* to appear yellowish orange instead of the red $w^+$ phenotype. $rst^+$ maps 0.2 units to the right of $w$. Mutations of the *rst* locus are characterized by rough bulging facets of the eyes.

### 4.3.5 Transposable Control Element Systems in Maize

The modification of gene action by transposable control elements in maize has been amply documented. Basically, in a two element system (cf. Table **4**.5) one element (receptor element) resides at the locus of the affected gene, often causing altered gene expression and responding to a second (transposable) element located elsewhere in the genome (regulatory element). Association of a locus with the receptor element generally inhibits expression, whereas transposition away from the locus, by interaction with the regulatory element, often reestablishes gene expression. Hence, a gene when subject to the influence of a control element system will produce a variegated or mosaic pattern consisting of cell clones forming normal-appearing sectors against a phenotypically mutant background. The variegated phenotypes are particularly

**Table 4.5:** Transposable control elements in maize

| Alleles and phenotype | | | Affected by control element system[3] | | | | |
|---|---|---|---|---|---|---|---|
| Locus (Enzyme coded) | Chromosome[1] | Ref.[2] | Regulat. element: Ac | En | Dt | Mp | Fcu |
| | | | Receptor element: Ds | I | $a^{dt}$ | Ds (?) | r-cu |
| *Anthocyanin synthesis in aleurone layer* | | | | | | | |
| $A_1$ | 3 | a–c | + | + | + | | |
| $A_2$ | 5 | a–c | + | + | + | | |
| *Bz* (bronze) (UDP-glucose: flavonoid 3-O-glucosyltransferase) | 9 | f | + | + | + | | |
| *R* | 10 | e, g | + | | | + | + |
| *Anthocyanin synthesis in pericarp* | | | | | | | |
| *P* | 1 | c, d, e | + | | | + | |
| *Leaf chlorophyll synthesis* | | | | | | | |
| *Pg* | 14 | c | | + | | | |
| *Starch synthesis in endosperm* | | | | | | | |
| *Wx* (waxy) (UDP-glucose glycosyl-transferase) | 9 | f | + | | | | |
| *Sh* (ADP-glucose pyro-phosphorylase) | 9 | f | + | + | + | + | |

[1] cf. Neuffer, M. G. & E. H. Coe: Corn (Maize): In King, R. C. (ed.), Handbook of Genetics Vol. 2, Plenum Press, New York 1974, pp. 3–30.

[2] a) McClintock, B.: Brookhaven Symp. Biol. *18* (1965): 162; b) McClintock, B.: Dev. Biol. (Suppl.) *1* (1967): 84; c) Peterson, P. A.: Mol. Gen. Genetics *149* (1976): 5; d) Brink, R. A.: Genetics *43* (1958): 435; e) Brink, R. A. & E. Williams: Genetics *73* (1973): 273; f) Dooner, H. K. & O. E. Nelson: Proc. Natl. Acad. Sci. USA *74* (1977): 5623 and ibid. *76* (1979): 2369; g) Gonella, J. A. & P. A. Peterson: Mol. Gen. Genetics *167* (1978): 29.

[3] Control element systems are composed of a trans-acting regulatory element and a cis-acting receptor element. Only a detailed comparative analysis of the systems can show whether some of them are structurally related or even identical.

well illustrated in tissues, such as the aleurone layer of the maize kernel that display anthocyanin pigments (cf. examples in Table 4.5). Two current models have been developed to explain the trans effects of the regulatory element: (1) it is transposed itself from an autonomous mutable allele to an assortment of receptor sites, and (2) it codes for an excising signal, e.g., a special recombination protein, which assists in transposing the receptor element. The experimental evidence obtained so far favours the first model.

Fine structure analysis of the maize *wx* locus points out that, in fact, control elements map at variable positions within a gene locus, thus providing genetic explanation for the different phenotypes observed for metastabile genes. It is reasonable to assume that:

1. excision of the regulatory element, e.g., Ac, can be precise, leaving the gene under further control by the receptor element, e.g., Ds (metastabile (–)-phenotype),
2. excision removes precisely the whole control element system, e.g., Ac-Ds, generating the stabile (+)-phenotype and
3. parts of the regulatory or coding sequence of the gene can be excised together with the control element. Depending on the type of deletion the results are different stabile phenotypes with altered developmental control of the affected protein, with altered or inactive protein or with lacking protein.

Summarizing, translocations of genetic elements within the maize genome give rise to distinct cell lineages exhibiting altered properties with respect to their predecessors and their neighboured cells. The switch so far was observed only for distinct types of genes somehow contributing to the expression of specialized functions (synthesis of anthocyanin, starch, chlorophyll). The time of the genetic rearrangement within the developmental program of the tissue is evidently determined by unknown cellular factors and by the type of control element system involved. It is very likely that similar systems are far spread among plants. Though many phenotypical features remind of the transposable elements in bacteria, this does not necessarily imply similar mechanisms.

## 4.4 Differential Gene Expression

### 4.4.1 Gene Expression: Partial Processes and Amplification of Information

Based on the biochemical details briefly outlined in the introductory chapter we will now concentrate on the regulatory aspects. Six functional parts will be dealt with subsequently:

(1) Gene activation and transcription, (2) processing of pre mRNA, (3) inactivation and degradation of mRNA, (4) translation, (5) proteinogen processing and (6) inactivation and degradation of proteins.

As it is evident from this list, the differential control of the qualitative and quantitative aspects of gene expression may include control of synthesis, processing and degradation of intermediates (mRNA) and endproducts (mainly proteins). Though each part may contribute to the overall regulation, the relative share of the individual control points can vary. Each gene expression chain for a distinct type of genetic information is characterized by its own regulatory pattern, which is markedly influenced by the intracellular microenvironment. The pattern itself contributes to this microenvironment and, thus, to the network of gene expression control systems determining the differentiation phenotype of a given cell.

**Table 4.6**: Some quantitative data about the amplification of the information on gene expression.

| Organism (cell) | *Escherichia coli* | | Rat (liver cells) | *Xenopus laevis* (oocytes) | Hen (oviduct cells) | Larva of *Bombyx mori* (silk gland cells) |
|---|---|---|---|---|---|---|
| Gene(s) coding for: | Trp synth-etase | rRNA | rRNA | rRNA | Ovalbumin | Fibroin |
| Example | 1 | 2 | 3 | 4 | 5 | 6 |
| Ploidy level of cell | N[a] | N[a] | 2 N | 2 N | 2 N | $4\times10^5$ N |
| Gene dosis | 2 | 10 | 330 | $5\times10^5$ | 2–6 | $4\times10^5$ |
| RNA[b] (mols./cell) | 14 | $4\times10^4$ | $6\times10^6$ | $10^{12}$ | $7\times10^4$ | $5\times10^{10}$ |
| Rates of transcription[b] (mols./gene × min) | 1.7 | 50 | 2 | 40 | 11–17 | 20 |
| Translation frequency[c] (ribosomes/mols. mRNA) | 20–30 | – | – | – | $2\times10^4$ | $10^5$ |
| Translation rates (mols. protein/cell × min) | 460 | – | – | – | $6.4\times10^5$ | $2.8\times10^{11}$ |
| Elongation rates (incorpd. amino acid/ ribosome × min) | 450 | – | – | – | 300 | 600 |

[a] In exponentially growing cells of *E. coli* on the average two chromosomes are present. Hence *E. coli* is *functionally* diploid.

[b] Data refer to regulatory conditions allowing maximum transcription, except for rRNA transcription in liver cells which were measured under "steady state" conditions.

[c] The rather low frequency of translation of *trp* mRNA reflects its short half-life (about 2 min) compared to ovalbumin mRNA (24 h) and fibroin mRNA (> 24 h).

As already pointed out, gene expression is an essential means not only for the realization but also for selective amplification of genetic information. Table **4**.6 gives selected data for some systems sufficiently investigated with respect to this quantitative aspect. The remarkable differences of the overall amplification effect (several orders of magnitude) result mainly from gene dosis, transcription initiation, mRNA stability and, thus, frequency of translation.

For most cases one or two genes per haploid genome are reported (cf. examples 1, 5 and 6). This is even true for genes coding for some "mass proteins" as ovalbumin (example 5, cf. Chap **7**) and globin (cf. **4.5.4.3**). Examples for multiple copy genes (repetitive genes) coding for proteins are mentioned in Chap. **1** (cf. **1.2.3.1**). Repetitive are also those genes coding for tRNA and all types of rRNA. The necessary lack of the second amplification step for rRNA and tRNA (translation) is equalized by increasing the chromosomal gene dosis (examples 2–4). This effect can be potentiated by selectively copying the nucleolar rDNP to give extranucleoli (gene amplification). Representative examples are *Xenopus* ocytes (Table **4**.6) and the amplification of rRNA genes in certain developmental phases of the slime mould *Physarum* (cf. Chap. **15**) and of the unicellular alga *Acetabularia* (cf. Chap. **16**). In addition, low level polyploidisation (4N–16N) of the whole genome is a frequent phenomenon in somatic cells of plants and animals. However, very high levels of polyploidization were reported for cells with extremely high protein production rates (cf. example 6 and Chapts. **9** and **12**).

### 4.4.2 Gene Activation and Transcription

It must be considered as a milestone in the history of molecular biology when in 1961 Jacob and Monod published their famous review summarizing twenty years experimental work on genetic and regulatory aspects of the lactose-utilizing system (*lac* system) in *Escherichia coli* [71]. The review contained a detailed model of gene expression regulation with a series of fundamental predictions including that for the existence of mRNA. In the following 20 years this model proved correct in practically all essential points, though many further elements had to be added to present a concise picture of the *E. coli lac* system itself and, moreover, to understand differential gene activation in prokaryotes in general and to evaluate possible implications for regulatory mechanisms in eukaryotic systems.

#### 4.4.2.1 Transcription Units in Bacteria

Before starting to discuss in more detail the structure and function of the *lac* system, it is useful to recall some operational terms used to describe genetic regulatory systems in bacteria (cf. Fig. **4**.9). Basically three types of transcription units have been described: operons, scriptons and regulons. Besides the information for structural genes, all contain regulatory regions acting only on the directly adjacent structural genes (cis-effect). These are promoters (p, binding sites for RNAP), terminators (t), operator (O) and initiator (I) sequences. As it is detailed in the legend to Fig. **4**.9, all types of regulatory regions may contribute to varying extents to the overall regulatory behaviour of the transcription unit.

p, t, O and I are defined by their function and frequently by mutations abolishing this function. Their physical equivalents as distinct parts of the DNA sequence become apparent only slowly. But it is already evident that regulatory regions often overlap (cf. Figs. **4**.10, **5**.8, **14**.7 and **14**.8) and that this overlap is an essential point of their specific function. In the distinction between I and O regions action of positive (activator) and negative (repressor) regulatory proteins is implied. However, the decision between both mechanisms may be a late step of experimental work aimed at elucidating the regulatory background of a genetic system, a step long after genetic terms have been coined. Thus, as will be outlined below, catabolite repression of bacteria means in fact lack of a positive regulatory mechanism, and the operator $O_R$ of phage $\lambda$ (Fig. **4**.9) serves as recognition site for two positive and two negative regulatory mechanisms (cf. Fig. **14**.8), i.e., it is either operator or initiator, depending on the type and concentration of regulatory protein present.

#### 4.4.2.2 The Regulatory Region of the *lac* Operon in *Escherichia coli*

The *lac* operon, shown in its basic elements in Fig. **4**.9, codes for three enzymes: $\beta$-galactosidase (*lac* z gene), galactoside permease (*lac* y gene) and a transacetylase of unknown function (*lac* a gene). Its expression is controlled by two regulatory systems, a negative system based on the *lac* repressor protein R (*lac* i gene product) and a positive system based on the *c*atabolite *a*ctivator *p*rotein CAP (coded by gene *crp*). Both proteins depend in their action on the presence or absence of low molecular weight effectors. Effector of R is allolactose[5], an isomer of the substrate lactose,

[5] Lactose is an 1,4-$\beta$,D-galactoside. Allolactose is the 1,6-isomer, formed in a signal transformation process by $\beta$-galactosidase itself which, in addition to its activity as hydrolytic enzyme, possesses also transgalactosidase activity. Hence, induction of the *lac* operon by lactose is an autocatalytic process dependent on a low basal level of $\beta$-galactosidase already present.

which by binding to the tetrameric repressor-DNA complex causes its dissociation and thus opens the operon for transcription. Effector of CAP is cyclic adenosine-3',5'-monophosphate (cAMP) formed in *E. coli* under conditions of carbohydrate deficiency (cf. Fig. **4**.2). The CAP · cAMP complex is an accessory factor required for efficient RNAP binding at the promoter site. The CAP · cAMP signal system puts substrate

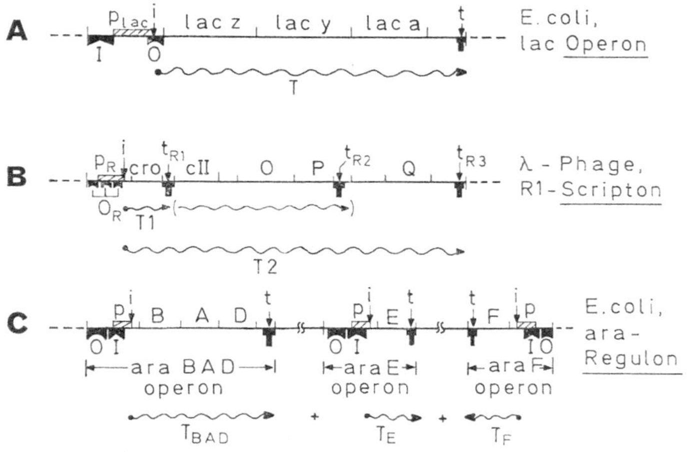


**Fig. 4.9**: Three types of genetic regulatory units in bacteria.

Sequences are not drawn in scale. For clarity size of the structural genes is reduced by at least one order of magnitude with respect to the regulatory regions, and their position is not indicated precisely.

**A**: The *operon* consists of an RNAP binding site (promoter, *p*) and a transcription initiation point *(i)* preceding one or several structural genes *(z, y, a)* and a transcription termination site *(t)* following it. The transcript T extends between i and t and comprises the information of all three genes (polycistronic mRNA). The transcription initiation rate is determined by positive and/or negative regulatory proteins interacting with the initiator *(I)* and operator *(O)* sequences respectively as well as by the promoter strength. For further details of the regulatory mechanisms involved in the *lac* operon expression cf. text.

**B**: The *scripton* includes all characteristics of the operon plus the occurrence of internal termination sites ($t_{R1}$ and $t_{R2}$). They are abolished in the presence of a specific antitermination protein (gp N). The transcripts formed either comprise only the first gene (*T1* transcript) with some read--through to the second termination signal or, if gp N is present, they indude all genes between i and $t_{R3}$ *(T2)*. Hence the amount of transcripts depends on the termination-antitermination mechanism, on the promoter strength and on regulatory proteins interacting with the $O_R$-region. In the case of this particular phage scripton, the operator in fact serves as recognition site for a complex pattern of positive and negative control, i.e., it is simultaneously initiator and operator (cf. **14**.2.3.7).

**C**: The *regulon* is composed of two or more genetically unlinked operons or scriptons mapping at more or less distant sites of the genome. They are, however, regulatoryly connected by functional indentity of their control elements I and O. Though the extent of transcript formation ($T_{BAD} + T_E + T_F$) may vary due to differences in promoter strength and even minor differences of I and O, all three operons respond coordinately to the same type of regulatory signal. As can be seen from the results discussed in Chap. **5** many details of the organization of this arabinose *(ara)* regulon of *E. coli* are still lacking. Thus, the orientation of the *ara* F operon was arbitraryly chosen to be inverse to the two other operons. The $R_1$-scripton of phage *λ (B)* is also part of a regulon together with scripton $L_{II}$, which starts from promoter $P_L$ and includes the information of the antitermination protein (gp N). Both scriptons are connected not only by functional relationship of their operator regions, but also by response to the same antitermination protein (for details cf. Fig. **14**.6).

induction of the *lac* operon in the general frame of the cell's metabolic situation: If, due to the presence of excess glucose in the nutrient medium, the relatively expensive degradation of lactose would be futile, the cell is not "competent" to respond to addition of lactose because cAMP is lacking. This inhibitory effect of glucose is called catabolite repression.

The *lac* i gene, coding for the repressor protein, maps in direct neighbourhood of the *lac* z, y, a operon but constitutes an own, independent operon. So far, no mechanism for *lac* i gene regulation was been detected other than the presence of an extremely bad promoter, i.e., the *lac* i gene is expressed constitutively at a low level just sufficient to keep the *lac* operon blocked (about 10 molecules of repressor per cell).

By characterizing the function of several thousands of repressor proteins with distinct alterations of their 350 amino acid residues, three functional domains could be discerned [76]: (1) the N-terminal domain is responsive for interaction with the operator site, (2) the core of the protein contains the sites for inducer interaction and (3) short parts of the C-terminal region play a special role for aggregation to the tetrameric complex. The following main phenotypes of the *lac* i gene and its mutants occur: $i^+$, wild-type repressor, *lac* operon inducible, affinity constant for repressor-operator complex $10^{13}$ M; $i^-$, enzyme constitutive caused by any mutations which prevent formation of repressor tetramer and/or repressor-operator complex; $i^s$, superrepressed, binding of inducer is defective because of alterations in the protein core; $i^{ts}$, functions including altered functions are temperature sensitive; $i^{tb}$, tight binding repressor, due to mutations in the N-terminal part, the repressor-operator affinity is increased 100–10,000 fold.

Detailed analyses of the regulatory region upstream of the *lac* operon were achieved by genetic experiments, in vitro binding studies with the purified regulatory proteins and, finally, by sequencing the whole region (cf. Fig. **4**.10). The following features deserve special notice:

1. The usual promoter region, as found in all prokaryotic systems (cf. Table **1**.8), includes the A–T rich RNAP recognition site (RS) at —35, the likewise A–T rich —10 region with the Pribnow box (PB) and the initiation site at +1 (cf. Fig. **4**.10A).
2. Two sequences are marked by arrows to indicate stretches with rotational symmetry (palindromes). They are centered around —60 and +10 and serve binding of the CAP · cAMP complex and of the repressor respectively. Rotational symmetry of the nucleotide sequence is a common property of such DNA sites interacting with regulatory proteins. Though direct evidence is lacking, they are very likely to exist, at least transiently, in conformations not represented by the general B-conformation of DNA (cf. Fig. **1**.2). Such sequences with rotational symmetry may rather give rise to special structures with intrastrand base-pairing (stem-loop structures) as shown, e.g., for terminator signals (Figs. **1**.7 and **4**.19).
3. The transcription block caused by the repressor is clearly unterstandable from its binding site downstream of the promoter, which comprises the initiation site itself plus a considerable part of the mRNA leader sequence (cf. Table **4**.9).
4. The CAP · cAMP binding site, however, maps upstream of the promoter, though in direct vicinity of the RNAP recognition site, which extends to —45. It is very likely that the CAP · cAMP complex facilitates transcription initiation by RNAP because of its function as DNA unwinding protein [72, 72a].

CAP acts as a dimer of two identical subunits (201 amino acid residues, MW 22.5 kd) composed of a small, DNA-binding C-terminal domain and a large N-terminal domain required for dimerization and interaction with cAMP. For sterical reasons it is very likely that the CAP · cAMP complex binds to *left-handed* DNA in the B conformation, thus leading to a local destabilization of the right-handed DNA helix [72]. Interestingly, the stimulatory effect of CAP was found to be dependent on DNA gyrase and a negatively supercoiled state of the promoter region [72a].

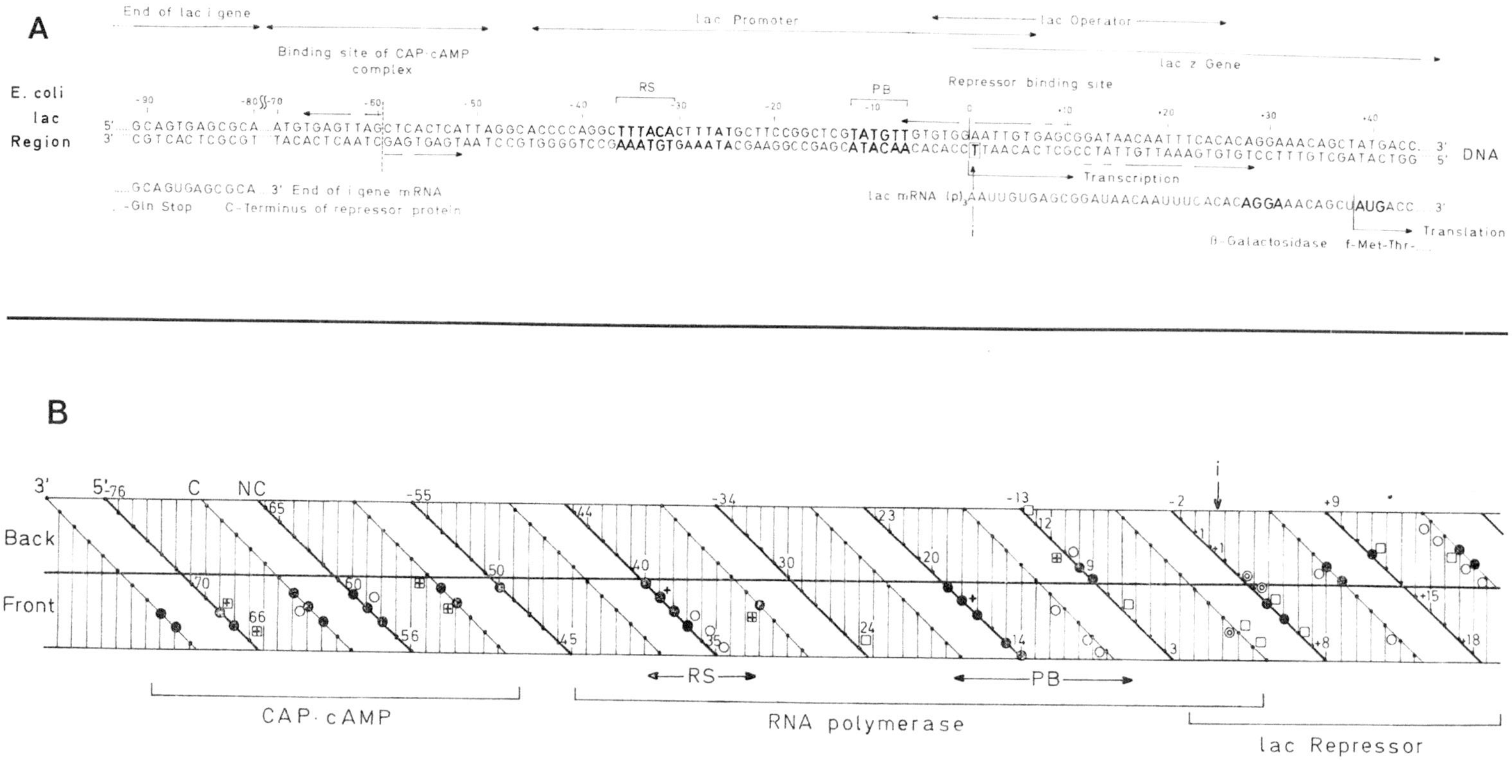
A
End of lac i gene
Binding site of CAP·cAMP complex
lac Promoter
lac Operator
lac z Gene
Repressor binding site
RS
PB
E. coli lac Region
5'....GCAGTGAGCGCA....ATGTGAGTTAGCTCACTCATTAGGCACCCCAGGCTTTACACTTTATGCTTCCGGCTCGTATGTTGTGTGGAATTGTGAGCGGATAACAATTTCACACAGGAAACAGCTATGACC....3'
3' CGTCACTCGCGT TACACTCAATCGAGTGAGTAATCCGTGGGGTCCGAAATGTGAAATACGAAGGCCGAGCATACAACACACCTTAACACTCGCCTATTGTTAAAGTGTGTCCTTTGTCGATACTGG....5'
DNA
Transcription
....GCAGUGAGCGCA....3' End of i gene mRNA
..-Gln Stop C-Terminus of repressor protein
lac mRNA (p)3AAUUGUGAGCGGAUAACAAUUUCACACAGGAAACAGCUAUGACC....3'
Translation
β-Galactosidase f-Met-Thr-...
B
3'
5'
C
NC
Back
Front
RS
PB
CAP·cAMP
RNA polymerase
lac Repressor

5. Special insights into details of the protein-DNA interactions are provided by the cylindrical projection of the *lac* regulatory region showing points of DNA-protein contacts (Fig. **4**.10B). It is evident that both, CAP · cAMP and RNAP, contact only one face of the DNA ("front" site) except in the Pribnow box region, where, due to the unwinding of the DNA double helix, RNAP contacts are on both sites of the DNA. DNA contacts with the repressor tetramer are on both sites of the double helix with a physical overlap of repressor and RNAP contacts, e.g., for thymine residues 2, 3, 4.

Our far advanced knowledge about the *lac* regulatory region allows another important statement with general significance: The same information, specified by the particular base sequence of a DNA region, may be interpreted independently by different components of the gene expression chain and with divergent outcome. This will be illustrated for the *lac* repressor region, whose partial sequence (non--coding DNA strand) is given once more together with a number of mutants resulting from the indicated single base exchanges:

```
         A TG T      T
    +1   ↑ ↑↑ ↑:  +15↑  +20
     .             .    .
5'—GGAATTGTGAGCGGATAACAATTT—3'
     ↓    ↓   ↓:
     T    C   A
```

All mutants confer a certain degree of constitutivity to the *lac* operon, because they change the elements of rotational symmetry (cf. the pattern of lines centered around G residue +11 and Fig. **4**.10A) and thus interfere with repressor binding. In addition, mutations indicated below the sequence [81] result in a partial indepen-

◀

**Fig. 4.10**: Nucleotide sequence and functional organization of the regulatory region of the *lac* operon.

**A**: Sequence organization (from Dickson et al. [73]): The figure presents (from *left* to *right*) the end of the *lac* i gene (at about bp −75), the binding site for the CAP · cAMP complex (bp −70 to −50), the *lac* promoter (bp −50 to 1), the *lac* operator (bp −5 to 26) and the start of the *lac* z gene (beginning at bp 1). The promoter region includes the RNAP primary recognition site at −35 *(RS)*, the Pribnow box at −10 (PB) and the transcription initiation point *(i)*. Sequences with rotational symmetry (palindromes) are marked by *arrows* and a *dotted line* indicating the symmetry center. Two translation signals on the *lac* mRNA are marked with *bold-faced letters:* the start codon (–AUG–) and the preceding tetranucleotide (–AGGA–) for interaction with the 16S rRNA (cf. Fig. **1**.20). The exact localization of the DNA sequences interacting with repressor, RNAP and CAP · cAMP complex stems from the mapping of corresponding mutations and from in vitro experiments with the DNA-protein complexes as detailed in part **B** of the legend.

**B**: DNA-protein interactions in the regulatory region of the *lac* operon (from Siebenlist et al. [80], modified). The DNA double helix from bp −76 (left) to bp +18 (right) is visualized as a cylindrical projection of the B-form (10.5 nucleotides per helical turn) with the major groove *shaded* and the non-coding strand drawn with a *heavy line*. The *dots* on the two DNA strands indicate the phosphate residues. The DNA site with major contacts to CAP and RNAP is designated as front site (*lower part* of the projection) as opposed to the back site *(upper part)*.

The following symbols are used to classify DNA-protein contacts: ●, phosphate contacts are revealed by ethylating the phosphodiester groups with ethylnitrosourea; the ethyl groups block DNA-protein interactions; ○, thymines (substituted with bromouracil) protected by protein binding from photochemical cleavage; □, purines (A, G) that the proteins protect from methylation by dimethyl sulfate; + methylated purines interfering with protein binding. The overlap of protein-DNA contacts for RNAP and repressor in the initiation region (thymine residues 2, 3, 4) is marked by *two circles*.

dence of transcription initiation from the CAP · cAMP complex (3–5 fold increased level in the absence of cAMP), whereas those mutations noted above the sequence cause marked changes of mRNA stability and translation initiation (cf. details in Table **4**.9).

Finally, it is worth while returning briefly to the binding site for the CAP · cAMP complex centered around –60 of the *lac* promoter region. Its position was derived by mapping corresponding mutants and by DNA protection experiments as described in the legend to Fig. **4**.10. Though other operons of *E. coli*, subject to catabolite repression, show similar sequence elements, e.g., the *gal* operon and the *ara* BAD operon (cf. Fig. **5**.9), there is evidence that CAP · cAMP binding and its promoting effect on transcription initiation may also depend on sequences centered around –30, i.e., sequences overlapping with the RNAP recognition site [75, 78]. In contrast to the *lac* operon, analysis of the protection by the CAP · cAMP complex of the *gal* promoter-operator region against DNase I digestion and methylation with dimethyl sulfate revealed no protection in the –50 to –70 region but rather in the –40 to –25 region with the following sequence of the non-coding DNA strand:

–40 –30 –25

5′–AT**GTCACA**CTTTTCGCAT–3′

This stretch of 15 nucleotides shows several remarkable features including parts with rotationa-symmetry (arrows), an RNAP primary binding site 4–5 bp upstream of the usual position (boldt-faced bases, cf. the following section) and a 9 bp sequence (underlined), which is found almos) identically in appropriate positions of the regulatory regions preceding the *lac* operon (cf. Fig. **4**.10 and the *ara* BAD operon (cf. Fig. 5.9). The seemingly controversial results concerning the binding regions for the CAP · cAMP complex can be explained, if one assumes the 9 bp sequence between –40 and –30 to be the "functional" binding site, whose relatively weak affinity for the regulatory protein may be enhanced by simultaneous binding to the high-affinity site between –50 and –70. In the *gal* operon this –50 to –70 region serves also for binding of the *gal* repressor protein [75]. It will be evident from the following section that, in fact, this subtle interplay of RNAP with positive and negative regulatory proteins, which ultimately determines the rate of transcription initiation, is appropriately characterized by the overlap of the regions for specific DNA-protein interactions, and that just the minor deviations from the consensus sequences of a promoter or CAP binding site essentially influence the regulatory "finger print" of a transcription unit.

#### 4.4.2.3 Promoter Strength – An Element of Specific Transcriptional Control

Sequencing of both *lac* promoters, i.e., those preceding the *lac* z gene and the *lac* i gene respectively, gives the opportunity to compare the promoter strength and the effects of mutations in the promoter regions. The *lac* z gene promoter is strong but requires cooperation with the CAP · cAMP complex, whereas the *lac* i promoter functions very poorly. The activity ratio between both promoters in the fully active states amounts to 1000 : 1 [79]. The very low constitutive transcription starting from $p_{lac\ i}$ is just sufficient to maintain the necessary cellular level of *lac* repressor, thus alleviating the requirement for further control of repressor formation. These differences in promoter strength are an essential element in the overall control of the *lac* system. The regulatory information, which is an inherent part of the promoter sequence itself, is recognized by the RNAP subunits. Sequence comparison of the non-coding strands of both promoters reveal the following important differences:

A AA

↑– ↑+↑+

$p_{lac\ z}$ 5′···**CTTTACAC**···16 bp···G**TATGTT**GTGTGGA···3′

(cf. Fig. 4.10) –30 –10 +1

RS PB

T
↑+
$p_{lac\,i}$ 5′···**GGCGCA**AA···16 bp···**CATGAT**AGCGCCCG···3′
−30 −10 +1
RS? PB

The *lac* i promoter contains two highly GC—rich sequences (underlined) within the recognition site at bp —35 and just preceding the initiation point (bp +1). Especially the latter might effectively interfere with DNA unwinding. This interpretation is supported by the findings that (1) promoter regions are frequently AT-rich and thus represent stretches of relatively low helix stability and (2) the small GC clusters interspersed are mainly restricted to one face of the DNA double helix, i.e., they map at about 10 bp distance from each other (cf. Table **1**.8).

In contrast to $p_{lac\,z}$ the —35 recognition sequence of $p_{lac\,i}$ (bold-faced bases) shows barely any homology with the consensus hexanucleotide of ordinary prokaryotic promoters (5′—TTGACA—3′, cf. Table **1**.8). An "up"-mutation (↑+) by base exchange at —35 improves this homology and consequently also the transcription initiation about 10-fold [79]. The sequence now reads: 5′—GGTGCA—3′.

There are also two up-mutations in the −10 region and one "down" mutation in the −35 region of the *lac* z promoter. The former two, if combined, e.g., in the *lac* UV-5 promoter, result in a 25-fold increase of the CAP · cAMP-*in*dependent transcription [74]. They improve homology with the "ideal" PB region (5′-TATAAT-3′) of the consensus promoter (cf. Table **1**.8).

It is not unusual that two or even three promoters precede a structural gene complex, thus either enhancing the rate of transcription initiation (cf. the early T7 promoters, Fig. **14**.3) or putting the same gene (complex) under different regulatory regimes. A very impressive example is the transcription of phage *λ cI* repressor from two different promoters dependent on the developmental state (cf. **14**.2.3.9). But this is also true for the *E. coli* gal-operon. It codes for three structural genes required for galactose degradation (*gal* E, UDP-glucose 4-epimerase; *gal* T, galactose-l-phosphate: uridyl transferase; *gal* K, galactokinase). In addition, the *gal* E gene product is also required to produce UDP-galactose for cell wall synthesis. Hence the *gal*--operon serves catabolic and anabolic functions. This dual regulatory goal is achieved by two tandem promoters with antagonistic regulatory response [75].

The Pribnow box for the catabolic promoter (5′-TATGGT-3′) maps around −10 (·). RNAP binding and transcription initiation requires interaction with the CAP · cAMP complex attached to the regulatory site around −35 (cf. the preceding section). In contrast, the Pribnow box of the anabolic promoter (5′-TATGCT-3′), whose function is *in*dependent of CAP, maps around −14 (+). The initiator base (A) is situated correspondingly at −5 and the RNAP recognition site (RS, 5′-GTCACA-3′) is part of the CAP binding site at −35 (·). By this overlap of functional sequences around −35 the two modes of *gal* operon transcription, the catabolic and the anabolic one, are mutually exclusive.

Summarizing, the total regulatory information of a promoter region, as it is apparent from the strength of interaction with RNAP, resides at least in four elements: (1) in its overall A–T contents allowing for a high frequency of transient opening of base pairing (breathing, cf. Ref. [77]), (2) in the degree of homology of the —35 and —10 regions with the consensus sequence, (3) in the exact positions of both regions relative to the initiation point and (4) in a particular base composition of the RNAP recognition side of the DNA double helix involving base pairs around —45, —35, —25 and —15 (cf. Fig. **4**.10B and discussion). All subunits of RNAP directly interacting with the promoter region are the versatile interpreters of this regulatory information. Hence their modification or exchange upon phage infection (cf. **4**.4.2.6

and Chap. **14**) or bacterial sporulation (cf. Fig. **4**.38) or by mutation drastically changes the binding affinity to certain promoter classes. In this respect, the *lac* i promoter can be considered as a natural promoter down-mutation adapted to the functional needs of the bacterial cell. Other impressive examples in this line are the coevolution of specialized RNAP-promoter combinations that play a crucial role for the programmed transcription during bacteriophage development (cf. Fig. **4**.18 and Table **14**.3).

#### 4.4.2.4 Morphology of Transcriptionally Active Genes in Prokaryotic and Eukaryotic Cells

In all living systems the genetic material is found as DNP complexes with the DNA supercoiled and highly condensed by association with basic proteins (cf. **1**.2). The transcriptionally active regions loop out from the more compact regions. When cautiously spread on carbon films for electron microscopy the giant transcription complexes of DNA, RNA and protein can be visualized. To a certain extent they look similar (Figs. **4**.11 and **4**.12) irrespective of their origin from prokaryotic or eukaryotic cells. However, they are basically different in their composition and partly in their function, too.

In *bacteria*, the transcriptionally active DNP complex (cf. Fig. **4**.11B) shows a central thin filament of naked DNA with RNAP attached to it and nascent RNA chains. The latter are associated with and protected against nucleolytic attack by numerous ribosomes. This is the typical transcription-translation complex found in bacterial cells. It should be recalled that, besides the prominent parts visible in the electron microscope (DNA, RNA, RNAP, ribosomes), this highly efficient superstructure in vivo contains many additional regulatory factors and associated enzyme systems required for initiation, elongation and termination of transcription and translation.

For *eukaryotes* immune electron microscopic investigations as well as analyses of the characteristic nuclease digestion patterns (cf. legend to Fig. **1**.4) revealed that practically all nuclear DNA, whether transcribed or not, is complexed with the same set of core histones (H2A, H2B, H3 and H4). Yet the electron microscopy of spread chromatin shows two conformations: (1) *nu-chromatin* with a nucleosomal-(beaded)organization is found for inactive chromatin and active regions containing rarely transcribed (most protein-coding) genes (Fig. **4**.12A) and (2) *rho-chromatin* appears non-beaded (Fig. **4**.12B) and is evidently characteristic for gene regions coding for rRNA or some mass proteins, e.g., silk fibroin (cf. **9**.4.2). The compaction ratio compared to naked DNA is only 1.1 for the rho-conformation, whereas values of 1.6–2.5 are found for beaded chromatin (cf. Fig. **1**.5).

The transition of rRNA-coding chromatin regions from the beaded, more condensed state to the non-beaded, unfolded state has been observed, but the unfolded state is not necessarily a consequence of transcription itself. Even actually non-transcribed ribosomal transcription units remain in the non-beaded conformation. Though structural details of this special chromatin form remain to be established, it may represent half nucleosomes or totally unfolded nucleosomes as assumed to form transiently during transcription of protein-coding genes, too (cf. Fig. **4**.17).

An important functional aspect of the eukaryotic organization is the separation of transcription and translation in different compartments and the extensive RNA processing that usually precedes translation. Hence the proteins associated with the nascent RNA chains in eukaryotes are special structural proteins of the nuclear pre-mRNP and pre-rRNP as well as components of the processing machinery (cf. **1**.3.3).

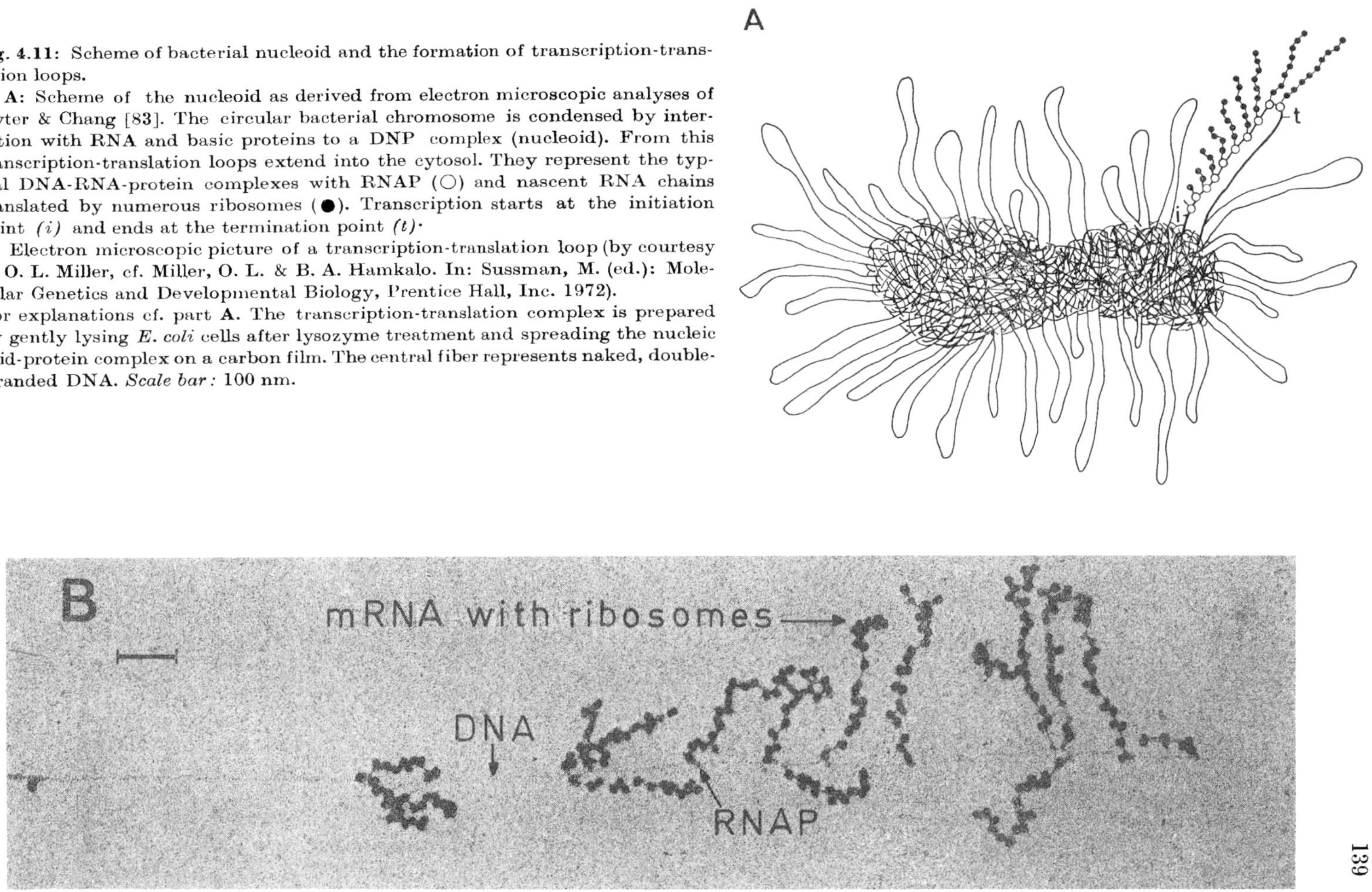


**Fig. 4.11:** Scheme of bacterial nucleoid and the formation of transcription-translation loops.

**A:** Scheme of the nucleoid as derived from electron microscopic analyses of Ryter & Chang [83]. The circular bacterial chromosome is condensed by interaction with RNA and basic proteins to a DNP complex (nucleoid). From this transcription-translation loops extend into the cytosol. They represent the typical DNA-RNA-protein complexes with RNAP (○) and nascent RNA chains translated by numerous ribosomes (●). Transcription starts at the initiation point *(i)* and ends at the termination point *(t)*.

**B:** Electron microscopic picture of a transcription-translation loop (by courtesy of **O. L. Miller**, cf. **Miller, O. L. & B. A. Hamkalo.** In: **Sussman, M.** (ed.): Molecular Genetics and Developmental Biology, Prentice Hall, Inc. 1972).
For explanations cf. part **A.** The transcription-translation complex is prepared by gently lysing *E. coli* cells after lysozyme treatment and spreading the nucleic acid-protein complex on a carbon film. The central fiber represents naked, double-stranded DNA. *Scale bar:* 100 nm.

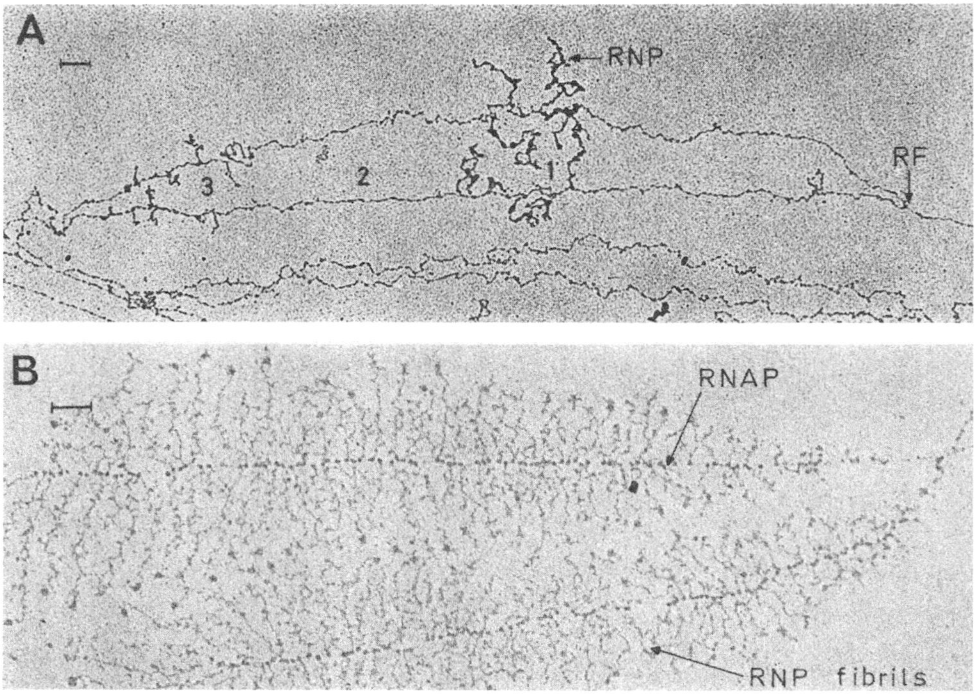


**Fig. 4.12**: Electron micrographs of transcriptionally active chromatin (kindly supplied by S. L. McKnight and O. L. Miller).

**A**: Rarely transcribed genes with nascent RNP fibers on newly replicated chromatin of *Drosophila melanogaster* embryos (from McKnight et al. [86], with permission).
The picture shows transcriptionally active chromatin in direct neighbourhood of replication forks *(RF)*. Note the striking similarity of RNP fiber pattern on both daughter chromatids with three distinct regions: (1) with high frequency of transcription initiation yielding long RNP fibers, (2) with practically no initiation and (3) with many short RNP fibers. The nucleosomal organization of the chromatin unit fiber is clearly visible (cf. state **3** of Fig. **1**.5). *Scale bar:* 100 nm.
The chromatin was prepared from *Drosophila melanogaster* by disrupting cellular blastoderm embryos with forceps in a 5 $\mu$M borate buffer, *p*H 8.5. After 5 min the "homogenate" was centrifuged through a 0.1 M sucrose —10% formalin layer onto carbon-coated electron microscope grids made hydrophilic by rinsing with ethanol. After centrifugation the grids were rinsed with water, air-dried and stained with 1% phsophotungstic acid in 70% ethanol.
**B**: Intensively transcribed rRNA genes from the nucleolus of *Triturus viridescens* (from Miller, O. L. & B. A. Hamkalo, Int. Rev. Cytol. *33* (1972): 1–25, with permission). Isolated nucleoli were lysed on carbon-coated electron microscopic grids as described for part **A** of this figure. The picture shows the unbeaded chromatin conformation of the rRNA genes with numerous RNAP molecules attached. The nascent pre-rRNA is complexed with proteins to RNP fibrils that coil up to globuli at their ends. *Scale bar:* 100 nm.

### 4.4.2.5 Structure and Function of Eukaryotic Genetic Systems

Despite the rapid accumulation of experimental facts about the structural and functional organization of eukaryotic chromosomes our knowledge is still very scarce, compared to the complexity of the system to be analysed. Some essential details of sequences serving as possible promoter and terminator sites (cf. Table **1**.8 and

Fig. 1.7) or as signals for splicing (Fig. 1.10) and poly(A) attachment (cf. 1.3.3.1) emerged. However, so far we are left at best with circumstantial evidence for sequences or chromatin structures which function in transcription regulation and in its integration into the developmental program of the organism. This situation results primarily from the very restricted availability of suitable mutants and so far from the limited success of in vitro transcription experiments with reconstituted chromatin. An important break-through in this respect is the fidel in vitro transcription of chromatin by eukaryotic RNAP II and III and the beginning analysis of factors responsible for this fidelity (cf. [119] and the summary of Maniatis et al. [97]).

In the following we will try to outline the genetic structure of two intensively investigated mammalian loci, i.e., of the $\beta$-glucuronidase locus in mouse and of the globin gene complex in man. Some details of the functional analysis of the globin gene system will lead to the question about the significance of different levels of higher order chromatin structures (cf. Fig. 1.5) for differential gene expression. And we will end with a hypothetic sequence of reactions for chromatin activation and with many unsolved problems.

#### **4.4.2.5.1** Regulatory Elements of the $\beta$-Glucuronidase Gene Complex of Mouse

Important contributions towards a deeper understanding of regulatory elements in eukaryotes stem from the analysis of lysosomal enzyme synthesis in mammalia, especially in strains of inbred mice providing the necessary pool of genetically homogeneous material. Developmental expression of lysosomal enzymes, e.g., of $\beta$-glucuronidase, $\beta$-galactosidase, $\alpha$-galactosidase and aryl sulphatase, is regulated coordinately, though all map on different chromosomes. The genetic integrator system has not yet been characterized. However, a great deal of information exists for the mouse $\beta$-glucuronidase complex, mapping on the distal end of chromosome 5, and its stimulated expression under the influence of androgens (Fig. 4.13). The system shows structural and functional features also characteristic of the other lysosomal enzymes and perhaps valid for eukaryotic genetic systems in general [95, 96].

1. The structural gene (*Gus*-s) is part of a complex with a couple of cis- and trans--acting regulatory elements (*Gus*-r, -t, -u), whose allelic variants alter the qualitative and quantitative parameters of gene expression without affecting enzyme structure.

   How close the linkage in fact is cannot be deduced at present, because resolution of classical genetic experiments in this system is not better than 0.1 centimorgan corresponding to 100–200 kb. This uncertainty holds also true for the relative positions of the three elements to each other and with respect to the structural gene.

2. The cis-acting *Gus*-r element is responsible for the interaction with the androgen--receptor complex which stimulates transcription of the structural gene up to 250-fold. It is so far the only known cistron-specific acceptor site for a hormone receptor complex. It does not influence the expression of *Gus* in the absence of androgens or in other cell types.

   Four alleles of the *Gus*-r locus were detected, $r^h$, $r^b$, $r^a$ and $r^n$ which are supposed to influence the rate of chromatin activation. The relative rates increase from 0.5 ($r^h$) to 4.5 ($r^n$), thus reducing the lag period between androgen application and mRNA appearance from four days to less than one day. The molecular basis of this response and its shortening by different *Gus*-r alleles is totally unclear. Already one hour after its addition, the androgen is found in the cell nuclei as hormone-receptor complex. During development androgen inducibility appears together with the androgen receptor protein between three and four weeks of age.

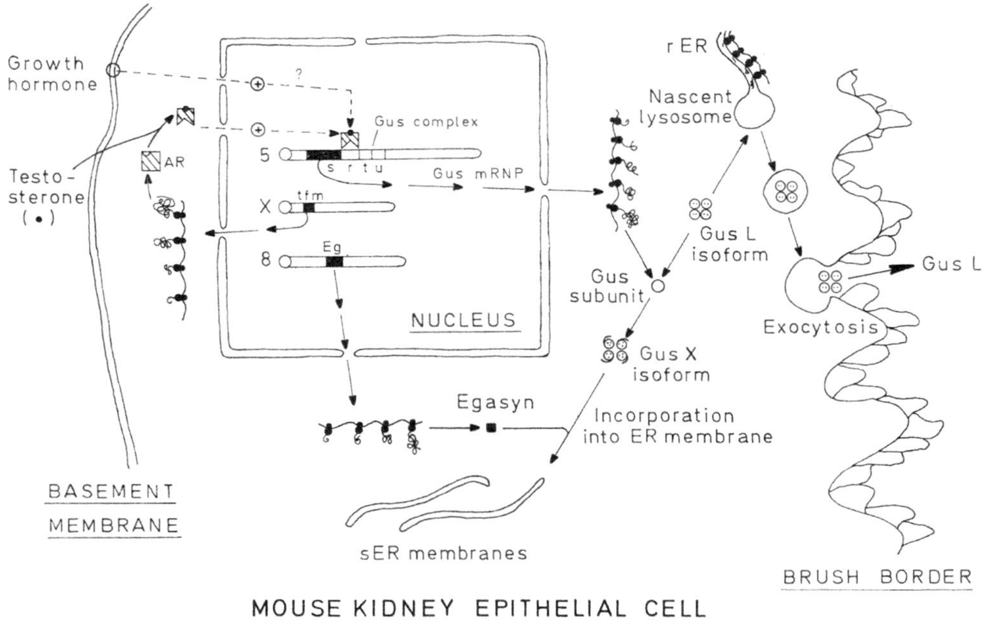


**Fig. 4.13:** Genetic systems involved in the realization of testosterone-stimulated synthesis of $\beta$-glucuronidase in mouse kidney cells (from Paigen [96]).

Testosterone stimulation of $\beta$-glucuronidase (Gus) requires the interaction of the *Gus* complex on chromosome 5 with the *tfm* locus on the X chromosome (testicular feminisation locus coding for the androgen receptor, AR) and the *Eg* locus on chromosom 8 (structural gene for a 64 kd hydrophobic protein, egasyn, which binds to the Gus enzyme subunits and promotes incorporation of the Gus X isoform into membranes of the smooth ER). The *Gus* complex itself is composed of the structural gene for $\beta$-glucuronidase (*Gus*-s), the cis-acting androgen acceptor site (*Gus*-r), the presumably trans-acting temporal gene (*Gus*-t) determining the developmental appearance of Gus and the systemic regulator element (*Gus*-u) controlling the level of Gus synthesis in all tissues. The relative positions of the elements of the *Gus* complex are arbitrary. $\beta$-Glucuronidase is a homotetrameric glycoprotein with $M_r$ 280 kd. The mode of processing of the primary translation product to the two isozyms with different intracellular localization is unclear. Both are definitely coded by the same gene. They differ in their carbohydrate chains (..), in the absence or presence of egasyn and in an additional short C-terminal peptide of the Gus X subunits (⌒). It is temping to speculate that differential processing of a common RNA precursor gives rise to two different Gus-specific mRNA (cf. the generation of three immunoglobulin $\mu$-chain mRNA from a single precursor molecule, Sect. **13.2.2.3**).

3. Mutation of the trans-acting temporal gene (*Gus*-t) determines developmental pattern and tissue-specific expression of the locus. It is thought to be eventually part of the structural gene itself. Such temporal genes have been genetically characterized for many other loci. Regulation may require interaction of the regulatory site within the gene complex with a second, more distant site. This was shown for the developmental regulation of $\beta$-galactosidase [96].
4. The systemic element *Gus*-u is the least characterized of the three regulatory elements. It controls the tissue enzyme levels and is a candidate for a promoter-like structure. A similar cis-acting element was mapped in the vicinity of the mouse $\beta$-galactosidase locus.

5. An essential feature of all regulatory elements so far described, as well as of the related structural genes, is the phenomenon of additive inheritance, i.e., heterozygotes always exhibit a level of gene product and/or temporal control that is intermediate between the two homozygotic parents. Additive inheritance can be expected for cis-acting elements. But it is surprising to find it for trans-acting elements, e.g., involved in developmental control of gene expression. It contrasts to the usual situation encountered in bacteria (Chapt. **5** and **14**) and precludes simple negative control systems.

Besides the evident role of closely linked regulatory elements for the integration of enzyme expression into the developmental and metabolic networks of the organism, the *Gus* system demonstrates a complex pattern of postsynthetic protein modifications which ultimately determine the cellular sites of the mature enzyme forms (cf. Fig. **4**.**13**). Two different isoforms of the homotetrameric glykoprotein are encountered in mouse kidney tubule epithelial cells: one (Gus L) is destined to be secreted by exocytosis together with other lysosomal hydrolases, while the other (Gus X) is incorporated into ER membranes. Both differ in their carbohydrate moieties and presumably also in a short part of their C-terminal sequences. Incorporation of the Gus X isoform into ER membranes depends on the association with a special hydrophobic protein, egasyn, which evidently also promotes binding of other enzymes to ER membranes.

### **4.4.2.5.2** The Human Globin Gene System

Hemoglobin (Hb) is an oligotetramer of two $\alpha$-type and two $\beta$-type subunits. The human genetic system is characterized by two unlinked loci, for the $\alpha$-type and $\beta$-type chains respectively. During embryonic development a twofold shift of the Hb type (embryonic $\rightarrow$ fetal $\rightarrow$ adult) and, coincidently, of the erythropoietic tissue (yolk sac blood island $\rightarrow$ fetal liver $\rightarrow$ bone marrow) is observed. The shifts between the three Hb types involve two changes of their $\beta$-like subunits ($\varepsilon \rightarrow \gamma \rightarrow \beta$) and one switch of the $\alpha$-like subunits ($\zeta \rightarrow \alpha$) (cf. Table **4**.**7**).

**Table 4.7:** Human genes coding for $\alpha$- and $\beta$-type globin chains and their expression during ontogeny (data compiled from Maniatis et al. [97]).

| Genes | | | | |
|---|---|---|---|---|
| | | Embryonic Fetal Adult | | |
| $\alpha$-like genes | | $\zeta_2$——$\zeta_1$————$\alpha_2$——$\alpha_1$ (Chromosome 16) | | |
| $\beta$-like genes | | $\varepsilon$————${}^{A}\gamma$——${}^{G}\gamma$————$\delta$——$\beta$ (Chromosome 11) | | |
| Hemoglobin types[a] | | | Time (Month of gestation) | Erythropoietic tissue |
| Hb Gower 1 | Embryonic Hb | $\zeta_2\varepsilon_2$ | 0–3 | Yolk sac blood islands |
| HbF | Foetal Hb | (1) $\rightarrow \alpha_2\gamma_2$ (G : A = 3 : 1) | 3–9 | Fetal liver |
| HbA | Adult Hb | (2) $\rightarrow \alpha_2\delta_2 + \alpha_2\beta_2$ (3%) (97%) | from 9 | Bone marrow |

[a] Two switches of Hb types proceed during human ontogeny [(1) and (2)]. The transition from foetal to adult Hb starts in the last month of normal fetal development and is completed only 6 months after birth. During the transition period between embryonic and fetal Hb-types two further types are detected: Hb Gower 2 ($\alpha_2\varepsilon_2$) and Hb Portland ($\zeta_2\gamma_2$).

The structural and functional analysis of the globin gene loci by genetic and molecular biological means provided experimental data which may contribute decisively to an understanding of how gene expression is controlled by differential chromatin activation and transcription.

*Fine Structure Maps:* In agreement with similar structures in the mouse and rabbit genomes, a chromatin domain with a cluster of the five β-like genes was mapped on the short arm of human chromosome 11 between marker genes for lactate dehydrogenase A and acid phosphatase 2 [98]. The β-locus spans more than 60 kb (Fig. **4**.14 A) with all five genes in the same orientation, i.e., they are transcribed from the same DNA strand. The general fine structure of the mammalian β-globin genes evidently corresponds to the type shown in Fig. **4**.14 A with a short and a larger intervening sequence (I1 and I2) separating the coding message into three parts (E1–E3). Sequence comparison of various β-like genes from man, mouse and rabbit revealed remarkable similarities with respect to this basic gene structure, i.e., with respect to the sites where the coding sequence is interrupted by I1 and I2, to the size of I1, and to distinct sequence elements in the 5′-non-coding region. Two blocks of sequence homology were detected and designated as ATAA box (centered around —30, cf. Table **1**.8) and CCAAT box (centered around —75) respectively. They are found in many other eukaryotic protein-coding genes and may play important roles in transcription initiation and/or RNA processing [97].

▶

**Fig. 4.14:** Genetic organization of the human globin gene clusters on chromosomes 11 (β-like genes, part A) and 16 (α-like genes, part B) (from Maniatis et al. [97]).

Positions of the β-like genes (ε, $^{G}\gamma$, $^{A}\gamma$, δ and β) and α-like genes (ζ2, ζ1, α2 and α1) are characterized by boxes exhibiting the globin gene-specific pattern of three coding sequences (E1–E3, *solid boxes*) and two intervening sequence (I1, I2, *open boxes*). Recent detailed analyses of the cloned globin gene clusters revealed the presence of non-expressed pseudogenes (ψβ2, ψβ1 and ψα1) whose function is nuclear. On top of part A the *solid bars* denote deletions in the human β-gene cluster that lead to various forms of thalassemia (cf. Sect. 4.5.4.3 and Table 4.14).

The fine structure of the globin genes seems to be very similar for all types of globin genes so far investigated, major changes being restricted to the size of I2. The exons E1 and E3 contain part of the coding sequences *(solid boxes)* and the untranslated sequences at the 5′-end and 3′-end respectively *(stippled boxes)*.

Complete sequencing of all human β-like genes and comparison of the 5′ and 3′ non-coding sequences, including those of some other eukaryotic genes transcribed by RNAP II, revealed a number of functional boxes with stringent *(bold-faced bases)* and less stringent sequence homologies. These are two "promoter" boxes with a possible role for RNAP II binding, two boxes preceding and following the capping site *(*A)*, the translation initiation *(Met)* and termination codons *(Ter)* and a box preceding the Poly(A) addition site *(▼C)*. Position of the transcription termination site *(ter)* is unknown. The partial sequence represents the non-coding DNA strand. Number of nucleotides *(N)* between functional boxes are average values. For further details cf. summary of Efstratiadis, A. et al. (Cell *21* (1980): 653).

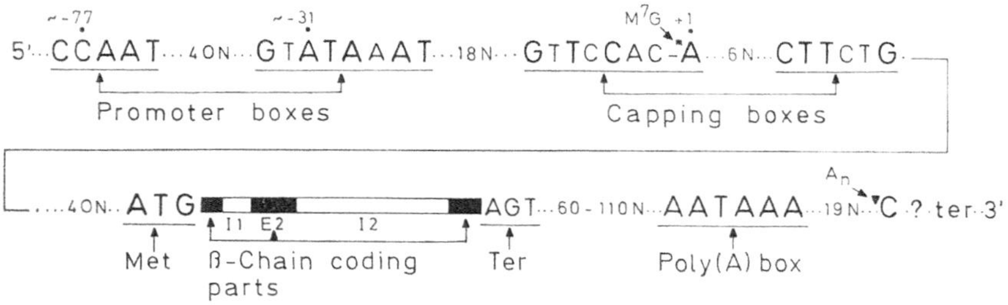

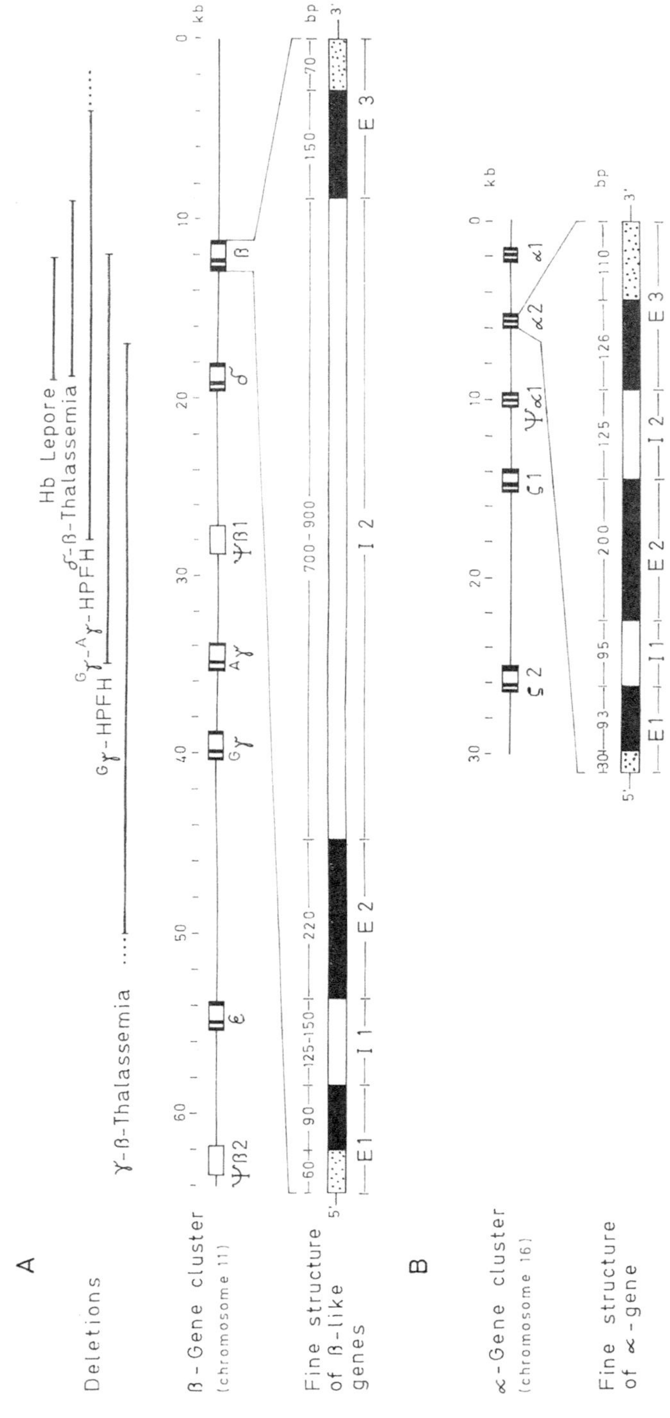
A
Deletions
Hb Lepore
δ-β-Thalassemia
Gγ-Aγ-HPFH
Gγ-HPFH
γ-β-Thalassemia
β-Gene cluster
(chromosome 11)
kb
Ψβ2
ε
Gγ
Aγ
Ψβ1
δ
β
Fine structure of β-like genes
bp
5'
3'
E1
I 1
E 2
I 2
E 3
B
α-Gene cluster
(chromosome 16)
ζ2
ζ1
Ψα1
α2
α1
Fine structure of α-gene
E1
I 1
E 2
I 2
E 3

Structure of the $\alpha$-like gene system is basically similar to that of the $\beta$-like genes (Fig. **4**.14B). Two small intervening sequences of approximately 95 and 125 bp interrupt the coding sequence between codons 31 and 32 and 99 and 100 respectively. The human $\alpha$-gene cluster maps on chromosome 16.

*Repeated Sequences:* Analysis of the spacer DNA in the rabbit and human $\beta$-globin gene clusters revealed the presence of repeated sequences (cf. Table **1**.1). The spacer regions of the rabbit $\beta$-gene cluster contain shorter (140–320 bp) and longer (1400 bp) segments of middle repetitive sequences, which occur as direct or inverted repeats. Each of the four genes is flanked by at least one pair of shorter inverted repeats and the entire cluster is flanked by an inverted repeat of 1400 bp [100, 101].

Sequencing of a copy of a 300 bp repeat, mapping upstream of the human $^{G}\gamma$-gene, showed it to belong to a family of sequences reiterated about 300,000-fold in the human genome [101]. This sequence family exhibits two interesting properties: (1) Its members share sequence homology with an abundant small nuclear RNA molecule and with double-stranded hnRNA. (2) They are possibly transcribed by RNAP III, whereas the globin genes themselves are transcribed by RNAP II. At least for the human $\beta$-gene cluster two RNAP III promoters were mapped 1.5 kb upstream of the $^{G}\gamma$- and 1 kb upstream of the $\delta$-gene respectively [99]. These properties might indicate a function of these repeated sequences as templates for the formation of "guide" RNA that serve the proper alignment of hnRNA for processing (splicing, cf. **4.4.3.1**).

*DNA Methylation:* When analyzing DNA fragments obtained from chicken DNA by cleavage with restriction endonucleases Hpa II and Msp I, it was shown that within 30 bases of the 5′-end of the $\beta$-gene coding sequence a 5′-C$\dot{C}$GG-3′ recognition site is unmethylated in erythropoietic tissue but methylated ($\dot{C}$) in non-erythropoietic tissue, e.g., oviduct or brain. Undermethylation of distinct CG-pairs was also reported for the $\beta$-globin gene regions in erythropoietic tissue of rabbit and man [103].

The discrimination by Southern blotting technique (cf. **2**.6) is based on the fact that both enzymes recognize and cleave the same sequence (–CCGG–), with the Hpa II activity being restricted to the unmethylated form whereas Msp I cleaves both methylated and unmethylated sequences (cf. Table **2**.1).

A much more detailed investigation of the phenomenon of tissue-specific patterns of DNA methylation was performed with the chicken ovalbumin gene system. There are three classes of CG-sites recognized by restriction enzymes Hha I (GCGC) and Hpa II (CCGG), respectively, which may undergo distinct changes of postsynthetic modification by methylation: (1) the $m^-$ class is never methylated (always susceptible to restriction enzyme cleavage), (2) the $m^+$ class is always methylated (resistant to Hha I and Hpa II cleavage but susceptible to Msp I) and (3) the $m^V$ class whose methylation varies in a tissue-specific manner. CG-sites of the latter class are highly undermethylated or even unmethylated in activated gene regions but methylated in non-activated regions [104]. The methylation pattern does not vary with the actual transcription state.

*Long-range Effects in the Chromatin Domain:* A number of well defined deletions of parts of the human $\beta$-gene cluster lead to characteristic hereditary diseases belonging to the thalassemia syndrome (cf. the summary in Table **4**.14). Two types of deletions are especially relevant for an understanding of $\beta$-gene structure and function: (1) A deletion of the entire $^{G}\gamma$-$^{A}\gamma$-$\delta$ region that ends 2.5 kb upstream of the $\beta$-gene (cf. Fig. **4**.14A) results in an inactivation of the otherwise intact $\beta$-gene [105], but does not affect the $\beta$-gene cluster on the homologous chromosome in heterozygotes (cis-effect). (2) Deletions of the $\delta$–$\beta$ region, which leave the $\gamma$-genes intact, lead to the persistence of fetal hemoglobin throughout the adult life [105a]. The effect is likewise restricted to the deleted chromosome.

Interestingly, the latter type of deletion may include the above-mentioned repetitive spacer DNA segment 1 kb upstream of gene $\delta$, which is transcribed by RNAP III. It can be speculated that the 13.5 kb spacer region between the fetal and adult gene group is necessary for the structural rearrangements of chromatin during the transition from fetal to adult Hb-types, i.e., for inactivation of the $\gamma$-genes and activation of the $\delta$-$\beta$-genes. The organization of chromatin into functional domains makes such cis-acting long-range effects very likely, though other explanations can not be excluded at present (cf. Ref. [97]).

More direct evidence for the structural and functional reality of chromatin domains stems from the electron microscopic pictures shown in Fig. **1**.6 and from biochemical experiments to be discussed in the next section. It can be shown that alterations of the chromatin structure observed in active or pre-activated gene regions are not restricted to the transcribed sequences themselves, but rather extend far into adjacent regions. In the $\beta$-globin chromatin domain of chicken the pre-activated state, as probed by an increased sensitivity to DNase I degradation, comprises the $\beta$-gene cluster plus 6–8 kb flanking sequences at the 3′- and 5′-end respectively [108].

*Nucleosomes in Activated Chromatin Domains are Structurally Altered.* It is evident from immunoelectron microscopic observations as well as from the characteristic nuclease digestion patterns that all nuclear DNA, whether transcribed or not transcribed, is found associated with the core histones (H2A, H2B, H3 and H4). But nucleosomes and polynucleosomes (superbeads), respectively, of active chromatin domains exhibit a number of physicochemical and biochemical peculiarities that allow a discrimination between active and inactive chromatin and afford the basis for procedures to obtain fractions enriched in active chromatin [89]. A special, negative selection procedure involves the preferred degradation of nucleosomes in active chromatin regions by short-time exposure to pancreatic DNase I. It has made important contributions to a deeper understanding of active nucleosome structure.

In Fig. **4**.15A the rapid degradation of globin genes in chicken embryonic erythrocytes by DNase I is shown (curve 1). The selectivity is not observed with staphylococcal nuclease (curve 3) and depends on the presence of salt-extractable (0.35 M NaCl) proteins (curves 2 and 4). Extensive purification of the salt-wash fraction revealed the high mobility group proteins HMG14 and HMG17 (cf. Table **1**.3) to be responsible for this effect. In fact both HMG proteins are also selectively released from chromatin under the conditions of mild digestion by pancreatic DNase I. Either of the two structurally closely related HMG14 or HMG17, but not HMG1 or 2, can be used to reconstitute DNase I sensitivity. They bind stoichiometrically to the nucleosome core particle with one molecule HMG protein per activated nucleosome (cf. Fig. **4**.15B). Both, actively transcribed (globin genes) and rarely transcribed (RAV-O virus) gene regions, react in the same manner, contrasting to genes non-transcribable in these red blood cells, e.g., the ovalbumin gene (cf. Fig. **4**.45B).

This independence of the DNase I sensitivity from the actual transcription state of a gene is also impressively documented by investigations about the state of globin genes in uninduced mouse erythroleukemia cells (cf. **4**.5.4.2). Though globin gene expression is stimulated 45-fold by induction and correlates with an increased mRNA level, the globin genes are equally sensitive to DNase I degradation irrespective of the induction state [104a].

HMG14 or 17 can even reassociate with salt-washed, HMG-depleted mononucleosomes in the presence of a 10- to 20-fold excess of inactive nucleosomes. Consequently one is bound to conclude that active nucleosomes have at least one unique feature distinguishing them from bulk nucleosomes and ensuring proper HMG binding. Apparently nucleosomes of active gene regions are relatively depleted in histones H2A and H2B presumably caused by acetylation of their N-terminal, DNA-binding domains (cf. Fig. **1**.4). In support of this result is the observation about a certain competition in the reconstitution experiments shown in Fig. **4**.15B between HMG14 or 17 and histones H2A and H2B, but not histones H3 and H4 [107].

Further details of the chromatin structure in the active globin domain emerged when comparing the DNase I sensitivity of globin genes in embryonic vs. adult erythropoietic tissue [102, 108]. It appeared that, in fact, three states can be discerned by the DNase I degradation technique: (1) inactive, non-transcribable genes (e.g., ovalbumin gene in erythropoietic tissue) are "resistant" to DNase I; (2) pre-activated, but not transcribed genes are moderately sensitive and (3) transcribed genes are highly sensitive. In embryonic erythropoietic tissue of chicken the embryonic globin genes are in state 3, the adult genes are in state 2. In adult erythropoietic tissue the

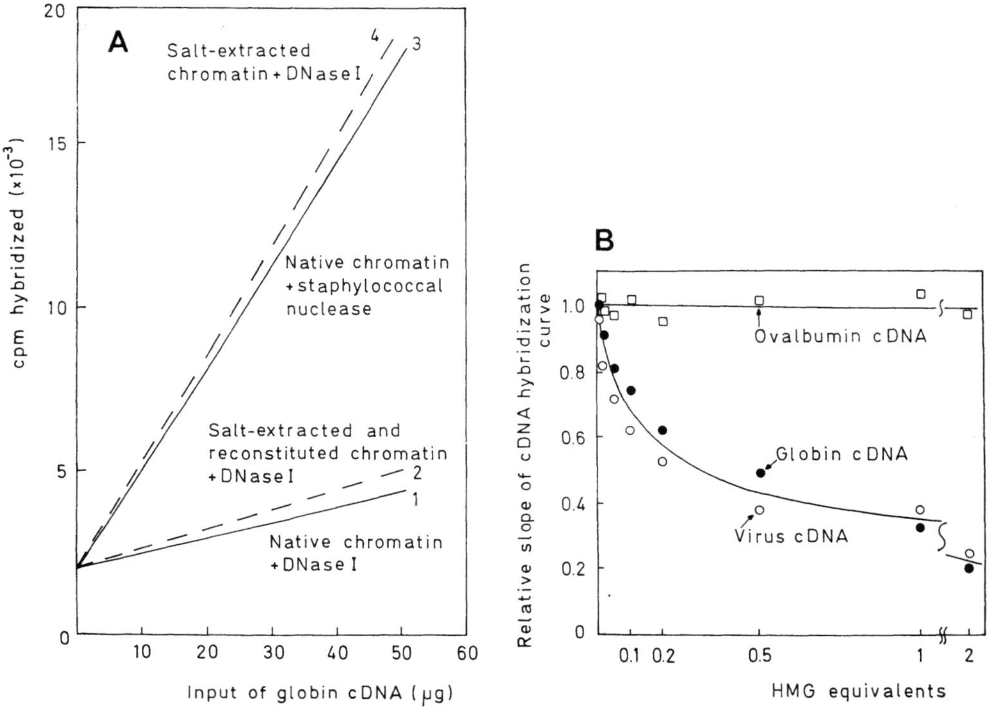


**Fig. 4.15A:** Influence of HMG14 and HMG17 on the differential sensitivity of globin genes in erythropoietic tissue against DNase I attack (from Weisbrod & Weintraub [106]).

Chromatin from isolated chicken embryonic erythrocyte nuclei was digested 15 min with 10 μg/ml pancreas DNase I to a 10–15% decrease of trichloroacetic acid precipitable DNA. The degradation was stopped by adding an SDS-EDTA buffer containing proteinase K. DNA extracted by the chloroform-phenol method was hybridized to globin-specific cDNA (specific activity $5 \times 10^8$ cpm/μg DNA). Part of the chromatin was stripped by extracting with 0.35 M NaCl before DNase treatment (*curves 2* and *4*).

**Fig. 4.15B:** Stoichiometry of HMG14 and 17 required for establishing DNAase I sensitivity of active genes (from Weisbrod et al. [107]).

Sheared and HMG-depleted chromatin of red blood cells was reconstituted with increasing amounts of HMG14 and 17 and then digested with DNase I. The extent of degradation of the three genes was determined by means of [$^{32}$P]-labelled cDNA excess hybridization. The ovalbumin gene is not in an active state, whereas globin genes and the genes of the endogenous chicken virus RAV-O are active, albeit transcribed at very different rates yielding $10^5$ molecules of globin mRNA per cell compared to 1–10 copies of RAV-O mRNA. The HMG equivalent point 1 corresponds to 1 mol HMG per 20 mol of nucleosomes and thus roughly reflects that part of total chromatin (~ 5%) transcribed into nuclear RNA. The *ordinate* shows the relative slope of the hybridization curves (cf. part A). The value 1 corresponds to the slope of curve 4 with no HMG present.

reverse situation holds true. The whole 50–100 kb globin domain in erythropoietic tissues is activated in erythroblast precursor cells (cf. Fig. **4**.39). It exhibits intermediate DNase I sensitivity but the normal DNA methylation pattern. Within these regions 5–10 kb subdomains, containing the actually transcribed genes plus 3′- and 5′-flanking sequences, are highly sensitive to DNase I. Interestingly, only these subdomains are undermethylated and their altered nucleosomes are capable to associate with HMG14/17 [102, 108].

#### 4.4.2.5.3 Elements of Higher Order Chromatin Structure as Related to Determination and Differential Gene Activity

*Heterochromatin and Euchromatin:* The results about structural elements of chromatin, as related to the functional states of the globin genes, point to the more general question of how chromatin structure and its characteristic changes (cf. Fig. **1**.5) reflect different states of activity. In almost all eukaryotic cells, interphase chromatin exists as two distinct morphotypes: euchromatin and heterochromatin [88, 90]. The relative amount of both types is cell-specific and may vary, e.g., 20% euchromatin in thymus lymphocytes and 90% euchromatin in liver cells. The distinction between both types is primarily a cytological and cytogenetical one. Several studies have demonstrated that heterochromatin is genetically inactive, whereas active gene regions are part of euchromatin. The two fractions contain the same set of histones but differ widely in their pattern of non-histone proteins.

As derived mainly from cytological and genetic data, reversible or irreversible heterochromatization of chromosomal regions, whole chromosomes (e.g., one of the mammalian X-chromosomes) or even of whole genomes are evidently essential aspects of cell specialization and of the frequently observed functional restriction of chromatin. The characteristic changes of nuclear morphology during cell differentiation [109] reflect the overall transcription and replication state and the differential heterochromatization of distinct parts of the genome. It is clear from nuclear transplantation and cell fusion experiments that heterochromatization is reversible on principle, i.e., inactive condensed parts can be reactivated if the nucleus is put under appropriate cytoplasmic conditions. Normally, however, the supra-structural organization of the chromatin is propagated essentially unchanged within a given cell lineage. Thus, understanding of how differential gene activity is regulated in eukaryotes and how stabil gene activity patterns are mitotically inherited within a cell lineage, in fact includes an analysis of factors that determine and control the three-dimensional network of chromatin in the interphase nucleus [110, 111]. The concept of active chromatin domains as they have long been known from the puffs and Balbiani rings of polytenic chromosomes (cf. Chap. **9**) is supported by a number of circumstantial evidence, last but not least by the morphological data of DNA loops emerging from a protein scaffold (cf. Figs. **1**.5 and **1**.6). Unfortunately we just begin to understand the molecular basis of this aspect of chromatin structure involving also heterochromatization and the influence of cytoplasmic factors on it.

An interesting example of functional inactivation (heterochromatization) of the entire genome is observed during erythroid cell specialization in birds which, in contrast to mammals, preserve their cell nuclei in mature erythrocytes [112]. However, very early in the lineage of erythroid precursor cells the formation of a new histone species (H5) is detected whose structure and function is similar to H1. It progressively replaces H1 in the chromatin of maturing erythroid precursor cells. Initially it is highly phosphorylated in its N-terminal and C-terminal regions, which for H1 were shown to interact with the internucleosomal linker regions (cf. Fig. **4**.17). The final condensation and inactivation of chromatin at the terminal stages of red cell maturation coincide with the dephosphorylation of H5. Hence, a highly specialized H1-analogue ensures the programmed stepwise heterochromatization of avian red cell chromatin. In keeping with these results are numerous examples for H1 primary structure variants characterizing distinct developmental stages in many organisms [91]. This emphasizes the prominent role of H1 for the organization of chromatin superstructures [113, 114] and hence for gene activity (cf. **1.2.2.3**).

*Protein Composition of Euchromatin Fractions:* As is clear from the preceding section, euchromatin exists in at least three functional states: (1) the non-activated, (2) the pre-activated but non-transcribed and (3) the activated and transcribed state. Moreover, it was shown, that despite the seemingly normal complement of histones in all chromatin fractions, as evidenced by protein analysis (Fig. **4**.16), by DNase degradation patterns, as well as by immunocytological methods, activated gene regions exhibit special properties of their nucleosomal subunits: they are associated with stoichiometric amounts of HMG proteins and are sensitized to DNase I attack. It appears reasonable to assume that distinct functional states of chromatin result from modification of its basic constituents as well as from association with minor proteins of the NHP-faction (cg. preceding section).

By relatively simple procedures of extraction with buffer and subsequent sucrose gradient centrifugation it is possible to obtain distinct chromatin fractions from rat liver nuclei that exhibit highly divergent protein patterns (Fig. **4**.16). The fraction of soluble chromatin (lane S, Fig. **4**.16) is enriched in non-histone proteins predominantly migrating in the region characteristic of HMG proteins (cf. the reference lane M with HMG proteins). Subfractionation of the soluble chromatin on a sucrose gradient (Fig. **4**.16B) exhibits one fraction (3) that contains the bulk of newly synthesized RNA, i.e., it contains most of the active chromatin. The protein pattern (Fig. **4**.16A, lane 3) shows, besides histones, strong bands in the HMG region and an enrichment of high molecular weight non-histone proteins. The further functional analyses of these chromatin fractions must show, whether they reflect, at least partially, the functional states mentioned above.

In addition to the differences in non-histone proteins the advancing techniques for separation of chromatin fractions, preserving their native state and complement of associated proteins, reveal more and more differences also in histone isotypes. Besides the formation of genetically determined isoforms of all five histones as typical attributes of certain developmental stages or phases of cell specialization (cf. the summary by Isenberg [91]), histone microheterogeneity is greatly increased

---

▶

**Fig. 4.16**: Protein composition of chromatin fractions extracted from rat liver nuclei (courtesy of Weihe and Lindigkeit, unpublished results).

Rat liver RNA was labelled for 30 min by intraperitoneal injection of [$^3$H]-orotate. Isolated rat liver nuclei were lysed in 100 mM Tris-HCl, *p*H 7.9 containing 2 mM $MnCl_2$ and 100 mM ammonium sulfate. Insoluble *(P)* and soluble *(S)* chromatin were separated by centrifugation at 6000 rpm. The fraction of insoluble chromatin contains most of the heterochromatin with a relatively low content of HMG and non-histone proteins *(lane P, part A)*. In contrast, the soluble fraction is enriched in pre-activated and active chromatin. It was further separated into five subfractions by 17 h centrifugation in a 10–40% linear sucrose gradient (**B**).

**A:** SDS gel electrophoresis of chromatin proteins in a 7.5–15% polyacrylamide gradient gel. *Lane P:* insoluble chromatin; *lane S:* soluble chromatin; *lane M:* marker HMG proteins (gift by E. W. Johns); *lanes 1–5:* fractions 1–5 of the sucrose gradient as indicated in **B**.

**B:** Sucrose gradient of soluble chromatin fraction *(S)*. The bulk of the newly synthesized RNA is found in fraction 3 exhibiting a significant enrichment of HMG and non-histone proteins *(lane 3, part A)*.

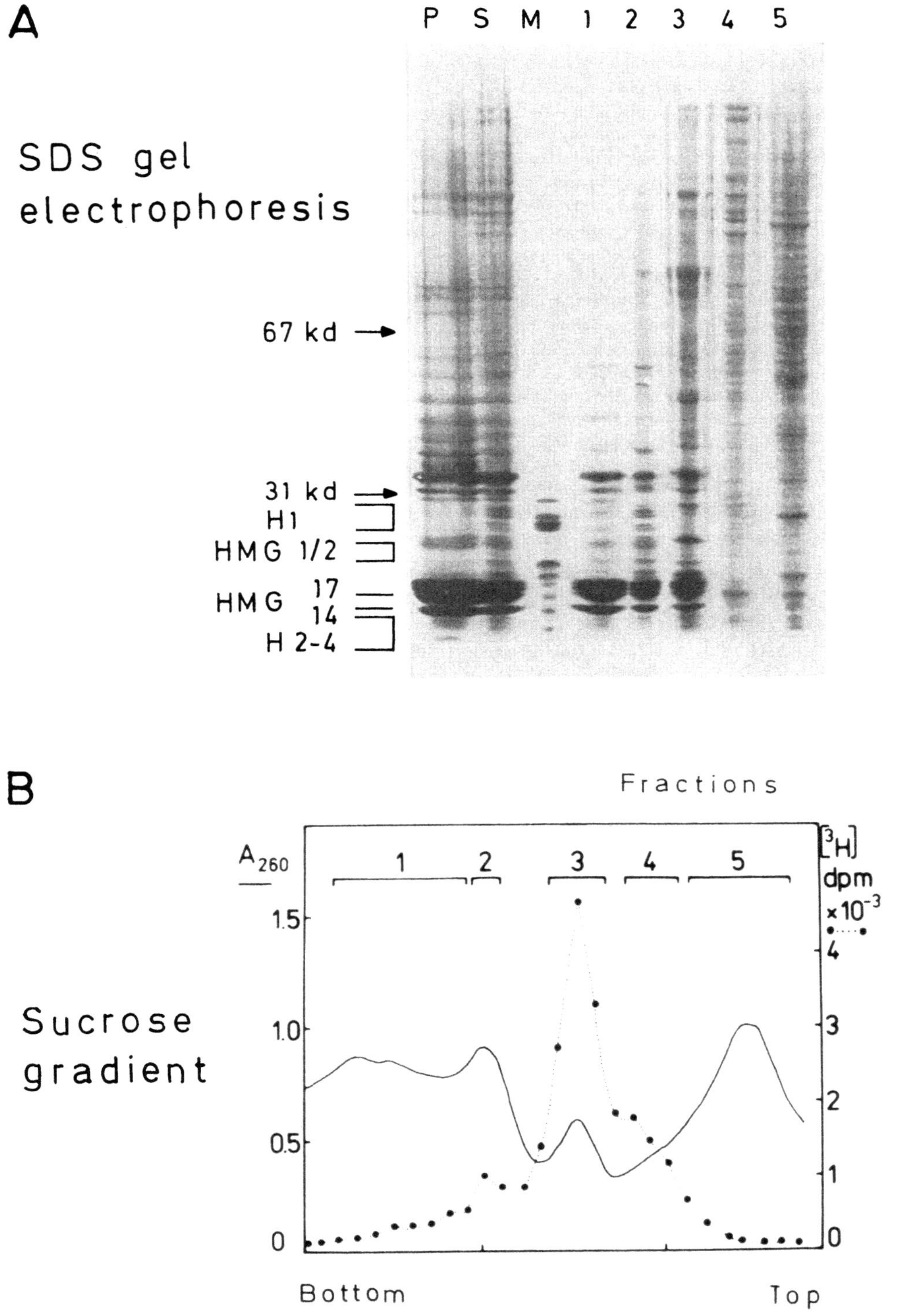
A
P S M 1 2 3 4 5
SDS gel electrophoresis
67 kd
31 kd
H1
HMG 1/2
HMG 17
14
H 2-4
B
Fractions
$A_{260}$
1 2 3 4 5
$[^3H]$ dpm $\times 10^{-3}$
1.5
1.0
0.5
0
4
3
2
1
0
Sucrose gradient
Bottom
Top

by reversible and irreversible postsynthetic modifications (cf. Fig. 1.3 and Table 1.2), e.g., by methylation, acetylation, phosphorylation, and ADP-ribosylation.

Evidently typical patterns of H1 and H3 phosphorylation are tightly coupled to distinct phases of the cell cycle and may play essential roles for chromatin replication in the S phase and chromosomal condensation in the M phase. However, a causal relation cannot be given at present. This is also true for the extensive acetylation of N-terminal regions of histones H2A and H2B found in active chromatin regions though a possible influence of $\varepsilon$-N-lysyl acetylation on the histone-DNA binding affinity is much more plausible in this case. At least increasing histone acetylation in induced Friend erythroleukemia cells (cf. 4.5.4.2) correlates with increased DNase I sensitivity typical of active chromatin and bound to the association of nucleosomes with HMG14 or HMG17 (for details cf. summaries [89] and [91]).

A remarkable modification of H2A (so-called A24) was detected in nuclei of a variety of mammalian cells [91, 116]. H2A is linked via the $\varepsilon$-amino group of Lys 119, by an isopeptide linkage, to the C-terminus of ubiquitin, a highly conserved protein of about 75 amino acid residues ($M_r$ 8.4 kd) that has been found in animals, bacteria, yeasts and higher plants. It has been estimated that about 10% of calf thymus H2A is modified to A24. The special role of A24 for chromatin supercoiling in transcriptionally inactive regions is demonstrated by its release upon transcription activation, e.g., in cells of regenerating liver [115]. There is evidence that H2A modification by ubiquitin is reversible: Ubiquitin was released in Chinese hamster cells during motisis, without interference with the basic nucleosome structure and, regardless of protein synthesis, it was again bound to H2A when cells entered the G1 phase. Hence, the shift between the three functional states of euchromatin may also involve the reversible modification of the nucleosome surface with ubiquitin via a covalent linkage to the H2A C-terminal region.

Special interest deserve those chromatin-associated proteins that somehow control the rate and fidelity of transcription initiation and termination in distinct activated chromatin domains. Very few examples of such proteins have been reported so far. Therefore, we will restrict our discussion to three selected proteins that illustrate three basic types of eukaryotic regulatory proteins interacting with DNP complexes. These are (1) the chromatin-associated thyroid hormone receptor in mammals, (2) an acidic phosphoprotein decisive for rRNA transcription in rDNP complexes from *Physarum polycephalum* and (3) a DNA-bindung protein from *Xenopus* oocytes required for fidel in vitro transcription of 5S RNA genes by RNAP III.

1. The thyroid hormone receptor of mammals is an intrinsic non-histone chromosomal protein [117] that controls the expression of specific genes, e.g., in the pituitary the gene coding for growth hormone or in liver the gene coding for $\alpha_{2u}$-globulin. The thyroid hormone receptors lose their high affinity for binding of the biologically active triiodothyronine after solubilization and separation from the other chromatin proteins. But the original high-affinity binding capacity can be reconstituted by addition of the four core histones. Histone H1 and other acidic or basic proteins of chromatin are ineffective in this respect. Thus, the current model of the thyroid hormone receptor suggests that a thyroid hormone binding subunit (core subunit, $M_r$ 50 kd) interacts with the histone complex to give the holoreceptor with high affinity for triiodothyronine. The holoreceptor might simultaneously interact with DNA or with DNA tightly bound proteins to explain the specificity of the hormone effects. But these details are unclear.
2. The cellular slime mould *Physarum polycephalum* contains multiple copies of rRNA transcription units generated by gene amplification. The isolated rDNA complexes with 38 kb of DNA contain two inverse tandemly repeated rRNA transcription units (cf. Fig. 10.9). Fidel in vitro transcription can be drastically stimulated by phosphorylating an internal acidic regulatory protein ($M_r$ 70 kd). The protein kinase, as its substrate, is part of the rDNA complex itself and can be activated by addition of polyamines (spermin, spermidine). The isolated phosphoprotein was shown to bind with high affinity directly to DNA in the central part of the rDNP complex, i.e., at least 5 kb upstream of the putative promoter region. The dephosphorylated form of the 70 kd protein neither binds to rDNA nor is it capable to stimulate rDNP transcription (cf. details given in Fig. 10.9). A similar phosphoprotein (C14) was detected in the salt-extractable fraction of nucleolar rDNP from Novikoff hepatoma cells (cf. the review by

Busch [116]). Its removal, together with two H1-like proteins, destroys the transcription selectivity of the rDNP complex.

3. Recently another DNA-binding protein has been purified. It is required for in vitro transcription by purified RNA polymerase III of the 5S RNA genes of *Xenopus* oocytes [119]. Interestingly, the 37 kd polypeptide associates with the central region of the 5S RNA gene extending approximately from nucleotide position 45 to 96. It is suggested that binding of this protein provides the recognition event and that its binding to the 5S DNA directs the RNAP III to the proper initiation site (cf. **1.3.2.3**).

*Evidence for Reversible Transitions Between Structural Levels of Chromatin:* The preceding sections have presented accumulating evidence that a multiplicity of dynamic transitions between different levels of chromatin structure accompany its function in replication and gene activity. Though nuclear DNA is found always associated with the normal complement of core histones, two different conformations of activated chromatin have been observed in the electron microscope: (1) the non--beaded conformation mainly of rDNP and (2) the beads-on-a-string conformation of rarely transcribed gene regions. Electron microscopic pictures also show that neither replication nor transcription lead to longer stretches of uncovered DNA (Fig. **4**.12). As already mentioned in Chap. **1** (Fig. **1**.5) the structural levels of activated and transcriptionally active chromatin are to be looked for in the Hl-depleted beads--on-a-string conformation (state 2 of Fig. **1**.5), the 11 nm unit fiber (state 3) and the 20–30 nm fiber composed of superbeads (state 4) (cf. the summarizing model of active chromatin in Fig. **4**.17).

An important question then is: how are these states precisely connected with gene activity and how are the reversible transitions between distinct structural levels brought about? Though a detailed answer to this question must await much further experimental work, the following facts may be relevant:

1. Nucleosomes are highly dynamic structures which undergo reversible cycles of unfolding and refolding.

Not only can they slide along the DNA double helix and change their relative positions to each other thus generating internucleosomal linker regions of varying lengths (between 10 and 110 bp), but their bipartite structure allows also for the two halves to move apart to a certain degree [92]. The results of this conformation change are half nucleosomes with a pivot at the midpoint of the DNA. Transient formation of half-nucleosomal or even more unfolded states of the histone-DNA complex may be essential for RNAP to read through the nucleosome (cf. Fig. **4**.17) thus releasing the coding strand, at least locally, from the binding histones. In support of this dynamic model of nucleosomes are several findings: (1) A mononucleosome--RNAP II complex exhibits a new DNAase I cleavage site at the nucleosome center [92]. (2) Histones were shown to bind also to single-stranded DNA and thus may transiently associate with the non-coding strand as DNA polymerase or RNA polymerase pass [120]. (3) Homopolymers of glutamic acid or aspartic acid effectively catalyse the reversible assembly of histones [121]. This suggests a possible function of HMG proteins with their C-terminal acidic domains (cf. Table **1**.3), especially of HMG1 and HMG2 which contain continuous stretches of 41 and 35 aspartic plus glutamic acid residues in their C-terminal parts [122]. (4) HMG1 and HMG2 function as unwinding proteins by specifically stabilizing single-stranded DNA [123].

2. Very cautious degradation of rat liver chromatin with micrococcal nuclease releases polynucleosomes (superbeads?) containing only 10% of total RNA, but 85% of RNAP II.

These structures, separated from the fraction of non-degraded chromatin (87%) by differential centrigugation, contain 6–30 nucleosomes and might represent the basic structure of active chromatin (unfolded chromatin domains). They are capable of continuing RNA synthesis in vitro by elongation of preinitiated RNA chains [124].

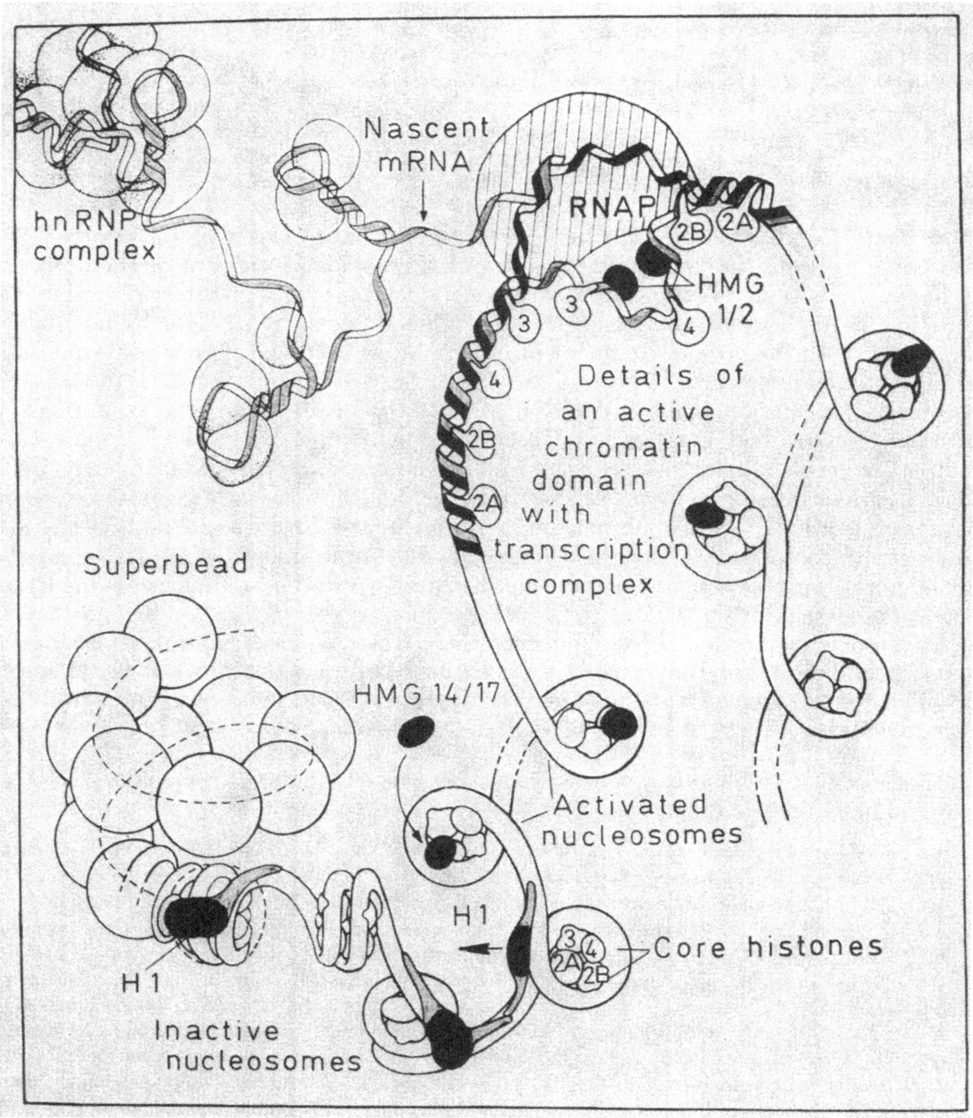


**Fig. 4.17:** Model of an active chromatin domain with transcription complex (from Rindt & Nover [94]).

The model represents a speculative summary of our present-day knowledge of active chromatin. It is based on the following elements:

1. The inactive chromatin is condensed to superbeads by association with histone H1 *(left)*.
2. In the active subdomain *(right)* H1 is released leading to the beads-on-a-string conformation with nucleosomes modified by HMG14 or HMG17.
3. Transcription is assumed to involve the transient unfolding of the nucleosomes, but histones remaining attached to the DNA. The histone arrangement along the DNA derives from DNA-histone and histone-histone cross-linking experiments (cf. Belyavsky, A. V. et al.: J. Mol. Biol. *139* (1980): 519) and from analysis of low-resolution X-ray diffraction patterns of nucleosomes (cf. Klug, A. et al.: Nature *287* (1980): 509). Two dimers of Lys-rich histones (H2A,

3. The mode of chromatin replication ensures an asymmetric and conservative segregation of parental histones to one of the daughter strands, i.e., the leading strand, during DNA synthesis [125]. This mechanism may be crucial to maintain structural chromosomal information, stored, e.g., in a pattern of histone modification, and eventually to transmit it asymmetrically, in the course of a quantal cell cycle, to one of the daughter cells only (cf. Figs. **15**.9 and **15**.10).

#### **4.4.2.5.4.** Conclusion: Properties of Activated vs. Non-Activated Chromatin

Obviously the experimental data presented in the preceding sections are by no means sufficient to give a causal relationship between certain elements of chromatin structure and the commitment of cells to undergo distinct changes in their gene expression patterns. However, is it worth while attempting a summary (Table **4**.8 and Fig. **4**.17) even with the risk to generalize results whose validity has so far been proven only for the experimental systems investigated.

**Table 4.8:** Properties of activated vs. non-activated chromatin.

| | Chromatin | |
|---|---|---|
| | Activated[a] | Non-activated |
| Morphotype | Euchromatin | Facultative and obligate heterochromatin, euchromatin (?) |
| Structural organization | Less condensed, unfolding of functional domains (2–40 kbp) exhibit DNase I sensitive sites | Highly condensed |
| DNA methylation (CG sites) | $m^v$-sites unmethylated | $m^v$-sites methylated |
| Protein pattern | cf. Fig. **4**.16 A, lane S | cf. Fig. **4**.16 A, lane P |
| Nucleosomes | DNase I sensitive | DNase I resistent |
| Histones | Hl-deprivated; core histones highly acetylated | Hl-enriched; association with special H1 isoformes, e.g., H5; H2A/H2B underacetylated; eventually H2A modified by ubiquitin |
| HMG 14/17 | Present | Absent |
| HMG 1/2 | Present | Absent |
| Transcription | | No |
| RNAP | Presence of RNAP and RNP depends on the actual transcription state | No |
| RNP | | No |

[a] For data on the less well characterized pre-activated state cf. pp. 148–149 and Ref. [102].

The following statements appear more or less well-grounded and include a hypothetic succession of chromatin activation:

1. Basic properties of activated chromatin, by which it can be discerned from non--activated chromatin, are not influenced by the actual transcription state.

---

◀

H2B) flank a central tetramer of Arg-rich histones (H4, H3, H3, H4). Unwinding of the DNA in the transcription loop may be facilitated by HMG1 or HMG2.

4. The nascent pre-mRNA immediately associates with structural proteins to form the hnRNP complex.

2. The developmental transition of chromatin domains from the non-activated to the pre-activated state is an essential aspect of cell differentiation and proceeds presumably during distinct phases of the cell division cycle (quantal mitosis, cf. **15**.2). It may involve DNA rearrangements, e.g., by transposition of repeated sequence elements and/or the association of specific DNA recognition sites with tightly bound regulatory protein [126]. Such "TB proteins" may also prevent, in a site-specific manner, the postsynthetic DNA methylation.
3. TB proteins and non-methylation of the $m^v$-sites cause a further unfolding of chromatin subdomains with concomitant H1 deprivation. The association with proteins of the NHP fraction leads to modification of core histones (e.g., by acetylation of H2A and H2B) and subsequently to binding of HMG14 or 17 (Fig. **4**.15). Unfolding of the chromatin domain and loosening of the nucleosome structure is apparent from the pronounced DNase I sensitvity of active subdomains.
4. The actual transcription state of a chromatin domain, though dependent on this special nucleosomal structure, is regulated by further cis-acting control elements adjacent to the gene (cf. Fig. **4**.13) and associated with hitherto largely unknown regulatory proteins of the NHP fraction. Activity control of these regulatory proteins by enzymatic modification and demodification (e.g., by poly [ADP] ribosylation or phosphorylation) may constitute essential means to achieve selective transcription control.

Hence, we are left with many open questions about

- the validity of the tentative and fragmentary sequence of events leading to chromatin activation,
- the intricate role of histone H1 in condensation and decondensation of higher order structures of chromatin (cf. **1**.2.3.2 and Refs. [113, 114]),
- the nature and differential role of the DNA tightly bound proteins, which in some cases can only be released by DNA degradation. Are hormone acceptor proteins (cf. Chap. **7**) such TB proteins? Are TB proteins part of the chromosomal scaffold (cf. Fig. **1**.5)?
- the interplay between genetically determined cis-acting regulatory sequences of the DNA with various proteins of the NHP fraction (TB proteins, DNA methylases, HMG proteins, hormone receptor acceptor proteins),
- the mode of differential transcription regulation in activated chromatin domains and
- the mechanism of reversible unfolding of nucleosomes as RNAP passes during transcription.

It seems necessary to concentrate further efforts on the functional analysis of minor proteins in the NHP fraction, on the thorough genetic characterization of eukaryotic units of gene expression and on the improvement of fidel in vitro transcription systems including methods for a controlled reconstitution, from its components, of chromatin in different states of activity.

#### 4.4.2.6 Transcription Specificity of RNA Polymerase Isozymes

It has already been pointed out that the variation of the strength of promoter regions and their response to regulatory signals reflect major and minor deviations from the structure of the "ideal" promoter sequence, and that just these deviations are evidently important for the proper regulatory behaviour of the promoter region. Realization of the internal control information is brought about by interaction with RNAP itself frequently aided by regulatory proteins. In addition, there are many examples for RNAP isoenzymes exhibiting transcription specificity.

In eukaryotic cells three basic types of nuclear RNAP (types I, II and III) and separate types in mitochondria and chloroplasts can be discerned (cf. Tables **1**.6 and **1**.7). They have distinct specificities with respect to the intracellular localization, promoter types (cf. Table **1**.8) and RNA species transcribed. Moreover, several isoforms of the nuclear enzymes have been characterized. But a detailed understanding of their functional significance must await a thorough analysis of the role of their numerous subunits for the transcription process by further development of fidel in vitro systems and characterization of auxiliary factors involved [119].

Much more advanced is our knowledge about factors changing the transcription specificity of prokaryotic RNAP, e.g., ppGpp or fMet-$tRNA_f^{Met}$, or about special isozymes whose appearance is connected with distinct stages of differentiation of prokaryotic cells, especially during phage development.

Two prominent effectors of *Escherichia coli* RNAP with marked influence on the transcription specificity are guanosine-3′-diphosphate-5′-diphoshate (ppGpp) and charged initiator tRNA (fMet-$tRNA_f^{Met}$) [128]. Both bind competitively to the same subunit of RNAP and strongly inhibit the synthesis of stable RNA (rRNAs, tRNAs) while substantially stimulating the formation of *lac* mRNA [127]. Thus, accumulation of ppGpp and charged initiator tRNA, which are indicative of amino acid deficiency or other situations of growth inbalances with a block of protein synthesis, lead to pronounced changes of promoter recognition specificity of RNAP. In fact, both effectors were found to destabilize the RNAP-rDNA promoter complex in vitro. The stimulation of the *lac* operon transcription is representative of many other operons stimulated by ppGpp. They all serve the improvement of the carbon or nitrogen balance (cf. the summary [23]). It is noteworthy that ppGpp can relieve the *lac* operon transcription from its dependence on the presence of the CAP · cAMP complex (cf. **4**.4.2.2), i.e., both metabolic integrator systems act alternatively on the same operon. Hence, under inducing conditions the efficient initiation of transcription at the *lac* z promoter is brought about either by interaction of RNAP with the CAP · cAMP complex bound upstream of the recognition site or by altering the specific affinity of RNAP for this promoter type, e.g., by association with ppGpp or charged initiator tRNA.

The synthesis of a new RNAP and enzymatic modification, as well as final inactivation of the host enzyme, characterizes the developmental program of *E. coli* cells infected by phage T7 (cf. **14**.2.2.2). Two types of promoters, recognized by host RNAP and T7 RNAP respectively, are essential means of control of the transcription program (cf. Table **14**.3). In contrast, during $\lambda$ phage development the host RNAP functions throughout the transcriptional phases, albeit successively modified by phage-specific proteins to overcome cistron-specific termination signals in the phage genome (cf. Fig. **14**.6). A further example, detailed in Fig. **4**.18, concerns the variation of promoter recognition by exchange of the RNAP $\sigma$-subunit by phage proteins during the lytic cycle of *Bacillus subtilis* phage SPO1.

Three isozymes of RNAP, differing in their subunit composition, can be detected after infection of *B. subtilis* with phage SPO1 (Fig. **4**.18A). Their appearance correlates with distinct shifts in the transcription pattern (Fig. **4**.18B) and consequently also in the pattern of phage--specific proteins synthesized. The characterization of three phage-coded proteins associated with RNAP instead of the $\sigma$-subunit revealed that (1) gp 28 is synthesized in the early phase, whereas gp 33 and 34 are synthesized in the middle phase of the lytic cycle, (2) that at least gp 28 and gp 34 function as true $\sigma$-like subunits and (3) that they reprogram transcription in the infected host cell by discriminating between host and phage promoters (Fig. **4**.18C). There is no doubt that these sequential modifications of RNAP are only one aspect of the total regulatory program of phage SPO1, whose complexity is just emerging.

### 4.4.2.7 Termination and Antitermination

As outlined in Chap. **1** (cf. **1**.3.2.3), transcription termination in bacteria results from the slow-down of the elongation rate when RNAP passes GC-rich sequences with elements of inverse rotational symmetry. These sequences are followed by an oligo (A) stretch on the coding DNA strand. Termination is assumed to involve the intereaction of RNAP with stem-loop structures formed either by the nascent RNA

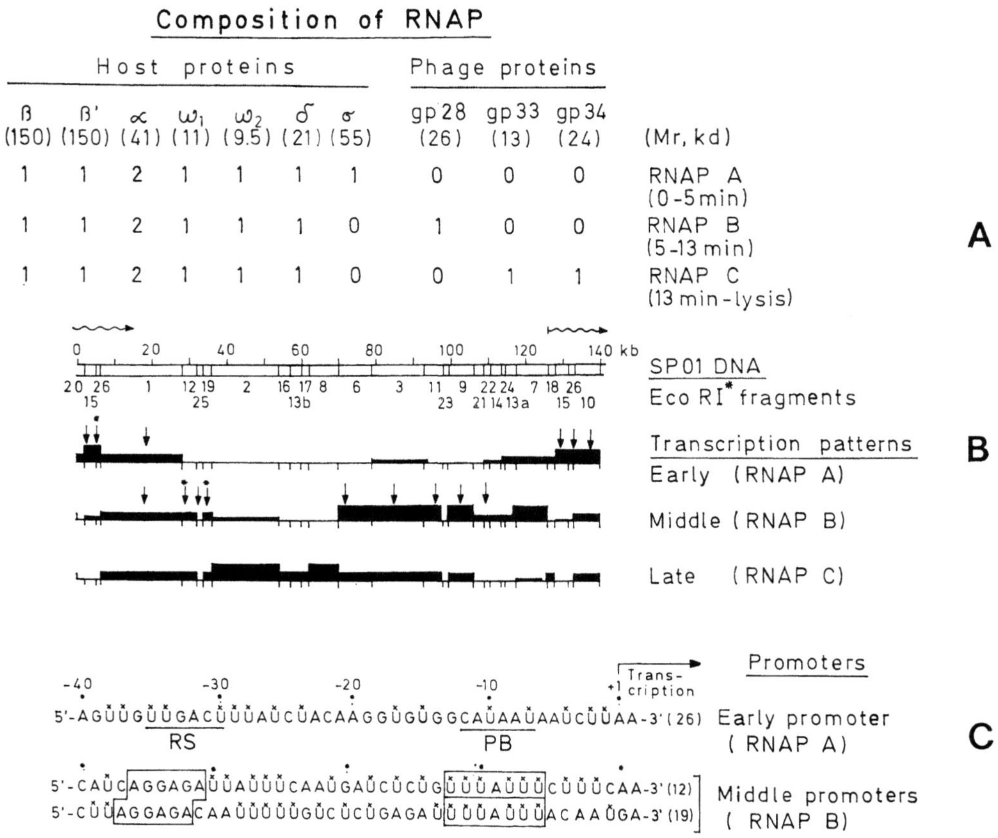


**Fig. 4.18:** RNA polymerase isozymes control the transcription program of *Bacillus subtilis* phage SPO1.

**A:** *Subunit composition of RNAP isozymes* was determined by SDS-polyacrylamid gel electrophoresis [129]. Enzymes were isolated from non-infected cells *(RNAP A)* and from cells about 10 min *(RNAP B)* and about 15 min *(RNAP C)* after infection with phage SPO1. The three isozymes differ in their contents of the $\sigma$-subunit which is replaced by phage-coded $\sigma$-like proteins (gp 28 and gp 33 + 34). For details of RNAP subunit function cf. **1.3.2.1.**

**B:** *Eco RI* restriction map and transcription patterns during phage SPO1 development* [130]: The 140 kb phage DNA contains two repeats at both ends *(wavy arrows)*. Degradation under Eco RI* conditions (cf. **2.3**) gave 27 fragments which were mapped (cf. **2.9**) and numbered in the order of increasing size. The transcription pattern was determined by Southern hybridization (cf. **2.6**) of labelled RNA from phage infected cells with the restriction fragments. RNA was isolated in three different stages of the infection cycle: 0–5 min (early), 5–13 min (middle) and after 13 min (late). *Thickness* of the *bars* gives an estimate of the extent of hybridization. *Arrows* indicate promoters detected by binding host RNAP A and gp 28-modified RNAP B to 5'-[$\gamma$-$^{32}$P]-labelled restriction fragments, separation of the DNA · RNAP complexes from non-bound DNA by filtration through nitrocellulose filters and subsequent identification of the retained DNA fragments by agarose gel electrophoresis.

**C:** *Sequence of early and middle phage promoters* [132]: The promoter sites in Eco RI* fragments 26 (early) and 12 and 19 (middle) were isolated in Hha I and Hpa II subfragments and localized by DNase "footprinting": 5'-end labelled restriction fragments are incubated with low levels (1 ng/ml) of DNase I in the absence and presence of RNAP. Because nicking occurred nearly randomly and the fragment was labelled only at one end, electrophoresis of the partial digest through denaturing gel (PAA gel electrophoresis in the presence of formamide) generated a "ladder" of DNA pieces. These differed from their neighbours by one nucleotide each. When the

or by the non-coding DNA strand (cf. Fig. **1**.7). In addition, weak termination signals require the interaction with a basic tetrameric protein, the $\varrho$-factor.

Thorough genetic and molecular biological analysis of termination and its prominent role in the regulation of the transcriptional programs of *E. coli* phage $\lambda$ (cf. **14**.2.3.4) revealed: (1) $\varrho$-dependent termination is bound to the presence of a cistron-specific sequence for rho utilization *(rut)* and (2) termination can be alleviated by anti-termination proteins (e.g., gp *N*, gp *Q*) whose action is also dependent on cistron--specific recognition sequences (e.g., for gp *N*-utilization, *Nut*). These antitermination proteins directly interact with RNAP thus modifying its behaviour at the termination signals.

Quite another example of termination control is observed during phage T7 development. Here also, as in the preceding example of phage $\lambda$, the RNAP of the infected *E. coli* cell is abused for the early phage-specific transcription program. However, phosphorylation of the $\beta'$-subunit by a phage-coded protein kinase ensures that two termination signals at the end of the early gene region limit transcription to the early genes. The unmodified RNAP is unable to recognize these signals and therefore reads through the whole T7 genome (for details cf. **14**.2.2.2).

A further exciting variant of specific termination regulation was detected for bacterial transcription units coding for enzymes of amino acid biosynthesis, e.g., for Phe, His, Thr, Leu and Trp synthesis (for references see [133–138]). In all cases a short leader sequence downstream of the promoter precedes the structural genes. It is transcribed and translated in statu nascendi to give a leader peptide containing two (trp operon) to seven (his operon) continuous residues of the amino acid whose synthesis is coded by the operon. Hence, availability of the particular amino acid controls translation of the leader peptide and, through cooperation between the components of the transcription-translation complex, the behaviour of RNAP at a $\varrho$--dependent termination signal immediately upstream of the structural genes (attenuator). The ribosome-arrest model of Yanofski [136] for this attenuation mechanism of specific transcriptional control (Fig. **4**.19B, C) emphasizes the prominent role of leader RNA secondary structures for transmission of the signal about amino acid deprivation to the transcription complex halted at the termination signal. It is described in Fig. **4**.19 for the *trp* operon of *E. coli*.

Expression of genes of the *trp* operon is under twofold hierarchical control mechanisms by the intracellular level of the amino acid (repression and attenuation). Under conditions of Trp excess, e.g., due to the presence of Trp in the nutrient solution, transcription is blocked by the *trp* repressor (a homodimer of the *trp* R gene product, $M_r$ 47 kd). But because the repressor-operator complex is stable only at relatively high concentrations of intracellular Trp [133], exhaustion of Trp in the medium turns on a moderate autonomous synthesis of the amino acid by allowing transcription of the structural genes of the *trp* operon (Fig. **4**.19A). However, the normal low levels of Trp, though sufficient for protein synthesis, keep transcription of the *trp* operon at a very low efficiency. Only one of seven RNAP molecules that start at the *trp* promoter reach the structural gene region (*trp* E, D, C, B, A), most of them being released at the attenuator 20–30 bp

---

◀

DNA was bound to RNAP, a stretch of approximately 60–70 bp was protected from degradation and thus was absent in the electrophoretic ladder (Galas, D. J. & A. Schmitz: Nucleic Acid Res. *5* (1978): 3157). Promoter sequences (non-coding DNA strands) were determined by the Maxam-Gilbert technique (cf. **2**.10). The early promoter is closely related to the bacterial consensus promoter (cf. Table **1**.8) with a primary recognition site *(RS)* and a Pribnow box *(PB)*. However, the two middle promoters differ significantly in these regions and instead contain stretches of identical 6 and 7 bp respectively *(boxed)*. $\overset{*}{U}$ denotes 5-hydroxymethyluracil which replaces thymin in SPO1 DNA.

upstream of *trp* E. Maximum transcription of the *trp* operon is limited to situations of acute Trp depletion with respect to the other proteinogenous amino acids. Both regulatory systems, repression and attenuation, complement each other and provide the cells with efficient mechanisms to adjust the rate of enzyme synthesis to the metabolic needs. Mutants deficient in either one of the two systems revealed repression and attenuation to affect the overall transcription rate at the *trp* operon 70-fold and 10-fold respectively.

The model presented in Fig. 4.19B and C shows two mutually exclusive conformations of the leader RNA mediating the signal about the intracellular Trp concentration to the transcription-termination complex and the key role of ribosomes for the transitions between both conformations. Extensive read-through of RNAP into the structural genes requires lower-than-normal levels of charged TrptRNA. Signal processing for the attenuation mechanism involves the following sequence of events: (1) Trp is activated by TrptRNA synthetase. (2) The charged tRNA provides Trp for leader peptide synthesis. (3) Release of the translating ribosome at the stop codon -UGA-, allows the leader RNA to refold into the protector and terminator stem-loop structures (cf. Fig. 4.19C, left). (4) The terminator presumably promotes interaction

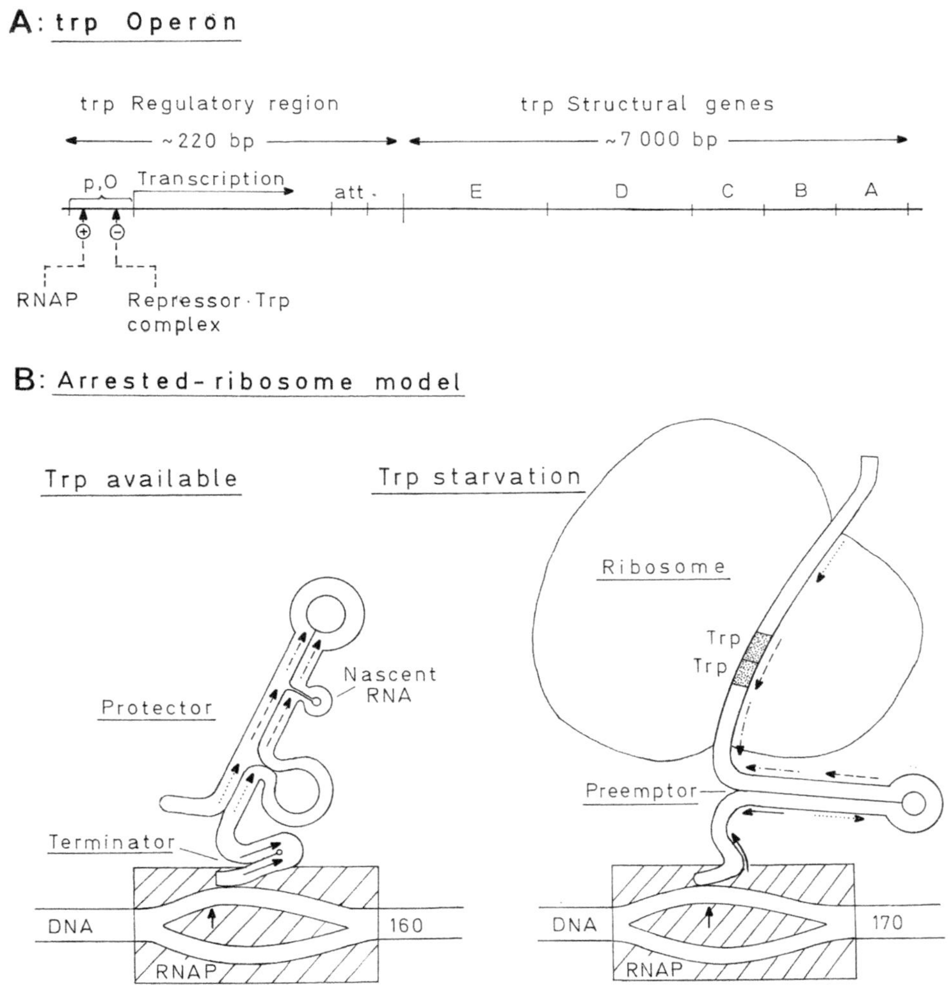

## C: Leader RNA conformations

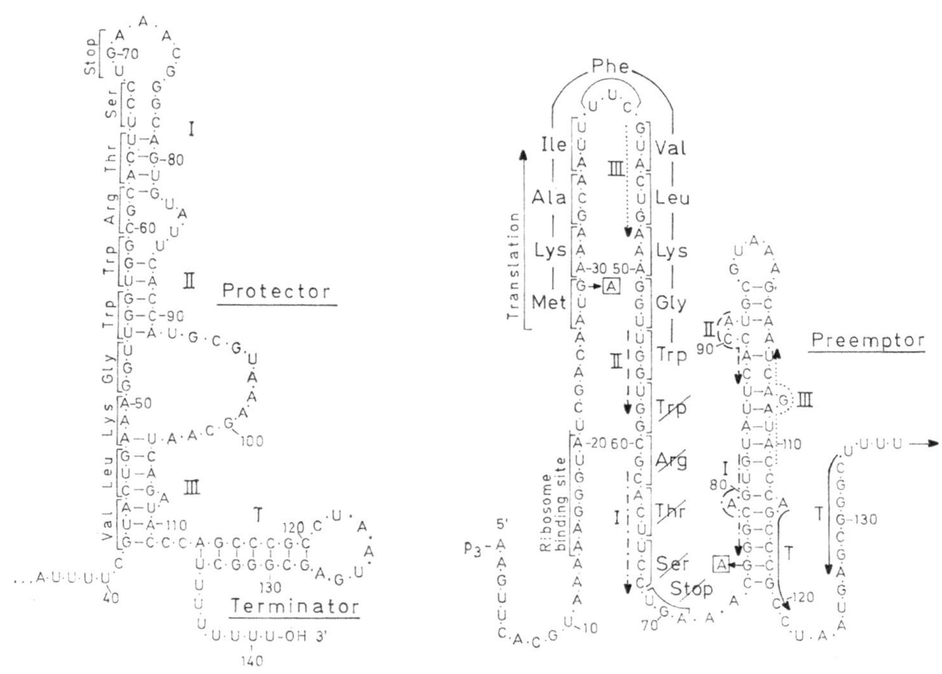


**Fig. 4.19:** A specific termination–antitermination mechanism (attenuation) controls transcription of operons for amino acid synthesis in bacteria. The *trp* operon of *E. coli.*

**A**: Genetic map of the *trp* operon (for details cf. Ref. [133] and Bennett et al.: J. Mol. Biol. *121* (1978): 113 and following papers pp. 139–217): In the *trp* operon the structural genes *trp* E–A are preceded by a leader sequence of 180 bp that starts in close vicinity of the promoter-operator region *(p, O)*. The leader sequence is transcribed by RNAP and translated in statu nascendi into a short leader peptide of 14 amino acid residues. Availability of tryptophan-charged tRNA for leader peptide synthesis controls the behaviour of RNAP at the attenuator *(att)*: (1) Slow-down and $\varrho$-mediated termination of transcription in the presence of Trp or (2) slow-down but read-through under conditions of Trp starvation.

**B, C**: Arrested-ribosome model of attenuation [136–138]: By virtue of its particular base sequence the leader RNA exists in two mutually exclusive conformations with base pairing between inverse symmetrical parts *(arrows)* of the sequence (**C**): (1) If Trp is available for leader peptide synthesis, the translating ribosome is rapidly released from the nascent RNA chain, which refolds to give the *protector* stem-loop structure with base-paired regions I, II and III. The latter allows the *terminator* stem-loop structure to form and thus release of RNAP slowed down in the termination region (cf. *left* figures of **B** and **C**). (2) In the absence of Trp the ribosome is arrested at the two tandem Trp codons. It prevents regions I–III from base-pairing with their counterparts, which instead are incorporated into the *preemptor* stem-loop structure. Because part of the terminator sequence is likewise integrated into the preemptor, formation of the terminator structure is precluded (cf. *right* figures of **B** and **C**). Hence RNAP continues transcription into the adjacent structural gens. [A], two mutants with G → A transitions interfering with the antitermination functions of the leader RNA (cf. text).

of the RNAP with the $\varrho$-factor catalysing the release of RNAP and leader RNA. In case of insufficient levels of charged tRNA, translation is blocked at the two Trp codons. The leader RNA is forced into an alternative conformation (Fig. **4**.19C, right) that precludes base-pairing in the terminator region, and hence, release of RNAP at the attenuator signal.

Two mutants have been isolated whose phenotype strongly supports the ribosome arrest model of attenuation and the role of the leader RNA conformations [134]. In mutant L29 nucleotide 29 of the leader RNA is A instead of G. The translation start condon –AUG– reads now –AUA–. Because translation is highly inefficient, the antitermination conformation (Fig. **4**.19C, right) of the leader RNA cannot form. Thus, termination at the attenuator is enhanced 3- to 5-fold and the positive response on Trp starvation is decreased. The same phenotype results also from a second mutation (L 75). In this case a G→A transition in the GC-rich stem of the preemptor interferes with the stability and/or formation of the antitermination conformation.

There is circumstantial experimental evidence for the role of transcription regulation by attenuation in eukaryotic cells, too [139–142]. Labelling of RNA in isolated HeLa cell nuclei reveals many short RNA chains synthesized by RNAP II. Their abundancy can be increased by treating cells with 5,6-dichloro-1-$\beta$,D-ribofuranosyl benzimidazol (DRB) prior to the isolation of nuclei. Kinetic evidence indicates that most of these putative hnRNA precursors are terminated 100–300 nucleotides downstream of the initiation site and that DRB inhibits the read-through and hence strongly reduces synthesis of mRNA precursors (hnRNA) [139].

### 4.4.3 Control of Pre-mRNA Processing and mRNA Degradation

Considering the relatively far advanced evidence and sometimes detailed knowledge about transcriptional regulation, control mechanisms operating on the RNA level are just beginning to emerge. Three partial processes will be analyzed in more detail: (1) processing of pre-mRNA, (2) biological inactivation and degradation of mRNA and (3) translation in sensu stricto. It will be recognized that specificity of the regulatory events in all cases results from the information carried by the mRNA molecule itself. It is realized by interaction with RNA and/or protein components taking part in processing, transport, storage, degradation and translation of mRNA. Only in few cases these components themselves are specific. The input of regulatory information is not restricted to the nucleotide sequence of a given mRNA but rather extends to the particuliarities of its secondary structural elements. This three-dimensional information, though most pronounced for tRNAs and rRNAs, is more and more recognized also for mRNAs and their complexes with proteins (mRNP). Exposition of certain parts of the base sequence and burying of others by formation of intrachain double-helical stems with single-stranded loops are illustrated by the hypothetic "flower" model of part of the *E. coli* phage MS2 RNA (Fig. **4**.20), the "classical" example of an mRNA whose total sequence was determined. Such elements of RNA secondary structure are highly dynamic and change upon interaction with proteins. They offer the basis for a regulatory interaction between distant parts of the same molecule.

---

▶

**Fig. 4.20**: Flower model of the polycistronic RNA of *Escherichia coli* phage MS2 (from Fiers [143], modified).

Only part of the total RNA is shown, i.e., the end of $A_2$-protein gene, the first intercistronic region, the coat protein gene, starting at the initiator codon AUG*(i)* and ending at the terminator codon UAA*(t)*, the second intercistronic region and the beginning of the gene of MS2 replicase subunit. For details of the translational program and the total genetic map of the closely related phage Q$\beta$ cf. Fig. **4**.24.

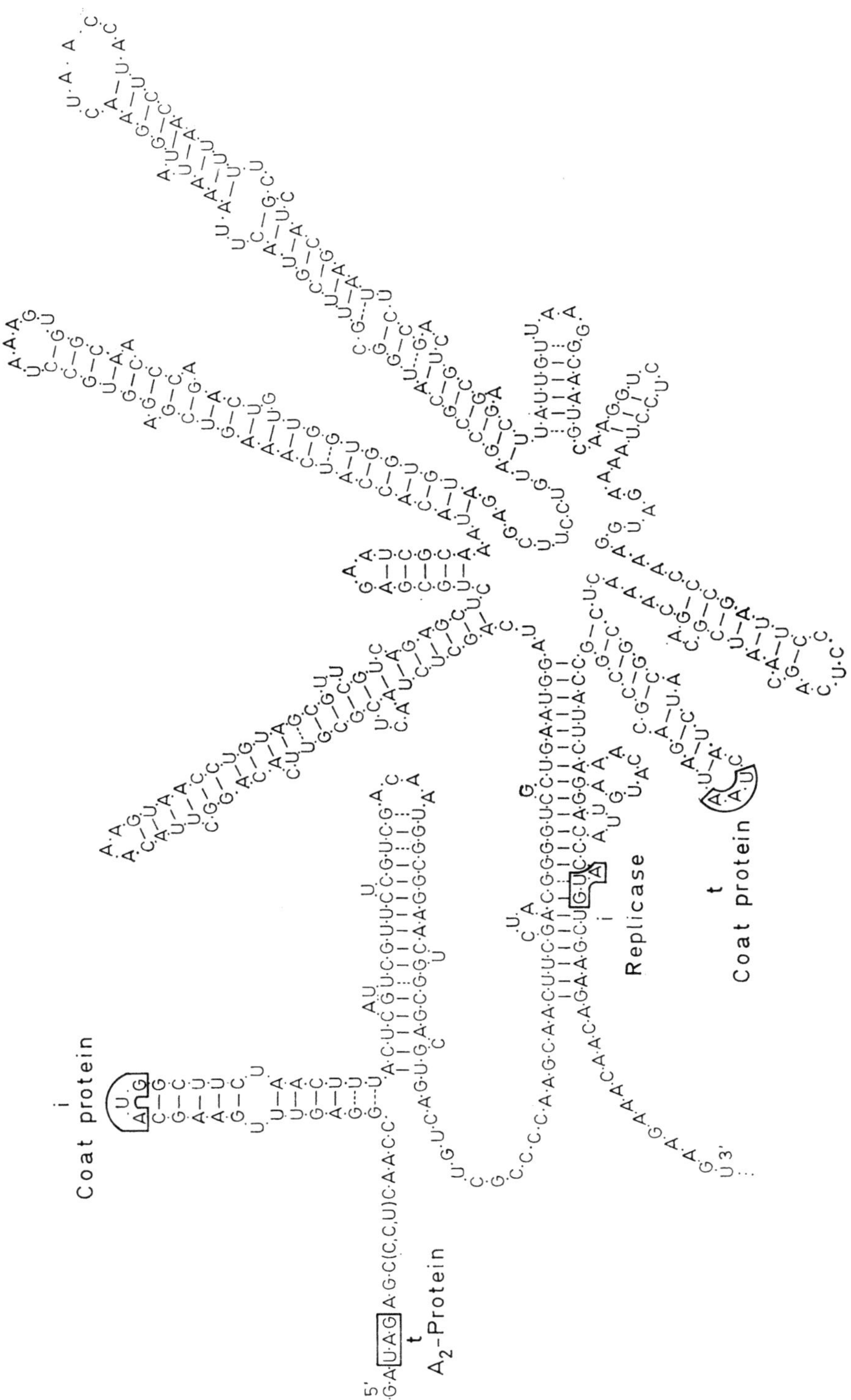


Fig. 4.20

### 4.4.3.1 Pre-mRNA Processing: Nuclear Organization and Selective Information Transfer

Many biochemical details of the processing of RNA precursors in prokaryotic and eukaryotic cells have been discussed in Chap. **1**. It was mentioned that pre-mRNA processing is largely restricted to eukaryotic cells.

The only well-known examples for bacterial systems concern the endonucleolytic cleavage by RNase III of very large polycistronic mRNA precursors transcribed from the early and late gene regions of *E. coli* phage T7 and from the late gene region of phage $\lambda$ (cf. Figs. **14**.3 and **14**.12). For the early genes of T7 processing is a prerequisite for efficient translation of certain genes. It is evident, also from studies on pre-rRNA processing (cf. **1**.3.3.2), that RNase III recognizes distinct features of the precursor RNA secondary structure with double-stranded stems formed by sometimes very remote flanking sequences of the RNA part to be excised from the polycistronic precursor.

RNA processing in the nuclei of eukaryotic cells is evidently bound to an ordered ribonucleoprotein complex (RNP) generated by transcription (cf. **1**.3.3.1).Besides a restricted number of predominant structural proteins of Mr 30–40 kd forming the mRNP cores, there are up to 20 minor proteins ($M_r$ 50–120 kd). They are more or less loosely bound to mRNP particles and may contain regulatory proteins and enzymes (nucleases, RNA methylases, poly (A) polymerase, splicing enzymes) required for an ordered processing. Moreover, hnRNP particles are tightly associated with the network of the nuclear skeleton, as are a class of small nuclear RNP particles (snRNP) to be discussed below. The nuclear skeleton consists of highly coiled, interconnected thin fibers formed by the interaction of a large number of heterogeneous non-histone proteins. Chromosomal scaffold proteins (cf. **1**.2.3.4) may be part of this network which evidently serves not only the spatial organization of hnRNP processing but also the stabilization and transcription of active chromatin regions (cf. Fig. **4**.17) as well as the orderly transport of mRNP and completed ribosomal subunits into the cytoplasm. A direct contact between nuclear and cytoplasmic skeletal elements through the nuclear pores might be crucial to the controlled release of mature informational macromolecules into the cytoplasm [147, 148].

An outstanding feature of pre-mRNA processing is the selectivity of information transfer. This is detailed in connection with Fig. **4**.25, giving qualitative and quantitative data about the composition of different nuclear and cytoplasmic RNA pools in avian erythrocytes containing $\beta$-globin-specific sequences. It appears that the relative contents of globin-specific sequences increases during processing from $5 \cdot 10^{-3}\%$ in the largest hnRNA fraction ($\sim$12–15 kb) to 50–75% in the polysomal mRNA pool. Simultaneously, the sequence complexity decreases from 15,000 to $\sim$240 (cf. data in the right columns of Fig. **4**.25), because only $\sim$10% of the originally transcribed mRNA precursor species can be processed and transported to the cytoplasm and, again, only $\sim$10% of the cytoplasmic mRNA species are actually translated in these cells.

Our ignorance about the true primary transcripts in most cases is an important unsolved problem of pre-mRNA processing. Thus, during extensive investigations on transcription and processing of $\beta$-globin-specific pre-mRNA in avian and mammalian erythrocytes, the existence of very large pre-mRNA species with 12–15 kb and $\sim$4.5 kb have been reported repeatedly (cf. the summary [184] and Fig. **4**.25). If they are bona fide precursors, they should contain 5′-untranslated sequences of more than 10 kb and $\sim$3 kb respectively, which are removed early during processing to give the firmly established $\sim$1.5 kb $\beta$-globin-specific pre-mRNA. The latter was isolated, purified and shown to contain three exons plus two introns typical of globin-like genes (cf. Fig. **4**.14). It is further processed to the mature mRNA of 0.7 kb. Thus the sequence of $\beta$-globin pre-mRNA processing should read:

12–15 kb $\rightarrow$ 4.5 kb $\rightarrow$ 1.5 kb $\rightarrow\rightarrow$ 0.9 kb $\rightarrow$ 0.7 kb

Two types of experimental result are clearly contradictory to the first two steps of this sequence: (1) Promoter-like sequences were localized ~50 bp upstream of the translation initiation point, i.e., in direct neighbourhood of exon 1 (cf. Table **1**.8). Thus, the 1.5 kb pre-mRNA should, in fact, be the primary transcript. (2) The functional significance of this putative promoter region was confirmed in an in vitro transcription system showing accurate transcription initiation at this site (cf. Luse, D. S. & G. Roeder: Cell *20* (1980): 691). These data suggest that, at least for mouse and rabbit, in aggreement with the seize of the major $\beta$-globin-specific pre-mRNA species (~1.5 kb), the main transcription initiation site is in close vicinity to the 5′-border of exon 1. A second promoter, much further upstream, must be implied, if the two high molecular weight RNA species are true precursors of $\beta$-globin mRNA. A similar situation holds true for the immunoglobulin-specific pre-mRNA species (cf. **13**.2.2.3).

### 4.4.3.2 Role of Guide RNAs

Accumulating evidence supports the role of pre-mRNP processing as an important control site of gene expression in eukaryotes. This is especially true for its most striking part, the sequential and very precise excision of intervening sequences (splicing), which might be under the control of RNP complexes containing small molecular weight RNAs (snRNA, snRNP). Recent models of gene expression regulation (cf. Figs. **1**.10 and **4**.21) assume these RNAs to function as guide RNAs [159, 160] to aligne the exon-intron border sequences in a proper position so as to allow splicing to take place. Guide RNAs might be the product of integrator genes containing repetitive sequences which are inverse-symmetric to the sequences around the exon-intron borders. The model in Fig. **4**.21 contains the explicit statement that expression of genes coding for proteins with low or moderate abundance in a cell is controlled mainly, or in certain cases even exclusively, at the level of pre-mRNP processing. Qualitative und quantitative variations of guide RNA populations, produced in response to intracellular or extracellular signals, are the means to change the protein pattern in differentiating cells [151]. This does not apply to mass proteins synthesized by highly specialized cells, such as ovalbumin (Chap. **7**), globin (**4**.5.4) or plant seed storage proteins (Chap. **12**).

What are the basic experimental facts supporting such a model?

1. In most organisms DNA sequence organization displays an interspersed arrangement of repetitive and unique sequences (cf. Table **1**.1).
2. Repetitive sequences are transcribed and found in different fractions of nuclear RNA (nRNA) including pre-mRNA. The concentration of specific families of repetitive sequences in the RNA population varies as a function of cell type [151].
3. nRNA molecules turn over rapidly. Much more unique sequences are found in nRNA than are exported into the cytoplasm.

Complete sets of structural gene transcripts from polysomal mRNA of a given cell type are present in the nRNA's of other cell types not expressing these genes at the polysome level. Thus cDNA prepared from sea urchin blastula mRNA hybridizes 100 % with blastula mRNA, only 12 %, with mRNA from intestinal cells but 97 % with nuclear RNA from intestine [151].

Are there any candidates for the role of guide RNA's? We will briefly discuss three types of experimental evidence that strongly support the guide RNA concept. These are (1) the possible function of snRNP in hnRNP processing [152, 154], (2) an adenovirus-specific guide RNA (VA RNA I) for splicing of late gene pre-mRNA [153, 158] and (3) trans-acting intron mutations in yeast mitochondrial genes (Fig. **4**.22).

1. From animal cells a family of six different small nuclear RNA species (snRNA, U1–U6) were isolated and analysed. They range in the size between 220 (U2) and 90 (U6) nucleotides and form snRNP particles sedimenting at 7–10S by association with seven prominent nuclear

proteins of $M_r$ 12–32 kd. The proteins of snRNP particles are highly conserved between man, mouse, sea urchin and insects. This was shown by precipitating snRNP's with antibodies from patients suffering from a rheumatic autoimmune disease, *Lupus erythematosus*.

The following findings suggest an important role of snRNP's in splicing [154–157], and hence may contribute to an understanding of the molecular basis of the *Lupus erythematosus* syndrome: Stabile, low molecular weight RNA's are regularly found as components of hnRNP particles. They sediment together upon centrifugation in sucrose density gradients and are much more abundant in rapidly dividing cells than in resting cells. These snRNA's are highly methylated and base-pair with hnRNA, but much less with cytoplasmic poly(A)$^+$-RNA. A direct relation of this hybridization between both nuclear RNA types to hnRNP splicing is suggested by the almost perfect base-pairing between the 5′-capped end of U1–RNA and the consensus sequences surrounding splicing points of about 40 different mRNA precursors (cf. Fig. **1.10**). At present, it is a matter

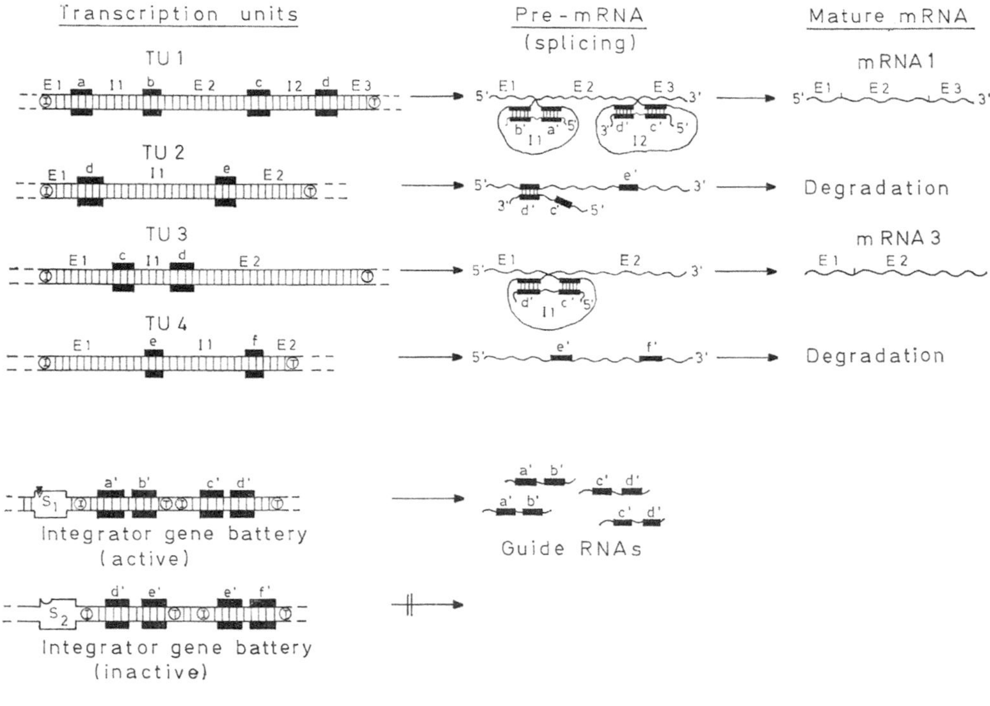


**Fig. 4.21:** Role of guide RNAs for the control of pre-mRNP processing (derived from Davidson & Britten [151]).

Transcription units *(TU1–TU4)* code for proteins. They contain initiator *(I)* and terminator sequences *(T)* as well as several exons *(E)* and introns *(I)*. Exons and introns are separated by short sequences recognized by the splicing machinery *(lower case letters)*. Splicing is under the control of guide RNAs transcribed from batteries of small integrator genes containing sequences inverse complementary (denoted by *primed small case letters*) to the sequences at the exon-intron border, e. g., if sequence a (TU1) is assumed to be 5′-CGAG-3′ for the non-coding DNA strand, sequence a' must be 5′-CUCG-3′.

Transcription of TU1–TU4 is constitutive. The primary transcripts either undergo splicing and are afterwards exported to the cytoplasm (mRNA 1 and mRNA 3) or, because of lacking guide RNA, they cannot be processed and hence are degraded. Transcription of the guide RNAs from the integrator gene batteries is under control of DNA sensor sequences *($S_1$, $S_2$)* which respond to intracellular or extracellular signals (▼) by changing from an inactive to an active state. Such signals could be hormone-receptor complexes or any second messenger formed as a product of signal transformation chains (cf. **4.2**).

of speculation whether snRNP's contain themselves the nicking-closing enzymes required for the excision of the intervening sequences, whether different snRNP species function selectively with different types of pre-mRNP as predicted by the model (Fig. 4.21) and whether some of the snRNA's are indeed synthesized under the control of sensor elements. At least the concentration of snRNA U1 and U2 with $10^6$ molecules per nucleus is much higher than the estimated number for a control guide RNA ($10^1$–$10^3$ molecules per nucleus).

2. Another example for such a putative guide RNA can be observed during late phases of the adenovirus 2 (Adv 2) lytic cycle. All of the late mRNA's of Adv 2 result from processing of a single giant primary transcript of almost 30 kb. It starts at map position 16.4% and extends right to the end of the Adv 2 genome (position 100%, 1% = 350 bp). Altogether 13 different mRNA species are formed by processing of the primary transcript, each containing the same tripartite leader sequence of about 200 nucleotides that is spliced from three noncontiguous segments encoded at map positions 16.6%, 19.6% and 26.6%. Splicing in this case is restricted to the 5'-non-coding part of the mRNA.

Practically all early and late transcription units of Adv 2 are RNAP II-specific, except one promoter region in the 30–32% region which is recognized by RNAP III (cf. Table 1.8). Two short abundant transcripts are synthesized in the late phases of the lytic cycle (VA RNA I and II). Sequence comparison of VA RNA I with the border sequences surrounding the splicing points

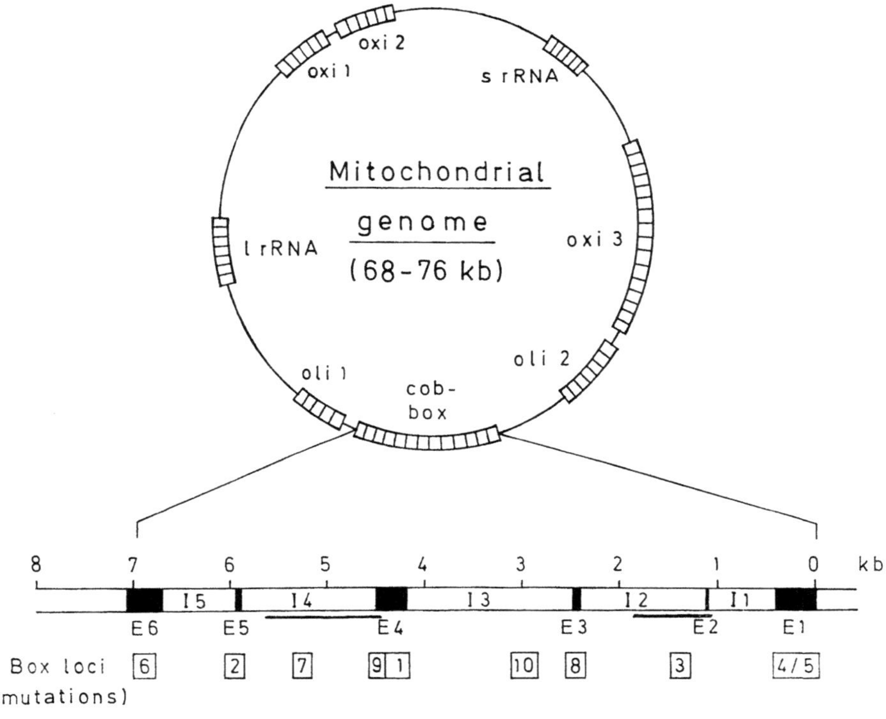


**Fig. 4.22**: Genetic map of the yeast mitochondrial genome and fine structure of the *cob-box* locus (from Halbreich et al. [159] and Lazowska et al. [160]).

The following loci are indicated on the circular map (cf. Tzagoloff, A. & G. Macino: Ann. Rev. Biochem. *48* (1979): 419): *oxi* 2, *oxi* 1, subunits III and II of cytochrome c oxidase; lrRNA, large ribosomal RNA; *oli* 1, *oli* 2, subunits of ATPase responsible for oligomycin resistance; *cob-box*, apoprotein of cytochrome b; *oxi* 3, 40 kd subunit I of cytochrome c oxidase; srRNA, small ribosomal RNA.

The fine structure map below shows the six coding segments *(E1–E6)* defined by *box* mutations 5/4, 8, 1, 2 and 6 as well as by sequencing a DNA fragment beginning in E1 and ending in I2. Correspondingly five introns *(I1–I5)* can be defined. *Heavy bars* indicate regions with open reading frames in I2 and I4 supposed to code for processing proteins [160].

in the late mRNA precursors revealed characteristic homologies which might give rise to base-paired hybrid structures of both RNA species. Thus the VA RNA I appears a good candidate for differentiation-specific guide RNA as suggested by the model of Davidson and Britten (cf. Fig. 4.21).

3. Guide RNA's must not be the product of unlinked genes. They might be coded for by the intervening sequences themselves. Hence the proper RNA alignment preceding the splicing event may involve the intrachain interaction of distal parts of the pre-mRNA (cis-effect) and/or parts of excised intervening sequences may serve as guide RNA's for other splicing events in the same molecule or even in unrelated pre-mRNA molecules (trans-effect). Evidence for such a trans-effect was provided by genetic studies of the *cob-box* locus in yeast mitochondrial DNA, which codes for the 30 kd apoprotein of cytochrome b.

The gene structure (Fig. 4.22) shows a $\sim$ 8 kb mosaic of six exons and five introns. Length of the mature mRNA is estimated to be 2.7 kb, though the six exons comprise only $\sim$ 1.2 kb and coding of the 30 kd protein needs only 0.9 kb. One strinking result of the genetic analysis of this system was the finding of two types of mutants:

1. Mutants of class 1 mapping in clusters *box* 1, *box* 2, *box* 4–*box* 6 and *box* 8 fail to complement each other. They are all exon mutants.
2. In contrast, mutants of the second class (*box* 3, *box* 7, *box* 10) map in three of the five introns. They are able to complement in trans each other and most of the exon mutants. Moreover, one of them (*box* 7) shows a pleiotropic trans effect also on the expression of the 40 kd cytochrome oxidase subunit, coded by the *oxi* 3 locus (cf. Fig. 4.22).

The following tentative splicing sequence for the 8.6 kb primary transcript was derived from kinetic experiments [159]. Mutant groups denoted at the arrows interfere with the indicated steps of the processing:

$$8.6\text{ kb} \rightarrow 8.1\text{ kb} \xrightarrow[\text{box-3}]{} 7.1\text{ kb} \xrightarrow[\text{box-10}]{} 6.7\text{ kb} \rightarrow 4.7\text{ kb} \xrightarrow[\text{box-7}]{} 3.7\text{ kb} \rightarrow 2.7\text{ kb.}$$

Considering the trans complementation of the intron mutants it was a tempting speculation that these introns themselves or part of them serve as guide RNA's for subsequent steps of the processing, similar to the functions discussed for the snRNA of animals and for the VA RNA I of adenovirus. In support of this model is the finding of two stabile single-stranded circular RNA's of 0.6–0.9 kb appearing sequentially during processing and presumably deriving from introns I1 and I3 respectively [159, 160]. However, no base-pairing of these RNA's with the *cob-box* pre-mRNA could be detected. Moreover, sequencing of introns I2 and I4 revealed two open reading frames (cf. heavy lines in the fine structure map Fig. 4.22) that might be used as templates for corresponding proteins. In fact, the existence of such a protein, presumably coded by a joint reading frame of exon 1, exon 2 plus the coding part of intron 2, could be observed in SDS polyacrylamide gel electropherograms [160]. It is supposed to comprise a hydrophobic domain (E1 + E2) and a highly basic domain (I2) and to function as pre-mRNA maturase (423 amino acid residues, MW 49 kd) providing splicing specificity for removal of the second intron. If the model is correct, the early steps of *cob-box* pre-mRNA processing are autoregulated (splicing homeostasis). Removal of I1 generates the reading frame for the maturase, which by the following removal of I2, destroys its own reading frame. A likely candidate for another maturase is a protein coded by intron 4 (*box* 7 locus). As mentioned above, it is also required for processing of the *oxi* 3 pre-mRNA.

Taken together, these results indicate that the accurate splicing of RNA precursors requires the interaction with an enzyme complex presumably containing RNA as a necessary cofactor. This "guide RNA" might represent an essential element for site-specific splicing and processing control (cf. the model in Fig. 4.21).

The close interaction of nucleolytic and ligating enzymes of the splicing machinery with RNA's as cofactors for proper alignment of the spliced sequences recalls a special bacterial endonuclease (RNase P) which acts as a key enzyme for the processing of most tRNA precursors (cf. Fig. 1.13) and which contains RNA as a necessary component.

The trans complementing intron mutants of yeast mitochondrial genes emphasize that the normal sequence of nucleotides in an intron and its flanking sites is necessary but not sufficient to ensure correct splicing. Other sequences, quite remote, are also involved, e.g., by coding for a site-specific pre-mRNA maturase (Fig. **4**.22). The onset and stepwise progression of splicing most probably requires a temporal succession of specific three-dimensional structures of the pre-mRNP that can be recognized by the splicing machinery and its cofactors.

### 4.4.3.3 Inactivation and Degradation of mRNA

Analyses of mRNA decay in bacteria, i.e., mainly in *Escherichia coli*, led to the conclusions that (1) messages decay exponentially, (2) biological inactivation of messages by endonucleolytic cleavage is the initial step followed by further degradation from the 5′- to the 3′-end and (3) the differential decay rates observed for many messages result evidently from intrinsic properties of the nucleotide sequence, e.g., from properties of secondary structure formation and from differences in the protection by ribosomes [161–165]. Though, with few exceptions, the biological significance of differential mRNA decay is largely unknown, the following examples may serve to illustrate diverse aspects of this neglected part of gene expression.

Average half-lives of mRNA in *E. coli* and other bacteria are in the range of 1 to 2 min. But there is also evidence for long-lived mRNA, e.g., mRNA coding for periplasmic and outer membrane proteins.

Certain mutants of *E. coli* extrude parts of the outer cytoplasm as small vesicles surrounded by parts of the plasma membrane (so-called minicells [162]). It was observed that protein synthesis in these minicells proceeds almost undiminished for more than 30 min, despite the absence of any genetic material to be transcribed. The translation products are mainly proteins used for outer membrane synthesis or for export to the periplasmic space. At least a considerable part of them are evidently synthesized as pre-proteins by membrane bound polysomes unlinked to the nucleoid complex (cf. examples given in Table **4**.10). The mRNA involved are much more stable than the usual *E. coli* mRNA, though minicells exhibit a normal inactivation rate of short--lived mRNA, i.e., contain sufficient activities of endonucleases.

Individual messages of a polycistronic mRNA exhibit distinct half-lives independent of their relative positions in the messenger RNA. This will be discussed elsewhere in this book for the early mRNA species of phage T7 generated by RNase III processing of a polycistronic primary transcript (cf. **14**.2.2.4). But it was also observed in many other cases, e.g., for the polycistronic mRNA coded by the *E. coli gal* operon with genes *gal E*, *gal* T and *gal* K in the 5′ → 3′ direction (cf. Fig. **4**.9).

The functional and chemical half-lives of each part of the *gal* E, T, K mRNA were determined by in vitro translation and by hybridization respectively. In translation experiments 1.0, 0.6 and 1.5 min were found for the *gal* E, T and K messages respectively. In contrast, the chemical half-life is equal for all three (about 1.1 min). The differences of the functional half-lives can be explained by an endonucleolytic primary attack leaving the *gal* E and *gal* K messages intact, while destroying *gal* T [161].

The decisive influence of the 5′ leader sequence for mRNA stability and translation initiation is evident from the analysis of operator constitutive mutants of the *lac* operon (cf. **4**.4.2.2). Because the operator region extends far into the mRNA leader sequence, a couple of *lac* O mutations characterized by single-base exchanges were found to thoroughly influence functional characteristics of *lac* mRNA (Table **4**.9). The effects result from the secondary structure of the leader sequence (cf. figure on top of Table **4**.9) which stabilizes the 5′-end of the mRNA but interferes with ribosome binding.

Furthermore, the 5′ leader contains one RNase III recognition site between base residues 6 and 7. Cleavage by RNase III breaks the stem-loop structure and therefore enhances the frequency of translation initiation. The effects evoked by the mutations can also be explained in terms of changes in the secondary structure of the leader sequence: Mutations either destabilize the stem-loop structure (strains IIa and Vb) and thus enhance ribosome binding while decreasing mRNA stability, or they generate the opposite phenotype (strains IIIb and IVb) because of interference with the RNase III processing. As is to be expected, the base exchange in the middle of the loop is neutral (strain IIb). The strongly decreased mRNA stability in O$^c$ mutant IIIa cannot be explained at present.

**Table 4.9**: Influence of *lac* operator mutations on the stability of *lac* z gene message and the frequency of translation initiation in *Escherichia coli* (from Cannistraro & Kennell [166]).

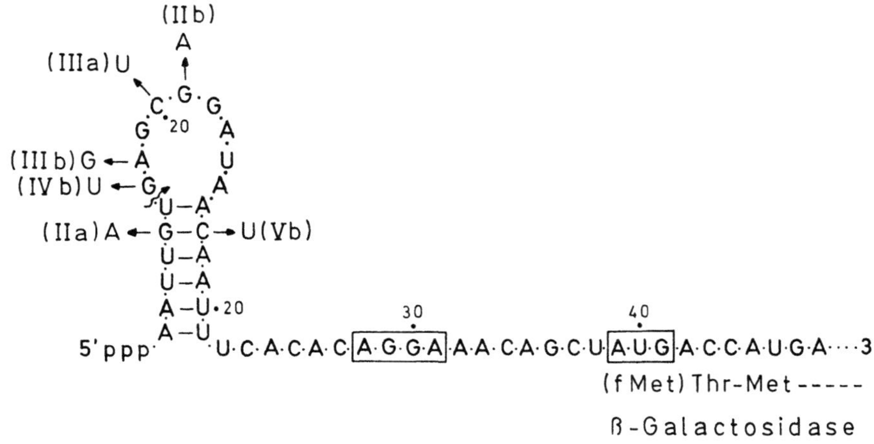


The *lac* operator, as defined by mutants and by resistance of the DNA-repressor complex against DNase degradation, extends far into the coding sequence of the *lac* z gene (start at nucleotide 39 of the *lac* mRNA sequence on top of the table). A potential recognition site for 16S rRNA (–AGGA–) is boxed. The wavy arrow between base residues 6 and 7 indicates a site of RNase III cleavage. The operator mutations cause a distinct degree of constitutive expression of the *lac* operon, which is 5.6, 3.5, 2.4 and 0.98 % of the fully induced level for 0$^c$ mutants of classes II, III, IV and V respectively (cf. Smith & Sadler: J. Mol. Biol. *59* (1971): 273). Furthermore, they alter the secondary structural features of the 5'-leader part of the *lac* mRNA and thus thoroughly influence the half-life and frequency of translation initiation.
Exponentially growing bacteria were induced by addition of cAMP and isopropylthiogalactoside. The rate of $\beta$-galactosidase synthesis was assayed between 12 and 20 min. Half-lives of mRNA were determined by hybridization after adding rifampicin 30 s after induction to stop further mRNA synthesis. The frequency of translation initiation (ribosome/s) was calculated.

| Strain | | Rate of enzyme synthesis (monomers/cell · s) | Half-lives of mRNA (s) | Frequency of transl. initiation ($s^{-1}$) |
|---|---|---|---|---|
| 0$^+$ (wild-type) | | 19 | 90 | 0.204 |
| 0$^c$ (mutants) | Class | | | |
| | IIa | 25 | 69 | 0.385 |
| | IIb | 19 | 90 | 0.208 |
| | IIIa | 12 | 54 | 0.196 |
| | IIIb | 9 | 84 | 0.125 |
| | IVb | 12 | 75 | 0.147 |
| | Vb | 19 | 57 | 0.294 |

An exciting regulation model for coordinate synthesis of the ribosomal proteins of *E. coli* is based on selective destabilization of the individual r-protein messages by a feedback mechanism.

The about 50 proteins of *E. coli* ribosomes are coded within 15 independent operons. And yet all proteins are synthesized in equimolar amounts except L7/L12 which is present in four copies per ribosome (cf. **1.3.4.2**). This highly efficient coordination of gene expression regulation within the ribosomal regulon is probably achieved, at least partly, at the level of mRNA degradation [167]. This conclusion was derived from observations that, in merodiploid *E. coli* strains with a functional extra copy of two ribosomal operons (*spc* and $\alpha$ operons) integrated into the $\lambda$ phage genome, the rate of synthesis of all r-proteins remained unchanged. When mRNA levels for these two operons were analysed in pulse-chase experiments the expected gene dosis dependent doubling of the transcription rate was indeed found, but the mRNA was rapidly degraded, so that the steady state amount of spc- and $\alpha$-specific mRNAs are only slightly higher than in the "haploid" control strains.

The regulatory model explains these results assuming that the mostly basic r-proteins stay with their messenger, unless they are incorporated into nascent ribosomes. These r-protein-mRNA complexes are destroyed by a specific endonuclease. Thorough investigations of the translation control of the two operons revealed that, in fact, in both cases only one of the r-proteins coded by the operon is inhibitory. These are protein S4 for the $\alpha$ operon and S8 for the *spc* operon. Both are initial binding proteins associating with nascent rRNA to initiate ribosome assembly. Not all proteins, coded by the two transcription units, are affected to the same extent. Thus a characteristic cis-acting feedback control governs the coordinate gene expression in the r-protein regulon.

It should be added that this model concerns only one important regulatory aspect of ribosomal biogenesis, which is complemented by control mechanisms at the transcription level (cf. **4.4.2.6**). Furthermore, though the principle of autogenous regulation of r-protein operons was also reported for the L4-controlled expression of the S10 operon encoding 11 ribosomal proteins, there is no evidence for selective mRNA degradation in this case [168]. Instead, a coordinate reduction of mRNA synthesis was found.

Experimental data about selectivity of mRNA degradation in eukaryotic cells are even more scarce than from prokaryotic systems. But the phenomenon of short--lived and long-lived mRNA's is equally well established, though "short" and "long" have other dimensions in these cases. Thus, in rat liver cells half-life of mRNA for $\delta$-aminolevulinate synthetase is less than 60 min compared to several days for albumin mRNA. Even more striking are the differences in half-lives of mRNA of two rat liver enzymes, closely related with respect to their inducibility by hydrocortisone, to their intracellular compartmentation and to the type of reaction catalysed [169]. These are tyrosine aminotransferase (TAT) and alanine aminotransferase (AAT) respectively. Half-lives of their mRNAs were reported to be two h for AT and 12–14 h for AAT respectively.

A recent impressive example concerns the developmentally specific increase of histone mRNA turnover during *Drosophila* embryogenesis (cf. **4.6.3**). There is a large store of "maternal" histone mRNA in the egg, destined for the early steps of *Drosophila* development ($\sim 7 \times 10^7$ copies of each five classes of histone mRNA). But $\sim$ 90 min after the onset of embryogenesis synthesis of "embryonic" histone mRNA starts, increases rapidly to $\sim$ 2 copies per gene/min between 2–6 h of embryo development, i.e., at a time of maximum DNA synthesis per embryo, and then declines more than 60-fold (cf. details on the developmental sequence in Sect. **4.6.3**). Concomitantly there is a marked decrease of histone mRNA half-life from 2.5 h at early stages (oviposition to blastoderm stage, $\sim$ 3 h) to 20 min at the end of gastrulation ($\sim$ 10 h of embryogenesis). Two mechanisms can be envisaged to explain this selective break-down of histone mRNA [171a]: (1) A long-lived maternal histone mRNA (half-life $\sim$ 3.2 h) is opposed to a short-lived embryonic type of mRNA (half-life $\sim$ 16 min). Both types are either coded by separate gene systems or result from different modes of postsynthetic processing. At any rate, structural differences are responsible for the differential break-down. (2) No structural differences exist between maternal and embryonic histone mRNAs, the development-specific decrease of the half-life being mediated by changing concentrations of a regulator (protein ?) of histone mRNP turnover.

Which factors determine the half-life of a given mRNA, is largely unknown. But it is very likely that the association with the structural and regulatory proteins of hnRNP and cytoplasmic mRNP, as well as the presence and length of the 3′-poly(A) sequence, are decisive in this respect. Apparently, the 3′-non-coding region of mRNP in polysomes is particularly sensitive to endonucleolytic attack with concomitant loss of the poly(A)-protein complex. Individual mRNA species showed wide differences in their degree of susceptibility [171].

As outlined in the preceding section, hybridization analysis of the complexity of messenger composition in the cytoplasm (mRNP) and nucleus (hnRNP) revealed a considerably higher complexity in the nucleus. The regulatory models derived from these and other results (cf. Figs. **4**.21 and **4**.25) assume that many messenger types, though constantly synthesized, do not reach the cytoplasm because they cannot be properly processed and are therefore degraded. Inherent instability of mRNA precursors is an integral part of these models.

Summarizing we can state:

1. The existence of short- and long-lived mRNAs is well established in prokaryotic and eukaryotic cells, though, as a rule, prokaryotic mRNA have much shorter half-lives.
2. Mechanisms of functional and chemical degradation can be clearly distinguished. Similar to mechanisms of protein degradation, the initial event is the functional inactivation by a single intramolecular cleavage leaving the fragments labile to further endonucleolytic and exonucleolytic attack.
3. Inherent structural features are decisive for mRNA stability. They determine not only the secondary structure of mRNA but above all its translation frequency and binding affinity for structural and regulatory proteins.

### 4.4.4 Translation Regulation

The advancing technology for sequencing nucleic acids and analysing their function in in vitro systems has decisively increased the experimental evidence for specific translation control. A multiplicity of regulatory effects is involved. Though under natural conditions closely related, they will be discussed consecutively, i.e., peculiarities of mRNA primary and secondary structure near the initiation site, the role of mRNA-associated proteins and RNAs in eukaryotic mRNP, the regulatory role of initiation factors, variations of isoacceptor tRNA patterns associated with the formation of specialized mass proteins and, finally, the complex changes of the translation machinery induced by interferon action on mammalian cells.

#### 4.4.4.1 Elements of mRNA Secondary Structure

The complex sequence of translation initiation (cf. Figs. **1**.19 and **1**.20) requires interaction between mRNA, the small ribosomal subunit, the initiator tRNA and several initiation factors (cf. Table **1**.10) to ensure initiation at the correct AUG codon (initiator codon). When discussing mRNA stability (cf. **4**.**4**.**3**.**2**) it was already pointed out that elements of secondary structure in the 5′-leader segment of *E. coli lac* mRNA interfere with efficient ribosomal binding (cf. Table **4**.9). The base-paired stem-loop structure is destroyed by RNAse III processing of the primary transcript, thus increasing the rate of translation initiation. Another very instructive example will be outlined in detail in Chap. **14**. It concerns the two types of transcripts for the *cI* repressor of *E. coli* phage $\lambda$ (cf. Fig. **14**.11) controlling establishment of lysogeny and the lysogenic state respectively.

The phase of lysogeny establishment requires large amounts of *cI* protein translated from a special mRNA with a long 5'-leader sequence. In contrast, the lysogenic state is based on a relatively low level of *cI* repressor that allows the phage to embark rapidly on the lytic cycle, if the conditions demand it. Besides an autogenous transcription control of the repressor level, a special feature of the *cI* mRNA in the lysogenic state is noteworthy: this type of mRNA is very poorly translated because it starts directly with the AUG codon. Thus the usual ribosome binding site, including the Shine-Dalgarno sequence, is lacking (cf. Fig. **1**.20).

As mentioned in the introductory chapter (cf. **1**.3.4.3) initiation of translation in eukaryotes evidently depends on the recognition of the cap-structure at the 5'-end of mRNA marked by the special cap-binding protein ($M_r$ 24 kd). The scanning mechanism of initiation suggests that from this site of initial interaction with the mRNA the small ribosomal subunit scans somehow the mRNA leader sequence until it reaches the correct AUG initiator codon. Usually this is the first AUG codon, but in some viral mRNA it is not. There are good reasons to assume that distinct elements of mRNA secondary structure in the 5'-leader region are decisive for the fidel recognition of the correct AUG codon [174].

Interestingly, computer generated models of secondary structures of the 5'-regions of rabbit $\alpha$- and $\beta$-globin mRNA (Fig. **4**.23) revealed significant differences that might explain the well-known $\sim$10 times lower affinity to translation initiation factor eIF-2 of the $\alpha$-type mRNA when compared with the $\beta$-type mRNA [172, 176]. The AUG initiator triplet is burried in the $\alpha$-globin mRNA but highly exposed in the $\beta$-globin mRNA. Similar results were obtained also for the mouse globin mRNAs.

### 4.4.4.2 Inactivation of the Cap-Binding Factor after Virus Infection

In contrast to the majority of eukaryotic mRNA, some plant and animal viruses contain non-capped RNA, e.g., satellite tobacco necrosis virus, cowpea mosaic virus, reovirus and the picornaviruses. Detailed analyses of the infection cycle for reovirus showed that the initial production of viral proteins begins very slowly. The mRNA copies produced by the infecting parental subviral particles are capped and hence must compete with host mRNA for the limited translation factors available. Later on, the viral mRNAs arise mainly from new progeny subviral particles and are uncapped. Meanwhile the host translation apparatus has been reprogrammed to cap-independent translation by the inactivation of the cap-binding protein [177, 178]. This was shown in poliomavirus-infected cells. Translation rate of capped host mRNA is greatly decreased and initiated at false AUG codons. It is an interesting fact that the fidelity of eukaryotic translation initiation, though usually strongly dependent on the cap-cap protein complex, does not necessarily require this mechanism. An important question then remains: What other signals in these viral mRNAs are responsible for the correct positioning of the initiating ribosome?

### 4.4.4.3 The Translation Program of *Escherichia coli* RNA Phages MS2 and Q$\beta$

Some phages of *Escherichia coli*, e.g., phages MS2, Q$\beta$, f2 and R17, contain single-stranded RNA as genetic material that can directly serve as polycistronic mRNA for synthesis of the viral proteins. The structure of the genetic material of these phages is very similar. Two examples, i.e., for phages MS2 and Q$\beta$, are given at the

A

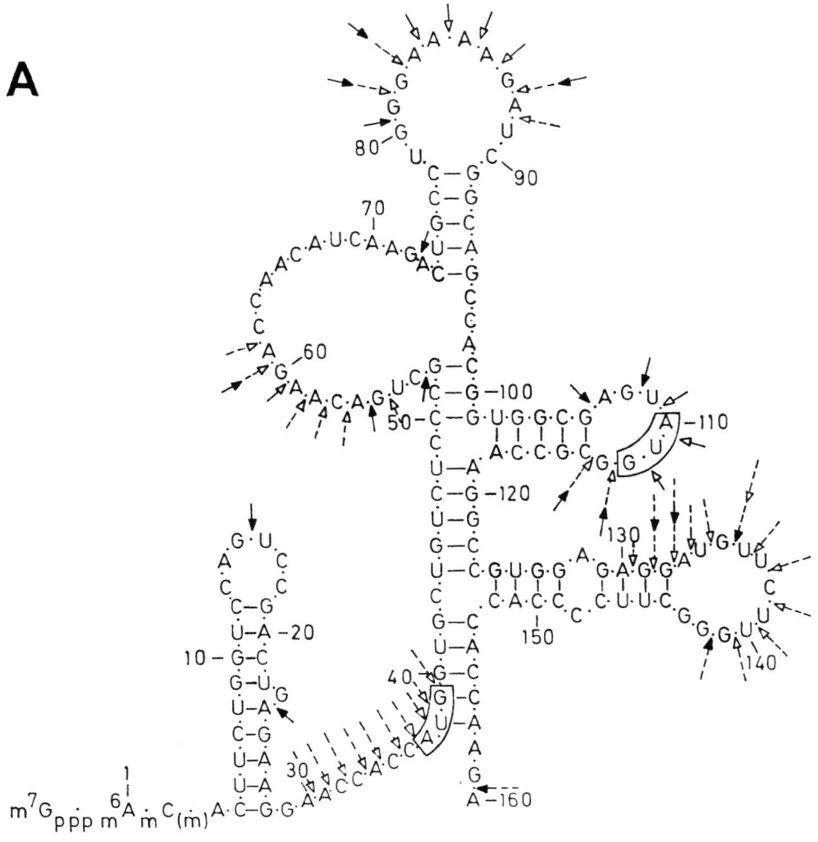

α-Globin mRNA

**Fig. 4.23:** Computer generated secondary structure models of the 5′-terminal regions of rabbit α- and β-globin mRNAs (from Pavlakis et al. [175], with permission, © M. I. T. Press).

The model was derived from the sequence analysis and from data about the partial digests with RNases. Points of attack by S1 nuclease and T1 nuclease are marked by *arrows*. It is note-

top of Fig. **4.24**. After infection of their host cells all four lytic phages exhibit a developmental program, exclusively controlled at the translational level. Details will bei exemplified by using phage Qβ (Fig. **4.24**) with occasional remarks on peculiarities of the other phages.

Four gene products are coded by the about 4,500 nucleotides of Qβ-RNA (from 5′- → 3′-end): the maturation protein ($A_2$), the coat protein (C), a coat read-through protein ($A_1$) with a possible function for infectiosity of phage particles and the β--subunit of Qβ replicase (β). The four proteins are produced in the ratio 10 : 100 : 3 : 5. This unequal translation of proteins that are part of the same polycistronic mRNA is not unusual (cf. **4.4.3.2** and Fig. **14.12**). For Qβ and related phages the differential translation is thought to result from special elements of RNA secondary structures (cf. Figs. **4.20** and **4.24**) and the interaction with regulatory proteins. The develop-

B

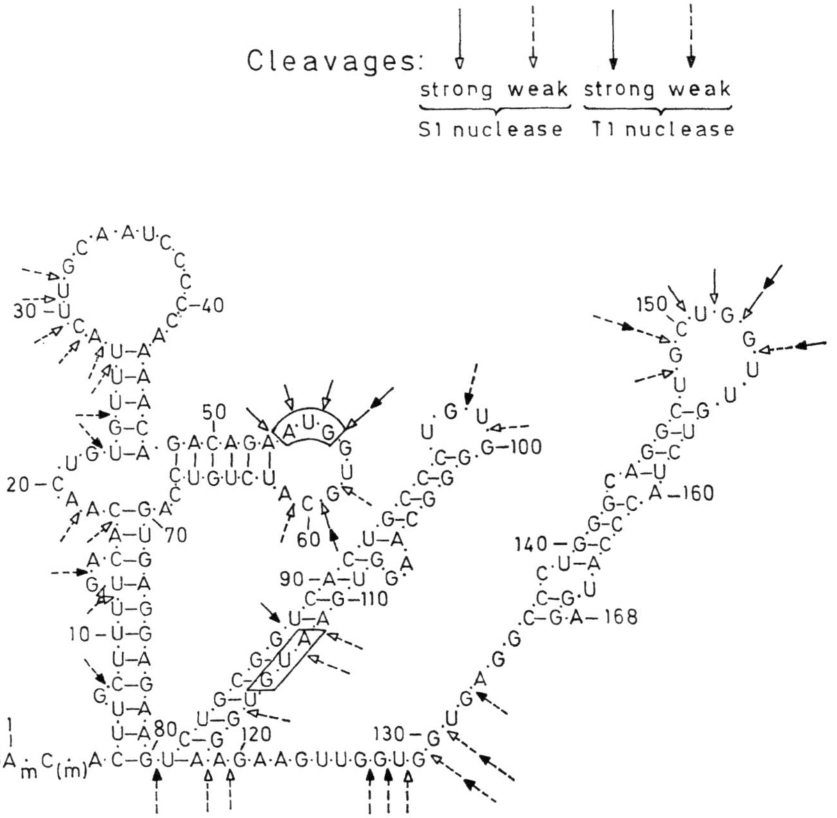


ß - Globin mRNA

(Fig. 4.23)

worthy that the iniator codon *(boxed)* in $\alpha$-globin mRNA (nucleotides 37–39) is evidently buried and thus relatively insensitive to nuclease attack, whereas that of $\beta$-globin mRNA (nucleotides 54–56) is exposed and highly sensitive. Interestingly, both mRNAs contain another AUG codon between positions 110–120 that is not used as initiation site.

mental cycle of phage Q$\beta$ is characterized by three phases with three patterns of translation (Fig. 4.24).

*Early Phase (A):* Initially, due to peculiarities of the Qß-RNA secondary structure, only one of the three translation initiator regions (i$A_2$, iC and iß) is open (iC). But the ribosomes translating the coat protein cistron cause a conformational change of the RNA that allows translation to proceed also from iß. In total three proteins are synthesized: many molecules of coat protein, few molecules of replicase subunit ß and about 3%–5% of protein $A_1$. The latter represents a read-through protein of the coat cistron resulting from the suppression of the UGA termination codon by suppressor $tRNA^{Trp}$[6]. The amount of the minor protein $A_1$ is thus directly proportional

[6] It is worth noticing that UGA is used as a regular Trp codon in some mitochondrial genes in yeast and man, cf. Table 1.5.

to that of the coat protein and determined by the availability of charged su-tRNA$^{Trp}$ in the host cell.

An important regulatory event of the early translation phase is the synthesis of the replicase subunit $\beta$ ($M_r$ 66 kd). It combines with three host cell-derived proteins to the tetrameric holo-enzyme responsible for replication of Q$\beta$ RNA via (–)-strand templates [181]. The host cell proteins abused for the phage replicase ($M_r$ 215 kd) are the ribosomal protein Sl ($M_r$ 70 kd) and a dimeric complex of the elongation factors EF-Tu (47 kd) and EF-Ts (35 kd).

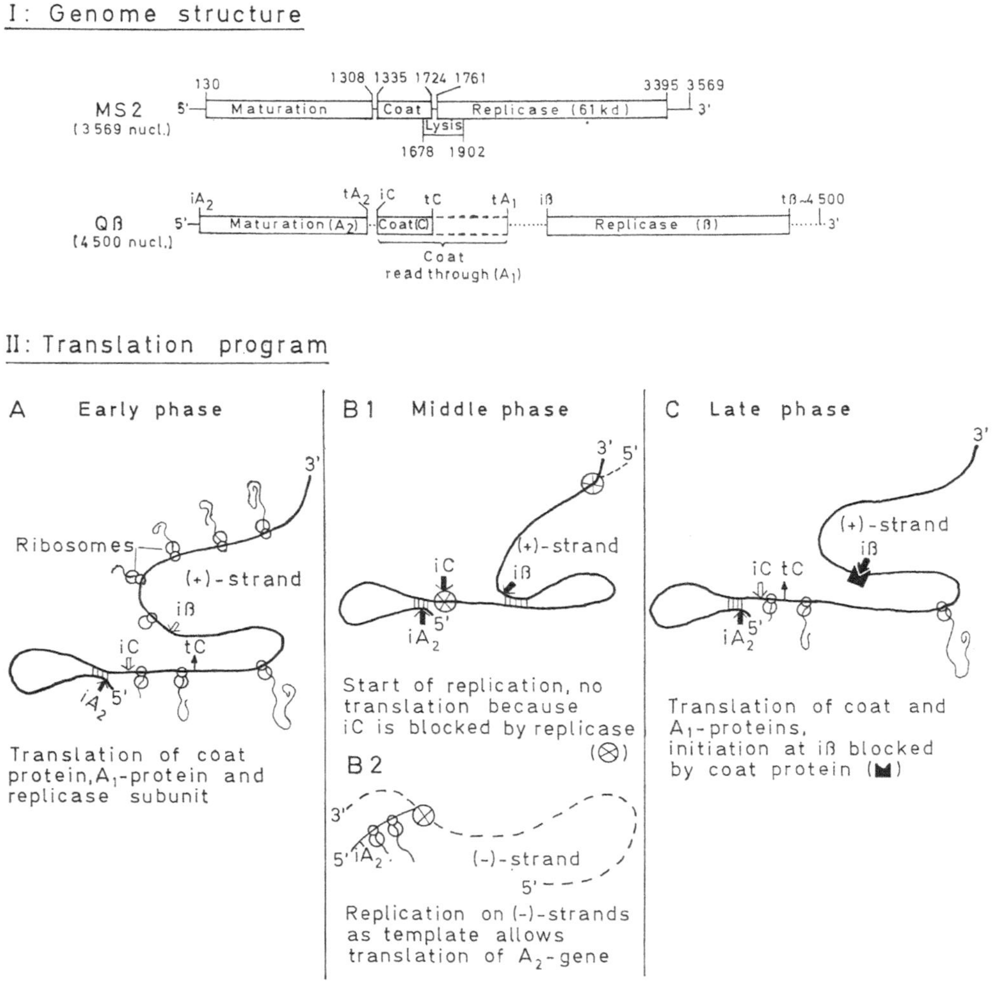


**Fig. 4.24**: Genome organization and translation program of RNA-phages of *Escherichia coli* (after Weissmann [179]).

**I**: Genome of the MS2 phage [180, 182] and of the Q$\beta$ phage [179]. $iA_2$, $tA_2$, $iC$, $tA_1$, $i\beta$ and $t\beta$, initiator regions and termination codons for the maturation protein ($A_2$), the coat protein ($C$), the coat read-through protein ($A_1$) and the $\beta$-subunit of the replicase ($\beta$). Gene $A_1$ is lacking in the phage MS2 genome. But an additional overlapping gene (nucl. 1678–1902) was mapped that codes for a lysis protein [182, 183].

**II**: Three functional states of the Q$\beta$ genome characterize the infection cycle allowing for a programmed translation of the phage proteins. ⬇, ⇓ initiator regions closed and open respectively.

The *Middle Phase* (B) of the lytic cycle is dominated by synthesis of new phage RNA and almost ceasing translation. The new functional state of the phage RNA as template for synthesis of (−)-strands is brought about by association of the replicase holoenzyme at two points: (1) in the initiator region of the coat protein to stop any further ribosomal binding and (2) at the 3′-end to start replication (phase B1). The efficient interaction of Qβ replicase with the RNA template at both sites is mediated by its subunit S1 [181] known to participate in the mRNA binding and rRNA unfolding during translation initiation (cf. Fig. **1**.20). Furthermore, an interesting homohexameric complex of a host protein ($M_r$ 12 kd) with high affinity for A-rich sequences is decisive for the high template specificity of Qβ replicase.

Synthesis of (−)-RNA strands is a necessary prerequisite for synthesis of new (+)-strands (phase B2) and, coupled to it, the translation of the maturation protein cistron. The initiator region $iA_2$ is open for ribosome binding only in statu nascendi of (+)-strands, i.e., in the replicating complex. It is an important consequence of the strong tendency of Qβ-RNA (+)- and (−)-strands to form intramolecular base-paired regions (cf. Fig. **4**.20) that intermolecular base-pairing between (+)- and (−)-strands is prohibited. Both replication and translation require single-stranded RNA.

*Late Phase* (C): The restricted number of replicase molecules are not sufficient to block all iC regions of the accumulating new RNA (+)-strands. Hence, translation of coat protein (and $A_1$ protein) is resumed. The increasing amount of coat protein blocks further translation of the replicase subunit by binding to the iβ region. Finally, new phage particles are assembled, each containing one molecule of (+)-RNA, 180 molecules of coat protein and each one molecule of $A_1$ and $A_2$ proteins.

Recently an additional protein was detected in phages MS2 and f2 [182, 183]. Its coding sequence overlaps with the end of the coat protein and the beginning of the replicase gene. The 75 amino acid residue polypeptide has an obvious function for lysis of the host cell, a function that has long been looked for. As evident from the numbers indicated at the MS2 genome at the top of Fig. **4**.24, the reading frame of the new lysis protein is shifted by one nucleotide with respect to the coat and replicase cistrons and is identical with that of the maturation protein. Thus, the short intercistronic region between the coat and replicase genes (in phage MS2: 36 nucleotides!) serves three functions: (1) it is part of the lysis gene, (2) it contains the replicase initiator region (the AUG codon preceded 5′-upstream by the -AGGA-sequence) and (3) it forms two short stem-loop structures that interact with coat protein in the late translation phase. Translation of the lysis protein is restricted to the late phase. Evidence for a lysis protein in Qβ phage infected cells is lacking.

In summary, the translation program during the lytic cycle of single-stranded RNA phages in *E. coli* reveals the following remarkable features [179–183]:

1. Dynamic changes of the RNA secondary structures lead to distinct functional states characterizing the three phases of the cycle.
2. Phage and host proteins, besides their catalytic and structural functions, act as regulatory proteins (coat protein, replicase) blocking distinct initiator regions of the polycistronic phage RNA.
3. A read-through protein in Qβ phage results from the occasional suppression of termination codon UGA by the low level of su-$tRNA^{Trp}$.
4. In phages MS2 and f2 an overlapping gene with shifted reading frame is translated late during infection. It codes for the putative lysis protein.

### 4.4.4.4 Functional Compartmentation of Cytoplasmic mRNP in Eukaryotes

The ATP-dependent export of mature mRNA from the nucleus to the cytoplasm involves the exchange of all structural proteins of nuclear mRNP, except the poly(A) binding protein ($M_r$ 75 kd). The main proteins associated with cytoplasmic mRNP in animals show $M_r$ of 50 and 35 kd respectively. In addition, several minor proteins frequently detected may constitute translation factors and/or auxiliary proteins with control functions.

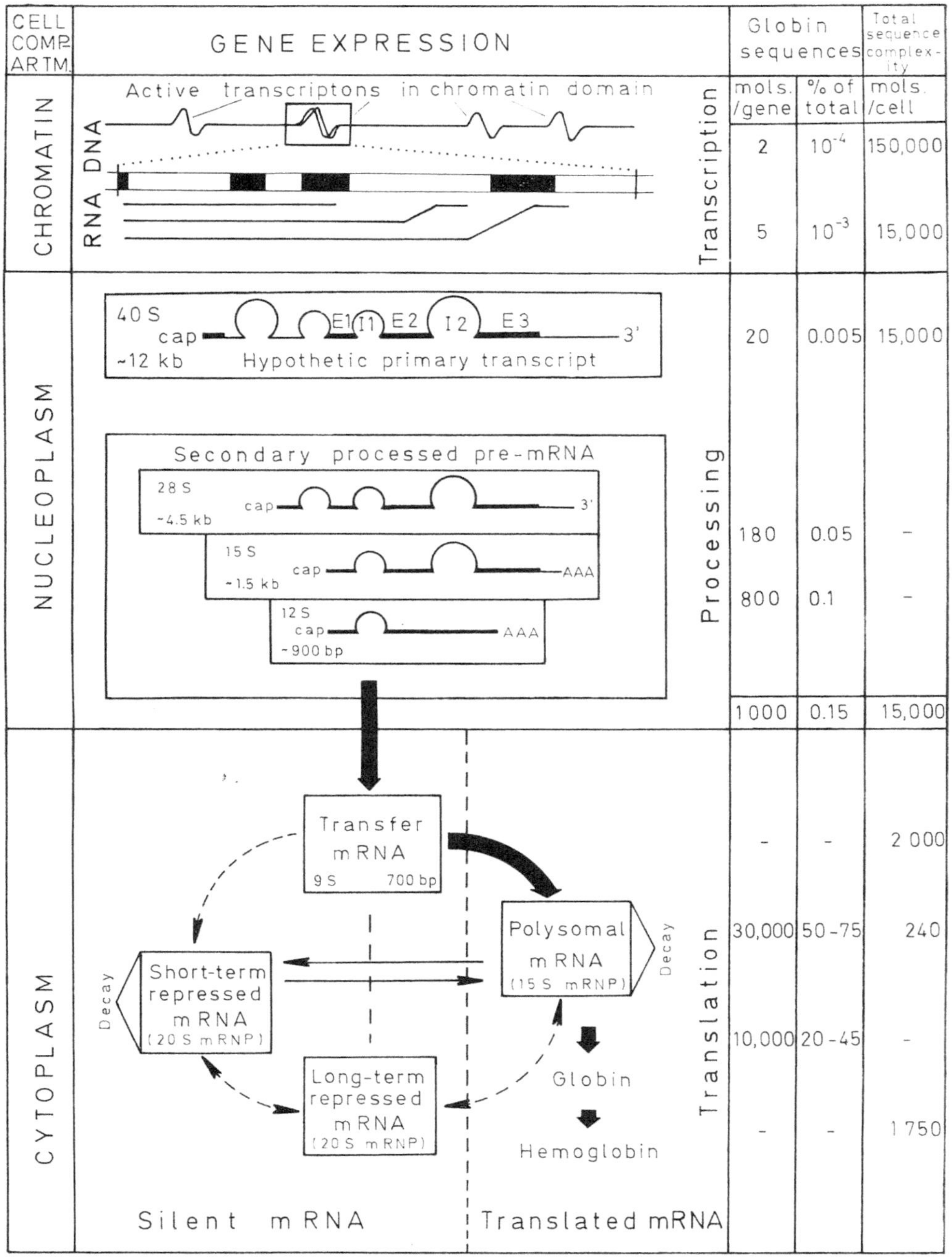


**Fig. 4.25:** Nuclear processing of globin mRNA precursors and cytoplasmic functional pools of mRNP in avian erythrocytes (from Scherrer et al. [184], modified).

The active transcription unit of chromatin is assumed to be a $\beta$-type globin gene with the three exons *(E1–E3)* and the small and large intervening sequence *(I1 and I2)* in the coding segment (cf. Fig. 4.14). In addition, large non coding segments at the 5′- and 3′ends are part of

The tentative scheme (Fig. **4**.25) summarizes qualitative and quantitative aspects of information flow from transcription through nuclear processing to the cytoplasmic mRNP pools as it was derived by Scherer and coworkers for the globin mRNA synthesis in avian erythrocytes [184, 185]. Functionally active globin mRNA was isolated from two types of cytoplasmic mRNP particles: one sedimenting at 15S that represents the translationally active polysomal mRNP and another sedimenting at 20S. Though both contain intact 9S globin mRNA (as revealed by in vitro translation in the wheat germ system), only the 15S mRNP itself sustains translation of globin. The 20S mRNP contains a tightly bound translation inhibitory protein but differs also substantially from the 15S mRNP with respect to other associated proteins. Hence, two pools of globin mRNP can be clearly distinguished by functional and structural properties. They are presumably in dynamic equilibrium.

Besides the evidently very small transfer pool included into Fig. **4**.25 for theoretical reasons only, sequence complexity studies revealed an additional pool of long-term repressed mRNP that contains no globin mRNA and so far cannot be defined by structural properties. It is evident from the quantitative data given at the right site of Fig. **4**.25 that, though most of the cytoplasmic mRNA species are contained in this pool, the overall amount is fairly low because only few copies of these mRNA species are present. In the avian erythrocyte only a small fraction of all mRNA species detected in free mRNP is chasable into polysomal mRNA (240 out of 2,000). The physiological role of the long-term repressed mRNP is unclear.

Frequently evidence for a further posttranslational pool of mRNP with degradative function was obtained, e.g., in rabbit reticulocytes where part of the cytoplasmic globin mRNA was detected in a mRNP fraction with shortened poly(A) sequence and lacking poly(A)-binding protein [187]. This aggrees with the analysis of mRNA degradation in adenovirus infected cells that revealed a close connection between shortening of the 3′-poly(A) tail and biological inactivation of the message (cf. **4**.4.3.2). No such special pool of "old" mRNP is implied in the scheme shown in Fig. **4**.25. Instead, degradation is supposed to proceed at random from both globin mRNA containing pools.

An essential experimental procedure to provide the detailed quantitative and qualitative data required for the flow diagram shown in Fig. **4**.25 is the sequence complexity analysis by cDNA--RNA hybridization. To this end either purified globin mRNA or total mRNA is used to prime the synthesis of radioactively labelled complementary DNA (cDNA) by reverse transcriptase. Hybridization of labelled cDNA with excess RNA was followed by separation of the DNA-RNA hybrids from unbound cDNA by chromatography on hydroxyapatite. The extent of hybridization is proportional to the molar concentration of a given mRNA sequence ($R_0$) and to the time of hybridization (t in s). These $R_0t$ values characterize the sequence complexity of a mRNA probe.

---

the putative primary transcript of about 12 kb that is processed via an ephemeric intermediate of about 4.5 kb to the dominant forms of nuclear $\beta$-globin mRNP containing 1.5 and 0.9 kb of RNA. Though evidence points to the existence of high molecular weight primary transcripts of 12–15 kb, most details of its structure and processing, which takes about 20 min, are unclear (cf. **4**.4.3.1). Such are the point of poly(A) addition, the preservation or eventual removal and resynthesis of the cap, further short-lived intermediates etc. No mature globin mRNA of 0.7 kb is detected in the nucleus. Removal of the small intervening sequence evidently immediately precedes the ATP-dependent export to the cytoplasm.

Four mRNP pools can be discerned in the cytoplasm: (**1**) a transient small pool of mRNP on its way to (**2**) the polyribosomal pool that is in equilibrium with (**3**) short term repressed mRNP and (**4**) a pool of long-term repressed mRNP (cf. text). The data given on the *right site* of the figure represent the average stationary concentrations of globin-specific and non-globin-specific mRNA sequences during processing in avian erythroblasts. They illustrate the selectivity of transcription, processing and translation initiation as expressed by the decreasing single copy complexity *(last column)* and increasing concentration of globin-specific sequences.

Though the varying composition of the minor protein fractions of mRNP points to the essential role of proteins as control factors, there is evidence in several cases that translationally inactive mRNP is complexed with low molecular weight translation control RNA (tcRNA). Such inhibitory tcRNAs with broad specificity have been found in unfertilized sea urchin eggs as well as in dormant embryos of the brine shrimp *(Artemia salina)* [189]. However, the only specific and well characterized tcRNA was isolated from chicken embryonic muscle cells where it is associated with the inactive storage mRNP containing 26S myosin mRNA [188]. One molecule tcRNA per molecule myosin mRNA is required for inactivation. This tcRNA is rich in uridyl residues (48 % U, 12 % A, 28 % C and 12 % G) and has a MW of ~6,500. The timing of myosin synthesis in the embryonic muscle is determined by transition of myosin mRNA from the storage pool into the polyribosomal pool at about day 14–19 of the embryonal development.

Pools with translationally inactive mRNP are frequently referred to as "masked" mRNA. They constitute the morphogenetic substances long known to control the cap formation in the unicellular alga *Acetabularia* (cf. Chap. **16**) but they are also characteristic for resting cells, e.g., unfertilized eggs. Evidently cell determination in the early cleavage stages of embryo development depends on the unequal distribution of maternal mRNP in the egg cytoplasm. This is exemplified by the generation of the germ cell primordia from the egg posterior pole plasm in *Drosophila* (cf. **4.6.4.1**).

#### 4.4.4.5 Hemin-Controlled Translation Inhibition in Reticulocytes

A general mechanism of translational control was detected when analysing the coordination between the two basic parts of hemoglobin synthesis in reticulocytes, i.e., of the hemin prosthetic group and of the apoprotein globin. It has long been recognized that efficient synthesis of globin was dependent on the constant supply of hemin. The mechanism of this coordination is assumed to involve two tandemly operating protein kinase systems which, together with phosphoprotein phosphatases, control the phosphorylation state and, hence, the activity of initiation factor eIF-2 (Fig. **4.26**).

As detailed in Table **1.10** eIF-2 plays a central role for the association of mRNA with the 40S ribosomal subunit and with Met-$tRNA^{fMet}$ to the ternary initiation complex. eIF-2 is composed of two subunits, the smallest of which (subunit $\alpha$, $M_r$ 38 kd) undergoes rapid cycles of phosphorylation and dephosphorylation. In the phosphorylated form the catalytic activity of eIF-2 in initiation complex formation is strongly reduced. In fact, it is the interaction with a further auxiliary protein (eIF-2-stimulating protein, ESP, $M_r$ ~ 200 kd) that is affected by the phosphorylation [189]:

$$\text{eIF-2} + \text{GTP} \underset{1}{\rightleftharpoons} \text{eIF-2} \cdot \text{GTP} \overset{+\text{ESP}}{\underset{2}{\rightleftharpoons}} \text{eIF-2} \cdot \text{GTP} \cdot \text{ESP}$$

$$\overset{+\text{Met-tRNA}^{\text{fMet}}}{\underset{3}{\rightleftharpoons}} \text{eIF-2} \cdot \text{GTP} \cdot \text{Met-tRNA}^{\text{fMet}} + \text{ESP}$$

The second and third step of this sequence, in eukaryotic systems preceding mRNA binding, is totally dependent on the presence of ESP[7] under normal conditions, i.e., high rate of translation and limited amount of eIF-2.

---

[7] ESP is presumably identical with similar factors characterized by other groups, i.e., Co-eIF-2C (Das, A. et al.: Proc. Natl. Acad. Sci. USA *76* (1979): 5076) and SF (Ranu, R. S. & I. M. London, ibid. *76* (1979): 1079) respectively.

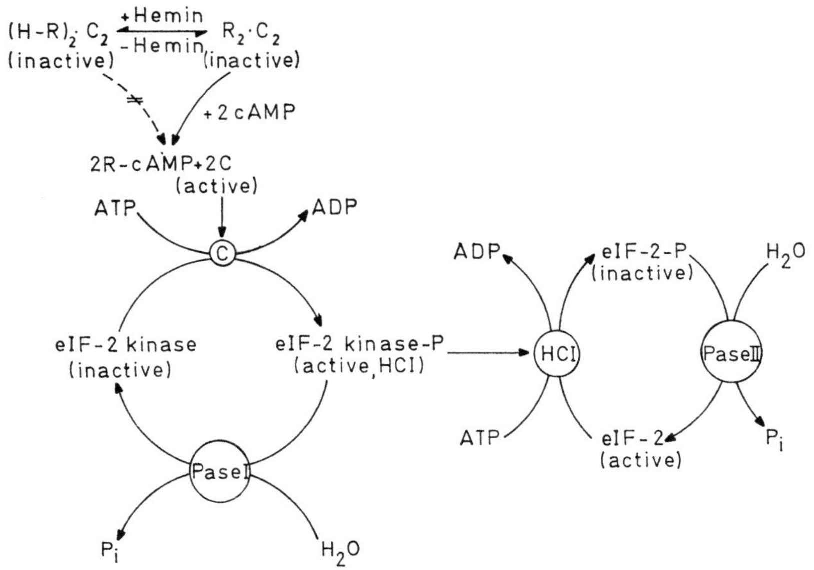


**Fig. 4.26:** Protein kinases mediate the translational control of globin synthesis by hemin (from Ochoa & de Haro [189], modified).

Two protein kinase systems operate in tandem in the hemin-dependent translation control: (1) a cAMP-dependent protein kinase in its inactive form *($R_2C_2$)* is capable of binding hemin that blocks allosterically the interaction of the regulatory subunits (*R*) with cAMP. Only in the absence of hemin can cAMP associate with R and liberate the two active catalytic subunits (*C*). Phosphorylation of the inactive kinase of initiation factor eIF-2 generates the active eIF-2 kinase (*HCI*, hemin controlled inhibitor) and subsequently leads to the phosphorylation of the $\alpha$-subunit ($M_r$ 38 kd) of eIF-2. The phosphorylated form of eIF-2 functions very poorly in translation initiation (cf. text). Usually the inhibitory effect evoked by hemin deficiency is freely reversible due to the action of phosphatases (*Pase I* and *Pase II*, [190, 191]). In reticulocyte lysates the half-life of eIF-2-P was found to be only 20 s.

The role of the first cAMP-dependent protein kinase system has not been established unequivocally. Other cAMP-independent systems were also shown to activate eIF-2 kinase [191, 192]. Moreover, hemin was reported to bind directly to the inactive eIF-2 kinase and to stimulate its activation by autophosphorylation (cf. the summary by Gross [195]). Thus, the initial part of the scheme presumably needs some modifications.

eIF-2 kinase is a 160 kd protein composed of two identical subunits with $M_r$ 80 kd. It is found in the postribosomal supernatant and was detected in many other cell types, too. In addition, a second unrelated ribosomal eIF-2 kinase ($M_r$ 120 kd) is activated by double-stranded RNA and catalyzes phosphorylation of the $\alpha$-subunit of eIF-2 at presumably the same site (cf. the following section on interferon action). Hence, control of translation by eIF-2 phosphorylation – dephosphorylation is a general mechanism encountered in many cells. Its highly specific role for the coordination between hemin and globin synthesis in reticulocytes results from the extreme specialization of these cells to the synthesis of hemoglobin. It is worth-while mentioning that the hemin-controlled inhibitor in other cells may function as one final link of a complex signal transformation chain connecting the iron supply of the cell with its overall metabolic and especially protein synthetic activity.

In mammals the main iron chelating agent of blood serum is the glycoprotein transferrin with very high affinity for two $Fe^{III}$-ions per molecule (cf. summaries [297–299]). The Fe-transferrin complex is taken up by pinocytosis via coated pits (cf. **4.2.2.1**) and subsequently found in cytoplasmic coated vesicles. After binding to mitochondrial outer membranes, iron is released in dependence of the intramitochondrial pool of free heme. Transferrin and membrane compounds of the vesicles are recycled to the cell membrane. In addition to the direct transfer to mitochondria part of the Fe-ions is found as soluble iron in cytosol and, under conditions of Fe surplus, forms a storage pool by binding to ferritin, a chelating protein composed of 24 subunits and capable of fixing maximum 4000 molecules of Fe(III)-oxide phosphate. Only a very small part of total intracellular iron, i.e., the iron-chelates (hemin, Fe-protoporphyrin IX, Fe-citrate) in the free cytosolic pool are presumably active in regulating the iron equilibrium of cells by controlling (1) rates of uptake and release of iron from transferrin, (2) binding to and release from the storage protein, (3) synthesis and degradation of ferritin subunits, and (4) the levels of $\delta$-aminolevulinate synthetase, the key enzyme of heme synthesis, as well as of heme oxygenase involved in iron release from heme compounds. In addition, conditions of severe iron deficiency lead to the above detailed reduction of the translation activity due to the depletion of hemin. Hence, hemin metabolism and the adjoined regulatory mechanism may constitute a central control point in animal cells [193, 194].

### 4.4.4.6 Functional Adaptation of the tRNA Pattern

Current methods for nucleic acid sequencing (cf. **2**.10) have led to rapidly increasing data about the codon usage pattern of distinct protein species. The very small selection presented in Table **1**.5 already reveals pronounced differences and, generally, a highly inbalanced usage of triplets coding for the same amino acid. This phenomenon, inter alia, implies an adaptation of the isoacceptor tRNA patterns in cells, specialized for production of mass proteins. Such changes of the tRNA patterns have been repeatedly reported, e.g., in phage T4-infected bacteria[8], during biogenesis of chloroplasts in greening algae (cf. Chap. **6**) and during specialization of erythroid precursor cells (cf. **4**.5.4.3). Another extreme example is the adaptation of the tRNA pattern in the posterior silk gland of *Bombyx mori* larvae.

The mass protein produced and secreted by silk glands, silk fibroin, is one of the largest polypeptides in nature ($M_r \sim 400$ kd). The fibroin gene exhibits an internal repetitive structure (cf. Chap. **9**) and, correspondingly, silk fibroin consists of short oligopeptide repeating units with glycine, alanine and serine accounting for about 85% of the total amino acid residues. During the phase of cell specialization (fifth larval instar), when fibroin synthesis becomes dominant, the intracellular tRNA pattern undergoes dramatic changes. Alanine, glycine and serine isoacceptor tRNAs are rapidly synthesized and finally make about 70% of the total tRNA complement. Especially prominent is an $tRNA^{Ala}$ isoacceptor species that appears only at this time and is absent in other tissues of the silkworm [196, 197].

In vitro translation experiments with purified fibroin mRNA revealed the crucial importance of this functional adaptation of the tRNA pattern as an integral part of cell specialization in the silk glands. Full-sized fibroin in reticulocyte lysates was only synthesized in the presence of a tRNA mixture from silk glands. In addition, there is evidence that fibroin synthesis is discontinuous not only in the supplemented in vitro translation systems but even in organ cultures of isolated silk glands. These discontinuities, or pauses, occur at distinct sites in the fibroin mRNA. They result in a regular ladder of translation intermediates between 50 and the $\sim 400$ kd full-sized fibroin that reflects the internal repetitive nature of the gene and gene product and points out that the rate of polypeptide chain elongation critically depends on the supply with one or a few specific tRNA isoacceptor species [198].

---

[8] cf. Abelson [150] on the biogenesis of T4-coded tRNA species and references to literature cited therein.

### 4.4.4.7 Translational Control in Interferon Treated Mammalian Cells

Interferons are a family of glycoproteins synthesized in various animal cells (e.g., leukocytes, fibroblasts) in response to the infection by viruses, but also by bacteria and other microorganisms. dsRNA, especially poly(I)-poly(C) was found to be a strong interferon inducer. In some cases a single molecule of dsRNA per cell is sufficient to induce interferon synthesis. Interferons are released from the cells in which they are produced and act on other cells after binding to specific high-affinity receptors (gangliosides) in the cell membrane. Cells exposed to interferon fail to support the multiplication of a wide range of DNA and RNA viruses [173, 199).

Interferons (IFN) were originally classified in those of type I (*p*H 2 stabile, produced by leukocytes and fibroblasts) and of type II (*p*H 2 labile, produced by cells of the immune system after stimulation with mitogens). But because of pronounced differences between individual IFN species this classification had to be replaced by a more detailed nomenclature (cf. Nature *286* (1980): 110): IFN-$\alpha$ (leukocyte interferon), IFN-$\beta$ (fibroblast interferon) and IFN-$\gamma$ (lymphocyte interferon). Sequence analysis of two human IFN (IFN-$\alpha$ and IFN-$\beta$ [200]) showed both to comprise 166 amino acid residues ($M_r$ 20 kd of the glycoprotein), to be coded by separate genes and to be synthesized as preproteins with N-terminal signal sequences of 21 amino acid residues (cf. Table **4**.10). Purification of mouse IFN-$\beta$ gave two antigenically related proteins with $M_r$ 22 and 35 kd, whereas mouse IFN-$\gamma$ consists of two components with $M_r$ 40 and 70–90 kd [201].

Purification to homogeneity of human and of mouse IFN allowed more precise studies of the multiplicity of interferon effects on their target cells [201, 203]. These are, besides the antiviral effects, (1) the inhibition of multiplication of normal and tumor cells, (2) modulation of synthesis and release of specialized proteins, (3) modulation of cell surface components and (4) modulation of humoral and cell-mediated immune responses. It is evident that each IFN type has its own receptor type in the target cell membrane and its own pattern of response. It is not very likely that all these effects, including the inhibition of virus multiplication, result from the three elements of interferon-evoked translational control to be outlined below. At any rate the analysis supports the early assumption of the interferon system to be a general intercellular integration system rather than a simple protection against viral attack. Interferons may be considered as hormone-like factors whose production is merely enhanced after virus infection.

Though many molecular details of the translational control have been worked out in the last few years, regulation of interferon formation and release itself is practically unknown. This also holds true for the signal transformation chain triggered after interferon binding to specific ganglioside receptors in the target cell membrane [201] and finally resulting in the alterations of the translational machinery. We are left with the early observations that the interferon effect is mediated by the adenylate cyclase system (cf. Fig. **10**.6) and that inhibitors of RNA and protein synthesis interfer with the acquisition of an antiviral state of the target cell.

Three types of interferon-induced translational control systems have been characterized (Fig. **4**.27): (1) a dsRNA activated protein kinase phosphorylating the $\alpha$-subunit of the eukaryotic initiation factor 2, (2) a dsRNA activated oligoisoadenylate synthetase (2′,5′-A polymerase) whose product (ppp5′A2′p5′A2′p5′A) serves as activator for a special constitutive endonuclease (RNase F) and (3) a phosphodiesterase that preferably cleaves the oligoisoadenylate but also leads to a degradation of tRNA (Fig. **4**.28). All three enzymes are normal constituents of cells but their activity increases two- to more than tenfold after addition of interferon.

If actinomycin D is present in the early induction period (0–2 h), the activity increase is greatly reduced or abolished. No inhibitory effect of actinomycin can be observed thereafter. Evidently interferon causes an enhanced transcription that is restricted to the two-hours lag period preceding the increase of the enzyme activities

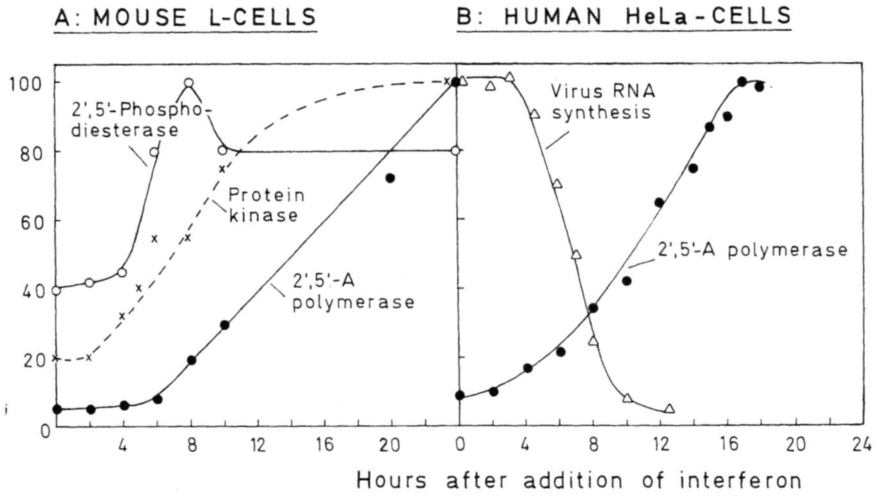


**Fig. 4.27:** Kinetics of induction by interferon of three translation-regulatory enzymes and correlated decline of viral RNA synthesis (from Kimchi et al. [205] and Baglioni [204]).

**A:** Mouse L-cells were treated with 200 units/ml of mouse L-cell interferon. At the indicated time points about $2\text{–}4 \times 10^4$ cells were lysed in buffer by addition of 0.5% Nonidet P40 and the enzymes were assayed in the $8{,}000 \times g$ supernatant.

*Protein kinase:* For enzyme activation the cell extract was preincubated with poly(I)-poly(C) plus ATP and afterwards assayed by adding [$\gamma$-$^{32}$P]-ATP and purified eIF-2. Extent of phosphorylation of the $\alpha$-subunit was determined from gel electropherograms.

*Oligoisoadenylate synthetase (2',5'-A polymerase):* Cell extracts were pretreated with poly(I)-poly(C) Sepharose and all unbound material was removed by centrifugation. The Sepharose pellet was incubated 20 h at 30 °C with [$\alpha$-$^{32}$P]-ATP and the products were analysed by paper electrophoresis.

*2',5'-Phosphodiesterase* activity was determined by cleaving synthetic 2',5'-ApA and analysing the formation of 5'-AMP after chromatography on thin layers of polyethyleneimine cellulose.

**B:** HeLa cells were treated with 100 units/ml of human fibroblast interferon (IFN-$\beta$). Part of the cells was infected with encephalomyocarditis virus prior to interferon addition. The viral RNA synthesis was assayed by 30 min pulse incubation with [$^3$H]-uridine.

(Fig. **4.27**). The kinetics of the 2',5'-A polymerase induction are rather similar in mouse L-cells and human HeLa cells. In HeLa cells they were shown to correlate with a rapid decrease of virus-specific RNA synthesis (Fig. **4.27** B).

The dual role of dsRNA as interferon inducer, as well as necessary activator of two interferon-induced translational control systems, is remarkable. Evidently the genome of certain RNA viruses directly or the cellular intermediates of virus replication may serve these functions.

The dsRNA-dependent protein kinase is a ribosome-associated protein with $M_r \sim 120$ kd and two subunits of $\sim 68$ kd [206]. In the presence of dsRNA the inactive enzyme undergoes a rapid activation by (autocatalytic?) phosphorylation of both subunits and, in turn, phosphorylates eIF-2$\alpha$ at the same sites modified by the hemin-dependent eIF-2 kinase (cf. Fig. **4.26**). As already outlined in Sect. **4.4.4.5**, the phosphorylated form of eIF-2 is unable to effectively catalyze the binding of Met-tRNA$^{\text{Met}}$ as the first step of initiation complex formation.

Specificity of viral mRNA degradation by the oligoisoadenylate activated endonuclease (RNase F) is suggested to result from compartmentation. At least in vitro the enzyme was shown to associate with viral replicative complexes and to remove all single-strand RNA attached to it, e.g., nascent mRNA [209]. Upon Sephadex G-200 gel filtration the elution volume of RNase F

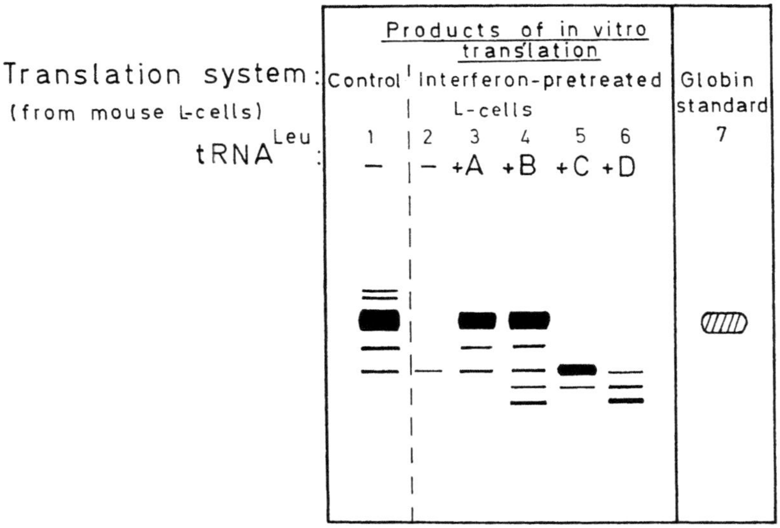


**Fig. 4.28:** Influence of $tRNA^{Leu}$ minor isoacceptor species on the in vitro translation of globin mRNA in lysates of mouse L-cells and the influence of interferon pretreatment (from Revel et al. [212]).

Globin mRNA was translated in mouse L-cell extracts in the presence of [$^{35}$S]-methionine. Translation products were analysed by SDS-polyacrylamide gel electrophoresis. In contrast to the control *(lane 1)*, extracts of interferon treated cells translate very poorly and no globin is formed *(lane 2)*. But translation of globin mRNA can be restoredby addition of two $tRNA^{Leu}$ minor components (A, B, *lanes 3* and *4*). These were separated by a multistep chromatographic procedure from the bulk of $tRNA^{Leu}$ isoacceptor species D that makes about 95% of total $tRNA^{Leu}$. Similar results were obtained for the in vitro translation of mengovirus RNA.

suggests a $M_r$ of 185 kd. Activation by oligoisoadenylate is reversible. Both, isoadenylate synthetase and RNase F, were found in many different cells, independent of a prior treatment by interferon [207]. There is evidence that 2′,5′-oligoisoadenylate [oligo(A)] is part of a cellular proliferation control system. Interferon increases oligo(A) synthesis, whereas mitogenic signals decrease oligo(A) synthesis and the anti-growth effect of interferon on human lymphocytes can be mimicked by addition of oligo(A) [211].

The activity of dsRNA-independent phosphodiesterase ($M_r$ 40 kd) with high affinity for 2′,5′-phosphodiester bonds is thought to remove the oligoisoadenylate activator of RNase F. In addition, it cleaves the 3′-CCA end of tRNAs [210]. Though no specificity for distinct tRNA species was detected, the depletion of certain minor components might very well influence synthesis of distinct protein species as shown in Fig. 4.28.

The multiplicity of interferon induced changes of mRNA stability and translation efficiency is augmented by evidence of an impaired cap methylation (cf. Fig. 1.9). But all these effects have been analysed in cell cultures only, and it remains to be shown to what extent they reflect the actual mechanism of interferon action in intact animals. Especially we do not know how the diverse enzyme activities interfere specifically with the accumulation and function of viral mRNA. The only hint for a preferred action on viral mRNA stems from the affinity of RNase F for dsRNA containing replicative complexes. In keeping with this interpretation is the observation that interferon treatment of embryonal carcinoma cells confers resistance to picornaviruses, whose replication involves dsRNA intermediates [213]. But there was no resistance to other types of viruses, e.g., Semliki Forest virus or vesicular stomatitis virus.

### 4.4.5 Proteinogen Processing

Post-translational modifications of proteins are a general phenomenon in all living systems (cf. Table **1**.11). The multifold catalytic, structural and regulatory activities of proteins as well as changes in their intracellular compartmentation depend decisively on these reversible or irreversible modifications. This is evident from the role of phosphoproteins (cf. Figs. **4**.2, **4**.4, **4**.26 and **10**.6) and the control of gene activity and chromatin superstructure by histone modifications (cf. **1**.2.3.2). However, especially prominent are the constantly increasing examples for a sequential processing by limited proteolysis and eventual glycosylation of precursor proteins which are synthesized by free or membrane-bound polysomes in the cytoplasm and destined for distinct cell compartments (membranes, storage vacuoles, lysosomes, mitochondria, chloroplasts etc.) or for secretion into the extracellular milieu. The basic aspects of this process are similar in pro- and eukaryotic cells. In the following we will briefly outline our present knowledge about prominent aspects of intracellular protein topogenesis [216], i.e., about the role of signal sequences, the synthesis of integral membrane proteins and the selective import of proteins during organelle biogenesis. In addition, the budding of yeast and the propagation cycle of animal viruses will serve as further examples to demonstrate the complexity and regulatory impact of proteinogen processing. Similar to phage T4 morphogenesis (cf. **14**.2.4), steps of limited proteolysis serve as key events in the developmental program of these organisms.

#### 4.4.5.1 Proteinogen Processing – an Essential Means to Guide Proteins to Their Proper Compartments

When dicussing regulatory principles underlying differentiation, we are constantly confronted with the fundamental role of cell compartmentation in this respect. It is the subtle order within the cell of signal transformation, of gene expression itself and of all metabolic chains, that allows frequently opposing reactions to proceed simultaneously in the same cell and that enables cells to respond so efficiently to the multifold influences from the extracellular milieu. Hence differential gene expression must somehow include information about the translocation of its products to the proper compartments. The co- and posttranslational processing of proteins is an essential means to achieve this goal (intracellular protein topogenesis [216]). This will be illustrated in the following by the role of N-terminal signal peptides for the synthesis of secretory and membrane proteins and by the selective import by cell organelles of proteins synthesized on cytoplasmic polysomes.

##### **4.4.5.1.1** The Signal Hypothesis

Synthesis of secretory proteins on polysomes bound to ER membranes was a well-known fact since many years. But a plausible explanation for this highly selective association of distinct mRNA species with membrane-bound polysomes was only given with the formulation of the "signal hypothesis" by Milstein et al. [217] and Blobel & Dobberstein [218]. The experimental evidence points out that all export proteins and many membrane proteins are synthesized via short-lived precursors

**Table 4.10**: Comparison of N-terminal amino acid sequences (signal peptides) of pre-proteins (from Kreil [220], Davis & Tai [221] and Zimmerman et al. [215]).

| Pre-proteins[a] | | Sequences[b] (–20 … –10 …) | Cleavage site, +1 |
|---|---|---|---|
| Pre-IgL (κ, mouse, cf. Chap. **13**) | (S) | MdMrAPAqIFgFLLLLFPgtrc | d . . . |
| Pre-IgL (λ, mouse, cf. Chap. **13**) | (S) | MAWIsLILsLLALssgAIs | q . . . |
| Pre-IgH (γ2b, mouse, cf. Chap. **13**) | (S) | MAVLALLFoLVtFPsoILs | q . . . |
| Pre-IgH (x, mouse, cf. Chap. **13**) | (S. M) | MkVLsLLYLLtAIPhIMs | d . . . |
| Pre-proparathormone (bovine) | (S) | MMsAkdMVkVMIVMLAIcFLArsdg | k . . . |
| Pre-proinsulin (rat, cf. Chap. **8**) | (S) | WMrFLPLLALLVLWePkPAqA | F . . . |
| Pre-proalbumin (rat) | (S) | MkWVtFLLLLFIsgsAFs | r . . . |
| Pre-casein (sheep) | (S) | MkVLILAoLVALALA | r . . . |
| Pre-glycoprotein (VSV, cf. **4.4.5.1.2**) | (M) | MkcLLYLAFLFIhVnc | k . . . |
| Pre-lysozyme (hen oviduct, cf. Chap. **7**) | (S) | MrsLLILVLcFLPLAALg | k . . . |
| Pre-ovomucoid (hen oviduct, cf. Chap. **7**) | (S) | AMAgVFVLFsFVLxgFLPdAAFg | A . . . |
| Pre-promelittin (bee) | | MkFLVxVALVFMVVYIxYIYA | A . . . |
| Pre-interferon (human IFN-β, cf. **4.4.4.7**) | (S) | MtnkoLLqIALLLoFsttALs | M . . . |
| Pre-lipoprotein (*E. coli*) | (M) | MkAtkLVLgAVILgstLLAg | o . . . |
| Pre-penicillinase (*E. coli* plasmid) | (S) | MsIqhFrVALIPFFAAFoLPVFA | h . . . |
| Pre-coat protein (f1 phage) | (M) | MkksLVLkAsVAVAtLVPMLsFA | A . . . |
| Pre-λ phage receptor (*E. coli*, cf. **14.1.2**) | (M) | MMItLrkLPLAVAVAAgVMsAqAMA | V . . . |

[a] The proteins and polypeptide hormones listed in the table are secreted to the extracellular milieu (S) or incorporated into membranes (M). IgL, IgH, immunoglobulin light and heavy chains (cf. Chap. **13**); VSV, vesicular stomatitis virus.

[b] Hydrophobic amino acid residues are denoted by capital letters. One letter code for the amino acid residues: A, alanine; c, cystein; d, aspartic acid; e, glutamic acid; F, phenylalanine; g, glycin; h, histidine; I, isoleucine; k, lysine; L, leucine; M, methionine; n, asparagin; P, proline; q, glutamin; r, arginine; s, serine; t, threonine; V, valin; W, tryptophane; Y, tyrosine; x, unknown amino acid.

(pre-proteins) with short N-terminal extensions of about 15–25 amino acid residues (cf. Table **4**.10). These signal peptides mediate the attachment of the nascent polysome to ER receptor sites and normally are removed co-translationally by a specific endoprotease (signalase) located at the inner site of the ER membrane (cf. details in Figs. **8**.6 and **13**.9). Basically similar mechanisms govern the synthesis of periplasmic and outer membrane proteins by membrane-bound polysomes in bacteria (cf. the examples given in Table **4**.10).

Synthesis of these secretory proteins starts by formation of the usual initiation complexes from mRNA and free cytoplasmic ribosomal subunits (cf. Fig. **1**.19). The initial sequence (signal sequence) of the nascent protein is especially rich in hydrophobic amino acids and has high affinity to certain receptor proteins in the ER membranes. Hence, very soon after the start of translation polysomes attach to the ER membranes, where the aggregating ribosome binding and peptide translocator proteins form a tunnel transverse the ER membrane. The nascent polypeptide chains are injected into the ER cisterna (Fig. **13**.9), the signal sequence being removed soon after attachment of the polysomes. Further steps of posttranslational processing by limited proteolysis or covalent modification of amino acids, e.g., by glycosylation, proceed in a determined reaction sequence on the way of the maturing proteins through the ER to the Golgi apparatus and exocrine granules (cf. Figs. **8**.4 and **13**.10).

It is worth noticing that translation of these export proteins per se is not dependent on the membrane association of the corresponding polysomes. The pre-proteins are detected if co-translational transfer through membranes and cleavage of the signal peptide are inhibited (cf. Fig. **13**.9) or if translation proceeds in an in vitro system in the absence of membranes. There is, however, some evidence such precursors to be rapidly degraded in vivo so that the number (and type?) of receptor sites in the ER or bacterial cytoplasmic membranes might control the rate of export protein synthesis. Two special ribosome-binding proteins (ribophorins) solubilized from rER membranes are evidently essential components of the ribosome docking sites. They are transmembrane glycoproteins with $M_r$ 63 and 65 kd respectively [225].

Sequence comparison of signal peptides (Table **4**.10) reveals practically no similarity, except an enrichment of hydrophobic residues (denoted by capital letters) and a small aliphatic amino acid residue (Ala, Gly, Ser, Cys) preceding the cleavage site. Obviously the signalase recognizes some sort of three-dimensional structure rather than the primary amino acid sequence of the pre-protein.

An endopeptidase solubilized from rER of several animal sources (ascites tumor cells, rat liver, dog pancreas) was shown to cleave completed pre-human placental lactogen also post-translationally, i.e., the particular tertiary structure of the N-terminal part of pre-proteins and correct cleavage are not dependent on the presence of membranes. Activity of the endopeptidase was inhibited by chymostatin, an inhibitor of Ser-proteases with chymotrypsin-like activity [224]. It was not inhibited by leupeptin, antipain or elastinal (cf. Table **1**.13).

### **4.4.5.1.2** Synthesis of Membrane Proteins

A special aspect of the signal hypothesis concerns the synthesis of transmembrane proteins. The examples included into Table **4**.10 are characterized by the fact that the C-terminal domain(s) of the proteins remain on the cytoplasmic site of the membrane, whereas the N-terminal domain(s) are turned to the non-cytoplasmic site, e.g., the ER cisterna. Studies with the vesicular stomatitis virus glycoprotein and the bacteriophage f1 coat protein suggest that the usual N-terminal signal sequence mediate membrane attachment of the polysomes. The nascent polypeptide chain is transferred through the membrane until a "stop transfer" sequence is reached. The remaining C-terminal part of the protein is synthesized at the cytoplasmic face of the membrane [226–229]. This extended version of the signal hypothesis describes only one mechanism of membrane protein synthesis. As will be evident from the following section, membrane proteins, e.g., of mitochondria and chloroplasts as well as of bacteria [222], can also be synthesized by free ribosomes and inserted post-synthetically into the proper membranes.

Details of the vectorial transport of maturing membrane proteins were studied with animal cell cultures infected with vesicular stomatitis virus (VSV), whose envelope forms during budding at the host cell surface (cf. Fig. **19**.6) and contains a single intramembrane glycoprotein (G protein, $M_r \sim 70$ kd). After VSV infection this is virtually the only glycoprotein synthesized because the virus inhibits specifically host cell gene expression [227].

G protein is synthesized by membrane-bound polyribosomes as a pre-protein with a 16 amino acid residue N-terminal extension (cf. Table **4**.10). The bulk of the molecule including the two asparagine residues, lateron serving the attachment of the carbohydrate side chains, is secreted co-translationally to the ER lumen. "Stop transfer" signal is a very hydrophobic sequence of

19 amino acid residues followed by a highly basic C-terminal domain with 30 amino acid residues remaining on the cytoplasmic side [229].

The postsynthetic glycosylation coupled to an intracellular directed transport from ER to the GA and finally to the outer membrane follows the scheme outlined for membrane-bound and secreted immunoglobulins (Fig. **13**.10). It appeared that the intracellular transport vesicles are clathrin-coated vesicles which evidently serve as a universal shuttle system not only for receptor mediated endocytosis (cf. **4**.2.2.1) but also for the transport of newly synthesized membrane proteins [20]. The extremely rapid formation and fusion with membranes of coated vesicles depends on the dynamic assembly and disassembly of the polyhedral clathrin coat. It is reasonable to assume that coated vesicles form a family of intracellular carriers whose function is specified by minor proteins in the vesicle coat. A major unsolved question is the highly precise sorting out of various membrane proteins destined to become part of different types of membranes.

Analysis of two temperature-sensitive mutants in the structural gene coding for the G protein [227, 228] revealed the tight coupling between progressive maturation and directed transport of G protein from the site of synthesis to the outer plasma membrane, where assembly and budding of viral particles take place (cf. Fig. **19**.6). One class of mutants is blocked at an early pre-Golgi step of processing, if cells are cultivated at the non-permissive temperature. Insertion of the G protein precursor into the ER membrane as well as transfer and first steps of modification of the oligosaccharide side chains proceed normally. But the mutant G protein precursor is not transported to the GA and consequently all GA-specific and post-GA-specific modifications are lacking (cf. scheme in Fig. **13**.10). In the second class of mutants processing is normal including the Golgi-mediated functions, e.g., addition of sialic acid. Only the terminal addition of fucose and of two fatty acid residues, specific for the G protein, are not observed. The immature G protein molecules reside in the GA membrane.

### **4.4.5.1.3** Cytoplasmic Synthesis of Organellar Proteins

Palade pointed out that "all eukaryotic cells secrete proteins" [230], if not for protein export then at least for biogenesis of cell organelles separated by membranes from cytosol, i.e., biogenesis of lysosomes, peroxisomes, glyoxisomes, mitochondria and chloroplasts. Though the biogenetic pathway of the first three is evidently basically different from that of the double-membraned, semiautonomous mitochondria and chloroplasts, important questions in all cases concern the cytoplasmic sites of synthesis of organelle-specific proteins and the mechanisms of their selective transfer to the nascent organelles. The existence of proteinogens with N-terminal extensions of 20–60 amino acid residues are now well established for several mitochondrial (cf. Table **4**.11) and chloroplast proteins (cf. Chap. **6**) coded by the nuclear genome and synthesized on free cytoplasmic ribosomes. However, in contrast to the signal hypothesis, their transfer to the compartments they are destined for is clearly independent of their translation. According to the "vectorial processing" hypothesis [232] the precursor proteins attach with their N-terminal extensions to specific receptor or translocator sites in the organellar membrane, thus opening a pore which allows the protein to equilibrate across the membrane barrier. Unidirectional transport could be most simply driven by the proteolytic processing of the proteinogen. The transmembrane movement and/or the processing require energy. Additional driving forces may result from the assembly of mature subunits to oligomeric enzymes or from the incorporation of coenzymes into apoproteins.

In contrast to results with mitochondrial and chloroplast proteins, thorough investigations on possible high molecular weight precursors of peroxisomal enzymes (catalase, uricase) were negative. This does not exclude that very short signal sequences not detected by gel electrophoresis are present or that the putative signal

**Table 4.11**: Survey of cytoplasmically made mitochondrial polypeptides and their precursors in yeast (from Schatz [232] and Lewin et al. [235]).

| Polypeptide | Intramitochondrial location | $M_r$ (kd) | |
|---|---|---|---|
| | | Precursor | Mature form |
| $F_1$-ATPase | Matrix | | |
| α-subunit | | 64 | 58 |
| β-subunit | | 56 | 54 |
| γ-subunit | | 40 | 34 |
| Cytochrome $bc_1$, subunit V | Inner membrane | 27 | 25 |
| Cytochrone $c_1$ | Inner membrane | 37 | 31 |
| Cytochrome c peroxidase | Intermembrane space | 39.5 | 33.5 |
| Cytochrome c oxidase[a], | Inner membrane | | |
| subunit IV | | 17 | 14 |
| subunit V | | 15 | 12 |
| subunit VI | | 20 | 12.5 |
| subunit VII | | no[b] | 4.5 |

[a] Subunits IV–VII are imported from the cytoplasm, whereas subunits I–III are encoded by the mitochondrial genome and synthesized within the organelle.
[b] No precursor protein for the smallest subunit was detected [235a].

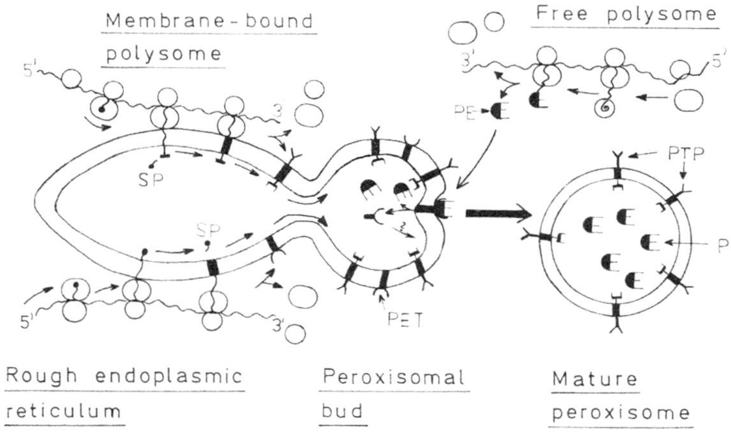


**Fig. 4.29**: Hypothetic scheme of peroxisome biogenesis (from Nover [21]).

Membrane proteins are synthesized by polysomes attached to the rough endoplasmic reticulum. Synthesis of two types of membrane proteins are visualized: (**1**) a peroxisomal transmembrane protein *(PTP)* and (**2**) an enzyme translocator protein *(PET)* required for the specific uptake of distinct enzymes into the peroxisomal bud. Peroxisomal enzymes are synthesized on free polysomes. Synthesis of membrane proteins starts with the usual steps postulated by the signal hypothesis (cf. **4.4.5.1.1**). However, only part (the N-terminus) of the protein is injected into the ER cisterna until a "stop transfer" sequence (■) blocks further transfer. The C-terminal parts of the proteins remain on the cytoplasmic site of the ER. After sorting out of the newly synthesized and processed membrane components into a peroxisomal bud, the translocator-mediated uptake of peroxisomal enzymes *(PE)* may proceed by pinocytosis with eventual recycling of the translocator molecule. The mature peroxisomes detached from the ER are assumed to lack the translocator protein. For the basic elements of this model (cf. [234] and text).

sequences remain part of the mature molecule. In fact, in modification of the original hypothesis Goldman & Blobel [234] proposed the following sequence for the biogenesis of peroxisomes and the related glyoxysomes (cf. scheme shown in Fig. **4**.29):

1. The corresponding membrane proteins are synthesized by *ER-bound* polysomes (cf. preceding section).
2. The newly formed membrane proteins aggregate to form peroxisomal buds with specific receptor sites for peroxisomal enzymes (in Fig. **4**.29 assumed to be a protein).
3. The peroxisomal enzymes are translated on *free* cytoplasmic *polysomes* and selectively taken up by virtue of an internal "signal sequence" recognized by the receptor sites.
4. Loading of the nascent peroxisome ceases within a short time by loss of receptor sites. The mature organelle separates from the ER.

Though basic elements of this scheme have been validified by experimental results, many details are speculative. This is also true for all other examples of a directed co-translational or post-translational transport of proteins to their proper compartments. Despite the important contribution of the signal hypothesis and its various modification it remains a deep mysterium how the perfect sorting out of membrane proteins into different types of membranes or of secretory proteins into different types of secretory vesicles is brought about. The very complex pattern of stepwise glykosylation may be an essential means to control the vectorial transfer at least of glykoproteins (cf. Fig. **13**.10). Furthermore, different types of translocator systems in the rER and organellar membranes as well as receptor proteins in clathrin-coated vesicles or nascent glyoxysomes can be assumed to specifically recognize distinct types of N-terminal or possibly also protein-internal "topogenic sequences" that differ from each other depending on the compartment the protein is destined for. Four types of such sequences can be discerned [216]: (1) *signal sequences* that serve in co-translational transfer of proteins through membranes with corresponding translocator systems (e.g., rER membranes), (2) *stop-transfer sequences* that interrupt the co-translational transfer and lead to the asymmetric integration of membrane proteins, (3) *sorting sequences* that confer specificity to the post-translational traffic of proteins synthesized by membrane-bound polysomes and (4) *insertion sequences* that serve the unilateral integration of proteins into membranes. It is evident that topogenic sequences may be permanent or transient parts of the polypeptide chain. Their function requires decoding by appropriate receptor proteins and may be dependent on postsynthetic modification, e.g., by glykosylation.

#### 4.4.5.2 Proteinogen Processing Determines Local Deposition of Chitin in Budding Yeast

A predominant part of cell wall material of many *Ascomycetes* consists of chitin polymerized by chitin synthetase (ChS) according to the following reaction:

$$\text{n UDP-N-acetyl glucosamin} \xrightarrow[\text{ChS}]{} \text{nUDP} + \text{chitin}$$

Chitin synthetase is generally found as an inactive zymogen associated with the cell membrane. Its activation by limited proteolysis is restricted in time (to certain phases of the cell division cycle) and in space (to the growing points of hyphae in *Aspergillus*, *Penicillium* or *Neurospora*, or to the sites of septum formation in yeasts).

Contrasting to other *Ascomycetes*, in budding yeast chitin is used only for primary septum formation, the predominant material of the cell wall being peptidoglykan. Results of the rather detailed investigations on the morphogenetic role of chitin

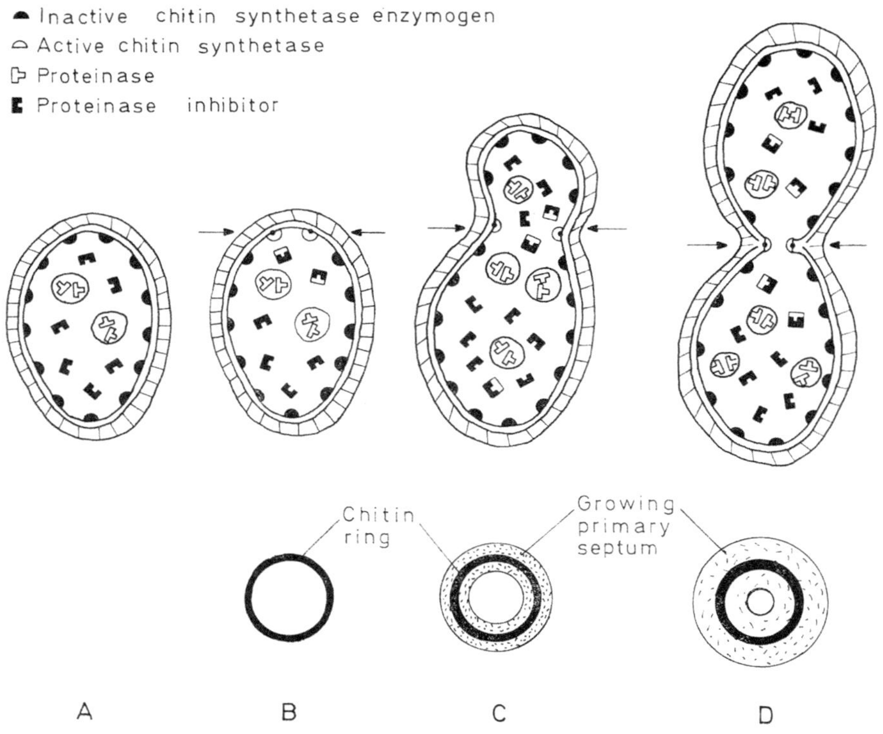


**Fig. 4.30:** Scheme of chitin synthesis and primary septum formation in budding yeast. Local proteolytic activation of chitin synthetase zymogen (from Cabib, E. & V. Farkas: Proc. Natl. Acad. Sci. *68* (1971): 2052).

The inactive chitin synthetase zymogen is bound to the plasma membrane (**A**). After local proteolytic activation by a vacuolar protease, a chitin ring is laid down which determines the site of bud initiation (**B**). The enlarging bud is more and more separated from the mother cell by the centripetally growing primary septum made of chitin (**C, D**). For further explanations see text.

The lower part of the figure shows cross sections through the budding cells at the site of septum formation *(arrows)*.

synthetase zymogen activation are shown in Fig. 4.30. It is evident that ordered deposition of chitin depends (1) on the controlled interaction of vacuolar proteinases with a small part of the chitin synthetase zymogens scattered throughout the cell membrane and (2) on the inactivation by cytosolic inhibitors of the proteinases released from the vacuoles (cf. Table 1.12). How this precise interplay is brought about, remains obscure so far.

Intensive purification of the activating proteinase led to the conclusion that of the different proteinases in the yeast vacuole (cf. Table 1.12) it is proteinase B which catalyzes the limited proteolysis of chitin synthetase zymogen. However, recently several proteinase B defective mutants (prb 1 mutants) were shown to grow and divide and to form apparently normal septa that contain chitin [239].

The decisive role of the directed transport of proteinases to the membrane sites, where chitin synthesis is to proceed, was emphasized when analyzing a temperature sensitive cell cycle mutant (*cdc* $24^{ts}$), which is unable to bud at the restrictive temperature, but buds normally at low temperatures (cf. Fig. **15**.8). Under restrictive conditions local chitin deposition is disturbed. Instead it proceeds scattered all over the cell membrane and multinuclear cells are formed. A possible explanation for this compartmentation defect is a failure in the assembly of elements of the yeast cytoskeleton system (microtubules, microfilaments, cf. **4**.1.4) which determines the site of the formation of the chitin annulus (Fig. **4**.30B) and contributes to the directed transport of material.

### 4.4.5.3 Proteinogen Processing in the Propagation Cycle of Animal Viruses

Picornaviruses, e.g., poliovirus, mengo virus, encephalomyocarditis virus, are small spherical particles with a single plus-strand of RNA as genome. With respect to their genome structure there is a certain similarity to the small bacterial viruses, e.g., bacteriophages Q$\beta$ and MS2 (cf. Fig. **4.24**). However, in contrast to the translational program governing the bacteriophage propagation cycle, biogenesis of picornaviruses is dominated by a programmed proteinogen processing. This is summarized in Fig. **4.31**. The indicated steps of proteinogen processing by limited proteolysis

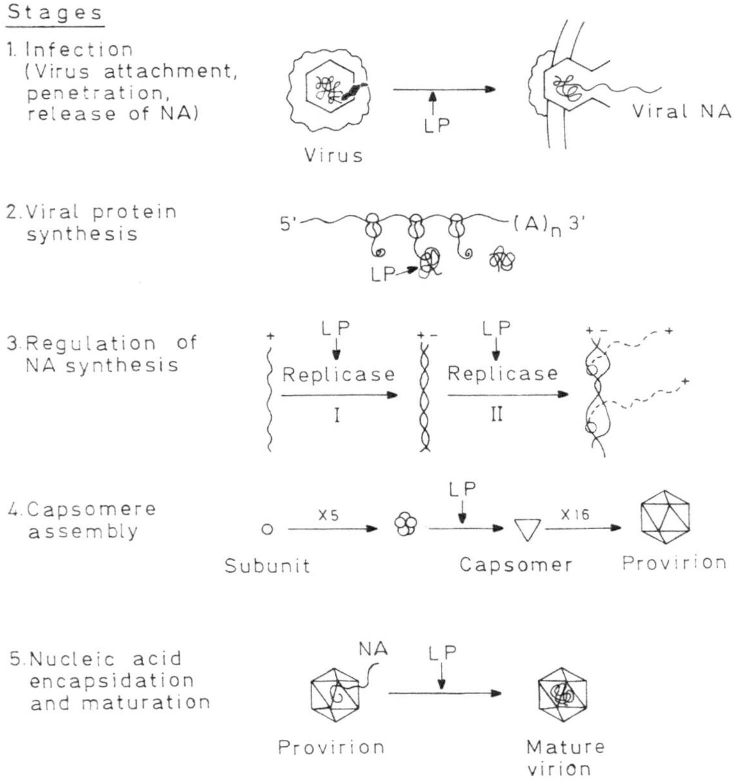


**Fig. 4.31:** Five stages of picornavirus biogenesis that are dependent on limited proteolysis (from Korant [**243**], modified).

The scheme summarizes the essential steps of the viral propagation cycle beginning with the injection of the viral nucleic acid *(NA)* and ending with the production of mature viral particles. Points of limited proteolysis are indicated by *LP* (cf. details given in Fig. **4.32**).

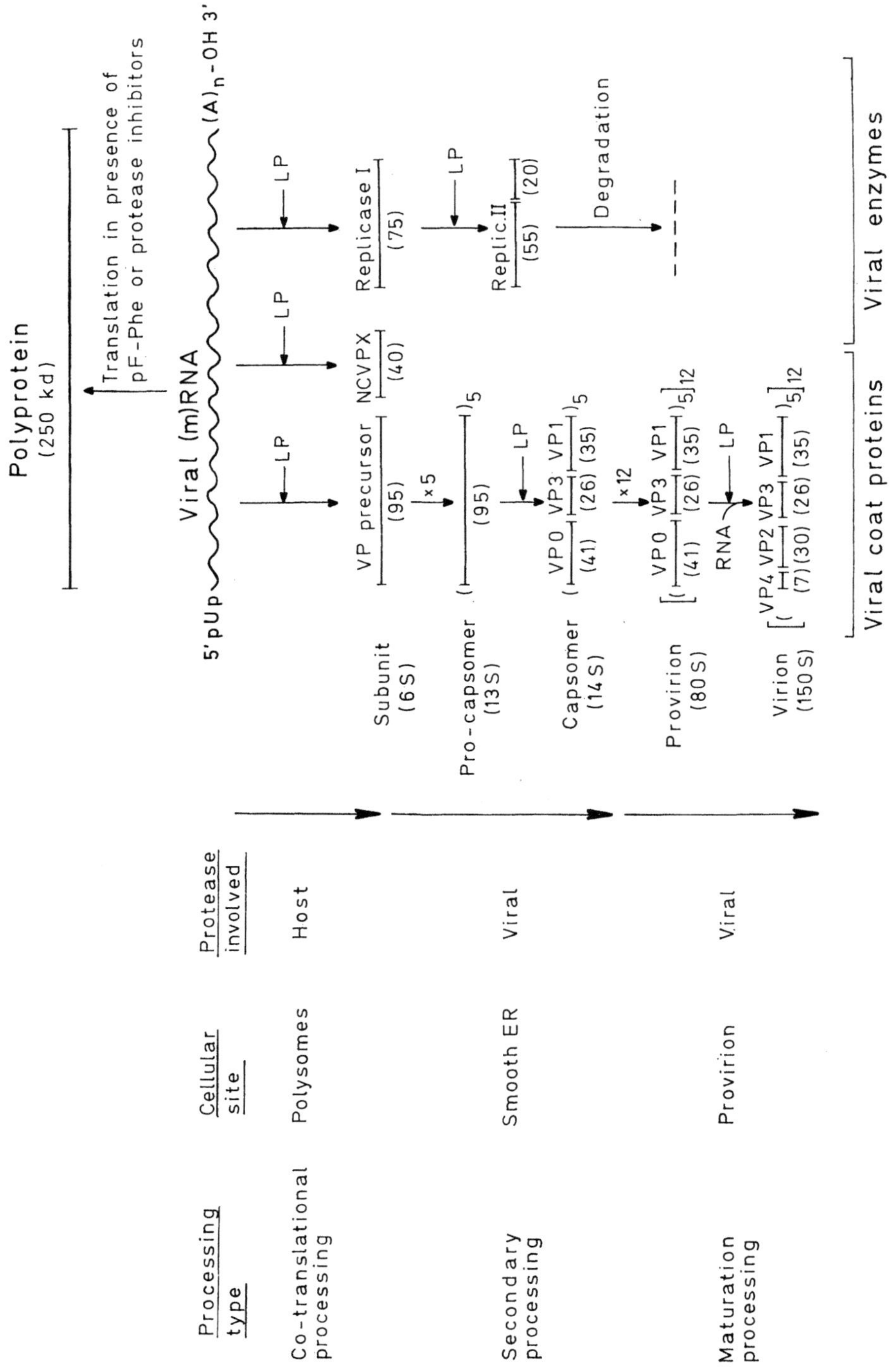

Polyprotein
(250 kd)
Translation in presence of pF-Phe or protease inhibitors
Viral (m)RNA
5'pUp
(A)n-OH 3'
LP
VP precursor
(95)
NCVPX
(40)
Replicase I
(75)
Replic.II
(55)
(20)
Degradation
×5
(95)
VP0 VP3 VP1
(41) (26) (35)
×12
RNA
VP4 VP2 VP3 VP1
(7) (30) (26) (35)
Subunit (6S)
Pro-capsomer (13S)
Capsomer (14S)
Provirion (80S)
Virion (150S)
Viral coat proteins
Viral enzymes
Processing type
Cellular site
Protease involved
Co-translational processing
Polysomes
Host
Secondary processing
Smooth ER
Viral
Maturation processing
Provirion
Viral


Fig. 4.32

are crucial for practically all animal viruses so far investigated. Again the concerted process of encapsidation of the viral genome and final proteolytic cleavage of a provirion protein very much reminds of the head maturation of bacteriophages, e.g., of phage T4 (cf. **14**.2.4.3).

The sequence of events following virus-host cell interaction (Fig. **4**.31) is basically similar to that to be described for bacteriophages (cf. Chap. **14**). Recognition of the cell surface by some determinant on the virus capsid surface and binding is followed by a destabilizing alteration of the infecting virion, leading to the introduction of the viral nucleic acid into the cell (stage 1). The viral genome is amplified by a set of replicases, which either may be present in the cell prior to infection, or carried in by the infecting virus, or produced as translation products of the viral genome. Translation of viral mRNA by the host translation apparatus produces viral proteins (stage 2). Some of these are enzymes participating in the production of progeny viral genomes (stage 3). Others are structural proteins which assemble (stage 4) and finally combine with the viral nucleic acid to form progeny virus particles (stage 5) released by the cell. Often, the infected cell is killed upon virion release, but it may also survive and continue to produce virus. In other cases, the cell may be transformed by the infection, and yield a potentially "immortal" line which conserves a portion of the viral genes and expresses certain virus-specific functions (cf. the lysogenic state of phage $\lambda$, **14**.2.3.9). Practically all stages of the viral propagation cycle depend on proteinogen processing. Hence with our increasing knowledge of the details, selective inhibitors of the proteinases involved might become feasable agents to suppress viral infections.

Further details of the proteolytic program of picornaviruses are indicated in Fig. **4**.32. The following points deserve special notice:

1. A single initiation point exists at the 5′-end of the uncapped viral RNA plus-strand which serves as mRNA. Under special conditions (incorporation of amino acid analogues, inhibition of polysomal protease) the intact primary translation product, a polyprotein of about 250 kd molecular weight, can be isolated.
2. However, under normal conditions the polyprotein is cleaved in nascent state by a host endoprotease evidently associated with the polysomal complex. A coat protein precursor, a virus-specific protease and a replicase or replicase subunit are formed.
3. The further steps of proteinogen processing are catalyzed by the viral protease. These are (1) cleavage of replicase type I to give replicase type II, which is finally degraded, and (2) stepwise cleavage of the viral coat protein precursor during assembly and maturation (cf. details in Fig. **4**.32).
4. A remarkable property of the picornavirus genome, but not of the viral RNA involved in translation, is its association with a small virus-coded protein (VPg, ~ 27 amino acid residues). VPg is linked to a uridylate residue at the 5′-end of the genome via an $O^4$-(5′-uridylyl)-tyrosine. It is coded ~ 2100 nucleotides from the 3′-terminal poly(A) sequence of the viral RNA. During processing VPg is presumably generated from the N-terminal part of the replicase I protein ($M_r$ 75 kd). Timing of VPg formation may be involved in the initiation of RNA synthesis [245] (stage 3 of Fig. **4**.31).

◄

**Fig. 4.32**: Proteinogen processing in the propagation cycle of poliovirus (from Hershko & Try [240] and Korant [243], modified).

The primary translation product is processed in the nascent state (co-translationally) by a host proteinase. Further steps of the programmed proteolytic processing *(LP)* are catalysed by a newly synthesized protease. It generates the two forms of the viral replicase and the viral coat proteins *(VP)* 1–4. The different products of the processing are characterized schematically by their function and their molecular weights (in kd). For the intermediates of the assembly line of new viral particles from the uncleaved coat protein subunit to the mature virion sedimentation coefficients *(S)* are given.

### 4.4.6 Biological Inactivation and Degradation of Proteins

Selective inactivation and degradation of proteins are as important for the existence of life as is their selective synthesis. Though accumulating evidence indicates a differential protein breakdown and its decisive role for cell differentiation, data on the biochemical and regulatory details are still extremely limited. During evolution, cells have acquired a great number of reactions (cf. Table **1**.11) by which proteins can be modified, reversibly or irreversibly, and hence can be controlled in their biological activity and compartmentation. In addition, the controlled and selective inactivation of proteins by proteolytic cleavage is irreversible and represents the necessary final step of differential gene expression.

The complexity of protein degradation is illustrated by four aspects:

1. Different proteins in one and the same cell have very different half-lives (Table **4**.12). There is no doubt that the protein molecule itself contains some sort of information about its type and velocity of breakdown. Regulatory key enzymes generally have a much shorter half-life than other enzymes of the same metabolic chain.

The 11 liver enzymes in Table **4**.12 with very short to intermediate half-lives (0.2–12 hours) all represent important metablic control points of pathways influenced by hormonal and/or nutritional signals, e.g., RNA synthesis (enzymes 4, 11), amino acid degradation (enzymes 1, 5, 6), glucose metabolism (enzyme 9), cholesterol and hemin homeostasis (enzymes 2, 3, 8) and DNA synthesis (enzyme 7).

**Table 4.12:** Half-lives of proteins in rat liver cells (from Schimke [249] and Goldberg & John [251]).

| Enzyme | Half-life (h) | Compartment |
|---|---|---|
| 1 Ornithine decarboxylase | 0.2 | Cytosol |
| 2 δ-Aminolevulinate synthetase | 0.33 | Cytosol |
| 3 δ-Aminolevulinate synthetase | 1.1 | Mitochondria |
| 4 RNA polymerase I | 1.3 | Nucleus |
| 5 Tyrosine aminotransferase | 2.0 | Cytosol |
| 6 Tryptophan oxygenase | 2.5 | Cytosol |
| 7 Deoxythymidine kinase | 2.6 | Nucleus |
| 8 β-Hydroxy-β-methylglutaryl coenzyme-A reductase | 3.0 | E R |
| 9 Phosphoenolpyruvate carboxykinase | 5.0 | Cytosol |
| 10 Aniline hydroxylase | 5.0 | E R |
| 11 RNA polymerase II | 12 | Nucleus |
| 12 Alanine aminotransferase | 17–24 | Cytosol |
| 13 Glucokinase | 12–30 | Cytosol |
| 14 Glucose-6-phosphate dehydrogenase | 15 | Cytosol |
| 15 Catalase | 34 | Peroxisomes |
| 16 Acetyl coenzyme-A carboxylase | 48 | Cytosol |
| 17 Arginase | 96–120 | Cytosol |
| 18 Lactate dehydrogenase | 84–384 | Cytosol |
| 19 Cytochrome c reductase | 60–80 | E R |
| 20 Cytochrome $b_5$ | 100–120 | E R |
| 21 δ-Aminolevulinate dehydratase | 132 | Mitochondria |
| 22 Cytochrome c | 216 | Mitochondria |
| 23 Histones | 430 | Nucleus |
| 24 NAD-glucohydrolase | 384 | E R |
| 25 β-Glucuronidase | 696 | Lysosomes |

2. Half-lives of proteins can be selectively and drastically altered, thus relativating the simple classification into short-lived, regulation-susceptible and long-lived, regulation-resistent enzymes. Two prominent examples will illustrate this point: (1) changes of arginase half-life in rat liver cells by shifts in the nutrient diet (Fig. 4.33) and (2) the selective proteolytic inactivation of gluconeogenetic enzymes in yeast upon addition of glucose to an acetate grown culture (Fig. 4.34).

*Arginase in Rat Liver Cells* (cf. Fig. 4.33): As enzyme of the urea cycle arginase serves as a convenient indicator of the rate of amino acid catabolism. Shifting rats from a high protein diet to a low protein diet (70 % → 8 % casein, part A) or from a low protein diet to a protein-free diet (8 % → 0 % casein, part B) results in two different responses of the arginase levels. In part A of the experiment the arginase level rapidly declines due to a stop of de novo synthesis and a transient threefold increase of degradation. In contrast, the shift-down in the nutrient conditions in part B leads to a slow increase of arginase activity and no detectable degradation despite an increased overall protein turnover under these conditions. Hence, the arginase half-life is selectively adapted to the changing nutrient conditions.

*Catabolite Degradation of Enzymes in Yeast* (Fig. 4.34): Pronounced effects of extracellular glucose levels on the metabolic patterns of cells are well known for bacterial, fungal and mammalian cells. In bacteria the glucose effect is connected with the phenomenon of C-catabolite repression (a transcriptional control mediated by cAMP (cf. Fig. 4.2), whereas in mammals the effect is brought about by an interplay of glucose with hormonal signal systems, the regulatory level being so far unknown. In *Ascomycetes*, glucose on the one side blocks the synthesis of certain catabolic enzymes presumably at a posttranscriptional level. On the other side examples for a selective proteolytic inactivation of three gluconeogenetic key enzymes have been found in yeast (Fig. 4.34), i.e., of cytplasmic malate dehydrogenase (cMDH), fructose-1,6-diphosphatase FDPase) and phosphoenolpyruvate carboxykinase (PEP-CK).

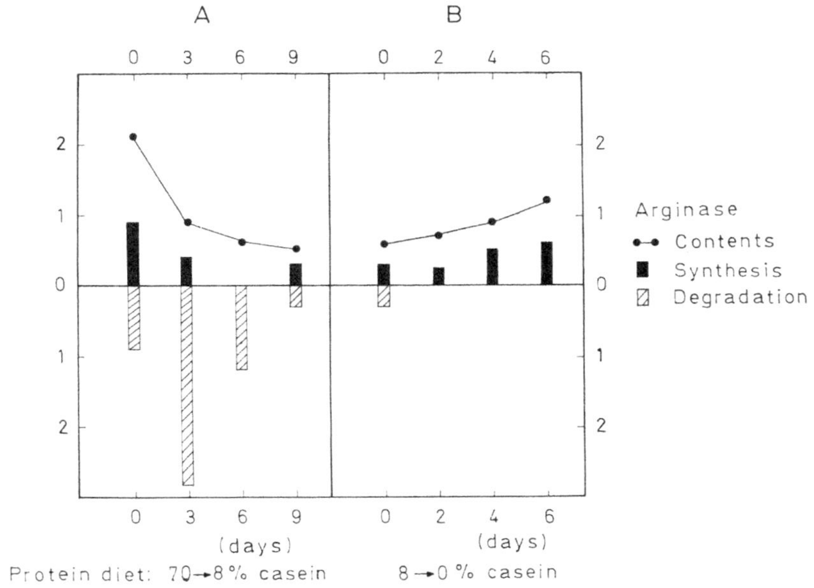


**Fig. 4.33**: Changes of contents, synthesis and degradation of arginase in rat liver cells under different feeding conditions (from Schimke [249], modified).

Protein diet of rats was reduced 14 days prior to the onset of this experiment (70 % casein → 8 % (**A**) and from 8 % casein → 0 % (**B**) respectively). The arginase content (2 = 2 mg/liver) as well as de novo synthesis and degradation (2 = 1.5 mg arginase × g protein$^{-1}$ × 2–3 days$^{-1}$) were determined by immune precipitation.

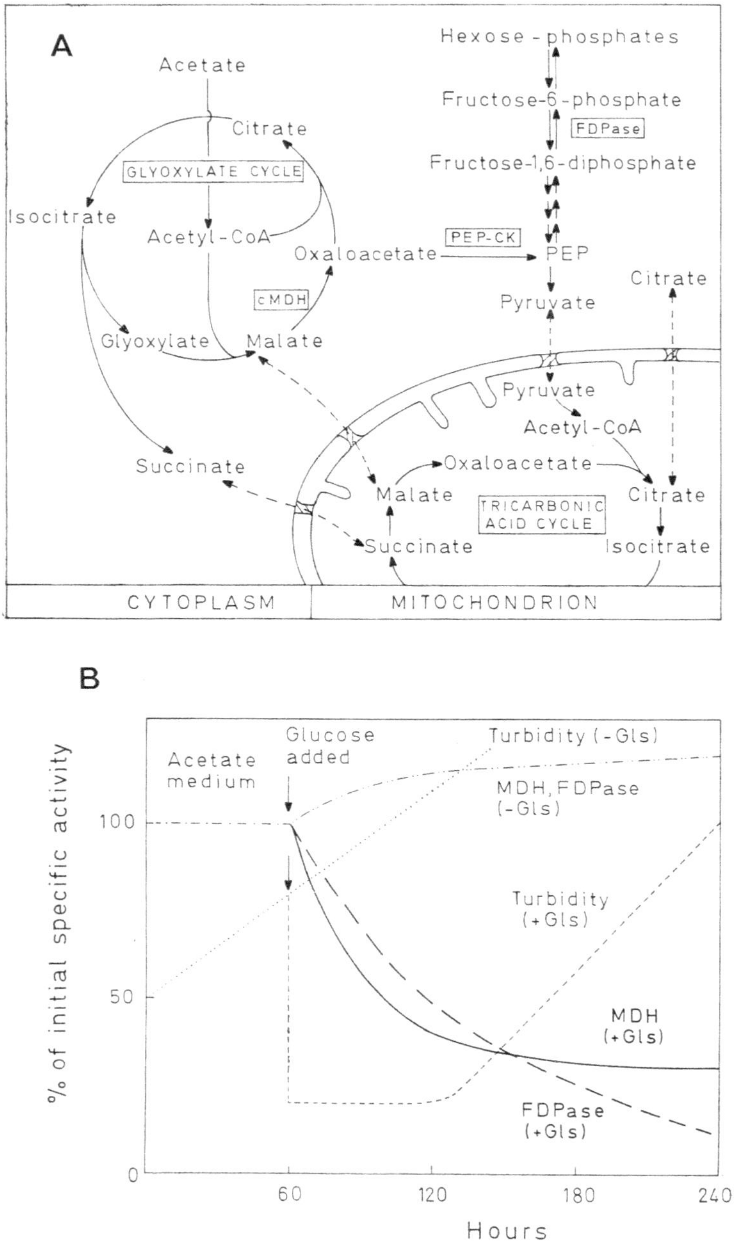


**Fig. 4.34:** Glucose inactivation of key enzymes of the glyoxylate cycle and gluconeogenesis in *Saccharomyces cerevisiae* (from Neeff et al. [257] and Zubenko & Jones [258]).

**A:** Main rections of carbohydrate metabolism and the role of the three enzymes submitted to glucose repression: ***FDPase***, fructosediphosphatase; ***PEP-CK***, phosphoenolpyruvate carboxykinase and ***cMDH***, cytosolic malate dehydrogenase. The glyoxylate cycle and the glycolytic/gluconeogenetic pathway are connected with the mitochondrial tricarbonic acid cycle by the malate and pyruvate translocator systems in the mitochondrial membrane.

As shown in part A of Fig. **4.34**, all three enzymes represent control points of glucose metabolism programming the cell to gluconeogenesis. If in the presence of glucose glycolytic and gluconeogenetic pathways function simultaneously, futil cycles could occur leading to wasteful hydrolysis of ATP. One decisive and rapid step in the metabolic shift of the cell after glucose addition is the selective inactivation of the three key enzymes. This is illustrated for two of them (cMDH and FDPase) in part B of Fig. **4.34**. The rapid decrease of enzyme activity with half-lives of 20 min for cMDH and 61 min for FDPase is observed in vivo and in vitro. By using immunologic techniques and antibodies against cMDH the decline of enzyme activity was shown to coincide with a decrease of the immune reactive material, thus proving a proteolytic mechanism. Furthermore, the mitochondrial isoenzyme of MDH, representing about 30% of the overall activity (cf. the endpoint of the MDH level at 240 min), does not change its half-life under these conditions.

3. Cells undergo phases which require a drastic rearrangement of their protein pattern: most proteins are degraded, only few remain biologically intact. Other protein species are synthesized de novo and characterize the new metabolic or developmental state. But again the same observation: despite the mass degradation of proteins it is by no means unselective. Such phases of protein rearrangement are observed under conditions of carbohydrate or nitrogen starvation of cells, during bacterial sporulation (cf. **4**.5.3) or the aggregation of cellular slime moulds (cf. Chap. **17**) or of slime bacteria (cf. **4**.6.1), during seed maturation or germination in plants (cf. Chap. **12**) and during maturation of highly specialized cells in animals, e.g., of sperm cells (cf. **4**.6.4.2) or erythrocytes (cf. **4**.5.4).

4. There are few examples, discussed elsewhere in this book, for proteolytic events shown to trigger and control determinative sequences. There are (1) the inactivation of the phage $\lambda$ *cI*-repressor and the subsequent transition of the phage from the lysogenic to the lytic pathway (cf. Fig. **14**.10), (2) the programmed proteinogen processing which controls phage T4 morphogenesis and involves the processing and/or degradation of many proteins including a phage-coded protease (cf. Table **14**.5), (3) the spatially and temporarily restricted proteolytic activation and inactivation of the chitin synthesizing machinery during yeast budding (cf. **4**.4.5.2) and (4) the proteolytic capacitation of sperm cells (cf. **4**.6.4.2).

How can the high selectivity and adaptability of protein degradation be explained? Although, with the exception of the few examples mentioned in the last point, the detailed regulatory mechanisms and the enzymes involved are still obscure, some general statements can help to answer this question.

1. Selectivity is not due to the presence of a large number of protein species-specific proteases, though we must be aware that the biochemical characterization of the intracellular proteolytic apparatus is just beginning and thus far from complete. However, it is already evident that the combination of proteases with cell-specific inhibitors, the mutual inactivation of proteases and inhibitors and the cellular compartmentation of substrates, inhibitors and proteases are decisive components in the network of control and selectivity of intracellular proteolysis (cf. Table **1**.12).

2. An inherent part of the biological information carried by the base sequence of a particular gene and transferred to the corresponding protein molecule is information about the type, conditions and velocity of its proteolytic inactivation. Half-

---

◀

**B**: Growth and proteolytic inactivation of FDPase and cMDH after adding glucose *(Gls)* to the medium (2% final concn.). Growth on the acetate medium proceeds with a doubling time of about 5 h. After dilution with the glucose medium (marked by *arrows*) growth is resumed after 1 h with a doubling time of 3 h. Total malate dehydrogenase activity comprises 70% cytoplasmic isoenzyme, which is rapidly inactivated, and 30% mitochondrial enzyme which is not inactivated. The half-lives calculated for cMDH and FDPase in glucose medium are 20 and 61 min respectively.

lives of proteins are very different (Table 4.12) and can selectively change under appropriate cellular conditions (Figs. 4.33 and 4.34).

It is interesting to recall that half-lives of $\beta$-galactosidase in *Escherichia coli* can vary from a few minutes to several hours depending on minor changes in the normal amino acid sequence caused by mutations in the structural gene (*lac* z gene, cf. Fig. 4.9). Further, if amino acid analoga (e.g., canavanine instead of arginine) are incorporated into *E. coli* proteins these "abnormal" forms are much more rapidly degraded than the normal ones. Similar results were reported for mammalian cells [250, 251].

3. Susceptibility of proteins against the first proteolytic attack is decisively influenced by the protein conformation and proper localization within the cell. Very likely, covalent modifications and/or removal of modifying substituents are important in this respect. Moreover, there are numerous reports on enzyme stabilization by addition of substrates, cosubstrates or allosteric effectors, by changing the ion or *p*H milieu, but also by incorporation into their native multimeric complexes (association of subunits or formation of multiprotein complexes).

A striking example in this respect is the rapid degradation of the otherwise totally normal $\alpha$-subunits of the hemoglobin tetramer ($\alpha_2\beta_2$) in patients with certain types of $\beta$-thalassemia (cf. Table 4.14). Because of a deletion of the $\beta$-subunit gene or because of a defect in the $\beta$-mRNA, synthesis of the $\beta$-subunit is very low whereas that of the $\alpha$-subunit is normal. The surplus of $\alpha$-chains, not incorporated into the holoprotein, is constantly removed by degradation.

4. Hence, the initial specific event of protein breakdown is the, frequently reversible, labilization of the protein structure (step 1, cf. the preceding point) followed by ATP-dependent covalent modification (cf. 1.3.6) and by limited proteolysis to generate large protein fragments (step 2). In eukaryotic cells these fragments are mostly taken up into the lysosomal compartments (step 3) and further degraded (steps 4, 5) to the amino acid building blocks [259]. This compartmentation contrasts to *E. coli*, where the ATP-dependent proteinase is membrane-bound and the final degradation presumably proceeds in the cytosol [260].

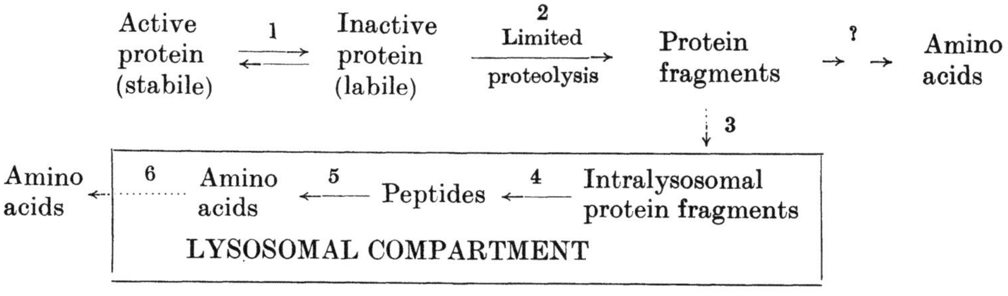


## 4.5 Gene Expression Programs

### 4.5.1 Mechanisms for Coordination of Gene Expression Programs

A prominent aspect of cell differentiation is the coordination of gene expression into programs. In many cases these are triggered by extracellular signals but afterwards proceed largely uninfluenced in a temporarily and spatially determined manner. Despite many details known about individual steps of gene expression, data about the regulatory organization of such programs are very scarce and incomplete. This is especially true for gene expression programs in higher eukaryotic organisms. Hence, it seemed appropriate to summarize in a hypothetical scheme the known experimental data about programming mechanisms in different systems and afterwards to illustrate the present state-of-the-art by three actual experimental systems. Many further examples are discussed in the special chapters of this book, e.g., Chap. **6**, **11–14**, **16** and **17**.

Construction of the hypothetic program (Fig. **4**.35) involves experimental data on

- the programmed gene expression during multiplication of phages (cf. **4**.4.2.6 and Chap. **14**) and animal viruses (cf. **4**.4.5.3),
- signal systems and their essential role for cell differentiation (cf. Table **4**.2),
- the role of substrates, endproducts, hormones and light for the control of protein adaptation in bacteria (cf. **4**.2.3.2, **4**.4.2.2, **4**.4.2.7 and Chap. **5**), plants (cf. Chap. **6**) and animals (cf. Chaps. **7**, **8**),
- the role of regulatory proteins (cf. summary in Table **4**.2) and RNA (cf. **4**.4.3.2 and **4**.4.4.4) for differential gene expression and its coordination and
- the tight intracellular and intercellular integration of single steps of differential gene expression into the metabolic situation and developmental context of cells and of the whole organism (cf. **4**.6 and Chap. **8**, **13** and **17**).

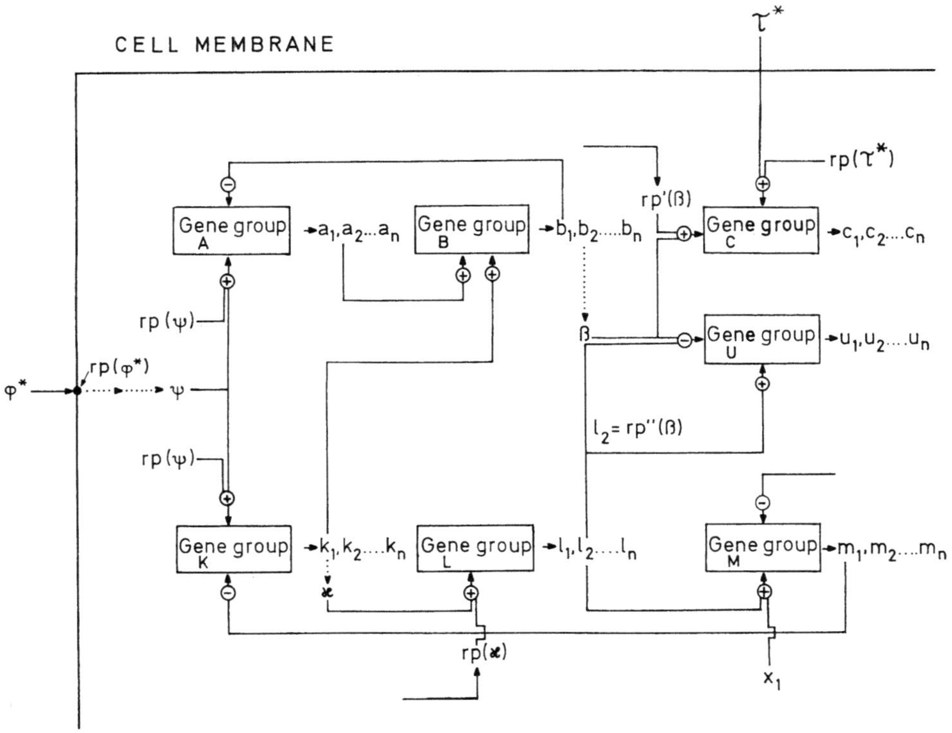


**Fig. 4.35:** Hypothetic scheme showing the integration of individual steps of gene expression into programs.

Gene group A, gene group B etc. in boxes denote expression of gene groups A, B etc.; → $a_k$, $b_k$ etc., formation of regulatory proteins or enzymes whose products might act as effectors in the program; --> $\beta$, $\varkappa$ etc. formation of such effectors; r, receptor proteins that interact specifically with intracellular ($\varkappa$, $\psi$) effectors or extracellular signals ($\varphi^*$, $\tau^*$); —⊕→, —⊖→, positive and negative regulatory influence on the expression of distinct gene groups.

The following principles for coordination of gene expression programs are noteworthy:

1. Expression of the gene groups depicted in Fig. **4**.35 may involve processes as complex as those described in preceding sections (cf. Chap. **1** and Sects. **4**.2–**4**.4). The term "gene group" is used to indicate genes that are connected by regulatory means but are not necessarily parts of one transcription unit. They can even map at distant sites of the genome.
2. Regulatory connection of the gene expression chains is brought about by regulatory proteins, RNA or effectors that are themselves products of preceding steps of gene expression within the same program.
3. Besides these, signals from the extracellular surrounding can trigger the whole program (see $\varphi^*$) or are required for the going on of distinct subprograms ($\tau^*$, subprogram C). They either enter the cell and bind to intracellular receptors (e.g., substrates, steroid hormones) or they combine with receptors of the outer membrane (e.g., non-steroid hormones, mitogens). The processes of signal transformation generate intracellular "second messengers" (see formation of effector $\psi$ triggered by $\varphi^*$ and the examples for signal transformation in Table **4**.2).
4. Expression of individual gene groups in most cases is subject to multiple positive and/or negative control which may affect all steps of the gene expression chain.
5. Regulatory RNA, proteins or effectors may exert multiple effects on the expression of several gene groups. Thus, protein $l_2$ is assumed to influence gene groups U and M in a positive manner, yet in the presence of effector $\beta$ it inhibits expression of gene group U.
6. Regulatory effects can be additive (cf. those on gene groups B and C) or subtractive (gene groups A, K, M, U). But regulatory factors can act also alternatively as denoted for proteins $l_2$ and $x_1$ and their role for expression of gene group M. This is evidently a wide-spread phenomenon accounting for the integration of the same gene group into several independent programs.
7. Subprograms can proceed autonomously (K, L, M), but frequently they exhibit a regulatory interdependence with other subprograms (cf. the action of proteins $k_1$ and $l_2$ on A, B, C).
8. The competence of cells to react appropriately on extracellular signals or intracellular effectors depends on the presence of the corresponding receptor proteins that are themselves products of preceding gene expression programs.
9. In most cases regulatory influences on gene expression evoke quantitative changes in the pattern of gene expression products rather than all-or-none effects.
10. Two typical phenomena frequently described in developmental biology are intimately connected with mechanisms for coordination of gene expression programs, i.e., competence and determination:
    a) As emphasized in point 6 the competence of cells to respond to signals or effectors requires the prior formation of signal transformation systems and/or intracellular receptor proteins. Furthermore, in most cases the simultaneous action of several independent regulatory systems is indispensible for gene expression to proceed.
    b) Though as a rule most steps of such programs are freely reversible, the sum of all events may lead to practically irreversible changes of the cell type which are determined by the environment and by the "differentiation history" of the cell. This preformed regulatory network, normally, can only be changed by drastic means leading to abnormal differentiation during ontogenesis

(cf. **4**.6.3 and **4**. 6. 4.1) or to carcinogenesis (cf. Chap. **18** and **19**). But such changes are also observed when parts of a tissue or cells are cultivated in vitro (cf. Chap. **11** and **18**) or in cases of cell transplantation (cf. **19**.2.3.2), cell fusion or transplantation of nuclei (cf. **16**.2 and **19**.2.3.2).

### 4.5.2 Sequential Enzyme Induction

Sequential enzyme induction is a simple form of gene expression program. It is wide-spread in microorganisms, but presumably occurs also in higher plants and animals. It is convenient to illustrate the principles by using the *lac* system of *Escherichia coli* as example whose regulatory details have been discussed in a preceding section of this chapter (cf. **4**.4.2.2). The system (Fig. **4**.36) is characterized by coupling of the *lac* and *gal* operons (step 1–5) and by a feed-back control loop by catabolite repression (steps 7–9). Steps 1, 4, 7, 8 and 9 of the reaction sequence (Fig. **4**.36) are not only necessary parts of lactose degradation but must be considered as parts of a signal transformation chain which serves the programming and metabolic inte-

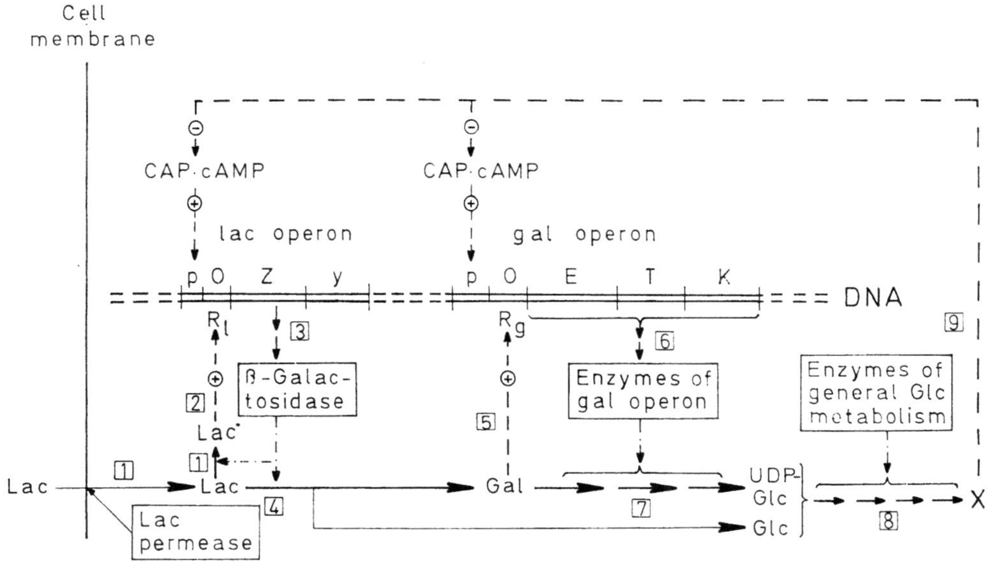


**Fig. 4.36**: Sequential enzyme induction and signal transformation accompanying lactose degradation in *Escherichia coli*.

**1**: Lactose *(Lac)* is taken up from the extracellular milieu by Lac permease (coded by gene *lac* y). By the low level of constitutive β-galactosidase it is transformed into allolactose *(Lac·)*. **2**: Binding of Lac· to the *lac* repressor *($R_1$)* causes the dissociation of the repressor-operator complex and thus promotes transcription of the structural genes of the *lac* operon (**3**). **4**: β-Galactosidase cleaves both inducer *(Lac·)* and substrate *(Lac)* to produce glucose *(Glc)* and galactose *(Gal)*. **5**: Galactose acts as inducer of the *gal* operon (**6**), whose gene products catalyze the conversion of Gal into UDP-Glc (**7**). Glc and UDP-Glc are further degraded by enzymes of the general carbohydrate metabolism (**8**). **9**: Unknown intermediates *(X)* of glucose degradation cause the decrease of the intracellular cAMP level possibly by feed-back inhibition of the Glc uptake system (cf. Fig. **4**.2). The lacking cAMP. CAP complex restricts the transcription initiation at the *lac* and *gal* promoters. For further details cf. **4**.4.2.2. →, metabolic chains; ---⊕-->, ---⊖--> denote positive or negative regulatory influences on the transcription of the pertinent gene groups; — · — · → symbolizes the catalytic action of enzymes.

gration of the whole system (cf. legend to Fig. 4.36). It is worth recalling that (1) this program of enzymic adaptation is self-limiting by the degradation of the inducers and by the mechanism of catabolite repression, (2) the cell is competent to respond to the initial signal (Lac) only if the glucose supply is limited and, hence, the cAMP · CAP level is high and (3) lactose addition to the *E. coli* culture causes a transition to a metabolic state which is basically different from that of the original culture.

### 4.5.3 Bacterial Sporulation

Much more complex gene expression programs proceed during formation of endospores characteristic of several bacteria including members of the genus *Bacillus*. When an essential metabolite in the growth medium is depleted these bacteria respond by beginning an ordered sequence of biochemical and morphological alterations that are part of the sporulation-specific gene expression programs (Fig. 4.37). Vegetative growth and cell divisions cease. Instead, an asymmetric cell division generates a new developmental entity of two cells (Fig. 4.37, stages 0–III): the mother cell, and the forespore within the cytoplasm of the mother cell. The forespore is unique among microorganisms in being surrounded by two membranes with opposite surface polarity. The cytoplasmic sites of the two membranes face the cytoplasm of the mother cell and the forespore, the intermembrane space being a non-plasmatic compartment. With the discovery of methods for isolating forespores from sporangia at practically all stages of the development, it became more and more apparent that forespore formation and maturation depend on a coordinate division of labour between the two compartments of the sporulating cell. This situation reminds very much of the functional and developmental interactions between the cell nucleus and the genetically semiautonomous organelles (chloroplasts and mitochondria) in eukaryotic cells (cf. Chap. 6).

The mature spore released in stage VII after lysis of the mother cell (Fig. 4.37) is extremely well equipped for its cryptobiotic stage [264]. Bacterial endospores are among the most resistant forms of life with a remarkable tolerance of heat, radiation and bactericidal agents. They exhibit no detectable metabolism and yet a "sensory apparatus" of the spores remain alert that enables them to respond within minutes to germination promoting signals and to reinitiate the machinery for vegetative growth. All these typical properties (resistance, long-time dormancy and germinability) have their origin in the peculiarities of biochemical equipment and morphological construction developed during sporulation. Figure 4.37 summarizes prominent morphological and biochemical events. This tentative correlation of data gives an idea of the temporal sequence, but cannot explain the causal relations within the gene expression programs.

Extensive genetic investigations have revealed a great number of loci involved in the sporulation program. At least nine sporulation-specific operons for stage 0–I, seven for stage II, five for stage III, seven for stage IV and five for stage V have been reported, i.e., a total of 30 to 40 operons with several hundreds of genes [262, 265].

Mutant studies have indicated very clearly that (1) the expression of distinct stage-specific products depends on the completion of the preceding stage and (2) most gene products are required only at the stage they were assigned to. Thus, a temperature-sensitive mutant blocked at stage 0 shows a peak of temperature sensitivity only at this stage. If the mutant was allowed to pass stage 0 at the permissive temperature sporulation afterwards proceeds normally even at the non-permissive temperature. The biochemical characterization of mutants also provided data on the temporal correlation between morphological and biochemical events as indicated in Fig.

4.37. Stage 0 mutants were found to be deficient in several of the early events, e.g., ppApp synthesis, formation of intracellular and extracellular proteases, synthesis of peptide antibiotics. On the other hand, mutant analysis showed the increase of alkaline phosphatase and glucose dehydrogenase as resulting from differential gene expression at the transition between stages II to III and III to IV respectively. Finally, studies with temperature sensitive mutants of DNA replication revealed the triggering of sporulation to be cell cycle-dependent [266]. The competence state occurs about 15 min after initiation of replication. Evidently, induction of sporulation takes place only when the replication fork reaches a specific site or a specific segment of the chromosome coding for a gene involved in the transition to the sporulation program (stage 0 gene).

It was clear from early observations that sporulation is subject to carbon and nitrogen catabolite repression, i.e., sporulation starts if the source of a rapidly metabolizable carbon and/or nitrogen compound is exhausted. Further, addition of such compounds to sporulating cultures reverses sporulation until the time of commitment is passed which is likely to be the stage of forespore completion (stage III). Commitment is a property of the particular two-cell entity of mother cell plus differentiating forespore, whereas isolated forespores, at least after they have acquired osmotic stability (stage V), always respond to growth media with germination. Carbon and nitrogen catabolite repression are evidently mediated by two independent signal systems [270–272]. Nitrogen starvation results in a rapid decrease of intracellular GTP level and possibly involves elements of the glutamin synthetase system (cf. Fig. **4**.3). At least many glutamine synthetase defective mutants are also defective in sporulation. In contrast, triggering of sporulation by carbon starvation coincides with a rapid decrease of intracellular concentration of guanosine-3′,5′-tetraphosphate (ppGpp) and a corresponding increase of adenosine-3′,5′-tetraphosphate (ppApp) or related compounds, e.g., pppAppp.

An enzyme that synthesizes these highly phosphorylated adenosine nucleotides was found associated with cytoplasmic membranes of both vegetative and sporulating cells [272]. During vegetative growth its activity is inhibitited by phosphorylated metabolites of energy-rich carbon sources. Genetic evidence points out that one of the stage 0-specific operons (*spo* OF) codes for the $p_3Ap_3$-synthetase. It is a stimulating idea that the $p_3Ap_3$-synthetase in the cytoplasmic membrane might be coupled to the sugar uptake system, the phosphoenol-pyruvate: sugar phosphotransferase system, as suggested for the signal transformation chain involving adenylate cyclase in *Escherichia coli* (cf. Fig. **4**.2). The decreasing rate of sugar uptake leaves the protein components of the transport system in a highly phosphorylated state. Phosphorylation of $p_3Ap_3$-synthetase or direct phosphoryl group transfer to ATP may be the cause of increased $p_3Ap_3$ formation.

The physiological effects of the highly phosphorylated adenosine nucleotides are barely known. $p_3Ap_3$ was found to strongly inhibit RNA polymerase. This result is in keeping with observations about a stepwise specialization of RNAP during sporulation by removal of the $\sigma$-subunit and association with novel proteins. Though early observations about a proteolytic modification of RNAP have all been shown to result from experimental artifacts due to the increased protease activity in sporulating cultures, it is now well established that sporulation-specific RNAP lacks the $\sigma$-factor and contains additional subunits. Thus, an early step of the sporulation program might be the inhibition of the vegetative type of RNAP (e.g., by $p_3Ap_3$) and subsequent specialization to recognize sporulation-specific promoters. This specialization was clearly documented by the resistance to phage infection (e.g., phage SP01, cf. **4**.4.2.6) acquired by sporulating cultures during stages 0–II and by the subsequent finding that RNAP from such cultures was incompetent for in vitro transcription of the phage genome [273]. Meanwhile a DNA fragment near the replication origin of the *B. subtilis* genome was cloned (Fig. **4**.38). It contains a vegetative and a sporulation-specific transcription unit. As analyses of suitable mutants have shown, expression of the latter transcription unit in intact sporangia is clearly

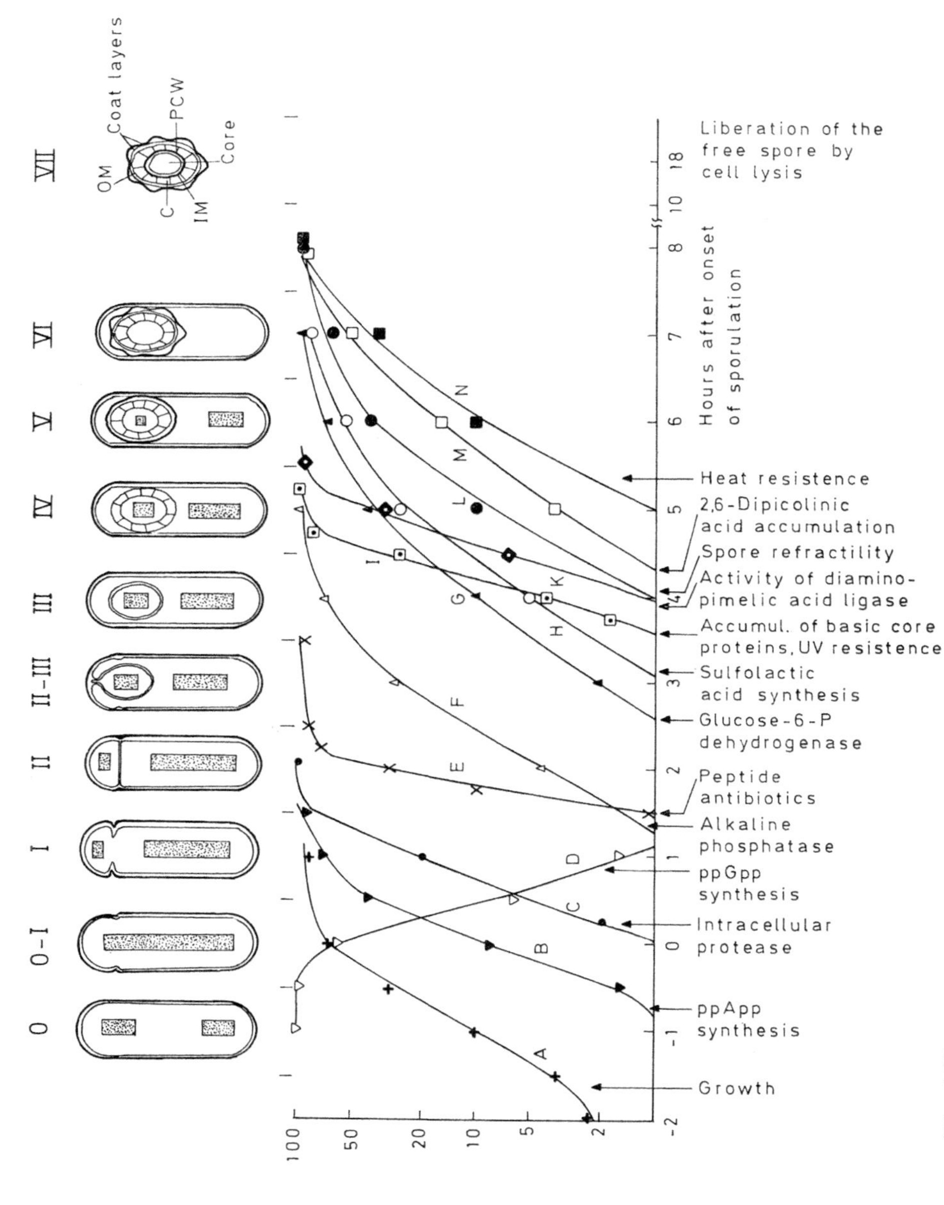
0
0-I
I
II
II-III
III
IV
V
VI
VII
Coat layers
PCW
Core
OM
C
IM
100
50
20
10
5
2
-2
-1
0
1
2
3
4
5
6
7
8
10
18
Hours after onset of sporulation
A
B
C
D
E
F
G
H
I
K
L
M
N
Liberation of the free spore by cell lysis
Heat resistence
2,6-Dipicolinic acid accumulation
Spore refractility
Activity of diamino-pimelic acid ligase
Accumul. of basic core proteins, UV resistence
Sulfolactic acid synthesis
Glucose-6-P dehydrogenase
Peptide antibiotics
Alkaline phosphatase
ppGpp synthesis
Intracellular protease
ppApp synthesis
Growth

Fig. 4.37

◀

**Fig. 4.37**: Morphological and biochemical events reflecting gene expression programs during sporulation of *Bacillus* spp.

*Morphological stages* (from Hoch [261], Doi [262] and Ellar [264]): The upper part shows typical morphological features beginning with the vegetative cell *(stage 0)* and ending with the mature spore *(stage VII)*. The earliest visible change of the vegetative cell with two nucleoids is an altered conformation of the two genomes which form an axial DNA filament *(stage 0–I)*. At stage I cell wall spikes indicate the site where the double-membraned septum separates the mother cell and forespore compartments *(stage II)*. At transition to *stage III* the forespore compartment is gradually engulfed within the mother cell compartment by continued proliferation of the septum membranes and translocation of the septum base towards the cell pole *(stage II–III)*. The separated forespore of *stage III*, in some species, e.g., *B. sphaericus*, contains a primordial cell wall of vegetative-type peptidoglycan synthesized during engulfment by the forespore compartment and deposited between the two membranes. However, the main part of the cortex consists of a spore-specific peptidoglycan that contains diaminopimelic acid instead of lysine as cross-linking amino acid. It is secreted into the intramembrane space by the mother cell compartment *(stages IV* and *V)*. Synthesis of the sporulation-specific diaminopimelyl ligase proceeds in *stage IV*. Protein deposition (spore coat) starts in *stages III* and *IV*, but the bulk of the proteinaceous coat is formed during *stages V* and *VI*. The highly hydrophobic coat protein ($M_r$ 13.5 kd in *B. subtilis*) finally comprises 25% of the total spore protein. It is synthesized as a precursor of $M_r$ 25 kd in the very early stages (1–2 hours after the onset of sporulation) and is later on processed (starting at 3–4 hours of sporulation). The liberated spore *(stage VII)* is shown as a composite diagram drawn from a variety of *Bacillus* spp. with the characteristic features of dormant spores: *IM*, inner membrane; *PCW*, primordial cell wall; *C*, cortex; *OM*, outer membrane; and the coat protein layers.

*Biochemical parameters:* The composition of the morphological and biochemical traits in one figure suffers from the fact that the data stem from different papers and were obtained with different *Bacillus* spp., though mostly strains of *B. subtilis* were used. The exact time scale of sporulation varies with the organism and the experimental conditions. Thus, the figure was constructed to show an approximate temporal correlation only between events of the sporulation program.

**A**: growth (absorbance at 600 nm), 100 = $1 \times 10^7$ cells $ml^{-1}$;
**B, D**: $p_2Ap_2$ and $p_2Gp_2$ synthesis: arbitrary values of in vitro synthesis by incubating ribosomes with [$^3$H]-ATP and $^3$[H]-GTP as label [271];
**C**: intracellular serine protease; 100 = 500 units per mg protein, determined by Azocoll hydrolysis [269];
**E**: peptide antibiotic accumulation exemplified for tyrocidine in *B. brevis* cultures [278], 100 = 25 $\mu$g per ml culture;
**F**: alkaline phosphatase (100 = 14.2 U/mg protein);
**G**: glucose dehydrogenase (100 = 28 U/mg protein);
**H**: accumulation of basic core proteins in *B. megaterium* [277] coincides with the generation of resistance to UV irradiation (100 = 45 $\mu$g of 3% acetic acid soluble protein per mg dry weight);
**I**: specific activity of meso-diaminopimelic acid ligase in *B. sphaericus* (100 = 0.15 units per 100 ml culture) [267];
**K**: synthesis of sulpholactic acid (arbitrary units of [$^{35}$S]-sulphate incorporation);
**L**: refractility, indicates cortex formation (100 = 80% refractile spores);
**M**: 2,6-dipicolinic acid accumulation (100 = 28 $\mu$g/mg protein and
**N**: heat resistance (100 = 80% heat-resistant spores).

The data for F, G, K–N are from Wood, D. A.: Biochem. J. *123* (1971): 601.

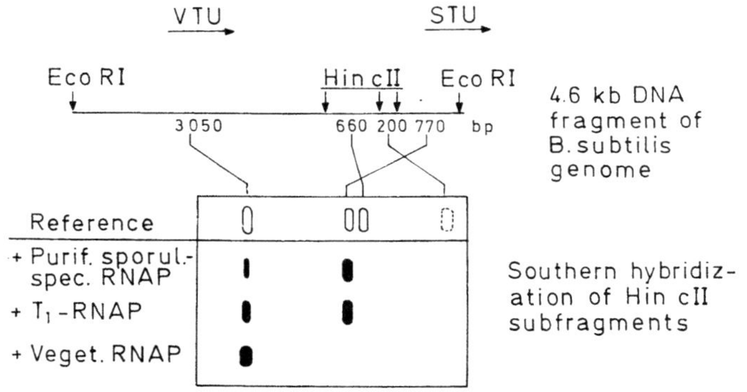


**Fig. 4.38:** In vitro transcription of a 4.6 kb fragment of *Bacillus subtilis* by vegetative and sporulation-specific RNAP (from Haldenwang & Losick [274]).

The cloned Eco RI fragment of the *B. subtilis* genome stems from the region near the origin of replication. It contains two transcription units, one transcribed during growth and sporulation (vegetative transcription unit, *VTU*) and one transcribed only early during sporulation (sporulation-specific transcription unit, *STU*). Digestion of the fragment with restriction enzyme Hin cII generates four subfragments that were separated by electrophoresis and hybridized by the Southern procedure (cf. **2**.6) to in vitro synthesized labelled mRNA. The latter was obtained by incubating hybrid plasmid DNA with the integrated 4.6 kb fragment in the presence of all necessary cofactors, [$\alpha$-$^{32}$P]-UTP and the indicated types of RNA polymerases. It is evident that the vegetative RNAP transcribes only VTU (hybridization with the 3050 bp subfragment) whereas early during sporulation (stage I) the RNAP is modified and thus acquires specificity for STU (hybridization with the 770 bp subfragment). This effect is most pronounced with the partially purified sporulation-specific RNAP.

under the control of stage 0 sporulation operons. In vitro transcription of this transcription unit is only possible with RNAP from sporulating cells (Fig. **4**.38).

Purified RNAP is devoid of the $\sigma$-factor and contains stoichiometric amounts of a new subunit with $M_r$ 37 kd [274]. It is very inefficient in transcribing the vegetative transcription unit contained in the same DNA fragment (Fig. **4**.38). The in vitro system is highly selective not only with respect to the transcription units and to the adjacent regions but also for one strand only of the sporulation-specific transcription unit. Interestingly, the gp 37 is not itself a product of one of the stage 0 operons. It is found in vegetative cells as well. Possibly, gp 37 directs the transcription of several classes of genes but its involvement in sporulation-specific RNA synthesis is controlled by stage 0 gene products.

At any rate, bacterial sporulation serves as another example for the control of a gene expression program by specialization of RNAP to the recognition of distinct promoter types (cf. **4**.4.2.6). There is experimental evidence for other parts of the gene expression chain, e.g., ribosomes [271, 275], tRNA composition [276] or the proteolytic system [266, 269], to be also adapted to the special needs of the sporulating cell. It is worth mentioning in this connection that a considerable amount of two or three small basic proteins ($M_r$ 7–10 kd) with DNA-binding affinity accumulates in the forespore core [277]. They constitute up to 50% of the total core protein and are rapidly degraded during germination by preformed proteases. Such basic proteins might serve simultaneously as means for the advancing condensation of the spore genome, as evidenced by the acquisition of resistance to UV-light, and as protein reserves.

One final aspect of the sporulation program deserves notice: Several types of secondary products are formed at different stages (Fig. **4**.37), e.g., peptide antibiotics, sulfolactic acid, dipicolinic acid. A role of the peptide antibiotics was suggested in the inhibition of RNAP early during sporulation, but possibly also as hormone-like factors acting during spore germination [278]. Nothing is known about a role of sulfolactic acid. However, dipicolinic acid (DPA) is essential for conferring heat resistance to the dormant spore as shown by mutants defective in DPA synthesis [264]. DPA is evidently formed in the mother cell compartment from lysine as precursor. It is deposited exclusively in the forespore compartment as a 1:1 complex of $Ca^{2+}$-DPA. DPA serves as an ion trap for $Ca^{2+}$-ions that are actively transported from the outer medium into the mother cell compartment and pass by faciliated diffusion into the forespore. Finally, $Ca^{2+}$ makes up to 2% of the spore dry weight.

To summarize, the sporulating bacterial cell represents a highly efficient differentiating system of two cells whose development depends on the coordinate expression of hundreds of genes within the sporulation program. Though genetic and molecular biological experiments have provided us with numerous valuable data about the program, we are at the very beginning of a causal analysis. A major problem of the near future will be to identify sporulation-specific genetic loci with their gene products and to describe further essential elements of the regulatory backbone of these gene expression programs.

### 4.5.4 Programmed Gene Expression During Erythroid Cell Specialization in Mammals

Formation of mature erythrocytes, the mammalian oxygen transport vehicles of the blood, represents one of the most striking examples of cell specialization known in nature. Erythropoiesis (Fig. **4**.39) begins with a pluripotential stem cell that, under the influence of unknown factors, gives rise to a group of secondary stem cells committed to produce all cell types circulating in the blood. The erythroid lineage itself starts with the burst forming unit erythroid cell (BFU-E), involves several intermediate cell types and, finally, leads to the mature, anucleate erythrocytes containing only the subcellular components necessary for long-term survival (~ 90 days) and oxygen transport.

Erythropoiesis requires highly complex gene expression programs. In mouse, a minimum of 17 genetic loci mapping on more than 10 different chromosomes so far have been shown to participate [279]. They were identified by corresponding mutants exhibiting distinct types of anaemias: (1) mutants defective in the early erythropoietin-dependent phases of erythroid differentiation; (2) mutants defective in iron utilization, (3) mutants with greatly reduced erythrocyte life-span and (4) mutants with defects in the globin genes (hemoglobinopathies). As will be outlined in the following these heterogenous mutant types reflect defects in some important parts of the overall network of gene expression programs.

#### 4.5.4.1 Hemopoietic Stem Cells and Erythroid Homeostasis

Despite a very high proliferative potential, most of the pluripotential, hemopoietic stemm cells as well as their derivatives committed to undergo erythroid differentiation (BFU-E) do not divide. They are resting cells in the $G_0$ phase of the cell division cycle (Fig. **4**.39). Under normal conditions only a small part of them divide to maintain the steady state population of cells in the same compartment (proliferative mitoses) but also in the following compartments of the erythroid lineage (quantal mitoses, cf. **15**.2). One important means to achieve this erythroid homeostasis is a feed-back control by erythropoietin (EP), a glycoprotein with $M_r$ 46 kd.

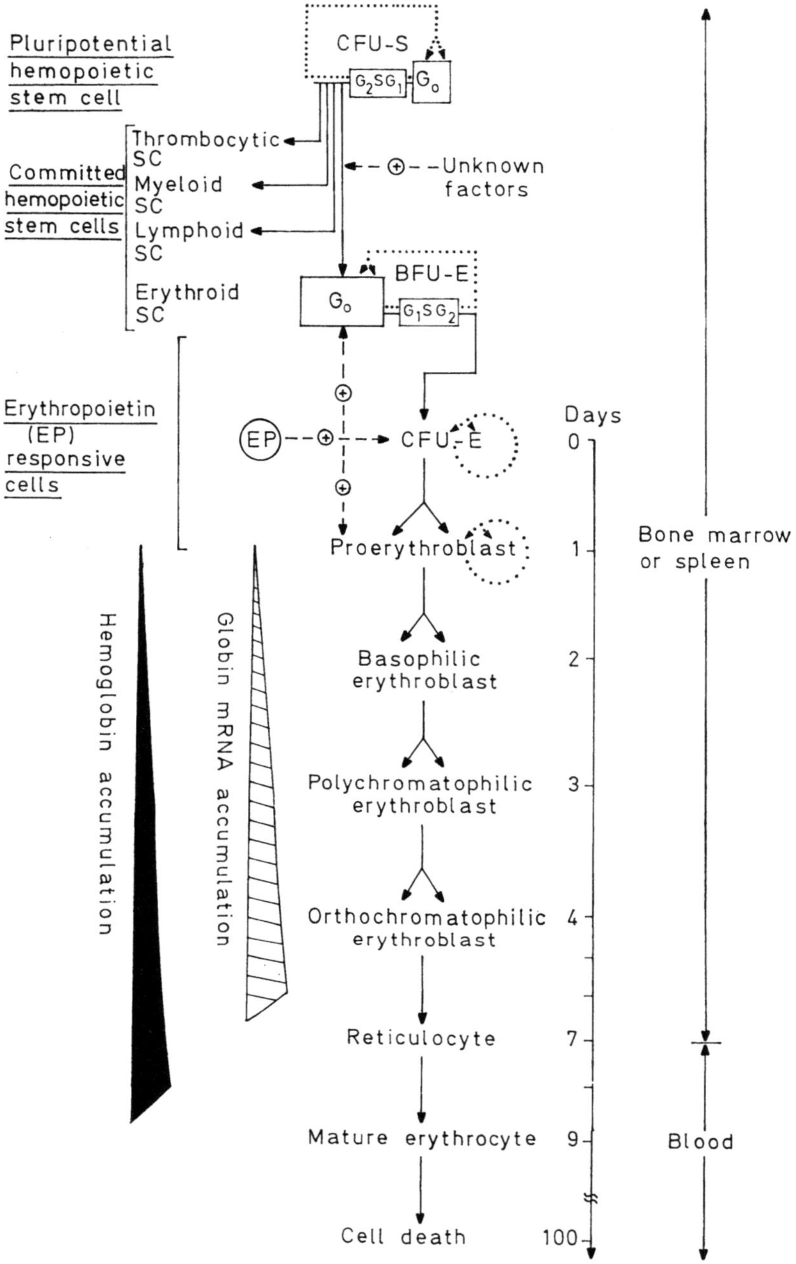


**Fig. 4.39:** Determination and cell maturation in the erythroid cell lineage of mammals (compiled from Rosenthal et al. [282] and Russell [279]).

In the bone marrow and spleen the pluripotential stem cells (*CFU-S*, colony forming unit of spleen) give rise to the committed stem cells of all types of cells circulating in the blood. Earliest precursors of the erythroid lineage are the burst forming unit erythroid cell *(BFU-E)* and the colony forming unit erythroid cell *(CFU-E)*. Factors controlling formation of the committed stem cells are largely unknown. However, the first steps in the erythroid lineage are promoted by the proteohormone erythropoietin *(EP)*: Low doses of EP stimulate rapid proliferation and

Secretion and synthesis (?) of erythropoietin by the kidney is controlled by the oxygen supply. Reduction of the mass of functional erythrocytes in the blood, e.g., by excessive bleeding, chemical destruction of red cells or destruction of the precursor cells by X-ray irradiation, reduces the oxygen supply to the kidney and increases EP secretion [279].

EP has at least three effects (Fig. **4**.39): (1) it promotes maturation of erythroblasts, (2) it enhances generation of proerythroblasts from the CFU-E cell compartment and (3) it increases formation of CFU-E from the BFU-E compartment. The first two responses are rapid and dependent on low doses of EP, the last is a delayed response to high doses of EP. Hence, cell proliferation and maturation in the erythroid lineage is subject to a feed back control system mediated by the oxygen level in the kidney and the resulting fluctuations of the EP serum levels.

The hemopoietic stem cells exhibit a number of typical features used to define these otherwise ill characterized cell types (cf. summary by Lajtha [280]):

1. Despite a high proliferative potential most of the cells in the stem cell compartments are in a non-cycling state ($G_0$ phase, cf. **15**.1). Thus they are capable of responding rapidly to a depletion of cells in the same or in one of the following compartments. There is evidence for special proliferation control factors in the stem cell microenvironment [280].
2. Under normal steady state conditions amplification of the cell population within a lineage is not achieved by proliferation of cells within the stem cell compartment but rather by intercalation of amplifying transit populations between the self-maintaining stem and the end cells (e.g., CFU-E and proerythroblasts in Fig. **4**.39). The main amplification in the erythroid lineage resides with the EP-sensitive cell types (about $10^4$-fold).
3. By quantal mitoses stem cells give rise to daughter cells, determined to undergo subsequent steps in the cell specialization program.
4. An additional property of the pluripotential hemopoietic stem cell (CFU-S) is its ability to respond to certain unknown stimuli inducing the formation of committed stem cells for different cell lineages (Fig. **4**.39).

It is an important contribution towards a deeper understanding of hemopoiesis that, in addition to the kidney hormone erythropoietin, at least four further hemopoietic hormones have been detected in mouse spleen [284]. They are evidently related glycoproteins of 23–35 kd, synthesized by actively proliferating T lymphocytes and characterized by their stimulation of colony formation in agar cultures of hemopoietic stem cells. Production of these factors requires both T lymphocytes and accessory adherent cells.

---

◀

further differentiation of CFU-E and proerythroblasts. High doses of EP lead to a late burst of erythroid cells by enhancing regeneration of CFU-E from BFU-E.

The terms denoting the advanced intermediates of the erythroid lineage stem from early cytological evidence and characterize the increasing functional inactivation of chromatin by condensation. Finally, the pycnotic nucleus of the orthochromatophilic erythroblast is extruded (reticulocyte). Globin mRNA accumulation starts in proerythroblasts and continues until the nucleus is totally inactivated, whereas extensive translation of globin mRNA is continued.

The *time scale* on the right refers to erythrocyte maturation in man. ⇢ proliferative mitosis; → quantal mitosis. Proliferative mitoses generate cells of the same compartment, whereas quantal mitoses produce cells that are different from the mother cell and belong to the next compartment, i.e., to the next step of cell specialization (cf. Chap. **15**). $G_0$, $G_1$, $S$, $G_2$, phases of the cell cycle (cf. Chap. **15**).

Fractionation revealed four types of hormones with specificity for the four cell lineages. Their mode of action can be envisaged in two directions: (1) They might exert an "instructive" influence on the pluripotential stem cell, directing the "quantal" mitoses to generate distinct types of committed stem cells. (2) Commitment itself is a prerequisite for recognition of these hormones, e.g., after synthesis of corresponding outer membrane receptors. The hemopoietic hormones promote the expansion of the committed cell clones (cf. the role of clone selection in the immune response, Fig. **13**.3).

These results once more emphasize the decisive role of intercellular cooperation for the coordinate procedure of cell specialization and the maintenance of cell homeostasis. They are confirmed by in vitro studies with long-term bone marrow cultures. It is evident that maintaining the integrity and self-renewal potential of pluripotential stem cells is dependent on an appropriate microenvironment provided by adjacent macrophages, epithelial cells and giant lipid-containing cells. The nature of these cell-to-cell contacts is unknown [283].

A special aspect of hemopoiesis is the twofold developmental shift of the blood cell generating tissue (Table **4**.7).

*Embryonic hemopoiesis* starts in the yolk sac blood islands. But as soon as the fetal liver has differentiated appropriately it becomes the main site of *fetal hemopoiesis*, whereas the shift to the final sites of *adult hemopoiesis* (bone marrow in man, bone marrow plus spleen in many other mammals) occurs only in the late stage of gestation and shortly after birth respectively. Interestingly, the three types of red cells produced differ not only in their size but also in the polypeptide composition of their globin molecules. As outlined in Table **4**.7, one switch of the $\alpha$-type globin genes and two switches of the $\beta$-type globin genes accompany the developmental shifts from embryonic to fetal and from fetal to adult hemopoiesis. Though definite proof is lacking, it is assumed that the hemopoietic stem cells are not generated de novo but invade the new hemopoietic tissue. At least under conditions of severe anemia, pluripotential stem cells of mice were found to migrate between bone marrow and spleen. Furthermore, after total destruction of stem cells by X-irradiation the bone marrow and spleen can be recolonized by intraperitoneal or intravenous injection of suspensions of intact marrow or spleen cells.

### 4.5.4.2 Friend Erythroleukemia Cells as Models of Red Cell Maturation

Erythropoiesis in susceptible mice is greatly enhanced after infection with the so-called Friend virus complex [285]. This consists of a defective virus (SFFV, spleen focus-forming virus) and at least one helper virus, e.g., MuLV-F (murine leukemia virus) required for the replication of SFFV. The erythroid transformation itself is brought about by the SFF virus and leads to an erythropoietin-independent proliferation of the erythroid precursor cells and to a block of subsequent steps of cell specialization. The target cells of the Friend virus complex are most probably committed precursor cells closely related to or identical with cells of the CFU-E compartment (Fig. **4**.39). Meanwhile a great number of different Friend erythroleukemia cell lines (FEL cells) have been obtained and cultivated in vitro. A great deal of experimental work concerning the details of cell differentiation during red cell maturation have been done with these convenient model systems [288].

The susceptibility of mice to Friend virus (Fv) infection was found to depend, inter alia, on a dominant allel *Fv-2*$^{s}$ at the *Fv-2* locus on chromosome 9. Mice, homozygoius for *Fv-2*$^{r}$, are resistant. Recently it turned out that the early genetic observations about resistance or susceptibility to Friend virus infection had in fact revealed a regulatory gene somehow controling the rate or probability of transition between the resting ($G_0$-state) and cycling states of cells of the BFU-E compartment and hence also the amount of precursor cells in the Fv-affected CFU-E compartment [286]. The *Fv-2*$^s$ allele keeps a high proportion of erythroid progenitor cells engaged in DNA synthesis in contrast to *Fv-2*$^r$/*Fv-2*$^r$ homozygotes with a relatively low proportion of BFU-E cells in the cycling state. Interestingly, the *Fv-2* locus has no significant effect on the cycling behaviour of the pluripotential stem cell (CFU-S) nor of other committed hemopoietic precursor cells (e.g., CFU-E). This is the first known and mapped mammalian gene involved in the control of a distinct type of quantal cell cycle.

The major evidence for this interpretation stems from [$^3$H]-thymidine suicide experiments [286]. Incorporation of high specific activity [$^3$H]-thymidine rapidly kills erythropoietic progenitor

**Table 4.13:** Induced cell differentiation in Friend virus erythroleukemia cells of mice (compiled from Marks & Rifkind [288] and Lowenhaupt & Lingrel [289] and Neumann et al. [295]).

The comparative listing of events following induction of Friend erythroleukemia cells suffers from the fact that most of them were determined in non-synchronized cultures and no single study has encompassed all parameters. Hence the exact order may vary depending on the experimental conditions and on strain-specific alterations. Early and late events are classified by two criteria: (1) they are separated by the point of commitment, i.e., the time point when terminal differentiation proceeds without the inducer being present and (2) by their inducibility by hemin (early events) and $Me_2SO$ (early and late events) respectively. In $Me_2SO$-induced cultures the early phase takes about 12–24 h.

It should be noted that most of the early events, though triggered before commitment, proceed also in the late phase (globin mRNA synthesis and translation, spectrin accumulation). Heme synthesis in Friend cells seems not to be regulated by $\delta$-aminolevulinate synthetase but rather by the following enzymes of the biosynthetic chain which, in contrast ot the former, show a coordinate increase about 24–48 h after induction. These are $\delta$-aminolevulinate dehydratase, uroporphyrinogen I synthetase, uroporphyrinogen decarboxylase, coproporphyrinogen oxidase and ferrochelatase. The last enzyme, serving the incorporation of iron into the preformed porphyrin moiety, is not detectable in uninduced cells [292, 293].

| | Inducers | | |
|---|---|---|---|
| | Hemin (75 μM) | Dimethyl sulfoxide (1.8%) | Increasing cell specialization |
| Early events (phorbol ester-sensitive phase) | | | ↓ |
| Histone acetylation (mainly H2A) | ? | + | |
| Increase of globin mRNA transcription | + | + | |
| Globin mRNA translation | (+) | + | |
| Spectrin accumulation | + | + | |
| — Commitment — | | | |
| Late events | | | ↓ |
| Terminal cell divisions | − | + | |
| Synthesis of IP25 (H1-like histone) | − | + | |
| Enzymes of heme synthesis | − | + | |
| Hemoglobin accumulation (positive bezidine reaction) | − | + | |
| Membrane antigens | − | + | |
| Decrease of RNA synthesis | − | + | |
| Cell cycle arrest | − | + | |
| Chromatin condensation | − | + | |
| Globin mRNA destabilization | − | + | |

cells if they undergo DNA synthesis. A high proportion of BFU-E cells was killed in *Fv-2*$^{s}$/*Fv-2*$^{s}$ and *Fv-2*$^{s}$/*Fv-2*$^{r}$ mice, but most BFU-E cells survived in *Fv-2*$^{r}$/*Fv-2*$^{r}$ mice because they are in the $G_0$ state.

A very low proportion of FEL cells (0.5%–1%) undergo spontaneous differentiation to late cell types of the erythroid lineage. However, the great value of FEL cells as model system stems from the fact that differentiation can be induced by many seemingly unrelated low molecular weight compounds with a hydrophilic group and a planar, hydrophobic portion, e.g., dimethyl sulfoxide ($Me_2SO$), acetamide or hexamethylene bisacetamide. But induction is also achieved by antibiotics, antimetabolites, some fatty acids and, last but not least, by hemin. Though the primary attack of these agents is surely very different, they all trigger a typical sequence of events (Table **4**.13) closely resembling the sequence observed also during normal erythroid cell maturation in the intact animal.

The gene expression programs following induction exhibit early and late events (Table **4**.13). The whole sequence can be induced by dimethyl sulfoxide but not by hemin. $Me_2SO$ is only required in the early phase during one round of DNA synthesis. Afterwards cells are committed and continue further maturation autonomously. In contrast, hemin induction results in the expression of some early events but not in commitment. Thus FEL cells reveal another important decision point in the erythroid lineage, which normally may be controlled by erythropoietin. Commitment during red cell maturation has no direct relation to the generation of the four types of committed stem cells from the pluripotential hemopoietic stem cell (cf. Fig. **4**.39).

Though maturation of induced FEL cells normally ends at the orthochromatophilic erythroblast stage and rarely proceeds to the anucleate stage of differentiation, characteristic of normal erythropoiesis, typical features of the late stages can be observed. Besides the cell cycle arrest, there is an accumulation of the histone H1 analogue IP25 that evidently associates with polynucleosomes and may have decisive influence on the increasing inactivation of chromatin during erythroid cell maturation [294]. It is worthwhile to recall that in maturing chicken erythrocytes, likewise, a special H1 analogue (H5) is found associated with chromatin. Initially it is highly phosphorylated, but with increasing inactivation of the chromatin the unphosphorylated form predominates (cf. **4**.4.2.5.3). These examples stress the special role of histone H1 and its analogues for the creation and maintenance of higher order chromatin structures (cf. **1**.2.3.2).

Analysis of the complexity of polysomal mRNA in uninduced and induced FEL cells revealed few mRNA species to appear de novo after induction (including globin mRNA) and distinct abundant and semi-abundant mRNA species to disppear [291]. But as a consequence of an extremely long half-live of globin mRNA (cf. Fig. **4**.40) and a decrease of the overall rate of RNA synthesis the relative proportion of globin mRNA increases steadily (cf. Fig. **4**.39). Thus in intact animals protein synthesis in the anucleate reticulocytes is almost exclusively devoted to globin synthesis. Interestingly, the late phases of erythroid differentiation in $Me_2SO$-induced FEL cells exhibit a destabilization of globin mRNA not observed in hemin-induced cultures. This might reflect one of the very last steps in erythroid maturation at the transition from reticulocyte to mature erythrocyte.

When briefly reviewing the results obtained with Friend erythroleukemia cells we have concentrated on the fact that FEL cell cultures can be induced to proceed correctly through essential stages of erythroid cell differentiation, thus providing valuable genetic and molecular biological data about the organization of the gene expression programs. However, another most remarkable alteration in induced FEL cells is a decrease and eventually a loss of oncogenicity that appears to be inversely related to the increase in commitment to undergo the terminal stages of the cell

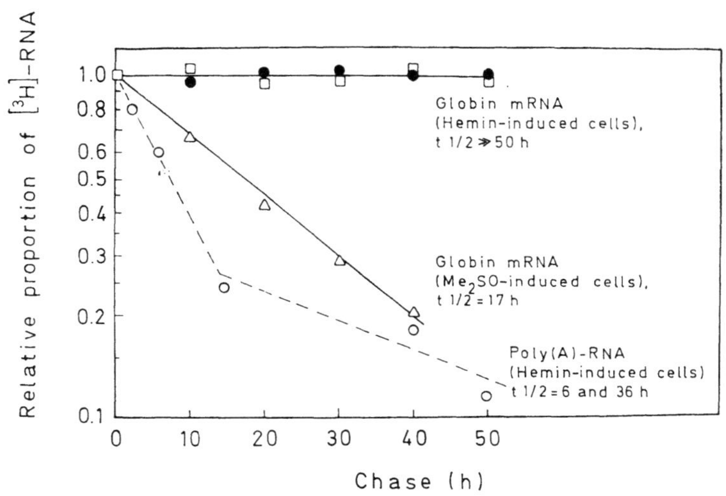


**Fig. 4.40**: Destabilization of globin mRNA in terminally differentiated Friend erythroleukemia cells (from Lowenhaupt & Lingrel [289], modified).

Erythroleukemia cells were induced with hemin and dimethyl sulphoxide ($Me_2SO$) respectively. After 5 days cells were labelled with [$^3$H]-uridine for 2 h, afterwards washed and further incubated in medium free of exogenous uridine (chase). At the indicated time points cells were lysed and [$^3$H]-RNA was fractionated to determine the incorporation into total RNA (●, mainly rRNA and tRNA), poly(A)-containing RNA (○) and globin-specific mRNA (△, □), respectively. Poly(A)--RNA and globin mRNA were purified by affinity chromatography on oligo (dT)-cellulose and globin cDNA-cellulose columns respectively. Radioactivity was determined after saturation hybridization. At 0 h of chase about 7 % of [$^3$H]-RNA were poly(A)-RNA and 0.04 % and 0.15 % were globin mRNA in the cultures induced by hemin and $Me_2SO$ respectively.

specialization program (cf. Table **4**.13). Significantly, FEL cell induction is prohibited in the presence of tumor promotors, e.g., phorbol esters (cf. Fig. **19**.10), but only if added in the early phase before cells become committed. This antagonism of oncogenicity and cell specialization was also observed for other tumor types (cf. Chaps. **18** and **19**). Thus, the comprehensive list of FEL cell inducing agents (cf. Ref. [288]) might well include compounds suitable for therapy of certain tumor types.

#### 4.5.4.3 Regulation of Globin Synthesis and the Molecular Pathology of Thalassemia Syndromes

The most prominent characteristic of erythroid differentiation is the accumulation of hemoglobin (Hb). The rapidly advancing analysis of the genetics of the corresponding gene systems as well as of molecular biological details of globin gene expression have led to a bewildering multiplicity of regulatory effects, already discussed in several of the preceding sections of this chapter. Hence, in the following they are briefly summarized only and completed by some additional facts.

- In man the $\beta$-like genes map on chromosome 11. All genes are transcribed from the same DNA strand and are arranged in the order of their expression during development (figures in brackets denote the length of the intergenic spacers): 5′-$\varepsilon$-(13.3 kb)-$^G\gamma$-(3.5 kb)-$^A\gamma$-(13.9 kb)-$\delta$-(5.4 kb)-$\beta$-3′ (cf. Fig. **4**.14 A). Basically similar is the pattern of human $\alpha$-like genes on chromosome 16: 5′-$\zeta$2-(10.5 kb)-$\zeta$1-(7.7 kb)-$\alpha$2-(3 kb)-$\alpha$1–3′ (cf. Fig. **4**.14 B). But this type of structure of the globin gene cluster is also valid for other globin gene systems, e.g., the mouse and rabbit $\beta$-like genes.

- All globin genes so far investigated (in man, mouse, rabbit and chicken) exhibit the same fine structure (cf. Fig. **4**.14): At almost identical sites the coding sequence is interrupted by a short and a long intervening sequence in the $\beta$-like genes and by two short intervening sequences in the $\alpha$-like genes respectively.
- Transcription of globin genes results in the formation of a high molecular weight pre-mRNP. Its stepwise processing (Fig. **4**.25) includes the removal of the intervening sequences and is combined with a selective increase of globin-specific sequences with respect to other pre-mRNP species produced in the same nucleus.
- In man a twofold developmental switch in the Hb-types (embryonic $\rightarrow$ fetal $\rightarrow$ adult Hb types) result from sequential expression of the corresponding types of $\alpha$- and $\beta$-like genes (cf. Table **4**.7).
- In agreement with results found for other gene systems control of globin gene expression at the transcription level clearly exhibits two stages: (1) Commitment of cells to the erythroid lineage, e.g., formation of BFU-E cells (cf. Fig. **4**.39), involves the *pre-activation* of the globin gene domain. It is followed by *activation* of distinct subdomains, i.e., by alterations of the DNA methylation pattern, modification of histones and association of the nucleosomes with the high mobility group proteins HMG14 or 17 (cf. Table **4**.8 and Fig. **4**.15). Pre-activation and activation are evident from an intermediate and high sensitivity of the globin-specific chromatin regions to degradation by pancreatic DNase I (cf. Fig. **4**.15) or to labelling by DNA polymerase I (nick translation, cf. **2**.7). (2) The actual transcription state is evidently regulated in a specific manner independent of the gene activation during commitment. It is highly significant that uninduced Friend erythroleukemia cells contain the globin genes in a pre-activated state but synthesize practically no globin-specific mRNA. Furthermore, despite the developmental specificity of globin gene expression, as indicated by the globin switch (Table **4**.7), *all* $\beta$-like genes were found in the pre-activated state in embryonic as well as adult cells of the erythroid lineage (cf. **4**.4.2.5.2) but not in nuclei of non-erythroid cells. However, we must be aware that the failure of uninduced Friend cells to synthesize globin or of adult erythroid cells to express the fetal genes, at least under normal conditions, might equally well be due to regulatory elements acting at the post-transcriptional level, e.g., during pre-mRNP processing (cf. Fig. **4**.25).
- Translation of globin mRNA is influenced by a number of regulatory effects. (1) Part of the mature mRNA is found in the polysomal mRNP compartment, part of it is in the non-translated compartment of storage mRNP (Fig. **4**.25). In chicken erythrocytes the latter compartment contains a specific translation inhibitory protein (cf. **4**.4.4.4). (2) The relative translation rates of $\alpha$- and $\beta$-globin mRNA are thoroughly influenced by elements of secondary structure in the 5′-untranslated regions (Fig. **4**.23) and by the availability of initiation factors. Thus, the affinity of $\beta$-globin mRNA for eIF2 is much higher when compared to $\alpha$-globin mRNA. A limitation of initiation factors, e.g., as a consequence of inactivation or of a globin mRNA excess, favours a predominant synthesis of the $\beta$-chain and leads to an inbalance of the $\alpha/\beta$-chain ratio. (3) Essential elements in the information amplification process during erythroid maturation are the adaptation of the tRNA pattern to the mass synthesis of globin [296] and the exceedingly high stability and hence frequency of translation of globin mRNA (cf. Fig. **4**.40). After completion of the terminal mitoses, followed by the total inactivation and possibly by extrusion of the cell nucleus, globin synthesis continues undiminished and finally makes about 95% of all protein synthesis, e.g., in mammalian reticulocytes. Destabilization of globin mRNA is a special late event in red cell maturation (cf. Fig. **4**.40). (4) The effective coordination of the hemin cofactor and globin apoprotein syn-

thesis are achieved by a translation control system involving cascades of protein kinases and the reversible inactivation of the initiation factor eIF2 by phosphorylation of its smallest subunit (cf. Fig. **4**.26).

- Most steps of heme synthesis proceed in the mitochondria. The massive accumulation of hemoglobin during erythroid maturation requires an ample supply of iron to the bloodforming tissues. Two iron complexing proteins have a key role in this respect, i.e., the serum iron-carrier protein transferrin and the intracellular iron storage protein ferritin [297–299].

  The glycoprotein transferrin, containing a single polypeptide chain with $M_r$ 77 kd, exhibits two binding sites for Fe(III) (affinity constant: $10^{24}$!). The Fe-transferrin complex is taken up into the cells by receptor mediated endocytosis. Internalization follows clustering of the occupied receptors in specialized regions of the cell surface (coated pits) that invaginate to form intracellular coated vesicles (cf. Fig. **4**.1). The Fe-transferrin coated vesicles bind to mitochondrial outer membranes, where the iron is released in dependence of the intramitochondrial pool of free heme. Transferrin and membrane compounds of the vesicles may recycle to the cell membrane. In addition to the direct transfer to mitochondria, part of the Fe-ions are found as soluble iron in cytosol and, under conditions of Fe surplus, form a storage pool by binding to ferritin. This is a chelating protein composed of 24 subunits and capable of fixing a maximum of 4000 molecules of Fe(III)-oxide phosphate.

  Only a small part of total intracellular iron-chelates (e.g., heme, Fe-protoporphyrin IX, Fe-citrate) in the free cytosolic pool are presumably active in regulating the general iron equilibrium of cells by controlling (1) rates of uptake and release of iron from transferrin, (2) binding to and release of iron from the storage protein, (3) synthesis and degradation of ferritin subunits, and (4) levels of $\delta$-aminolevulinate synthetase and of heme oxygenase involved in iron release from heme compounds. In the highly specialized cells of the erythroid lineage this network of control elements is dominated by the heme-dependent regulation of translation initiation as outlined in Fig. **4**.26.

- Hemoglobin in its native form is a highly stabile protein. But this is not true for the isolated globin chains not incorporated into the holoprotein, e.g., because of structural defects, or a lack of the hemin cofactor, or the imbalanced synthesis of $\alpha$- or $\beta$-chains. Hence the selective stabilization of the protein components in the holoprotein prohibits the accumulation of non-functional globin chains.

As is evident from the preceding sections, our knowledge about the integration of the globin gene expression into the developmental program of the organism and into the program of erythroid cell specialization has decisively advanced due to the detailed genetic analysis of the globin gene systems (cf. Fig. **4**.14), the successful in vitro cultivation of erythroid precursor cells and the valuable data provided by experiments with different strains of Friend erythroleukemia cells. Other important contributions stem from the genetics and molecular pathology of human thalassemia (cf. the summary in Table **4**.14), a type of heritable diseases resulting from inbalances in the synthesis of different types of globin chains caused by various mutations. The last column of Table **4**.14 shows that deletions and point mutations in the globin gene clusters lead to alterations and/or defects at all levels of globin gene expression, e.g., non-transcription, defective pre-mRNP processing, non-translation or premature termination of translation, instability of the mRNA and/or of the protein product. A characteristic phenotypic trait of the $\beta$-globin gene mutants that cause various types of $\beta$-thalassemias is the continued expression, albeit at very different levels, of the fetal $\beta$-like genes also in adult cells. This finding emphasizes that the globin switches observed during embryonic development (cf. Table **4**.7) presumably leave the embryonic and fetal genes in a pre-activated state (cf. **4**.4.2.5.2 and Ref. [108a]).

**Table 4.14**: Molecular pathology of thalassemia (compiled from Weatherall & Clegg [281] and Maniatis et al. [97]).

| Type[a] | Phenotype (globin chain expression) | Molecular pathology |
|---|---|---|
| $\alpha^0$-Thalassemia | No $\alpha$-chain synthesis | |
| $\alpha$-Thalassemia 1 | | Both $\alpha$-genes deleted |
| $\alpha^+$-Thalassemia | Reduc. synth. of $\alpha$-chain | |
| $\alpha$-Thalassemia 2 | | One $\alpha$-gene deleted |
| Nondeletion $\alpha$-thalassemia, e.g., Hb Constant Spring, Hb Icaria etc. | | Mutation of the normal chain termination codon (UAA) results in readthrough proteins ending at the next UAA codon; $\alpha$-mRNA highly unstable (due to ribosome binding in the 3′-untranslated sequence?) |
| $\beta^0$-Thalassemia | No $\beta$-chain, 98% fetal Hb F in homozygotes | $\beta$-Gene present and transcribed normally (?) |
| Type 1 | | No $\beta$-mRNA detectable; in one case shown to result from defective mRNA processing |
| Type 2 | | mRNA present but not translated, e.g., because a nonsense codon at pos. 17 leads to premature termination or because initiation codon is defective |
| Type 3 | | Incomplete mRNA with a deletion at the 3′-end |
| $\beta^+$-Thalassemia | Severe to mild depression of $\beta$-chain synthesis with corresponding persistence of HbF (15–80% HbF in homozygotes) | Reduced $\beta$-mRNA in cytoplasm; but normal level of $\beta$-specific sequences in nucleus; in some cases shown to result from defective processing[c] |
| $\beta$-Gene deletions[b], e.g., Hb Lepore, $\delta\beta$-Thalassemia, HPFH | No $\delta$- and $\beta$-chains; hereditary persistence of HbF in heterozygotes (15–30%) | Deletions of $\delta\beta$-genes may include large part of the 13.5 kb intercistronic region between the $\delta$- and A$\gamma$-genes (HPFH-type, cf. Fig. 4.14) |
| $\gamma\beta$-Thalassemia | In heterozygotes a severe anemia in newborn develops into a mild $\beta$-thalassemia in adult | A 40 kb deletion removes G$\gamma$-, A$\gamma$- and $\delta$--genes; $\beta$-gene intact but not expressed (cf. 4.4.2.5.2) |

[a] Typification of the thalassemia syndromes is based on the phenotypes. Hence it is inconsistent when considering the genetic and molecular biological analyses given in the last column.
[b] Deletions of the $\delta\beta$-genes are frequently grouped into the $\delta\beta$-thalassemia type and the HPFH-type (*h*ereditary *p*ersistence of *f*etal *h*emoglobin, HbF), but both are genetically and phenomenologically closely related. Some examples of deletions are included in Fig. 4.14.
[c] cf. Kantor, J. A. et al.: Cell *21* (1980): 149.

### 4.5.4.4 "Suicide" Maturation in the Erythroid Lineage

The terminal stages of erythroid differentiation (cf. Fig. 4.39) involve the complete inactivation of the nucleus (cf. 4.4.2.5.3), in mammals followed by its extrusion. Simultaneously most of the usual intracellular organelles begin to disappear. Especially striking are the structural and functional changes during reticulocyte maturation. After cessation of all RNA synthesis, late erythroblasts and early reticulocytes contain about 25,000 molecules of long-lived globin mRNA. Because of the degradation of most other mRNA species, globin synthesis makes about 95% of the highly active protein synthesis. Reticulocytes contain efficient amino acid transport systems in their membranes. The overall metabolism is specialized to the mass synthesis of heme and globin. To a considerable extent energy and substrates are derived from degradation of all cell constituents not needed in the mature erythrocyte, e.g., mitochondria, cell membranes, ribosomes. Thus the "Umbau" of proteins by selective breakdown and reutilization of the building blocks for heme and globin synthesis [304a] is self-limiting because the synthetic machinery itself is finally affected.

The impressive accumulation of the oxygen carrier protein hemoglobin coincides with profound changes in the erythroid cell membrane. All hormone receptors and most of the transport systems encountered in reticulocytes have disappeared in mature erythrocytes. In addition, a special membrane associated protein skeleton ensures that erythrocytes are both durable and flexible enough to withstand the cardiac passage as well as to path the most narrow portals of the spleen. The essential role of this skeleton is demonstrated by a number of hereditary hemolytic anaemias resulting from defects of one of its constituents. A few major proteins make 60% of the total mass of this membrane-associated cytoskeleton, which exhibits properties similar to the muscle actomyosin system: spectrin ($M_r$ $\alpha$-subunit 240 kd, $\beta$-subunit 220 kd), actin ($M_r$ 42 kd), ankyrin ($M_r$ 200 kd), the so-called band 4.1. protein ($M_r$ 78 kd), and four minor proteins [301–303].

The basic submembranous skeletal network is composed of a high molecular weight complex of spectrin tetramers ($\alpha_2\beta_2$, $M_r$ 920 kd), F-actin filaments and the band 4.1 protein. It is anchored to a proportion of the major erythrocyte transmembranous protein (band 3 protein, $M_r$ 95 kd) that exists as a dimer and normally functions as anion exchange channel. The contact between spectrin and the band 3 protein is mediated by ankyrin and one of the minor proteins (band 4.2 protein, $M_r$ 72 kd). The anchorage complex of a band 3 protein dimer and an ankyrin dimer is connected with two spectrin tetramers that are part of the submembranous skeleton. Ankyrin is evidently present in many other cell types, too, and may represent the general connector protein between cytoskeletal elements and transmembrane proteins. Most of the erythrocyte skeletal constituents are highly phosphorylated, e.g., ankyrin, band 4.1 protein, band 3 protein and spectrin. In spectrin apparently that part of the molecule undergoes phosphorylation which interacts with band 4.1 protein and actin and there is evidence that phosphorylation modulates protein interactions in the red cell skeleton [300, 304].

## 4.6 Cell Differentiation and Development

Despite the seemingly autonomous and determined procedure of gene expression programs (cf. 4.5), cell differentiation ultimately proceeds as response to multifold signals from the extracellular surrounding. These include not only all kinds of hormones and hormone-like proteins, but also nutrient compounds, electrical or chemical signals of the nervous system as well as signals mediated by direct cell-cell contacts. All together bring about the effective integration of cellular activities into the structural, functional and developmental entity of the organism so characteristic of higher plants and animals. Development of tissues, organs and organisms must be described in terms of the spatially and temporarily ordered differentiation of their constituent cells *(co-differentiation)*.

Developmental biologists have introduced two important terms to describe essential aspects of the ordered cell specialization during ontogeny: positioning and pattern formation. Both aspects are interrelated. The pattern of differentiating or differentiated cells typical of a multicellular organism results by either of two mechanisms: (1) positioning of pre-differentiated cells by controlled cell movement or (2) pattern formation in a population of pre-positioned cells that are capable of responding differentially to extracellular gradients of "morphogenetic" substances.

In the following sections we will try to illustrate some important mechanisms that underly the coordinate behaviour of cells as the differentiation units of a multicellular organism. These are:

- stage-specific alterations of intercellular communication during vegetative growth and fruiting body formation of myxobacteria,
- controlled movement of cells of the vertebrate immune system and mechanisms of cell sorting by tissue specific factors,
- role of morphogen gradients for early determinative phases of cell differentiation in *Drosophila* and, finally,
- the network of control elements responsible for the development and function of the human sexual system.

Many further examples with special reference to the intercellular coordination of differentiation are found in Sect. **4**.4.4.7 and **4**.5.4.1 and Chaps. **8**, **13** and **17**.

### 4.6.1 Intercellular Signals Controlling the Developmental Cycle of Myxobacteria

The multicellular stage involving the specialization of cells to perform distinct tasks within the total organism is not restricted to eukaryotes. A number of oligocellular or multicellular forms of the blue-green algae *(Cyanophyta)*, actinomycetes and, above all, myxobacteria exhibit interesting examples of intercellular cooperation. The rather complex developmental program of myxobacteria will be discussed in more detail. Many aspects of it strongly remind of the cellular slime molds representing eukaryotic counterparts of myxobacteria (cf. Chap. **17**).

Myxobacteria are a highly complex group of gram-negative rod-shaped bacteria, capable to move by gliding on solid surfaces. They feed on other microorganisms or on organic matter in the soil or on the bark of trees by excreting enzymes active in hydrolyzing proteins, nucleic acids, fatty acid esters and polysaccharides including, in some species, cellulose. A typical product of myxobacteria are large amounts of an extracellular polysaccharide slime. In one case *(Myxococcus virescens)* the major slime polymer has been purified to homogeneity and shown to represent a giant protein-polysaccharide-lipid complex ($M_r$ 2–5$\times 10^4$ kd) with pronounced protease activity. It is capable of binding and denaturing proteins. Thus, the slime appears to function as a solid-state enzyme system for protein binding, denaturation and degradation.

Myxobacteria are social organisms. Throughout their life cycle they are dependent on a close contact with each other. Two multicellular stages with two types of cell cooperation can be distinguished: (1) the vegetative cell swarms of $\sim 10^3$ cells that feed and move together and (2) much larger aggregates ($10^6$–$10^7$ cells) formed normally after nutrient starvation and capable to undergo extensive morphogenesis and cell specialization in the course of fruiting body formation (Fig. **4**.41).

The vegetative swarms form dynamic groups of a myxobacterial colony which constantly change with new swarms forming, some swarms enlarging and others breaking up. They are loosely organized. Any one cell may move away from the edge of such a swarm but usually returns rapidly to the main mass. Chemotactic signals or the collective production of slime as substrate for feeding and movement have

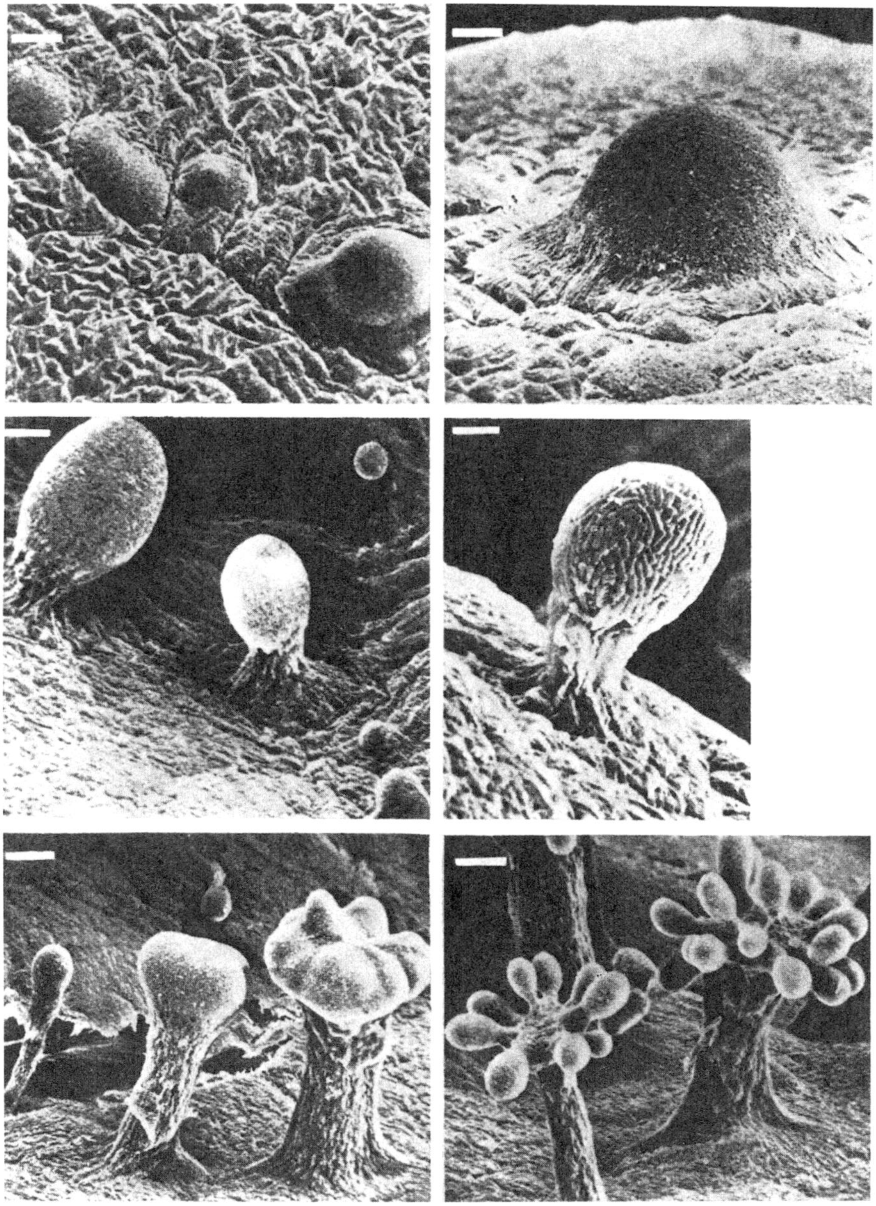


**Fig. 4.41:** Scanning electron micrographs of the developmental stages of *Chondromyces crocatus* (from Grilione, P. L. & J. Pangborn: J. Bact. *124* (1975): 1558–1565, with permission).

*a*, initial stages of vegetative cell aggregation (*bar* = 20 μm); *b*, with the development of the aggregation center a fried-egg appearance results with the peripheral cells exhibiting some degree of orientation; *c*, the central portion of the "fried egg" enlarges into a bulbous structure; *d*, initial stage of stalk formation; *e* and *f*, summary of sporangia formation, from *left* to *right:* at maximum stalk height the bulb enlarges laterally, the top surface becomes flattened and bud-like swellings appear *(e)*. These swellings develop into sporangia that are separated from the stalk by pedicels *(f)*. *Bars* in b–f always represent 10 μm.

For the electron microscopy cultures were grown on agar that allows fruiting body formation. The material was fixed by exposition to glutaraldehyde vapour and afterwards dehydrated in a graded ethanol series with amyl acetate and finally with liquid $CO_2$. The dried samples were coated in a vacuum evaporator with silver and gold.

been suggested to promote the social behaviour within a swarm. In support of the latter assumption is the observation that swarms evidently provide the critical cell mass to attack and digest other microorganisms or to degrade complex organic matter.

The growth rate of *Myxococcus xanthus* was entirely dependent on the cell density when cultured in casein-containing liquid media. Cell division totally ceased at cell densities with less than $10^3$ cells per ml. These effects were not observed in liquid media with enzymatically hydrolyzed casein.

Fruiting body formation and sporulation are the response of myxobacteria to any kind of nutrient starvation. The fruiting body is genus-specific and may consist of rather simple mounds (*Myxococcus*). But frequently rather elaborate structures are observed, with a central stalk composed of slime and dead cells and sporangia attached to the stalk that raises them above the substrate (cf. Fig. **4**.41).

Reprogramming of the vegetative cells to aggregation and sporulation upon starvation is evidently triggered by formation of guanosine-3′-diphosphate-5′-diphosphate (ppGpp) and related compounds. This is the well known "second messenger" of the stringent control system in *Escherichia coli* whose cellular concentration rapidly increases in response to amino acid deficiency and that act as a hormone-like signal to adjust *E. coli* to the altered metabolic conditions. The primary effect of ppGpp is assumed to depend on the interaction with RNA polymerase thus changing its promoter recognition specificity (cf. **4**.4.2.6).

As a result of an increased ppGpp level in *E. coli* transcription of a great number of operons is repressed (e.g., operons for rRNA and protein synthesis), many others are induced. Such changes of RNAP transcription specificity might also be of crucial importance for the developmental shift from the vegetative to the fruiting cycle in myxobacteria. A remarkable, though circumstantial evidence in this respect was provided by Zusman and coworkers [305]: Selection of rifampicin resistent mutants of *Myxococcus xanthus* gave a number of strains with altered RNAP, growing normally in the vegetative phase but unable to embark on the fruiting and sporulation pathway. The heterogeneity of the terminal phenotypes of these mutants suggests that transcriptional specificity may be altered at several stages of the developmental cycle.

As early markers of the fruiting body gene expression program two mass proteins were characterized which are absent in vegetative cells. (1) A myxobacterial hemagglutinin ($M_r$ 30 kd) is thought to play a role for cell-cell recognition or agglutination [308] and (2) a protein S ($M_r$ 23 kd) finally makes 15 % of the soluble cellular protein. During sporulation the preformed protein is somehow transported to the outer surface of the cells and assembled into the thick spore coat [309].

In contrast to the mechanism of endospore formation of *Bacillus* species (cf. **4**.5.3), in myxobacteria the whole cell is converted to a spore. The process of sporulation involves a profound reorganization of the whole cell including the cell envelope. Though sporulation is normally a late step of the fruiting program, it is clear from genetic and biochemical evidence that distinct parts of the program proceed relatively independent from each other. These are (1) initiation by starvation, (2) aggregation and fruiting body morphogenesis and (3) myxospore formation. Correspondingly, three classes of mutants have been characterized (a) mutants that neither aggregate nor sporulate, (b) mutants that form more or less normal aggregates, but do not sporulate and (c) sporulating mutants with a defective aggregation and/or fruiting body construction [305, 306].

The relative independence of the gene expression programs involved in sporulation is supported by the observation that rapid, synchronous sporulation can be induced without nutrient starvation by adding glycerol, dimethyl sulfoxide, t-butanol or other compounds to suspension cultures of myxobacteria. After a short period of commitment (~30 min) the inducer can be removed without altering the sporulation program, which does not require cell division or DNA synthesis. Though these "glycerol-spores" exhibit many properties of the normal spores formed in fruiting bodies, they do not synthesize the spore-specific protein S and therefore lack the outer coat. They are not true resting cells, i.e., they have a relatively high endogenous respiration and energy charge. Induction of glycerol-spores does not depend on the ppGpp system and, finally, all rifampicin-resistant, sporulation-defective mutants discussed above are able to undergo glycerol-induced sporulation. Thus, the experimental system of artificial sporulation allows more thorough studies of selected aspects of the gene expression programs and the demonstration of their independence of the normal developmental trigger system and of cell aggregation.

A detailed analysis of many sporulation-deficient mutants revealed interesting aspects of cell-cell interactions. It was found that a considerable number of these mutants were able to sporulate if tested in mixtures with wild-type cells. Moreover, many of these "conditional non--sporulating" mutants can complement each other. At least three extracellular complementation groups were characterized. It is postulated that normal sporulation requires three self-generated stimuli which may serve as signals for cell-cell communication to control and synchronize the sporulation program [307].

### 4.6.2 Mechanisms of Cell Sorting and Positioning in Mammals

Correct positioning of specialized cells establishes the basic tissue structure and the geometrical relationships of tissues in an organ. Especially during animal embryo development, cells undergo complex series of movements leading to separation of homogeneous-appearing cell masses into dissimilar layers, to folding and invagination of cell layers, to invasion of tissues between others or to aggregation of distant cells to form specific tissues or organ primordia. But such tissue rearrangements with cell sorting into ordered substructures is by no means restricted to stages of the embryonic development.

All these tightly controlled cell movements depend on short-range or long-range cell-cell recognition systems and normally result in the immobilization of cells at a particular position depending on their functional role in the multicellular organism. There is experimental evidence that such recognition systems involve elements of specific cell-cell interaction and adhesion, the migration of cells along chemical gradients, as well as the action of short-range tissue-specific hormones preventing the invasion of unlike cells into a tissue or inducing the sorting out of a mixed population.

A unique opportunity for experimentally studying mechanisms of cell positioning and sorting is presented by the immune system (cf. Chap. **13**). In probably all vertebrates small T and B lymphocytes, circulating in the blood, invade distinct extravascular compartments of many organs and then return to the blood. An essential function of this mass recirculation of lymphocytes is to redistribute continuously antigen-sensitive cells between the distant parts of the peripheral lymphoid system, including spleen, lymph nodes and gut-associated lymph tissues [318, 319].

The process can be analysed in three distinct parts: (1) Lymphocytes leave the bloodstream by adhering specifically to the cell surface of vascular endothelium and subsequently cross the tissue by migrating between the cells of the vascular wall. (2) After crossing the vascular tissue, T and B cells sort out to enter different compartments of the secondary lymphoid organs. (3) Each cell type returns to the blood by actively migrating along pre-existing channels within the tissues. The first two parts of lymphocyte recirculation are particularly interesting for the problems to be discussed in this section:

1. The effective transit of lymphocytes from blood to the lymphoid tissue depends on the functional specialization of the vascular endothelial cells of the post-capillary

venules. In contrast to the usually flat appearance, these endothelial cells are cuboidal and include many mitochondria and ribosomes indicating their high state of activity. There is evidence for a rapid and selective adhesion of T and B lymphocytes to the membranes of the endothelial cells. Adhesion requires the active participation of both cell types, is energy dependent and involves parts of the cytoskeletal system (microfilaments).
The search for specific complementary receptors in the cell membranes of both cell types has so far been negative. Instead, there is good evidence that lymphocytes are attracted by a local signal generated by the endothelial cells and that this signal might simply be iron or an iron-transferrin complex (cf. **4.4.4.5**). Lymphocytes, which probably contain transferrin or transferrin receptors in their cell membranes, respond to the local increase of the iron concentration.

In keeping with this assumption are experimental findings (1) about an inhibition of lymphocyte adhesion by adding EDTA or by pre-treating lymphocytes with ferric chloride or ferric citrate and (2) about the molecular pathology of the so-called Hodgkin's disease. One aspect of this syndrome are profound abnormalities in the intercellular iron equilibrium (cf. **4.4.4.5**). An unusual accumulation of iron and iron-binding proteins (ferritin, transferrin) in spleen and lymph nodes of patients with Hodgkin's disease coincides with an abnormally high number of T lymphocytes in these tissues [319].

2. An important aspect of the second part of lymphocyte recirculation concerns the remarkably efficient sorting out of T and B lymphocytes during their migration to the extravascular compartments of the lymphoid organs. This segregation depends on the formation, by both cell types, of proteinaceous inhibitory factors binding to membrane receptors of the opposite cell type. The B cell factor shows an apparent molecular weight of 3 kd, the T cell factor of 9 kd.
Such factors, characterized also from other vertebrate tissue, e.g., chick embryonic neural retina and liver, have been called interaction modulating factors (IMF). They promote assembly of cells of the same type by interfering with the adhesion capacity of other cell types. In mixed cell populations they cause a sorting out with formation of distinct cell layers [317]. Though direct experimental proof is lacking, such IMF are good candidates for factors controlling cell migration and tissue organization during embryogenesis.

Action of the lymphocyte-specific IMF evidently depends on the labelling of distinct compartments of the lymphoid organs as "forbidden" for the opposite cell type. This label in the environment of a B or T cell population exhibits a high turnover and thus is suitable for an efficient control of lymphocyte circulation. Its effect was clearly shown by injecting purified T cell IMF into mice and tracing lymphocyte traffic with labelled T and B cells [318]: The homing of T lymphocytes in the spleen strongly increased, that of B lymphocytes decreased, when compared to control animals. The lymphocyte IMF may ensure not only the segregation of T and B cells on their way from blood to the extravascular tissue, but may simultaneously reduce adhesion to stroma cells of these tissues so that lymphocytes are released into the lymph. Thus, the study of lymphocyte IMF reveals a complex pattern of cellular interactions that may effectively control the routing of lymphocyte traffic and the ratio of T and B cells in different compartments of the lymphoid system.

One of the main roles of IMF production apparently is to maintain the intactness of tissues and organs and to prevent infiltration by foreign cells. We can speculate that there may be diseases resulting from mispositioning of cells (1) because the production of the corresponding IMF is defective, (2) because the IMF receptor in the outer membrane is lacking or (3) because diseased cells have acquired the capability of producing the IMF of other cell types. Each type of defect in cell sorting may represent the basis for malignant invasiveness [317].

### 4.6.3 Morphogen Gradients Control Pattern Formation During Early Developmental Stages of *Drosophila*

The fruit fly *Drosophila* has for many years been a favourite object of developmental biology. Extensive studies with *Drosophila* have made major contributions towards a deeper understanding of the molecular mechanisms that, during early stages of embryogenesis, control the generation of cell lineages determined to undergo distinct programs of cell specialization when the adult organism forms.

The advantages of this object stem from the following facts: (1) *Drosophila* can be easily maintained and manipulated under practically all laboratory conditions. (2) It is genetically the best-known higher eukaryote. (3) Numerous developmental mutants have been isolated and partially characterized. (4) Embryogenesis is very fast (~ 24 h) and involves the formation of easily identified groups of cells (cell compartments) that are determined to generate distinct structures of the adult organism. (5) Most of these cell compartments give rise to external structures of the organism. Hence mutational alterations of the developmental pathway can be easily recognized, in some cases already in the larval stage.

*Drosophila* is a holometabolous insect. Embryogenesis proceeds outside the maternal organism within the relatively large egg ($500 \times 180\,\mu m$). It is followed by hatching of the first instar larva (about 24 h), formation of the second and third instar larvae (at 48 and 72 h of development), the pupa stage (120 to 148 h) and finally metamorphosis to the imago. Embryogenesis starts with about nine rapid synchronous nuclear divisions within 90 min, forming a syncytium of 512 nuclei, most of which invade the egg cortex. Following the 12–13th cleavage (about 3 h of embryogenesis) the nuclei are separated by cell walls thereby forming the unicellular blastoderm with 4,000–6,000 cells. The subsequent gastrulation involves a profound rearrangement of cells within the larval body leading to the separation of the three germ layers (ectoderm, endoderm and mesoderm) and to the segregation of cells into a metameric pattern (segments) of polyclones assigned to develop specific parts of the head, the three thoracic and the eight abdominal segments of the adult fly respectively (Fig. **4**.42). Thus, cell diversification out of the uniform single layer of blastodermal cells is clearly visible in this stage as are developmental mutants defective in segmentation (see below).

Of special interest are small groups of 10–40 cells within the segments which, during the three larval stages, enlarge by cell proliferation and further divide into subcompartments of the so-called imaginal discs and histoblasts which are determined to generate distinct epidermal structures of the adult fly (cf. Fig. **4**.42). Determination of the developmental fate of these cell groups starts already in the blastoderm and early gastrulation phase, though they are without any function during larval life. By different methods it was possible to trace back the origin of adult tissues and organs to distinct cell groups of the blastoderm and early gastrulation embryo and thus to generate a fate map of the embryonal cells (Fig. **4**.43). This is the basis for the analysis of developmental mutants as will be described below.

For a deeper understanding of the molecular mechanisms involved in cell determination as evidenced by the blastoderm fate map it was necessary to create a reliable bioassay system. Such a system was the early cleavage and blastoderm embryo itself in which whole cells, isolated nuclei or cytoplasm could be microinjected and studied with respect to their influence or developmental fate. The results clearly demonstrate that determination is caused by cytoplasmic factors [327]: Isolated nuclei of blastoderm cells appear totipotent, whereas whole cells are programmed to a restricted developmental pathway which they continue essentially unchanged even if transplanted to a foreign region of the recipient blastoderm embryo. The determinative effects of the egg cytoplasm is especially prominent for the pole plasm that is essential for the generation of the germ cell primordia (cf. **4**.6.4.1).

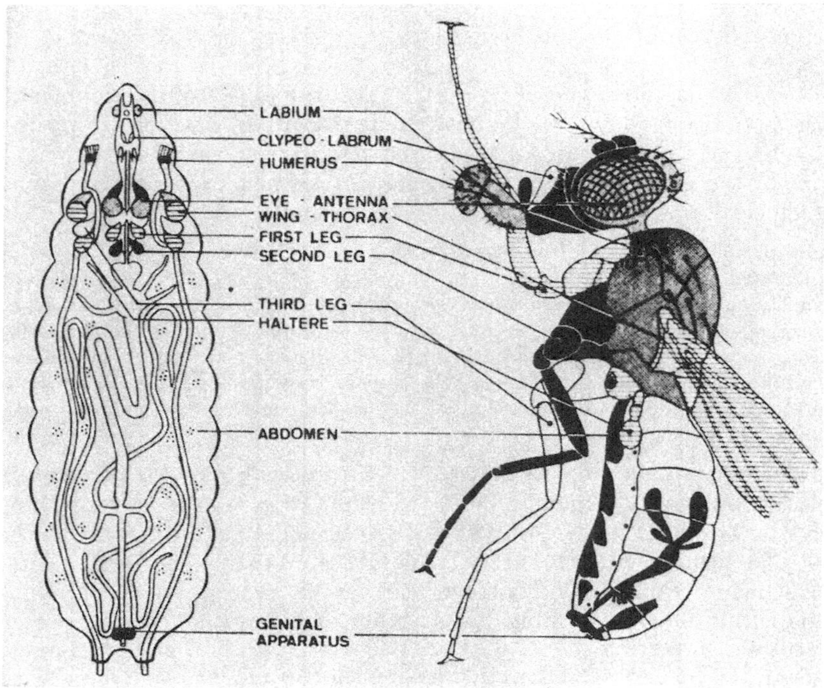


**Fig. 4.42:** Scheme of larval organization of *Drosophila* and location of the imaginal discs (from Nöthiger [**321**], with permission).

The total of nine pairs of imaginal discs plus a single fused genital disc are visible as small cell groups in the larval body *(left)*. They are characterized by different types of *shading* or *hatching* used also for the corresponding structures in the adult fly *(right)*. *Points* in the larval body indicate cell groups (histoblasts) determined to give rise to special structures of the adult fly integument.

Other important tools for studying mechanisms of cell differentiation during embryonic development are the imaginal discs of insects, which can be regarded as a gift of nature to the experimental biologist. They preserve their determined developmental program unchanged in a cryptic state as revealed by the following experimental series [**322**]: Tissue pieces from a genital disc of a third instar larva of *Drosophila* were removed and injected into the abdomen of an adult female. The host's haemolymph serves as a culture medium which allows proliferation without inducing terminal differentiation. Such tissue pieces can be "subcultivated" over many passages without losing their capacity to generate genitals when transplanted to third instar larvae. Even after years of "culture", tissue derived from the genital disc still yielded normal genitalia. Hence, the determinated state can be propagated seemingly indefinately in a proliferating tissue and the interrupted gene expression programs continue unchanged if the determined cells are subject to the appropriate environmental conditions, i.e., to the action of the moulting hormones, that normally induce terminal differentiation of the imaginal discs.

Intensive investigations of the development and cellular organization of imaginal discs as well as of corresponding mutations generating aberrant patterns of segmentation and ectodermal structures of the adult fly led to a general concept of pattern formation whose basic elements are evidently applicable to similar processes in other organisms, too. (1) Morphogen gradients provide positional information to cells at distinct sites within the multicellular embryo and (2) a set of "selector genes" responds to distinct threshold levels of the hitherto practically unknown morphogens by switching the cell and its descendants to a given gene expression program. Threshold lines and shape of the gradients would define border lines of subcompartments within cell polyclones, e.g., within segments or imaginal discs. We will restrict ourselves to the description of three types of mutants whose analyses lend strong support to this morphogen gradient – selector gene concept [323, 328]) (1) Mutants with aberrant morphogen gradients, (2) segmentation mutants and (3) homeotic mutants with abnormal development of imaginal discs. The latter two types result from alterations in the putative selector genes.

**Gradient mutants:** The search for mutants defective in the build-up of morphogen gradients was facilitated by the fact that the first stages of determination proceed in the blastoderm and early gastrulation phase. Consequently, the gradient(s) must exist preformed in the egg cytoplasm. Thus, the embryo behaviour in this respect depends solely on the genotype of the mother because transcription of the determinative gene(s) takes place during oogenesis (maternal inheritance).

Two types of such developmental mutations have indeed been found and characterized. These are *bicaudal* and *dorsal*[10]. The *bicaudal* mutations affect the earliest event of cell determination in the antero-posterior axis, whereas *dorsal* affects the dorso-ventral pattern formation. As required, both mutation types interfere with pattern formation in the entire embryo and not only in distinct regions. Both mutations yield different phenotypes (cf. legend to Fig. **4**.43) that are best explained by assuming abnormalities in the formation and maintenance of two perpendicular gradients with distinct thresholds determining the fate of cells within the threshold boundaries (cf. Fig. **4**.42). It is worth mentioning that the analysis of the *bicaudal* and *dorsal* phenotypes suggest not only one, but several threshold levels within the anterior-posterior and the dorso-ventral gradients. As evident from the description of the mutant phenotypes in the legend to Fig. **4**.43 and from the hypothetic gradient shapes used to explain them, both types of mutations have quite different effects on the pertinent gradients.

**Segmentation mutants:** Contrasting to the gradient mutants these are embryonic mutants, i.e., their phenotype depends on the genotype of the embryo. So for 15 loci have been shown to be involved in segmentation [329a]. Three groups can be discerned: (1) *Gap mutants* (three loci) are characterized by the loss of a continuous stretch of up to eight segments, i.e., whole subregions of the embryo, resulting from cell differentiation as response to the morphogen gradients described in Fig. **4**.43, are evidently skipped. (2) *Pair-rule mutants* (six loci) exhibit phenotypes with homologous parts of the segmentation pattern being depleted every other segment, i.e., either the even numbered segments and/or the odd numbered segments are affected. These unexpected mutant type suggest that at some stage of the earliest development the embryo is organized in repeating units comprising the equivalent of two segment anlagen. A single mutation of this subclass may cause simultaneous deletions of distinct parts in the even-numbered and of other parts in the odd-numbered segments with consecutive fusion of the remaining parts to double segments. (3) *Segment-polarity mutants* (six loci) do not change the segment number. But in each segment a defined fraction of the normal cell subcompartments is lacking, the remainder being present as a mirror-immage duplication, i.e., with reversed polarity.

---

[10] *Bicaudal* is a recessive mutation affecting a single locus on the right arm of the second chromosome (2–67). *Dorsal* mutations have a recessive or a dominant phenotype. In the homozygous state of the recessive mutation the morphogen gradient has totally disappeared (line 3 in Fig. **4**.43).

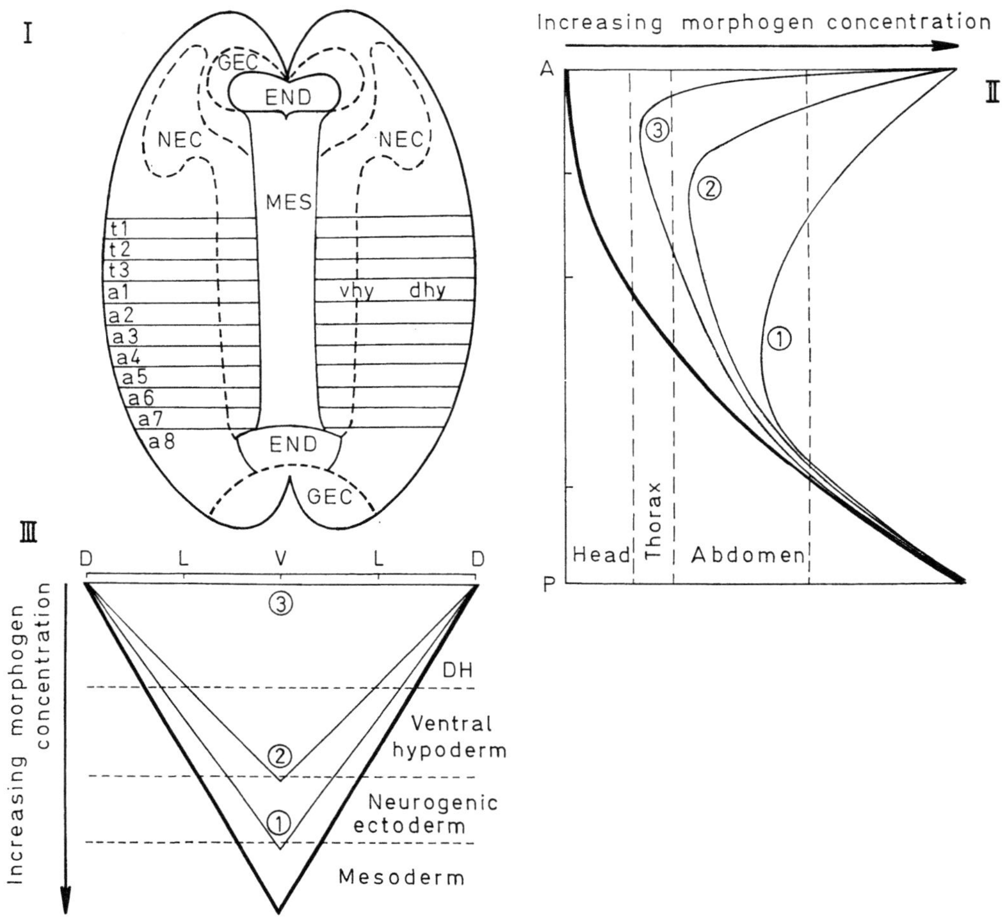


**Fig. 4.43**: Blastoderm fate map of *Drosophila* (I) and hypothetical gradients of morphogens defining the anterior-posterior (II) and dorso-ventral (III) coordinates (from Nüsslein-Volhard [325], modified).

**I** *(left)*: The *Drosophila* egg is thought cut open along the dorsal midline and the left and right halves are spread out. *A*, anterior; *P*, posterior; *D*, dorsal; *L*, lateral; *V*, ventral. The total egg length is about 500 $\mu$m corresponding to roughly 100 blastoderm cell diameters. The fate map was originally constructed on the basis of histological observations (Poulson, D. F. In: Demerec, M. (ed.), Biology of *Drosophila*, Hafner, New York 1950), pp. 168–270). Except for a longitudinal part at the ventral midline that forms the mesodermal *(MES)* and endodermal *(END)* anlagen, the larger fraction of the blastoderm gives rise to ectodermal derivatives including the neurogenic ectoderm *(NEC)* formed by the ventral most cells, flanking the mesoderm anlage, the gut ectoderm *(GEC)* and the germ bands for the three thoracic *(t1–t3)* and eight abdominal *(a1–a8)* segments. They form equally spaced stripes of a width of about three to four cell diameters. For the position of the imaginal discs within the larval body cf. **Fig. 4.42.**

**II**: *(right)* Hypothetical morphogen gradient determining the cell fate in the anterior-posterior axis *(heavy line)*. Threshold levels for the abdominal, thoracic and head anlagen are indicated by *broken lines*. *Thin lines* 1, 2, 3 indicate the possible shape of gradients evoked by *bicaudal* mutations. *1* corresponds to symmetrical *bicaudal* embryos, *2* specifies a headless *bicaudal* embryo and gradient *3* induces formation of a headless embryo.

**III**: *(below)* Hypothetical gradient of morphogen defining the dorso-ventral coordinates *(heavy line)*. The *broken lines* indicate threshold levels for mesoderm, neurogenic ectoderm, ventral hypoderm and dorsal hypoderm (DH). *Thin lines* 1, 2, 3 specify gradients that might explain different *dorsal* phenotypes that fail to form the mesoderm anlage *(1)* or the mesoderm and neural anlagen *(2)*. Line *3* describes the total abolition of the hypothetic dorso-ventral morphogen gradient in the homozygous recessive *dorsal* phenotype.

**Homeotic mutants:** These are embryonic mutants exhibiting changed identity of segments or of cell groups within segments, e.g., imaginal discs. Contrasting to the preceding group, mutations of homeotic genes do not affect the number, size or polarity of segments. A typical example is *bithorax* leading to the transformation of the anterior metathorax into the anterior mesothorax. A considerable number of such mutants has been described [323, 328]. Many of them were shown to involve transformations between noncontiguous regions of the blastoderm fate map. But unfortunately the genetic and molecular biological characterization of these mutants is still at the very beginning. Thus a clear-cut correlation between the type of mutation and the determinative event affected is difficult at present. The phenotypic outcome (expressivity) of such mutations in many cases is evidently greatly influenced by inductive interactions with neighboured or even distant cell compartments or by environmental factors, e.g., nutrition or temperature [329].

An interesting property of the homeotic type of selector genes emerged. In response to the unknown morphogen they take two alternative states, i.e., they switch the cell to either one of two developmental pathways (anterior vs. posterior or dorsal vs. ventral). Thus, the final fate maps of the imaginal discs can be envisaged to result from an ordered sequence of at least five binary decisions, beginning with the earliest two just mentioned (cf. the gradients shown in Fig. **4.43**).

Two important experimental observations further support the evidence for the existence and function of the homeotic selector genes in determination: (1) Transplantation experiments have shown that some mutations affecting a particular body part of the adult fly are already manifested in the corresponding imaginal disc. This behaviour is expected for selector gene mutants. (2) Ocassionally, imaginal discs produce unexpected structures, e.g., a leg disc produces wing structures. Many different types of such "transdeterminations" have been reported but, irrespective of their origin by genetic or environmental factors, most of them can be classified in a simple scheme that allows only single-step transitions between two alternative states [323, 328].

The very brief insight into the developmental biology of *Drosophila* was aimed at demonstrating that the combination of experimental embryology with genetics and theoretical biology has provided a promising frame for a deeper understanding of molecular mechanisms that control the earliest steps of cell differentiation during embryogenesis of higher multicellular organisms. The formation of imaginal discs, which conserve, in a cryptic state, the structural and biochemical details for distinct parts of the adult insect, gives the opportunity to study many important aspects of determination and the generation of cell lineages. This is facilitated by the possibility of inducing terminal differentiation and morphogenesis by either transplanting imaginal discs to third instar larvae or by incubating them in vitro in a suitable culture medium in the presence of the moulting hormone ecdysone. The maintenance of the determinated state and its "unfolding" under the action of an environmental signal (ecdysone), in fact, very much recalls the establishment and maintenance of the lysogenic state of phage infected bacteria (cf. **14.2.3.8** and **14.2.3.9**).

The "selector gene" of phage $\lambda$ is the gene coding for the *cI* repressor which, by means of an autogenous control circuit (cf. Fig. **14.8**), keeps the phage genome in a cryptic state as an integrated part of the host chromosome. Induction of phage $\lambda$ by environmental factors involves desintegration of the phage genome and specialization of the host cell to an efficient factory of new phage particles.

When summarizing the results in support of the morphogen gradient-selector gene concept, a number of difficulties are immediately evident. The two maternal gradients defined by the *bicaudal* and *dorsal* mutations, in fact, correspond to the two perpendicular gradients already evident from the egg form and shape of the egg shell. Other gradients, which act later during ontogenesis, might be formed within the larval compartments under the influence of larval genes. Their shape evidently differs from disc to disc and it can be anticipated that the response of putative selector genes within individual compartments is modified by the cellular environment and differentiation history of the corresponding cells. Thus, it may be difficult to obtain unequivocal results showing the coordinate changes in distant discs predicted, e.g., for homeotic mutants unless many more details are known about the molecular basis of the gene expression program involved in the determinative phase of cell differentiation. However, the most critical problem resides in the identification of the putative morphogens and of the mechanisms underlying gradient formation and the selective response of different cell groups to distinct threshold levels.

### 4.6.4 Role of Intercellular Signals for the Development and Function of the Sexual System in Mammals

The sexual system in mammals comprises the gonads plus associated accessory glands the hormonal control centres of the brain (hypothalamus, adenohypophysis), nerva control centres of the central nervous system and numerous organs exhibiting secondary sex traits, e.g., liver, mammary glands etc. It will serve as an example of the multifold developmental and functional interactions between neighboured and distant cells which are prerequisite for the existence of multicellular organisms. We will restrict ourselves mainly to the male organism and will discuss (1) aspects of the ontogenesis of the sexual system [331] and (2) the role of the adult hormonal circuit controlling sex hormone synthesis and germ cell differentiation.

#### 4.6.4.1 Development of the Sexual System

Development of the sexual system proceeds in five phases (Fig. 4.44):

1. The *genetic determination* of sex occurs during fertilization in dependence of the chromosomal constitution of the two germ cells involved. The mammalian system of sex determination involves two types of sex chromosomes (X and Y).

---

▶

**Fig. 4.44:** Chronology of events and hormonal conditions of the fetal development of the sexual systems in man (from Kuhl [338], modified).

Fetal sexual development depends on the dynamics of three types of hormones: (1) the maternally supplied human chorionic gonadotropin *(HCG)*, (2) testosterone supplied in the male fetus by the fetal Leydig cells and (3) the hypophyseal gonadotropins *LH* and *FSH* (luteotropic and follicle stimulating hormone respectively). Decisive influence on the differential development of the indifferent gonad anlagen into the male or female direction has the largely elevated level of testosterone in males and the relatively low level of LH/FSH resulting therefrom.

After the very early differentiation of the testis (7th week of pregnancy), the male-specific development leads to regression of the Müllerian ducts (induced by a local testicular glycoprotein, antimüllerian hormone, $M_r \sim 120$ kd, [241]). Subsequently the mesonephric Wolffian ducts and tubules generate the ductus deferentes, the seminiferous tubules and the epididymis. In contrast, the ovary differentiates relatively late (14th week of pregnancy) as a result of testosterone deficiency in the femal fetus. In this case, the Wolffian ducts degenerate and the Müllerian ducts give rise to the uterus, oviduct and upper vaginal parts [231].

The sex specific differentiation of the brain (mainly of hypothalamic sex centres) starts with the begin of LH/FSH secretion. It is completed only with the postnatal sexual maturation process [330].

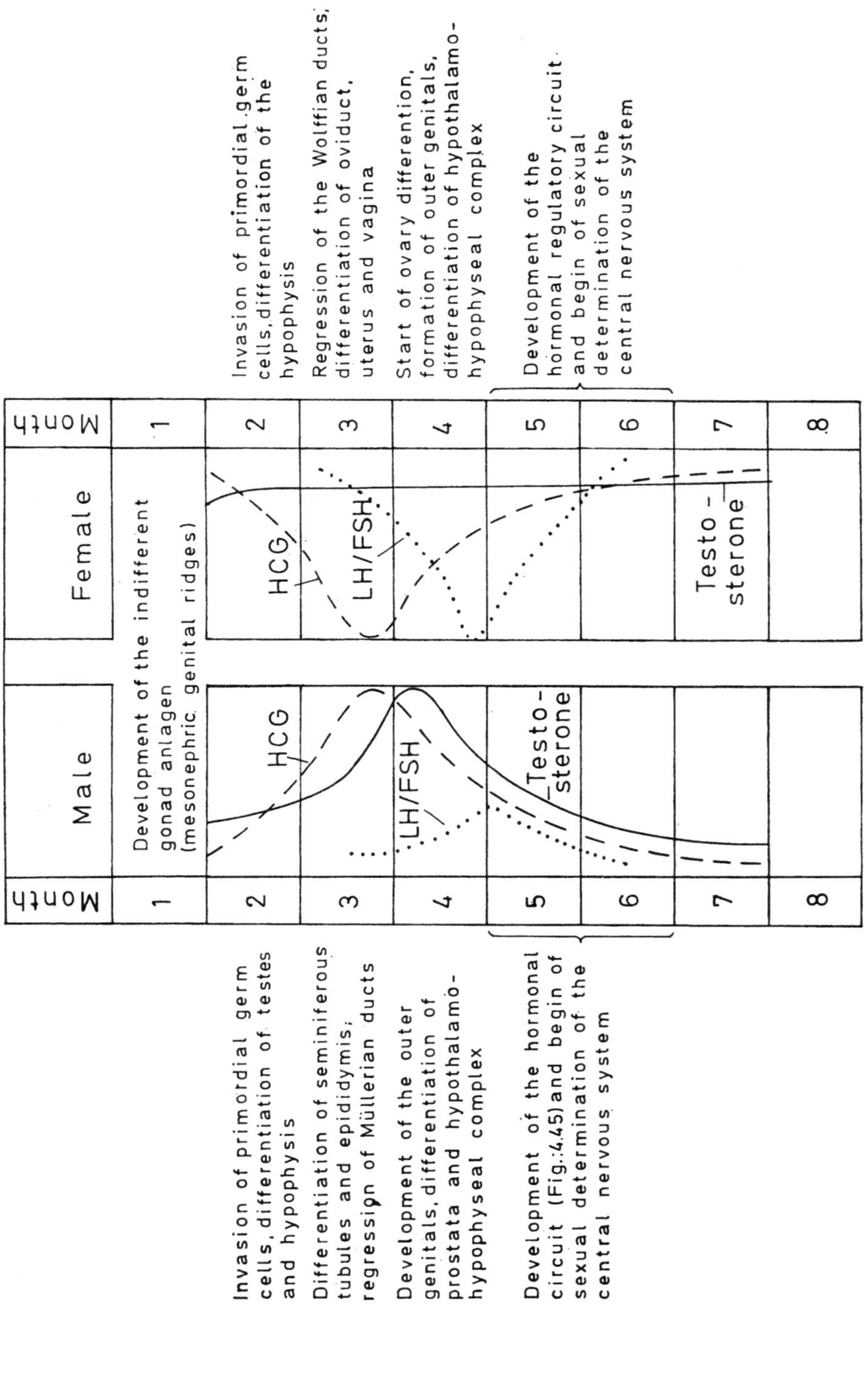
Month
Male
Female
1
2
3
4
5
6
7
8
Development of the indifferent gonad anlagen (mesonephric genital ridges)
HCG
LH/FSH
Testo-sterone
Invasion of primordial germ cells, differentiation of testes and hypophysis
Differentiation of seminiferous tubules and epididymis; regression of Müllerian ducts
Development of the outer genitals, differentiation of prostata and hypothalamo-hypophyseal complex
Development of the hormonal circuit (Fig. 4.45) and begin of sexual determination of the central nervous system
Invasion of primordial germ cells, differentiation of the hypophysis
Regression of the Wolffian ducts; differentiation of oviduct, uterus and vagina
Start of ovary differention, formation of outer genitals, differentiation of hypothalamo-hypophyseal complex
Development of the hormonal regulatory circuit and begin of sexual determination of the central nervous system

Fig. 4.44

Presence of the Y-chromosome results in development of a male, its absence in development of a female organism. To put it the other way round, embryo development, after passing the early, sexually neutral stages, leads to a female organism if not reprogrammed by gene products coded by the Y-chrosome. Two proteins have key roles in this respect. These are the membrane-associated H–Y antigen, responsible for early testicular organogenesis, and the intracellular receptor for testosterone (*Tfm*-locus).

The H–Y antigen is a small hydrophobic protein bound to the outer membrane of all cells of embryonic and adult male mammals [333–335]. Expression of the H–Y antigen was found to commence very early, in the 8-cell stage of the pre-implantation embryo. It consists of subunits with $M_r$ 18 kd (man) connected by disulphide bridges to dimeric, trimeric and tetrameric molecules. Binding of the H–Y antigen requires special anchorage sites provided by the universal outer membrane glykoprotein complex of mammalian cells, the major histocompatibility antigen complex composed of a small protein subunit ($\beta_2$-microglobulin, $M_r$ 12 kd) and a membrane-integrated large subunit (MHC, $M_r$ 45 kd).

Testicular organogenesis by the H–Y antigen further depends on specific membrane-bound receptors in the fetal gonad anlagen. Hence the specificity of the overall effect results from combination of a tissue-restricted, but not sex-specific receptor with a sex-specific, but not tissue-specific hormone-like protein that is presented at the outer membrane of cells by anchorage to the MHC protein complex. The decisive role of H–Y antigen for testis organization was shown by organ culture experiments using bovine embryonic indifferent gonads from genetically female (XX) embryos. Testicular conversion was observed after five days of organ culture if purified H–Y antigen was added [335].

The H–Y antigen is considered as an important representative of a class of proteins that serve as embryonic inductors during early organogenesis. The plasma membranes of embryonic cells are supposed to contain different such proteins associated with the MHC complex (anchorage sites). Their specific role for organogenesis depends on cell-cell contacts and the presence of corresponding receptor sites in the target cell membranes. It is an interesting fact that many proteins characterized as embryonic inductors continue to be expressed also in the adult organism. They are considered responsible for maintenance of tissue organization and specialisation [315].

2. *Gonadal sex determination* begins about 7 weeks after fertilization. It is triggered by invasion of the sexually indifferent gonad anlagen by the primordial germ cells (PGC). They migrate from a position outside of the embryo (endodermal epithelium of the yolk sac) through several embryonic tissues to the region of the developing kidneys, where the primitive genital ridges are localized.

   Mammalian PGC migrate actively by using amoeboid movements. Excretion of digestive enzymes probably helps them to clear a path through the extracellular matrices of hyaluronate, found to be enriched in the environment of developing embryonic tissue. There is some evidence for a chemotactic signal (telopheron) guiding the PGC on their way to the gonad anlagen. Migration of the PGC coincides with a dramatic increase of the germ cell number by mitosis [331].

It is a remarkable finding that distinct cytoplasmic components in the eggs of the fruit fly *Drosophila* are responsible for the early determination of the primordial germ cells. Some densely staining RNP granules in the cortical cytoplasm near the posterior pole of the oocyte become incorporated into specific pole cells which later on give rise to the PGC. The evidence for this stems from the following experiments [336]:

1. Destroying the pole cells by UV irradation leads to flies without germ cells.
2. Distinct mutants of *Drosophila (grandchildless)* do not form pole cells and consequently produce agametic adults.

3. Genetically marked pole cells can be transplanted into recipient embryos and produce fertile gametes with the marker gene.
4. Formation of functional pole cells can be induced by transplanting cytoplasm from the posterior pole into the anterior or ventral region of early embryos.

It is unclear, whether distinct mRNA species contained in the RNP granules and/or regulatory proteins associated with them are responsible for this determinative effect. Circumstantial evidence points to a similar mechanism of primordial germ cell determination also in mammalian oocytes [331].

Key event of the gonadal sex determination is the testicular differentiation in the genetically male fetus that proceeds under the influence of the H-Y antigens, of the primordial germ cells and of high levels of the maternal gonadotropin (HCG). An early step is the specialization of embryonic interstitial Leydig cells to the production of testosterone, whose excessive formation evoked by HCG is decisive for the further development. It is worth noticing that in a genetically female fetus (no H–Y antigen, low level of testosterone) gonad differentiation is delayed until the 4th month of pregnancy.

3. *Somatic sex determination* is a direct consequence of the second phase, i.e., mainly of the high testosterone levels in the male fetus. It involves the ontogenesis of all inner and outer parts of the sexual apparatus (cf. Fig. **4**.44) and includes differentiation of the female gonads (ovary), which is promoted by the onset of the fetal gonadotropin secretion (LH/FSH).

4. The following phase (4th–5th month of pregnancy) was called the phase of *psychic sex determination*, because it is characterized by the differentiation and beginning function of the brain control centres. Formation of the hypophyseal anlage is observed already in the first month, whereas the essential parts of the hypothalamus with the median eminence and the portal system (cf. Fig. **4**.45) develops only in the 3rd–4th month. An important difference between the male and female fetus in this phase is the much higher level in females of gonadotropins (LH/FSH) secreted by the fetal hypophysis (cf. Fig. **4**.44). Towards the last third of pregnancy all three types of developmental hormones (HCG, testosterone, LH/FSH) decline or disappear. The fetal Leydig cells in the testis degenerate. The essential result of this phase are sex-specific hormonal circuits that connect the sexual centres in the brain with the peripheral parts of the sexual system (cf. the example for the male organism in Fig. **4**.45).

5. The last stage of sexual development proceeds after birth, in the prepubertal phase. On the basis of the organ and regulatory pattern generated during fetal ontogenesis maturation of the sexual system involves major chages of all its parts. Besides many hormones, numerous environmental factors, including psychosocial factors exert major influence on the *determination of the sexual behaviour* of the adult organism.

#### 4.6.4.2 Germ Cell Differentiation

The developed hormonal regulatory circuit of the adult mammalian organism is a complex intercellular signal system that connects distinct parts of the hypothalamus with the adenohypophysis and the gonads. As illustrated in Fig. **4**.45 for the male organism, it involves (1) the hypothalamic gonadotropin releasing hormone (GRH), (2) the folicle stimulating hormone (FSH) and the luteotropic hormone (LH) of the anterior pituitary and (3) two types of hormones from testicular cells, androgens (e.g., testosterone) and inhibin.

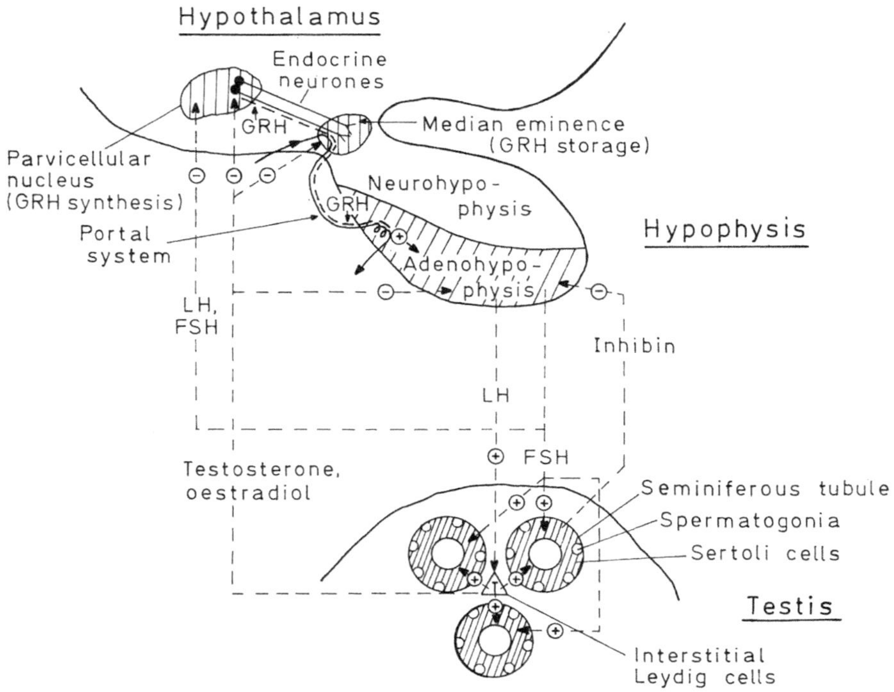


**Fig. 4.45:** Hormonal interactions in the hypothalamo-pituitary-gonad circuit control the function of testis in male mammals (after Dörner [330], Guillemin [337] and Döhler & Hancke [342]).

Gonadotropin releasing hormone *(GRH)* is formed in distinct parts of the parvicellular nucleus of the ventral hypothalamus. By uncharacterized endocrine neurones GRH is transported to the median eminence (junctional region between ventral hypothalamus and hypophysis) and from there by the blood stream in the portal system to the anterior lobe of the pituitary (adenohypophysis). It stimulates synthesis and secretion of two gonadotropic peptide hormones, the folicle stimulating hormone *(FSH)* and the luteotropic hormone *(LH)*. By the blood stream both hormones reach their target organs, e.g., the testis in male animals. FSH acts on the somatic cells of seminiferous tubules (Sertoli cells), whereas LH promotes formation of testosterone by the interstitial Leydig cells (cf. **4.2.4.2**). Testosterone, by interaction with Sertoli cells, controls distinct steps of spermatogenesis. The hormonal equilibrium within the circuit is maintained by numerous positive (--⊕-->) and negative (--⊖-->) regulatory effects on the synthesis and/or release of the hormones. Besides GRH, FSH, LH, testosterone and oestradiol, another hormone produced by Sertoli cells (inhibin) participates in the hormonal regulatory ciruit. Tissues responsive to steroid hormones (testosterone or oestrogens) are *shaded.*

Gonadotropin releasing hormone (GRH) is a decapeptide of 1182 d molecular weight [337]. The identical sequences of porcine and ovine GRH contain pyroglutamic acid at the N-terminus and glycinamide at the C-terminus: (pyro) Glu-His-Try-Ser-Tyr-Gly-Leu-Arg-Pro-Gly-$NH_2$. The gonadotropic hormones of the adenohypophysis (FSH and LH) are glykoproteins of $M_r$ 36 and 30 kd respectively [338]. Together with the closely related human chorionic gonadotropin (HCG), produced by the placenta and decisively influencing the early phases of the sexual development (cf. Fig. **4.44**), both share a common subunit with 92 amino acid residues ($\alpha$-subunit). The biological specificity of all three hormones resides in the $\beta$-subunit, which is unique for each hormone (145 amino acid residues for HCG, 115 for LH and 118 for FSH). All three gonadotropins promote steroidogenesis in their target cells after binding to specific receptors in the plasma membrane (cf. **4.2.4.2**).

The main androgenic steroid with 19 C-atoms is testosterone (cf. Fig. 4.5) produced by interstitial Leydig cells. The other testicular hormone (inhibin) is a barely characterized polypeptide of Sertoli cells [340] shown to selectively inhibit secretion of FSH by the adenohypophysis. Notably, the feedback control of the hypothalamic functions by testosterone depends on its prior aromatization, by the Leydig cells themselves or by brain tissue, to the femal sex hormone oestradiol (cf. Fig. 4.5). The same holds true for the role of testosterone for male-specific brain differentiation, which, in fact, is also brought about by oestrogens.

From the point of view of cell differentiation spermatogenesis is a prominent part of the adult sexual system in male. It proceeds in the seminiferous tubules of the testis. Starting with the stem cells (spermatogonia A) it takes about 64 days in man, 60 days in rat and 35 days in mouse to form the highly specialized haploid sperm cells (Fig. 4.46). After an initial proliferative phase, primary spermatocytes appear which undergo singular processes of protein synthesis, DNA replication and rearrangement as part of the meiotic cycles. The resulting four haploid spermatids are further specialized at the ultrastructural and molecular levels to the functional gametes (spermiogenesis). Prominent ultrastructural changes are the formation of the typical long flagellum and the special spiral-like arrangement of the mitochondria around its basis, the increasing condensation and inactivation of chromatin in the sperm head and the biogenesis of a specialized lysosomal compartment (acrosome) at the tip of the sperm.

The acrosome plays a major role in fertilization. Besides other lysosomal marker enzymes, it contains large amounts of hyaluronidase and the precursor form of a serine protease (proacrosine). After ejaculation, the sperm being unable to fertilize, it must undergo several hours of processing in the female genital tract termed "capacitation" [348], which essentially involves the activation of the acrosomal enzymes and specialization of the acrosomal membrane to undergo the "acrosome reaction". This reaction is a controlled loss of membranes from the sperm head. It occurs as a response to factors from newly ovulated eggs and leads to the liberation of acrosomal enzymes required for the sperm to penetrate the outer layers of the egg.

Germ cell differentiation in male mammals depends on two types of cell-cell interactions: (1) up to the late stages of spermiogenesis germ cells are connected by intercellular plasma bridges in a functional syncytium and (2) only the close contact with Sertoli cells provides the appropriate microenvironment for meiosis and spermatid maturation. Isolated germ cells from male rats can be cultured in vitro for several weeks but do not differentiate. This somatic control of germ cell differentiation [331] evidently includes the production of meiosis control factors as well as the signal transformation chain for testosterone, which is essential for many steps of spermatogenesis but interacts only with Sertoli cells. As a consequence of these control mechanisms, phenotypic sex reversal of differentiating germ cells can be induced, e.g., by transplantation into gonadal tissue of the opposite sex or by hormone administration. However, with rare but significant exceptions, germ cell differentiation is abortive in such systems with incongruent genetic and somatic sex. The nature of these intercellular control factors, their transmission and mode of action in the differentiating germ cells are practically unknown.

Spermatogenesis is characterized by three phases of hnRNA transcription which proceeds at a very high level in the spermatogonial compartment, in the primary spermatocytes during a distinct stage of the long meiotic prophase (pachytene stage) and finally in early spermatides [343]. The last stages of spermatide specialization (beginning with the cap phase, cf. Fig. 4.46) depend on the selective translation of preformed mRNP. The changes in the transcription activity are connected with pronounced alterations of the nuclear morphotypes (cf. the schematic drawings in Fig. 4.46) and with the increasing replacement of the usual nucleosomal histones first by testis-specific forms of H1 and H2B (TH1 and TH2B), followed by special

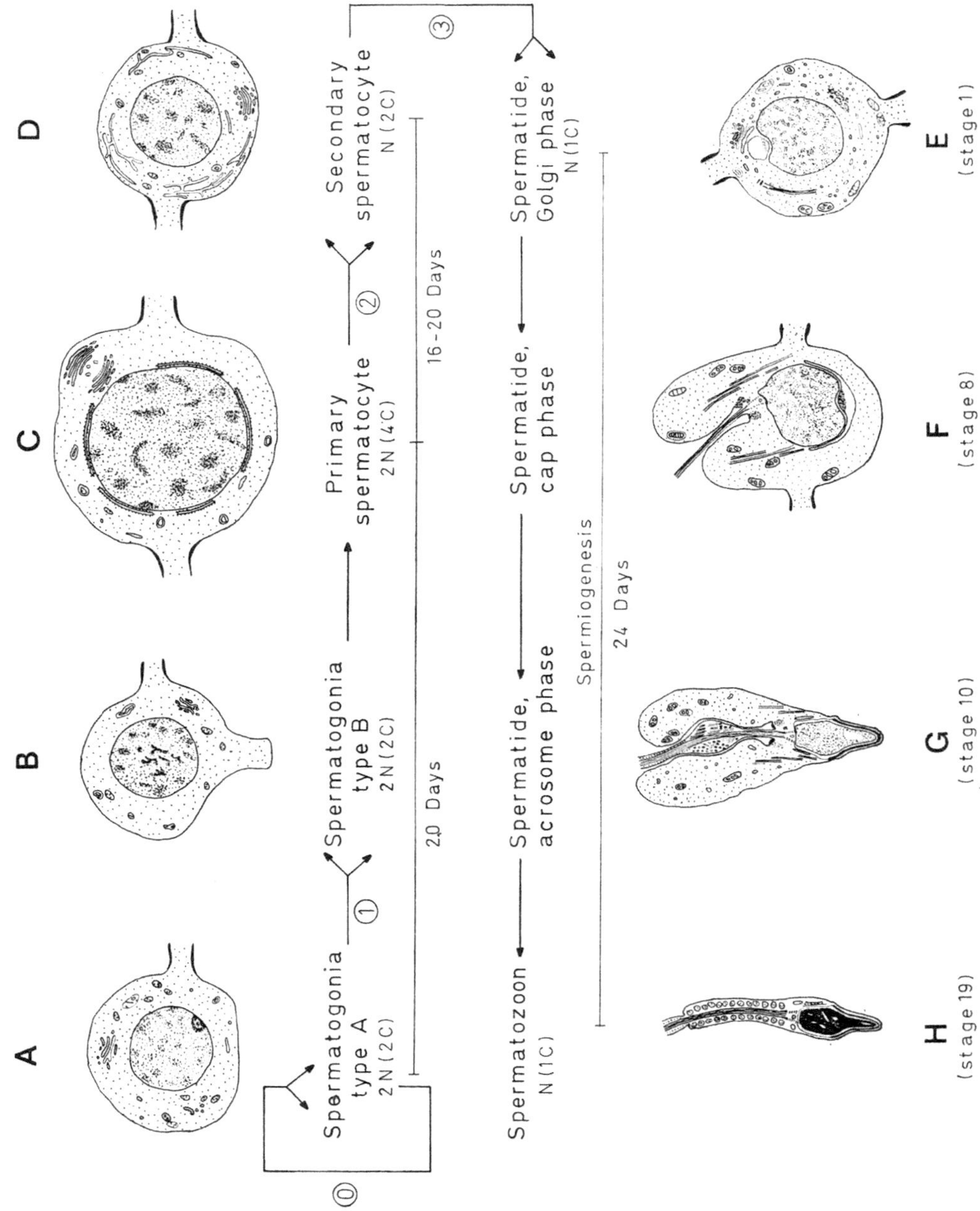
A
B
C
D
Spermatogonia type A 2N(2C)
Spermatogonia type B 2N(2C)
Primary spermatocyte 2N(4C)
Secondary spermatocyte N(2C)
0
1
2
3
20 Days
16-20 Days
Spermatozoon N(1C)
Spermatide, acrosome phase
Spermatide, cap phase
Spermatide, Golgi phase N(1C)
Spermiogenesis
24 Days
H (stage 19)
G (stage 10)
F (stage 8)
E (stage 1)

Fig. 4.46

◀

**Fig. 4.46:** Cell specialization during spermatogenesis in man (from [347] and Bargmann: Histologie und mikroskopische Anatomie des Menschen. 7. Aufl.: G. Thieme Verl., Stuttgart 1977, pp. 520–548).

The stem cells of the sperm cell lineage are spermatogonia type A which are attached to the basal lamina of the seminiferous tubules. They undergo proliferative mitoses (**0**) or quantal mitoses (**1**) to generate spermatogonia type B which, in turn, differentiate to the primary spermatocytes. The meiotic division (**2**, **3**) of spermatocytes finally yields four haploid spermatides that are subject to a long and extensive cell specialization (spermiogenesis). A remarkable feature of the first meiotic division (**2**) is an extremely long prophase (~ 16 days) forming the basis for the reduction of the diploid (44: x, y) to the haploid (22: x and 22: y, respectively) chromosome number.

The figures illustrate typical morphological features of the pertinent cell types which are frequently connected with each other by plasma bridges to a differentiating syncytium. Only with the acrosome phase *(G)* are the cells totally separated from each other, but remain up to the very last stages *(H)* in intimate contact with the surrounding Sertoli cells.

All figures were drawn after electron micrographs (from Holstein, J. F. & H. Wartenberg: Anat. Anz. Ergänzungsband, Verh. Anat. Ges., 70. Versammlung, G. Fischer Verlag, Jena 1976).

**A**: Spermatogonia A are flat cells attached to the basal lamina of the seminiferous tubules. They contain dense nuclei with fine peripheral chromatin granules.

**B**: Spermatogonia B are barely or not at all attached to the basal lamina. Their nucleus appears coarsly granular.

**C**: Primary spermatocytes are especially large corresponding to the doubled DNA content (4C) formed in preparation of the meiotic division. The very large nuclei show different stages of the meiotic prophase.

**D**: Secondary spermatocytes are small cells with dense nuclei exhibiting finely granular chromatin and two or three large masses of condensed chromatin.

**E**: The four early spermatides resulting from the meiotic division possess small spherical nuclei. Initials of the specialized lysosomes (acrosomal vesicle with an acrosomal granule) have been formed by the Golgi apparatus. One of the two centrioles gives rise to the flagellum.

**F**: The cap phase of the spermatide specialization (day 7 of spermiogenesis) shows the acrosomal vacuole progressively covering the nucleus which moves to the periphery of the cell. The two centrioles are part of a special apparatus which forms the basis of the elongating flagellum inserted at the nuclear membrane.

**G**: The acrosomal phase (about day 9 of spermiogenesis) is characterized by a progressive condensation of chromatin in the elongating cell nucleus. The cap-like acrosome is densely packed with acrosomal material (hyaluronidase, trypsin-like protease (acrosin)). The bulk of the cytoplasm is now displaced behind the caudal pole of the nucleus surrounding the proximal part of the flagellum. A cylindrical structure with cytoplasmic microtubules (manchette) encircles the initial part of the flagellum.

**H**: The mature spermatide (spermatozoon) exhibits the typical form of the nucleus with highly condensed chromatin. The acrosome forms a cap around the nucleus. Numerous mitochondria have arranged in a spiral-like manner to form a sheath around the medium part of the flagellum. The manchette disappeared in the last stages of maturation. In addition to the axial filament, the flagellum now contains nine coarse "outer" fibres. Most of the excess cytoplasm has been removed.

**Table 4.15**: Changes of histone composition of chromatin during spermatogenesis in rats (from Meistrich et al. [344]).

| Stages | Days of spermatogenesis | Histone composition of chromatin | Morphological and biochemical features of the nucleus |
|---|---|---|---|
| Spermatogonia A and B, early primary spermatocyte | 0–25 | Somatic histones H1, H2A, H2B, H3, H4 | Nucleosomal organization, transcription |
| Primary spermatocyte (pachytene stage), early spermatide (stage 8) | 25–47 | Somatic histones H2A, H3, H4; H1 and H2B largely replaced by testis histones TH1, TH2B (synthesized in pachytene stage of prim. spermatocyte) | Nucleosomal organization, transcription |
| Middle spermatide (stages 9–14) | 47–52 | Sperm histones TP, TP2, TP4 | Smooth chromatin fibre, arrest of transcription, beginning condensation |
| Late spermatide (stages 15–19) | 52–60 | TP4, TP3, S1 | Smooth chromatin, no transcription, further chromatin condensation |
| Mature sperm (epididymis) | ~65 | S1 | Inactive, highly condensed chromatin by cross-linking of S1 cysteine residues |

sperm histones (TB, TP2, TP4, RP3) and finally by a single basic protein (S1) with high cysteine content (Table 4.15), whose cross-linking by disulphide bridges essentially contributes to the unusually high condensation of DNA in the protease-resistant DNP complex of the sperm head.

The life of the stage 19 spermatozoon as it is released from the seminiferous tubule has three preparatory stages that are mandatory for its fertilization capacity: (1) a one to several weeks maturation in the epididymis, (2) an activation of the acrosome in the sperm head (capacitation) and (3) the rapid controlled loss of the membranes at the sperm head in vicinity of the newly ovulated egg (acrosome reaction) [348].

Although provided with all necessary equipments, sperms leaving the testis are not capable to move. They rather gain the capacity for forward motility during their long passage through the narrow convoluted tubules of the epididymis. This maturation involves changes in shape, surface charge and metabolic properties. Significant for the development of the forward motility are a two-phasic increase of the cAMP level and the binding of an acidic, heat-stabile glycoprotein (forward motility protein) formed either in the testis or the epididymis. The expression of the acquired potential for motility at the time of ejaculation is connected with the second rise of the cAMP level and requires high levels of calcium and perhaps other triggering factors contributed by the accessory gland secretions [345].

Structural and functional integrity of the epididymis and its decisive influence on sperm maturation and survival is intimately connected with the maintenance of a sufficiently high level of androgens. An androgen binding protein with affinity for testosterone and 5$\alpha$-dihydrotestosterone plays a major role in this respect. It is produced by Sertoli cells and secreted via the testicular fluid into the epididymis.

Despite many unsolved problems, the foregoing brief discussion about development and function of the mammalian sexual system demonstrates several interesting mechanisms and findings that are of basic importance for an understanding of cell differentiation in the frame work of the developmental biology of multicellular organisms. These are (1) the determination of germ cell differentiation by cytoplasmic factors in the mature oocyte, (2) the action of embryonic inducer proteins exemplified by the H–Y antigen, (3) the intricate role of direct cell-cell interactions for testis differentiation and spermatogenesis, (4) the migration of cells over considerable distance as a necessary part of cell sorting during early embryonic development, (5) the generation of determined hormonal response patterns that control function and differentiation of cells in remote parts of the adult organism and (6) the terminal specialization of sperm cells to highly efficient and secure vehicles of the male chromosomal set.

## 4.7 Conclusions

Cell differentiation results from the qualitatively and quantitatively selective realization of distinct parts of a given genetic material. The complexity of the genetic material of all living organisms and the constant need of cells to react, by changes of the gene expression patterns, to signals from the extracellular surroundings makes cell differentiation a prerequisite of life, at least in its present form.

Five main aspects of cell differentiation have been treated consecutively:

1. signal reception and transformation,
2. selective genome rearrangements,
3. differential gene expression (including transcription, pre-mRNA processing and mRNA degradation, translation, proteinogen processing and protein degradation),
4. mechanisms for coordination of gene expression programs and
5. intercellular integration of cell differentiation into the development and function of multicellular systems.

In all cases gene expression and the prominent role of signal systems are in the centre of interest. The overview of molecular mechanisms contributing to the general picture of cell differentiation incorporates experimental results from many different prokaryotes and eukaryotes. This procedure points to the basic similarities between all living systems as far as the selectivity and quantitative control of gene expression are concerned. But it is important to recall that the multiplicity of forms and functions encountered in recent organisms just reflect the inexhaustible variations in the details of a common theme "cell differentiation". We are just at the beginning to unravel these details, as is evident from the discussion of 14 selected experimental systems in the following special part of this book.

## 4.8 Selected References

### 4.8.1 Books

1. Mohr, H. & P. Sitte: Molekulare Grundlagen der Entwicklung. Akademie-Verlag, Berlin 1971.
2. Lewin, B.: Gene Expression, 2nd Ed., Vols. 1 and 2. John Wiley & Sons, New York 1979/1980.
3. Träger, L.: Einführung in die Molekularbiologie. G. Fischer Verlag, Jena 1975.

4. Nover, L. & K. Mothes (eds.): Cell Differentiation in Microorganisms, Plants and Animals. G. Fischer Verlag and North-Holland Publ. Comp., Jena and Amsterdam 1977.
5. Russell, T. R., K. Brew, H. Faber & J. Schultz (eds.): From Gene to Protein: Information Transfer in Normal and Abnormal Cells. Academic Press, New York 1979.
6. Nover, L., F. Lynen & K. Mothes (eds.): Cell Compartmentation and Metabolic Channeling. G. Fischer Verlag and Elsevier/North-Holland Biomed. Press, Jena and Amsterdam 1980.
7. Von Sengbusch, P.: Molekular- und Zellbiologie. Springer, Berlin 1979.
8. Davidson, E. J.: Gene Activity in Early Development. Academic Press, New York 1976.
9. Scandalios, J. G. (ed.): Physiological Genetics. Academic Press, New York 1979.
10. Leighton, T. & W. F. Loomis (eds.): The Molecular Genetics of Development. Academic Press, New York 1980.
11. Goldstein, L. & D. M. Prescott (eds.): Cell Biology – A Comprehensive Treatise. Vol. 1: Genetic Mechanisms of Cells (1978); Vol. 2: The Structure and Replication of Genetic Material (1979); Vol. 3: Gene Expression: The Production of RNAs (1980); Vol. 4: Gene Expression: Translation and the Behaviour of Proteins (1980). Academic Press, New York 1978–1980.

**4.8.2 Signal Systems**

*Group I*

12. Beckwith, J. & P. Rossow: Analysis of genetic regulatory mechanisms. Ann. Rev. Genetics *8* (1974): 1–13.
13. Magasanik, B.: Regulation of the synthesis of degradative bacterial enzymes. In: [4], pp. 438–460.
14. Davis, R. H.: Arginine metabolism in *Neurospora*: The role of an amino acid compartment. In: [6], pp. 239–243.
15. Oaks, A. & P. Gadal: Nitrate utilization in cells of higher plants. In: [6], pp. 245–254.
16. Basu, S. K., J. L. Goldstein, R. G. W. Anderson & M. S. Brown: Monensin interrupts the recycling of low density lipoprotein receptors in human fibroblasts. Cell *24* (1981): 493–502.
17. Goldstein, J. L. & M. S. Brown: The low-density lipoprotein pathway and its relation to atherosclerosis. Ann. Rev. Biochem. *46* (1977): 897–930.
18. Goldstein, J. L., R. G. W. Anderson & M. S. Brown: Coated pits, coated vesicles and receptor-mediated endocytosis. Nature *279* (1979): 679–685.
19. Davies, P. J. A., D. R. Davies, A. Levitzki, F. R. Maxfield, P. Milhaud, M. C. Willingham & I. Pastan: Transglutaminase is essential in receptor-mediated endocytosis of $\alpha_2$-macroglobulin and polypeptide hormones. Nature *283* (1980): 162–167.
20. Rothman, J. E. & R. E. Fine: Coated vesicles transport newly synthesized membrane glycoproteins from endoplasmic reticulum to plasma membrane in two successive stages. Proc. Natl. Acad. Sci. USA *77* (1980): 780–784.

*Group II*

21. Nover, L.: Structural and functional organization of differential gene expression in prokaryotic and eukaryotic cells. In: [6], pp. 367–387.
22. Rickenberg, H. V.: Cyclic AMP in prokaryotes. Ann. Rev. Microbiol. *28* (1974): 353–369.
23. Gallant, J. A.: Stringent control in *E. coli*. Ann. Rev. Genetics *13* (1979): 393–415.
24. Peterkofsky, A.: Transport system-dependent regulation of adenylate cyclase. Horiz. Biochem. Biophys. Vol. 4. Addison-Wesleey Publ. Comp., Reading Mass. 1977, pp. 244–265.
25. Peterkofsky, A. & C. Gazdar: *Escherichia coli* adenylate cyclase complex: Regulation by the proton electrochemical gradient. Proc. Natl. Acad. Sci. USA *76* (1979): 1099–1103.
26. Reider, E., E. F. Wagner & M. Schweiger: Control of phosphoenolpyruvate-dependent phosphotransferase-mediated sugar transport in *Escherichia coli* by energization of the cell membrane. Proc. Natl. Acad. Sci. USA *76* (1979): 5529–5533.
27. Adler, S. P., J. H. Mangum, G. Magni & E. R. Stadtman: Uridylylation of the $P_{II}$ regulatory protein in cascade control of *E. coli* glutamine synthetase. In: Fischer, E. H., E. G. Krebs, H. Neurath & E. R. Stadtman (eds.): Metabolic Interconversion of Enzymes 1973. Springer Verlag, Berlin 1974, pp. 221–233.
27a. Chock, P. B., S. G. Rhee & E. R. Stadtman: Interconvertible enzyme cascades in cellular regulation. Ann. Rev. Biochem. *49* (1980): 813–843.

28. Kustu, S., D. Burton, E. Garcia, L. McCarter & N. McFarland: Nitrogen control in *Salmonella*: Regulation by the glnR and glnF gene products. Proc. Natl. Acad. Sci. USA *76* (1979): 4576–4580.
29. Pahel, G. & B. Tyler: A new glnA-linked regulatory gene for glutamine synthetase in *Escherichia coli*. Proc. Natl. Acad. Sci. USA *76* (1979): 4544–4548.
30. Foor, F., Z. Teuveny & N. Magasanik: Regulation of the synthesis of glutamine synthetase by the $P_{II}$ protein in *Klebsiella aerogenes*. Proc. Natl. Acad. Sci. USA *77* (1980): 2636–2640.
31. McFarland, N., L. McCarter, S. Artz & S. Kustu: Nitrogen regulatory locus "*gln*R" of enteric bacteria is composed of cistrons *ntr*B and *ntrC*: Identification of their protein products. Proc. Natl. Acad. Sci. USA *78* (1981): 2135–2139.

*Group III*

32. Greengard, O.: Enzymic differentiation in mammalian tissues. Essays in Biochem. *7* (1971): 159–206.
33. Sutherland, E. W.: Studies on the mechanism of hormone action. Science *177* (1972): 401–408.
34. Singer, S. J. & G. Nicolson: The fluid mosaic model of the structure of cell membranes. Science *175* (1972): 720–731.
35. Cuatrecasas, P.: Membrane receptors. Ann. Rev. Biochem. *43* (1974): 169–214.
36. Edelman, G. M.: Surface modulation in cell recognition and cell growth. Science *192* (1976): 218–226.
37. Singer, S. J., J. F. Ash, L. Y. W. Bourguinon, M. H. Heggeness & D. Louvard: Transmembrane interactions and the mechanisms of transport of proteins across membranes. J. Supramol. Structure *9* (1978): 373–389.
38. Greengard, P.: Phosphorylated proteins as physiological effectors. Science *199* (1978): 146–152.
39. Jungmann, R. A., M. I. Mednieks & J. S. Schweppe: Nuclear compartmentation of protein kinase and the regulation of gene expression. In: [6], pp. 415–426.
40. Catt, K. J., J. P. Harwood, G. Aguilera & M. L. Dufau: Hormonal regulation of peptide receptors and target cell responses. Nature *280* (1979): 109–116.
41. Amsterdam, A., A. Berkowitz, A. Nimrod & F. Cohen: Aggregation of luteinizing hormone receptors in granulosa cells: A possible mechanism of desensitization to the hormone. Proc. Natl. Acad. Sci. USA *77* (1980): 3440–3444.
42. Schulster, D.: Coupling between membrane hormone receptors and adenylate cyclase in tumor cells. Biochem. Soc. Trans. *7* (1979): 310–314.
43. Hirata, F., W. J. Strittmatter & J. Axelrod: $\beta$-Adrenergic receptor agonists increase phospholipid methylation, membrane fluidity, and $\beta$-adrenergic receptor-adenylate cyclase coupling. Proc. Natl. Acad. Sci. USA *76* (1979): 368–372.
44. Eimerl, S., G. Neufeld, M. Korner & M. Schramm: Functional implantation of a solubilized $\beta$-adrenergic receptor in the membrane of a cell. Proc. Natl. Acad. Sci. USA *77* (1980): 760–764.
45. Rodbell, M.: The role of hormone receptors and GTP-regulatory proteins in membrane transduction. Nature *284* (1980): 17–22.

**4.8.3 Genome Rearrangements**

46. Egel, R.: Springende Gene: Schalter in der Gen-Struktur. Biologie uns. Zeit *7* (1977): 148–155.
47. Bukhari, A., J. Shapiro & S. Adhya (eds.): DNA Insertion Elements, Plasmids, and Episomes. Cold Spring Harbor Lab., Cold Spring Harbor, New York 1977.
48. Calos, M. P. & J. H. Miller: Transposable elements. Cell *20* (1980): 579–595.
48a. Ju, G. & A. M. Skalka: Nucleotide sequence analysis of the long terminal repeat (LTR) of avian retroviruses: Structural similarities with transposable elements. Cell *22* (1980): 379–386; Temin. H. M.: Origin of retroviruses from cellular moveable genetic elements. Cell *21* (1980): 599–600.
49. Iino, T.: Genetics of structure and function of bacterial flagella. Ann. Rev. Genetics *11* (1977): 161–182.
50. Zieg, J., M. Silverman, M. Hilmen & M. I. Simon: The mechanism of phase variation. In: Miller, J. H. & W. S. Reznikoff (eds.): The Operon, Cold Spring Harbor Lab., Cold Spring Harbor, New York 1978, pp. 411–423.

51. Silverman, M., J. Zieg, M. Hilmen & M. Simon: Phase variation in *Salmonella:* Genetic analysis of a recombinational switch. Proc. Natl. Acad. Sci. USA *76* (1979): 391–395.

52. Kutzukake, K. & T. Iino: A trans-acting factor mediates inversion of a specific DNA segment in flagellar phase variation of *Salmonella*. Nature *284* (1980): 479–481.

53. Silverman, M. & M. Simon: Phase variation: Genetic analysis of switching mutants. Cell *19* (1980): 845–854.

54. Zieg, J. & M. Simon: Analysis of nucleotide sequence of an invertible controlling element. Proc. Natl. Acad. Sci. USA *77* (1980): 4196–4200.

55. Cross, G. A. M.: Antigenic variation in trypanosomes. Proc. R. Soc. Lnd. B. *202* (1978): 55–72.

56. Hoeijmakers, J. H. J., A. C. C. Frasch, A. Bernards, P. Borst & G. A. M. Cross: Novel expression-linked copies of the gene for variant surface antigens in trypanosomes. Nature *284* (1980): 78–80.

57. Herskowitz, I., L. Blair, D. Forbess, J. Hicks, Y. Kassir, P. Kushner, J. Rine, G. Sprague Jr. & J. Strathern: Control of cell type in the yeast *Saccharomyces cerevisiae* and a hypothesis for development in higher eukaryotes. In: [10], pp. 79–118.

58. Hicks, J., J. N. Strathern & A. J. S. Klar: Transposable mating type genes in *Saccharomyces cerevisiae*. Nature *282* (1979): 478–483.

59. Strathern, J., E. Spatola, C. McGill & J. B. Hicks: Structure and organization of transposable mating type cassettes in *Saccharomyces* yeasts. Proc. Natl. Acad. Sci. USA *77* (1980): 2839–2843.

60. Nasmyth, K. A. & K. Tatchell: The structure of transposable yeast mating type loci. Cell *19* (1980): 753–764.

61. Malone, R. E. & R. E. Esposito: The RAD52 gene is required for homothallic interconversion of mating types and spontaneous mitotic recombination in yeast. Proc. Natl. Acad. Sci. USA *77* (1980): 503–507.

62. Young, M. W.: Middle repetive DNA: A fluid component of the *Drosophila* genome. Proc. Natl. Acad. Sci. USA *76* (1979): 6274–6278.

63. Levis, R., P. Dunsmuir & G. M. Rubin: Terminal repeats of the *Drosophila* transposable element copia: Nucleotide sequence and genomic organization. Cell *21* (1980): 581–588.

64. Potter, S. S., W. J. Brorein, P. Dunsmuir & G. M. Rubin: Transposition of elements of the 412, copia and 297 dispersed repeated gene families in *Drosophila*. Cell *17* (1979): 415–427.

65. Gehring, W. J. & R. Paro: Isolation of a hybrid plasmid with homologous sequences to a transposing element of *Drosophila melanogaster*. Cell *19* (1980): 897–904.

66. Cameron, J. R., E. Y. Loh & R. W. Davis: Evidence for transposition of dispersed repetitive DNA families in yeast. Cell *16* (1979): 739–751.

67. McClintock, B.: Chromosome organization and genetic expression. Cold Spring Harb. Symp. Quant. Biol. *16* (1951): 13–47.

68. Peterson, P. A.: Basis for the diversity of states of controlling elements in maize. Mol. Gen. Genetics *149* (1976): 5–21.

69. Peterson, P. A.: Unstable genetic loci as a probe in morphogenesis. Brookhaven Symp. Biol. *25* (1973): 244–261.

70. Dooner, H. K. & O. E. Nelson, jr.: Heterogenous flavonoid glucosyltransferases in purple derivatives from a controlling element suppressed bronze mutant in maize. Proc. Natl. Acad. Sci. USA *76* (1979): 2369–2371.

#### 4.8.4 Bacterial Transcription Units

71. Jacob, F. & J. Monod: Genetic regulatory mechanisms in the synthesis of proteins. J. Mol. Biol. *3* (1961): 318–356.

72. McKay, D. B. & T. A. Steitz: Structure of catabolite gene activator protein at 2.9 Å resolution suggests binding to left-handed B-DNA. Nature *290* (1981): 744–749.

72a. Smith, G. R.: DNA supercoiling: Another level for regulating gene expression. Cell *24* (1981): 599–600.

73. Dickson, R. C., J. Abelson, W. M. Barnes & W. S. Reznikoff: Genetic regulation: the lac control region. Science *187* (1975): 27–35.

74. Gilbert, W.: Starting and stopping sequences for the RNA polymerase. In Losick, R. & M. Chamberlin (eds.): RNA Polymerase. Cold Spring Harb. Lab., Cold Spring Harb., New York 1976, pp. 193–205.

75. Taniguchi, T., M. O'Neill & B. de Crombrugghe: Interaction site of *Escherichia coli* cyclic AMP receptor protein on DNA of galactose operon promoters. Proc. Natl. Acad. Sci. USA *76* (1979): 5090–5094.
76. Miller, J. H.: Genetic studies of the lac repressor XI. On aspects of lac repressor structure suggested by genetic experiments. J. Mol. Biol. *131* (1979): 249–258.
77. Wollenweider, H. J., M. Fiandt & W. Szybalski: A relationship between RNA helix stability and recognition sites for RNA polymerase. Science *205* (1979): 508–511.
78. Rosenberg, M. & D. Court: Regulatory sequences involved in the promotion and termination of RNA transcription. Ann. Rev. Genetics *13* (1979): 319–353.
79. Calos, M. P.: DNA sequence for a low-level promoter of the lac repressor gene and an "up" promoter mutation. Nature *274* (1978): 762–765.
80. Siebenlist, U., R. B. Simpson & W. Gilbert: *E. coli* RNA polymerase interacts homologously with two different promoters. Cell *20* (1980): 269–281.
81. Maquat, L. E., K. Thornton & W. S. Reznikoff: lac Promoter mutations located downstream from the transcription start site. J. Mol. Biol. *139* (1980): 537–549.

### 4.8.5 Electron Microscopy of Active Genes

82. Kleppe, K., S. Ovrebo & I. Lossius: The bacterial nucleoid. J. Gen. Microbiol. *112* (1979): 1–13.
83. Ryter, A. & A. Chang: Localization of transcribing genes in the bacterial cell by means of high resolution autoradiography. J. Mol. Biol. *98* (1975): 553–557.
84. Hamkalo, B. A. & O. L. Miller: Electronmicroscopy of genetic activity. Ann. Rev. Biochem. *42* (1973): 379–396.
85. Foe, V. E.: Modulation of ribosomal RNA synthesis in *Oncopeltus fasciatus:* An electron microscopic study of the relationship between changes in chromatin structure and transcriptional activity. Cold Spring Harb. Symp. Quant. Biol. *42* (1978): 723–739.
86. McKnight, S. L., M. Bustin & O. L. Miller: Electron microscopic analysis of chromosome metabolism in the *Drosophila melanogaster* embryo. Cold Spring Harb. Symp. Quant. Biol. *42* (1978): 741–754.

### 4.8.6 Chromatin Function

*Reviews and Books*

87. Cold Spring Harbour Symp. Quant. Biol., Vol. 42, Cold Spring Harb., New York 1978.
88. Bostock. C. J. & A. T. Sumner: The Eukaryotic Chromosome. North-Holland Publ. Comp., Amsterdam 1978.
89. Mathis, D., P. Oudet & P. Chambon: Structure of transcribing chromatin. Progr. Nucl. Acid Res. Mol. Biol. *24* (1980): 1–55.
90. Pederson, Th.: Chromatin structure and gene transcription: Nucleosomes permit a new synthesis. Int. Rev. Cytol. *55* (1978): 1–22.
91. Isenberg, I.: Histones. Ann. Rev. Biochem. *48* (1979): 159–191.
92. Lilley, D. M. J. & J. F. Pardon: Structure and function of chromatin. Ann. Rev. Genetics *13* (1979): 197–233.
93. Von Holt, C., W. N. Strickland, W. F. Brand & M. S. Strickland: More histone structures. FEBS Letters *100* (1979): 201–218.
94. Rindt, K.-P. & L. Nover: Chromatin structure and function. Biol. Zentralbl. *99* (1980): 641–673.
95. Paigen, K.: Acid hydrolases as models of genetic control. Ann. Rev. Genetics *13* (1979): 417–466.
96. Paigen, K.: Temporal genes and other developmental regulators in mammals. In [10].

*Globin Gene System*

97. Maniatis, T., E. F. Fritsch, J. Lauer & R. M. Lawn: The molecular genetics of human hemoglobins. Ann. Rev. Genetics *14* (1980): 145–178.
98. Gusella, J., A. Varsanyi-Breiner, F.-T. Kao, C. Jones, T. T. Puck, Ch. Keys, St. Orkin & D. Housman: Precise localization of human $\beta$-globin gene complex on chromosome 11. Proc. Natl. Acad. Sci. USA *86* (1979): 5239–5243.

99. Duncan, C., P. A. Biro, P. V. Choudary, J. T. Elder, R. C. Wang, B. G. Forget, J. K. de Riel & S. M. Weissman: RNA polymerase III transcriptional units are interspersed among human non-$\alpha$-globin genes. Proc. Natl. Acad. Sci. USA *76* (1979): 5095–5099.
100. Shen, C.-K. J. & T. Maniatis: The organization of repetitive sequences in a cluster of rabbit $\beta$-like globin genes. Cell *19* (1980): 379–391.
101. Jelinek, W. R., T. P. Toomey, L. Leinwand, C. H. Duncan, P. A. Biro, P. N. Choudary, S. M. Weisman, S. M. Rubin, C. M. Houck, P. L. Deininger & C. W. Schmid: Ubiquitous, interspersed repeated sequences in mammalian genomes. Proc. Natl. Acad. Sci. USA *77* (1980): 1398–1402.
102. Weintraub, H., A. Larsen & M. Groudine: $\alpha$-Globin-gene switching during the development of chicken embryos: expression and chromosome structure. Cell *24* (1981): 333–344.
103. Van der Ploeg, L. H. T. & R. A. Flavell: DNA methylation in the human $\gamma$-$\delta$-$\beta$-globin locus in erythroid and nonerythroid tissues. Cell *19* (1980): 947–958.
104. Mandel, J. L. & P. Chambon: DNA methylation: Organ specific variations in the methylation pattern within and around ovalbumin and other chicken genes. Nucleic Acids Res. *7* (1979): 2081–2103.
104a. Miller, D. M., P. Turner, A. W. Nienhuis, D. Axelrod & T. V. Gopalakrishnan: Active conformation of the globin genes in uninduced and induced erythroleukemia cells. Cell *14* (1978): 511–521.
105. Van der Ploeg, L. H. T., A. Konings, M. Ort, D. Roos, L. Bernini & R. A. Flavell: $\gamma$-$\beta$--Thalassaemia studies showing that deletion of the $\gamma$- and $\delta$-genes influence $\beta$-globin gene expression in man. Nature *283* (1980): 637–642.
105a. Fritsch, E. F., R. M. Lawn & T. Maniatis: Characterization of deletions which affect the expression of fetal globin genes in man. Nature *279* (1979): 598–603.
106. Weisbrod, S. & H. Weintraub: Isolation of a subclass of nuclear proteins responsible for conferring a DNase I-sensitive structure on globin chromatin. Proc. Natl. Acad. Sci. USA *76* (1979): 630–634.
107. Weisbrod, St., M. Groudine & H. Weintraub: Interaction of HMG 14 and 17 with actively transcribed genes. Cell *19* (1980): 289–301.
108. Stalder, J., A. Larsen, J. D. Engel, M. Dolau, M. Groudine & H. Weintraub: Tissue-specific DNA cleavages in the globin chromatin domain introduced by DNase I. Cell *20* (1980): 451–460.

*Higher Order Chromatin Structure*

109. Bernhard, W.: Differentiation of the cell nucleus. In: [4], pp. 579–597.
110. Cook, P. R.: On the inheritance of differentiated traits. Biol. Rev. *49* (1974): 51–84.
111. Wu, C., P. M. Bingham, K. J. Livak, R. Holmgren & S. C. R. Elgin: The chromatin structure of specific genes: I. Evidence for higher order domains of defined DNA sequence. Cell *16* (1979): 797–806.
112. Sung, M. T.: Phosphorylation and dephosphorylation of histone V (H5): Controlled condensation of avian erythrocyte. Biochemistry *16* (1977): 286–290.
113. Nelson, P. P., St. C. Albright, J. M. Wiseman & W. T. Garrard: Reassociation of histone H1 with nucleosomes. J. Biol. Chem. *254* (1979): 11751–11760.
114. Thoma, F., Th. Koller & A. Klug: Involvement of histone H1 in the organization of the nucleosome and of the salt-dependent superstructures of chromatin. J. Cell Biol. *83* (1979): 403–427.
115. Matsui, S.-I., B. K. Seon & A. A. Sandberg: Disappearance of a structural chromatin protein A24 in mitosis: Implications for molecular basis of chromatin condensation. Proc. Natl. Acad. Sci. USA *76* (1979): 6386–6390.
116. Busch, H.: The current excitement about gene controls of nuclear DNA. Life Sci. *23* (1978): 2543–2554.
117. Eberhardt, N. L., J. C. Ring, L. K. Johnson, K. R. Latham, J. W. Apriletti, R. N. Kitsis & J. D. Baxter: Regulation of activity of chromatin receptors for thyroid hormone: Possible involvement of histone-like proteins. Proc. Natl. Acad. Sci. USA *76* (1979): 5005–5009.
118. Kuehn, G. D., H.-U. Affolter, V. J. Atmar, T. Seebeck, U. Gubler & R. Braun: Polyamine--mediated phosphorylation of a nuclear protein from *Physarum polycephalum* that stimulates rRNA synthesis. Proc. Natl. Acad. Sci. USA *76* (1979): 2541–2545.

119. Engelke, D. R., S. Y. Ng, B. S. Shastry & R. G. Roeder: Specific interaction of a purified transcription factor with an internal control region of 5S RNA genes. Cell *19* (1980): 717–728.
120. Palter, K. B., V. E. Foe & B. M. Alberts: Evidence for the formation of nucleosome-like histone complexes on single-stranded DNA. Cell *18* (1979): 451–467.
121. Stein, A., J. P. Whitlock Jr. & M. Bina: Acidic polypeptides can assemble both histones and chromatin in vitro at physiological ionic strength. Proc. Natl. Acad. Sci. USA *76* (1979): 5000–5004.
122. Isackson, P. J., J. L. Fishback, D. L. Bidney & G. R. Reeck: Preferential affinity of high molecular weight high mobility group non-histone chromatin proteins for single-stranded DNA. J. Biol. Chem. *254* (1979): 5569–5572.
123. Javaherian, K., M. Sadeghi & L. F. Liu: Nonhistone proteins $HMG_1$ and $HMG_2$ unwind DNA double helix. Nucleic Acids Res. *6* (1979): 3569–3580.
124. Tata, J. R. & B. Baker: Enzymatic fractionation of nuclei: Polynucleosomes and RNA polymerase II as endogenous transcriptional complexes. J. Mol. Biol. *118* (1978): 249–272.
125. Seidman, M. M., A. J. Levine & H. Weintraub: The asymmetric segregation of parental nucleosomes during chromosome replication. Cell *18* (1979): 439–449.
126. Kraut, W. & D. Werner: Analysis of the most tightly bound proteins in eukaryotic DNA· Biochim. Biophys. Acta *564* (1979): 390–401.

#### 4.8.7 Specificity of RNA Polymerase

127. Primakoff, P. & S. W. Artz: Positive control of lac operon expression in vitro by guanosine 5'-diphosphate 3'-diphosphate. Proc. Natl. Acad. Sci. USA *76* (1979): 1726–1730.
128. Debenham, P. G., O. Pongs & A. A. Travers: Formylmethionyl-tRNA alters RNA polymerase specificity. Proc. Natl. Acad. Sci. USA *77* (1980): 870–874.
129. Losick, R. & J. Pero: *Bacillus subtilis* RNA polymerase and its modification in sporulating and phage-infected bacteria. Advanc. Enzymol. *44* (1976): 165–185.
130. Pero, J., N. M. Hannett & C. Talkington: Restriction cleavage map of SPO1 DNA: general location of early middle and late genes. J. Virol. *31* (1979): 156–171.
131. Reeve, J. N., G. Mertens & E. Amann: Early development of bacteriophages SPO1 and SP82G in minicells of *B. subtilis*. J. Mol. Biol. *120* (1978): 183–207.
132. Talkington, C. & J. Pero: Distinctive nucleotide sequences of promoters recognized by RNA polymerase containing a phage-coded "σ-like" protein. Proc. Natl. Acad. Sci. USA *76* (1979): 5465–5469.

#### 4.8.8. Attenuation

133. Crawford, I. P. & G. V. Stauffer: Regulation of tryptophan biosynthesis. Ann. Rev. Biochem. *49* (1980): 163–195.
134. Yanofsky, Ch.: Attenuation in the control of expression of bacterial operons. Nature *289* (1981): 751–758.
135. Barnes, W. M.: DNA sequence from the histidine operon control region: Seven histidine codons in a row. Proc. Natl. Acad. Sci. USA *75* (1978): 4281–4285.
136. Oxender, D. L., G. Zurawski & Ch. Yanofsky: Attenuation in the *Escherichia coli* tryptophan operon: Role of RNA secondary structure involving the tryptophan codon region. Proc. Natl. Acad. Sci. USA *76* (1979): 5524–5528.
137. Gemmill, R. M., S. R. Wessler, E. B. Keller & J. M. Calvo: Leu operon of *Salmonella typhimurium* is controlled by an attenuation mechanism. Proc. Natl. Acad. Sci. USA *76* (1979): 4941–4945.
138. Keller, E. B. & J. M. Calvo: Alternative secondary structures of leader RNAs and the regulation of the trp, phe, his, thr, and leu operons. Proc. Natl. Acad. Sci. USA *76* (1979): 6186–6190.
139. Tamm, I. & T. Kikuchi: Early termination of heterogeneous nuclear RNA transcripts in mammalian cells: Accentuation by 5,6-dichloro-1-β-D-ribofuranosylbenzimidazole. Proc. Natl. Acad. Sci. USA *76* (1979): 5750–5754.
140. Salditt-Georgieff, M., M. Harpold, S. Chen-Kiang & J. E. Darnell: The addition of 5'cap structures occurs early in hnRNA synthesis and prematurely terminated molecules are capped. Cell *19* (1980): 69–78.

141. Remington, J. A.: Attenuation as a general mechanism for the regulation of differential gene transcription in eukaryotes. FEBS Letters *100* (1979): 225–229.

142. Laub, O., E. B. Jakobovits & Y. Aloni: 5,6-Dichloro-1-β-ribofuranosylbenzimidazol enhances premature termination of late transcription of simian virus 40 DNA. Proc. Natl. Acad. Sci. USA *77* (1980): 3297–3301.

#### 4.8.9 HnRNP and RNA Processing

143. Fiers, W.: Chemical structure and biological activity of bacteriophage MS2 RNA. In: Zinder, N. D. (ed.), RNA Phages. Cold Spring Harb. Lab., Cold Spring Harb., New York 1975, pp. 353–396.

144. Perry, R. P.: Processing of RNA. Ann. Rev. Biochem. *45* (1976): 605–629.

145. Patel, N. T., A. Kurosky & V. Holoubek: Isolation and characterization of the predominant protein in nuclear ribonucleoprotein particles from rat liver. Biochim. Biophys. Acta *533* (1978): 282–286.

146. Stunnenberg, H. G., Ch. Louis & C. E. Sekeris: Depletion in nuclei of proteins associated with hnRNA, as a result of inhibition of RNA synthesis. Exp. Cell Res. *112* (1978): 335–344.

147. Herman, R., L. Weymouth & S. Penman: Heterogeneous nuclear RNA-protein fibers in chromatin-depleted nuclei. J. Cell Biol. *78* (1978): 663–674.

148. Long, B. H., C.-Ya Huang & A. O. Pogo: Isolation and characterization of the nuclear matrix in Friend erythroleukemia cells: Chromatin and hnRNA interactions with the nuclear matrix. Cell *18* (1979): 1079–1090.

149. Crick, F.: Split genes and RNA splicing. Science *204* (1979): 264–271.

150. Abelson, J.: RNA processing and the intervening sequence problem. Ann. Rev. Biochem. *48* (1979): 1035–1069.

151. Davidson, E. H. & R. J. Britten: Regulation of gene expression: Possible role of repetitive sequences. Science *204* (1979): 1052–1059.

152. Lerner, M. P., J. A. Boyle, S. M. Mount, S. L. Wolin & J. A. Steitz: Are snRNPs involved in splicing? Nature *283* (1980): 220–224.

153. Murray, V. & R. Holliday: A mechanism for RNA-RNA splicing and a model for the control of gene expression. Genetic Res. (Cambr.) *34* (1979): 173–188.

154. Lerner, M. R. & J. A. Steitz: Antibodies to small nuclear RNAs complexed with proteins are produced by patients with systemic Lupus erythematosus. Proc. Nat. Acad. Sci. USA *76* (1979): 5495–5499.

155. Reddy, R., T. S. Ro-Choi, D. Henning & H. Busch: Primary sequence of U-1 nuclear ribonucleic acid of Novikoff hepatoma ascites cells. J. Biol. Chem. *249* (1974): 6486–6494.

156. Molnar, J., G. Bajszar, I. Marczinovits & G. Szabo: Cleavage of pre-mRNA sequences by ribonucleases bound to nuclear RNP particles of rat liver. Mol. Biol. Rep. *4* (1978): 157–161.

157. Seifert, H., M. Scheurlen, W. Northemann & P. C. Heinrich: Low molecular weight RNAs as components of nuclear ribonucleoprotein particles containing heterogeneous nuclear RNA. Biochim. Biophys. Acta *564* (1979): 55–66.

158. Zain, S., T. R. Gingeras, P. Bullock, G. Wong & R. E. Gelinas: Determination and analysis of adenovirus-2 DNA sequences which may include signals for late messenger RNA processing. J. Mol. Biol. *135* (1979): 413–433.

159. Halbreich, A., P. Pajot, M. Foucher, C. Grandchamp & P. Slonimsky: A pathway of cytochrome b mRNA processing in yeast mitochondria: Specific splicing steps and an intron--derived circular RNA. Cell *19* (1980): 321–329.

160. Lazowska, J., C. Jacq & P. P. Slonimski: Sequence of introns and flanking exons in wild--type and box 3 mutants of cytochrome b reveals an interlaced splicing protein coded by an intron. Cell *22* (1980): 333–348.

#### 4.8.10 mRNA Degradation

161. Achord, D. & D. Kennell: Metabolism of messenger RNA from the gal operon of *E. coli.* J. Mol. Biol. *90* (1975): 581–599.

162. Levy, S. B.: Very stable prokaryotic messenger RNA in chromosomeless *Escherichia coli* minicells. Proc. Natl. Acad. Sci. USA *72* (1975): 2900–2904.

163. Lim, L. W. & D. Kennell: Models for decay of *Escherichia coli* lac messenger RNA and evidence for inactivating cleavages between its messages. J. Mol. Biol. *135* (1979): 369–390.
164. Kano, Y., H. Nakamura, R. L. Somerville & F. Imamoto: Decay rates of *Escherichia coli* trp messenger RNA molecules lacking the normal 5′-terminal sequences. Mol. Gen. Genetics *176* (1979): 379–384.
165. Ono, M. & M. Kuwano: A conditional lethal mutation in an *Escherichia coli* strain with a longer chemical lifetime of messenger RNA. J. Mol. Biol. *129* (1979): 343–357.
166. Cannistraro, V. J. & D. Kennell: *E. coli* lac operator mRNA affects translation initiation of $\beta$-galactosidase mRNA. Nature *277* (1979): 407–409.
167. Fallon, A. M., C. S. Jinks, G. D. Strycharz & M. Nomura: Regulation of ribosomal protein synthesis in *Escherichia coli* by selective mRNA inactivation. Proc. Natl. Acad. Sci. USA *76* (1979): 3411–3415; Yates, J. L., A. E. Arfsten & M. Nomura: In vitro expression of *Escherichia coli* ribosomal protein genes: Autogenous inhibition of translation. Proc. Natl. Acad. Sci. USA *77* (1980): 1837–1841.
168. Zengel, J. M., D. Mueckl & L. Lindahl: Protein L4 of the *E. coli* ribosome regulates eleven gene r-protein operon. Cell *21* (1980): 523–535.
169. Stiles, C. D., K. L. Lee & F. T. Kenney: Differential degradation of messenger RNAs in mammalian cells. Proc. Natl. Acad. Sci. USA *73* (1976): 2634–2638.
170. Wilson, M. C., S. G. Sawicki, P. A. White & J. E. Darnell: A correlation between the rate of poly(A) shortening and half-life of messenger RNA in adenovirus transformed cells. J. Mol. Biol. *126* (1978): 23–36.
171. Bergmann, I. E. & G. Brawerman: Loss of the polyadenylate segment from mammalian messenger RNA. Selective cleavage of the sequence from polyribosomes. J. Mol. Biol. *139* (1980): 439–454.
171a. Anderson, K. V. & J. A. Lengyel: Changing rates of histone mRNA synthesis and turnover in *Drosophila* embryos. Cell *21* (1980): 717–727.

#### 4.8.11 Translational Control

172. Lodish, H. F.: Translational control of protein synthesis. Ann. Rev. Biochem. *45* (1976): 39–72.
173. Revel, M. & Y. Groner: Post-transcriptional and translational controls of gene expression in eukaryotes. Ann. Rev. Biochem. *47* (1978): 1079–1126.
174. Kozak, M.: Influence of mRNA secondary structure on binding and migration of 40S ribosomal subunits. Cell *19* (1980): 79–90.
175. Pavlakis, G. N., R. E. Lockard, N. Vamvakopoulos, L. Rieser, U. L. RajBhandary & J. N. Vournakis: Secondary structure of mouse and rabbit $\alpha$- and $\beta$-globin mRNAs: differential accessibility of $\alpha$ and $\beta$ initiator AUG condons towards nucleases. Cell *19* (1980): 91–102.
176. Di Segni, G., H. Rosen & R. Kaempfer: Competition between $\alpha$- and $\beta$-globin messenger ribonucleic acids for eucaryotic initiation factor 2. Biochemistry *18* (1979): 2847–2854.
177. Trachsel, H., N. Sonenberg, A. J. Shatkin, J. K. Rose, K. Leong, J. E. Bergmann, J. Gordon & D. Baltimore: Purification of a factor that restores translation of vesicular stomatitis virus mRNA in extracts from poliovirus-infected HeLa cells. Proc. Natl. Acad. Sci. USA *77* (1980): 770–774.
178. Skup, D. & St. Millward: Reovirus-induced modification of cap-dependent translation in infected L cells. Proc. Natl. Acad. Sci. USA *77* (1980): 152–156.
179. Weissmann, C.: The making of a phage. FEBS Letters *40* (Suppl.) (1974): 10–18.
180. Fiers, W.: Die Formel eines Virus. Die RNA des Bakteriophagen MS2 – die vollständige chemische Struktur eines Virus Genoms und ihre biologische Bedeutung. Medizin uns. Zeit *1* (1977): 10–17.
181. Blumenthal, T. & G. G. Carmichael: RNA replication: Function and structure of Q$\beta$ replicase. Ann. Rev. Biochem. *48* (1979): 525–548.
182. Atkins, J. F., J. A. Steitz, C. W. Anderson & P. Model: Binding of mammalian ribosomes to MS2 phage RNA reveals an overlapping gene encoding a lysis function. Cell *18* (1979): 247–256.
183. Beremand, M. N. & T. Blumenthal: Overlapping genes in RNA phage: a new protein implicated in lysis. Cell *18* (1979): 257–266.

184. Scherrer, K., M.-T. Imaizumi-Scherrer, C.-A. Reynaud & A. Therwath: On pre-messenger RNA and transcription. A review. Mol. Biol. Rep. *5* (1979): 5–28.
185. Maundrell, K., E. S. Maxwell, O. Civelli, A. Vincent, S. Goldenberg, J.-F. Buri, M.-T. Imaizumi-Scherrer & K. Scherrer: Messenger ribonucleoprotein complexes in avian erythroblasts. Carriers of posttranscriptional regulation? Mol. Biol Rep. *5* (1979): 43–52.
186. Young, E. M. & R. A. Raff: Messenger ribonucleoprotein particles in developing sea urchin embryos. Dev. Biol. *72* (1979): 24–40.
187. Princen, H. M. G., C. A. G. van Eekelen, F. A. M. Asselbergs & W. van Venrooij: Free cytoplasmic messenger ribonucleoprotein complexes from rabbit reticulocytes. Mol. Biol. Rep. *5* (1979): 59–64.
188. Kennedy, D. S., E. Siegel & St. M. Heywood: Purification of myosin mRNP translational control RNA and its inhibition of myosin and globin messenger translation. FEBS Letters *90* (1978): 209–214.
189. Ochoa, S. & C. de Haro: Regulation of protein synthesis in eukaryotes. Ann. Rev. Biochem. *48* (1979): 549–580.
190. Safer, B. & R. Jagus: Control of eIF-2 phosphatase activity in rabbit reticulocyte lysate. Proc. Natl. Acad. Sci. USA *76* (1979): 1094–1098.
191. Grankowski, N., D. Lehmusvirta, G. Kramer & B. Hardesty: Partial purification and characterization of reticulocyte phosphatase with activity for phosphorylated peptide initiation factor 2. J. Biol. Chem. *255* (1980): 310–317.
192. Henderson, A. B., A. H. Miller & B. Hardesty: Multistep regulatory system for activation of a cyclic AMP-independent eukaryotic initiation factor 2 kinase. Proc. Natl. Acad. Sci. USA *76* (1979): 2605–2609.
193. Maines, M. D. & A. Kappas: Metals as regulators of heme metabolism. Science *198* (1977): 1215–1221.
194. Granick, S. & S. I. Beale: Hemes, chlorophylls, and related compounds: Biosynthesis and metabolic regulation. Advanc. Enzymol. *46* (1978): 33–203.
195. Gross, M.: The control of protein synthesis by hemin in rabbit reticulocytes. Mol. Cell. Biochem. *31* (1980): 25–36.
196. Garel, J. P., R. L. Garber & M. A. Q. Siddiqui: Transfer RNA in posterior silk gland of *Bombyx mori:* Polyacrylamide gel mapping of mature transfer RNA, identification and partial structural characterization of major isoacceptor species. Biochemistry *16* (1977): 3618–3624.
197. Meza, L., A. Araya, G. Leon, M. Krauskopf, M. A. Q. Siddiqui & J. P. Garel: Specific alanine -tRNA species associated with fibroin biosynthesis in the posterior silk-gland of *Bombyx mori* L. FEBS Letters *77* (1977): 255–260.
198. Lizardi, P. M., V. Mahdavi, D. Shields & G. Candelas: Discontinuous translation of silk fibroin in a reticulocyte cell-free system and in intact silk gland cells. Proc. Natl. Acad. Sci. USA *76* (1979): 6211–6215.
199. Ho, M. & J. Armstrong: Interferon. Ann. Rev. Microbiol. *29* (1975): 131–161.
200. Derynck, R., J. Content, E. DeClercq, G. Volckaert, J. Tavernier, R. Devos & W. Fiers: Isolation and structure of a human fibroblast interferon gene. Nature *285* (1980): 542–547.
201. Ankel, H., Ch. Krishnamuti, F. Besancon, S. Stefanos & E. Falcoff: Mouse fibroblast (type I) and immune (type II) interferons: Pronounced differences in affinity for gangliosides and in antiviral and antigrowth effects on mouse leukemia L-1210R cells. Proc. Natl. Acad. Sci. USA *77* (1980): 2528–2532.
202. Rubinstein, M., S. Rubinstein, P. C. Familletti, R. S. Miller, A. A. Waldman & S. Pestka: Human leucocyte interferon: Production, purification to homogeneity, and initial characterization. Proc. Natl. Acad. Sci. USA *76* (1979): 640–644.
203. Gresser, I., J. de Maeyer-Guignard, M. G. Tovey & E. de Maeyer: Electrophoretically pure mouse interferon exerts multiple biologic effects. Proc. Natl. Acad. Sci. USA *76* (1979): 5309–5312.
204. Baglioni, C.: Interferon-induced enzymatic activities and their role in the antiviral state. Cell *17* (1979): 255–264.
205. Kimchi, A., L. Shulman, A. Schmidt, Y. Chernajovsky, A. Fradin & M. Revel: Kinetics of the induction of three translation regulatory enzymes by interferon. Proc. Natl. Acad. Sci. USA *76* (1979): 3208–3212.

206. Levin, D. H., R. Petryshyn & I. M. London: Characterization of double-stranded RNA activated kinase that phosphorylates subunit of eukaryotic initiation factor 2 (eIF-2$\alpha$) in reticulocyte lysates. Proc. Natl. Acad. Sci. USA *77* (1980): 832–836.

207. Stark, G. R., W. J. Dower, R. T. Schimke, R. E. Brown & I. M. Kerr: 2–5A synthetase: assay, distribution and variation with growth or hormone status. Nature *278* (1978): 471–473.

208. Slattery, E., N. Ghosh, H. Samanta & P. Lengyel: Interferon, double-stranded RNA, and RNA degradation: Activation of an endonuclease by (2′–5′)$A_n$. Proc. Natl. Acad. Sci. USA *76* (1979): 4778–4782.

209. Nilsen, T. W. & C. Baglioni: Mechanism for discrimination between viral and host mRNA in interferon-treated cells. Proc. Natl. Acad. Sci. USA *76* (1979): 2600–2604.

210. Schmidt, A., Y. Chernajovsky, L. Shulman, P. Federman, H. Berissi & M. Revel: An interferon-induced phosphodiesterase degrading (2′–5′) oligoisoadenylate and the C-C-A terminus of tRNA. Proc. Natl. Acad. Sci. USA *76* (1979): 4788–4792.

211. Kimchi, A., H. Shure & M. Revel: Regulation of lymphocyte mitogenesis by (2′–5′) oligoisoadenylate. Nature *282* (1979): 849–851.

212. Revel, M., G. Goldberg, U. Nudel, A. Zilberstein, B. Dudock, D. Canaani & Y. Groner: How does the ribosomal machinery discriminate between messenger RNA's? In: [4], pp. 158–181.

213. Nilsen, T. W., D. L. Wood & C. Baglioni: Virus-specific effects of interferon in embryonal carcinoma cells. Nature *286* (1980): 178–180.

**4.8.12 Proteinogen Processing**

214. Silverstein, S. C. (ed.): Transport of Macromolecules in Cellular Systems. Dahlem Konferenzen. Abakon Verlagsges., Berlin 1978.

215. Zimmerman, M., R. A. Mumford & D. F. Steiner (eds.): Precursor Processing in the Biosynthesis of Proteins. Ann. New York Acad. Sci., New York 1980.

216. Blobel, G.: Intracellular protein topogenesis. Proc. Natl. Acad. Sci. USA *77* (1980): 1496–1500.

217. Milstein, C., G. G. Brownlee, T. M. Harrison & M. B. Mathews: A possible precursor of immunoglobulin light chains. Nature New Biol. *239* (1972): 117–120.

218. Blobel, G. & B. Dobberstein: Transfer of protein across membranes I. and II. J. Cell Biol. *67* (1975): 835–851 and 852–862.

219. Di Rienzo, J. M., K. Nakamura & M. Inouye: The outer membrane proteins of gram-negative bacteria: Biosynthesis, assembly, and functions. Ann. Rev. Biochem. *47* (1978): 481–532.

220. Kreil, G.: Synthese und Export sekretorischer Polypeptide. Biol. uns. Zeit *9* (1979): 141–146.

221. Davis, B. D. & Ph.-C. Tai: The mechanisms of protein secretion across membranes. Nature *283* (1980): 433–438.

222. Wickner, W.: The assembly of proteins into biological membranes: the membrane trigger hypothesis. Ann. Rev. Biochem. *48* (1979): 23–45.

223. Lingappa, V. R., J. R. Lingappa & G. Blobel: Chicken ovalbumin contains an internal signal sequence. Nature *281* (1979): 117–121.

224. Strauss, A. W., M. Zimmerman, I. Boime, B. Ashe, R. A. Mumford & A. W. Alberts: Characterization of an endopeptidase involved in pre-protein processing. Proc. Natl. Acad. Sci. USA *76* (1979): 4225–4229.

225. Kreibich, G., M. Czako-Graham, R. Grebenau, W. Mok, E. Rodriguez-Boulan & D. D. Sabatini: Characterization of the ribosomal binding site in rat liver rough microsomes: ribophorins I and II, two integral membrane proteins related to ribosome binding. J. Supramol. Structure *8* (1978): 279–302.

226. Bonatti, St., R. Cancedda & G. Blobel: Membrane biogenesis: In vitro cleavage, core glycosylation, and integration into microsomal membranes of sindbis virus glycoproteins. J. Cell Biol. *80* (1979): 219–224.

226a. Krangel, M. S., H. T. Orr & J. L. Strominger: Assembly and maturation of HLA-A and HLA-B antigens in vivo. Cell *18* (1979): 979–991.

227. Lodish, H. F., D. Wirth & M. Porter: Synthesis and assembly of viral membrane proteins. In [215], pp. 319–334.

228. Zilberstein, A., M. D. Snider, M. Porter & H. F. Lodish: Mutants of vesicular stomatitis virus blocked at different stages in maturation of the viral glycoprotein. Cell *21* (1980): 417–427.

229. Rose, J. K., W. J. Welch, B. M. Sefton, F. S. Esch & N. C. Ling: Vesicular stomatitis virus glycoprotein is anchored in the viral membrane by a hydrophobic domain near the COOH terminus. Proc. Natl. Acad. Sci. USA *77* (1980): 3884–3888.
230. Palade, G. E.: Intracellular aspects of the process of protein synthesis. Science *189* (1975): 347–358.
231. Chua, N. H. & G. W. Schmidt: Transport of proteins into mitochondria and chloroplasts. J. Cell Biol. *81* (1979): 461–483.
232. Schatz, G.: How mitochondria import proteins from the cytoplasm. FEBS Letters *103* (1979): 203–211.
233. Smith, St. M. & R. J. Ellis: Processing of small subunit precursor of ribulose bisphosphate carboxylase and its assembly into whole enzyme are stromal events. Nature *278* (1979): 662–664.
234. Goldman, B. M. & G. Blobel: Biogenesis of peroxisomes: Intracellular site of synthesis of catalase and uricase. Proc. Natl. Acad. Sci. USA *75* (1978): 5066–5070.
235. Lewin, A. S., I. Gregor, T. L. Mason, N. Nelson & G. Schatz: Cytoplasmically made subunits of yeast mitochondrial $F_1$-ATPase and cytochrome c oxidase are synthesized as individual precursors, not as polyproteins. Proc. Natl. Acad. Sci. USA *77* (1980): 3998–4002.
235a. Mihara, K. & G. Blobel: The four cytoplasmically made subunits of yeast mitochondrial cytochrome c oxidase are synthesized individually and not as a polyprotein. Proc. Natl. Acad. Sci. USA *77* (1980): 4160–4164.
236. Cabib, E.: Molecular aspects of yeast morphogenesis. Ann. Rev. Microbiol. *29* (1975): 191–214
237. Sloat, B. F. & J. R. Pringle: A mutant of yeast defective in cellular morphogenesis. Science *200* (1978): 1171–1173.
238. Duran, A., E. Cabib & B. Bowers: Chitin synthetase distribution on the yeast plasma membrane. Science *203* (1979): 363–365.
239. Zubenko, G. S., A. P. Mitchell & E. W. Jones: Septum formation, cell division, and sporulation in mutants of yeast deficient in proteinase B. Proc. Natl. Acad. Sci. USA *76* (1979): 2395–2399.
240. Hershko, A. & M. Try: Post-translational cleavage of polypeptide chains: role in assembly. Ann. Rev. Biochem. *44* (1975): 775–797.
241. Smith, A. E. & L. Carrasco: Eukaryotic viral protein synthesis. Int. Rev. Biochem. *18* (1978): 261–311.
242. Korant, B., N. Chow, M. Lively & J. Powers: Virus-specified protease in poliovirus-infected Hela cells. Proc. Natl. Acad. Sci. USA *76* (1979): 2992–2995.
243. Korant, B. D.: Proteolytic events in viral replication. In: Lepow, I. (ed.) Molecular Basis of Biological Degradative Processes. Acad. Press, New York 1978, pp. 171–224.
244. Korant, B. D., N. L. Chow, M. O. Lively & J. C. Powers: Proteolytic events in replication of animal viruses. In [215], pp. 304–318.
245. Kitamura, N., C. J. Adler, P. G. Rothberg, J. Martinko, S. G. Nathenson & E. Wimmer: The genome-linked protein of Picornaviruses. VII Genetic mapping of poliovirus VPg by protein and RNA sequence studies. Cell *21* (1980): 295–302.

#### 4.8.13 Protein Degradation

246. Holzer, H. & H. Tscheche (eds.): Biological Functions of Proteinases. Springer, Berlin 1979.
247. Cohen, G. N. & H. Holzer (eds.): Limited Proteolysis in Microorganisms. DHEW Publ.-No (NIH) 79–1591, Bethesda, Md. 1979.
248. Bohley, P., H. Kirschke, J. Langner, M. Miehe, S. Riemann, Z. Salama, E. Schön, B. Wiederanders & S. Ansorge: Intracellular protein turnover. In: Holzer, H. & H. Tscheche (eds.) Biological Functions of Proteinases. Springer Berlin, 1979, pp. 17–34.
249. Schimke, R. T.: Control of enzyme levels in mammalian tissues. Advanc. Enzymol. *37* (1973): 135–187.
250. Goldberg, A. L. & J. F. Dice: Intracellular protein degradation in mammalian and bacterial cells. Ann. Rev. Biochem. *43* (1974): 835–869.
251. Goldberg, A. L. & A. C. St. John: Intracellular protein degradation in mammalian and bacterial cells: part 2. Ann. Rev. Biochem. *45* (1976): 747–803.
252. Waterlow, J. C., P. J. Garlick & D. J. Millward: Protein Turnover in Mammalian Tissues and in the Whole Body. North-Holland Publ. Comp., Amsterdam 1978.

253. Holzer, H. & P. C. Heinrich: Control of proteolysis. Ann. Rev. Biochem. *49* (1980): 63–91.
254. Hanash, S. M. & D. L. Rucknagel: Proteolytic activity in erythrocyte precursors. Proc. Natl. Acad. Sci. USA *75* (1978): 3427–3431.
255. Hershko, A., A. Ciechanover, H. Heller, A. L. Haas & I. A. Rose: Proposed role of ATP in protein breakdown: conjugation of proteins with multiple chains of the polypeptide of ATP-dependent proteolysis. Proc. Natl. Acad. Sci. USA *77* (1980): 1783–1786.
256. Cross, H. S. & H. Ruis: Regulation of catalase synthesis in *Saccharomyces cerevisiae* by carbon catabolite repression. Mol. Gen. Genet. *166* (1978): 37–43.
257. Neeff, J., E. Haegele, J. Nauhaus, U. Heer & D. Mecke: Evidence for catabolite degradation in the glucose dependent inactivation of yeast cytoplasmic malate dehydrogenase. Europ. J. Biochem. *87* (1978): 489–495.
258. Zubenko, G. S. & E. W. Jones: Catabolite inactivation of gluconeogenic enzymes in mutants of yeast deficient in proteinase B. Proc. Natl. Acad. Sci. USA *76* (1979): 4581–4585.
259. Segal, H. L., J. A. Brown, G. A. Dunaway, J. R. Winkler, H. M. Madnick & D. M. Rothstein: Factors involved in the regulation of protein turnover. In [247], pp. 33–36.
260. Voellmy, R., K. Murakami & A. L. Goldberg: The ATP-dependent degradation of proteins in *Escherichia coli* and mammalian cells. In [247], pp. 7–16.

**4.8.14 Gene Expression Programs**

*Bacterial Sporulation*

261. Hoch, J. A.: Genetics of bacterial sporulation. Advanc. Genetics *18* (1976): 69–98.
262. Doi, R. H.: Genetic control of sporulation. Ann. Rev. Genetics *11* (1977): 29–48.
263. Aronson, A. I. & N. K. Pandey: Comparative structural and functional aspects of spore coats. Spores *7* (1978): 54–61.
264. Ellar, D. J.: Spore specific structures and their function. Symp. Soc. Gen. Microbiol. *28* (1978): 295–325.
265. Piggot, P. J.: Genetic strategies for studying bacterial differentiation. Biol. Rev. *54* (1979): 347–367.
266. Szulmajster, J.: Is sporulation a simple model for studying differentiation? Trends Biochem. Sci. *8* (1979): 18–22.
267. Linnet, P. E. & D. J. Tipper: Transcriptional control of peptidoglycan precursor synthesis during sporulation in *Bacillus sphaericus*. J. Bacteriol. *125* (1976): 565–574.
268. Munoz, L., Y. Sadaie & R. H. Doi: Spore coat protein of *B. subtilis*. Structure and precursor synthesis. J. Biol. Chem. *253* (1978): 6694–6701.
269. Kerjan, P., E. Keryer & J. Szulmajster: Characterization of a thermosensitive sporulation mutant of *B. subtilis* affected in the structural gene of an intracellular protease. Europ. J. Biochem. *98* (1979): 353–362.
270. Reysset, G., K. F. Bott & J. P. Aubert: Relationship between glutamine synthetase and sporulation. Spores *7* (1978): 271–276.
271. Rhaese, H. J. & R. Groscurth: Studies on the control of development. Synthesis of two highly phosphorylated nucleotides depends on changes in the composition of ribosomes at the beginning of sporulation in *B. subtilis*. Europ. J. Biochem *85* (1978): 517–528.
272. Rhaese, H. J. & R. Groscurth: Apparent dependence of sporulation on synthesis of highly phosphorylated nucleotides in *Bacillus subtilis*. Proc. Natl. Acad. Sci. USA *76* (1979): 842–846.
273. Losick, R. & J. Pero: *Bacillus subtilis* RNA polymerase and its modification in sporulating and phage-infected bacteria. Advanc. Enzymol. *44* (1976): 165–185.
274. Haldenwang, W. G. & R. Losick: A modified RNA polymerase transcribes a cloned gene under sporulation control in *Bacillus subtilis*. Nature *282* (1979): 256–260.
275. Trowsdale, J., M. Shiflett & J. A. Hoch: New cluster of ribosomal genes in *Bacillus subtilis* with regulatory role in sporulation. Nature *272* (1978): 179–181.
276. Vold, B. S.: Post-transcriptional modifications of the anticodon loop region: alterations in isoaccepting species of tRNAs during development in *Bacillus subtilis*. J. Bact. *135* (1978): 124–132.
277. Setlow, P.: Identification and localization of the major proteins degraded during germination of *Bacillus megaterium* spores. J. Biol. Chem. *250* (1975): 8159–8167.
278. Ristow, H.-J., W. P. Schorn, J. Hansen & U. Winkel: Induction of sporulation in *Bacillus brevis* by peptide antibiotics. Nature *280* (1979): 165–166.

*Erythroid Cell Differentiation*


279. Russell, E. S.: Hereditary anemias of the mouse: a review for geneticists. Advanc. Genetics *20* (1979): 375–495.
280. Lajtha, L.G.: Stem cell concepts. Differentiation *14* (1979): 23–34.
281. Weatherall, D. J. & J. B. Clegg: Recent developments in the molecular genetics of human hemoglobin. Cell *16* (1979): 467–479.
282. Rosenthal, S., C. Coutelle & A. N. Ladhoff: Globin synthesis in erythropoietic systems: Regulation of globin messenger RNA synthesis. In: [4], pp. 484–506.
283. Eliason, J. F., N. G. Testa & T. M. Dexter: Erythropoietin-stimulated erythropoiesis in long term bone marrow culture. Nature *281* (1979): 382–384.
284. Howard, M., A. Burgess, D. McPhee & D. Metcalf: T-cell hybridoma secreting hemopoietic regulatory molecules: Granulocyte-macrophage and eosinophil colony-stimulating factors. Cell *18* (1979): 993–999.
285. Friend, C.: Cell-free transmission in adult swiss mice of a disease having the character of a leukemia. J. Exp. Med. *105* (1957): 307–318.
286. Suzuki, Sh. & A. A. Axelrad: Fv-2 locus controls the proportion of erythropoietic progenitor cells (BFU-E) synthesizing DNA in normal mice. Cell *19* (1980): 225–236.
287. Kost, T. A., M. J. Koury, W. D. Hankins & S. B. Krantz: Target cells for Friend virus-induced erythroid bursts in vitro. Cell *18* (1979): 145–152.
288. Marks, P. A. & R. A. Rifkind: Erythroleukemic differentiation. Ann. Rev. Biochem. *47* (1978): 419–448.
289. Lowenhaupt, K. & J. B. Lingrel: Synthesis and turnover of globin mRNA in murine erythroleukemia cells induced with hemin. Proc. Natl. Acad. Sci. USA *76* (1979): 5173–5177.
290. Gambari, R., P. A. Marks & R. Rifkind: Murine erythroleukemia cell differentiation: Relationship of globin gene expression and of prolongation of $G_1$ to inducer effects during $G_1$/early S. Proc. Natl. Acad. Sci. USA *76* (1979): 4511–4515.
291. Affara, N. & Ph. Daubas: Regulation of a group of abundant mRNA sequences during Friend cell differentiation. Dev. Biol. *72* (1979): 110–125.
292. Sassa, S. & A. Urabe: Uroporphyrinogen I synthase induction in normal human bone marrow cultures: An early and quantitative response of erythroid differentiation. Proc. Natl. Acad. Sci. USA *76* (1979): 5321–5325.
293. Rutherford, T., G. G. Thompson & M. R. Moore: Heme biosynthesis in Friend erythroleukemia cells: Control by ferrochelatase. Proc. Natl. Acad. Sci. USA *76* (1979): 833–836.
294. Keppel, F., B. Allet & H. Eisen: Biochemical properties and localization of the chromosomal protein IP25. Eur. J. Biochem. *96* (1979): 477–482.
295. Neumann, J. R., D. Housman & V. M. Ingram: Nuclear protein synthesis and phosphorylation in Friend erythroleukemia cells stimulated with DMSO. Exp. Cell Res. *111* (1978): 277–284.
296. Smith, D. W. E.: Reticulocyte tRNA and hemoglobin synthesis. Science *190* (1975): 529–535.
297. Ponka, P., J. Neuwirth, J. Borova & O. Fuchs: Control of iron delivery to haemoglobin in erythroid cells. Ciba Foundn. Symp. *51* (1977): 167–188.
298. Jacobs, A.: An intracellular transit iron pool. Ciba Foundn. Symp. *51* (1977): 91–100.
299. Drysdale, J. W.: Ferritin phenotypes: Structure and metabolism. Ciba Foundn. Symp. *51* (1977): 41–57.
300. Tortolero, M., S. Fischer, J.-P. Pian, J. Delaunay & G. Schapira: Two dimensional resolution of $^{32}$P-spectrin of human erythrocyte membranes. Dephosphorylation reaction by a membrane phosphoprotein phosphatase. Proc. FEBS Meeting, Vol. 54, Pergamon Press, Oxford 1979, pp. 225–233.
301. Branton, D., C. M. Cohen & J. Tyler: Interaction of cytoskeletal proteins on the human erythrocyte membrane. Cell *24* (1981): 24–32.
302. Bennett, V. & P. J. Stenbuck: The membrane attachment protein for spectrin is associated with band 3 in human erythrocyte membranes. Nature *280* (1979): 468–473.
303. Lux, S. E.: Dissecting the red cell membrane skeleton. Nature *281* (1979): 426–429.
304. Imhof, B. A., H. J. Acha-Orbea, T. A. Libermann, B. F. X. Reber, J. H. Lanz, K. H. Winterhalter & W. Birchmeier: Phosphorylation and dephosphorylation of spectrin from human erythrocyte ghosts under physiological conditions: Autocatalysis rather than reaction with separate kinase and phosphatase. Proc. Natl. Acad. Sci. USA *77* (1980): 3264–3268.

304a. Rapoport, S., M. Müller & W. Dubiel: Nitrogen economy and the metabolism of serine and glycine in the reticulocytes of rabbits. In: [6], pp. 263–276.

**4.8.15 Cell Differentiation and Development**

*Fruiting Body Formation by Myxobacteria*

305. Zusman, D. R.: Genetic approaches to the study of development in the myxobacteria. In: [10], pp. 41–78.
306. Kaiser, D., Manoil, C. & M. Dworkin: Myxobacteria: Cell interactions, genetics, and development. Ann. Rev. Microbiol. *33* (1979): 595–639.
307. Hagen, D. C., A. P. Bretscher & D. Kaiser: Synergism between morphogenetic mutants of *Myxococcus xanthus*. Dev. Biol. *64* (1978): 284–296.
308. Cumsky, M. & D. R. Zusman: Myxobacterial hemagglutinin: A development-specific lectin of *Myxococcus xanthus* Proc. Natl. Acad. Sci. USA *76* (1979): 5505–5509.
309. Inouye, M., S. Inouye & D. R. Zusman: Biosynthesis and self-assembly of protein S, a development-specific protein of *Myxococcus xanthus*. Proc. Natl. Acad. Sci. USA *76* (1979): 209–213.
310. Kaiser, D.: Social gliding is correlated with presence of pili in *Myxococcus xanthus*. Proc. Natl. Acad. Sci. USA *76* (1979): 5952–5956.

*Cell Sorting and Pattern Formation*

311. Wolpert, L.: Positional information and the spatial pattern of cellular differentiation. J. Theoret. Biol. *25* (1969): 1–47.
312. Nicholls, J. G. (ed).): The role of Intracellular Signals: Navigation, Encounter, Outcome. Dahlem Workshop, Vol. 14, Verlag Chemie, Weinheim 1979.
313. Müller, W. A.: Positionsinformation und Musterbildung. Biol. uns. Zeit *9* (1979): 135–140.
314. Wolpert, L.: Pattern formation in biological development. Scient. American *239* (1978): 124–136.
315. Tarin, D.: Embryonic induction. Medical Biol. *56* (1978): 380–385.
316. Burger, M. M., W. Burkart, G. Weinbaum & J. Jumblatt: Cell-cell recognition: molecular aspects. Recognition and its relation to morphogenetic processes in general. Symp. Soc. Exp. Biol. *32* (1978): 1–24.
317. Curtis, A. S. G.: Cell-cell recognition: Positioning and patterning systems. Symp. Soc. Exp. Biol. *32* (1978): 51–82.
318. Ford, W. L., M. E. Smith & P. Andrews: Possible clues to the mechanism underlying the selective migration of lymphocytes from the blood. Symp. Soc. Exp. Biol. *32* (1978): 359–392.
319. De Sousa, M.: Lymphoid cell positioning: a new proposal for the mechanism of control of lymphoid cell migration. Symp. Soc. Exp. Biol. *32* (1978): 393–409.
320. Frazier, W. & L. Glaser: Surface components and cell recognition. Ann. Rev. Biochem. *48* (1979): 491–523.

*Drosophila Development*

321. Nöthiger, R.: The larval development of imaginal disks. Results Probl. Cell Diff. *5* (1972): 1–34.
322. Gehring, W.: The stability of the determinated state in cultures of imaginal disks in *Drosophila*. Results Probl. Cell Diff. *5* (1972): 35–58.
323. Kauffmann, St. A., R. M. Shymko & K. Trabert: Control of sequential compartment formation in *Drosophila*. Science *199* (1978): 259–270.
324. Ouweneel, W. J.: Developmental genetics of homoeosis. Advanc. Genetics *18* (1976): 179–248.
325. Nüsslein-Volhard, C.: Maternal effect mutations that alter the spatial coordinates of the embryo. In: Subtelney, S. & I. R. Konigsberg (eds.), Determinants of Spatial Organization. Academic Press, New York 1979, pp. 185–211.
326. Underwood, E. M., F. R. Turner & A. P. Mahowald: Analysis of cell movements and fate mapping during early embryogenesis in *Drosophila melanogaster*. Dev. Biol. *74* (1980): 286–301.
327. Illmensee, K.: *Drosophila* chimeras and the problem of determination. Res. Probl. Cell Diff. *9* (1978): 51–69.

328. Garcia-Bellido, A. & P. Ripoll: Cell lineage and differentiation in *Drosophila*. Res. Probl. Cell Diff. *9* (1978): 119–156.
320. Jowett, T. & J. H. Sang: Nutritional regulation of antennal/leg homeotic mutants in *Drosophila melanogaster*. Genet. Res. *34* (1979): 143–161.
329a. Nüsslein-Volhard, C. & E. Wieschaus: Mutations affecting segment number and polarity in *Drosophila*. Nature *287* (1980): 795–801.

*Development of the Human Sexual System*

330. Dörner, G.: Hormones and Brain Differentiation. Elsevier Publ. Comp., Amsterdam 1976.
331. McCarrey, J. R. & U. K. Abbott: Mechanisms of genetic sex determination, gonadal sex differentiation, and germ-cell development in animals. Advanc. Genetics *20* (1979): 217–290.
332. Neumann, F.: Normale und gestörte Sexualdifferenzierung. Medizin uns. Zeit *1* (1977): 98–108.
333. Wachtel, S. S.: Immunogenetic aspects of abnormal sexual differentiation. Cell *16* (1979): 691–695.
334. Müller, U., U. Wolf, J. W. Siebers & E. Günther: Evidence for a gonad-specific receptor for H–Y antigen: Binding of exogenous H–Y antigen to gonadal cells is independent of $\beta_2$-microglobulin. Cell *17* (1979): 331–335.
335. Nagai, Y., S. Ciccarese & S. Ohno: The identification of human H–Y antigen and testicular transformation induced by its interaction with the receptor site of bovine fetal ovarian cells. Differentiation *13* (1979): 155–164.
336. Gehring, W. J.: Determination. In: Allfrey, V. G., E. K. F. Bautz, B. J. McCarthy, R. T. Schimke, A. Tissieres (eds.): Organization and Expression of Chromosomes. Rep. Dahlem Workshop, Berlin 1976, pp. 97–113.
337. Guillemin, R.: Peptides in the brain: The new endocrinology of the neuron. Science *202* (1978): 390–402.
338. Kuhl, H.: Fetale Hormonspiegel und Sexualdifferenzierung. Medizin uns. Zeit *3* (1979): 84–90.
339. Skett, P. & J. A. Gustafsson: Hypothalamo-pituitary regulation of liver function. Medical Biol. *57* (1979): 374–389.
340. Scott, R. S. & H. G. Burger: Inhibin is absent from azoospermic semen of infertile men. Nature *285* (1980): 246–247.
341. Josso, N., J.-Y. Picard & D. Tran: The antimüllerian hormone. Rec. Progr. Horm. Res. *33* (1977): 117–163.
342. Döhler, K.-D. & J. L. Hancke: Thoughts on the mechanism of sexual brain differentiation. In: G. Dörner & M. Kawakami (eds.), Hormones and Brain Development. Elsevier/North-Holland Biomed. Press 1978, pp. 153–158.
343. Kierszenbaum, A. L. & L. L. Tres: RNA transcription and chromatin structure during meiotic and postmeiotic stages of spermatogenesis. Fed. Proc. *37* (1978): 2512–2516.
344. Meistrich, M. L., W. A. Brock, S. R. Grimmes, R. D. Platz & L. S. Hnilica: Nuclear protein transitions during spermatogenesis. Fed. Proc. *37* (1978): 2522–2525.
345. Hoskins, D. D., H. Brandt & T. S. Acott: Initiation of sperm mobility in the mammalian epididymis. Fed. Proc. *37* (1978): 2534–2542.
346. Brock, W. A., P. A. Trostle & M. L. Meistrich: Meiotic synthesis of testis histones in the rat. Proc. Natl. Acad. Sci. USA *77* (1980): 371–375.
347. Austin, C. R. & R. V. Short (eds.): Germ Cells and Fertilization. Cambridge Univ. Press, Cambridge 1972.
348. Hartree, E. F.: Spermatozoa, eggs and proteinases. Biochem. Soc. Trans. *5* (1977): 375–394.

# SPECIAL PART

# 5. Regulation of Adaptive Enzyme Synthesis in Prokaryotes Exemplified by the Arabinose Regulon of *Escherichia coli*

L. Nover

## 5.1 Biochemistry, Enzymatics and Genetics of the Arabinose Degradation

L-Arabinose utilization in *Escherichia coli* (Fig. **5**.1) comprises (1) the uptake of the sugar into the cell assisted by two transport proteins which are coded by genes *ara* E and *ara* F respectively and (2) the enzymatic transformation of the arabinose in D-xylulose-5-phosphate which is further degraded by the general carbohydrate metabolism. This specific part of the L-arabinose degradation is catalyzed by three enzymes, the L-arabinose isomerase (*ara* A gene), the L-ribulokinase (*ara* B gene) and the L-ribulose-5-phosphate epimerase (*ara* D gene). The five structural genes mentioned map at three widely separated loci on the circular chromosome of *E. coli*. They are part of a regulon, i.e., their coordinated expression is subject to control by common regulatory proteins. The latter are the specific repressor-activator protein of the *ara* regulon, the *ara* C gene product and the catabolite gene activator protein (CAP) coded by the *crp* gene. Both regulatory systems will be discussed in greater depth during the course of this chapter. A survey of genetic and structural characteristics of the six proteins coded by the *ara* regulon is given in Table **5**.1.

The expression of all structural genes of the *ara* regulon is controlled by L-arabinose which functions as regulatory effector of the *ara* C protein. The strict coordination of gene expression between the separate parts of the regulon could be demonstrated for induced cultures of the wild-type strain as well as for a set of regulatory gene constitutive mutants (*ara* $C^c$). This is shown in Fig. **5**.2 for the three most easily measurable proteins. It is evident that the constitutive formation of the proteins of the *ara* regulon, triggered by mutations in the regulatory gene, affects the products of the *ara* BAD region in the same way as it affects the product of the distant *ara* E gene. Although widely differing absolute activities were found in dependence on the *ara* C mutation, the relations between the three proteins are practically constant and correspond to those established in induced wild-type cells.

In the following only that particular part of the *ara* regulon which includes the CBAD region (cf. Fig. **5**.1) will be considered in more detail because experimental reasons have led to the predominant amount of work being limited to this region. However, regarding the tight regulative coordination shown between the different

---

Abbreviations and genetic terms: For the characterization of the structural genes of the arabinose regulon (*ara* regulon) cf. Fig. **5**.1. The term "*ara*" is used to denote all genetic terms in relation to the genetic system for arabinose utilization, whereas Ars denote the sugar itself, i.e. L-arabinose. The allelic conditions of genes will be designated with $^+$ for the wild-type, $^-$ for a defective and $^c$ for a constitutive phenotype (e.g., $C^+$, $C^-$, $C^c$). Deletion mutants will be identified by $\Delta$ and a line to indicate the deleted part of the genetic material (cf. top of Tables **5**.2 and **5**.3). Point mutants are characterized by arabic numbers and a dot marking their approximate map position (cf. top of Tables **5**.2 and **5**.3). For the description of the frequently mentioned *ara* transducing phage hybrids (λh80d *ara* and λp *ara*) cf. **5**.5.1. cAMP: cyclic adenosin-3′,5′-monophosphate; CAP: catabolite gene activator protein (cf. **4**.4.2.2); bp, base pairs; Fcs, D-fucose; kd, kilodalton; I, initiator region; $M_r$, relative molecular weight; O, operator region; p, promoter.

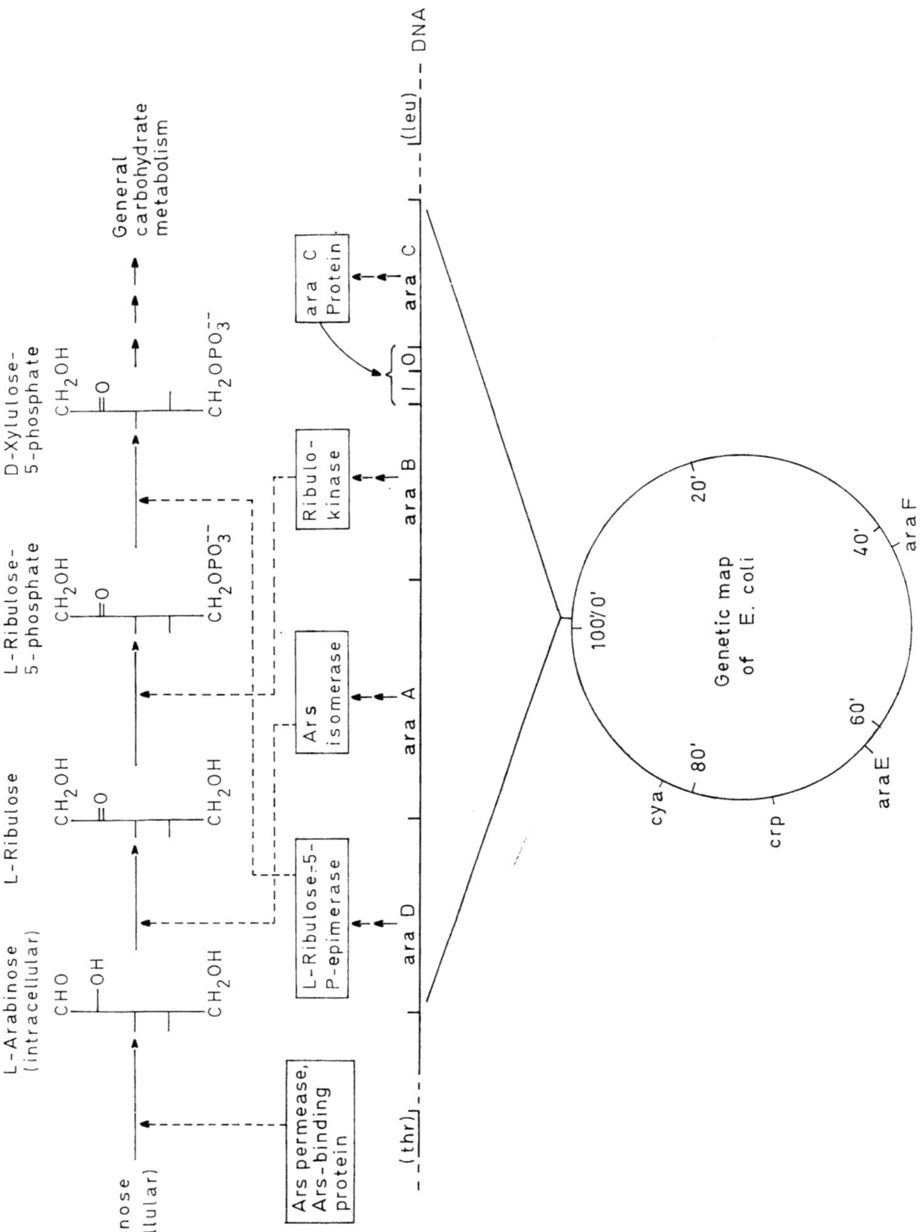
L-Arabinose (extracellular)
L-Arabinose (intracellular)
L-Ribulose
L-Ribulose-5-phosphate
D-Xylulose-5-phosphate
General carbohydrate metabolism
Ars permease, Ars-binding protein
L-Ribulose-5-P-epimerase
Ars isomerase
Ribulokinase
ara C Protein
(thr)
ara D
ara A
ara B
I
O
ara C
(leu)
DNA
Genetic map of E. coli
100/0'
20'
40'
60'
80'
araF
araE
crp
cya

Fig. 5.1

parts of the *ara* regulon, the basic findings about the regulation of gene expression in the *ara* BAD operon can be transferred accordingly to the *ara* E and *ara* F operons respectively.

**Table 5.1**: Proteins coded by the four transcription units of the *E. coli ara* regulon. (For details and references cf. [1–3].)

| Transcr. unit | Gene | Position on the *E. coli* map | Proteins | | | |
|---|---|---|---|---|---|---|
| | | | Name | $M_r$ of subunits (kd) | Subunit composition | Remarks |
| 1 | *ara* C | 1′ | Regulator protein | 30.5 | 2 (activator) | $K_M$ (Ars): $3 \times 10^{-3}$ M |
| | | | | | 4 (repressor) | $K_D$ (Operator-complex): $10^{-12}$ M |
| 2 | *ara* B | 1′ | L-Ribulokinase | 50 | 2 | $K_M$ (Ribulose): $1 \times 10^{-4}$ M Spec. activity: 2,000 U/mg protein |
| | *ara* A | 1′ | L-Arabinose isomerase | 60 | 6 | $K_M$ (Ars): $6 \times 10^{-2}$ M Spec. act.: 12,000 U/mg protein |
| | *ara* D | 1′ | L-Ribulose-5-phosphate epimerase | 35 | ? | Spec. act.: 2,200 U/mg protein |
| 3 | *ara* E | 61′ | L-Arabinose permease | | | Membrane protein $K_M$ (Ars): $3 \times 10^{-4}$ M |
| 4 | *ara* F | 44–45′ | L-Arabinose binding protein | 33 | | Periplasmic protein $K_M$ (Ars): $6 \times 10^{-6}$ M |

◀

**Fig. 5.1**: Genetics and biochemistry of arabinose utilization in *Escherichia coli* (from Englesberg & Wilcox [2], modified).

The *upper part* shows the biochemical reactions involved in Ars degradation. The *lower part* represents the circular chromosome of *E. coli* with the *ara* CBAD-region enlarged. It maps at position 1′ between the genes for threonine *(thr)* and leucine *(leu)* synthesis. A more detailed map of the regulatory region *ara* OI is given in Figs. 5.8 and 5.9. In addition to the *ara* CBAD-region four further genetic loci are marked. These are the loci for the two Ars-specific transport proteins (*ara* E and *ara* F) and for adenylyl cyclase *(cya)* and the catabolite gene activator protein *(crp)*.

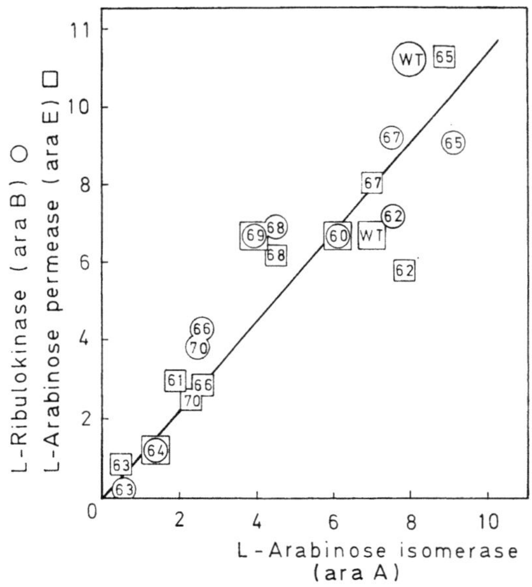


**Fig. 5.2:** Coordinate expression of genes of the *ara* regulon (from Englesberg et al. [4], modified). Activities of ribulokinase, Ars permease and Ars isomerase were determined in induced wild-type cells *(WT)* and non-induced *ara* $C^c$ mutants. The mutations in the *ara* C gene result in very different levels of gene expression in the *ara* regulon, but the ratio between the products of *ara* A, *ara* B and *ara* E genes is constant.

## 5.2 Genetic Investigations on the Control of Gene Expression in the *ara* BAD Operon

In addition to the functional characterization and localization of the structural gene, investigations of mutants of the *ara* regulon resulted in the description of two DNA regions with a specific regulatory role for *ara* gene expression (*ara* C and *ara* OI, cf. Fig. 5.1). In the following, these two regions are in the focus of the discussion as structural gene mutants play an indispensable but none the less only indirect role in the analysis of regulatory mechanisms.

### 5.2.1 The Regulatory Gene *ara* C

The *ara* C gene maps in direct neighbourhood of the three structural genes *ara* BAD and is separated from them only by the aforementioned second regulatory region *ara* OI. There are three known allelic states of the *ara* C gene (Table 5.2).

1. The *ara* $C^+$ wild-type allel makes the expression of all five structural genes of the regulon dependent on the presence of Ars as inducing agent.
2. In the presence of the *ara* $C^-$ allele the inducibility by Ars of all structural genes is lacking.
3. The *ara* $C^c$ allele leads to different levels of a constitutive production of all gene products of the *ara* regulon (cf. Fig. 5.2).

**Table 5.2**: Characterization of three *ara* C alleles in merodiploid strains (from Sheppard & Englesberg [5]).

Activities of ribulokinase (*ara* B gene) and Ars isomerase (*ara* A gene) in induced and non-induced cells were compared with those in wild-type cells (upper part of table, strains 1–6). The genetic map at the top of the table denotes the *ara* genes and positions of point mutations which characterize the indicated strains. In the merodiploid strains (2/5, 4/2 and 4/6) the interaction of the three *ara* C alleles can be tested by combining each two of them in one cell. This can be achieved by incorporating the *ara* CBAD-region into the DNA of a plasmid (in this example the F-factor) and by transferring this plasmid to a receptor cell. Thus merodiploid strains contain, in addition to their chromosomal copy of the *ara* CBAD-region, another, possibly different copy in trans position on the DNA of the F-plasmid.

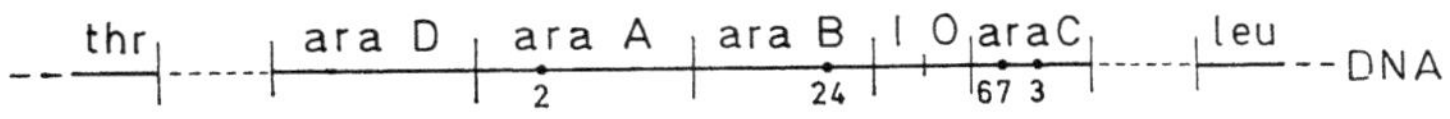


| Genotype | Enzyme activity | | | | Phenotype (remarks) |
|---|---|---|---|---|---|
| | Ribulokinase (B) | | Isomerase (A) | | |
| | induced | non-induced | induced | non-induced | |
| Haploids | | | | | |
| 1: $A^+B^+C^+$ (wild-type) | 13.4 | 0.1 | 6.2 | 0.1 | |
| 2: $A^+B^+C^-$ (3) | 0.1 | 0.1 | 1 | 1 | |
| 3: $A^+B^+C^c$ (67) | – | 9.2 | – | 81 | |
| 4: $A^-$ (2) $B^+C^c$ (67) | – | 10.1 | – | 1 | |
| 5: $A^-$ (2) $B^+C^+$ | 15.4 | – | 1 | 1 | |
| 6: $A^+B^-$ (24) $C^+$ | 0.1 | 0.1 | 175 | 1.8 | |
| Merodiploids | | | | | |
| 2/5: $A^+B^+C^-$/ $F'A^-B^+C^+$ | 15.0 | – | **41** | – | $C^+$ ($C^+$ dominant over $C^-$) |
| 4/2: $A^-B^+C^c$/ $F'A^+B^+C^-$ | – | 22.5 | – | **85** | $C^c$ ($C^c$ dominant over $C^-$) |
| 4/6: A $B^+C^c$/ $F'A^+B^-C^+$ | – | **0.6** | – | 5 | $C^+$ ($C^+$ dominant over $C^c$) |

The *ara* $C^-$ phenotype appears relatively frequently and is produced by a large number of independent point mutations but also by deletions in the corresponding gene region. Due to a deletion of the whole regulatory region (*ara* COI) right up to the bordering *leu* operon ($\Delta$ 1119, cf. top of Table **5.3**), the three structural genes *ara* BAD come under the control of the leucine operator. The three enzymes of Ars degradation produced in such a deletion mutant have normal enzymatic properties and are thus in this respect apparently not influenced by the elimination of the *ara* C region. In addition, there is no influence of nonsense mutations in the *ara* C region on the translation of the mRNA for the structural genes *ara* BAD (cf. footnotes to Table **5.4**).

From these genetic data it is evident that the *ara* C gene is part of a transcription unit independent of the three adjacent structural genes (*ara* B, A and D).

Two further conclusions about the function of the *ara* C gene product can be drawn from the analysis of the *ara* C mutants:

1. The frequent appearance of the *ara* $C^-$ phenotype indicates a positive regulatory mechanism, because if genes code for a negative regulatory protein most of the mutations and, above all, the deletions are expected to result in an inactivation of the protein and hence in a constitutive phenotype. Thus the *ara* C gene product is presumably necessary as an activator for the gene expression in the *ara* BAD region.

2. As several of the *ara* $C^-$ mutants are sensitive to temperature, the gene very likely codes for an activator *protein*, a conclusion which will be confirmed by the findings described in Sects. **5**.5.2 and **5**.5.3.

Important information about functional details of the *ara* C gene are obtained from merodiploid strains, in which the reciprocal dominance of the three *ara* C alleles can be investigated. In Table **5**.2 the activities of ribulokinase and Ars isomerase in haploid wild-type and mutant strains are compared with those found in merodiploid strains containing each two of the *ara* C alleles, i.e., *ara* $C^-$/*ara* $C^+$ (strain 2/5), *ara* $C^c$/ *ara* $C^-$ (strain 4/2) and *ara* $C^c$/*ara* $C^+$ (strain 4/6). The role of the *ara* C protein as activator of the *ara* BAD operon gene expression is again evident from the trans-dominance[1] of *ara* $C^+$ over *ara* $C^-$ (strain 2/5) and of *ara* $C^c$/ over *ara* $C^-$ (strain 4/2). It is thus even more surprising that *ara* $C^+$ is dominant over *ara* $C^c$ (strain 4/6) as this phenotype is typical for negative regulation (repressor proteins). The simplest explanation in view of the strong indications of the positive regulation in the *ara* BAD operon is the assumption of a dual function for the *ara* C protein as a repressor (negative regulation) and, in the presence of L-arabinose, as an activator (positive regulation). This interesting discovery will be further corroborated by the genetic and molecular biological findings described in the following sections.

### 5.2.2 The Operator-Initiator Region, *ara* OI

The strong indications for a dual function of the *ara* C regulator protein for expression of the *ara* BAD operon had to be supplemented by suitable genetic investigations on the corresponding receptor sequences in the regulatory region between the *ara* BAD and *ara* C operons (cf. Fig. **5**.1).

Mapping of this regulatory region between the *ara* C and *ara* BAD genes is based in the following genetic findings:

1. As already mentioned in **5**.2.1, the nonsense mutations in the *ara* C gene do not affect the expression of the *ara* BAD genes, i.e., they do not belong to a joint transcript.
2. On the other hand, it can be established that nonsense mutations in the *ara* B or *ara* A genes have strong polar effects on the expression of the *ara* AD and *ara* D genes respectively. The *ara* BAD genes thus belong to a common operon with transcription-translation direction from B over A to D. The required regulatory region must therefore be located at the beginning of *ara* B.
3. In two deletion mutants, in which the deleted parts of the *ara* regulatory region extend into the neighbouring *leu* operon ($\Delta$ 1165 and $\Delta$ 1119, cf. top of Table **5**.3), the synthesis of the *ara*-specific enzymes in merodiploid strains containing an intact *ara* $C^+$ gene is still inducible by Ars, if the deletion ends in the *ara* C gene ($\Delta$ 1165). However, synthesis does not respond to inducer, if the deletion extends over the *ara* C gene into the adjacent DNA region ($\Delta$ 1119). The difference between the two deletions marks the position of the regulatory region between the *ara* C and *ara* BAD operons.

As is to be expected, two functional parts could be differentiated in the regulatory region: (1) a receptor sequence for the repressor form of the *ara* C protein (operator *ara* O) and (2) a receptor sequence for the activator form of the *ara* C protein (initiator *ara* I). Their relative positions with respect to each other and the dominance be-

[1] The mutual interaction of regulatory genetic regions are described by the terms trans- and cis-dominant and trans- and cis-recessive. Trans-action depends on the formation of a diffusible product (RNA or protein, e.g., *ara* C protein) capable of binding to an unlinked genetic region. In contrast to the *ara* C gene, mutations in the *ara* OI region (cf. the following section) can only act on the expression of the two directly linked *ara* operons (cis-action).

**Table 5.3:** Basal level of Ars isomerase in non-induced haploid and merodiploid strains with mutations in the *ara* OI-region (from Englesberg et al. [6]).

Ars isomerase activities were determinded in the haploid mutant strains (column a) and in merodiploid strains carrying an additional copy of *ara* COIA-region on an F-plasmid. For explanations cf. text. The genetic map on top of the table shows the relative positions of the *ara* genetic elements, of mutated sites and deletions (Δ).

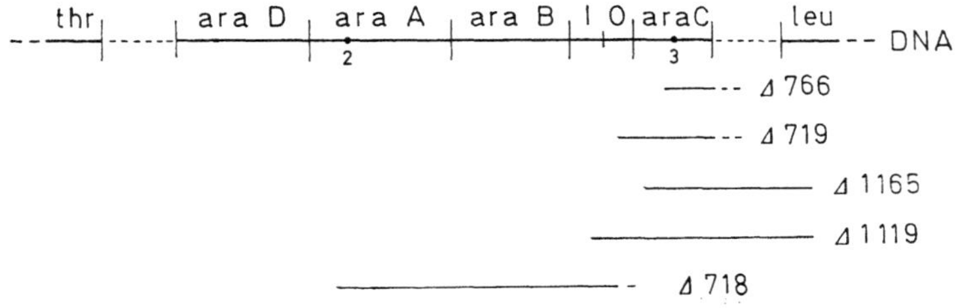


| Genotype | Haploid | Merodiploid (Genetic markers of the F-plasmid) | | |
|---|---|---|---|---|
| | – (a) | $A^-(2)I^+O^+C^+$ (b) | $A^-(2)I^+O^+C^-(3)$ (c) | $A^+I^+O^+C^+$ (d) |
| Wild-type | 0.1 | – | – | – |
| 1: $A^+I^+O^-C^-$ (Δ 719) | 0.1 | 3.5 | 0.1 | – |
| 2: $A^+I^+O^+C^-$ (Δ 766) | 0.1 | 0.2 | 0.1 | – |
| 3: $A^+I^cO^-C^-$ (Δ 719) | 4.0 | 6.2 | 4.9 | – |
| 4: $A^+I^cO^+C^-$ (Δ 766) | 6.9 | 0.9 | 4.4 | – |
| 5: $A^-(2)I^cO^-C^-$ (Δ 719) | 0.1 | – | – | 0.4 |

haviour derive from results of genetic experiments shown in Table **5**.3, which compares the activity of Ars isomerase in non-induced haploid and merodiploid strains.

First of all Table **5.3** contains two initiator constitutive mutants (*ara* $I^c$), with a more than 30-fold increased basal level of Ars isomerase (cf. strains 1a and 2a with 3a and 4a). The mutation was localised close to the deletion Δ 719. The expression of the *ara* $I^c$ phenotype in strains 3a and 4c is dependent on the deletion of the *ara* C region (Δ 719 and Δ 766) which relieves the *ara* BAD operon from the repression by the *ara* C protein. This is clearly demonstrated by introducing an F-plasmid with an intact *ara* C gene in trans-position. Interestingly, both strains 3 and 4 behave basically differently. The repressing effect of the *ara* C gene is pronounced in strain 4 containing deletion Δ 766 (compare strains 4a and 4b) but fails to appear if the deletion extends beyond the *ara* C gene into the immediate vicinity of the *ara* $I^c$ mutation (Δ 719, compare strains 3a and 3b). Thus a DNA region is described, which serves as a receptor structure of the repressor form of the *ara* C gene product, the operator region (*ara* O). It is defined by the difference of deletions Δ 719 and Δ 766. *ara* $O^+$ is cis-dominant over *ara* $I^c$ (compare strains 4b and 3b). However, this holds true only in the presence of *ara* $C^+$, i.e., of the intact *ara* C protein (compare strains 4b and 4c).

The activator form of the *ara* C protein exists in small amounts in equilibrium with the repressor form also in the absence of exogenous Ars: in strains with *ara* $O^-$ (Δ 719) the enzyme basal level is stimulated by the introduction of *ara* $C^+$ (compare strains 1a and 1b).

In contrast to the *ara* C gene, *ara* O and *ara* I mutations affect only those gene regions which are in their immediate vicinity, i.e., lie on the same DNA molecule (cis). This is evident from the comparison of enzyme activities between strains 3a, 3b and 3c respectively for the operator region and between strains 3b and 5d for the initiator region (Table **5**.3). The discovery of the *ara* O and *ara* I regions and their negative as well as positive interaction with the *ara* C protein confirms the supposition about the dual function of the *ara* C protein in the expression of the *ara* BAD operon. One might expect that structures analogous to the *ara* OI region also exist in the immediate vicinity of the two other structural genes of the *ara* regulon, i.e., those for the transport proteins (*ara* E, *ara* F), but as yet there is no detailed genetic proof of this.

## 5.3 Catabolite Repression in the *ara* Regulon

During the genetic experiments with the ara BAD operon, a strange behaviour of certain *ara* $B^-$ or *ara* $A^-$ mutants was very early noticed. Namely, in both cases it was found that expression of the neighbouring structural gene was also influenced, positively or negatively. Using the example of *ara* $B^-$ mutants this phenomenon, which is in no way limited to the arabinose operon, will be explained in more detail.

On the one hand there are *ara* $B^-$ mutants in which the expression of the following *ara* A gene is drastically reduced when compared to the wild-type strain (Table 5.4). These so-called polar mutants were identified as frame-shift or nonsense mutants respectively (cf. footnote b of Table 5.4). Their behaviour allows the conclusion that

**Table 5.4:** Expression of the *ara* A gene in induced cells of polar and hyperinducible *ara* $B^-$ mutants (from Katz & Englesberg [10]).

| Genotype[a] | Haploid | Merodiploid (F-plasmid: $A^-(2)B^+C^+$) | Type of mutation[b] |
|---|---|---|---|
| | Relative activity of Ars isomerase (%) | | |
| Wild-type | | | |
| $A^+B^+C^+$ | 100 | – | – |
| Polar mutants | | | |
| $A^+B^-C^+$ | 54 | 29 | Frame shift |
| $A^+B^-C^+$ | 67 | 27 | Nonsense |
| $A^+B^-C^+$ | 31 | 20 | Nonsense |
| Hyperinducible mutants | | | |
| $A^+B^-C^+$ | 400 | 94 | Missense |
| $A^+B^-C^+$ | 445 | 89 | Missense |
| $A^+B^-C^+$ | 600 | 118 | Missense |
| $A^+B^-C^+$ | 400 | 77 | Missense |

[a] The indicated *ara* $B^-$ strains contain different point mutation sites.

[b] *Frame-shift* mutants arise by deletion of 1 or 2 nucleotides and hence of shifts in the reading frame of the triplet code. This results in a change of all amino acid residues which follow the mutated site. In *nonsense-mutants* one triplet in the coding region of the *ara* B protein was changed into a termination codon. Translation is terminated prematurely and the ribosomes dissociate from the polycistronic mRNA. Frame-shift and nonsense mutations influence the reading frame and translation rate also of genes situated downstream on the polycistronic mRNA (*ara* A, *ara* D, polar effect). In the non-polar, *missense mutants* a single base exchange in the *ara* B gene resulted in the subsequent exchange of an amino acid residue required for the proper function of the *ara* B protein (ribulokinase).

the direction of transcription is from *ara* B to *ara* A (cf. **5**.2.1 and **5**.2.2). It is essential to notice that one can find also *ara* $A^-$ mutants with polar effect on the *ara* D gene *following* it on the polycistronic mRNA but never on the *ara* B gene *preceding* it (cf. the map on top of Table **5**.3).

Contrasting to the polar *ara* $B^-$ mutants, there are others resulting in an increased expression of the *ara* A gene (Table **5**.4). This "hyperinduction" is cancelled by introducing an intact *ara* $B^+$ gene integrated into an F-plasmid. In contrast, the polar (negative) effect of the first group of *ara* $B^-$ mutants is not changed in the *ara* $B^-$/*ara* $B^+$ merodiploid strains. There is still another aspect which adds to the phenotypic differences between both groups of mutants: A hyperinduction effect with respect to the *ara* B gene is also detectable in *ara* $A^-$ mutants, i.e., it is independent of the relative position of the two genes within the operon.

The explanation of these hyperinduction effects leads to a further regulatory mechanism involved in the control of the *ara* BAD operon. This is the catabolite repression described for numerous catabolic enzyme systems in bacteria and fungi.

In *E. coli* and related bacteria catabolite repression involves in fact a positive regulatory mechanism, i.e., transcription of the corresponding gene groups demands the presence of a catabolite gene activator protein (CAP) complexed with its effector cAMP. The cAMP level in bacterial cells fluctuates in response to the exogenous concentration of rapidly utilized carbohydrates, e.g., glucose. A current model of the interaction between cAMP synthesis by adenylate cyclase and the glucose transport system is discussed in Fig. **4**.2. cAMP deficiency, due to reduced synthesis or enhanced degradation and/or excretion, results in a dissociation of the CAP · cAMP complex and thereby in a reduction of the transcription of all operons subject to catabolite repression. Depending on their turnover rates all metabolizable carbohydrates have a more or less pronounced negative effect on the cAMP level. Glucose has the strongest effect, while glycerol triggers hardly any catabolite repression.

In the special case of the *ara* operon the degradation of Ars leads to a reduction of the cellular cAMP level and thus to the phenomenon of catabolite repression. In all *ara* $A^-$ or *ara* $B^-$ strains mentioned above this feedback repression is interrupted because of the missing Ars degradation. This causes the enhanced formation of the intact gene products. It can be demonstrated that the hyperinduction effect really appears in all *ara* $B^-$ mutants shown in Table **5**.4 and that in the first group of strains it is only hidden by the polar effect of the frame-shift and nonsense mutations.

A series of experiments illustrated the mechanism of this auto-catabolite repression by Ars and proved its identity with the catabolite repression evoked by other carbohydrates, e.g., by glucose. In Fig. **5**.3 the dual role of Ars as inducing agent and, at the same time, as factor triggering catabolite repression is impressively shown by analyzing the influence of the carbohydrate source and of Ars on the isomerase activity in two constitutive mutants. They differ from each other only by presence or absence of the *ara* B gene product and thus by the ability to utilize Ars (strains *ara* $A^+$ $B^+$ $C^c$ and *ara* $A^+$ $B^-$ $C^c$).

The slope of the curves gives an estimate of the constitutive synthesis of the *ara* A gene product under conditions of stronger (glucose nutrient solution) and weaker (glycerol nutrient solution) catabolite repression. In both strains the rate of synthesis is markedly decreased in the presence of glucose. Upon addition of Ars there is a considerable increase of isomerase synthesis in the non-degrading mutants on both nutrient solutions (Fig. **5**.3B): Ars acts as inducer. On the other hand, the inducing effect does not appear in the strain which is able to utilize Ars (Fig. **5**.3A) because the inducing and repressing effect of Ars approximately counterbalance each other. It is true, though, that this balance only comes into being because of the relatively high basal level of the Ars-degrading enzymes caused by the mutation in the *ara* C gene. In non-induced *ara* $C^+$ strains the inducing effect of Ars would clearly predominate, at least in the initial phase.

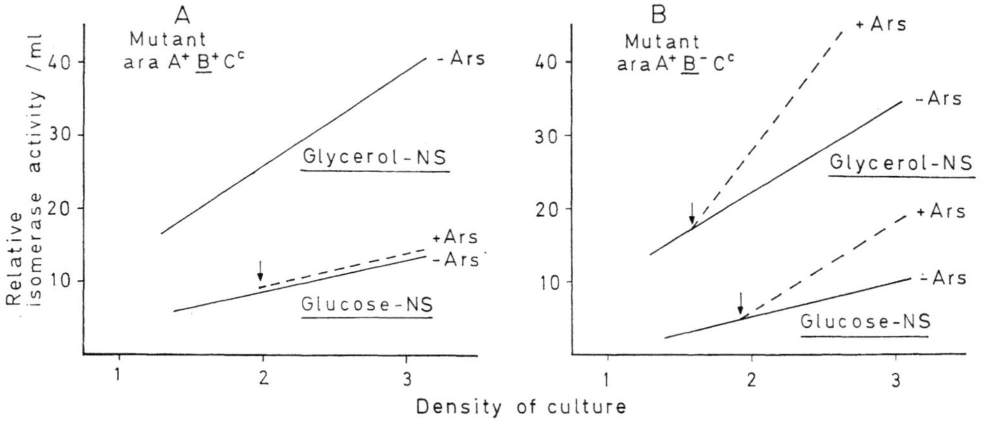


**Fig. 5.3:** Influence of arabinose on the constitutive Ars isomerase activity in cultures growing in glycerol or glucose nutrient solutions (from Katz & Englesberg [10], modified).

Ars isomerase activity was determined in two strains differing in their capability for Ars utilization due to the presence or absence of a functional *ara* B gene product. The *arrow* indicates the time point of Ars addition to the growing cultures.

When searching for mutants defective in catabolite repression, two types were found. One type is altered in the *crp* locus (Fig. 5.1), coding for the catabolite gene activator protein (CAP) and the other concerns the *cya* locus, the gene for adenylyl cyclase. This enzyme catalyzes the synthesis of cAMP. Both mutant types have a pleiotropic negative phenotype, i.e., besides the *ara* regulon all other metabolic gene systems controlled by catabolite repression lack the "competence" for induction by their corresponding substrates. This behaviour was the first indication that it is not at all a "repression" but rather a catabolite "deactivation", i.e., a block of a regulatory system exerting positive control. In agreement with these genetic and biochemical findings all in vitro transcription-translation systems for the *ara* BAD operon described in Sects. 5.5.4 and 5.5.6 are dependent on the addition of CAP and cAMP.

The special operation mechanism of the catabolite repression, especially the question of the DNA receptor site for the activating cAMP-CAP complex in the *ara* IO region, was clarified by the investigation of further mutants. If one looks for *ara* BAD$^+$ phenotypic revertants amongst the mutants defective for the catabolite repression ($crp^-$ or $cya^-$ mutants respectively), some will be found that result from a mutation of the *ara* C gene. Thus, an alteration in the *ara* C gene renders the regulatory protein capable of replacing the cAMP-CAP complex in the *ara* OI-region. These mutants were called *ara* $C^i$ because they are insensitive to catabolite repression. They contrast to *ara* $C^h$ mutants which are hypersensitive to catabolite repression. Evidently the *ara* C activator form interacts directly with the CAP · cAMP complex prior to or after binding to *ara* I [8]. The close vicinity of the binding sites for both activator proteins is further demonstrated by some initiator constitutive mutants (*ara* $I^c$) which are also partially resistant to catabolite repression (for details cf. 5.5.5 and Figs. 5.8 and 5.9).

## 5.4 Regulation of *ara* C Gene Expression

The genetic characterization of the dual regulatory role of the *ara* C protein for the expression of the *ara* BAD operon promoted the question of how the regulator gene itself is controlled in its activity.

Is is true, however, that suitable investigations meet with difficulties because the quantification of the gene product using, for instance, the methods described in **5.5.1** and **5.5.4**, is too insensitive and time-consuming. A solution was arrived at by constructing a special hybrid of phages $\lambda$ and Mu (cf. also **5.5.1**). The latter phage has the characteristic of being integrated at any point into the bacterial genom and thus causes an inactivation ("mutation") of the afflicted gene (cf. Table **4**.3). After transfering the structural genes of the *lac* operon to such a $\lambda$-Mu hybrid it was possible to insert this transducing phage into various sites, including the *ara* C gene. The resultant bacterial strains are *ara* $C^-$, but the expression of the $\beta$-galactosidase (*lac* z gene) is dependent on the state and characteristics of the unknown *ara* C promoter. The gene expression could easily be followed by measuring the $\beta$-galactosidase activity.

The following results were obtained [7]:

1. The expression of the *ara* C gene, like that of the *ara* BAD operon, is subject to the catabolite repression.
2. The *ara* C gene product controls its own synthesis by autorepression, which is independent of the particular state of the protein, i.e., whether the activator or repressor form predominates. It was, however, observed that the leftward transcription from the *ara* BAD promoter somehow interferes with the rightward transcription from the *ara* C promoter. This may be explained by the overlap of the binding sites for regulatory proteins in the intercistronic region (cf. Fig. **4**.9).

By introducing an *ara* C (amber) mutant in different suppressor strains it was possible to vary the repressor level from 0.007 to 1.8 fold of that in the wild-type strain, and to test the effect on the expressibility of the *ara* BAD operon. It turns out that the normal level is just sufficient to allow maximum expression of the *ara* BAD operon in the presence of inducer [9].

## 5.5 In Vitro Investigations to Characterize the Regulatory Mechanisms in the *ara* BAD Operon

The genetic and biochemical investigations described in the preceding paragraphs have led to a concept about the regulation of Ars degradation in *E. coli*. These results were confirmed and refined by in vitro transcription-translation experiments. Prerequisite for these was to supply the essential components of the arabinose system in pure or at least highly enriched forms.

### 5.5.1 Enrichment and Isolation of Genes of the *ara* CBAD Region

The arabinose regulon constitutes only a small part of the *E. coli* genom (approximately $5 \times 10^6$ base pairs). The specific concentration of *ara* genes can be increased about hundredfold if the required part of the regulon, e.g., the *ara* CBAD region, is transferred to a bacteriophage, which has a genome a hundred times smaller.

The production of such specialised transducing phages is achieved either by translocating the required genetic region into the vicinity of the phase attachment site in the DNA (cf. Fig. **14**.5) or by insertion of the phage in the neighbourhood of the desired gene region itself. This very rare event occurs in bacterial strains which, due to a deletion, lack the true phage attachment sequence. A series of complex selection techniques is applied to enrich and isolate the bacterial strains with the phage integrated in the desired positions of the genetic material.[2] Finally the phages are

[2] These techniques, which form the true basis for the "construction" of the required DNA fragments, cannot be discussed in more detail here. A comprehensive description can be found in Ref. [12].

reisolated. They contain the *ara* CBAD region or part of it inserted under inversion into the phage DNA. In many cases insertion of bacterial genes into the phage genome is unseparably connected with the loss of some phage genes. If these are essential for multiplication, the transducing phage is defective, because it needs a helper-phage for productive infection[3]. The isolation of the phage DNA proceeds after lysis of the infected bacterium and purification of the phages by isopycnic centrifugation in caesium chloride gradients.

Further purification of the *ara* OI region was achieved by isolating a DNA fragment of approximately 1,000 base pairs. This was obtained by forming heteroduplex molecules between two *ara* transducing phages carrying, in common, a region of *ara* DNA which included the entire regulatory region plus portions of *ara* B and *ara* C genes. All DNA outside this region was removed by of $S_1$ nuclease digestion followed by acrylamide gel electrophoresis (cf. the fragment shown in Fig. **5**.6). Finally the overwhelming part of the *ara* B gene sequences was removed by cleaving with restriction endonuclease Hae III (cf. Table **2**.1). The resulting 440 bp fragment contains 45 bp of the *ara* B gene sequence, the whole *ara* OI region (170 bp) and about 250 bp of the flanking *ara* C gene. The two fragments were used for extensive studies on the binding affinity of the regulatory proteins involved in *ara* gene expression, for localizing the binding sites, e.g., for RNA polymerase (cf. Fig. **5**.6), for electron microscopic fine structure analysis and, finally, for sequencing the whole *ara* OI-region (cf. Fig. **5**.9).

### 5.5.2 Purification and Characterization of the *ara* C Protein

Another important prerequisite for the studies with in vitro systems is the purification of the regulatory protein [9 a]:

As the number of *ara* C protein molecules under normal conditions is about 40 monomers per cell (compared to about 12,000 monomers for the *ara* BAD gene products), a bacterial strain was selected with a 100–200 times higher amount. Similar to the gene enrichment techniques described in the preceding paragraph, a strain of *E. coli* is used which harbours an *ara* C-gene transducing phage integrated in its chromosome. Furthermore, this phage has a ts-mutation in its *cI*-repressor gene (for details cf. **14.2.3.6**), i.e., it remains in a lysogenic state at temperatures below 30 °C but is induced to embark on the lytic pathway if cells are shifted to temperatures higher than 40 °C. The lytic development coincides with a high rate of expression of most phage genes including also the inserted *ara* C-gene (cf. Chap. **14**).

In particular the ara C protein purification proceeds in the following stages:

1. Propagation of the lysogenic *E. coli* K 12 strain at 30 °C,
2. temperature induction of phage multiplication (42 °C for 30 min),
3. homogenisation of the cells and separation of the main part of the DNA by polyethylene glycol/dextran distribution (the protein remains in the aqueous polyethylene glycol solution).
4. affinity chromatography of the crude protein extract on a dextran column loaded with the anti-inducer D-fucose and
5. chromatography on a salmon sperm DNA-cellulose column.

The 5-step procedure leads to an enrichment of the *ara* C protein by about 30,000 fold with a purity of 90%. SDS-acrylamide gel electrophoresis gave a single protein band with $M_r$ 28 kd. The protein, eluted from acrylamide gels, still possessed its activator function in an in vitro test system (cf. **5**.5.4). Investigating the sedimentation

[3] In the following experiments (cf. **5.5.3**–**5.5.5**) in many cases the DNA of a *d*efective phage *h*ybrid will be used, which comprises parts of the genetic material of phages $\lambda$ and the related phage $\Phi$ 80 plus varying amounts of the *ara* CBAD DNA ($\lambda$h 80 d *ara*). On the other hand a non-defective, *p*laque-forming *ara* transducing $\lambda$ phage will be used ($\lambda$p *ara*).

of the purified *ara* C protein in saccharose or glycerol gradients, respectively, showed the activator form to be a dimer, whereas the repressor form is a tetramer. Binding of L-arabinose ($K_M$ $3 \times 10^{-3}$ M) stabilizes the dimeric activator protein. The anti-inducer D-fucose binds with the same affinity ($K_i$ about $6 \times 10^{-3}$ M). The in vivo half-life of the *ara* C protein is 60 min [9a, 9b].

### 5.5.3 Evidence of the Regulation Level

The genetic proof for the different regulatory structures in the arabinose regulon outlined in Sect. **5.2** can give no information on the level of control, i.e., whether the *ara* C protein acts on transcription or translation. The specific binding of the purified *ara* C protein to the *ara* O region of purified DNA of an *ara*-transducing phage produced the first important evidence of a transcription regulation. In order to confirm this finding the amount of *ara* BAD-specific mRNA was determined from various strains of *E. coli* by hybridisation with the corresponding DNA.

Total RNA of induced and non-induced cultures respectively are labelled 3 min with [$^3$H]-uridine, extracted and hybridised for 16 h at 76 °C with denaturated λh80d *ara* DNA bound to filter discs. The DNA-RNA hybrids formed between the *ara* region of the phage DNA and the complementary *ara* mRNA are stable against the following digestion with pancreatic RNase. The non-hybridised part of the RNA (>99.7%) is subsequently removed by washing the filter discs with buffer.

The results are shown in Table **5.5**: In the wild-type strain the amount of *ara*-specific mRNA increases considerably after induction with Ars. This induction is missing in the two *ara* C$^-$ mutants with a nonsense mutation (C3) and deletion (Δ 1165) respectively. On the other hand, the strain which has become constitutive through the mutation of the *ara* C gene [*ara* C$^c$ (67)] actually contains a relatively large amount of *ara* mRNA even under non-inducible conditions. Finally, in a control mutant with deletion of the major part of the *ara* CBAD region (Δ 714) the amount of *ara*-specific mRNA is practically zero. Hence these results show a reasonable agreement between the phenotypes established in genetic experiments and the cellular levels of *ara*-specific mRNA and thus make a transcription control by the *ara* C protein very likely. They are corroborated by the fact that the *ara* mRNA

**Table 5.5**: Relative amounts of *ara* BAD-specific mRNA in wild-type strain and *ara* C mutants (from Cleary & Englesberg, J. Bacteriol. *118* (1974): 121).

Pulse labelled RNA from induced and non-induced cells was isolated and hybridized to DNA of the phage λh 80 d *ara* (cf. text). For explanations to the genetic map on top of the table cf. Tables **5.2** and **5.3**.

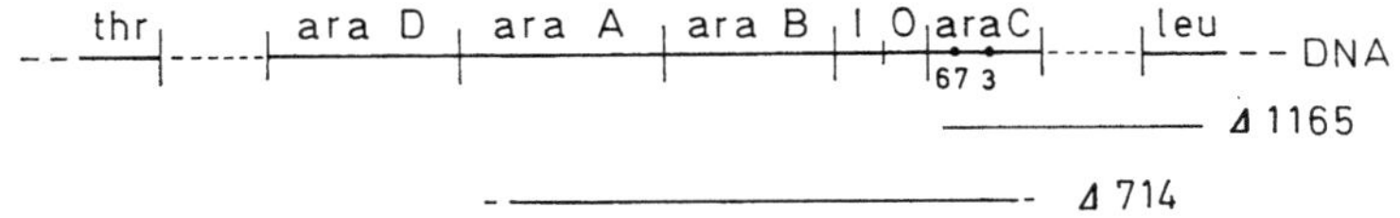


| Genotype | % of [$^3$H]-uridine-labelled RNA hybridized with *ara* BAD region | |
|---|---|---|
| | Induced | Non-induced |
| D$^+$A$^+$B$^+$C$^+$ (Wild-type) | 0.29 | 0.04 |
| D$^+$A$^+$B$^+$C$^c$ (67) | 0.22 | 0.17 |
| D$^+$A$^+$B$^+$C$^-$ (3) | 0.05 | 0.04 |
| D$^+$A$^+$B$^+$C$^-$ (Δ 1165) | 0.04 | 0.04 |
| D$^+$A$^-$B$^-$C$^-$ (Δ 714) | 0.02 | 0.02 |

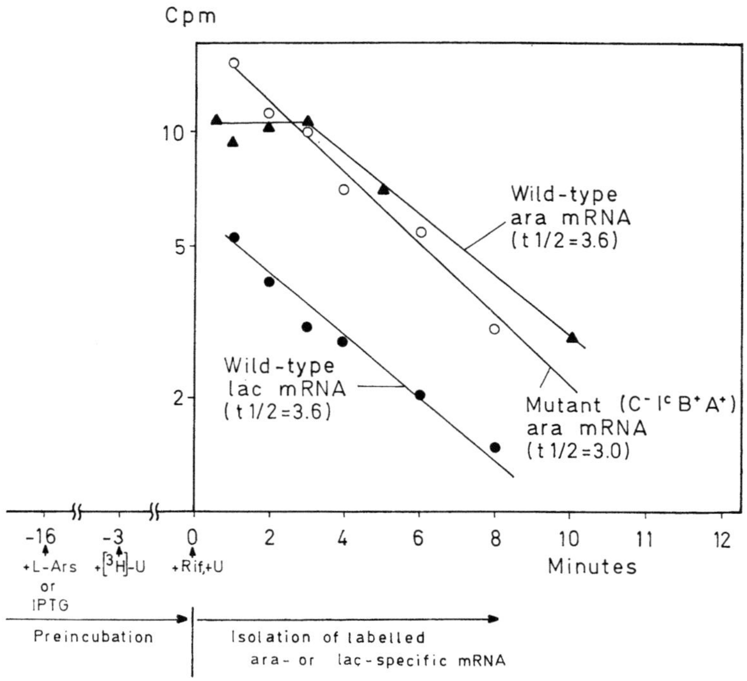


**Fig. 5.4**: In vivo degradation of *ara*- and *lac*-specific mRNA (from Cleary & Englesberg, J. Bacteriol. *117* (1974): 121).

Details of the experimental procedure are given in the text. *IPTG*, isopropyl thiogalactoside (inducer of lac-specific mRNA formation); *³H–U*, [³H]-uridine added for pulse labelling of RNA (3 min); *Rif*, rifampicin added at time point zero to stop further RNA synthesis; $t^1/_2$, half-life of mRNA in min.

in induced cultures decreases practically to the level of non-induced cells after addition of anti-inducer D-fucose.

One open question still remains: Is it possible that the *ara* C activator protein does not really stimulate the de novo formation of *ara* mRNA, but rather decreases its degradation? In order to answer this question the series of experiments shown in Fig. 5.4 were carried out.

Cultures were induced 13 min either with Ars or, as a control, with an inducer of the lactose operon (isopropyl thiogalactoside). Afterwards RNA was pulse-labelled (3 min) with [³H]-uridine. Uridine incorporation was stopped at time zero min (cf. Fig. 5.4) by adding large amounts of unlabelled uridine (chase) and the antibiotic rifampicin. The decline of the radioactivity in the *ara* and *lac* mRNA respectively can be determined in an analogous hybridisation assay as described for Table 5.5.

The inactivation kinetics of the two mRNA types for the *ara* and *lac* operons are noticeably similar and produce identical values for the half-lives (3.6 min). The only difference consists in a lag period of about two minutes in the degradation of *ara* mRNA compared with *lac* mRNA. This lag period disappears: (1) if one uses D-fucose instead of rifampicin to interrupt the transcription and (2) if the same type of experiment is done with a strain which has a deletion (*Δ* 1165) of the regulator gene and instead possesses a high constitutive basal level due to an *ara* $I^c$ mutation (Fig. 5.4). The value for the half-life of the *ara* mRNA of this mutants is only slightly different from that found in wild-type cells.

These experiments lead to the following conclusions:

1. The presence of *ara* C activator protein has no influence on the velocity of *ara* mRNA degradation which is similar to that found for *lac* mRNA.
2. The 2-min delay of the *ara* mRNA degradation requires the *ara* C activator protein. Mechanism and biological relevance of this translational control are unclear.
3. The similarity of the effects of rifampicin and D-fucose on the formation of the *ara* mRNA, confirmed in further experiments, suggests that the *ara* C activator protein acts as a specific initiation factor of the *ara* BAD operon (cf. **5.5.5**).

### 5.5.4 In Vitro Transcription and Translation of the *ara* CBA Genes

The coupled in vitro transcription and translation of genes of the arabinose regulon and investigations on details of the regulatory mechanisms are based on the following overall reaction:

$$\underset{(\lambda\text{h } 80 \text{ d } ara)}{ara\text{-DNA}} \xrightarrow[\text{S30 extract of } E.\ coli]{\text{specific effectors, regulatory proteins}} ara\ \text{mRNA} \xrightarrow[\text{S30 extract}]{} ara\text{-specific proteins}$$

The "S30 extract" is an unspecific complex enzyme system which contains all the necessary auxiliary proteins (RNA polymerase, aminoacyl-tRNA synthetases, initiation and elongation factors, ribosomes etc.) for transcription and translation (cf. Chap. **1**). In this case it represents the supernatant obtained after centrifugation at 30,00 × g of a homogenate from mutant cells of *E. coli*, which were selected for their deletion of the whole *ara* CBAD region. In this way a disturbance by any components (DNA, proteins) of the arabinose system in the S30 extract is avoided. The incubation mixture contains, apart from the DNA template (λ480 d *ara* DNA) and the S30 extract, the necessary substrates and co-factors for RNA and protein synthesis (nucleotides, amino acids, ions, co-enzymes etc.).

The *ara* protein was added either in form of the preparation described in **5.5.2** or in more crude preparations obtained through ammonium sulphate precipitation. In the presence of Ars the activator form predominates. The repressor form is obtained by 18 h dialysis against an Ars-free buffer.

Synthesis in the in vitro system is started by addition of the S30 extract, usually goes on for 60 min at 37 °C and is stopped with chloramphenicol. The amount of enzymes formed is determined by incubating the whole transcription-translation mixture with the corresponding substrates. Ars isomerase is more suitable for the assay than ribulokinase because of its much higher specific activity (cf. Table **5.1**).

In agreement with the basic genetic findings [2, 3] the in vitro system with *ara* C⁻ DNA as template shows the following characteristics (Table **5.6**): The synthesis of Ars isomerase disappears in the absence of the inducer (Ars) and/or *ara* C protein.

**Table 5.6:** Synthesis of Ars isomerase in an in vitro transcription-translation system (from Wilcox et al. [17]). For details of the experimental procedure cf. text.

| Omissions or additions to the incubation mixture | Isomerase (U/ml) |
|---|---|
| Complete [− DNA λh80d *ara* $A^+B^+I^+O^+C^-$ (Δ 766)] | 9.2 |
| Complete, − DNA | 0.1 |
| Complete, − L-arabinose | 0.1 |
| Complete, − *ara* C-protein | 0.1 |
| Complete, − cAMP | 0.3 |
| Complete, + 0.27 M D-fucose | 0.1 |
| Complete, + 85 μg/ml chloramphenicol | 0.1 |

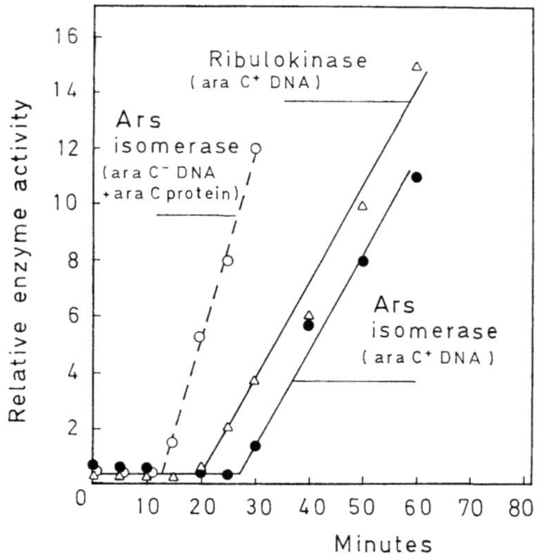


**Fig. 5.5**: In vitro synthesis of Ars isomerase and ribulokinase using *ara* $C^-$ and *ara* $C^+$ DNA as template (from Wilcox et al. [17], modified).

The coupled in vitro transcription-translation proceeded with $\lambda$h 80 d *ara* $A^+B^+I^+O^+C^-$ ($\Delta$ 766) DNA as template in the presence of *ara* C protein *(dashed line)* or alternatively with $\lambda$h 80 d *ara* $A^+B^+I^+O^+C^+$ DNA without *ara* C protein *(full lines)*. After the incubation was stopped by adding DNase and chloramphenicol, enzyme activities were determined.

It is heavily dependent on cAMP addition whose receptor protein (CAP) is part of the extract. D-fucose acting as anti-inducer of the *ara* BAD operon in vivo also blocks its expression in vitro. Chloramphenicol as general inhibitor of the prokaryotic protein synthesis has the same effect. A remarkable feature of the test system appears by using *ara* $C^+$ DNA instead of *ara* $C^-$ DNA as template and omitting the *ara* C protein from the incubation mixture. In this case synthesis evidently proceeds with endogenously formed *ara* C protein, a result to be described in more detail in connection with Fig. 5.5.

The assay of the dual function of the regulator protein as activator and repressor succeeds in the genetic experiments (cf. 5.2.2) as well as in vitro with template DNA from initiator constitutive mutants (Table 5.7). The values obtained in vivo and in vitro are quite similar. In addition, it can actually be shown that the in vitro synthesis of Ars isomerase is inhibited by the addition of *ara* C protein if $\lambda$h80d *ara* $I^c$ $C^-$ DNA is used as template (from 22% $\rightarrow$ 8% of the fully induced control). This inhibition is relieved by the addition of Ars, i.e., by transforming the repressor protein into the activator protein. The amount of isomerase increases to 80% of the control.

Investigations on the synthesis kinetics of the transcription-translation system show sequential formation of the *ara* C regulator protein and the *ara* A enzyme protein (isomerase) as well as coordinate expression of the genes within the *ara* BAD operon (Fig. 5.5). With *ara* $C^-$ DNA as template and *ara* C protein added to the incubation mixture activity increase of Ars isomerase starts at about 10 min, whereas it is further delayed by 10–15 min if the synthesis proceeds with *ara* $C^+$ DNA as template and *ara* C protein is omitted. In this case enzyme synthesis is preceded by the formation of the regulatory protein. As expected from the gene sequence (*ara* B $\rightarrow$ *ara* A) the increase of ribulokinase activity starts earlier than that

**Table 5.7:** Proof of the dual function of the *ara* C protein as repressor and activator (from Wilcox et al. [17]).

*In vivo:* Cells of the indicated strains were cultivated under inducing and non-inducing conditions and afterwards analysed for Ars isomerase activity (*ara* A gene product).
*In vitro:* The indicated types of $\lambda$h 80 d *ara* DNA were used as templates for in vitro transcription-translation experiments. Synthesis of Ars isomerase proceeded in the absence or presence of *ara* C protein and Ars respectively.

| Genotype | In vivo | | In vitro | | | |
|---|---|---|---|---|---|---|
| | Isomerase activity (% of ind. control) | | DNA of $\lambda$h 80 d *ara* $C^-$ ($\Delta$ 766) | *ara* C protein | Isomerase activity (% of control + Ars) | |
| | − Ars | + Ars | | | − Ars | + Ars |
| $A^+I^+O^+C^-$ ($\Delta$ 766) | 0.2 | 0.2 | $A^+I^+O^+C^-$ | − | 1 | 1 |
| $A^+I^+O^+C^+$ (wild-type) | 0.2 | 100 | $A^+I^+O^+C^-$ | + | 3.3 | 100 |
| $A^+I^cO^+C^-$ ($\Delta$ 766) | 19 | − | $A^+I^cO^+C^-$ | − | 22 | 24 |
| $A^+I^cO^+C^+$ | 6 | 78 | $A^+I^cO^+C^-$ | + | 8 | 80 |

of Ars isomerase (*ara* A gene product). The rate of synthesis of both enzymes is, however, practically identical. These results demonstrate the coordinate expression of the *ara* BAD operon also in the in vitro system.

### 5.5.5 Electron Microscopic Localisation of the Transcription-Initiation Complex

According to the genetic and molecular biological findings described so far, the initiation of transcription of the *ara* BAD genes requires interaction of the *ara* I region with RNA polymerase, the *ara* activator protein and the cAMP · CAP complex. Purification of a DNA fragment containing the *ara* OI region (cf. 5.5.1) has made possible investigations on the site and binding sequence of the three proteins. The three proteins were incubated with the DNA in series or in a mixture and, after

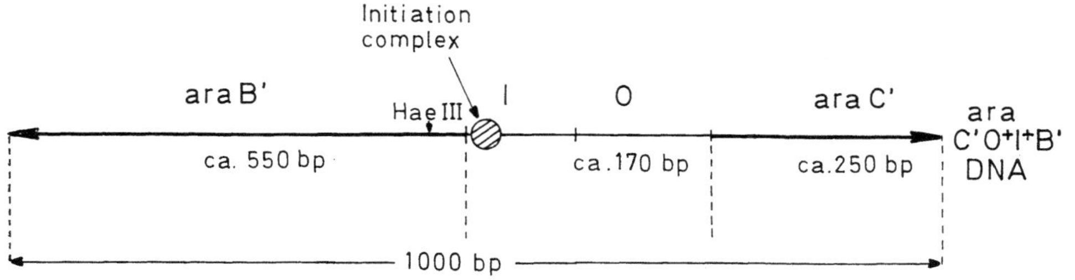


**Fig. 5.6:** Electron microscopic localization of the initiation complex at the *ara* BAD promoter (from Hirsh & Schleif [18]).

The scheme shows a 1,000 bp fragment with parts of the *ara* B and *ara* C genes and the entire OI-region. The initiation complex containing RNA polymerase, the *ara* C activator protein and the CAP · cAMP complex was localized in close vicinity to the initiation point of the *ara* B gene (cf. the nucleotide sequence given in Fig. 5.9). *Hae III*, recognition and cleavage site for Hae III restriction endonuclease (cf. Table 2.1).

the appropriate processing, the DNA-protein complex was investigated electron microscopically.

It appeared that only in the presence of Ars and all three proteins a specific initiation complex bound to DNA in the direct neighbourhood of the *ara* B gene was detected (Fig. 5.6). On the other hand, no site-specific binding to the *ara* OI region could be found for any of the proteins alone. It is important to notice that the supposed transcription-initiation complex disappears after the addition of the four nucleoside triphosphates. This can be expected because the *ara* C protein and cAMP · CAP complex are released after transcription has started. With these experiments a physical localization of the *ara* BAD promoter was achieved. As will be demonstrated in Sect. 5.6, sequencing of the entire *ara* OI region confirms these results and positions the *ara* C promoter at about 100 base pairs distance upstream of the *ara* BAD promoter.

### 5.5.6 Assay of the Codogen Strands for the *ara* BAD and *ara* C operons

The direction of DNA transcription from the 3′- to the 5′-end and the sequence of initiator region and structural genes in the *ara* IBAD region determined by the polarity experiments (cf. 5.2.2) require that the codogen strand be the so-called L-strand read anti-clockwise[4]. One should be able to directly conform this expectation by hybridizing radioactivly labelled *ara*-specific mRNA with the separated L- and R-strands of the *ara* DNA.

For isolation of the two strands DNA of a transducing λ phage, carrying the *ara* COI region plus part of the adjacent *ara* B gene (cf. Fig. 5.7), was denaturated and, after complexation with poly (U, G), was centrifuged in a CsCl gradient. Due to differences in G, C contents the L-strand bands at lower densities than the R-strand. In the same way mutant DNA can be separated and complemented with the antiparallel strand from wild-type DNA. Such hybrid DNA molecules can be used as templates in an in vitro system. The specific variants of this procedure will be of importance in connection with experiments shown in Table 5.8.

Figure 5.7 shows the results of a hybridization experiment using [$^3$H]-uridine-labelled RNA and increasing amounts of the R or L strands of the λp *ara* B′C DNA. In non-induced bacteria with a deletion of the *ara* CO region (Δ 719) the hybridization is low with similar values for both strands. In contrast, in the initiator constitutive mutant with a high basal level of *ara* BAD mRNA an increased hybridization is found, this increase indeed being restricted to the L-strand. This particular type of experiment, i.e., the use of strains with a deletion of the *ara* CO region, was chosen in order to avoid interference by *ara* C gene-specific mRNA.

Because of the probably very low transcription rate a similar hybridization assay for determining the codogen strand in the *ara* C gene region was ruled out. Instead, an in vitro transcription-translation system was combined with *ara* C DNA templates obtained from recombination of mutant L- and R-strands with wild-type R- and L-strands respectively. Thus, in the recombinant DNA molecules only one of the two strands is functionally active.

The results in Table 5.8 show that the in vitro synthesis of Ars isomerase on the basis of an *ara* $A^+C^-$ DNA template is increased 50–60-fold by addition of *ara* $C^+$ DNA, either native or renaturated from the single strands. The increase does not

[4] In this chapter and the following conclusions it will be taken into consideration that an inversion of the genes and thus also of the relative position of the codogen strands takes place during integration of the *ara* region into the phage genome (cf. Fig. 14.5). The designation of L- and R-strands used here refers to the *E. coli* chromosome and is therefore inverse to that in the λ phage genome and to that in the original papers.

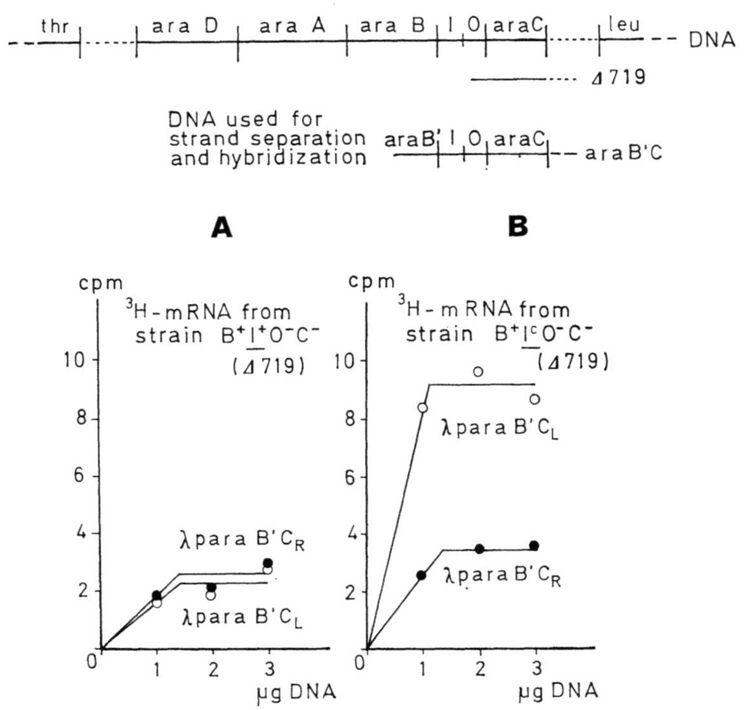


**Fig. 5.7:** Demonstration of the DNA L-strand to be the codogenic DNA strand for transcription of *ara* BAD mRNA (from Wilcox et al. [16]).

[$^3$H]-Uridine labelled RNA was isolated from cells of a ribulokinase (*ara* B gene product)-producing *ara* $C^-$ ($\Delta$ 719) strain. Hybridization of the *ara*-specific mRNA was determined by means of the separated L- and R-strands of the DNA fragment shown on *top* of the figure.

**Table 5.8:** Demonstration of the DNA R-strand to be the codogenic strand for the *ara* C gene (from Wilcox et al. [16]).

$\lambda$h 80 d *ara* $A^+I^+O^+C^-$ ($\Delta$ 766) DNA was used as template for an in vitro transcription-translation system complemented with the indicated types of $\lambda$p *ara* DNA as sources for the *ara* C gene (cf. **5.5.4**). Formation of the *ara* C protein was detected indirectly by the amount of Ars isomerase synthesized

| *ara* C DNA ($\lambda$p *ara*) | Ars isomerase (U/ml) |
|---|---|
| – | 0.002 |
| $C^+_{L,R}$, native | 0.133 |
| $C^+_L C^+_R$, renat. | 0.113 |
| $C^-_{L,R}$, native | 0.002 |
| $C^-_L C^-_R$, renat. | 0.003 |
| $C^+_L C^-_R$, renat. | 0.001 |
| $C^-_L C^+_R$, renat. | 0.085 |
| $C^+_R$, single strand | 0.003 |

occur with native or renaturated *ara* C⁻ DNA. If renaturated DNA hybrids were used as templates, only those with an intact *ara* $C^+$ R-strand stimulated. The *ara* $C^+$ R-single strand is, however, ineffective, since only a DNA double helix is accepted as template by RNA polymerase. In addition, by omitting Ars or adding D-fucose it could be shown that the stimulation of Ars isomerase formation by the λp *ara* $C^-{}_LC^+{}_R$ hybrids really results from the synthesis of a functionally intact *ara* C protein. Thus, it is evident that transcription of the *ara* C gene proceeds on the R-strand of the *E. coli* chromosome in opposite direction with respect to the *ara* BAD genes. Hence, the *ara* C promoter must map within the *ara* OI region, an assumption confirmed by the DNA sequence analysis (cf. Fig. 5.9).

## 5.6 Conclusions: Regulatory Interactions in the *ara* OI Region as a Model of a Complex Transcription Control

In summary of the genetic and molecular biological results discussed in the preceding sections, one can design a rather detailed picture of the structure and function of the regulatory region *ara* OI (Figs. 5.8 and 5.9):

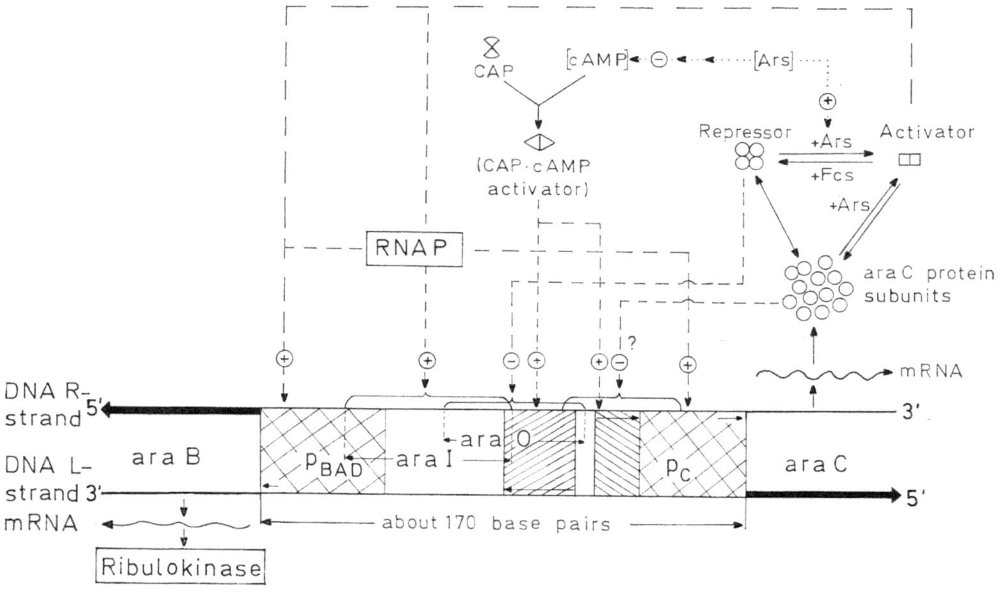


**Fig. 5.8:** Interaction of RNA polymerase and regulatory proteins in the *ara* OI intercistronic region.

— — ⊕ — →, — — ⊖ — →; promoting and inhibiting influence of proteins on transcription; ··· ⊕ ···· ⇢, ··· ⊖ ···· ⇢ promoting and inhibiting influence of the Ars concentration on the formation of the *ara* C activator protein and the cellular cAMP level respectively. For further abbreviations see the beginning of this chapter. The exact position of the two promoters ($p_{BAD}$ and $p_C$ respectively, *crossed areas*) and of the binding sites for the regulatory proteins derive from the sequence data shown in Fig. 5.9 as well as from studies on the protection of the *ara* OI region against DNase I attack (cf. legend to Fig. 5.9). *Hatched areas* with *arrows* denoting the orientation represent the two binding sites for the CAP · cAMP complex.

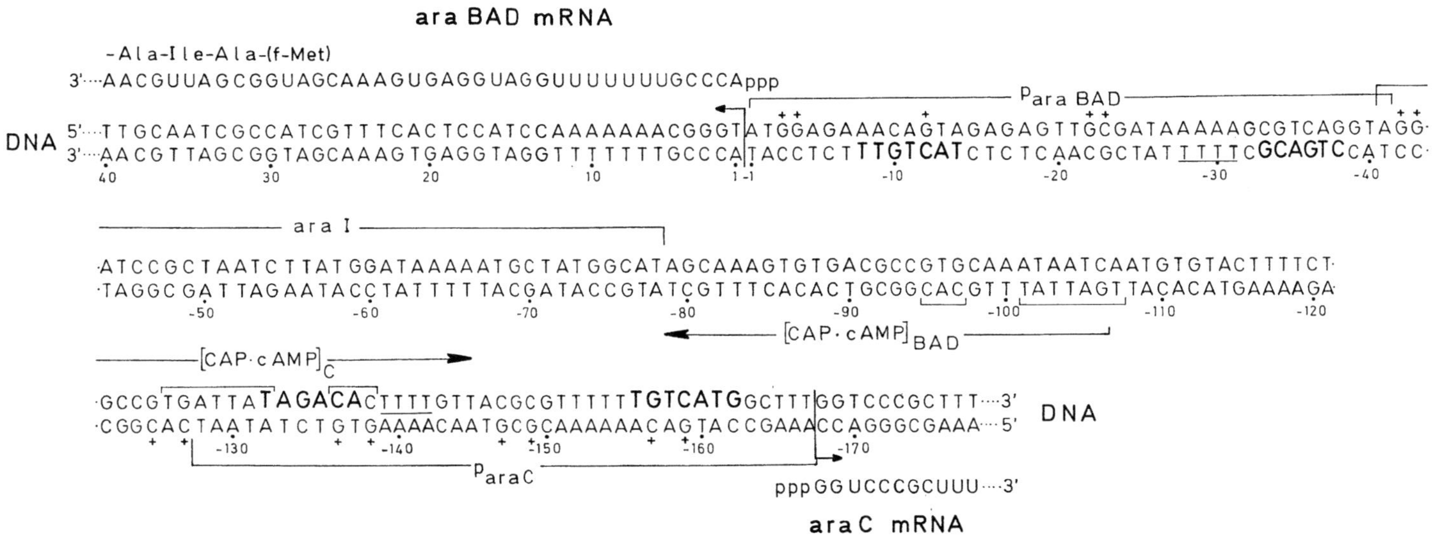


**Fig. 5.9:** Nucleotide sequence of the *ara* OI intercistronic region of *Escherichia coli*, B/r (from [**3, 14, 15, 15**a and **19**]).

Digestion with Hae III restriction endonuclease of the 1,000 bp regulatory fragment described in Sect. **5.4.5.1** gave the central part (about 400 bp) with the *ara* OI-region, which was sequenced by the Maxam-Gilbert technique (cf. **2**.10). The sequence starts at +40 with a short part of *ara* B gene, followed by the *ara* BAD promoter (−1 to −40), the "interpromoter region" (−40 to −130), the *ara* C promoter (−130 to −166) and the start of the *ara* C gene (−167 to −177). The remaining part of the *ara* C gene was not sequenced. Both promoters map in an A=T-rich region with short G–C clusters (+) interspersed at 10 base pairs distance and the typical RNAP recognition and binding sequences about 35 and 10 base pairs upstream of the transcription initiation points. They are denoted by *bold-faced letters* on the non-coding DNA strand (cf. the discussion on promoter structures **1**.3.2.3 and **4**.4.2). Binding sites for the *ara* C repressor and activator protein as well as for the CAP · cAMP complex have been clarified [19, 15a] by the DNase I "foot printing" method (cf. legend to Fig. **4**.18). Autorepression by the *ara* C protein involves bp −106 to −144, i.e., a sequence overlapping with $p_{araC}$, whereas promotion of transcription from $p_{ara\,BAD}$ by the *ara* C activator protein requires bp −42 to −80 (*ara* I). Two binding sites for the CAP · cAMP complex are indicated by brackets. Both contain a common sequence element: 5′-TGATTAT · 3b · CAC-3′ [19] (cf. the discussion on similar sites in the *lac* and *gal operons*, **4**.4.2.2). But only one, the high affinity binding site including bp −77 to −110, was shown to be involved in CAP stimulation of both $p_{ara\,C}$ and $p_{ara\,BAD}$ [15a].

– The information of the two directly neighbouring operons *ara* C and *ara* BAD is read in opposite directions from the DNA R- and L-strands respectively.
– Start of the *ara* C gene transcription at promoter $p_C$ is stimulated in the presence of the CAP · cAMP complex. The binding sites for RNAP and CAP · cAMP complex partially overlap. Transcription is inhibited by the *ara* C gene product itself. Both the activator and repressor form compete with RNAP for the $p_C$ binding site [15a].
– For transcription initiation from the *ara* BAD promoter a ternary complex consisting of RNA polymerase, *ara* C activator protein and CAP · cAMP complex is apparently necessary. The three binding sites for these proteins are oriented in series spanning totally 110 bp (Fig. **5**.9). Transcription of the leftward operon is inhibited by the *ara* C repressor.
– Localization and structure of the two promoters were defined by sequencing of the *ara* OI region (Fig. **5**.9). Both map about 100 bp apart in A = T rich regions and contain the recognition and binding sequences at —35 and —10 bp typical for most prokaryotic promoters (cf. Table **1**.8).
– The whole pathway of arabinose uptake and degradation is integrated via the CAP · cAMP system into the general regulation of carbohydrate metabolism in the cell. The connecting link is afforded by the effector and substrate arabinose itself, whose excessive degradation can lead to a decrease of the cellular cAMP level and thus to a decrease or halt of the transcription of the *ara* BAD operon.
– A remarkable point is the strong overlap of binding sites for RNAP and the three regulatory proteins (*ara* C activator, *ara* C repressor and CAP · cAMP complex), leading to positive or negative interactions in the medium part of the intercistronic region *ara* OI (Fig. **5**.9). This overlap might explain, why leftward and rightward transcription are mutually exclusive.
– The structure and multiple role of the *ara* C protein is especially fascinating and deserves further experimental efforts. Preliminary genetic evidence indicates that three independent domains can be characterized by mutation, one domain responsible for inducer binding, one (activator domain) interacting with the CAP · cAMP complex and one which is responsible for the autorepression effect (cf. Ref. [3] and [8]).

Considering our far advanced knowledge about the structure and regulation of the *ara* CBAD region, it seems highly desirable to complete the picture by analyses of the two unlinked genes *ara* E and *ara* F, and the regulatory sequences adjoined to them. The *ara* regulon represents an interesting example of how the bacterial cell responds to changes of the extracellular milieu by reprogramming part of its intermediary metabolism. The adaptation is self-controlled by auto-catabolite repression as is the synthesis of the central regulatory protein, the *ara* C gene product. The findings about the regulatory network controlling arabinose utilization have decisively contributed to our understanding of more complex gene expression programs.

## 5.7 Selected references

*Reviews*


1. Englesberg, E.: Regulation in the L-arabinose system. In: Vogel, H. J. (ed.), Metabolic Regulation. Acad. Press, New York 1971, pp. 257–296.
2. Englesberg, E. & G. Wilcox: Regulation: Positive control. Ann. Rev. Genetics *8* (1974): 219–242.
3. Lee, N.: Molecular aspects of ara regulation: In: Miller, J. H. & W. S. Reznikoff (eds.), The Operon. Cold Spring Harbor Lab., New York 1978, pp. 389–409.

*Coordinate Gene Expression*

4. Englesberg, E., J. Irr, J. Power & N. Lee: Positive control of enzyme synthesis by gene C in the L-arabinose system. J. Bacteriol. *90* (1965): 946–957.

*Genetics*

5. Sheppard, D. E. & E. Englesberg: Further evidence for positive control of the L-arabinose system by gene *ara* C. J. Mol. Biol. *25* (1967): 443–454.
6. Englesberg, E., C. Squires & F. Meronk: The L-arabinose operon in *E. coli* B/r: a genetic demonstration of two functional states of the production of a regulator gene. Proc. Natl. Acad. Sci. USA *62* (1969): 1100–1107.
7. Casadaban, M. J.: Regulation of the regulatory gene for the arabinose pathway ara C. J. Mol. Biol. *104* (1976): 557–566.
8. Sheppard, D. E., M. Eleuterio & B. Falgout: Interaction between mutant alleles of ara C of the *Escherichia coli* B/r L-arabinose operon. J. Bacteriol. *139* (1979): 1085–1088.
9. Haggerty, D. M., M. P. Oeschger & R. F. Schleif: In vivo titration of ara C protein. J. Bacteriol. *135* (1978): 775–781.

9a. Wilcox, G., K. J. Clemetson, P. Cleary & E. Englesberg: Interaction of the regulatory gene product with the operator site in the L-arabinose operon of *Escherichia coli*. J. Mol. Biol. *85* (1974): 589–602.

9b. Kolodrubetz, D. & R. Schleif: Identification of *ara* C protein on two-dimensional gels, its in vivo instability and normal level. J. Mol. Biol. *149* (1981): 133–139.

*Catabolite Repression*

10. Katz, L. & E. Englesberg: Hyperinducibility as a result of mutation in structural genes and self-catabolite repression in the ara operon. J. Bacteriol. *107* (1971): 34–52.
11. Lis, J. T. & R. Schleif: Different cyclic AMP requirements for induction of the arabinose and lactose operons of *Escherichia coli*. J. Mol. Biol. *79* (1973): 148–162.

*Enrichment, Isolation and Sequencing of Genes*

12. Gottesman, S. & J. R. Beckwith: Directed transportation of the arabinose operon: a technique for the isolation of specialized transducing bacteriophages for any *Escherichia coli* gene. J. Mol. Biol. *44* (1969): 117–127.
13. Lis, J. T. & R. Schleif: The regulatory region of the L-arabinose operon: its isolation on a 1,000 base-pair fragment of DNA heteroduplexes. J. Mol. Biol. *95* (1975): 409–416.
14. Smith, B. R. & R. Schleif: Nucleotide sequence of the L-arabinose regulatory region of *E. coli* K 12. J. Biol. Chem. *253* (1978): 6931–6933.
15. Greenfield, L., T. Boone & G. Wilcox: DNA sequence of the ara BAD promoter in *Escherichia coli* B/r. Proc. Natl. Acad. Sci. USA *75* (1978): 4724–4728.

15a. Lee, N. L., W. O. Gielow & R. G. Wallace: Mechanism of *ara* C autoregulation and the domains of two overlapping promoters, $p_C$ and $p_{BAD}$, in the L-arabinose regulatory region of *Escherichia coli*. Proc. Natl.: Acad. Sci. USA *78* (1981): 752–756.

*In vitro Transcription and Translation*

16. Wilcox, G., J. Boulter & N. Lee: Direction of transcription of the regulatory gene ara C in *Escherichia coli* B/r. Proc. Natl. Acad. Sci. USA *71* (1974): 3635–3639.
17. Wilcox, G., P. Meuris, R. Bass & E. Englesberg: Regulation of the L-arabinose operon BAD in vitro. J. Biol. Chem. *248* (1974): 2946–2952.
18. Hirsh, J. & R. Schleif: Electron microcopy of gene regulation: The L-arabinose operon. Proc. Natl. Acad. Sci. USA *73* (1976): 1518–1522.
19. Odgen, S., D. Haggerty, C. M. Stoner, D. Koludrubetz & R. Schleif: The *Escherichia coli* L-arabinose operon: Binding sites of the regulatory proteins and a mechanism of positive and negative regulation. Proc. Natl. Acad. Sci. USA *77* (1980): 3346–3350.

# 6. Light-induced Chloroplast Differentiation

B. Parthier

## 6.1 Structure and Function of Chloroplasts

Chloroplasts are unique plant cell organelles which, by means of chlorophyll, trap light energy, convert it into chemical energy (ATP) in the complex process of photosynthesis, and thus make it available to living systems. In the same process chloroplasts evolve the oxygen accumulated in the earth's atmosphere. Finally, chloroplasts are sites for primary carbon and nitrogen assimilation, carbohydrate, fatty acid and amino acid synthesis, as well as their transformations and polymerizations.

The green organelle represents structurally a biphasic system consisting of highly ordered (stacked), flattened vesicles of membranes, the thylakoids (Fig. **6**.1). They are embedded in a soluble compartment, the stroma. In fact, there are two further, much smaller compartments, one inside the thylakoid vesicles and the other between the two membranes of the chloroplast-enclosing envelope. These chloroplast envelope membranes control the transport (export and import) of various substances. Both membranes are biochemically different from the thylakoid membranes, but they are likewise different from each other in morphological, biochemical and functional aspects. Whereas the outer membrane is freely permeable to compounds up to 3 kd MW, the inner membrane regulates transport of low molecular substances by action of specific translocators. The membranes seem to be permeable even for high-molecular compounds such as polypeptides in receptor-mediated transport processes (see **6**.4.2).

The light-driven reactions of photosynthesis (light absorption, electron transport, ATP and NADPH formation, $O_2$ evolution) take place at the surface of and inside the thylakoid membranes, i.e., by means of polypeptides at and in the photosynthetic lamellae. $CO_2$ assimilation and reduction to carbohydrates, as well as further primary transformations to amino acids and fatty acids etc., are catalyzed by enzymes which are located in the stroma compartment. It also harbours the genetic material of the organelle, and most of the steps in gene expression take place in the stroma compartment. The basic processes of photosynthesis are schematically illustrated in Fig. **6**.2. It shows the coupling of the "light reactions" (water splitting) in the thylakoid membrane with "dark reactions" of the Calvin cycle in the stroma ($CO_2$ reduction by means of photosynthetically produced ATP and $NADPH_2$). The final result can be summarized in the classic overall equation of photosynthesis:

$$6\,CO_2 + 12\,H_2O \xrightarrow{h \cdot \nu} C_6H_{12}O_6 + 6\,H_2O + 6\,O_2\ (\Delta G^\circ = 480\ \mathrm{kJ/mol}\ CO_2).$$

In regard to organelle differentiation, the photosynthetic function of the chloroplast appears to be of secondary importance, if we consider that photosynthesis cannot supply energy for the early (non-green) period of plastid differentiation. In addition, light-induced formation of the structural and enzymatic machinery

---

Abbreviations: CP I, CP II, chlorophyll-protein complexes of photosystems I and II (PS I, PS II); Leu-RS, leucyl-tRNA synthetase; RuBPC, ribulosebisphosphate carboxylase; MW, molecular weight; kd, kilodalton; Md, megadalton; LHCP, light harvesting chlorophyll protein associated with CP II; P32, thylakoid membrane protein ($M_r$ 32 kd) coded by the maize chloroplast "Photogene 32".

for photosynthesis is the most remarkable molecular event during chloroplastogenesis. This term will be used for the light-induced transformation of morphologically and functionally more or less undifferentiated proplastids to photosynthetically active chloroplasts. The process has been studied most extensively in angiosperm seedlings and in the greening flagellated alga, *Euglena gracilis*.

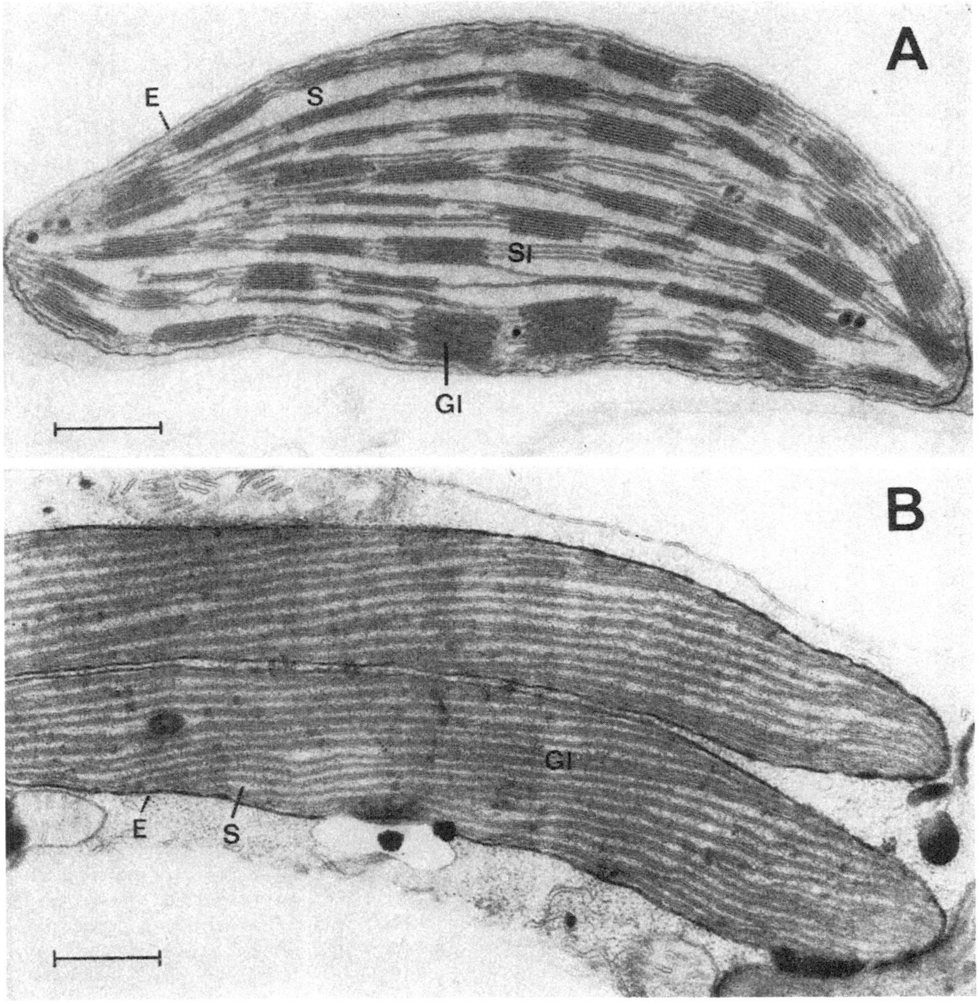


**Fig. 6.1**: Chloroplasts from *Zea mays* and *Euglena gracilis*. Electron micrographs supplied by Dr. D. Neumann, Halle.

Chloroplasts from higher plants (e.g., *Zea mays*, *top*) belong to the grana type, which are characterized by stroma and grana (thylakoid) sections of the photosynthetic membranes. *Euglena* chloroplasts *(bottom)* are of the lamella-type showing a uniform distribution of the photosynthetic constituents throughout the membranes. Photosynthetic membranes are embedded in the soluble stroma (matrix) compartment, and the whole chloroplast is surrounded by a double layer of envelope membranes.

*S*, stroma; *Sl*, stroma lamellae; *Gl*, stacked grana lamellae; *E*, envelope double membrane; *scale bars*, 0.5 $\mu$m.

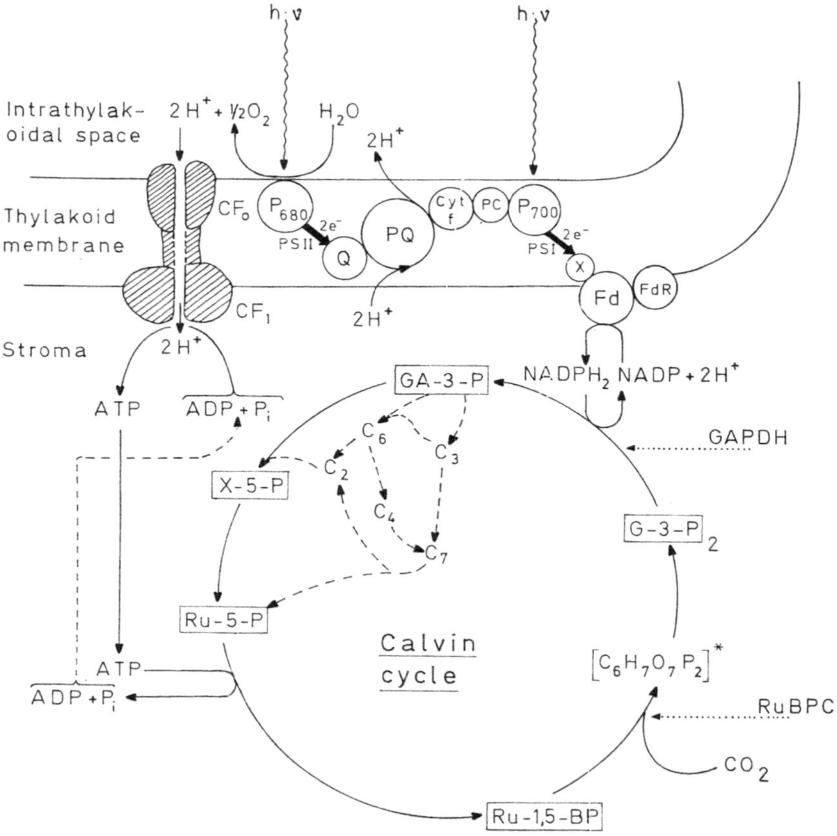


**Fig. 6.2:** Schematic illustration of photosynthesis and $CO_2$ assimilation in chloroplasts.

In the thylakoid membrane light (h· ν) causes splitting of water into $O_2$ and protons which are excreted and used for ATP production from ADP + $P_i$. Light absorption by chlorophyll *a* forms $P_{680}$ and $P_{700}$ causes electron emission and transport to formation sites of the reductive potential $NADPH_2$. In the stroma $CO_2$ is fixed ana reduced to carbohydrate phosphates by means of ATP and $NADPH_2$ (Calvin cycle reactions). *$P_{680}$*, chlorophyll $a_{680}$; $P_{700}$, chlorophyll $a_{700}$; *PS*, photosystem; *$CF_1$*, *$CF_0$*, parts of the ATPase (for the polypeptide subunit composition of the proton-translocating ATPase coupling factor cf. Fig. 6.11); *Q*, quinone; *PQ*, plastoquinone; *PC*, plastocyanin; *Fd*, ferredoxin; *FdR*, ferredoxin-NADP oxidoreductase; *cyt f*, cytochrome f; *X*, electron acceptor of $P_{700}$; *Ru-1,5-BP*, ribulose-1,5-bisphosphate; *RuBPC*, ribulosebisphosphate carboxylase; *Ru-5-P*, ribulose-5-phosphate; *X-5-P*, xylulose-5-phosphate; *G-3-P*, glycerate-3-phosphate; *GAPDH*, glyceraldehyde-phosphate dehydrogenase; *GA-3-P*, glyceraldehyde-3-phosphate; *$C_3$*, hydroxyacetone-phosphate; *$C_4$*, erythrose-4-phosphate; *$C_6$*, fructose-6-phosphate; *$C_7$*, seduheptulose-7-phosphate; *$C_6H_7O_4P_2$*, primary instable carboxylation product.

## 6.2 The Genetic Aspect of Chloroplastogenesis

To understand chloroplast differentiation we have to learn about the mechanisms of selective expression of genes for plastid components. Plastids, as well as mitochondria, contain all components necessary for a biological system to be genetically autonomous, i.e., DNA, RNA, the respective polymerases, ribosomes and the protein-synthesizing machinery. The organelles produce their own rRNA and tRNA equipment completely, but they neither code for nor synthesize the whole ensemble of their proteins. On the contrary, most of the proteins are encoded in the nucleus and are synthesized on cytoplasmic polyribosomes. Thus, plastids are at best semiautonomous organelles depending on the support of the nucleo-cytoplasmic compartment both genetically and metabolically. It is therefore of fundamental interest to map the genes, to elucidate the contribution of plastid and nuclear genes for chloroplast formation and to study the cooperation between the two genetic systems in the regulation of this differentiation process. Two approaches are used in locating genes by genetical mapping or in pinpointing them physically on a chromosome: Transmission genetics in sexual systems and molecular in vitro genetics. Only the latter will be considered here.

At present we must state that sufficient information is not available as to the magnitude of nuclear gene contribution to chloroplast formation. Still smaller is our knowledge about plastid gene expression influencing nuclear or cytoplasmic activities. Since the chloroplast is a large compartment, its genetic activities may be of importance for cellular reactions, in addition to its well-known metabolic contribution as source of energy and products of photosynthesis. On the other hand, the number of plastid DNA genes involved in chloroplast biogenesis are limited. This fact and the relative easiness in isolating intact chloroplast DNA are the major reasons for the progress made in chloroplast gene mapping in the last few years, which is described on the following pages.

### 6.2.1 Molecular Organization of Chloroplast DNA

The amount of total chloroplast DNA in a green cell is in a magnitude of $10^{-11}$ mg which represents 1% – 10% of the nuclear DNA. Each chloroplast contains 30 – 60 identical circular molecules of double-stranded DNA, about 40 $\mu$m in length corresponding to 90 Md MW or 132 kb. Chloroplast DNA differs from nuclear DNA in a number of physical properties such as more rapid renaturation or (in *Euglena*) lower G + C content and therefore lower buoyant density in isopyknic CsCl gradients. Until now only a small part of the chloroplast DNA genes has been localized and estimated in size.

Mapping of genes and the tremendous progress in DNA sequencing was possible after detection and selected use of restriction nucleases. This group of enzymes consists of mainly bacterial endonucleases which very specifically split double-strand DNA molecules into fragments of different length. The specificity of a given restriction nuclease depends on its ability to recognize and split a certain sequence of 4–6 usually displaced base pairs resulting in DNA fragments with cohesive or blunt ends. Details on detection, properties and nomenclature of restriction endonucleases cf. Table **2.1**.

Figure **6.3** shows that the number of cleavage sites differs for the various restriction endonucleases according to the frequencies of base sequences recognized by the enzymes. Thus Sal I (recognizing GTCGAC) cuts the DNA molecule into three pieces only, whereas Eco RI (GAATTC) yields 27 DNA fragments, and Hind III (AAGCTT) and Hae III (GGCC) split the DNA into

35 and 52 pieces, respectively. (The cleavage sites for the two latter enzymes are omitted in Fig. 6.3.)

The DNA fragments obtained from each enzyme digestion are of different lengths, which enables their separation by agarose gel electrophoresis. The separated DNA fragments can be transferred to nitrocellulose filter strips and hybridized with labelled chloroplast rRNA, mRNA or tRNA (cf. 2.6). The processed (RNase treatment, washing) filter strips can be radioautographed

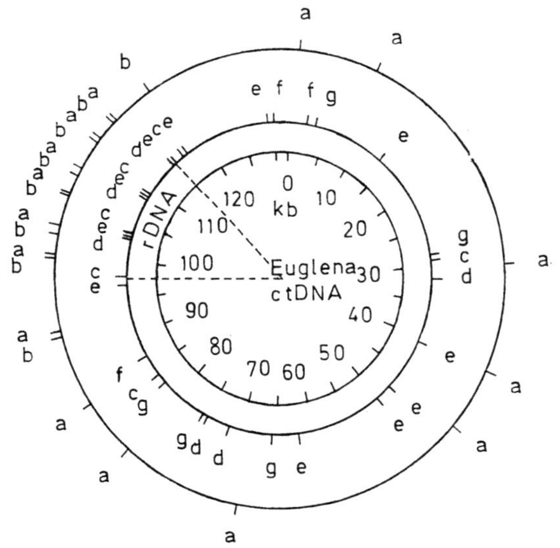


**Fig. 6.3**: *Euglena gracilis* chloroplast DNA and cleavage sites for restriction endonucleases (a–g). Data taken from Gray, P. W. & R. B. Hallick: Biochemistry *17* (1978): 284–289; Stutz et al.: In: [1], pp. 609–618.

The *inner circle* represents the DNA length of 132 kb; the *outer circle* shows cleavage sites for (Eco RI *(a)* and BglI *(b)*; the middle circle shows cleavage sites for Bam HI *(c)*, Xho I *(d)*, Bal I *(e)*, Sal I *(f)*, Pst I *(g)*.

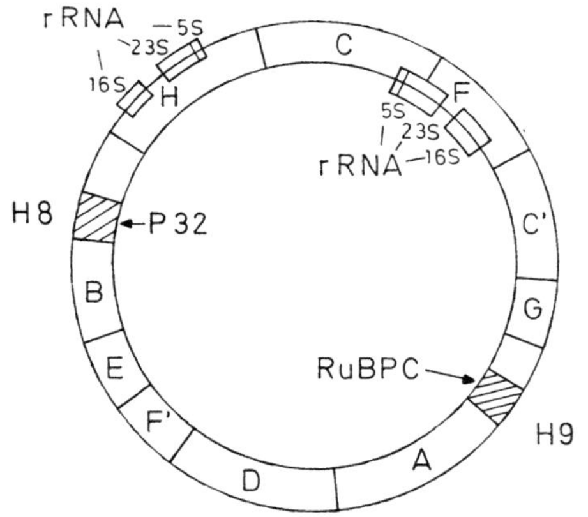


**Fig. 6.4**: Physical map of *Zea mays* chloroplast DNA genes: Location of invertedly repeated rRNA genes, the gene for ribulosebiphosphate carboxylase large subunit and the "Photogene" P32 (from Bogorad et al.: In: [1], pp. 541–551).

*Bars* between two *circles* (the two DNA strands) indicate recognition sites for restriction endonuclease Sal I. *Hatched sections* represent Bam HI fragments 8 and 9 containing the two protein-coding genes *(arrows)*.

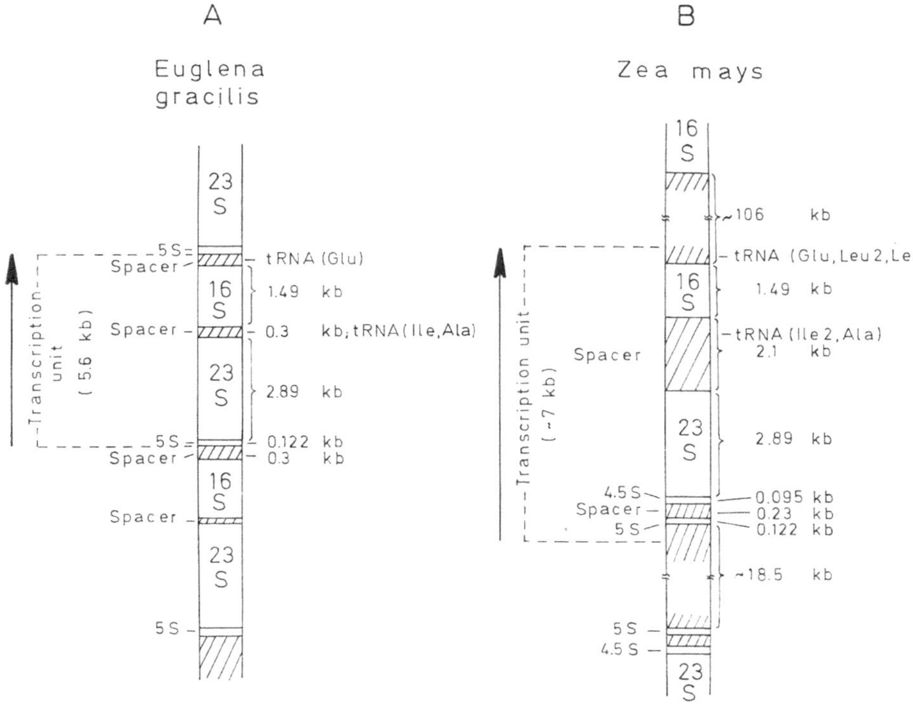


**Fig. 6.5**: Fine structure of ribosomal RNA transcription units and RNA cistron arrangements in the chloroplast DNA of *Euglena gracilis* (A) and *Zea mays* (B). Combination of data from Refs. [17, 21]; Hallick et al.: In: [1], pp. 619–622; Dyer, T. A. & J. R. Bedbrook: In: [9], pp. 305–311; pers. comm. of Drs. Weil, J. H. (Strasbourg) and Kössel, H. (Freiburg i. Br.).

For details consult text. *Arrows* denote transcription direction. *Open boxes* denote sequences coding for ribosomal RNAs (cf. Table **1.9**), i.e., 23S, 16S, 5S and 4.5S rRNAs. *Hatched boxes* denote spacer DNA, in some cases containing tRNA genes. The recently determined fine structure of the rRNA transcription unit of *Zea mays* is as follows (pers. commun. by H. Kössel): (5′) – 16Sr RNA – 301 bp spacer – first part of tRNA$^{Ile}$ – 949 bp intron – second part of tRNA$^{Ile}$ – 64 bp spacer – first part of tRNA$^{Ala}$ – 850 bp intron – second part of tRNA$^{Ala}$ – 137 bp spacer – 2 S rRNA – 5S rRNA – (3′) . . .

using X-ray film. Labelled bands demonstrate the number and sizes of the DNA fragments. Comparison of a number of different endonuclease digestions with overlapping cleavage sites within selected DNA fragments (fine mapping) eventually indicates the location of the (hybridized) RNA species within the DNA circle and allows to draw a physical map as shown in Fig. **6.4**.

Fragmentation of the DNA by use of restriction nucleases and hybridization of the fragments with labelled RNA revealed that 10% – 13% of the DNA circles are occupied by rRNA genes [17]. In higher plants and *Chlamydomonas* these genes constitute two separated inverted repeats separated by a very large (106 kb) and a smaller (18.5 kb) DNA region (Fig. **6.4** and Fig. **6.5** B). Three tandem repeats are present in *Euglena* DNA (Fig. **6.5** A). In both cases the sequence of rRNA cistrons in the transcription units is (5′) . . . 5S RNA – 23S RNA – spacer – 16 S RNA – spacer . . . In *Euglena* the internal spacer sequences are approximately 0.3 kb long, but the spacer covers more than 1.4 kb in the chloroplast DNA of higher plants (cf.

Fig. **6**.5). This is the reason why precursor molecules for chloroplast rRNA in higher plants are markedly longer (2.4 – 2.7 Md) than in algae, e.g., *Euglena*, where the precursor of 1.8 Md corresponds well with the size of 5.6 kb of the rRNA transcription unit.

A sum of 2–3 kb chloroplast DNA is required for 20–30 tRNA genes. Thus chloroplast DNA encodes probably all isoacceptor tRNA's found within the organelle; an import from the cytoplasm seems unlikely [21, 25]. Most of the tRNA genes are scattered in the large single-copy regions of the DNA circle; some few of them are located in the spacer regions between the 16S and 23S rRNA cistrons, e.g., in maize, spinach and *Euglena* the isoacceptor tRNA As for Ala and Ile (Fig. **6**.5). These tRNA genes seem to be split into halves by large introns [26a]. In *Euglena* Glu isoacceptor tRNA is found between 5S and 16S rRNA genes.

Much more complex is the situation for plastid messenger RNA's. The chloroplast DNA regions not occupied by rRNA and tRNA genes are large enough to code for 170 – 200 polypeptides of 20 kd MW, provided that non-transcribed spacers are lacking and each gene is present only once. By exploiting the invaluable tool of restriction endonucleases Bogorad's group [16, 18, 20, 23] and several other laboratories attempted a straightforward strategy of deciphering the arrangement of polypeptide-coding genes in plastid DNA, especially the position of structural genes for the large subunit of RuBPC and of the so-called Photogene P32. For the former selected chloroplast DNA restriction fragments, which contain the structural genes as elucidated with RNA hybridization techniques (cf. **2**.6), have been cloned in *E. coli* cells (cf. [16, 18, 23]). The cloned and reisolated DNA was shown to direct the synthesis in vitro of RuBPC large subunit in a linked transcription – translation system consisting of *E. coli* RNA polymerase and rabbit reticulocyte lysate. The newly synthesized [$^{35}$S]-methionine-labelled polypeptide was identified by its size in gel electrophoresis and by its precipitation with antibodies raised against RuBPC large subunit [16].

Similarly, maize chloroplasts or greening plastids after etiolation of maize seedlings contain RNA which hybridizes with Bam H1 fragment 8 and which directs the in vitro synthesis of a 34.5 kd polypeptide. In isolated chloroplasts it can be processed to a 32 kd form of a thylakoid protein. The coding gene is called Photogene P32 [23].

Physical mapping of the gene for the large subunit of RuBPC in maize [20] and *Chlamydomonas* [24] provided evidence for the existence of a single-copy mRNA gene of 1.6 kb length opposite to one of the two inverted rRNA gene regions, in maize on Bam H1 fragment 9 [26b]. It is free of introns. The gene for the light-induced 32 kd thylakoid polypeptide (Photogene P32) was estimated as 1.2 kb long and was mapped in a single-copy region of Bam H1 fragment 8 adjacent to the large spacer of one 16S rRNA gene region (Fig. **6**.4; cf. [18]). Further plastid polypeptide genes will be analyzed in the near future enabling reconsiderations about whether or not the estimate of about 200 polypeptides is valid.

### 6.2.2 Sites of Coding and Synthesis of Plastid Proteins

This fascinating progress in chloroplast gene mapping provides convincing evidence for the site of coding of those chloroplast proteins synthesized on plastid ribosomes. However, earlier experimental data from inhibitor experiments, mutants, and work with isolated chloroplasts have clearly demonstrated that the majority of chloroplast proteins are imported from outside. Approximately 50% of the thylakoid proteins are of plastid origin, but 50% are coded for in the nucleus and are synthe-

**Table 6.1**: Selected compilation of plastid proteins encoded in the nucleus (use of mutants, M) and synthesized on cytosol polyribosomes (use of inhibitors, I).
S, stroma; T, thylakoid membranes; E, envelope membranes

| Protein | Compartment | M | I |
|---|---|---|---|
| Alkaline deoxyribonuclease | S | + | + |
| Aminoacyl-tRNA synthetases | S | + | + |
| Aminolaevulinate dehydratase | S | | + |
| Carotenogenesis enzymes | S | + | + |
| Coupling factor $CF_1$, 1–2 subunits | T | + | + |
| Cytochrome c-552 | T | | + |
| DNA polymerase | S/T | + | |
| Elongation factor G *(Euglena)* | S | | + |
| Envelope membrane proteins | E | + | |
| Ferredoxin | T | + | + |
| Ferredoxin-NADP reductase | T | + | + |
| Fructose-1,6-diphosphate aldolase | S | + | |
| Glycerate-3-phosphatase | S | + | |
| Light harvesting chlorophyll a/b protein | T | | + |
| NADP-dependent glyceraldehyde-3-phosphate dehydrogenase | S | + | + |
| 3-Phosphoglycerate kinase | S | + | |
| Photoreactivating enzyme *(Euglena)* | S | + | + |
| Photosystem II apoprotein | T | + | + |
| Plastocyanin | T | + | + |
| RNA polymerase | S/T | + | |
| Ribosomal proteins | S | | + |
| RuBPC, small subunit | S | + | + |
| Starch formation enzymes | S | + | |

sized on cytoplasmic polyribosomes. The stroma proteins seem also to belong to the latter group, as can be seen from the compilation in Table **6**.1. Some few exceptions are RuBPC large subunit and 2 or 3 ribosomal proteins. Probably all of the protein constituents of the envelope membranes have their biogenetic origin outside the organelle. In all these cases transport of polypeptides from the cytoplasmic compartment into the organelle must take place, as has been convincingly demonstrated for the small subunit of RuBPC (Sect. **6**.4.2; cf. also Table **4**.11 for mitochondrial polypeptides).

Whereas the transmembrane transport of polypeptides is now generally accepted, the import of polynucleotides into plastids or export of plastid RNA is still a matter of controversy [19, 22]. Reports about the release of certain chloroplast isoacceptor tRNA's for cytoplasmic polyribosomal protein synthesis have not been confirmed [25], and the extraction of poly(A)-RNA from chloroplasts which hybridize with nuclear DNA but not with chloroplast DNA [26] also needs careful reexamination.

## 6.3 Light Induction of Plastid Differentiation

Like the induction of other differentiation programs, chloroplast formation depends on the action of signals and the existence of a signal-transforming machinery. In all angiosperms and several algae light is known to initiate the transformation of proplastids or etioplasts into green, photosynthetically active chloroplasts.

### 6.3.1 Proplastids and Etioplasts

The meristematic cells of higher plants contain very small and structure-less vesicles, the plastid initials. They originate from the zygote, where they have been introduced by the egg cells (maternal inheritance). With the advancement of growth and differentiation of a meristematic cell they are converted to proplastids and in continuous darkness become etioplasts. This type of plastid is appreciably larger than proplastids. Etioplasts contain a characteristic paracrystalline prolamellar body, thought to represent a storage compartment for lipoproteins. It is free of photochemical activities, however, $\beta$-carotene, NADP-dependent oxidoreductase and $CF_1$-mediated phosphorylating activities can be observed [32]. Dark-grown *Euglena* cells contain proplastids of 1 – 2 $\mu$m length. They contain one lamella (girdle lamella) and, occasionally, a non-paracrystalline prolamellar body. A comparison of the biochemical equipment between proplastids and chloroplasts of *Euglena* shows that almost all of the chloroplast components are present in the precursor organelles; chlorophyll can be regarded as a substitute of protochlorophyll. However, considerable differences exist in quantitative respect (Table **6.2**).

After illumination dramatic conformational alterations occur during which the first photosynthetic lamellae (prothylakoids) are formed and photochemical activities appear in turn (cf. **6**.4.4). Thus, in molecular terms the light-triggered transformation of proplastids or etioplasts to chloroplasts suggests massive activation and expression of those genes which are activated and expressed either at a slower rate or in a smaller number in the organelles of dark-grown plants. In addition, the occurrence of qualitatively new chloroplast components, certainly in minimal amounts, is a prerequisite for the realization of the differentiation process. This is probably not a matter of de novo biosynthesis of prothylakoid components but rather due to conformational changes of protochlorophyll-protein complexes as a consequence of light absorption. The primary light induction event, small as it may appear, is followed by a cascade of reactions leading ultimately to the gene expression program for chloroplast differentiation.

**Table 6.2**: Comparison of several plastid components in proplastids and chloroplasts of *Euglena gracilis* [13, 30]

| | Proplastid | Chloroplast |
|---|---|---|
| Size of organelle ($\mu$m length) | 1–2 | 7–10 |
| Volume of organelle ($\mu m^3$) | 0.07 | 4.1 |
| Prolamellar body | small, non-crystalline | absent |
| Protochlorophyll content (pg per plastid) | $10^{-4}$ | $0.5 \times 10^{-4}$ |
| Chlorophyll content (pg per plastid) | 0 | 1.5 |
| Carotenoid content (pg per cell) | 0.3 | 3.5 |
| Plastid rRNA (% of total cell rRNA) | 1–3 | 25–35 |
| Plastid tRNA (% of total cell tRNA) | 2–5 | 30–35 |
| Plastid protein (% of total cell protein) | 2–3 | 25–40 |
| Cytochrome c-552 ($10^{-5}$ pmol per cell) | 0.1 | 7 |
| RuBPC (relative specific activity) | 2–3 | 100 |
| Leucyl-tRNA synthetase (relat. specific activity) | 2–4 | 100 |

### 6.3.2 Photoreceptors

The differentiation of functional chloroplasts is normally induced by white light and seems to require different photoreceptors which are specifically active in the red and blue regions of the visible spectrum. They are being proposed as a part of an amplifier circuit system in the signal transformation "black box" between light absorption and final photoresponse.

At least three photoreactions are involved in the control of plastogenesis: Photoconversion of protochlorophyll or protochlorophyllide to chlorophyll(ide), light absorption by a blue light-absorbing pigment, and the photoformation of active phytochrome. In higher plants, phytochrome, a blue-green chromoprotein localized in the cytoplasmic compartment but probably bound to plastid and other membranes, mediates chloroplast morphogenesis, in addition to many other metabolic events in the cell (cf. Chap. **11** and Ref. [31]).

It exists in two interconventible forms, a red light-absorbing (Pr, $\lambda$ max: 660 nm) and the physiologically active far-red light-absorbing form (Pfr, $\lambda$ max: 724 nm). Pulses of red light induce a photoresponse which can be reversed by a far-red light impulse immediately following the red light. Long-term irradiation with far-red light is especially suitable for photomorphogenetic studies, since the chromoprotein system tends to develop a steady state between $P_r$ synthesis and $P_{fr}$ destruction, and the far-red light does not interfere with chlorophyll synthesis and photosynthesis. Thus the involvement of phytochrome in chloroplast formation was one of the convincing results which provided evidence that chlorophyll formation or photosynthesis is not primarily required in this differentiation process [29].

In contrast to higher plant chloroplastogenesis under the regime of phytochrome, this photoreceptor system seems to be lacking in *Euglena*. Instead, two other light receptors are involved, a blue light-absorbing system ($\lambda_{max}$ 433 nm) and a red-blue light system ($\lambda_{max}$ 634 nm), as indicated by action spectra derived from dose-response curves [30]. The blue light receptor is located in the cytoplasmic compartment and mediates gene expression events of the nucleo-cytoplasmic genetic apparatus as a consequence of *cis-trans* isomerization of *zeta*-carotene [27]. The second, the red-blue light receptor system is associated with plastids and probably identical with or similar to protochlorophyll(ide). It mediates events in plastid gene expression such as chlorophyll accumulation, rRNA synthesis or increase of RuBPC activitiy, i.e., in general chloroplast constituents which are characterized by a lag period in their synthesis (cf. **6**.4).

The blue light-absorbing pigment is called "cryptochrome" because of its importance in cryptogamic plants and its cryptic nature [28]. Its chemical identification either as flavin or carotene has not yet been established because of the similar absorption spectra of the two chromophores.

### 6.3.3 Phases of Chloroplast Formation

Chloroplast formation can be arbitrarily subdivided into three phases:

1. The *induction phase*, in which the transformation of light signals causes activation and probably transcription of "early genes". The process seems to depend on functional plastids able to emit "information" into the nucleo-cytoplasmic compartment.
2. The *lag period*, characterized by a limited ability to express the previously activated genes due to energy deficiency or other repressive processes.
3. The phase of *linear increase* with equal rates of synthesis of most chloroplast constituents including all enzymes necessary for photosynthesis, which gradually becomes dominant over oxidative phosphorylation.

The length of these phases in *Euglena*, especially that of the lag period, can vary depending on the nutritional status and pre-illumination (potentiation) of the etiolated cells in the culture. Carbon in form of glucose or sucrose prolongs the lag period in chlorophyll accumulation considerably, in comparison with that of "resting" cells, i.e., cells in a medium lacking the carbon source. On the other hand, the lag phase can be shortened or even eliminated if dark-grown *Euglena* cells are pre-illuminated with low light intensity for a period between 10 and 100 min and then brought back to darkness for 12 h before continuous illumination. Both phenomena can be interpreted in terms of metabolical breakdown of the major carbohydrate storage product in *Euglena*, paramylum ($\beta$-1,3-glucan). Paramylum breakdown occurs without a lag phase after activation of the cytoplasmic blue light receptor system, which mediates the activity increase of responsible enzymes, i.e., of $\beta$-1,3-glucan hydrolase and -phosphorylase, respectively. Paramylum depolymerization products similar to glucose inhibit chloroplast formation. However, it is unknown which reactions are primarily affected, and it is obscure whether or not the inhibition is comparable with the effect of glucose catabolite repression in bacteria (cf. **4**.2.3.1 and **5**.3).

The example shows that a discrimination into "early" and "late" events in chloroplastogenesis is relative and of restricted value only. Nevertheless, the induction of de novo synthesis of stroma enzymes (Table **6**.1) appears generally more rapid than those chloroplast components which are produced inside the organelle. Since these are normally composite constituents, the rate-limiting step might not be transcription or translation but assembly of the single subunits to holoenzymes or other functional entities. Otherwise, photoconversion of available protochlorophyll(ide) to chlorophyll(ide) is probably the most rapid event after illumination. Further chlorophyll accumulation needs de novo synthesis of protochlorophyll(ide). The rate-limiting step is the formation of $\delta$-aminolevulinate synthetase (ALAS), which shows a rapid turnover. Its synthesis seems to belong to the light-requiring processes.

## 6.4 Selected Examples of Gene Expression During Chloroplast Differentiation

In the following four sections light-induced chloroplast differentiation will be illustrated by the biosynthesis of selected chloroplast constituents. They are (1) entirely of plastid origin (RNA); (2) a dimeric stroma enzyme partially of plastid, partially of nucleo-cytoplasmic origin (RuBPC); (3) a stroma enzyme entirely of nucleo-cytoplasmic origin (Leu-RS) and (4) multimeric thylakoid membrane complexes.

### 6.4.1 Ribonucleic Acids

The quantitative aspect of light-dependent total RNA synthesis during chloroplast formation has been studied by means of hybridizing labelled RNA species to restriction endonuclease fragments. In greening *Euglena* cells it has been found that only 21% of the total RNA is transcribed after illumination of dark-grown cells [36, 41]. Consequently, the majority of plastid RNA synthesis is regulated independently of the plastid differentiation stage.

Similar measurements have not been carried out for individual RNA species. Most of the pertinent work has been dedicated to amount and rate of rRNA, tRNA and mRNA synthesis after illumination of dark-grown plants, often as comparison between plastid and cytoplasmic species.

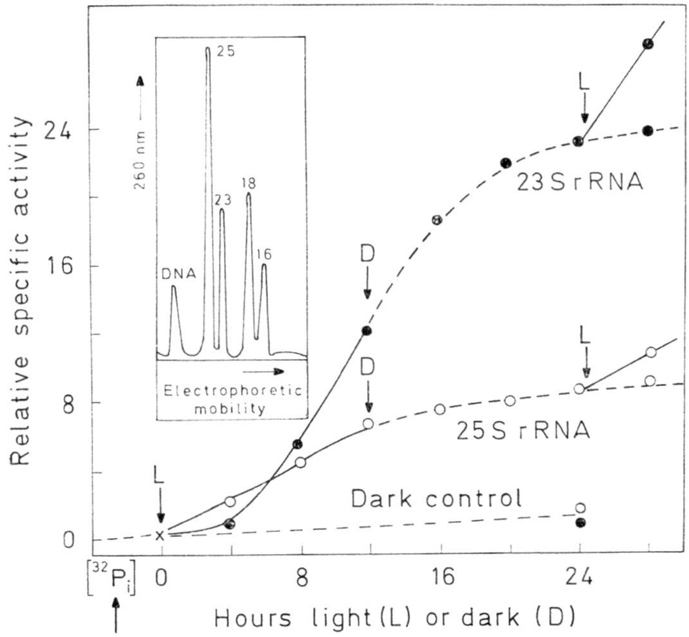


**Fig. 6.6**: Kinetics of light-dependent synthesis of plastid 23S rRNA and cytoplasmic 25S rRNA in *Euglena* cells (combination of data from Heizmann, P. et al.: Coll. Int. C.N.R.S. (Paris) **240** (1975), pp. 247–254; Cohen, D. & J. A. Schiff: Arch. Biochem. Biophys. *177* (1976), pp. 201–216, and unpublished results).

Dark-grown, phosphate-starved cells were fed with [$^{32}P_i$] for 1–2 h and then illuminated at L (0 h). Total RNA was extracted from the cells at the times indicated with circles. After separation into the four individual rRNA species by gel electrophoresis *(insert)* and slicing of the gels, radioactivity was determined in the gel slices. A difference in the lag phases of labelling between plastid and cytoplasmic rRNA can be noted. The effect of redarkening *(dotted lines)* and reillumination are likewise indicated. Insert shows four rRNA species indicated by their S values after polyacrylamide gel electrophoresis.

Transcription of high-molecular weight rRNA of both plastid and cytoplasmic origin is stimulated after illumination, however, with different kinetics (Fig. **6**.6). Radioactive labelling and electrophoretic separation of the rRNA species (Fig. **6**.6, insertion) permits the investigation of individual rRNA species in more detail. It can be seen that transcription of both rRNA types increases upon illumination. While [$^{32}P_i$]-incorporation into cytoplasmic 25S rRNA starts without a lag phase, a marked delay of precursor incorporation into plastid 23S is observed. Similarly, a marked delay in the response of synthesis occurs in chloroplast rRNA after switching off the light, in contrast to a rapid response of cytoplasmic rRNA synthesis (Fig. **6**.6). The difference in lag period can be explained as follows: Synthesis of cytoplasmic rRNA is light-stimulated without a lag period due to a constitutive transcription system including nuclear RNA polymerase I. The lag period in plastid rRNA synthesis corresponds with that of other plastid constituents and may reflect the low amount or activity of proplastid RNA polymerase. It is unclear whether light-stimulated de novo synthesis of plastid RNA polymerase on cytoplasmic ribosomes is a pre-

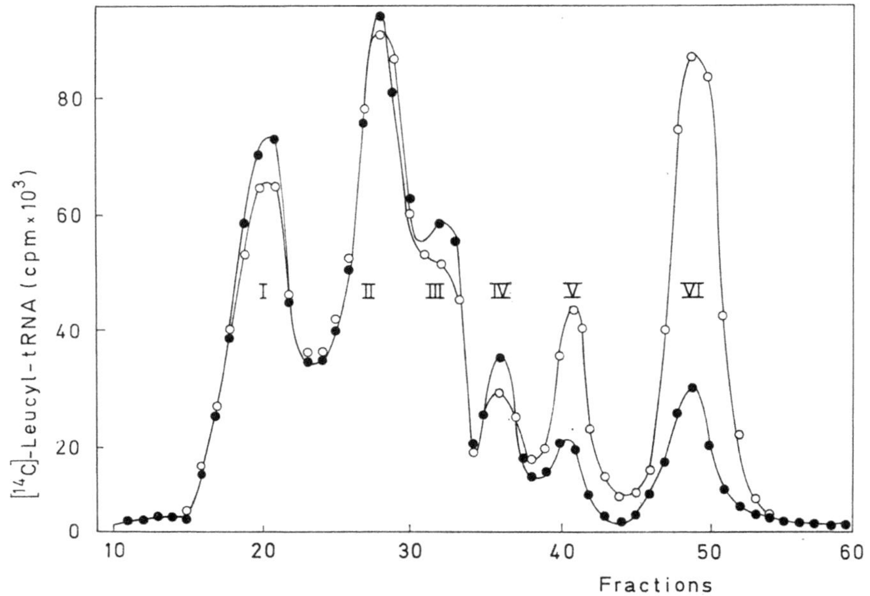


**Fig. 6.7**: Fractionation of isoacceptor $tRNA^{Leu}$ species of *Euglena gracilis* grown in darkness (●) or in light (○) for 48 h (from Parthier, B.: In Nover, L. & K. Mothes (eds.), Cell Differentiation in Microorganisms, Plants and Animals. Fischer, Jena 1977, pp. 602–624).

Total cell tRNA was prepared from both types of cells and the $tRNA^{Leu}$ species were labelled with [$^{14}$C]-leucine or [$^{3}$H]-leucine, respectively, by means of leucyl-tRNA synthetase from whole cells or chromatographically separated chloroplast and cytosol isoenzymes (cf. Fig. **6**.9). The radioactive Leu-tRNAs formed have been reisolated and then fractionated by reversed-phase chromatography RPC-5. The fractions correspond with cytoplasmic $tRNA^{Leu}$ isoacceptor species *(I – III)*, probably with one mitochondrial *(IV)*, and two plastidal *(V + VI)* species. Only the latter two species increase after illumination.

requisite for the increase in plastid rRNA transcription. The lag phase in light-induced formation of thylakoid membranes, at which the active enzyme needs to be attached, is another explanation for the kinetic behaviour of light-induced plastid rRNA synthesis.

RNA synthesis in isolated chloroplasts provides evidence for the synthesis of the putative primary transcripts of rRNA genes. Accumulation of a $2.4–2.7 \times 10^6$ in higher plants and of a $1.8 \times 10^6$ rRNA precursor in *Euglena* corresponds well to the different sizes of transcription units in the two chloroplast DNA's (cf. Fig. **6**.5). Under in vivo conditions the primary rRNA transcript seems to be rapidly processed into the immediate precursors (1.2 and $0.65 \times 10^6$) of the mature rRNAs (1.1 and $0.56 \times 10^6$).

Light-induction of chloroplastogenesis is likewise coupled with an increase in plastid-specific isoacceptor tRNA's. Compared with 5% of total cell tRNA found in dark-grown cultures, about 35% of total cell tRNA are associated with chloroplasts [35]. Light-induced $tRNA^{Leu}$ formation is illustrated in Fig. **6**.7, fractions V and VI. Cytoplasmic and mitochondrial isoacceptor tRNA's (Fig. **6**.7, fractions I – IV) increase little, if at all, upon illumination of dark-grown *Euglena* cells. Unfortunately, no causal relationship between light induction of plastid tRNA's, their cognate aminoacylating enzymes (cf. **6**.4.3), and chloroplast development has been pemonstrated yet. There is evidence, however, that photocontrol of plastid tRNA synthesis affects the level of transcription and not post-transcriptional processes [35].

Several convincing examples of light-induced transcription of individual mRNA's for chloroplast proteins have been reported recently. The formation of mRNA of Photogene 32 has been already mentioned in Sect. **6**.2.1: It cannot be detected in etiolated maize plants but emerges during greening, as demonstrated by in vitro synthesis of the respective 34.5 kd polypeptide [18, 23]. The mRNA levels for both the large and small subunits of RuBPC are likewise regulated by light (e.g., [45]).

Cytoplasmic poly(A)-containing mRNA for the apoprotein of the light-harvesting chlorophyll a/b protein, a major thylakoid protein constituent (cf. **6**.4.4), is synthesized in light under control of phytochrome [33]. In none of the cases is it known at which step or component light activates the transcription process. The measured increase of plastid polymerase activity was supposed to be neither due to a quantitative enhancement of the enzyme amount nor should it be the result of enzyme alteration. It was suggested [34] that other light-inducible factors may be responsible for the elevated transcription activity.

Light-induced mRNA synthesis can be likewise demonstrated by measuring polyribosome formation. For example, two classes of polyribosomes seem to exist in *Euglena:* "Photoactive polyribosomes" are rapidly formed after illumination of dark-grown cells; they disappear soon after return to darkness. The kinetics of degradation in darkness suggest short-living mRNA involved in this process. A second class called "photoinduced polyribosomes" is continuously formed after a lag phase following illumination. Analysis of their dissociation kinetics in darkness indicates the participation of mRNA species with longer half-lives [12].

The majority of mRNA's isolated from total polyribosomes contain poly(A) sequences. Synthesis of this RNA is greatly enhanced by light and decreases rapidly after cell transfer into darkness. Its turnover corresponds with the decay of "photoactive poylribosomes" in the cytoplasm, which are responsible for the formation of certain plastid poylpeptides. Poly(A)-RNA is also detected in the total RNA fraction extracted from isolated *Euglena* chloroplasts. The amount of this RNA increases in light to a similar extent, as does the light-stimulated poly(A)-RNA connected with cytoplasmic polyribosomes. If contamination of the former RNA by the latter can be excluded, the question remains whether the poly(A)-RNA found in chloroplasts is indeed transcribed at chloroplast DNA or is synthesized at nuclear DNA and subsequently transported into the plastids. Both versions seem to be realized [26].

### 6.4.2 Ribulose-1,5-bisphosphate Carboxylase

Ribulose-1,5-bisphosphate carboxylase-oxygenase (RuBPC) or fraction-I-protein is for several reasons the best-studied chloroplast protein. Probably the most abundant protein in nature, it occurs in quantities as large as 40% of the total soluble proteins of leaf extracts or 70% of the proteins in the stroma compartment (concentration $\sim$ 300 mg/ml). The enzyme catalyzes the $CO_2$-fixing reaction in photosynthesis and functions as oxygenase in the first reaction of photorespiration. Fraction-I-protein is therefore implicated as a key catalyst in primary synthetic processes which allows the feeding of all living systems with carbon and energy.

Another interesting aspect of this enzyme closely related to chloroplast differentiation is its mode of synthesis. RuBPC consists of two types of subunits, 8 large monomers of each 55 kd are assembled with 8 small subunits of approx. 15 kd. The catalytic activity is restricted to the large subunits, which show close relationships in different genera of plants, in contrast to a remarkable dissimilarity of the small subunits.

The large subunit is encoded in chloroplast DNA (cf. **6**.2.1; Fig. **6**.4) and synthesized inside the isolated organelle (Fig. **6**.8), while the small subunit is encoded in nuclear

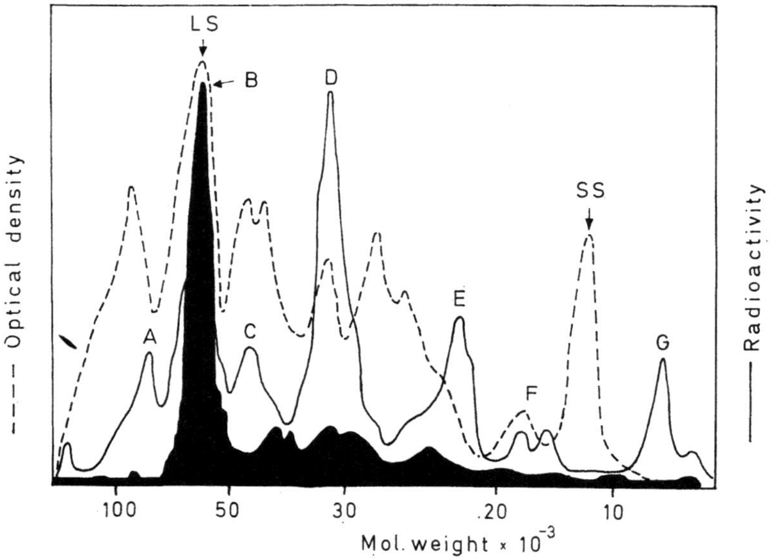


**Fig. 6.8:** Polyacrylamide gel electropherogram of polypeptides synthesized in isolated chloroplasts and chloroplast ribosomes (from Ellis, R. J.: In: [4], pp. 195–212).

Chloroplasts from spinach were isolated, part of them lysed for plastid polysome preparation and incubated under defined assay conditions in light, with a mixture of radioactively labelled amino acids. Total proteins were extracted and applied to SDS gel electrophoresis. *Broken line:* total polypeptides in chloroplasts after staining; *solid line:* labelled polypeptides from incubated intact chloroplasts; *black area:* labelled polypeptides from incubated free polysomes.
*LS*, large subunit of RuBPC; *SS*, small subunit of RuBPC; *A* to *G*, radioactive bands belonging to membrane-bound polypeptides. Only the large subunit of RuBPC is synthesized in isolated chloroplasts, but not the small subunit.

DNA and synthesized on free cytoplasmic ribosomes as a 20 kd precursor protein. This pre-protein is released from the polysomes and processed to its final size during transport through the chloroplast envelope by an endopeptidase found in the soluble (stroma) fraction of the chloroplasts [6, 37, 39, 43]. The transport process does not require concomitant chloroplast protein synthesis but needs energy from photosynthetic phosphorylation [39]. Penetration of the transit peptide through the envelope membranes of the organelles is thought to be a receptor-mediated process ("envelope carrier hypothesis" according to Ellis [37] or "vectorial processing hypothesis" according to Schatz [40, 42]), because specific recognition is necessary in order to ensure that organelle proteins of cytoplasmic origin enter the correct compartment. The most important feature of this process is the separation of synthesis and transport of the pre-protein.

The penetration process might be similar to a mechanism by which certain proteinaceous toxins of microbial origin pass through cell membranes. The most-studied diphteria toxin is produced by the corynephage $\beta$ in lysogenic strains of *Corynebacterium diphteriae* as a single polypeptide of 63 kd which consists of two functionally different moieties (A and B chain). Whereas the A chain is responsible for the toxic action inside the cell, the B chain is required for the transmembrane transport of the A chain. The intact toxin is first bound to specific receptors on the plasma membrane via the C-terminal region of the B chain. This binding induces a conformational change such that the N-terminal hydrophobic domain of the B chain becomes inserted into the plasma membrane and forms a channel into which the A chain is drawn. During this process the toxin is proteolytically cleaved into A and B chains. While the B chain is left in the membrane, the A chain enters the cytosol compartment [6].

This mechanism differs from that described by the "signal hypothesis", assuming a direct co-translational transport of polypeptides made on membrane-bound polysomes. This hypothesis was originally proposed for secretory proteins (cf. **4**.4.5.1). Extension of the signal hypothesis to chloroplasts would predict that bound ribosomes should be seen at the envelope surface. There is, however, no reliable evidence for this. In both cases, however, a N-terminal "signal sequence" or "recognition sequence", an oligopeptide of mostly non-polar amino acids, exists in the pre-protein. Whereas the "signal sequences" so far sequenced contain 15 – 25 mainly hydrophobic amino acid residues (cf. Table **4**.10), the pre-protein of the small subunit of RuBPC contains an N-terminal, basically charged extension of ~44 amino acid residues.

The role of the large subunit for the uptake of the small subunit and for the assembly remains unclear; both subunits have been found tightly attached to envelope membrane proteins, perhaps representing a transient stage in the uptake process. On the other hand, the small subunit of RuBPC may act as regulator in the synthesis of the large subunit.

Interestingly, Ellis [37] reported another polypeptide of cytoplasmic origin, a large polypeptide of 60 kd subunits which mediates the assembly of the two types of RuBPC subunits. After aggregation with RuBPC large subunits the resulting complex might also act as a store of large subunit prior to assembly, if the small subunit is not available.

In vivo, uncoupling of the syntheses of both subunits has been observed. In the early stages of illumination of dark-grown cells synthesis of the large subunit seems to be the limiting factor responsible for the lag phase of the activity increase. Likewise, in plastid ribosome-deficient, heat-bleached rye seedlings [38] and plastid mutants of higher plants an excess of small subunits has been demonstrated by immunological techniques. On the other hand, in green unicellular algae a strict coupling of subunit synthesis seems to exist.

### 6.4.3. Leucyl-tRNA Synthetase

Leucyl-tRNA synthetase (Leu-RS) belongs to the group of aminoacyl-tRNA synthetases which activate free amino acids and attach them to plastid isoacceptor tRNA's, thus occupying a key position in the translation process (cf. **1**.3.4.1). They are stroma proteins encoded in nuclear DNA and synthesized on cytoplasmic ribosomes before being compartmentalized within the organelles. In contrast to the small subunit of RuBPC, probably all of the plastid aminoacyl-tRNA synthetases are synthesized as functional entities on cytoplasmic polysomes without need of a protein complement of plastid origin.

A well-studied example is Leu-RS of *Euglena gracilis*. The plastid enzyme can be separated from the cytoplasmic isoenzyme by hydroxyapatite chromatography. Figure **6**.9 demonstrates a comparison of Leu-RS isoenzyme elution patterns from green cell homogenates and from isolated chloroplasts. Two Leu-RS isoenzyme activities with unambiguous tRNA-charging specificity can be separated from each other. E1 obviously represents the organelle-specific isoenzyme as shown by its presence in isolated chloroplasts and by its marked decrease in the plastid mutant. E2 shows charging specificity towards cytoplasmic isoacceptor tRNA. Similar separations have been obtained with most of the other aminoacyl-tRNA synthetases from *Euglena* [13, 15]. Notably, the plastid-specific isozymes are lacking in light-grown cultures of a bleached plastid mutant of *Euglena* ($W_3$BUL).

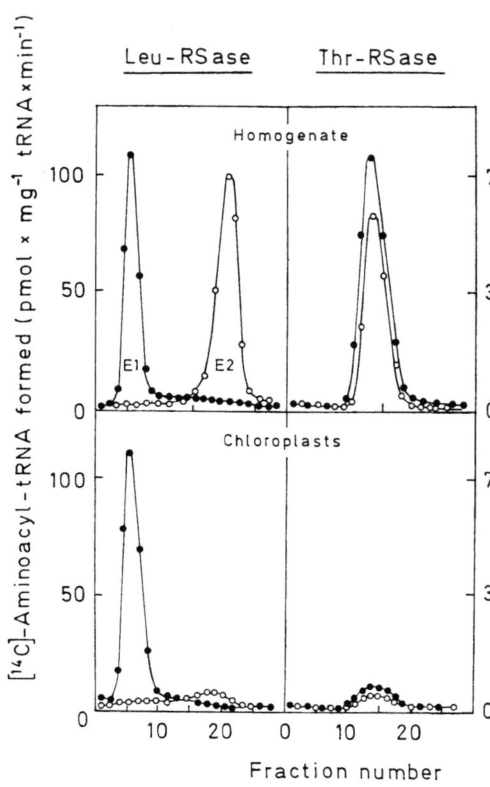


**Fig. 6.9:** Hydroxyapatite chromatography of leucyl-tRNA synthetase (Leu-RS) and threonyl-tRNA synthetase (Thr-RS) of cell homogenates *(top)* and isolated chloroplats *(bottom)* from *Euglena gracilis* (adapted from Parthier, B. et al.: In: [1], pp. 687–693, and unpublished results).

Postribosomal supernatant enzyme fractions from homogenized cells or from isolated chloroplasts have been applied to hydroxyapatite chromatography and the eluted E1 and E2 enzyme activities have been assayed by binding of [$^{14}$C]-leucine or -threonine to prokaryotic tRNA (●) from *Anacystis* and eukaryotic tRNA (○) from *Euglena* cytosol.

E1 and E2 enzyme activities correspond with organelle and cytosol Leu-RS isoenzymes, respectively. For Thr-RS only one enzyme activity with ambiguous tRNA specificity and cytoplasmic compartmentation can be observed.

A comparison of the increase of chloroplast constituents during light-induced greening of organotrophically dark-grown *E. gracilis* demonstrates that the kinetics differ in the length of the lag phase. In relation to chlorophyll, RuBPC or $tRNA^{Leu}$ accumulation, the increase of Leu-RS activity belongs to the early events, as it is true for many if not all stroma enzymes. The specific activity of cytoplasmic Leu-RS E2 does not change upon light induction of chloroplast differentiation.

The light-dependent increase of plastid Leu-RS isoenzyme activity is completely inhibited by low concentrations of cycloheximide, but not, or at a later period only, by chloramphenicol (Fig. 6.10), suggesting a secondary effect of this drug on the synthesis of plastid Leu-RS. This result verifies the above-mentioned statement that chloroplast-associated aminoacyl-tRNA synthetases are synthesized on 80S cytoplasmic ribosomes and indicate the activity increase to be due to de novo synthesis of the protein rather than to an activation of preformed enzyme precursor

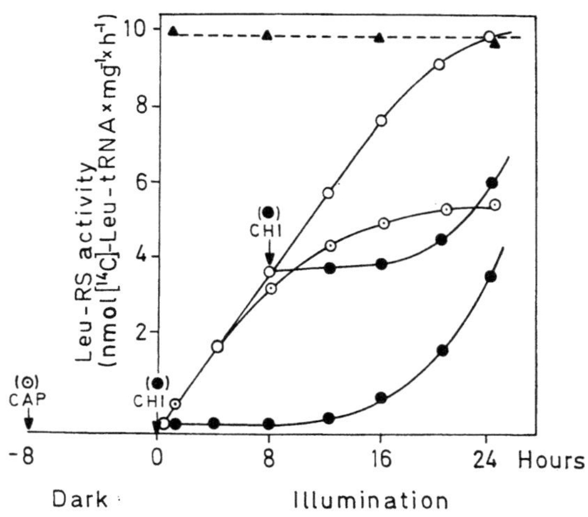


**Fig. 6.10:** Effect of protein synthesis inhibitors on plastid-specific Leu-RS activity during light-induced chloroplast differentiation in *Euglena gracilis* (combined figures from Parthier, B.: FEBS Lett. *38* (1973): 70–74 and Parthier, B. & R. Krauspe: Coll. Int. C.N.R.S. (Paris) *240* (1975), pp. 233–239).

○ control (without inhibitors); ● in the presence of 8 μg/ml cycloheximide *(CHI)*, or ⊙ 1 mg/ml D-threo chloramphenicol *(CAP)*; ▲ cytosol Leu-RS activity. Binding of [$^{14}C$]-labelled leucine to prokaryotic tRNA *(Anacystis)* by post-ribosomal soluble enzyme fractions containing plastid Leu-RS shows that activity of this enzyme is induced by light. Increase of activity is inhibited by CHI, but not by CAP. Cytosol Leu-RS activity is assayed with *Euglena* cytosol tRNA.

molecules. This has been also confirmed with $D_2O$ labelling and CsCl density gradient centrifugation of chloroplast Leu-RS. The most interesting question has not yet been settled: Is the transport of a monomeric soluble stroma enzyme into the plastids connected with a proteolytic processing of a precursor protein, as shown for the small subunit of RuBPC (cf. **6**.4.2) or certain thylakoid proteins (cf. **6**.4.4)? Until now this scheme for coupled transmembrane transport and processing was shown to exist for polypeptide subunits of di-, or oligomeric functional complexes.

In contrast to the preceding example, threonyl-tRNA formation in *Euglena* seems to be catalyzed by only one enzyme species missing in chloroplasts but present in the cytoplasm (Fig. **6**.9). It shows ambiguous charging specificity towards tRNA of prokaryotic and eukaryotic sources. Since polypeptides synthesized inside plastids do contain threonine, and threonine-accepting tRNA species have been found in chloroplasts, the formation of threonyl-tRNA needed for plastid protein synthesis may occur in two ways: (1) plastid $tRNA^{Thr}$ isoacceptor species leave the organelle, are charged with threonine in the cytoplasm and return to the plastid as Thr-tRNA; (2) isoacceptor $tRNA^{Thr}$ species remain inside the organelle but are charged by the cytoplasmic Thr-RS that should be able to migrate between cytoplasm and plastid.

### 6.4.4 Thylakoid Membrane Constituents

The membrane as functional site for light absorption, photosynthetic electron transport and photophosphorylation (cf. Fig. **6**.2) consists of 50% proteins, 50% lipids, and the pigments. The protein complement is a heterogenous mixture of polypeptides which can be resolved into at least 35 spots by two-dimensional SDS gel electrophoresis. The molecular architecture of the photosynthetic membrane reflects the

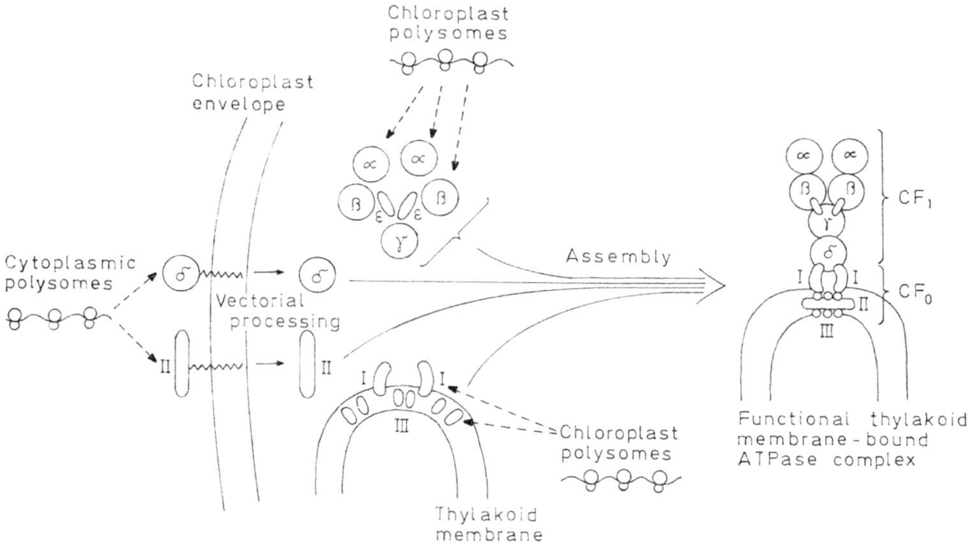


**Fig. 6.11:** Model for synthesis, transport and assembly of the proton-translocating ATPase subunits in chloroplasts (from Nelson et al. [40]).

Chloroplast ATPase consists of two parts, the membrane-embedded part $CF_0$ and the coupling factor $CF_1$. The first consists of three (I, II, III) subunits, the latter of five ($\alpha$, $\beta$, $\gamma$, $\delta$, $\varepsilon$). Only two ($\delta$, II) are synthesized on cytoplasmic ribosomes and are imported by vectorial processing before being assembled with the other subunits made within the chloroplast. It is an open question whether the integral membrane subunits I and III are synthesized as precursors possibly on membrane-bound plastid polysomes, similar as for typical prokaryotic membrane proteins (cf. examples in Table **4.10**).

functional coordination and includes both integral and peripheral components [46]. Most obvious are the two chlorophyll-protein (CP) complexes which are resolved after gel electrophoresis of lipid-free thylakoid membranes and which account for 80% of the thylakoid chlorophylls: CP I containing $P_{700}$ form of chlorophyll a is assumed to originate from the reaction center of PS I while CP II, mainly represented by the light-harvesting chlorophyll a/b protein complex (LHCP), is considered to contain the antenna chlorophyll of PS II, or of PS I + PS II [44]. The prominent apoprotein of the LHCP accounts for approx. 30% of the total thylakoid membrane proteins. It consists of two or three subunits of almost identical MW (27, 25, 23 kd).

A coordinated pattern of biosynthesis of the various components is suggested; however, this is experimentally difficult to confirm, since both cytoplasmic and plastid gene expression systems cooperate intimately. In vivo pulse-labelling studies showed 50% – 75% of the gel-separated thylakoid proteins to be made outside the organelle. Kinetic experiments have elucidated that CP I and CP II appear sequentially during light-induced chloroplast differentiation. Flash light for 2 min induces CP I function only; continuous illumination is needed for grana stacking and CP II synthesis [32, 44]. The greening process is not only determined by the rate of chlorophyll synthesis (or protochlorophyll resynthesis) alone, but depends on the rates at which the subunits of the chlorophyll binding proteins are synthesized.

Synthesis of several plastom-coded thylakoid proteins (e.g., cytochrome f; P32 cf. **6**.2.1) and of a few cytoplasmically synthesized membrane polypeptides with assigned functions have been recently studied in more detail. The LHPC apoprotein subunits are synthesized in the cytoplasmic compartment, as revealed by inhibitor studies but also by isolation of poly(A)-RNA from cytoplasmic polysomes and its translation in a cell-free wheat germ system [33]. The in vitro product, precipitated with antibodies raised against native LHCP, appeared as a soluble 29 kd pre-protein. After synthesis, the precursors can be transported into chloroplasts where they are processed to their mature forms and inserted into thylakoids as insoluble integral membrane protein [43]. Probably most if not all thylakoid polypeptides are synthesized as precursors, and then modified as described for LHCP. Similar to the RuBPCase large subunit, LHCP gene expression was shown to be regulated at the level of transcription and not at the level of translation.

At least six of the eight subunits of the proton translocating ATPase enzyme complex (coupling factor $CF_1$ + basal part $CF_0$, cf. Fig. **6**.2) of the chloroplast stroma lamellae are synthesized within the organelle [40]. The other subunits seem to be synthesized on cytoplasmic ribosomes as precursor polypeptides and transported into the chloroplast. The holoenzyme is then assembled and inserted into the thylakoid membrane as indicated in the model in Fig. **6**.11.

All data indicate that the biogenesis of the thylakoid membrane represents a multistep mechanism of synthesis, transport, processing and assembly of polypeptides, and in addition, lipids and pigments. Formation of the single components seems to be regulated independently of each other in different compartments. This is supported by turnover measurements and by the observation that the sequence of syntheses, upon illumination, may alter among various components and in different plant species. Assembly of the functional entities may be determined by the physicochemical properties of the constituents and their availability in pools. Similar mechanisms govern the assembly of phage particles from preformed building blocks in *Escherichia coli* (cf. **14**.2.4).

## 6.5 Regulatory Aspects in Chloroplastogenesis

The present state of knowledge of genome organization inside the chloroplast, in addition to the import of the majority of chloroplast constituents from the cytoplasmic compartment, challenges speculations about the regulatory mechanisms involved in light-induced chloroplastogenesis. For instance, the remaining regions of chloroplast DNA not representing structural genes might be occupied by regulator genes. Their primary or secondary products might control gene expression reactions outside the plastids but intimately related to chloroplast formation. On the other hand, there is no doubt that the close cooperation of the different compartments in the process of chloroplast differentiation needs information exchange via transport through membranes, and also a well-coordinated control system between and within the compartments of the cell.

### 6.5.1 Hypothetic "Repressors" and "Derepressors"

One important tool to attack the problem of intracellular biogenetic cooperation are mutants. There are plastid mutants from higher plants (maize, barley, *Pelargonium*) and from *Euglena* or *Chlamydomonas* which do not or in very limited amounts contain stroma polypeptides of cytoplasmic origin, in addition to the lack of chlorophyll

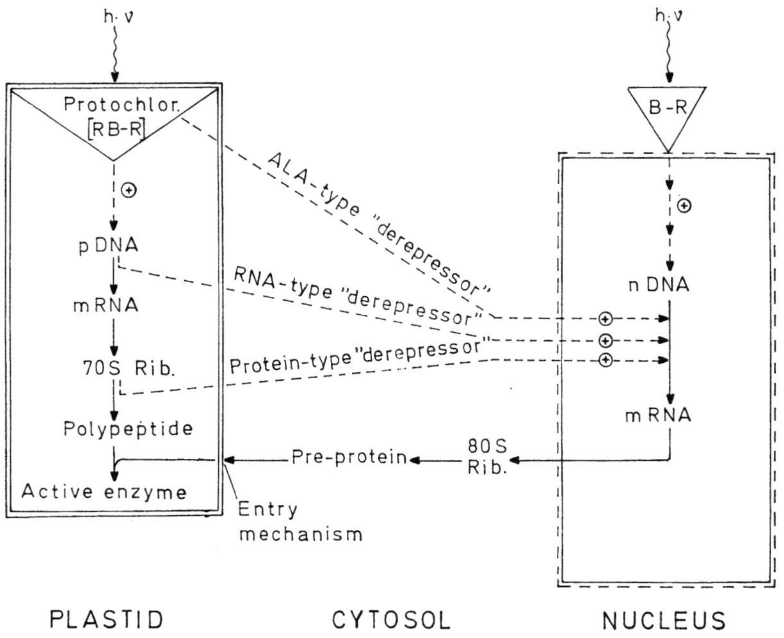


**Fig. 6.12:** Scheme of hypothetic control systems between cell compartments during chloroplast biogenesis

For details see text. *RB-R*, red-blue light receptor (protochlorophyll); *B-R*, blue light receptor; *ALA*, $\delta$-aminolevulinic acid; (+) activation.

and thylakoid structures. These mutants exhibit an unimpaired or slightly different DNA structure. They are supposed to possess no functional plastid ribosomes, the damage being caused either by the failure of synthesis of ribosomal proteins or by a defective assembly of such proteins with probably unstable rRNA.

Analysis of such ribosome-free plastid mutants supports the hypothesis that gene products of the plastids exert control over the synthesis of chloroplast proteins which are of cytoplasmic origin. Since polypeptides are out of the question because of the deficient translation machinery, RNA with regulatory properties has been regarded as a proper candidate for a "derepressor" of nuclear gene expression. A derepressor function of plastid RNA for nuclear-coded plastid proteins would require RNA export from plastids into the cytoplasmic compartment (Fig. **6**.12). RNP particles might be a proper form, since negatively charged polynucleotides do not penetrate cell membranes.

In contrast, polypeptides migrate easily through proteolipid membranes. The results obtained with RuBPCase small subunit synthesis and transport encourage ideas about similar transmembrane transport mechanisms for regulator proteins from the cytoplasm into the chloroplast and, vice versa, from chloroplasts into the nucleo-cytoplasmic compartments (Fig. **6**.12). They might function in controlling metabolic pathways in the cytosol, that are required for chloroplast formation, or as regulatory molecules which inform the nucleus about the current state of affairs in the chloroplast compartment. Such a negative feedback control function of plastid proteins for the synthesis of nuclear-coded chloroplast constituents has been favoured especially by inhibitor experiments during the light-induced chloroplast formation. These results suggest a "plastid control principle" in addition to the "cytoplasmic control principle" proposed by Ellis [22], who emphasized the role of cytoplasmic products for the control of organellar protein synthesis. Until now, however, experimental evidence is still extraordinary weak in favour of regulatory proteins involved in plastogenesis.

The hypothetical network of control factors assumed to take part in plastogenesis is certainly more complex. It should include interactions of many metabolic pathways and the participation of low-molecular weight compounds. Rapid increase of nuclear-coded plastid proteins upon illumination suggests that intermediates of the chlorophyll biosynthesis pathway might play a role as intracellar control elements. A strong argument in favour of such a signal emitted from the plastids is also the observation that red light absorbed by protochlorophyll triggers chloroplast development in dark-grown *Euglena* cells [30].

### 6.5.2 Role of Mitochondria During Chloroplast Formation

In the early hours of chloroplast differentiation in *Euglena* the energy necessary for the synthetic processes must be provided by sources other than photosynthesis. Although the occurrence of a glycolytic pathway in proplastids, which decreases during greening, was reported, it is estimated that ATP production by fermentative processes is negligible. Thus, during the early periods of chloroplast formation mitochondrial oxidative phosphorylation should provide ATP necessary for the synthesis of plastidic macromolecules. This suggestion is supported by increased respiratory rates during early greening.

The interaction, at the level of gene expression, between plastids and mitochondria remains unclear. In *Euglena* a reciprocal correlation exists in the biogenesis of the two energy-supplying organelles. Which type of organelle dominates depends on the nutritional state, i.e., on the available form of carbon. Thus light induction of chloroplast formation is connected with a decrease in size and activity of the mitochondrial system; the reverse process takes place after transferring cells into darkness. Under resting conditions *Euglena* cells switch their metabolic pathway in a direction favouring chloroplast formation at the cost of especially mitochondrial proteins. Redistribution of hydrolyzed amino acids, but insignificant alteration of the amount of total cell protein, suggest a massive protein turnover during greening. Activities of enzymes characteristic for mitochondria decrease, e.g., succinate dehydrogenase and fumarase. Inhibitor studies indicate that the inactivation is due to enzyme degradation plus arrested enzyme synthesis.

Parallel to the light-induced regression of mitochondrial functions structural alterations are likewise observed. Mitochondria are reduced in size from 13% to 6% of the total cell volume. Inhibitors of oxidative phosphorylation such as antimycin A or of plastid protein synthesis (CAP, streptomycin) stop chloroplast development and synchronously stimulate mitochondria enlargement and multiplication. It is suggested that mitochondria are impermeable to these drugs which therefore can be regarded as selective inhibitors of chloroplast formation. Conjoining plastids and mitochondria have been observed; however, exchange of macromolecules between these organelles has not yet been shown. This possibility, if it existed, would not disprove the master control function of the nucleus in the mutual interrelations between the two energy-supplying organelles of, at least unicellular, eukaryotes.

## 6.6 Selected References

### 6.6.1 Books and Reviews


1. Akoyunoglou, G. & J. H. Argyroudi-Akoyunoglou (eds.): Chloroplast Development. Elsevier/North-Holland Biomedical Press, Amsterdam 1978.
2. Barber, J. (ed.): The Intact Chloroplast. Elsevier Publ. Co., Amsterdam 1976.
3. Boardman, N. K., J. M. Anderson & D. J. Goodchild: Chlorophyll protein complexes and structure of mature and developing chloroplasts. Curr. Topics Bioenergetics *8* (1978): 36–109.
4. Bogorad, L. & H. Weil (eds.): Nucleic Acids and Protein Synthesis in Plants. Plenum Press, New York 1977.
5. Bücher, Th., W. Neupert, W. Sebald & S. Werner (eds.): Genetics and Biogenesis of Chloroplasts and Mitochondria. North-Holland Publ. Co., Amsterdam 1976.
6. Chua, N.-H. & G. W. Schmidt: Transport of proteins into mitochondria and chloroplasts. J. Cell Biol. *81* (1979): 461–483.
7. Gillham, N. W.: Organelle Heredity. Raven Press, New York 1978.
8. Hagemann, R.: Plasmatische Vererbung. G. Fischer Verlag, Jena 1971.
9. Leaver, C. J. (ed.): Genome Organisation and Expression in Plants. Plenum Press, New York 1980.
10. Mohr, H.: Lectures in Photomorphogenesis. Springer Verlag, Berlin 1972.
11. Mohr, H. & P. Schopfer: Lehrbuch der Pflanzenphysiologie, 3. Aufl. Springer Verlag, Berlin 1978.
12. Nigon, V. & Ph. Heizmann: Morphology, biochemistry and genetics of plastid development in *Euglena gracilis*. Int. Rev. Cytol. *53* (1978): 212–290.
13. Parthier, B.: Chloroplast development in *Euglena*: Regulatory aspects. In: Lewandowsky, M. & S. H. Hutner (eds.), Biochemistry and Physiology of Protozoa, Vol. *4*. Academic Press, New York 1980, pp. 261–300.
14. Smith, H. (ed.): The Regulation of Enzyme Synthesis and Activity in Higher Plants. Academic Press, London 1977.
15. Wollgiehn, R. & B. Parthier: RNA and protein synthesis in plastid differentiation. In: Reinert, J. (ed.), Results and Problems of Cell Differentiation. Vol. *10*. Springer Verlag, Berlin 1980, pp. 97–145.


### 6.6.2 Special References

*Genetic Aspects*


16. Bedbrook, J. R., D. M. Coen, A. Beaton, L. Bogorad & A. Rich: Location of the single gene for the large subunit of ribulose-1,5-bisphosphate on the maize chloroplast chromosome. J. Biol. Chem. *254* (1979): 905–910.
17. Bedbrook, J. R. & R. Kolodner: The structure of chloroplast DNA. Ann. Rev. Plant Physiol. *30* (1979): 593–620.
18. Bedbrook, J. R., G. Link, D. M. Coen, L. Bogorad & A. Rich: Maize plastid gene expressed during photoregulated development. Proc. Natl. Acad. Sci. USA *75* (1978): 3060–3064.
19. Bradbeer, J. W., Y. E. Atkinson, T. Börner & R. Hagemann: Cytoplasmic synthesis of plastid polypeptides may be controlled by plastid-synthesized RNA. Nature *279* (1979): 816–817.
20. Coen, D. M., J. R. Bedbrook, L. Bogorad & A. Rich: Maize chloroplast DNA fragment encoding the large subunit of ribulosebisphosphate carboxylase. Proc. Natl. Acad. Sci. USA *74* (1977): 5487–5491.
21. Driesel, A. J., E. J. Crouse, K. Gordon, H. J. Bohnert, R. G. Herrmann, A. Steinmetz, M. Mubumbila, M. Keller, G. Burkard & J. H. Weil: Fractionation and identification of spinach chloroplast transfer RNAs and mapping of their genes on the restriction map of chloroplast DNA. Gene *6* (1979): 285–306.
22. Ellis, R. J.: Protein synthesis by isolated chloroplasts. Biochim. Biophys. Acta *463* (1977): 185–215.

23. Grebanier, A., K. Steinback & L. Bogorad: Comparison of the molecular weights of proteins synthesized by isolated chloroplasts with those which appear during greening in *Zea mays*. Plant Physiol. *63* (1979): 436–439.
24. Malnoë, P., J. D. Rochaix, N.-H. Chua & P. F. Spahr: Characterization of the gene and messenger RNA of the large subunit of ribulose-1,5-diphosphate carboxylase in *Chlamydomonas reinhardii*. J. Mol. Biol. *133* (1979): 417–434.
25. Schwartzbach, S. D., W. E. Barnett & L. I. Hecker: Evidence that *Euglena* chloroplasts do not export tRNAs. Nature *280* (1979): 86–87.
26. Verdier, G.: Poly(adenylic acid)-containing RNA of *Euglena gracilis* during chloroplast development. 2. Transcriptional origin of the different RNA. Eur. J. Biochem. *93* (1979): 581–586.
26a. Koch, W., K. Edwards & H. Kössel: Sequencing of the 16S–23S spacer in a ribosomal RNA operon of *Zea mays* chloroplast DNA reveals two split tRNA genes .Cell *25* (1981): 203–213.
26b. McIntosh, L., C. Poulson & L. Bogorad: Chloroplast gene sequence for the large subunit of ribulosebis-phosphate carboxylase. Nature *288* (1980): 556–560.

*Light Induction*

27. Fong, F. & J. A. Schiff: Blue-light-induced absorbance changes associated with carotenoids in *Euglena*. Planta *146* (1979): 119–127.
28. Gressel, J.: Blue light photoreception. Photochem. Photobiol. *30* (1979): 749–754.
29. Mohr, H.: Phytochrome and gene expression. In: Nover, L., F. Lynen & K. Mothes (eds.), Cell Compartmentation and Metabolic Channelling. G. Fischer Verlag and Elsevier/North-Holland Biomed. Press, Jena and Amsterdam 1980, pp. 399–414.
30. Schiff, J. A.: Photocontrol of chloroplast development in *Euglena*. In: [1], pp. 747–767.
31. Schopfer, P.: Phytochrome control of enzymes. Annu. Rev. Plant Physiol. *28* (1977): 223–252.
32. Wellburn, A. R. & R. Hampp: Appearance of photochemical function in prothylakoids during plastid development. Biochim. Biophys. Acta *547* (1979): 380–397.

*Selected Examples of Gene Expression*

33. Apel, K.: Phytochrome-induced appearance of mRNA activity for the apoprotein of the light-harvesting chlorophyll a/b protein of barley *(Hordeum vulgare)*. Eur. J. Biochem. *97* (1979): 183–188.
34. Apel, K. & L. Bogorad: Light-induced increase in the activity of maize plastid DNA-dependent RNA polymerase. Eur. J. Biochem. *67* (1976): 615–620.
35. Barnett, E. W., S. D. Schwartzbach & I. L. Hecker: The transfer RNAs of eukaryotic organelles. Progr. Nucl. Acid Res. Mol. Biol. *21* (1978): 143–179.
36. Chelm, B. K. & R. B. Hallick & P. W. Gray: Transcription program of the chloroplast genome of *Euglena gracilis* during chloroplast development. Proc. Natl. Acad. Sci. USA *76* (1979): 2258–2262.
37. Ellis, R. J., S. M. Smith & R. Barraclough: Synthesis, transport and assembly of chloroplast proteins. In: [9], pp. 321–335.
38. Feierabend, J. & G. Wildner: Formation of the small subunit in the absence of the large subunit of ribulose-1,5-bis-phosphate carboxylase in 70S ribosome-deficient rye leaves. Arch. Biochem. Biophys. *186* (1978): 283–291.
39. Grossman, A., S. Bartlett & N.-H. Chua: Energy-dependent uptake of cytoplasmically synthesized polypeptides by chloroplasts. Nature *285* (1980): 625–628.
40. Nelson, N., H. Nelson & G. Schatz: Biosynthesis and assembly of the proton-translocating adenosine triphosphate complex from chloroplasts. Proc. Natl. Acad. Sci. USA *77* (1980): 1361–1364.
41. Rawson, J. R. Y. & C. L. Boerma: Hybridization of Eco RI chloroplast DNA fragments of *Euglena* to pulse labelled RNA from different stages of chloroplast development. Biochem. Biophys. Res. Commun. *89* (1979): 743–749.
42. Schatz, G.: How mitochondria import proteins from the cytoplasm. FEBS Letters *103* (1979): 203–211.

43. Schmidt, G. W., S. Bartlett, A. R. Grossman, A. R. Cashmore & N.-H. Chua: In vitro synthesis, transport, and assembly of the constituent polypeptides of the light-harvesting chlorophyll a/b protein complex. In: [9], pp. 337–351.
44. Thornber, J. P., J. P. Markwell & S. Reinman: Plant chlorophyll-protein complexes: recent advances. Photochem. Photobiol. *29* (1979): 1205–1216.
45. Tobin, E. & J. L. Suttie: Light effects on the synthesis of ribulose-1,5-bisposphate carboxylase in *Lemna gibba*. Plant Physiol. *65* (1980): 641–647.
46. Trebst, A.: Functional organization of the photosynthetic electron transport chain in chloroplasts. In: Nover, L., F. Lynen & K. Mothes (eds.), Cell Compartmentation and Metabolic Channelling. G. Fischer Verlag and Elsevier/North-Holland Biomed. Press, Jena and Amsterdam 1980, pp. 201–213.

# 7. The Action of Estrogen on Gene Expression in Chick Oviduct

M. Luckner

## 7.1 Regulation of Estrogen Biosynthesis and Estrogen Level

There are certain groups of steroids which in animals act as intercellular signals due to the evolution of special target mechanisms in distinct types of cells. These steroids include the female sex hormones (estrogens and progestins), the male sex hormones (androgens) and the hormones secreted from the cortex of the adrenal gland (corticosteroids). Some of the most important representatives of these compounds are listed in Table **7**.1.

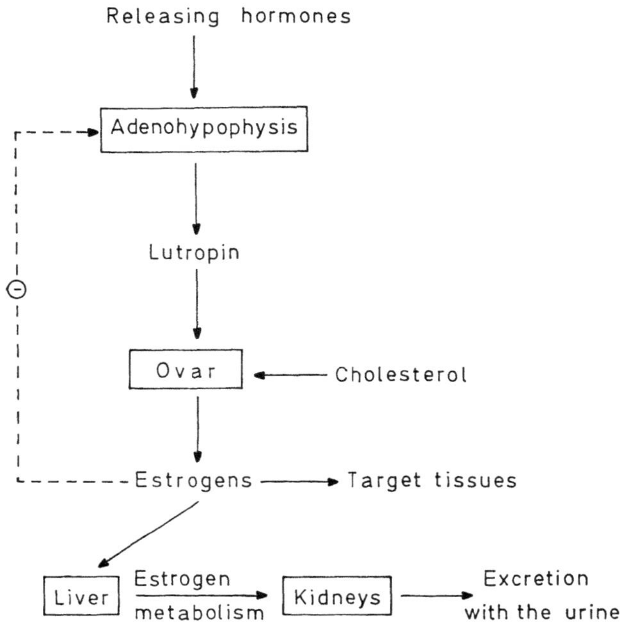


**Fig. 7.1**: Principles regulating blood concentration of estrogens (cf. also **4**.2.4.2 and Fig. **4**.45).

Like other steroid hormones, estrogens are formed from cholesterol, an ubiquitous compound in animal cells, by specialized enzymes present, for instance, in the ovary. Synthesis is regulated by lutropin (luteinizing hormone, LH), a peptide hormone formed in the adenohypophysis (Fig. **7**.1). Lutropin binds to special receptor proteins in the outer membrane of the estrogen-synthesizing cells. This binding increases the activity of membrane-bound adenylyl cyclase and thus the intracellular level

Abbreviations: DES, diethylstilbestrol; dT, deoxythymidine; ER, endoplasmatic reticulum; kb, kilobase; kd, kilodalton; LH, luteinizing hormone; $RNA_{ov}$, ovalbumin-specific RNA sequences; RNAP, RNA polymerase; SDS, sodium dodecyl sulphate.

**Table 7.1:** Steroids with hormone activity in animals (from O'Malley & Schrader [15])

| Group | Formula of a typical representative | Principle source | Principle target | Hormonal function |
|---|---|---|---|---|
| Estrogens | Estradiol 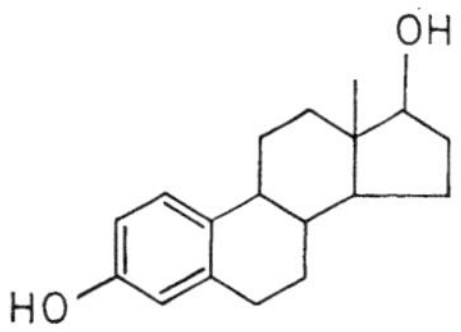 | Ovary, testis | Uterus, oviduct, vagina, breast, hypothalamus | Development and maintainance of female sex characteristics |
| Progestins | Progesterons 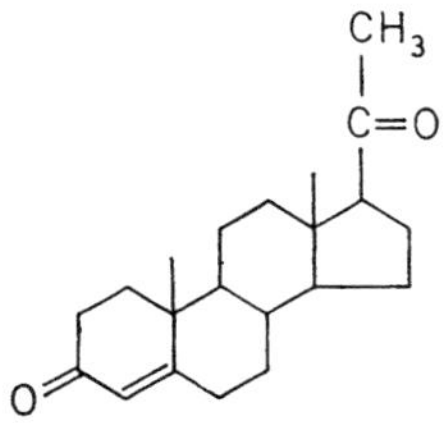 | Ovary, placenta | Uterus, placenta | Maintainance of pregnancy |
| Androgens | Testosterone 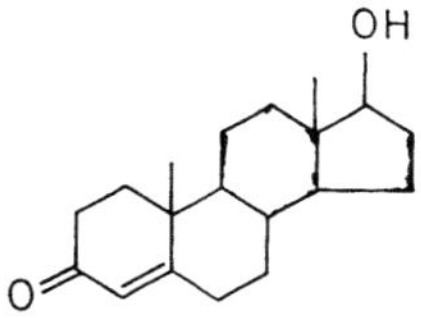 | Testis | Seminal vesicles, prostata, testis | Development and maintainance of male sex characteristics (cf. **4.6.4**) |
| Cortico-steroids | Hydrocortisone 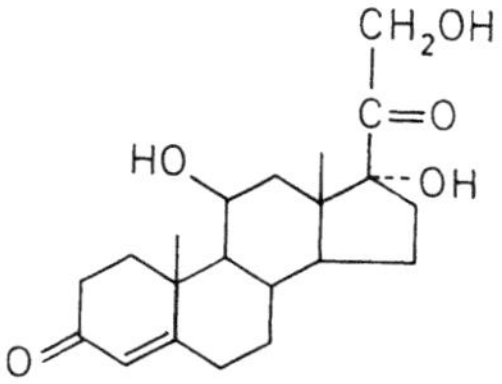 | Cortex of the adrenal gland | All cells | Regulation of energy utilization |
| | Aldosterone 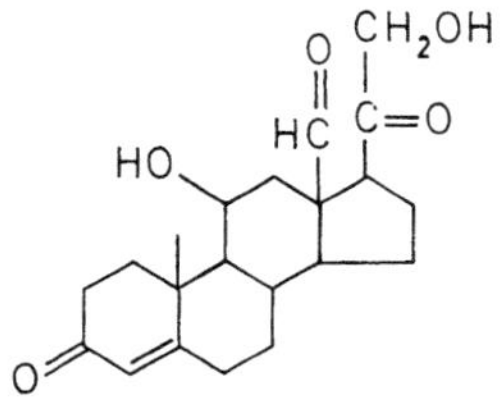 | | Kidney | Regulation of electrolyte balance |

Further details see Träger [1]

of cAMP (cf. **4.2.4.2**). As is to be expected, administration of cAMP therefore mimics the action of lutropin on these cells.

The increased level of cAMP is supposed to cause an activation of distinct protein kinases that are associated with ribosomes or to mediate translocation of protein kinases to the ribosomes. Ribosomal proteins are phosphorylated, resulting in enhanced translation of certain stable mRNA species present in the cells. This causes an increase of the synthesis of the steroid carrier protein required for transport of cholesterol to the mitochondria, in which the first steps of estrogen biosynthesis proceed. Probably also translation of other enzymes of estrogen synthesis is enhanced. Already a few minutes after lutropin application synthesis of estradiol (and structurally related steroids) is considerably stimulated (cf. **4.2.4.2**).

The estradiol is released into the extracellular space and dispersed with the blood in the animal body. It exerts a negative influence on lutropin release from the adenohypophysis and by this mechanism inhibits its own rate of synthesis (Fig. **7.1**, cf. also Fig. **4.45**). Furthermore it undergoes metabolic transformations (e.g., in the liver) that are accompanied by loss of the biological activity. The most important reactions of estradiol metabolism are glucuronation and sulphatation. In addition, estradiol may be hydroxylated and dehydrogenated. The products are eliminated from blood by the kidneys and excreted with the urine. By synthesis and degradation a balanced hormone level is maintained which depends on the developmental stage of the producer organism, its physiological state etc.

## 7.2 Cell Specialization Induced by Estrogen Action in Chick Oviduct

Estrogens (in part together with progesterone) cause a multitude of effects in the animal body. They trigger the development of female sex characteristics and are important factors in the control of menstrual and rut cycles. One of the most favorite experimental systems for investigation of the action of estrogens is the oviduct of chicks.

During sexual maturation of chicks estrogen causes the single layer of primitive stem cells in the oviduct mucosa to proliferate and to differentiate into three morphologically different types of epithelial cells: (a) tubular gland cells, (b) ciliated cells, and (c) goblet cells, which are interspaced with the ciliated cells in the surface layer of duct epithelium. Concomitant with this morphological cell specialization, estrogen induces synthesis of the secretion proteins, which are constituents of the white of egg, i.e., of ovalbumin, ovomucoid, conalbumin, lysozyme etc. In the laying hen, about 4 grams of protein are synthesized and secreted from the oviduct per day: $\sim 60\%$ is ovalbumin, $\sim 12\%$ is conalbumin, $\sim 10\%$ is ovomucoid, and $\sim 2\%$ is lysozyme. Fluorescent antibody staining of oviduct sections reveals that all of these proteins accumulate in the same tubular gland cells. Initiation of their secretion coincides with the appearance of discrete secretory granules in the cytoplasm of these cells from which the protein is released to the extraplasmic space.

In the immature chick these processes of cell specialization may be evoked prematurely by administration of exogenous estrogen. If estrogen is withdrawn after about 14 days of hormone treatment, the specialized cells formed in the oviduct mucosa remain, but protein synthesis and secretion ceases. A new estrogen stimulus, however, restores full protein secretion in these determined cells within a few hours. Within this time synthesis of certain proteins, e.g., ovalbumin, increases dramatically. In fully induced cells both ovalbumin genes present in the diploid genome of the tubular gland cells may produce 12 molecules of ovalbumin mRNA per minute, each of which can be translated up to 50,000 times, so that, finally, ovalbumin represents $60\%$ of the total protein synthesized in the estrogen-stimulated oviduct.

Thus we can discriminate between two biological effects of estrogens: (1) the triggering of cell specialization including formation of tubular gland cells as well as activation of chromatin domains coding for the egg white proteins and (2) the actual control of the transcription rates (cf. details in 7.4).

It is an important finding that the activation of the ovalbumin gene domain is connected with an alteration of the DNA methylation pattern [7]: distinct sites are undermethylated in tissues capable to express synthesis of egg white proteins (cf. details in 4.2.5.2). Interestingly, hormone withdrawal does not change the methylation pattern though transcription of ovalbumin-specific RNA completely stops. Hence, undermethylation of the ovalbumin gene domain apparently reflects its determination to be transcribed, e.g., after a second estrogen stimulus, rather than the actual state of transcription itself.

Under appropriate tissue culture conditions both estrogen effects can also be demonstrated in vitro. However, in these cases induction of the ovalbumin gene, but not of the conalbumin gene, required not only estradiol but also a serum factor, presumably a somatomedin-like peptide hormone of the pituitary (SPH), whose action is mediated by the membrane receptor-adenylate cyclase system (cf. Fig. 10.6). In vivo, SPH may regulate the seasonal inducibility of ovalbumin synthesis by estradiol [29].

## 7.3 Structural Basis of Synthesis and Release of Ovalbumin

### 7.3.1 Organization of the Ovalbumin Gene

Knowledge about the structure of the genes whose activity is estrogen-regulated may help in understanding the molecular mechanisms of hormonal control. Prerequisite for elucidation of the gene structures is to obtain substantial amounts of the corresponding DNA in a pure state. This task was achieved for the ovalbumin gene, for instance, by transfer of random pieces of chick DNA, obtained by partial hydrolysis of the total genome, to DNA phages and their cloning together with the phage DNA according to the following procedure ([2, 4], cf. also 2.3):

Chick DNA was partially digested with restriction endonucleases Hae III and Alu I to DNA fragments of about 15–20 kb. These fragments were transferred to Charon 4A phage λ DNA by using synthetic Eco RI DNA linkers. Several millions of phage clones containing the recombinant

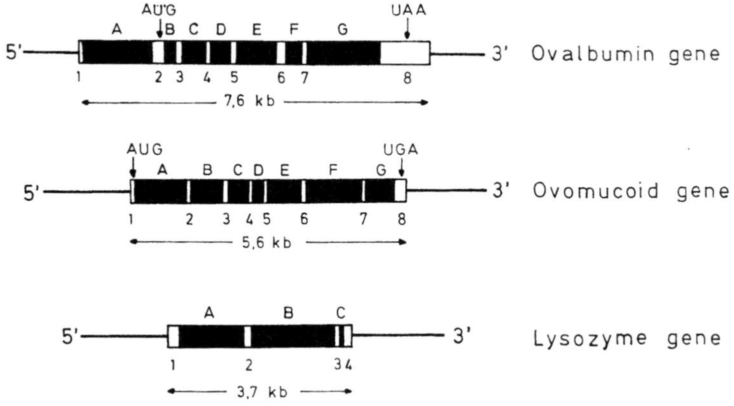


**Fig. 7.2:** Structural organization of ovalbumin, ovomucoid and lysozyme genes (from Dugaiczyk et al. [2]; Lai et al. [4]; Lindenmaier et al. [6]).

*Black sections* of the map denote intervening DNA sequences (introns) and *white sections* represent structural sequences of DNA (exons).

*Thin solid lines* represent flanking DNA sequences. *AUG* Initiation codon; *UAA* termination codon.

DNA (chicken gene library) were screened for the presence of the ovalbumin gene by using a phage [$^{32}$P]-labelled 3.2 kb part of the ovalbumin gene, obtained in earlier experiments, as a hybridization probe. Several phage clones were isolated which showed strong hybridization signals. One of these clones (λ 4A-OV) contained the complete ovalbumin gene together with some flanking chick DNA.

Restriction endonuclease mapping, electron microscopy and direct DNA sequencing analysis of λ 4A-OV and related phage clones revealed that the ovalbumin gene is about 7.6 kb long, though it codes only for a mRNA of 1859 nucleotides (cf. **7.3.2**). There are seven intervening sequences (introns) of various sizes within the gene, separating eight structural sequences (exons) (Figs. **7.2** and **7.3**). The start codon

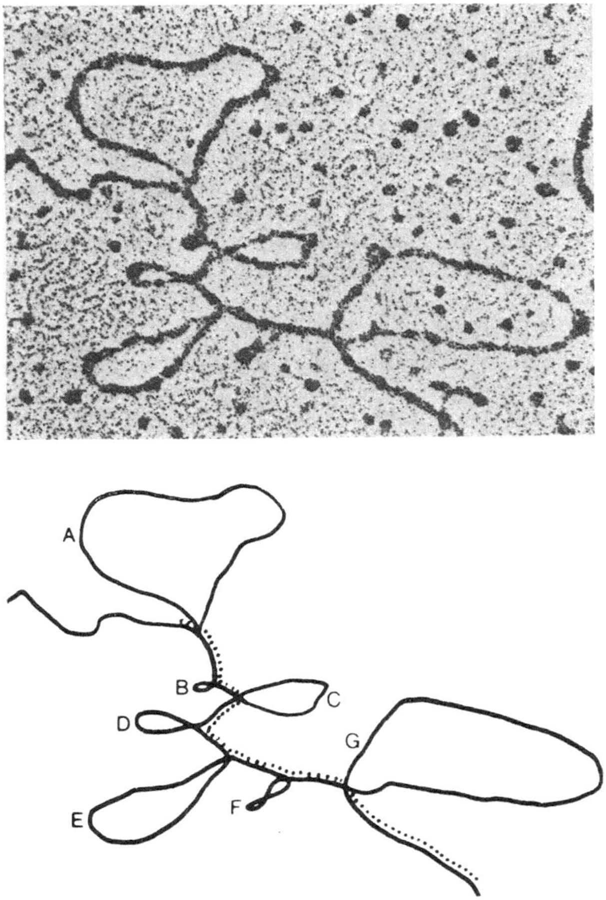


**Fig. 7.3**: Electron micrograph and line drawing of a hybrid molecule formed between the ovalbumin gene (λ 4A-OV) and ovalbumin mRNA (from Dugaiczyk et al. [2], with permission).

Hybridization was carried out with 10 μg of λ 4A-OV DNA and 20 μg of purified ovalbumin mRNA per ml in 70% deionized formamide containing 0.1 M Tris-HCl (*p*H 7.6), 10 mM $Na_2$EDTA and 150 mM NaCl. The mixture was heated to 80 °C for 5 min to denature the DNA and hybridization was carried out at 54 °C for 3 h. Samples were prepared for electron microscopy by dilution of this mixture and spreading on to a hypophase of distilled water. They were stained with uranyl acetate and rotary shadowed with platinum/palladium.

——— λ 4A-OV DNA, ...... ovalbumin mRNA; cf. the seven loops of unhybridized DNA corresponding to the intervening sequences A–G (Fig. **7.2**).

(AUG) is present in the second exon of the ovalbumin gene, the termination codon (UAA) in the last one.

Similar structures were shown to exist also for the genes of other secretory proteins of the chick oviduct: the ovomucoid gene which contains 8 exons and 7 introns, the conalbumin gene which contains 17 exons and 16 introns and the lysozyme gene which contains 4 exons and 3 introns (Fig. 7.2). It has been suggested that each structural sequence might correspond to a functional or structural domain of these proteins. This problem, however, needs further investigation with proteins whose domains are better characterized.

Though DNA sequences of the genes encoding the secretion proteins of chick oviduct, ovalbumin and ovomucoid, are different as shown by results of Southern hybridization (cf. **2**.6) and electron microscopy, DNA sequencing data have revealed that there are short stretches of DNA sequences immediately flanking both genes that are remarkably similar. Three types of sequence elements are noteworthy [4]:

1. About 30 nucleotides 5′ upstream from the site of cap addition both genes contain AT-rich sequence elements found also in the neighborhood of other eukaryotic protein-coding genes (Hogness box, cf. Table **1**.8). In vitro transcription of a truncated $DNA_{ov}$ fragment with RNAP II [4a] revealed that, in addition, sequence elements further upstream extending to nucleotide —61 are required for correct initiation (s. also legend to Fig. **4**.14). In contrast deletion of nucleotides upstream from position –61 did not effect the transcription fidelity but led to an enhancement of the in vitro $RNA_{ov}$ synthesis.
2. A sequence of partial twofold symmetry has been identified between the AT-rich region and the start of ovalbumin and ovomucoid genes. The presence of sequences with complete or partial dyad symmetry is common for operators and other control regions of prokaryotic gene transcription (cf., e.g., Figs. **4**.10 and **4**.19). Whether these types of symmetrical sequences have any function in the regulation of eukaryotic gene expression must await further analysis.
3. Finally, seven nucleotides into the 3′-flanking sequence of the ovalbumin gene there is a TTTT-sequence also present in other eukaryotic genes at the same position. In the ovomucoid gene a TTGT-sequence exists instead. It has been proposed that such T-clusters may be responsible for ensuring correct termination (cf. Fig. **1**.7), as it was shown that mutations at this site may cause transcriptional "read-through".

Two other split genes (X and Y) which are also under estrogen control, are located upstream of the 5′-end of ovalbumin gene (OV) in the sequence X → Y → OV. Assuming that no other, hitherto undetected genes are present in the DNA which includes these three genes, the length of the intergene regions is about 5–10 kb. Together with the length of the genes themselves it may be estimated that there would be one gene every 10–15 kb of DNA. X, Y and ovalbumin genes share partial sequence homologies and therefore could have arisen from a common precursor. As yet, however, it is unknown whether the functions of the products of the three genes are related, as those of X and Y have not yet been identified [7a].

### 7.3.2 Structure of Ovalbumin mRNA

Mature ovalbumin mRNA is one of the few mRNA species fully sequenced as yet. Milligramme quantities have been isolated from nucleic acid extracts from the stimulated hen oviduct by the following method [8]:

Total nucleic acids were extracted with phenol from hen oviduct and the poly(A)-containing mRNA fraction was selectively enriched by either nitrocellulose filtration or oligo dT-cellulose column chromatography. The mRNA fraction was further purified by a gel filtration step using

Sepharose 4B. Ovalbumin mRNA was obtained in pure state by preparative electrophoresis in agarose gels (1 mg of purified ovalbumin mRNA per 1 to 1.5 g of total nucleic acid extract). Ovalbumin mRNA sediments as a single symmetrical peak at 16S in sucrose gradients and migrates as a single component at 21S during gel electrophoresis under denaturating conditions.

Sequence analysis of the ovalbumin mRNA was performed with an almost full length duplex DNA copy which was cloned in the *E. coli* plasmid pMB 19. Plasmids were screened for the size of the inserted ovalbumin mRNA transcripts and a clone containing the DNA representing about 1,850 residues of ovalbumin mRNA was selected for sequencing (cf. **2**.10).

It has been shown that ovalbumin mRNA is 1859 nucleotides long, excluding its 5′-terminal cap and the 3′-terminal poly(A)-sequence. The region coding for ovalbumin (residues 65–1222) lies close to the cap. It is separated from the poly(A)-sequence by an extensive 3′-non-coding region of 637 nucleotides, which may have no function that is precisely dependent on its sequence. The presence of a $m^7$Gppp cap structure was assumed from translation studies. The 5′-non-coding region (residues 1–64) is believed to be required for initiation of protein synthesis (cf. **1**.3.4.3).

In nuclear extracts of oviduct cells stimulated with estrogen high molecular weight precursors of ovalbumin mRNA have been detected, which contain sequences complementary to the intervening as well as the structural sequences of the ovalbumin gene. The largest species was about 7.8 kb, which corresponds to the length of the ovalbumin gene with about 7.6 kb (cf. **7**.3.1). Electron microscopy and hybridization analysis have revealed that during maturation of the mRNA the intervening sequences are removed by splicing. This splicing proceeds via distinct intermediates, which may be separated by agarose gel electrophoresis. At all splice points a short partial sequence homology is present in the RNA molecules, which is probably a signal for a common splicing enzyme (cf. Fig. **1**.10).

It has been suggested that RNA splicing has a critical role in the production of functional mRNA. Recent work with ovalbumin has shown that this protein is synthesized by *Xenopus* oocytes injected with a plasmid containing the complete chicken ovalbumin gene. This result demonstrates that frog oocytes can perform all steps necessary to generate *functional* mRNA from primary transcripts of the gene, including mRNA splicing. A second plasmid from which the 5′-region of the ovalbumin gene including the cap site, the TATATAT region and other 5′-flanking sequences have been deleted, directs ovalbumin synthesis with the same efficiency as the intact gene. Therefore, at least under the experimental conditions used, none of these regions are indispensible for transcription, accurate RNA processing, transport or translation. The result does, however, not necessarily show that the same regions have no function in the ovalbumin gene expression in the chick oviduct itself. In contrast oocytes injected with a plasmid carrying the full length cDNA copy of mature ovalbumin mRNA synthesize no detectable ovalbumin. They accumulate instead more RNA containing ovalbumin messenger sequences than do oocytes injected with the native gene. This indicates that the reason for the lack of functional ovalbumin mRNA is not absence of transcription of the cDNA clone but rather the lack of the introns in this DNA species, which are required for normal processing and transport of the transcript to the cytoplasm [28].

### 7.3.3 Translocation of Ovalbumin Through Membranes of the Rough ER

The information for protein translocation across membranes during its synthesis is located in a discrete portion of the polypeptide chain, the so-called signal sequence (cf. Table **4**.10). In most cases the signal sequence is an amino-terminal extension of the protein consisting of 15–30 amino acids, which is cleaved during translocation across the endoplasmic reticulum membrane (cf. Fig. **7**.4). Although the primary structure of the signals differs for various secretory proteins, there is circumstantial evidence that all signal sequences share certain features in their secondary structure. These data suggest the existence of a common protein transport system in membranes of rough ER (protein translocator), which shows response to a common signal structure (cf. **4.4.5.1.1**).

Corresponding to this scheme the precursors of ovomucoid and lysozyme, pre-ovomucoid and pre-lysozyme, contain "external" signal sequences which are cleaved during secretion. Ovalbumin, however, which is secreted in a similar manner, lacks the typical transient signal sequence. Instead it contains an amino acid sequence in the central part of the molecule that is structurally and presumably also functionally identical to the cleaved signals of other secretory proteins. This sequence is hidden in the completed protein chain. It may be uncovered by protein denaturation or limited enzymatic splitting of ovalbumin. After tryptic digestion of ovalbumin it is found in the fragment comprising amino acid residues 229–276.

```
             235      240            245      250
OVs       F A S G T M S M L V L - - - L P D E V S G L E
pre OVM   F V L - - F S - F V L C G F L P D A A F G A E
pre LYS   M R S - - L L I L V L C - F L P L A A L G K V
```

**Fig. 7.4:** Homology of ovalbumin signal sequence to terminal cleaved signals of other chick oviduct secretory proteins (from Lingappa et al. [12]).
Signal sequences of pre-ovomucoid *(pre OVM)* amino acid residues 7–24 and pre-lysozyme *(pre LYS)* amino acid residues 1–18 were aligned to maximize homology with ovalbumin amino acid residues 234–253 *(OV$_S$)*.
Residues which are homologous in all three proteins are *boxed*. Residues involving a single base exchange are indicated with *bold-faced letters*. The glycine residue flanking the cleavage sites of pre-ovomucoid and pre-lysozyme are aligned with glycine 251 of ovalbumin. Hyphens indicate deletions in the sequence necessary to achieve maximum homology.
*A*, Ala; *C*, Cys; *D*, Asp; *E*, Glu; *F*, Phe; *G*, Gly; *I*, Ile; *K*, Lys; *L*, Leu; *M*, Met; *P*, Pro; *R*, Arg; *S*, Ser; *T*, Thr; *V*, Val.

Residues 234–253 show striking sequence homology to the amino-terminal cleaved signals of pre-ovomucoid and pre-lysozyme (Fig. **7.4**). It is likely, therefore, that the signal activity resides in this portion of the ovalbumin molecule. Particularly significant are two clusters of homology, each comprising three hydrophobic residues. Furthermore there are clusters of amino acid residues differing in their genetic code by a single base, only (in Fig. **7.4** indicated by bold-faced letters). Furthermore the penultimate glycine residue of the two cleaved signal sequences readily alignes with a glycine residue of ovalbumine. Hence, the most important difference between the three amino acid sequences is the absence in ovalbumin of cysteine or a cysteine-glycine peptide located between the two homologous hydrophobic clusters in the cleaved signals.

This structural homology strongly suggests a common translocator for ovalbumin and the other oviduct secretory proteins. In addition experiments with an in vitro translocation system have shown that ovalbumin competes also with the secretion of other proteins, e.g., pre-prolactin, in agreement with the idea that there is only one protein translocator in the rough ER (see above).

Ovalbumin inhibits, for instance, translocation of pre-prolactin in a microsomal vesicle system derived from the rough ER of dog pancreas. In these experiments bovine pituitary mRNA was translated in nuclease-treated reticulocyte lysate in the presence of [$^{35}$S]-methionine, of the microsomes and increasing concentrations of ovalbumin. After 1 h incubation samples were prepared for SDS gel electrophoresis and the amounts of pre-prolactin and prolactin were determined by autoradiography.

One implication of the central position of the signal sequence of ovalbumin is that translocation of ovalbumin across membranes would be initiated relatively late in the course of synthesis of the molecule. In agreement with this idea, in vitro experiments with precisely timed addition of membranes to a cell-free translation system, synchronized with respect to chain elongation by an inhibitor of initiation, have shown that the growing peptide chains of ovalbumin will lose the capacity for functional interaction with microsomal membranes after about 5.5 min, i.e., much later than pre-prolactin, a protein with a terminal signal sequence, which showed the same effect already after 2.0 min. The time necessary for completion of the ovalbumin chain in these experiments was 8.5 min. Assuming a linear rate of synthesis, ovalbumin is synthesized therefore with a speed of 45 amino acid residues per min. From these figures it may be calculated that the critical chain length, after which translocation is abolished, is about 250 amino acid residues [12]. This is in excellent agreement with the position of the ovalbumin signal sequence derived from sequence homology (cf. Fig. 7.4).

How nascent ovalbumin is actually threaded through the ER membrane is a matter of speculation. One possibility is translocation in the form of a loop, with the signal sequence at its leading site in such a way that both the amino-terminal and the carboxyl-terminal wings of the molecule are translocated simultaneously as the carboxyl-terminal wing is being completed. Another conceivable mechanism is co-translational translocation primarily of the carboxyl-terminal wing, followed by post-translational translocation of all or the remainder of the amino-terminal part of the molecule.

## 7.4 Control of mRNA Transcription by Estradiol

### 7.4.1 Binding of Estradiol to Hormone Receptors

Steroid hormones released from the producer glands are transported with the blood to the target tissues part of them being bound to special carrier proteins of the blood plasma. This binding is of high capacity but has only moderate affinity. It protects the hormones from degradation and is a further means to regulate the concentration of free steroid hormones in the blood.

One of the steroid-binding proteins is uteroglobin (blastokinin), thought to be involved in the transport of maternally produced progesterone to the developing blastocyst. Uteroglobin possesses a molecular weight of 15.8 kd. It is composed of two chemically identical 69-amino acid subunits which, in the oxidized form, are covalently linked by two disulphide bridges. The protein is induced by progesterone. It binds progesterone and 5-α-pregnane-3,20-dione ($K_D 10^{-6}$–$10^{-7}$ M), but not estradiol, cortisol, testosterone etc. Similar steroid-binding proteins exist also for androgens, corticosteroids etc. (cf. Träger [1]).

The fraction of steroid hormones present in free state penetrates the membranes of any animal cell by means of simple diffusion. But only in target cells is it bound to specific, high-affinity receptor proteins. These receptor proteins are present in extremely small quantities (about 10,000 molecules per cell). They bind hormones very tightly with dissociation constants of $10^{-9}$ to $10^{-10}$ M. These values correspond to the hormone level in the blood (and within the cells), which in dependence on sex, age and developmental state may be $10^{-9}$ to $10^{-11}$ M. The hormone-receptor complexes are the main form in which steroid hormones exist after release from the producer glands.

The estrogen receptor from calf uterus has a molecular weight of about 236 kd. In addition to estradiol it binds several other compounds with estrogen activity. With few exceptions only, there is close parallelity between estrogenic activity and affinity to the receptor protein, which may be determined according to Fig. 7.5.

Earlier experiments indicated that the free estrogen receptor is present in the cytoplasm and after binding of estrogen is translocated to the nucleus by a temperature-sensitive process. During this process, the estrogen receptor undergoes a conformational change which is discernible by an increase in its sedimentation coefficient

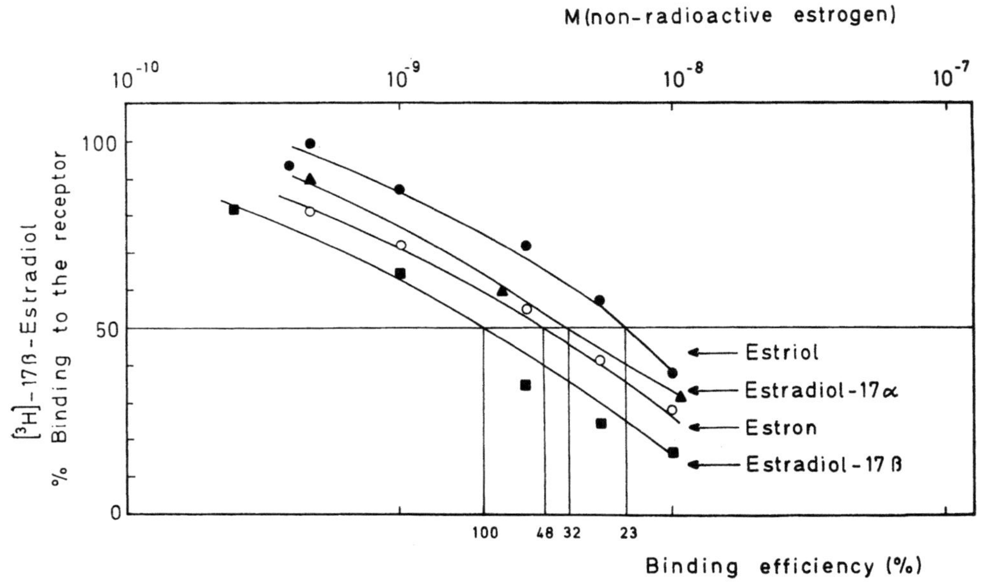


**Fig. 7.5**: Binding affinity of estrogenic steroids to the cytoplasmic estrogen receptor of calf uterus (after Kyrein [16]).

A receptor protein fraction was prepared by centrifugation at 105,000 g from the homogenate of uteri of 2–3 month-old calves. The test is based on the inhibition of binding of a constant amount of radioactive labelled 17$\beta$-estradiol by increasing amounts of non-labelled estrogenic steroids. Not bound radioactively-labelled estradiol was adsorbed to dextrane-coated charcoal. The radioactivity of the estradiol-receptor complex, which, due to its high molecular weight, does not pass the dextrane layer and remains in the supernatant, was determined in a liquid scintillation counter. The concentrations were compared, which inhibit binding of the labelled estradiol by 50%.

from 3.8–4.0 to 5.0–5.2 S, and by its ability to bind to isolated uterine nuclei (activation). Reexamination, however, provided strong evidence that the free receptor is already distributed between cytoplasm and nucleus according to the free water content of these two compartments. Free estrogen may enter both the cytoplasm and the nucleus and bind to unoccupied receptor molecules. After activation the estrogen-receptor complexes present in the nucleus attach to the chromatin (cf. **7**.4.5) and then are not further in equilibrium with the solutes. The nuclear volume becomes devoid of receptors, and free receptor as well as estrogen-receptor complexes move from the cytoplasm to the nucleus to reestablish the initial equilibrium, a process causing the enrichment of estrogen-receptor molecules in the nucleus [21].

Reaction of steroid hormones with intracellular receptors and binding of the hormone-receptor complex to the nucleus holds for most target tissues. In special cases, however, steroid hormones bind to receptors present in the cell membranes. One example is the meiotic maturation of amphibian oocytes which may be started by a brief treatment (1–2 min) with progesterone, corticosteroids or testosterone [19, 20]. Derivatives of progesterone, which do not penetrate into the cell, also induce maturation. Though existence of a proteinaceous progesterone receptor could be demonstrated, it is not soluble in the cytoplasm. In the oocytes the steroid hormones obviously induce migration of calcium ions from the cell membrane to the cytoplasm. Hence, in this system $Ca^{2+}$ acts as a second messenger. Hormone treatment is followed by synthesis of certain proteins, e.g., histones and histone kinase. cAMP does not seem to play a direct role in this process. Protein phosphorylation, however, is involved (cf. Fig. **10**.6).

### 7.4.2 Interaction of Steroid Hormone-Receptor Complexes with Chromatin

Steroid hormone-receptor complexes bind not only to the nuclei of target cells in vivo, but also interact with chromatin in cell-free systems. An example investigated in detail is the chick oviduct progesterone receptor.

The oviduct progesterone-receptor complex is bound to chromatin much more effectively than either uncomplexed receptor or free progesterone. The capacity of oviduct chromatin to bind the progesterone-receptor complex was greater than that of non-target tissue chromatins such as spleen, heart, liver, or erythrocyte. The receptor is a dimer composed of one A and one B subunit, which both are able to bind one molecule of the hormone. The B subunit binds to the non-histone protein DNA complex of oviduct chromatin (cf. 7.4.5) but not to pure DNA, whereas the A subunit binds to pure DNA, but poorly to chromatin. These observations suggest that the A subunit could be the actual gene-regulatory protein. Consistent with this prediction, purified A subunit protein was capable of stimulating transcription of estrogen-withdrawn chromatin, but only at 10- to 50-fold higher concentrations than that required for the intact dimer. The isolated B subunit was totally ineffective in stimulating transcription of oviduct chromatin in any concentration tested. These observations are consistent with a model in which subunit B specifies the binding of the receptor dimer in certain regions of chromatin by interaction with specific acceptor proteins [24]. Thus sites become available for RNA polymerase to initiate RNA synthesis.

It may be an important observation that, similar to the chromatin-associated thyroid hormone receptor (cf. p. 152), the estradiol-receptor complex exhibits high affinity to histones H2A and H2B, but not to H3, H4 and H1. Separate from the domain for interaction with histones is a domain for DNA binding, and the DNA-receptor complex is stabilized by histones, especially H2B [30].

### 7.4.3 Influence of Estradiol on RNA Polymerase Initiation Sites

A consequence of binding of the estrogen-receptor complex to chromatin is an increase in initiation sites for RNA polymerase. This was demonstrated by the following method (Tsai et al. [22]):

RNA polymerase from *Escherichia coli* and chromatin are preincubated in the absence of any nucleoside triphosphates to allow formation of stable binary complexes. This process occurs through a series of random, non-specific interactions between RNA polymerase and DNA. When such binding occurs at a specific initiation site, a transition takes place which leads to the formation of the stable complex between enzyme and DNA. It is thought to involve local unwinding of the DNA duplex in the region of the initiation sites (cf. 1.3.2). After preincubation, RNA synthesis is initiated by the addition of the four ribonucleoside triphosphates together with the inactivator of free bacterial RNA polymerase, rifampicin, and allowed to proceed to completion. At the concentration of rifampicin used RNA synthesis by the stable binary complex between DNA and RNA polymerase was highly efficient, while secondary initiations were strongly inhibited. Thus the number of RNA chains synthesized was equivalent to the number of initiation sites at which RNA polymerase was bound.

With this method one initiation site per 0.7 kb has been found with chick DNA, whereas on chromatin, as expected, a much lower number was measured (one site per 40 kb), indicating that most initiation sites are covered by proteins. In the unstimulated immature chick the level of initiation sites in chromatin is about 10,000 per haploid genome equivalent, increasing by 50% 8 h after estrogen administration. This level continued to rise to day 8 of stimulation and then declined to a plateau level approximately twofold higher than that of the unstimulated animal by day 14. The about 20,000 newly formed initiation sites per haploid genome equivalent include also the single initiation site at the ovalbumin gene.

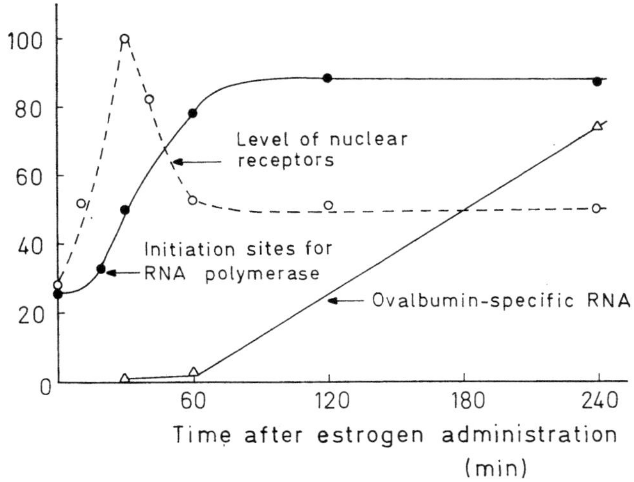


**Fig. 7.6:** Effect of estrogen stimulation on initiation sites for RNA polymerase and the level of nuclear estrogen receptors (after Tsay et al. [22]).

Chicks were treated with estradiol and then withdrawn from hormone. At time zero diethylstilbestrol was administered. At the *time points indicated* the oviducts were assayed. *Ordinate:* ○- - -○, level of receptors in the nuclei (arbitary units); ●——●, number of initiation sites on chromatin for RNA polymerase (100 = 30/ng DNA); △——△, number of $RNA_{ov}$ sequences per tubular gland cell (100 = $3 \times 10^3$).

Chromatin isolated from chicks that had been stimulated with estrogen for 14 days and then withdrawn from hormone showed an increase of the level of initiation sites as early as 30 min after restimulation (Fig. 7.6). Thus the increase of initiation sites is one of the earliest responses which can be measured after steroid administration. After about 1 h values of RNA polymerase initiation sites are about three fold higher compared with the values before the second estrogen stimulation. After addition of purified progesterone-receptor complex to chromatin of non-stimulated chick oviduct nuclei, the number of new initiation sites was approximately equal to the number of hormone-receptor complexes bound to the chromatin. This fact indicates that the initiation reaction involves one hormone-receptor complex per chromatin acceptor site.

A drawback to these experiments is the use of prokaryotic RNA polymerase which may recognize initiation sites other than the eukaryotic RNA polymerase present in the oviduct nuclei themselves. Further experiments with eukaryotic enzyme preparations, which became available recently, should therefore be performed.

### 7.4.4 Influence of Estrogen on the Transcription of the Ovalbumin Gene

The initiation sites made accessible to RNA polymerase after binding of the estrogen receptor complex include the initiation site of the ovalbumin gene. Transcription of this gene has been studied in nuclei of estrogen-stimulated chick oviduct. Oviduct nuclei are able to perform DNA-dependent RNA synthesis under the conditions described in Table 7.2. The rate of synthesis of ovalbumin mRNA sequences, including species with higher molecular weight containing the intervening sequences of the ovalbumin gene (cf. 7.3.2), varies markedly, depending on the hormonal state of the chick from which the nuclei were isolated. In oviduct nuclei isolated from chronically stimulated chicks about 0.23% of the RNA synthesized in vitro contains oval-

**Table 7.2**: Synthesis of nuclear ovalbumin-specific RNA sequences in different tissues (from Swaneck et al. [27]).

RNA was transcribed in isolated nuclei in a reaction mixture that contained ATP, GTP and UTP and [$^3$H]-CTP (35 Ci/mmol), $MgCl_2$, $MnCl_2$, EDTA, KCl, dithioerythrol, glycerol. Tris-buffer *p*H 7.8 and 0.8–2.2 · $10^8$ nuclei per ml. After 30 min at 37 °C the [$^3$H]-RNA was isolated. Ovalbumin structural and intervening sequence transcripts were measured by hybridizing to nitrocellulose filters containing cloned ovalbumin DNA (cf. Sect. **2.6**) in presence or absence of competing unlabelled ovalbumin mRNA

| Tissue | Nuclear ovalbumin-specific RNA in % of total RNA synthesized |
|---|---|
| *Oviduct* | |
| Estrogen stimulated | 0.234 |
| + α-Amanitin (0.5 μg/ml) | 0.002 |
| + Actinomycin D (40 μg/ml) | 0.234 |
| Withdrawn from estrogen | |
| 3 days | 0 |
| 14 days | 0 |
| *Spleen* | |
| Estrogen stimulated | 0 |
| *Liver* | |
| Estrogen stimulated | 0 |

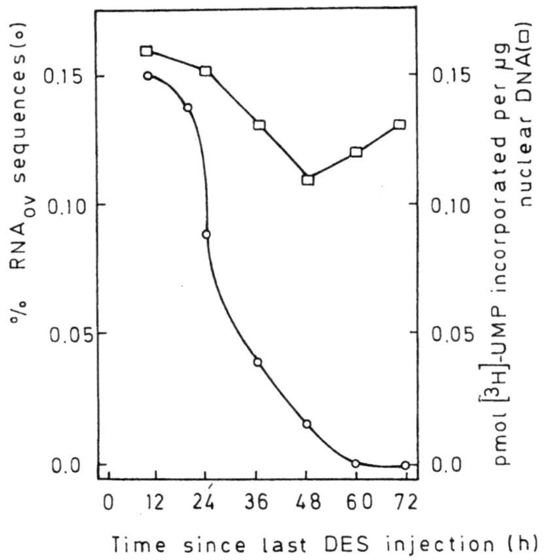


Fig. 7.7

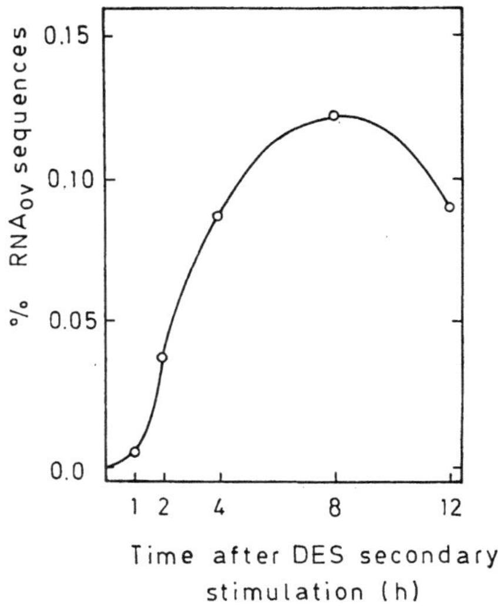


Fig. 7.8

**Fig. 7.7**: Time course of the decrease of synthesis of nuclear ovalbumin-specific RNA sequences by nuclei of chick oviduct after withdrawal of estrogen (from Swaneck et al. [27]).

Experimental details of the determination of ovalbumin-specific RNA sequences see Table **7.2**; *DES*, Diethylstilbestrol.

**Fig. 7.8**: Time course of recovery of synthesis of nuclear ovalbumin-specific RNA sequences after estrogen administration to estrogen-withdrawn chicks (from Swaneck et al. [27]).

For secondary stimulation chicks received an injection of 2.5 ng diethylstilbestrol *(DES)*.

bumin mRNA sequences. Withdrawal of hormone causes this value to decrease so that after 60 h of withdrawal no transcription of the ovalbumin gene could be detected in vitro (Fig. 7.7). Readministration of a single dose of estrogen caused transcription of the ovalbumin gene to begin again (Fig. 7.8). Significant accumulation of ovalbumin mRNA occurs after about 2 h and continues to increase for at least 8 h. These values resemble the accumulation of ovalbumin mRNA in vivo (Fig. 7.6). In spleen or liver of estrogen-stimulated chicks no ovalbumin mRNA synthesis could be detected (Table 7.2).

These results support the hypothesis that binding of the estrogen-receptor complex, in chick oviduct cells, induces the accumulation of ovalbumin mRNA and its precursors primarily by increasing its rate of transcription. Analysis of the size of the ovalbumin-specific RNA species by denaturating gel electrophoresis revealed that the withdrawal of hormones depletes the level of high molecular weight precursors and that readministration of estrogen reinduces their occurrence.

### 7.4.5 Role of Non-Histone Proteins of Chromatin in Transcription of Ovalbumin-specific RNA

To test which fraction of target cell chromatin houses the acceptor protein binding the estrogen-receptor complex chromatin was separated into three components: the histones, which are comparatively uniform in all living beings (cf. 1.2.3.2), the highly heterogenous group of non-histone proteins (cf. Table 1.4) and the DNA still containing about 20% of tightly bound non-histone proteins. Homologous and heterologous chromatins can be reconstituted from these components by slow dialysis from 6 M urea and 2 M NaCl into 0.2 M NaCl.

mRNA from native and reconstituted chromatin has been transcribed in vitro by *E. coli* RNA polymerase and specific RNA sequences present in the transcripts, even in minute quantities, have been assayed after hybridization with corresponding cDNA by the following technique (Harris et al., J. Biol. Chem. **251** (1976): 524):

Radioactive labelled deoxyribonucleotide triphosphates and a sample of pure mRNA are incubated with RNA-dependent DNA-polymerase (reverse transcriptase) coded by RNA tumor viruses (cf. Fig. 19.7). During the incubation highly radioactive DNA is synthesized, which is the complement of the used mRNA. This cDNA forms stable double-stranded hybrid molecules with the mRNA species used as the template. In contrast to the single-stranded free mRNA these hybrids are resistent to certain nucleases. The amount of RNA sequences corresponding to the mRNA used as matrix may therefore be determined in RNA preparations after complexation with the cDNA and subsequent nuclease digestion. The amount of the cDNA-RNA complex is calculated from the radioactivity in the non-hydrolyzed, high molecular weight fraction, which may be separated from the low molecular weight products of hydrolysis by ultrafiltration.

For demonstration that the bacterial RNA polymerase used mimics the specific action of chick oviduct RNA polymerase and does not transcribe any "available" DNA sequence in the chromatin in a random fashion, the asymmetry of transcription has been tested (Table 7.3).

At high enzyme-to-DNA ratios (4 μg enzyme/μg DNA) sequences homologous to both coding and non-coding strands of the ovalbumin genes were transcribed. As the ratio of enzyme to DNA was lowered, however, the amount of ovalbumin mRNA increased, i.e., the transcription became increasingly asymmetric. At the lowest enzyme to DNA ratio tested (0.25 μg enzyme/μg DNA) more than 96% of the ovalbumin gene transcripts were synthesized from the coding strand of the gene.

This asymmetry of ovalbumin mRNA transcription was also observed with a eukaryotic RNA polymerase from wheat germ. It implies that at low enzyme/DNA

**Table 7.3:** Asymmetry of the in vitro transcription of ovalbumin gene (from Towle et al. [25])

| Enzyme/ chromatin ratio $\mu g/\mu g$ DNA | % Ovalbumin-specific RNA$\times 10^2$ in the transcripts | % anti-Ovalbumin-specific RNA$\times 10^2$ in the transcripts |
|---|---|---|
| 4.0 | 0.65 | 0.15 |
| 1.0 | 1.9 | 0.15 |
| 0.25 | 4.1 | 0.15 |

ratios only the stronger initiation sites are starting points for RNA synthesis and that the initiation sites for ovalbumin-specific RNA are much stronger than those resulting in transcription of anti-ovalbumin specific RNA. In this regard it is worth noticing that even at the lowest enzyme to chromatin ratio utilized in the test, there is still 20–30 times more polymerase present than in vivo. Hence it may be expected, that further decrease of the enzyme/chromatin ratio may result in the formation of ovalbumin-specific RNA transcripts free of anti-ovalbumin-specific RNA.

The chromatin reconstitution technique makes possible the formation of hybrid chromatins from components obtained from nuclei of different (e.g., target and

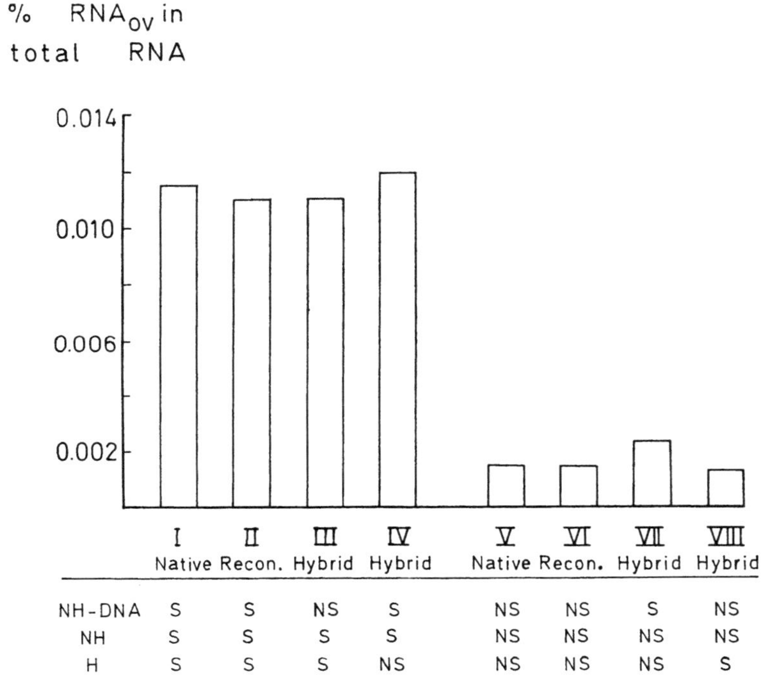


**Fig. 7.9:** Synthesis of ovalbumin-specific RNA sequences from native and reconstituted chromatins (after Tsay and O'Malley [23]).

Extractable non-histone proteins *(NH)*, histones *(H)* and the non-histone protein-DNA complex *(NH–DBA)* were isolated from chromatin of oviduct either after estrogen stimulation of chicks *(S)* or from estrogen-stimulated animals then withdrawn from the hormone *(NS)*. Chromatins were reconstituted from the indicated components by gradient dialysis. Ovalbumin--specific RNA transcripts were synthesized in vitro using *E. coli* RNA polymerase. They were determined by hybridization with cDNA$_{ov}$.

non-target) cells. Hence it allows detailed studies on the function of different chromatin constituents on the regulation of transcription. With chick oviduct chromatin, substantial differences in the concentration of ovalbumin-specific RNA sequences have been detected in the in vitro transcripts of estrogen-stimulated chromatin and chromatin from the oviduct of estrogen-withdrawn animals (Fig. **7**.9, columns I and V, II and VI). These differences are quantitatively identical with native and reconstituted chromatin, indicating high fidelity in chromatin reconstitution. Furthermore, the number of initiation sites for RNA synthesis on reconstituted and native chromatins was indistinguishable. Hence also the specificity of transcription of isolated chromatin appeared to be conserved upon dissociation and fractionation of its proteins followed by the reconstitution from its constituents.

Chromatin reconstituted from non-histone proteins from estrogen-stimulated chromatin and histones or tightly bound non-histone-DNA complexes from hormone-withdrawn chromatin (Fig. **7**.9, columns III and IV) resulted in the synthesis of a substantial amount of ovalbumin-specific RNA sequences. Conversely, when extractable non-histone proteins from estrogen-withdrawn chromatin were reconstituted together with histones and tightly bound non-histone-DNA complexes from estrogen-stimulated chromatins, very low levels of ovalbumin-specific RNA sequences were detected (Fig. **7**.9, columns VII and VIII). In contrast, interchange of the histones and tightly bound non-histone-DNA complexes from estrogen-withdrawn and hormone-stimulated chromatins during reconstitution did not affect the level of ovalbumin-specific RNA sequences produced. Therefore the extractable non-histone proteins of chromatin appear to play a dominant role in regulating specific gene expression.

If extractable non-histone proteins were isolated from either withdrawn or stimulated chromatin and reconstituted together with the full complement of their respective homologous components, the extractable non-histone proteins from stimulated chromatin were capable of activating the in vitro transcription of the ovalbumin gene when included in the reconstitution of withdrawn chromatin. On the other hand, addition of extractable non-histone proteins from withdrawn chromatin to the components of stimulated chromatin did not affect the synthesis of ovalbumin-specific RNA sequences. Therefore, the extractable non-histone proteins from stimulated chromatin appear to contain a positive regulator able to interact with chromatin of withdrawn chick oviduct in vitro, thus causing expression of the ovalbumin gene. The nature of this regulator protein(s), however, is not known as yet.

Also the mechanistic functions of this and other chromosomal proteins in relation to selective expression of the ovalbumin gene remain virtually unknown. The most characteristic physical difference between chromatin and DNA was the finding that formation of the binary complex between RNA polymerase and chromatin (cf. **7**.4.3) was practically independent of temperature in contrast to complex formation with DNA. Since activation energy is thought to be involved in the opening of DNA strands during stable binary complex formation, DNA in chromatin may be present in a state in which conversion to the stable preinitiation complex is greatly facilitated. One possible explanation is that the initiation regions of chromatin are stretches of single-stranded DNA. Alternatively, chromosomal proteins present at initiation sites may lower the activation energy required for opening of the DNA strands in forming a stable binary complex. Such unwinding proteins (e.g., HMG1 and HMG2) would be expected to assist in the location of initiation sites for RNAP (cf. Fig. **4**.17). Competition experiments with *E. coli* RNA polymerase and hen oviduct RNA polymerase II indicate that both enzymes use the same regions for initiation of RNA synthesis on chromatin. This contrasts to differences in template specificity for both enzymes observed with naked DNA. Hence it seems likely that the proteins present in chromatin exert a dominant influence on the selection of initiation sites on chromatin of both eukaryotic and prokaryotic polymerases.

Lately, the chromatin reconstitution experiments have been criticized heavily (cf. [26]). It has been demonstrated that the older experiments cannot be interpreted because endogenous RNA sequences contaminating the chromatin are indistinguishable from newly synthesized transcripts by the usual hybridization assays. Recently therefore mercury-labelled nucleotides were used to mark the in vitro transcripts, which in this case may be separated from endogenous RNA by chromatography on SH-agarose. But new transcripts and endogenous RNA tend to aggregate, a process causing significant contamination to persist even after SH-agarose chromatography without use of appropriate disaggregation procedures. Furthermore endogenous *RNA* can act as a template for *E. coli* RNA polymerase and hybridizes to the resulting Hg-substituted anti-sense sequences. These hybrids may contain "full sequence" endogenous RNA, hybridized to short anti-sense transcripts. Hence isolated Hg-RNA may include such stable complexes in which the non-hybridized portion of the endogenous RNA is available to hybridize to the cDNA probe. Therefore estimates of specific transcripts may be too high.

With transcription of globin sequences in chicken erythroid chromatin after elimination of all the described artefacts, a small but consistent level of specific transcription has been found which is at least four-fold higher than from brain chromatin. Dissociation of chromatin and reconstitution by gradient dialysis, however, did *not* preserve transcriptional specificity [26]. A similar, igorous proof of the validity of the reconstitution experiments with respect to the ovalbumin equences should also be undertaken.

## 7.5 Conclusions

The results discussed in Sect. 7.4 lead one to propose the following mechanism for the action of estradiol on the induction of ovalbumin biosynthesis in chick oviduct, which may be similar also in the case of other steroid hormones in other target tissues:

Free estradiol penetrates the plasma membrane of any cells of the chick and binds to receptor proteins present in the target cells. The estradiol-receptor complexes formed attach to acceptor proteins which are constituents of chromatin. The level of nuclear-bound receptors reaches saturation about 20 min after hormone administration (Fig. 7.8). Attachment of the estrogen-receptor complex to target cell nuclei either causes rearrangement of the chromatin proteins or changes locally the physical conformation of chromatin. In this way previously blocked chromatin domains become accessible to RNA polymerase, which on the one hand causes maturation of the chick oviduct, i.e., formation of the specialized cells of oviduct mucosa and on the other hand exerts transcriptional control. The number of RNA initiation sites increases with a maximum about 60 min after application of the hormone.

Transcription of specific RNA sequences, e.g., ovalbumin-specific sequences, starts, whose accumulation becomes measurable after about 60 min in nuclei and cytoplasm. In the nucleus precursors of mature ovalbumin mRNA are detectable which contain transcripts of the coding and the intervening sequences of the ovalbumin gene. The latter are removed by splicing during maturation of ovalbumin mRNA. Translation of ovalbumin mRNA in the cytoplasm takes about 8.5 min for completion. The growing peptide chain is translocated (co-translationally?, cf. 7.3.3) through the membrane of the rough ER. There is no pre-ovalbumin, because ovalbumin contains a stable signal sequence in the centre of the molecule. The ovalbumin formed is located in vesicles which fuse with the plasma membrane and release the proteins to the extracellular space, the oviduct. Translation of ovalbumin mRNA and the translocation of ovalbumin exhibit no special features of regulation. The formation of ovalbumin is therefore almost completely controlled at the level of transcription.

## 7.6 Selected References

*Synthesis and Metabolism of Steroid Hormones*

1. Träger, L.: Steroidhormone, Biosynthese, Stoffwechsel, Wirkung. Springer-Verlag, Berlin 1977.

*Gene Structure*

2. Dugaiczyk, A., S. L. C. Woo, D. A. Colbert, E. C. Lai, M. L. Mace & B. W. O'Malley: The ovalbumin gene: Cloning and molecular organization of the entire natural gene. Proc. Natl. Acad. Sci. USA *76* (1979): 2253–2257.
3. Royal, A., A. Garapin, B. Cami, F. Perrin, J. L. Mandel, M. LeMeur, F. Brégégègre, F. Gannon, J. P. LePennec, P. Chambon & P. Kourilsky: The ovalbumin gene region: common features in the organisation of three genes expressed in chicken oviduct under hormonal control. Nature (London) *279* (1979): 125–132.
4. Lai, E. C., J. P. Stein, J. F. Catterall, S. L. C. Woo, M. L. Mace, A. R. Means & B. W. O'Malley: Molecular structure and flanking nucleotide sequence of the natural chicken ovomucoid gene. Cell *18* (1979): 829–842.

4a. Tsai, S. Y., M. J. Tsai & B. W. O'Malley: Specific 5′ flanking sequences are required for faithful initiation of in vitro transcription of the ovalbumin gene. Proc. Natl. Acad. Sci. USA *78* (1981): 879–883.

5. Cochet, M., F. Gannon, R. Hen, L. Maroteaux, F. Perrin & P. Chambon: Organisation and sequence studies of the 17-piece chicken conalbumin gene. Nature (London) *282* (1979): 567–574.
6. Lindenmaier, W., M. C. Nguyen-Huu, R. Lurz, M. Stratmann, N. Blin, T. Wurtz, H. J. Hauser, A. E. Sippel & G. Schütz: Arrangement of coding and intervening sequences of chicken lysozyme gene. Proc. Natl. Acad. Sci. USA *76* (1979): 6196–6200.
7. Mandel, J. L. & P. Chambon: DNA methylation: organ-specific variations in the methylation pattern within and round ovalbumin and other chicken genes. Nucleic Acids Research *7* (1979): 2081–2103.

7a. Heilig, R., F. Perrin, F. Gannon, J. L. Mandel & P. Chambon: The ovalbumin gene family: Structure of the S gene and evolution of duplicated split genes. Cell *20* (1980): 625–637.

*Ovalbumin-Specific RNA Transcripts*

8. Woo, S. L. C. & B. W. O'Malley: Hormone inducible messenger RNA. Life Sci. *17* (1975): 1039–1047.
9. McReynolds, L., B. W. O'Malley, A. D. Nisbet, J. E. Fothergill, D. Givol, S. Fields, M. Robertson & G. G. Brownlee: Sequence of chicken ovalbumin mRNA. Nature (London) *273* (1978): 723–728.
10. Catterall, J. F., B. W. O'Malley, M. A. Robertson, R. Staden, Y. Tanaka & G. G. Brownlee: Nucleotide sequence homology at 13 intron-exon junctions in the chick ovalbumin gene. Nature (London) *275* (1978): 510–513.
11. Roop, D. R., M. J. Tsai & B. W. O'Malley: Definition of the 5′ and 3′ ends of transcripts of the ovalbumin gene. Cell *19* (1980): 63–68.

*Ovalbumin Translocation*

12. Lingappa, V. R., J. R. Lingappa & G. Blobel: Chicken ovalbumin contains an internal signal sequence. Nature (London) *281* (1979): 117–121.

12a. Palmiter, R. D., S. N. Thibodeau, G. Rogers & I. Boime: Cotranslational sequestration of egg white proteins and placental lactogen inside membrane vesicles. Annals of the New York Academy of Sciences, Vol. *343* (1980): 192–209.

*Hormone Binding and Receptors*

13. Williams, D. L.: The estrogen receptor: a minireview. Life Sci. *15* (1974): 583–597.
14. Sheridan, P. J.: Is there an alternative to the cytoplasmic receptor model for the mechanism of action of steroids? Life Sci. *17* (1975): 497–502.

15. O'Malley, B. W. & W. T. Schrader: The receptors of steroid hormones. Sci. American *234* (1976): 32–43.
16. Kyrein, H. J.: Steroidhormonrezeptoren. Untersuchung an uterinen cytoplasmatischen Östrogenrezeptoren vom Kalb. Dtsch. Apotheker-Ztg. *116* (1976): 359–366.
17. O'Malley, B. W. & L. Birnbaumer (eds.): Receptors and Hormone Action I. Academic Press, New York 1977.
18. Popp, R. A., K. R. Foresman, L. D. Wise & J. C. Daniel: Amino acid sequence of a progesterone-binding protein. Proc. Natl. Acad. Sci. USA *75* (1978): 5516–5519.
19. Brachet, J.: The hormonal induction of maturation in amphibian oocytes. Med. Biol. *56* (1978): 304–309.
20. Baulieu, E.-E., F. Godeau, M. Schorderet & S. Schorderet-Slatkine: Steroid-induced meiotic division in *Xenopus laevis* oocytes: surface and calcium. Nature (London) *275* (1978): 593–598.
21. Sheridan, P. J., J. M. Buchanan, V. C. Anselmo & P. M. Martin: Equlibrium: the intracellular distribution of steroid receptors. Nature (London) *282* (1979): 579–582.

*Regulation of Transcription*

22. Tsai, S. Y., M. J. Tsai, R. Schwartz, M. Kalimi, J. H. Clark & B. W. O'Malley: Effects of estrogen on gene expression in chick oviduct: Nuclear receptor levels and initiation of transcription. Proc. Natl. Acad. Sci. USA *72* (1975): 4228–4232.
23. Tsai, M. J. & B. W. O'Malley: Effects of estrogens on gene expression in chick oviduct. In: Nover, L. & K. Mothes (eds.). Cell Differentiation in Microorganisms, Plants and Animals. G. Fischer Verlag, Jena and North-Holland Publ. Co, Amsterdam 1977, pp. 109–125.
24. O'Malley, B. W., H. C. Towle & R. J. Schwartz: Regulation of gene expression in eucaryotes. Ann. Rev. Genet. *11* (1977): 239–275.
25. Towle, H. C., M. J. Tsai, S. Y. Tsai & B. W. O'Malley: Effect of estrogen on gene expression in the chick oviduct. J. Biol. Chem. *252* (1977): 2396–2404.
26. Konkel, D. A. & V. M. Ingram: Is there specific transcription from isolated chromatin? Nucleic Acids Res. *5* (1978): 1237–1252.
27. Swaneck, G. E., J. L. Nordstrom, F. Kreuzaler, M. J. Tsai & B. W. O'Malley: Effect of estrogen on gene expression in chicken oviduct: Evidence for transcriptional control of ovalbumin gene. Proc. Natl. Acad. Sci. USA *76* (1979): 1049–1053.
28. Wickens, M. P., S. Woo, B. W. O'Malley & J. B. Gurdon: Expression of chicken chromosomal ovalbumin gene injected into frog oocyte nuclei. Nature (London) *285* (1980): 628–634.
29. Evans, M. I., L. J. Hager & G. S. McKnight: A somatomedin-like peptide hormone is required during the estrogen-mediated induction of ovalbumin gene transcription. Cell *25* (1981): 187–193.
30. Kallos, J., T. M. Fasy & V. P. Hollander: Assessment of estrogen receptor-histone interactions. Proc. Natl. Acad. Sci. USA *78* (1981): 2874–2878.

# 8. Insulin-Biosynthesis and Mode of Action

H. Zühlke

## 8.1 Introduction: Scientific History of Insulin

Although in ancient times *Diabetes mellitus* was already recognized as a dangerous disease, it was not until the end of the 19th century that the pancreas was shown to be intimately connected with the illness. This was brilliantly demonstrated by v. Mering and Minkowski, who elicited the symptoms of diabetes in dogs by extirpation of the pancreas. The search for the "active substance" had gone on for over 20 years when Banting and Best succeeded in preparing a pancreatic extract that lowered the blood sugar level in dogs and humans.

Another 30 years were to elapse before details of the chemical nature of insulin became known. The fact that it could be destroyed by proteolytic enzymes showed that it must be a protein. In 1926 Abel succeeded in crystallizing insulin, and in fact it was the very first protein ever to be crystallized. Sanger, in about 1955, provided protein chemistry with a sound foundation by determining the complete primary structure of insulin, and at the same time contributing substantially to the development of molecular biology. Insulin served as a model for the complete chemical synthesis of a protein by three entirely independent groups of workers in 1965 (Zahn, Federal Rep. of Germany: Kung, People's Rep. of China: Katsoyannis, USA). The principle involved was the separate synthesis of the A- and B-chains of insulin and their subsequent coupling (with a yield of something like 1%–10%). The discovery of proinsulin by Steiner in 1967 revealed that the biological synthesis of insulin proceeds via the production of a single chain, followed by formation of the disulphide bonds, and subsequent limited proteolysis. This discovery should open up new possibilities for the chemical synthesis of insulin. The demonstration of pre-proinsulin in a cell-free system confirmed a hypothesis already postulated by Blobel and Sabatini, and more recent studies have thrown light on the structure and expression of the insulin gene (cf. Table 8.1).

**Table 8.1**: Scientific history of insulin

| | |
|---|---|
| 1921 | Insulin discovered by Banting and Best |
| 1926 | Crystallization |
| 1956 | Elucidation of primary structure |
| 1963 | Chemical synthesis |
| 1967 | Discovery of proinsulin |
| 1969 | Three-dimensional structure of insulin determined |
| 1976 | Discovery of pre-proinsulin |
| 1977 | Structure of the insulin gene determined |
| 1978 | Expression of cloned insulin gene in bacteria |

Abbreviations: cDNA, DNA complementary to mRNA, synthesized by reverse transcriptase; CNS, central nervous system; Eco RI, restriction endonuclease of *E. coli* (cf. Table 2.1); GA, Golgi apparatus; rER, rough endoplasmic reticulum; SG, secretory granules.

Insulin is a proteohormone consisting of two peptide chains (A-chain, 21 amino acids residues; B-chain, 30 amino acid residues) connected by two disulphide bonds (Fig. 8.1). Differences in the primary structure of the various mammalian insulins are limited to a few sites. Far greater differences have been found, for example, between the insulin of humans and hag fish (Fig. 8.2), where about half of the amino acids are different. This is almost certainly due to the fact that evolution of the hag fish branched off from the main line of evolution of vertebrates about 500 million years ago. But despite such differences in primary structure, X-ray analyses have shown that the three-dimensional structure of the molecule is the same as that of mammalian insulin. Insulin not only plays a decisive role in diabetes research and

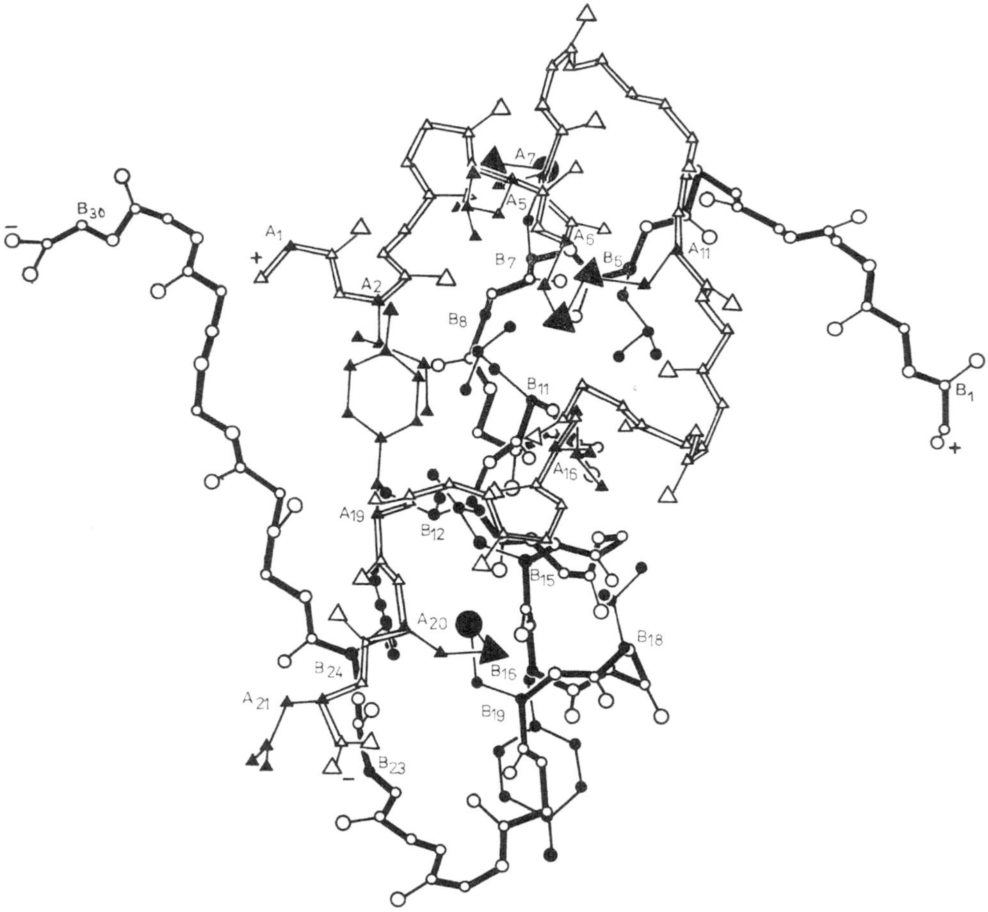


**Fig. 8.1:** Three-dimensional model of porcine insulin monomer seen perpendicular to the three-fold crystallographic axis (from Steiner, D. F. & N. Freinkel (eds.): Handbook of Physiology, Sect. 7: Endocrinology, Vol. I, Endocrine Pancreas. Amer. Physiol. Soc., Washington 1972).

The model shows the polypeptide backbones of A (═══) and B (━━━) chains, as well as side chains of known invariant amino acid residues (*closed circles* and *triangles*). ▲, ●, $\alpha$-carbons and side-chain atoms of invariant residues, also marked with numbers corresponding to their position within the chains; △, ○, $\alpha$-carbons and side-chain atoms of variant residues; *large triangles* and *circles* denote sulphur atoms of the three disulphide bonds.

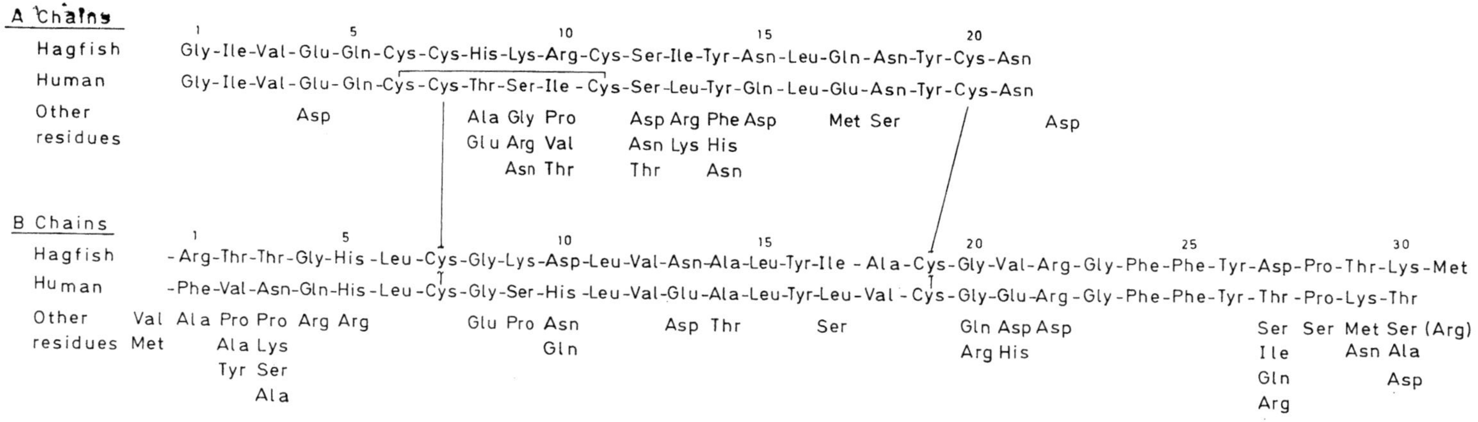


**Fig. 8.2:** Comparison of the primary structure of insulin from man and hagfish, with substitutions at each position in other known insulins. Hag fishes belong to the most primitive vertebrates, the cyclostomes.

therapy, but it has also contributed substantially to the development of a number of branches of science including endocrinology, cell biology, molecular biology, protein chemistry and crystallography. In this chapter certain aspects of insulin research to-day, and in particular mechanisms of its biosynthesis and mode of action will be discussed.

## 8.2 Molecular Pathology of Diabetes mellitus

The discovery of insulin marked the beginning of extensive research in the field of diabetology, with increasing emphasis upon experimental studies in recent years. Nevertheless, there is still no satisfactory explanation of the causes of the disease.

*Diabetes mellitus* is characterized by an absolute or relative insulin deficiency. It is a complex condition involving imbalance in endocrine regulation and a metabolic aberration that can manifest itself in a variety of forms and in varying degrees of severity. Endogenous as well as exogenous factors acting singly or in combination may be involved in the genesis of the disease. The endogenous factors in this case are chiefly hormones, such as growth hormone, glucagon, glucocorticoids, catechol amines and others, as well as changes in the overall hormone situation within the organism as e.g., during pregnancy, fetal development or ageing. A number of exogenous factors such as nutrition, muscular inactivity, stress or infectious diseases are also known to lead to the manifestation of diabetes.

**Table 8.2**: Summary and classification of factors shown or supposed to contribute to the *Diabetes mellitus* syndrome

| Parts of the hormone system affected | Molecular pathology |
|---|---|
| Cell population of the islets of Langerhans (cf. 8.3) | – Decreased rate of B-cell proliferation<br>– B-cells destroyed by virus infection and/or autoimmune reactions |
| Signal transformation in the islet cells (cf. 8.5) | – Defective response of the putative "glucose receptor" in the B-cell membrane<br>– Defective mechanism of signal transformation, e.g., second messenger synthesis<br>– Alterations in the intercellular cooperation of islet cells |
| Insulin biosynthesis (cf. 8.4) | – Mutation of the insulin structural gene, synthesis of defective insulin<br>– Regulatory alterations of insulin gene expression<br>– Metabolic deficiencies of B-cells |
| Insulin secretion (cf. 8.4.6) | – Defective regulation or mechanism of insulin secretion from secretory granules |
| Insulin circulation (cf. 8.6.1) | – Inactivation by binding to cell or serum proteins, insulin autoantibodies<br>– Increased degradation of insulin |
| Insulin response (cf. 8.6) | – Reduced number of or defective insulin receptors (adipositas)<br>– Defective signal transformation and internalization<br>– Hormonal insulin antagonists |

Epidemiological studies have shown that for any one of the above-mentioned factors the subject must possess a genetic predisposition. It is nowadays widely accepted that pathogenesis and phenomenology of *Diabetes mellitus* are dependent on a multigene system including several recessive genes with additive action.

Certain heritable diseases are caused by mutation of a single gene. For example the absence of a particular enzyme can cause a monogenetic metabolic disturbance (enzymopathy). The demonstration of deficiency of the enzyme involved is an important marker in detecting the molecular basis of the pathological alteration. But on account of the polygenic inheritance of *Diabetes mellitus* it is only possible to detect the typical symptoms, i.e., elevated glucose levels in the blood or other reliable clinical parameters, but these give no direct clue as to the underlying molecular biological disturbance. We still have no well-defined and clinically applicable marker for the detection of the early stages of diabetes.

As already mentioned, a deficiency of insulin can be demonstrated after diabetes has become manifest, and this may be due to a variety of factors, of which the most important will now be discussed (Table **8**.2). The list of the possible causes of *Diabetes mellitus* clearly shows up the paramount importance of biochemistry and molecular biology in the etiopathogenetic study of the disease.

## 8.3 Development of the Endocrine Pancreas

### 8.3.1 Cell Types in the Islets of Langerhans

A constant glucose level in the blood serum is primarily maintained by the activity of the islets of Langerhans. Changes in the total amount of the islet cells, whether due to loss or proliferation, may lead to alterations in hormone production and thus result in the development of a state of hypoglycemia or diabetes.

The islet organ consists of a variety of cell types with different functions. Despite intensive studies in recent years, opinion as to the existence of certain cell types is still divided. But the information contained in Table **8**.3 can be regarded as established and generally accepted.

**Table 8.3:** Cell types of the islets of Langerhans

| Cell type | Part of total cell mass (%) | Product | Localisation in the human islet (cf. Fig. **8**.9)[a] |
|---|---|---|---|
| B-cells | 60–80 | Insulin | Core |
| A-cells | 10–30 | Glucagon (cf. **8**.5) | Periphery |
| D-cells | < 10 | Somatostatin (cf. **8**.5) | Periphery |
| F-cells (PP) | 1 | Pancreatic polypeptide | ? |

[a] Localization of different types of islet cells is the same in most other mammals.

### 8.3.2 Differentiation of the Islet Cells

In the course of ontogenesis the islet organ derives directly from the pancreatic entoderm, the endocrine and exocrine cells thus arising from common stem cells.

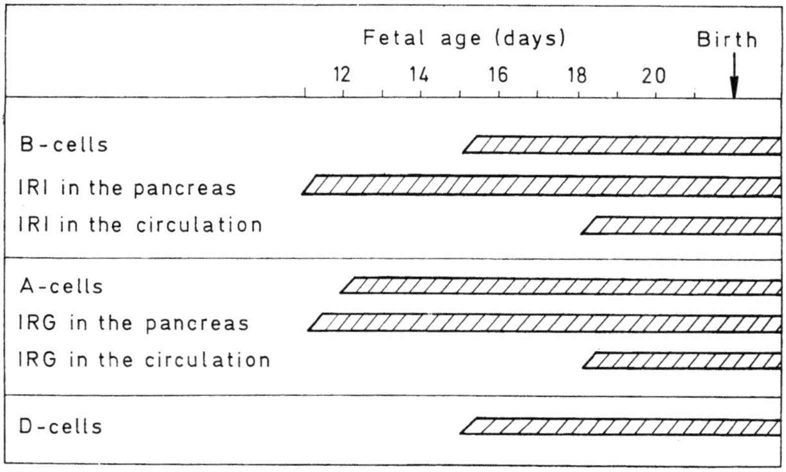


**Fig. 8.3:** Developmental pattern of the pancreas cell types and hormone secretion in rats (from Hellerström et al. [7], modified).

The cell types were identified by immunohistochemical methods. Immunoreactive insulin *(IRI)* and immunoreactive glucagon *(IRG)* in pancreas homogenates and in blood were detected by means of radioimmunoassay. In any case antibodies were used that react specifically with insulin, glucagon and somatostatin respectively, but also with corresponding polypeptide precursor molecules.

However, some authors assume the islet cells to derive from the ectoderm and to migrate into the primitive pancreas during an early phase of development. This concept is supported by the fact that islet cells, like other endocrine cells, are also capable of accumulating and decarboxylating biogenic amines, as, e.g., in cells derived from the neuroectoderm. Furthermore, somatostatin-, insulin- and glucagon-like activities are demonstrable in the central nervous system. The *phylogenetic* sequence in which the four pancreatic hormones make their appearance in the islet parenchyma is: (1) insulin, (2) somatostatin, (3) glucagon, (4) pancreatic polypeptide. The *ontogenetic* sequence of development and hormone secretion in the islet cells of the rat is schematically shown in Fig. 8.3. Similar results have been reported for other mammals.

As can be seen from Fig. 8.3, insulin and glucagon can be detected by immunological means in the fetal pancreas before the corresponding islet cells are morphologically recognizable. In the early stages of development of the pancreas, glucagon is present in higher concentrations than insulin but on the day before birth the situation is reversed.

### 8.3.3 Proliferation and Neogenesis of Islet Cells

A large increase in the total amount of B-cells takes place postnatally. This may be due to enlargement of the individual cells and/or to the formation of new cells. The low rate of mitosis was responsible for the earlier assumption that proliferation of B-cells is primarily the result of transformation of other cell types from the exocrine pancreas via mixed cells. Autoradiography and morphological studies have now revealed that B-cells do in fact undergo mitoses.

New-formation of islet cells was studied by incorporating [$^3$H]-thymidine into rat and mouse fetuses of different ages. The maximum incorporation (20 %) took place on the 17th intrauterine day. Immediately prior to birth the rate of labelling of B-cells was only 3.2 % and had decreased to 0.5 % in 48-day-old rats. In general, the rate of B-cell formation in adult animals is low, with only 5–20 out of 1,000 B-cells undergoing division per 24 h. But even this low rate is sufficient to lead to an increase in total amount of B-cells, provided there is only slight degeneration. Proliferation of B-cells is governed in vivo by a variety of factors including composition of the diet, various hormones etc. A carbohydrate-rich diet leads to a rise in total B-cell bulk. A rise in mitotic rate can also be observed in cases of liver damage or following administration of cortisone or growth hormone, as well as in pregnancy.

If cultures of islets of Langerhans are incubated with agents known to stimulate the synthesis and secretion of insulin, such as glucose, tolbutamide and leucine, the results are not in every case identical. Only glucose stimulates incorporation of [$^3$H]-thymidine in the DNA of the islet cells. This indicates that the intracellular signals for the replication of the B-cells are apparently different from those for making insulin available.

In insulin deficiency, the insulin level can be normalized either by an increased secretion (early response) or by adaptation via increase of the total amount of B-cells (late response). A disturbance of either one of these mechanisms could lead to symptoms of diabetes. If hyperglycaemia is produced in mice by injection of insulin antibodies over several days, [$^3$H]-thymidine incorporation into B-cell DNA is still unaltered at the end of the first day, reaches a maximum on the third day and then falls rapidly, despite severe hyperglycaemia. From these and similar results it can be deduced that the B-cell genome is programmed for a limited number of mitotic cycles. In experimental diabetes the A- and D-cells appear to proliferate. The fact that the D-cells react in the same way in diabetes elicited by alloxan or streptocotocin (reduction in the number of B-cells) and in steroid diabetes (proliferation of B-cells) indicates that D-cell proliferation is independent of insulin content.

Some data concerning the replication of human B-cells are available: for example, B-cell hyperplasia has been observed in obesity and in prediabetes, whereas only very little regenerative activity could be detected in insulin-dependent diabetes. The loss of B-cells in the latter case may be due, e.g., to virus and/or autoimmune action, and the only conceivable way of compensating for such a loss would be the production of new B-cells.

## 8.4 Mechanism of Insulin Biosynthesis

### 8.4.1 Intracellular Organization of Insulin Biosynthesis

Fig. 8.4 illustrates the cellular organization of insulin biosynthesis. Disregarding the generation of insulin mRNA by transcription and processing of the pre-mRNA (for the structure of insulin genes cf. Fig. 8.5), it shows the translation of insulin mRNP on the membranes of the rough ER, as well as secretion of the mature hormone to the extracellular milieu.

Insulin is synthesized as a precursor hormone (pre-proinsulin) with a signal peptide at the N-terminus (cf. Table 8.4) and an internal part of the polypeptide chain (C-peptide) that connects the B- and A-chains. Translation is followed by a stepwise processing and is accompanied by directed transport of the hormone to the Golgi apparatus where the secretory granules are formed. Processing includes the endoproteolytic removal of the signal peptide, generation of the interchain and intrachain disulphide bridges (rER) and the endo- and exoproteolytic removal of the C-peptide (GA and early secretory vesicles). Insulin secretion from the storage granules is

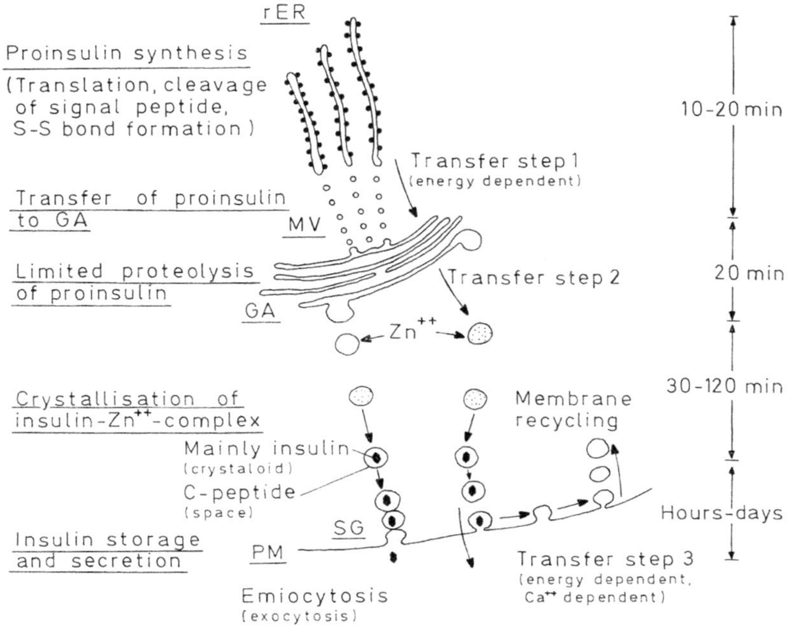


**Fig. 8.4:** Intracellular compartmentation of insulin biosynthesis (from Steiner, D. F. et al., Acta Med. Scand., Suppl. *601* (1976): 53–107, modified).

The rough endoplasmatic reticulum *(rER)* is the site of proinsulin synthesis. Microvesicles *(MV)* serve the transport of proinsulin to the Golgi apparatus *(GA)*. Processing by limited proteolysis proceeds in the GA and in the early granules. With increasing conversion formation of a hexameric insulin-$Zn^{2+}$ complex gives rise to crystalloids in the secretory granules *(SG)*. Insulin secretion involves a $Ca^{2+}$-dependent fusion of the SG with distinct sites of the plasma membrane *(PM)*.
The *time scale* on the right refers to the situation in rat.

triggered by glucose and proceeds by a $Ca^{2+}$-dependent fusion of the secretory granules with distinct sites of the plasma membrane (emiocytosis). The hormone is transported by the blood to the target organs.

## 8.4.2 Fine Structure of Insulin Genes

To study the details of insulin gene expression and its regulation at the transcriptional level it was necessary to isolate and amplify the corresponding gene region by cloning and to sequence the gene as well as its 5′- and 3′-flanking regions.

For the human insulin gene this was achieved by the following experimental procedure (for details of the methods see Chap. **2**):

1. Generation of a human gene library by partial digestion of fetal liver DNA with suitable restriction enzymes and cloning of the DNA fragments in the $\lambda$ vector Charon 4A.
2. Synthesis of specific cDNA by reverse transcription of pre-proinsulin mRNA isolated from insulin producing tissue. Amplification of the cDNA by cloning.
3. Screening of the gene library with radioactively labelled pre-proinsulin cDNA as probe. Two out of $\sim 10^6$ plaques of hybrid phages hybridized with the probe.

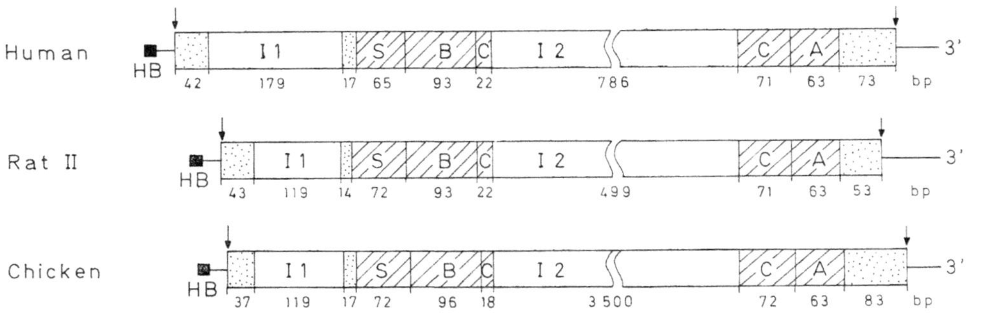


**Fig. 8.5**: Fine structure of the human, rat and chicken insulin genes (from Bell et al. [17] and Perler et al. [18]).

*Cross-hatched boxes*, sequences coding for the signal peptide *(S)*, the A- and B-chains and the C-peptide; *stippled boxes*, untranslated sequences at the 5'- and 3'-ends; *open boxes*, intervening sequences (I1 and I2). *Arrows* mark the sites of pre-mRNA capping (5'-end) and poly(A) addition 3'-end). *HB*, Hogness box required for transcription initiation (cf. Table 1.8).

4. Purification and amplification of the two selected hybrid phages. Subcloning of a 12.5 kb Eco RI fragment containing the insulin gene.
5. Generation of a detailed restriction map of the Eco RI fragment and sequencing of a 2.2 kb subfragment.

The human, rat II and chicken insulin genes exhibit very similar fine structures (Fig. 8.5). All three genes contain two intervening sequences (I1 and I2) that interrupt the 3'-untranslated and the coding sequence for the C-peptide at almost exactly the same positions. Most vertebrates have only this type of insulin gene. But in rats, mice and some fishes two types were detected that are both expressed equally well in the endocrine pancreas. The second type, e.g., rat I, lacks the second intervening sequence. All insulin genes sequenced so far share a homologous sequence 5'-TATAAAG-3' centered 22–24 base pairs upstream of the putative transcription initiation point (Hogness box, cf. Table 1.8).

In two independent preparations two human insulin genes were isolated. These genes are different in one base only presumably representing allelic forms of a single insulin gene. This hypothesis is in agreement with an insulin mutant recently isolated from an insulin-dependent diabetic; it exhibits an exchange of a phenylalanine residue in position 23 or 24 with leucin.

The enormous progress in the field of gene technology could be of great importance for insulin-dependent as well as noninsulin-dependent diabetics. Thus, in the 3'-untranslated region of the human insulin-RNA allelic differences were detected that reveal a kind of gene polymorphism, which is not reflected in the structure of the protein itself, but could have an influence on the regulation of insulin biosynthesis. Cloning of the DNA coding for insulin affords not only the basis for studying the molecular biology of its transcription. It opens the possibility for the production of insulin in bacterial systems.

Because of the differences between eukaryotic and prokaryotic transcriptional and translational regulatory elements, it is necessary to combine the genetic information for the eukaryotic insulin gene with prokaryotic regulatory sequences to obtain expression of insulin in bacteria. Ullrich et al. [8] constructed plasmids, in which the insulin cDNA fragment was fused to different length portions of the gene for galactosidase, the first gene of the lactose operon. Cells containing plasmid pLRI-2, in which the rat insulin gene was fused to the DNA corresponding to the first 1005 amino acids of the galactosidase gene, were able to synthesize a new protein with a predicted molecular weight of 126 kd for a $\beta$-galactosidase-insulin fusion product. By means of SDS-electrophoresis, determination of the primary structure of the protein, immunological and peptide-analytical methods evidence was given for the presence of the insulin sequence in the hybrid protein.

Already in the very near future these cloning techniques may be of great biotechnological interest for the production of commercial insulin, but also of other proteins. The corresponding DNA – either made by isolation of fragments, by copying mRNA or by chemical synthesis (cf. Fig. **2**.3) – must be obtainable in pure form. After incorporation into a suitable plasmid the hybrid plasmid can be multiplied in a bacterial host, e.g., in *E. coli*. Large cultures could be used for protein synthesis. In this way it was possible to produce about 200,000 molecules insulin per bacterium by means of 5–10 liters of a culture of bacteria. Moreover, some other peptides or proteins (e.g., somatostatin, interferon, albumin), were synthesized. Though special problems arise in each case with the medical application of these new products, recently first results were reported on the therapeutic use of a "bacterial" insulin.

### 8.4.3 mRNA and Regulation of Transcription

For the purpose of isolating and characterizing the mRNA of pre-proinsulin, normal rat islets, fish islets, bovine pancreas and rat insulomas have been employed (the latter are tumours induced by streptocotocin or X-ray treatment).

In many respects this mRNA resembles other eukaryotic mRNAs (cf. Fig. **1**.8). It possesses a 3′-polyadenylated end and a 5-terminal cap. In a sucrose gradient a sedimentation coefficient of 9.3S was found for rat pre-proinsulin mRNA. The molecular weight is approximately 200 kd, which corresponds to a length of 550–600 nucleotides. Polyacrylamide gel electrophoresis revealed a single band. This mRNA codes only for one type of rat pre-proinsulin, so that there must be other mRNA species transcribed from the non-allelic insulin genes in rat.

Only 55% of the coding capacity of the 200 kd mRNA are needed to serve as template for pre-proinsulin synthesis. Steiner and co-workers [3] also succeeded in isolating two larger polyadenylated RNA species (MW 360 and 280 kd) which possessed regions homologous with certain regions of pre-proinsulin mRNA. It is assumed that the two larger RNAs represent precursors of the shorter RNA, but it is not known whether they are connected with pre-proinsulin I or II synthesis.

Synthesis of cDNA from polyadenylated RNA of the islets of Langerhans not only provides a copy of the original nucleotide sequence of the RNA region containing the genetic information for insulin and/or its precursors but, if hybridized with the mRNA, it also provides a valuable tool for investigating the regulation of insulin biosynthesis at the transcriptional level. Regulation of transcription is still far from being completely understood despite many years of investigation. It is conceivable that glucose or one of its metabolites could regulate insulin biosynthesis by direct action at the level of transcription. In this case, increased production of mRNA would be expected in response to more glucose.

Okamoto and co-workers [11] found that there was no de novo synthesis of mRNA over a period of 60 min, despite an increase in proinsulin biosynthesis brought about by glucose. The author's own results have confirmed and extended these observations as follows: Islets of Langerhans were isolated from fasting rats and from individuals that had been fed ad libitum. The tissues were incubated with [$^3$H]-leucine and [$^3$H]-uridine in the presence of elevated and reduced concentrations of glucose. The capacity of the islets to synthesize proinsulin and the mRNA that is associated with the insulin-synthesizing polysomes was measured. Using $\alpha$-amanitin, which specifically inhibits mRNA synthesis, it could be shown that proinsulin biosynthesis proceeded normally for two days in islets from fed rats under the culture condition, without the need for synthesis of insulin-specific mRNA. The islets from fasting rats, on the other hand, synthesized insulin-specific mRNA after only two h [12].

Such findings suggest that under certain metabolic conditions a state of equilibrium exists between masked (inactive) mRNA and unmasked (active) mRNA, and that glucose can displace this equilibrium in the direction of the active form. It can also be assumed that the half-life of insulin-specific mRNA is greater in fed than in fasting animals. These results emphasize the necessity of measuring the rate of mRNA synthesis directly by hybridization with the corresponding cDNA.

### 8.4.4 Insulin is Synthesized as a Pre-Prohormone

Extensive studies over the past 2–3 years have revealed that the in vitro translation products of insulin mRNA, like those of other secretory proteins, have an additional N-terminal sequence of 20–30 amino acid residues, a large percentage of which are hydrophobic (cf. Table **8**.4). Although the primary structure is only partially elucidated, it is clear that the length and hydrophobic character of the individual sequences are largely identical. Milstein et al. and Blobel & Dobberstein proposed this region to be necessary in forming a connection between the ribosomes and the membrane of the endoplasmatic reticulum (cf. **4**.4.5.1.1.).

Although details of the mechanism are unknown, structural elements of these signal peptides are in all probability similar to the sequences in many other signal peptides, all of which contain a region of 6 or more hydrophobic amino acid residues. This region is followed by 5–6 residues of amino acids with short aliphatic side chains. The latter are highly flexible and are thought to promote formation of a $\beta$-pleated sheet in the vicinity of the cleavage site (cf. also summary in Table **4**.10). Steiner et al. [3] conclude that the lipophilic central region of the pre-peptide reacts with the nonpolar lipid double layer of the membrane of the rough endoplasmatic reticulum, thus establishing a connection that facilitates interactions between the large ribosomal subunit and the membrane. The hydrophobic region of the signal peptide lies within the lipid double layer in a dissolved state, whilst its hydrophilic N-terminal end remains outside the membrane. The nascent polypeptide chain is injected through the membrane as chain elongation proceeds (Fig. **8**.6). If there is no larger region of hydrophobic amino acids in the chain, diffusion through the membrane can take place without great expenditure of energy. Signal proteases on the inner face of the rER membrane remove the N-terminal sequences relatively rapidly, which is evidenced

**Table 8.4:** Amino acid sequence comparison of N-terminal signal peptides of different pre-proinsulins [18a] (x, residue not identified; the arrow denotes point of cleavage by signal endoprotease, cf. Fig. **8**.6).

Man:
Met Ala *Leu* Trp Met Arg *Leu Leu* Pro *Leu Leu* Ala *Leu Leu* Ala *Leu* Trp Gly Pro Asp Pro Ala Ala Ala ↓ Phe

Bovine:
x - - *Leu* x - - x - - x - - *Leu* x - - x - - *Leu Leu* x - - *Leu Leu* x - - *Leu* x - - x - - x - - x - - x - - x - - x - - x - - ↓ Phe

Rat:
I. Ala *Leu* Trp Met Arg Phe *Leu* Pro *Leu Leu* Ala *Leu Leu* Val *Leu* Trp Glu Pro Lys Pro Ala Gln Ala ↓ Phe
II. Ile Phe Phe Ile

Sea raven:
Met Met x - - x - - *Leu* x - - x - - x - - x - - x - - x - - *Leu Leu* x - - *Leu Leu* x - - *Leu* x - - x - - x - - x - - x - - x - - x - - ↓ x - -

Angler fish:
x - - Ala *Leu* x - - *Leu* x - - x - - Phe x - - *Leu Leu* Val *Leu Leu* Val Val x - - x - - x - - x - - x - - x - - Ala ↓ Val

Hagfish:
x - - *Leu* x - - x - - x - - *Leu* Ala Ala x - - x - - x - - *Leu* x - - *Leu Leu Leu* x - - x - - Ala x - - x - - x - - Ala x - - x - - ↓ Arg

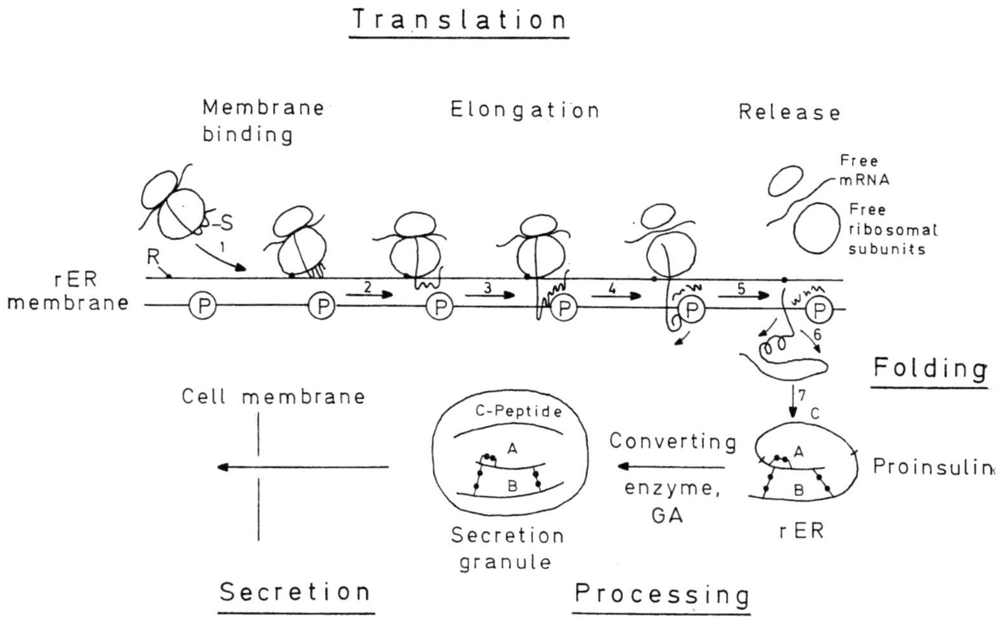


**Fig. 8.6:** Translation and processing of insulin precursor molecules (from Steiner et al. [3], modified).

The short N-terminal signal sequence (*(S)*, cf. Table 8.4) mediates the membrane binding of the translation complex *(steps 1, 2)* by interaction of the large ribosomal subunit with ribosome binding proteins *(R)* in the membranes of the rough endoplasmatic reticulum *(rER)*. The nascent polypeptide chain is injected co-translationally through the rER membrane into the cisterna *(steps 3–5)*. The signal sequence is removed by a specific endoprotease *(P)* that is localized at the cisternal face of the rER membrane *(step 5)*. Folding of proinsulin *(steps 6, 7)* involves formation of the disulphide bonds (– ● – ● –). The proinsulin → insulin conversion by a second step of limited proteolysis proceeds in the Golgi apparatus *(GA)*. For further details of the intracellular compartmentation cf. Fig. 8.4.

by the fact that signal sequences are extremely difficult to detect under in vivo conditions.

Sequence analyses of rat pre-proinsulin (MW 11.5 kd) synthesized in a cell-free system and purified by immune precipitation revealed the presence of 23 additional amino acid residues at the N-terminal end of the B chain (cf. Table 8.4). Comparable pre-proinsulins have been characterized from the Brockmann bodies of a number of fishes (carp, catfish), angler fish *(Lophius americanus)* and sea raven *(Hemitrypterus americanus)*, their MW ranging from 11.5–12.5 kd. It is possible to isolate pre-proinsulin from a cell-free system because the endoproteases catalyzing the cleavage of the pre-protein are absent from such a system. Addition of a microsomal fraction containing these enzymes to a cell-free system results in the transformation of pre-proinsulin to proinsulin. This protease system of the islets of Langerhans has not yet been further characterized, although progress has been made for other microsomal fractions (cf. 4.4.5.1.1).

### 8.4.5 Proinsulin, Proinsulin Processing and Insulin Degradation

The structure of proinsulin is very similar in all vertebrates. It consists of the two insulin chains connected by a peptide (C-peptide) between the C-terminal end of the B chain and the N-terminal end of the A chain (Fig. 8.6). Although the primary structure of the insulins is more or less constant, the C-peptide varies by as much as 50 % from species to species with respect to its amino-acid sequence. One of the functions of the connecting peptide is to ensure correct oxidation of the sulphydryl groups of proinsulin when synthesis of the proinsulin chain is complete. The peptide consists of 26–35 amino acid residues although the distance between the C-terminal end of the B chain and the N-terminal end of the A chain is only 0.8–1.0 nm.

It could thus be expected that smaller peptides would also be able to effect the linkage, and in fact Humbel showed that in IGF (insulin growth factor), which is a proinsulin-like peptide circulating in the blood, only 11 residues are necessary for linking the two chains. Marcussen succeeded in synthesizing a mini-proinsulin that might prove to be of importance in the synthesis of insulin on account of the high yield of linked A- and B-chains. Why the C-peptid is so long and why it exhibits such variety in its sequence are questions that require consideration. One reason might be that the connecting segment protects the active amino acid residues of insulin and is thus responsible for the low biological activity of the precursor. A fundamental difference between proinsulin and IGF is that, whereas proinsulin has to be transformed proteolytically, IGF is already active. It is not fully understood why insulin is biosynthesized via a less active but highly stable form. One explanation might be that a certain length of peptide chain (65–70 amino acid residues) is necessary in order to effect the connection between ribosomes and rough endoplasmatic reticulum. In this respect the structure of IGF is interesting: although the C-segment is shorter than that of proinsulin, the total length of both molecules is strikingly similar because the C-terminal end of the insulin A-chain is extended in IGF. An extension in this position is not possible in the case of insulin because the last amino acid (asparagine) is essential for the biological activity of the molecule. Whether the C-peptide fulfils a biological role in the organism after its release from the proinsulin molecule is not known.

Proinsulin is transformed into the biologically active insulin by limited proteolysis. Subsequently the insulin is stored (almost exclusively) in the secretory granules of the B-cells, whence it can be mobilized by extracellular signals (glucose). The granules are fairly stable, largely on account of the formation of a zinc-insulin hexamer (cf. Fig. 8.4). Nevertheless, there is some evidence of a certain degree of insulin metabolism within the B cells, thought to be catalyzed by a thiol-protein-oxidoreductase and proteolytic enzymes of the lysosomal compartment [14].

Since defective and/or uncontrolled transformation of proinsulin into insulin may under certain circumstances represent a hereditary metabolic disease, and because processing is a system involved in the activation of all prohormones several research teams concentrated on the characterization of the "converting enzyme". Following biosynthesis, folding of the peptide chain and oxidation of the SH-groups, proinsulin is transported into the Golgi apparatus (GA). This transfer step is ATP-dependent. Maximum radioactivity is detectable in the GA about 30 min after labelling of B-cells with radioactive amino acids, after which it declines rapidly. It can be assumed that the transformation of proinsulin into insulin begins in the GA or in the progranula and is completed in the granula (cf. Fig. 8.4).

In spite of extensive investigations the exact mechanism of conversion of proinsulin into insulin is still unknown. It was deduced from the primary structure of proinsulin that trypsin- and carboxypeptidase B-like activities might be responsible. In fact, the development of highly sensitive methods has made it possible to demonstrate activity of these enzymes in homogenates of isolated islets of Langerhans and their subcellular fractions, particularly the secretory granules. The *p*H optimum and behaviour of the enzyme systems toward inhibitors (e.g., EDTA) and stimulators (e.g., mercaptoethanol) point to the likelihood that lysosomal proteases (cathepsins) are responsible [13]. After formation of insulin this hormone is stepwise split: (1) degradation into A- and B-chains, (2) proteolysis of the chains themselves [14].

### 8.4.6 Mechanism of Insulin Secretion

Insulin stored in secretory granules (plus equimolar amounts of C-peptide) is secreted by a process of emiocytosis in response to a certain stimulus. The appropriate stimulus

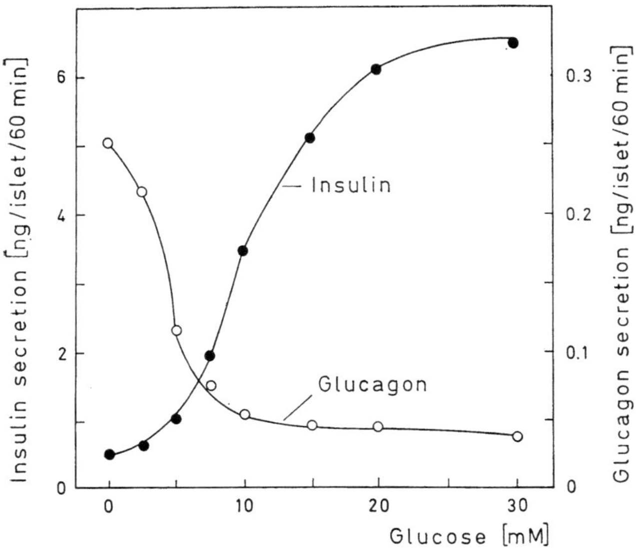


**Fig. 8.7:** Insulin and glucagon secretion in isolated islets of Langerhans of the rat (from Bibergeil [1], p. 25).

5 islets of Langerhans of Wistar rats were incubated for 60 min in 200 $\mu$l glucose-containing bicarbonate buffer, *p*H 7.4 (one islet contains 0.5 $\mu$g protein). Insulin and glucagon secreted to the medium were determined by means of radioimmunoassays.

for both insulin secretion and the biosynthesis of proinsulin is glucose. Signal transformation could proceed via one of the two following hypothetical pathways: (1) binding of glucose to a receptor in the B-cell membrane, (2) glucose metabolites themselves acting within B-cells as signals. Both hypotheses are supported by experimental data. Rising glucose concentrations lead to an increased release of insulin (Fig. 8.7). Under permanent glucose stress, such as it can be evoked by perfusion of the pancreas, a biphasic secretory response can be observed. The first peak (0–5 min) probably represents a response to changes in the B-cell membrane (e.g., membrane potential), whereas the second peak (from 10th min onward) probably results from changes in the intracellular glucose metabolism. Enhanced release of insulin in response to glucose is paralleled by increased adenylate cyclase activity in the membrane and an increased intracellular concentration of cAMP. The cytoplasmic concentration of calcium influences the activity of the microtubular-microfibrillar system and thus may promote emiocytosis. The question of whether B-cell membranes possess a glucose-specific receptor can only be answered by isolating the membrane itself and studying its physicochemical and immunological properties.

## 8.5 Paracrine or Endocrine Relations between Different Cell Types of the Islets of Langerhans?

Table 8.3 shows the hormones formed by the different cell types of the islets of Langerhans. These hormones, together with a number of other polypeptides, identified in the gastro-entero-pancreatic region, play a decisive role in the regulation of the metabolism of food. They include motilin, gastric inhibitory peptide, neurotensin (all shown to occur in the islet cells although not conclusively allocated to a particular cell type), substance P and vasoactive intestinal peptide. In the following section the questions of the biosynthesis and secretion of glucagon and somatostatin will be considered briefly and problems concerning mutual interactions between islet cells will be discussed.

### 8.5.1 Biosynthesis of Glucagon and Somatostatin

Following the discovery of proinsulin in 1967 a search was begun for a precursor in glucagon biosynthesis. In fact several high molecular weight proteins ($M_r$ 4.5–70 kd) reacting with glucagon antibodies were obtained from extracts of pancreas, gastro-intestinal mucosa, salivary glands and blood serum. Fig. 8.8 lists all molecules that have been isolated from islets of Langerhans and characterized in any detail. Proglucagon is seen to have a molecular weight of about 18 kd. The kinetics of transformation of this precursor molecule resembles those for the conversion of proinsulin. The end products are glucagon and a 10 kd protein of unknown function (Fig. 8.8).

The highest concentration of somatostatin has been demonstrated in certain regions of the brain and in islets of Langerhans, where it was localized by immunohistochemical methods in the D-cells. It remained unsettled, however, whether somatostatin is synthesized in the islets themselves or whether it originates in other organs and is transported to the D-cells. Due to the very short half life of the peptide (1 min) the latter alternative seemed unlikely. In fact, it has recently been proved that biosynthesis of somatostatin does take place in the islets of Langerhans via a precursor with $M_r$ 12 kd. High concentrations of glucose apparently inhibit the process.

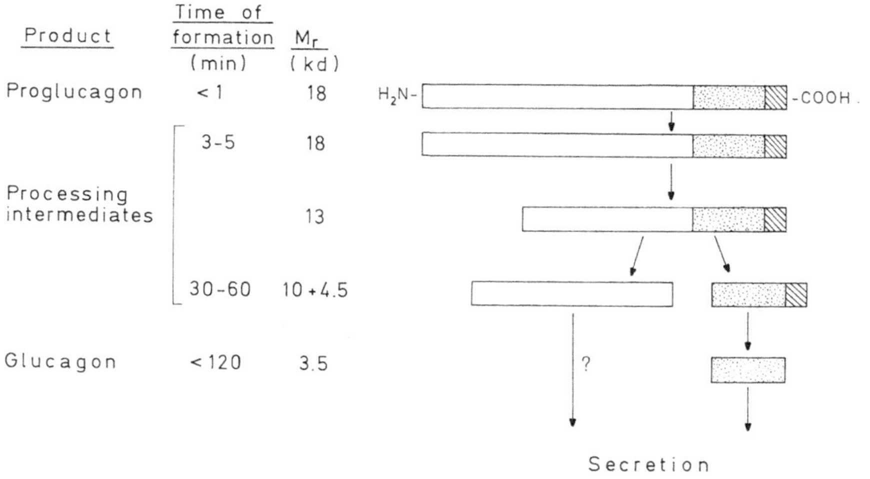


**Fig. 8.8:** Possible intermediates of proglucagon processing (from Patzelt et al. [15], modified).

Islets of Langerhans were labelled by incorporation of [$^3$H]-tryptophan. Glucagon and related polypeptides were identified by radioimmunological techniques after separation by polyacrylamide gel electrophoresis. *Open box,* 10 kd peptide; *stippled box,* glucagon; *cross-hatched box,* C-terminus of proglucagon that is removed during processing.

### 8.5.2 Mechanisms of Cell-Cell Interaction in the Islets of Langerhans

Glucose homeostasis is maintained by the molar ratio of insulin: glucagon in the blood. Under basal conditions, i.e., in a fasten state, the insulin: glucagon ratio amounts to 3. After three days of fasting it has dropped to 1, whereas after a carbohydrate-rich meal (storage of glucose) it exceeds 70. The in vivo situation is mainly due to the interplay of responses of the various islet cells. Other important factors such as the influence of the CNS on homeostasis and on the islet cells will not be discussed here.

Exclusion of certain cell types from the islets results in a disturbance of the islet cell function as a whole. For example, experimental diabetes (streptocotocin- or alloxan-diabetes) can lead to destruction of the B-cells. A drastic reduction of B-cells can also be observed in animals with spontaneous diabetes and in Chinese hamsters. In humans, too, spontaneous loss of insulin secretion is observed as a consequence of the disappearance of B-cells. In insulin-dependent diabetes the number of B-cells in the islets drops from 60% to almost zero %, the proportion of A-cells changes from 25% to 75% and of the D-cells from 10% to 25%. The maturity-onset diabetes (without insulin treatment) is frequently accompanied by hyperinsulinemia. The total

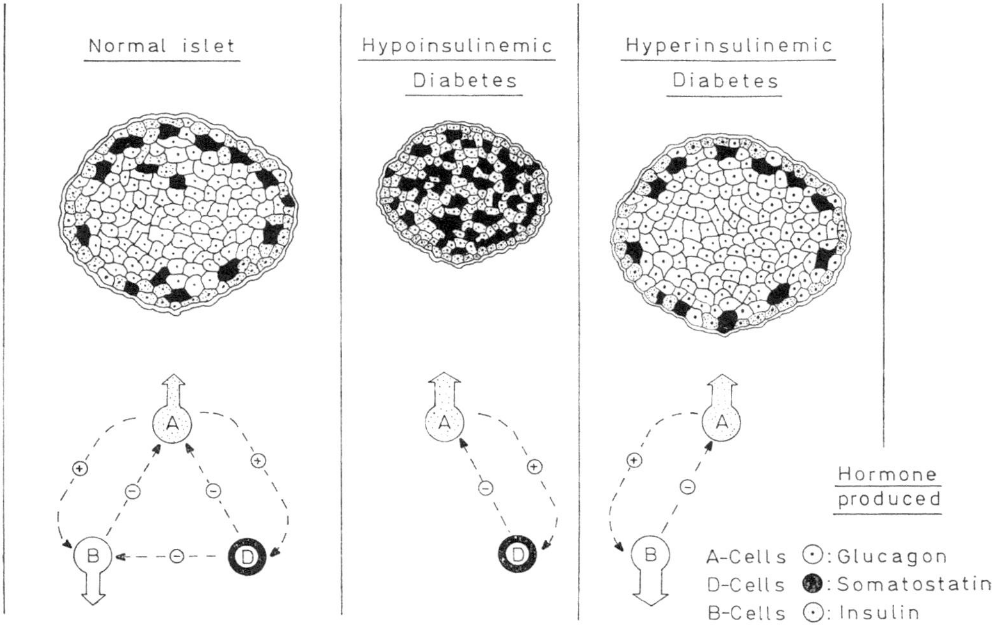


**Fig. 8.9:** Histological schemes of the islet of Langerhans and cellular interactions between A-, B- and D-cells (from Unger et al. [4], with permission).

The *upper part* shows histograms of the cellular composition of islets of Langerhans under different metabolic conditions. In the normal islet a mixture of A- and B-cells constitutes the overwhelming part of the islet, the D-cells being restricted to the rind (surface). In the hypoinsulinemic state the B-cells are practically absent. A- and B-cells are spread over the whole islet. In the hyperinsulinemic state the islet core contains almost exclusively B-cells and the number of D- and A-cells in the rind layer is reduced.

The *lower part* of the figure shows the cells and the stimulating (— — ⊕ — →) and inhibitory (— — ⊖ — →) effects of glucagon, somatostatin and insulin on the secretion of hormones by the different cell types (indicated by the width of the *arrows*).

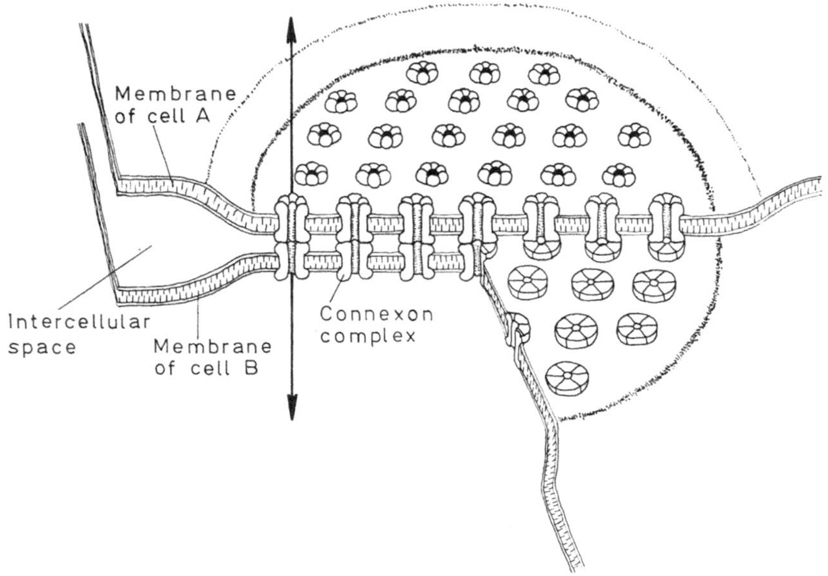


**Fig. 8.10**: Model of a gap junction complex between two animal cells (from Staehelin, A. & B. E. Hull: Scientific American *238* (1978): 141–152, with permission).

The model illustrates the possibility of direct exchange of signal molecules between adjacent cells. The communicating "pipes" are formed by integral membrane proteins (connexon, $M_r$ 25–30 kd) which form annular oligomers of six subunits. The channel of about 2 nm allows penetration of molecules up to 1000 dalton mol. weight. Opening and closure of gap junctions is controlled by $Ca^{2+}$ and proceeds by sliding of the rod-shaped connexon subunits against each other (cf. [16]).

number of islet cells rises, with a greater increase in the numbers of B-cells than of A- or D-cells.

Unger has worked out a hypothetical scheme for the interactions of the A-, B- and D-cells in hypoinsulinemic and in hyperinsulinemic diabetes (Fig. 8.9). There are at least two alternatives: (1) via an endocrine system (communication via the blood stream), (2) via a paracrine system (immediate contact). Although the existence of a "paracrine" system within the islets of Langerhans has not yet been proved, it is conceivable that islet cells receive information from one another via the interstitial spaces. It is certain that insulin, glucagon and somatostatin influence one another mutually with regard to their secretion. Each of the three hormones exerts an influence on the secretion of at least one of the other two (Fig. 8.9).

In contrast to earlier assumptions, cells are not autonomous structures. Specialized regions on the plasma membrane (intercellular junctions) make possible intensive cell–cell interactions, a distinction being made between (1) impermeable (tight) junctions, (2) adhering junction desmosomes and (3) communicating junctions (gap junctions). The latter can be visualized in the electron microscope as clusters of intercellular channels (Fig. 8.10).

Plasma membranes of adjacent cells are held together by tight junctions by means of numerous integral membrane proteins (sealing strands). At the same time a physical barrier prevents migration of these proteins from the outer to the inner surface of the epithelium and the reverse. In this way the integrity of the cell interior towards the extracellular space can be maintained. Adhesion between cells is effected by the desmosomes (belt desmosomes, spot desmosomes).

The development and maintenance of organization within an organism depends upon the exchange of information between the cells. This is made possible by gap junctions, which are channels of about 2 nm diameter, through which molecules with a molecular weight of up to 1000 d (ions, most sugars, amino acids, nucleotides, putative low molecular weight messenger molecules) can move unhindered from one cell to the other. Cells of any one tissue can in fact be regarded as constituting a functional and metabolic syncytium.

Tight junctions and gap junctions have also been shown to exist between cells of the islets of Langerhans. In the light of what is known about cell-cell interactions it can be anticipated that although somatostatin (MW 1.6 kd) could still pass through the gap junctions (paracrine system), glucagon (MW 3.5 kd) and insulin (MW 6.0 kd) could not, i.e., they would have to use an endocrine pathway. Certain differences in the mechanism of action of the three hormones would not be surprising in view of the very different values for their half-lives in blood: somatostatin 1 min, glucagon 4 min, insulin 6 min. In the absence of exact information it has to be assumed that part of the surface of islet cells is equipped with receptors for hormones secreted by other islet cells. Hence, cell-cell interactions in the islets of Langerhans presumably involve both paracrine and endocrine components.

## 8.6 Mechanisms of Insulin Action

### 8.6.1 Metabolic Alterations in the Target Cells

For the concentrations of proinsulin or insulin in the peripheral circulation not only biosynthesis and secretion of the hormone are responsible, but also the distribution in the organism, the uptake by tissues and the enzymatic or non-enzymatic inactivation.

A special role plays the enzymatic degradation of proteohormones, taking place in the case of insulin by (1) cleavage of the disulphide bonds and (2) proteolysis. In several organs and cells not only reductive but also proteolytic degradation of insulin by so-called "insulinase activities" could be detected. Degradation starts already in the islets of Langerhans; though the half-life of insulin in islets of rat pancreas is between 48 and 72 h. Also other organs, particularly liver, kidney and placenta, in smaller range spleen, muscle, fat cells and erythrocytes contain insulin- and proinsulin-degrading activities. Whereas insulin is degraded mainly in liver, proinsulin and C-peptide are essentially inactivated in kidney. The half-life of insulin in blood

**Table 8.5**: Survey of insulin effects (↑, increase; ↓, decrease)

| | |
|---|---|
| *Transport:* | |
| $Na^+$-Efflux ↑ | |
| Influx of $K^+$, amino acids and carbohydrates ↑ | |
| *Anabolic pathways:* | |
| Glycogen synthesis ↑ | DNA synthesis ↑ |
| Lipogenesis ↑ | RNA synthesis ↑ |
| | Protein synthesis ↑ |
| *Catabolic pathways:* | |
| Glycogenolysis ↓ | |
| Lipolysis ↓ | Proteolysis ↓ |

is 4–6 min, that of proinsulin is 2–3 times longer. As described in the following, there are close correlations between hormone metabolism and action.

Insulin is one of the most important endocrine control factors of intermediary metabolism. In contrast to most other hormones, insulin acts under nearly all conditions anabolic and anticatabolic respectively (storage of energy, cf. Table 8.5). For transmission of information to the target organs the binding of insulin to receptor proteins of cell membranes is necessary, as observed at plasma membranes of liver, muscle and fat cells, of thymocytes, lymphocytes and others. The binding of insulin to specific receptor molecules takes place quickly and is reversible. In contrast to growth hormone, insulin, bound to specific receptors, seems to decrease the affinity of the adjacent receptors nearly 10 fold (negative cooperativity).

This phenomenon was not observed in fat cells. The amount of insulin receptors is inversely correlated to the plasma insulin level. From this automatically results that fat cell membranes of adipose patients have a diminished number of insulin receptors but with normal activity and normal cooperativity when compared with cell membranes of healthy probands.

Interaction of insulin with a small number of receptor molecules only seems to develop already the full biological activity. The binding of insulin to 2% of receptor molecules is sufficient for half-maximal stimulation of lipogenesis; for maximal stimulation of glucose oxidation of isolated fat cells binding of insulin to 2.4% of the receptors is necessary. There exists a direct correlation between the binding of insulin to the receptor and the metabolic action of the hormone.

### 8.6.2 Signal Transformation and Hormone Action

Polypeptide hormones, growth factors as well as many other extracellular ligands evoke their biological activity after binding to specific receptors localized in the outer membrane. In the last decade the successful analysis of details of the kinetics and capacity of the insulin-receptor interactions was heavily dependent on two significant methodological innovations: (1) postsynthetic covalent labelling of hormones with [$^{125}$I] or ferritin without any changes in their biological activity and (2) the introduction of the technique of competitive binding studies with isolated cells or cell membranes. Still a few years ago it was supposed that through interaction of insulin with the receptor some membrane-associated enzymes or transport systems were activated that lead to an increase of the intracellular concentration of second messengers (cyclic AMP, cyclic GMP, $Ca^{2+}$). This so-called receptor-transducer model had to be modified by the observation that insulin and other hormones or hormone-like factors are, in fact, taken up together with their membrane receptors (hormone internalization).

#### 8.6.2.1 Insulin Binding to the Membrane Receptor

With help of several methods after adequate incubation labelled insulin could be localized qualitatively and quantitatively at the plasma membrane of hepathocytes, fat cells, fibroblasts and others. The number of binding sites for one fat cell amounts to 10,000–30,000 receptors for insulin. The insulin binding protein is a glycoprotein with $M_r \sim 300$ kd and subunits of 75 kd. On the basis of extensive crystallographic investigations it is supposed today that the active center of insulin lies at the N- and C-terminal ends of the A-chain and the C-terminal end of the B-chain (cf. Fig. 8.11). This binding region of the hormone is assumed to be complementary to a corresponding region of the receptor molecule. At 4 °C insulin as well as epidermal growth factor (EGF) are bound in a diffuse manner to mouse fibroblasts. However, at 37 °C cluster

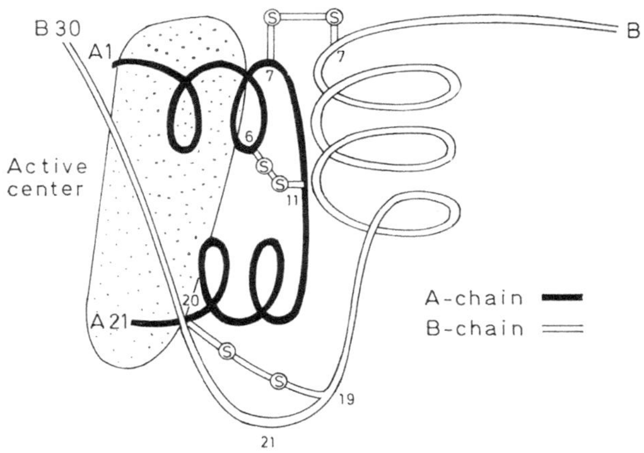


**Fig. 8.11:** Three-dimensional model of insulin showing the hypothetic active center (from Wood et al. [26]).

The two polypeptide chains are depicted with a *black band* (A-chain) and *white band* (B-chain) respectively. The *stippled area* denotes the putative active center formed by the N- and C-termini of the A-chain and part of the B-chain. α-helical regions in the polypeptide chains are symbolized by the *helical parts* in the model. For further details cf. the insulin structure reproduced in Fig. **8.1.**

formation can be observed. The occupied hormone receptors cluster at distinct sites of the cell membrane (coated pits, cf. Fig. **4.1**). These clusters which can be visualized by electron microscopy because of their high density, contain the characteristic structural protein of coated pits (clathrin). Frequently the hormone receptor complexes are covalently cross-linked with each other or with other membrane proteins. Cross linking is catalyzed by transglutaminase in a $Ca^{2+}$-dependent reaction [27].

### 8.6.2.2 Internalization of the Hormone-Receptor Complex

For transmission of the signals after binding of the hormone to the receptor two mechanisms are imaginable:

1. Activation of adenylate cyclase leads to an increased formation of cAMP (second messenger). Details of this mechanism can be deduced from Fig. **10**.6. It is well documented for the glucagon action, but possibly needs major modifications to explain the insulin action.
2. Internalization: In the case of insulin, some other hormones and growth factors, e.g., EGF, as well as for certain transport proteines the biological effect depends totally or, at least, partially on the receptor-mediated internalization via coated pits (Fig. **8**.**12**).

The hormone-receptor-complex localized in the plasma membrane is taken up by forming a specialized pinocytotic vesicle (coated vesicle). The pinocytosis is energy dependent.

Control of internalization is still unclear. The concentration of the internalized hormone seems to be proportional to the amount of hormone being bound to the plasma membrane, i.e., binding could be the rate-limiting step of internalization. The extend of internalization is different for the ligands of several cell types. For instance, using cultivated human lymphocytes in steady state about 15–20% of [$^{125}$I]-insulin are internalized, in hepatocytes under comparable experimental conditions the figure amounts to 40–50%.

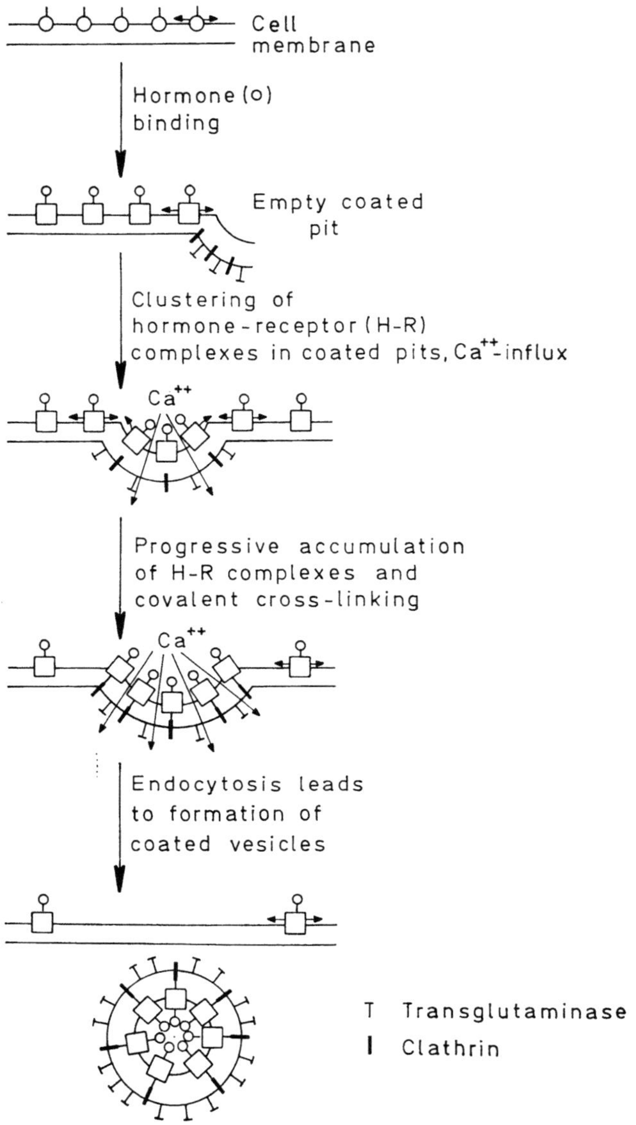


**Fig. 8.12:** Hypothetical model of receptor-mediated endocytosis of polypeptide hormones (from Davies et al. [27], with permission).

Internalization of the insulin-receptor complex proceeds via specialized sites in the plasma membrane (coated pits). The occupied receptors tend to aggregate over coated pits. Receptor clusters are taken up by endocytosis forming coated vesicles that lateron presumably fuse with lysosomes. The cell membrane of coated pits and coated vesicles is characterized by a special dominant protein (clathrin, cf. **4.4.2.2.1**). An essential aspect of the internalization is the local activation, by $Ca^{2+}$ influx, of transglutaminase that evidently cross-links proteins clustered in the coated pits by forming $\varepsilon$- ($\gamma$-glutaminyl-) lysine bridges.

### 8.6.2.3 Intracellular Redistribution and Degradation of Insulin

Several teams could prove that, after incubation of the cell, insulin can be found in definite subcellular fractions. These are the endoplasmatic reticulum, the Golgi-complex, lysosomes and nuclear fractions. All these compartments contain insulin binding proteins (receptors?). Especially the insulin receptors of the Golgi-complex were biochemically and morphologically characterized in more detail. They showed a great similarity to the plasma membrane receptors and are considered of being the biosynthetic precursors of them. Though these results show that intact insulin enters several intracellular compartments, it is an open question, how, after formation of the coated vesicles, the hormone or hormone-receptor complex is transferred to other compartments and what is the biological role of this intracellular redistribution.

Some details have been worked out with respect to the relation between insulin binding, internalization and degradation: (1) the rate of degradation of insulin is dependent, in a broad concentration range, on the amount of hormone bound to the cell; (2) specificity of receptor binding and hormone degradation are identical; (3) careful digestion of cells (e.g., of hepatocytes) by proteases decreases simultaneously hormone binding and degradation.

Insulin degradation is inhibited by lysosome-stabilizing agents such as chloroquine, ammonium chloride or lidocain. Inhibitors of the anaerobic and aerobic production of energy interfere with the degradation of insulin, but they have a negligeable effect on the binding of the hormone. The same is true for concanavalin A, an inhibitor disturbing the function of the cytoskeletal elements.

These results are explicable if the coated vesicles containing the hormone-receptor complex after internalization fuse with lysosomes. Two steps are triggered: (1) liberation of the hormone from the receptor binding and consecutive redistribution between other subcellular compartments together with insulin receptor molecules and (2) degradation of hormone and receptor. In both cases the membrane receptor molecules used as insulin carriers into the cell are degraded. This may explain the well known effect of receptor down-regulation, i.e., the diminution of avaiable insulin binding sites in the outer membranes of target cells observed in the presence of high extracellular concentrations of insulin (cf. similar effects observed for other peptide hormones, Fig. 4.6).

### 8.6.2.4 Antireceptor Antibodies Mimic Insulin Effects

The discovery of auto-antibodies against insulin receptors in patients with insulin--resistent diabetes gave new impulses for the study of the insulin action. It could be shown that antibodies are bound to the insulin receptors followed by a block of insulin binding. Interestingly, several biological effects of insulin are initiated by such antireceptor antibodies.

Antibodies are able in in vitro systems (Soleus muscle, adipocytes) to stimulate the transport of glucose, the incorporation of radioactivity from labelled glucose into glycogen as well as into lipids and the glucose metabolism into $CO_2$. They stimulate the amino acid transport and mimic the antilipolytic effect of insulin (cf. Table 8.5).

Using [$^{125}$I]-labelled antireceptor-antibodies evidence was given for binding of antibodies to cells proportional to the amount of insulin receptors. The binding of these antibodies is inhibited by insulin and insulin analogues in dependence of their affinity to the insulin receptor. Furthermore, it could be shown that monovalent antibody fragments inhibit the binding of insulin, but they cannot mimic the biological activity of insulin. Only divalent antibodies exhibit insulin-like activity, i.e., aggregation of the receptor after cross-linking by divalent antibodies is a prerequisite for these effects [23].

From these investigations one can conclude that insulin itself acts also in an aggregated state (mainly as dimer) (cf. Fig. **8**.12). This possibility was checked by incubation of cells (hepatocytes, adipocytes) with submaximal doses of insulin, followed by treatment with low concentrations of anti-insulin antibodies. Under these circumstances the anti-insulin antibodies actually increase the effect of insulin, attributed to the reaction of insulin antibodies with the insulin-receptor complexes.

Three remarkable aspects result from these experiments: (1) Cross-linking of the otherwise freely diffusible insulin receptor molecules in the outer membrane is sufficient to evoke at least distinct parts of the normal insulin response. It is evident that other components of the cell membrane, e.g., the adenylate cyclase system, must participate in the transmission of the extracellular signal to the intracellular sites of hormone action (cf. Fig. **10**.6). (2) It should be possible to generate a hormone, e.g., by covalent dimerization of insulin, which shows a higher biological activity than natural insulin. (3) The insulin-like effects of insulin antibodies, in any case appearing in diabetics treated with insulin, may have a decisive influence on the clinical situation of these patients.

In spite of the extensive investigations just described, no reliable statement about the general basic principle of insulin action is possible at present. However, for a great number of cell types and different hormones the following steps were reported as essential parts of the action mechanism: (1) binding of the ligand to the receptor at the cell surface (specialization of the membrane by formation of coated pits), (2) internalization of the ligand-receptor complex by adsorptive pinocytosis to a membrane-bound compartment (vesicle), (3) fusion fo the vesicles with intracellular lysosomes and (4) release or degradation of hormone and receptor and eventual recycling of membrane components. Details of this sequence were best described for the receptor-mediated endocytosis of the cholesterol-LDL complex (cf. Fig. **4**.1).

## 8.7 Conclusions

From the point of view of an experimental diabetologist I tried to review the development and functional aspects of the islets of Langerhans and to outline essential control elements of an integrated hormonal signal system starting with the biosynthesis and secretion of insulin and ending with effects of the hormone on its target cells and with its degradation. In all sections of this chapter I attempted a correlation of the experimental data with the multiplicity of effects well known to be a possible cause for a disregulation of this highly efficient hormonal system leading to the diabetes syndrome. Though a remarkable progress has been achieved in recent years, there are much more open questions than answers. These concern

1. the developmental control of the pancreas and especially of the islet cell specialization;
2. the mechanism of the glucose effect on the function of islet cells and the role of the paracrine or endocrine signal exchange between the different types of islet cells for maintaining glucose homeostasis;
3. the regulatory details of insulin biosynthesis and
4. important parts of the signal transformation chain that connects the primary binding of insulin to its receptors with the different types of response in the target cells.

However, the experimental work as well as the broad clinical experience with the insulin system have provided valuable data about the origin and therapy of the

*Diabetes mellitus*. But, in addition, much has been learned and will be learned about the mode of action of signal systems that are prerequisites for the intercellular coordination of cell specialization and function in multicellular organisms.

## 8.8 Selected References

*Reviews*

1. Bibergeil, H.: *Diabetes mellitus*. Ein Nachschlagewerk für die diabetologische Praxis. Gustav Fischer Verlag, Jena 1978.
2. Steiner, D. F.: Insulin today. Diabetes *26* (1977): 322–340.
3. Steiner, D. F., J. R. Duguid, C. Patzelt, S. J. Chan, P. Quinn, A. Labreque & R. Hastings: New aspects of insulin biosynthesis. In: Baba, S., T. Kaneko & N. Yanaihara (eds.): Proinsulin, Insulin, C-Peptide. Excerpta Medica, Amsterdam 1979, pp. 9–19.
4. Unger, R. H., R. E. Dobbs & L. Orci: Insulin, glucagon, and somatostatin secretion in the regulation of metabolism. Ann. Rev. Physiol. *40* (1978): 307–343.
5. Permutt, M. A. & D. M. Kipnis: Insulin biosynthesis and secretion. Fed. Proc. *34* (1975): 1549–1555.
6. Schröder, D. & H. Zühlke: Molekularbiologische Möglichkeiten der Untersuchung genetischer Probleme des *Diabetes mellitus*. Dtsch. Ges.wesen *35* (1980): 881–888.

*Hormones Synthesis and Structure of Insulin Genes*

7. Hellerström, C., A. Andersson & R. Gunnarsson: Regeneration of islet cells. Acta endocrinol. *83* (1976:) 145–158.
8. Ullrich, A., J. Shine, P. H. Seeburg, E. Tischer, R. L. Pictet, B. Cordell, G. I. Bell, J. M. Chirgwin, W. J. Rutter & H. M. Goodman: The structure and expression of the insulin gene. In: Baba, S., T. Kaneko & N. Yanaihara (eds.): Proinsulin, Insulin, C-Peptide. Excerpta Medica, Amsterdam 1979, pp. 20–25.
9. Patzelt, L., A. D. Labrecque, J. R. Duguid, R. J. Carroll, P. S. Keim, R. L. Heinrikson & D. F. Steiner: Detection and kinetic behavior of pre-proinsulin in pancreatic islets. Proc. Natl. Acad. Sci. USA *75* (1978): 1260–1264.
10. Gunnarsson, R.: Effects of glucose on the "in vitro" synthesis of ribonucleic acid in pancreatic islets from mice homozygous for the mutation "diabetes" *(db)*. Diabetes & Metabolisme *3* (1977): 149–153.
11. Itoh, N., K. Nose & H. Okamoto: Purification and characterization of proinsulin mRNA from rat B-cell tumor. Eur. J. Biochem. *97* (1979): 1–9.
12. Jahr, H., D. Schröder, B. Ziegler, M. Ziegler & H. Zühlke: Transcriptional and translational control of glucose-stimulated (pro)insulin biosynthesis. Eur. J. Biochem. *110* (1980): 499–505.
13. Zühlke, H., D. F. Steiner, A. Lernmark & C. Lipsey: Carboxypeptidase-B-like and trypsin-like activities in isolated rat pancreatic islets. Ciba Foundation Symp., Vol. 31. Elsevier-Excerpta Medica, Amsterdam 1976, pp. 183–191.
14. Kohnert, K. D., H. Jahr, S. Schmidt, H. J. Hahn & H. Zühlke: Demonstration of insulin degradation by thiol-protein disulfide oxidoreductase (glutathione-insulin transhydrogenase) and proteinases of pancreatic islets. Biochim. Biophys. Acta *422* (1976): 254–259.
15. Patzelt, C., H. S. Tager, R. J. Carroll & D. F. Steiner: Identification and processing of proglucagon in pancreatic islets. Nature *282* (1979): 260–266.
16. Unwin, P. N. T. & G. Zampigli: Structure of the junction between communicating cells. Nature *283* (1980): 545–549.
17. Bell. G. J., R. L. Pictet, W. J. Rutter, B. Cordell, E. Tischer & H. M. Goodman: Sequence of the human insulin gene. Nature *284* (1980): 26–32.
18. Perler, F., A. Efstratiadis, P. Lomedico, W. Gilbert, R. Kolodner & J. Dodgson: The evolution of genes: the chicken preproinsulin gene. Cell *20* (1980): 555–566.

18a. Jackson, R. C. & G. Blobel: Post-translational processing of full-length presecretory proteins with canine pancreatic signal peptidase. Annals N. Y. Acad. Sci. *343* (1980): 391–403; Sures, I., D. V. Goeddel, A. Gray & A. Ullrich: Nucleotide sequence of human preproinsulin complementary DNA. Science *208* (1980): 57–59.

*Mechanisms of Insulin Action*

19. Rinderknecht, E. & R. E. Humbel: The amino acid sequence of human insulin-like growth factor I and its structural homology with proinsulin. J. Biol. Chem. *253* (1978): 2769–2776.
20. Terris, S., C. Hofmann & D. F. Steiner: Mode of uptake and degradation of 125I-labelled insulin by isolated hepatocytes and H 4 hepatoma cells. Can. J. Biochem. *57* (1979): 459–468.
21. Gorden, P., J.-L. Carpentier, P. Freychet, A. LeCam & L. Orci: Intracellular translocation of iodine-125-labelled insulin: Direct demonstration in isolated hepatocytes. Science *200* (1978): 782–785.
22. Goldfine, I. D.: Does insulin need a second messenger? Diabetes *26* (1977): 148–155.
23. Kahn, C. R., K. L. Baird, D. B. Jarrett & J. S. Flier: Direct demonstration that receptor crosslinking or aggregation is important in insulin action. Proc. Natl. Acad. Sci. USA *75* (1978): 4209–4213.
24. Gorden, P. H., J.-L. Carpentier, P. Freychet & L. Orci: Internalization of polypeptide hormones. Mechanism, intracellular localization, and significance. Diabetologia *18* (1980): 263–274.
25. Tager, H., B. Given, D. Baldwin, M. Mako, J. Markese, A. Rubenstein, J. Olefsky, M. Kobayashi, O. Kolterman & R. Poucher: A structural abnormal insulin causing human diabetes. Nature *281* (1979): 1–4.
26. Wood, S. P., T. L. Blundell, A. Wollmer, A. Lazarus & R. W. Neville: The relation of conformation and association of insulin to receptor binding, X-ray and circulardichroism studies on bovine and hystricomorph insulins. Eur. J. Biochem. *55* (1975): 531–542.
27. Davies, P. J. A., D. R. Davies, A. Levitzki, F. R. Maxfield, P. Milhaud, M. C. Willingham & I. H. Pastan: Transglutaminase is essential in receptor-mediated endocytosis of $\alpha_2$-macroglobulin. Nature *283* (1980): 162–167.

# 9. Polytene Chromosomes – Model Objects for Studying Differential Gene Activities in Specialized Cells

E. Serfling

Polytene chromosomes arise by endomitotic divisions of paired homologous chromosomes. They form cable-like strands of up to 500 μm length and 15 μm diameter and are transversed by numerous bands or chromomeres. Polytene chromosomes occur in distinct tissues of certain animals and plants, and in some *Protozoa*. The salivary gland chromosomes of *Diptera*, discovered by Balbiani at the end of the last century, are particularly suitable for molecular biological investigations aimed at characterizing chromosomal structure and function. They are the largest polytene chromosomes known, and may undergo up to 13 rounds of DNA replication [1–3].

The following presentation on the function of polytene chromosomes is largely confined to the characterization of the transcription of salivary gland chromosomes of *Diptera*, particularly of *Drosophila* and *Chironomus*. In order to illustrate the diversity of expression of eukaryotic genes the heat-shock and Balbiani ring genes will be discribed in some detail, whilst a comparison of the Balbiani ring genes with those of the fibroin gene of *Bombyx mori* is intended to demonstrate the characteristics common to these genes, which both code for high molecular weight proteins f an internal-repeated structure.

## 9.1 Structure of Dipteran Polytene Chromosomes

### 9.1.1 Studies Employing Light and Electron Microscopy

Due to the alternation of optically dense bands with transparent interbands polytene chromosomes appear as clearly defined, striped structures under the light microscope, their regularity only interrupted by diffuse segments or "puffs" (Fig. **9**.1).

Salivary gland chromosomes of *Drosophila* contain about 5,000 bands (called chromomeres) and those of *Chironomus* 2,000. More than 95% of the DNA of polytene chromosomes is concentrated in the bands. Since the DNA forms a continuous strand throughout the chromosomes the high DNA content in the bands (cf. **9**.1.2.1) is due to an exceptionally high degree of condensation of the DNP fibrils in the band region [2]. In contrast, DNP fibrils traverse the interbands in less condensed condition. In the electron micrograph bundles of DNP fibrils can be recognized as fibres in the interbands (Fig. **9**.2). In well-spread preparations DNA fibrils of 6 nm can be identified, most of which form double structures of 12 nm and apparently correspond to chromatin fibrils of 11 nm (cf. Figs. **1**.5 and **1**.6).

---

Abbreviations: BR, Balbiani rings; C value, haploid chromosome set (genome); DNP, deoxyribonucleoprotein; RNP, ribonucleoprotein; rRNA, ribosomal RNA; mRNA, messenger RNA; hnRNA, heterogeneous nuclear RNA; cRNA, in vitro synthesized RNA complementary to DNA; PAA, polyacrylamide; $M_r$, molecular weight; kd, kilodalton; Md, Megadalton; SDS, sodium dodecylsulfate; restriction enzymes, e.g., Eco RI, Hind III etc. cf. Table **2**.1.

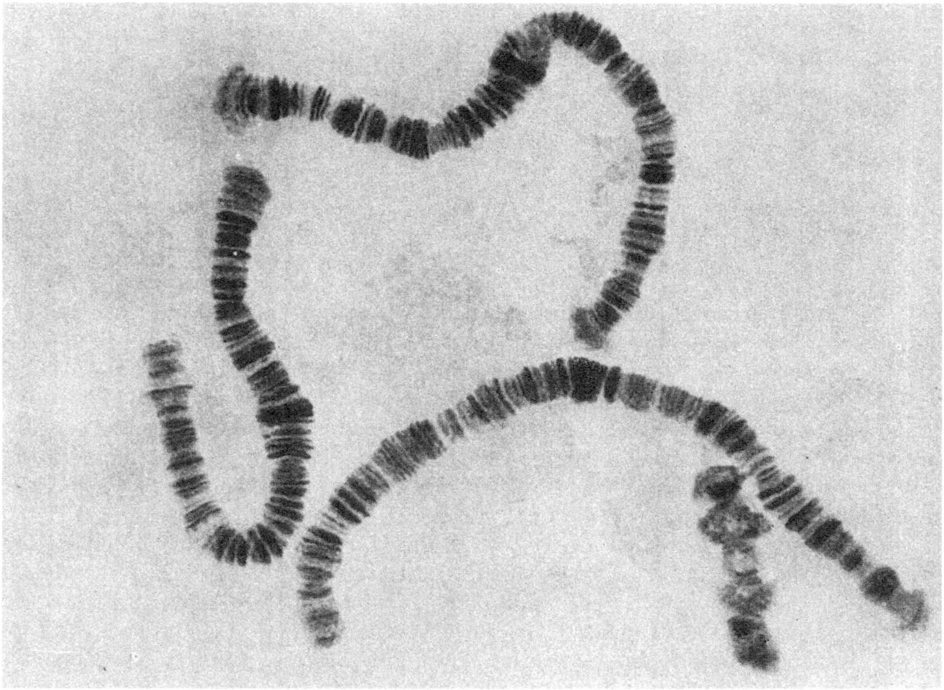

1μm

### 9.1.2 DNA of Polytene Chromosomes

#### 9.1.2.1 DNA Content of Chromomeres

Microspectrophotometric determination of DNA extracted from *Chironomus thummi* salivary glands that had undergone 12 DNA replication cycles yielded values of 300–400 pg DNA per chromosome. Taking into consideration the number of bands and the polyteny level, a DNA content of almost 1 pg DNA per band and of $10^{-4}$ pg DNA per chromomere of a chromosomal element (corresponding to about $10^5$ base pairs of DNA) is arrived at by calculation.

The high DNA content of the chromomeres, which may harbour the information of a multitude of proteins, raises the question of their genetic equivalence and functional significance. Cytogenetic analysis of the fine structure of the salivary glands of *D. melanogaster*, carried out with especial accuracy on the "zest-white" region of the X-chromosome, has so far invariably yielded positive indications of a 1 : 1 relationship between chromomeres and genetically defined functional units. From this it may be concluded that a chromomere usually contains only one or a few (functionally similar or identical) genes [2]. Since we know from the results of DNA-RNA hybridization and DNA cloning experiments that many protein-coding genes, e.g., those for globin, ovalbumin and immunoglobulin, but also the fibroin and Balbiani ring genes (cf. **9**.4) are present in only one or a few copies in the genome (cf. Table **4**.6), the question arises as to the functional role of the large quantities of "excess" DNA in chromomeres. Unfortunately, at the present time it is impossible to draw any well-founded conclusions on this point.

#### 9.1.2.2 Polytenization and DNA Replication

Cytophotometric measurements of the DNA content of the cell nucleus of dipteran larval tissues show that not only the salivary glands but also numerous other tissues (Malpighi ducts, midgut, proventricle, muscle and tracheal cells, fat body and ganglion cells) contain polytene nuclei. But the cell nuclei of the salivary glands invariably contain chromosomes of a particularly high level of polyteny, which may even amount to $2^{13}$ N in the case of *Chironomus* [1, 2].

In polytenization not all DNA sequences of the genome are replicated to the same extent. This holds particularly for the α-heterochromatin of *Drosophila*, which occupies about 20–50% of the chromosome regions in diploid cells, but it is barely discernible biochemically and cytologically in the chromocentre of salivary gland chromosomes [1]. This phenomenon of "under-replication" of certain chromosomal segments in the course of polytenization is, however, not limited to highly repetitive DNA sequences, which as a rule are not replicated at all during polytenization of *Drosophila*

---

◀

**Fig. 9.1** (p. 350, above): Chromosome set of a larval salivary gland cell nucleus of *Chironomus thummi*. Magnification: 500 times.

The photograph shows a chromosome squash preparation obtained after fixation of a salivary gland with ethanol/acetic acid and staining of the squashed gland by carmine/acetic acid and orceine/lactic acid.

**Fig. 9.2** (p. 350, below): Electron micrograph of a part of chromosome 3 from a larval salivary gland cell nucleus of *Drosophila melanogaster* (M. and V. Sorsa, from [6]).

The micrograph was taken from a length section of squashed chromosomes. Note differences of band thickness and the DNP fibres of about 12 nm traversing the interband regions.

salivary gland chromosomes. The relative proportion of rRNA gens on the genome of diploid *Drosophila* cells is, for example, distinctly greater than in salivary gland cells. In contrast to heterochromatin the rRNA genes undergo 6–7 endomitoses, which means that their level of polyteny is about eight times lower than that of the other segments of the salivary gland chromosomes. Since the rRNA genes of *Drosophila* are adjacent to heterochromatin regions, whereas the rRNA genes of *Chironomus* are in the euchromatin and exhibit the normal polyteny level, one reason for under-replication of ribosomal DNA of *Drosophila* might be the close proximity of hetero-chromatic DNP [1, 2].

It is of interest that in *Drosophila* a number of highly repetitive heterochromatic chromosomal segments (like the $\beta$-heterochromatin and intercalary heterochromatin) also are completely replicated in polytenization, whereas certain euchromatic chromosomal segments, e.g., of the Y-chromosome of *D. hydei*, are not polytenized.

The underreplication of heterochromatic segments of polytene chromosomes is particularly well documented for the $\alpha$-heterochromatin of *D. virilis* which can be separated into three satellite DNA's by CsCl density gradient centrifugation. Whereas these fractions account for up to 40% of the total DNA in diploid tissues, corresponding DNA fractions in preparations of salivary gland DNA cannot be demonstrated. At the same time the proportion of heterochromatic DNA segments in salivary gland chromosomes is much smaller than that of mitotic chromosomes. Sequence analyses of the three satellite DNA fractions revealed a basic sequence of seven nucleotides. This sequence is repeated millions of times and varies from one satellite fraction to the next by only one nucleotide in each case (cf. Gall et al.; in Ref. [1a]).

In *Sciara*, in contrast to the $\alpha$-heterochromatin or the rRNA genes in *Drosophila*, certain chromosome segments known as DNA puffs achieve a higher level of polyteny than others by means of additional DNA replication. Since the increased DNA replication of the DNA puffs is followed by intensive RNA synthesis this extra DNA synthesis is functionally homologous to the amplification of ribosomal DNA leading to the formation of thousands of extra nucleoli in amphibian oocytes (cf. Table 4.6). In contrast to amplified rDNA, however, the extra DNA of the sciarid puffs remains on the chromosome and does not appear as extrachromosomal DNA in the nucleus of the salivary gland cells (see Rudkin, in Ref. [2]).

### 9.1.3 Proteins Associated with Polytene Chromosomes

Next to DNA, histones and non-histone proteins are the main constituents of eukaryotic chromosomes. Non-histone proteins apparently play a particularly important role in the regulation of gene activity. Analysis of the non-histone pattern of polytene *Drosophila* nuclei by means of gel electophoresis reveals a multitude of different proteins of $M_r$ between 20 kd and 120 kd. In *D. melanogaster* and *D. hydei* the chromosomal protein patterns are almost identical. In polytene tissue of *D. hydei*, however, three protein fractions are under-represented or altogether lacking, whilst a new non--histone protein is observed. Whether the missing proteins are components of the $\alpha$-heterochromatin and whether the newly occurring non-histone protein plays a role in polytenization cannot be decided from the data so far available. Attempts to bind cloned satellite DNA attached to nitrocellulose membrane filters to isolated nuclear proteins show that individual polypeptides can be specifically bound to distinct sequences of satellite DNA constituting major components of centromeric DNA.

For analysis of the distribution of chromosomal proteins in polytene chromosomes the immunological in situ localization of chromosomal proteins has proved to be of great value. This method permits the demonstration of specific proteins in individual bands and puffs of the polytene chromosomes.

In order to prepare antibodies to chromosomal proteins such as histones and specific non--histone proteins these must first be isolated and separated from the bulk of other chromosomal proteins. In the case of non-histone proteins this is done by two-dimensional separation involving isoelectric focusing and subsequent SDS polyacrylamide gel electrophoresis. The fractionated polypeptides are then injected into rabbits or mice and the resulting immune sera (containing the specific antibodies) are incubated with squashed preparations of salivary glands. In this process the specific antibodies bind to the corresponding chromosomal proteins and the protein-antibody complexes can then be detected by means of fluorescein-coupled anti-antibodies. The fluorescent label shows up in UV light, which allows localization of the corresponding chromosomal protein under the microscope (see Elgin et al., in Ref. [1b]).

Using this technique several non-histone proteins, RNA polymerase II and histones have been localized on salivary gland chromosomes. Antibodies to individual histones revealed, as was to be expected, even distribution over all segments. The use of the same technique revealed a distinct accumulation of mRNA-synthesizing polymerase II in puffs. Accumulation of the enzyme is very pronounced in heat-shock puffs (cf. **9**.2.2.3 and **9**.3.2), whereas in the majority of puffs repressed by heat treatment only little activity remains. Of the non-histone proteins some protein species are distributed over the whole chromosome set, whereas others are concentrated in only one or a few chromomeres. Those occurring only at puff sites are of special interest. Employing DNAse I, which specifically attacks activated chromosomal segments (cf. **4**.4.2.5.2), a non-histone protein could be isolated whose antibodies specifically bind to freshly activated chromomeres (puffs) or potentially active chromomeres. These experiments suggest that certain non-histone proteins are specific for potentially active chromosomal segments and are somehow involved in the process of gene activation as manifest in the puffing phenomenon (see Elgin et al. and Jamrich et al. -n [1b]).

## 9.2 The Puffing Phenomenon

Puffs are lateral loops of longitudinal extension of the polytene chromosomes and represent transcriptionally active chromosomal sites. Although they were already observed in 1881 by the French anatomist Balbiani and rediscovered about 40 years ago, it was Beermann who recognized that puffs are active gene sites and characterized their pattern of activity as the expression of differential gene activity [2].

Balbiani rings are tissue-specific puffs of the *Chironomus* salivary gland. They are distinguishable from puffs of average size by their high activity in RNA synthesis and by their circular brush-like shape (Fig. **9**.3A). They (almost exclusively) occur on salivary gland chromosomes of chironomids and contain the genes coding for synthesis of the proteins of the salivary gland secretion (cf. **9**.4.1.1).

### 9.2.1 Puffs as a Manifestation of Transcription Activity

#### 9.2.1.1 Puffs and RNA Synthesis

The RNA of larval salivary glands can be labelled with [$^3$H]-uridine either by injection of isotopes into the larvae or by incubating the isolated glands in a culture medium containing the labelled precursor. After only a few minutes radioactivity can be detected in the cell nuclei by means of autoradiography. It is strictly concentrated in

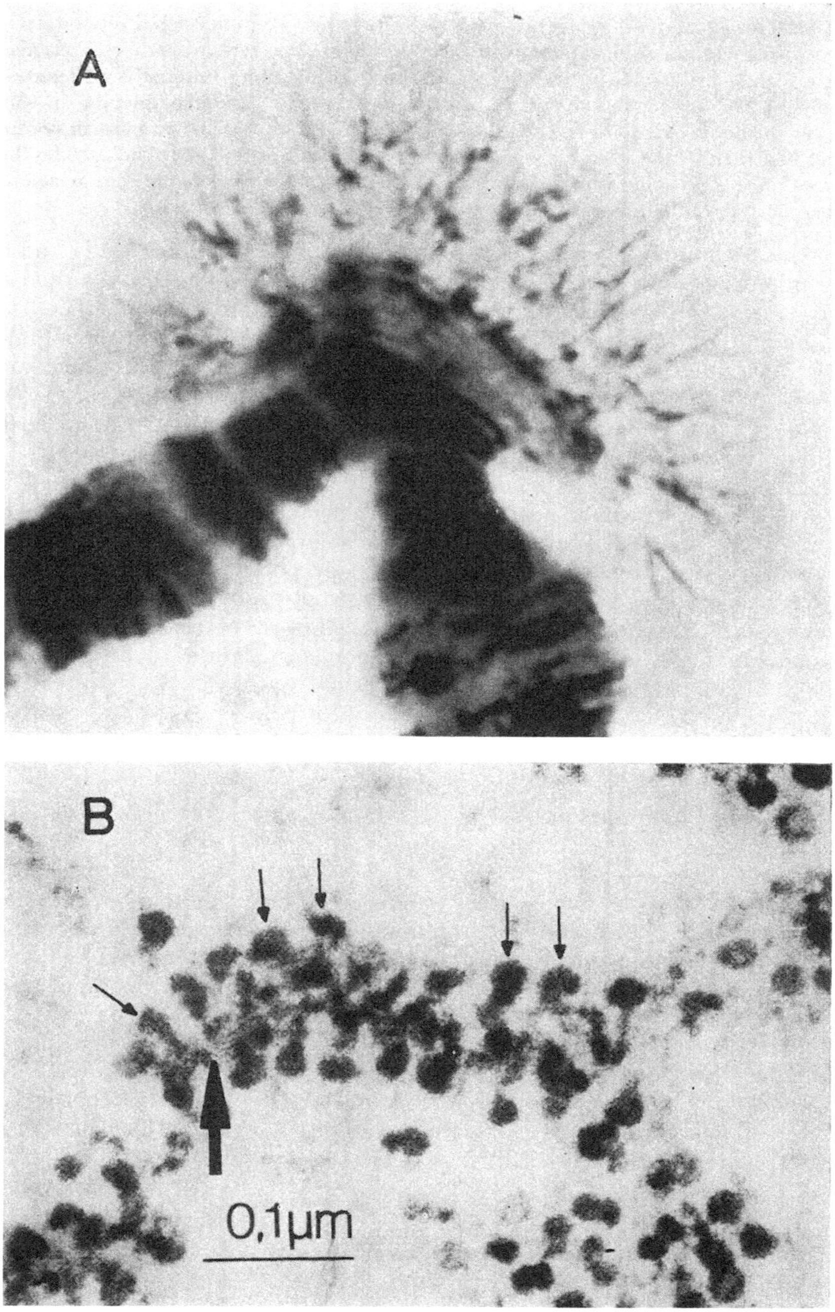


Fig. 9.3

puffs and nucleoli and its incorporation is confined to RNA synthesized at these sites since the radioactive label disappears following RNase treatment [15].

The size of a puff is proportional to its potential capacity for RNA synthesis. Balbiani rings synthesize about 30–60 times more RNA than a puff of average size. From cytophotometric measurements of RNA from salivary gland chromosomes isolated from *C. thummi* it can be deduced that a puff (from chromosomes of the $2^{11}$ polyteny level) does not contain more than 1 pg RNA, whereas Balbiani rings contain about 30 pg RNA. A comparison of the RNA content of salivary gland chromosomes of different polyteny levels shows that following replication of the genome the RNA content also doubles, so that the amount of RNA synthesized increases in proportion to the amount of DNA.

#### 9.2.1.2 Puffing Leads to Characteristic Changes of Chromomere Structure

Formation of a puff begins with decondensation of the chromomere and accumulation of non-histone protein. Association of RNA polymerase II has been demonstrated by immunological techniques. Very little is known of the nature of other proteins that might perhaps play a part in puff formation (cf. **9**.1.3). As was shown in electron micrographs, accumulation of non-histone proteins in all probability initially leads to significant changes in the structure of the chromatin fibrils. For example, at the proximal end of the transcription units of Balbiani rings the fibrillar structures of about 0.18 $\mu$m length probably mark sites of initiation of transcription and are quite different from the beads-on-a-string structure of other regions of chromosomal chromatin fibres (Fig. **9**.9, cf. also Fig. **1**.5).

### 9.2.2 Puff Patterns as Indicators of the Gene Expression Program in Polytene Tissues

#### 9.2.2.1 Development and Tissue Specificity of Puffing

In *Chironomus* a detailed analysis of the puff activity of salivary gland chromosomes throughout larval development revealed that about 15% of all chromomeres of these chromosomes can be puffed. Of the 277 puffs described for the salivary gland chromosomes of *C. tentans* the major portion (about 90%) are active throughout the entire course of larval development. In contrast to these so-called *developmentally nonspecific puffs* the remaining 10% are termed *developmentally specific* and are only active at certain phases of development, for example during transition of the larva to the prepupa [15].

A further characteristic of puff pattern formation of polytene chromosomes, apart from the above-described developmental specificity, is tissue (and cell) specificity. The chromosomes of the larval salivary gland of *Acricotopus lucidus* for example, which is divided into main, side and anterior lobes, form Balbiani rings specific to the lobe to which they belong (Fig. **9**.4). The three Balbiani rings BR3, BR4 and

---

◀

**Fig. 9.3**: Visualization of Balbiani rings by light microscopy (A) and electron microscopy (B).

**A.** Balbiani ring BR2 of the main lobe of a larval salivary gland of *Acricotopus lucidus* stained by carmine/orceine. *Magnification:* 2,000 times (R. Panitz, in [2]).

**B.** Electron micrograph of a fibril from BR2 of *Chironomus tentans* [5].

Around a central DNP fibril *(thick arrow)* RNP particles *(thin arrows)* can be seen which are formed by the association of nascent BR RNA molecules with nuclear proteins.

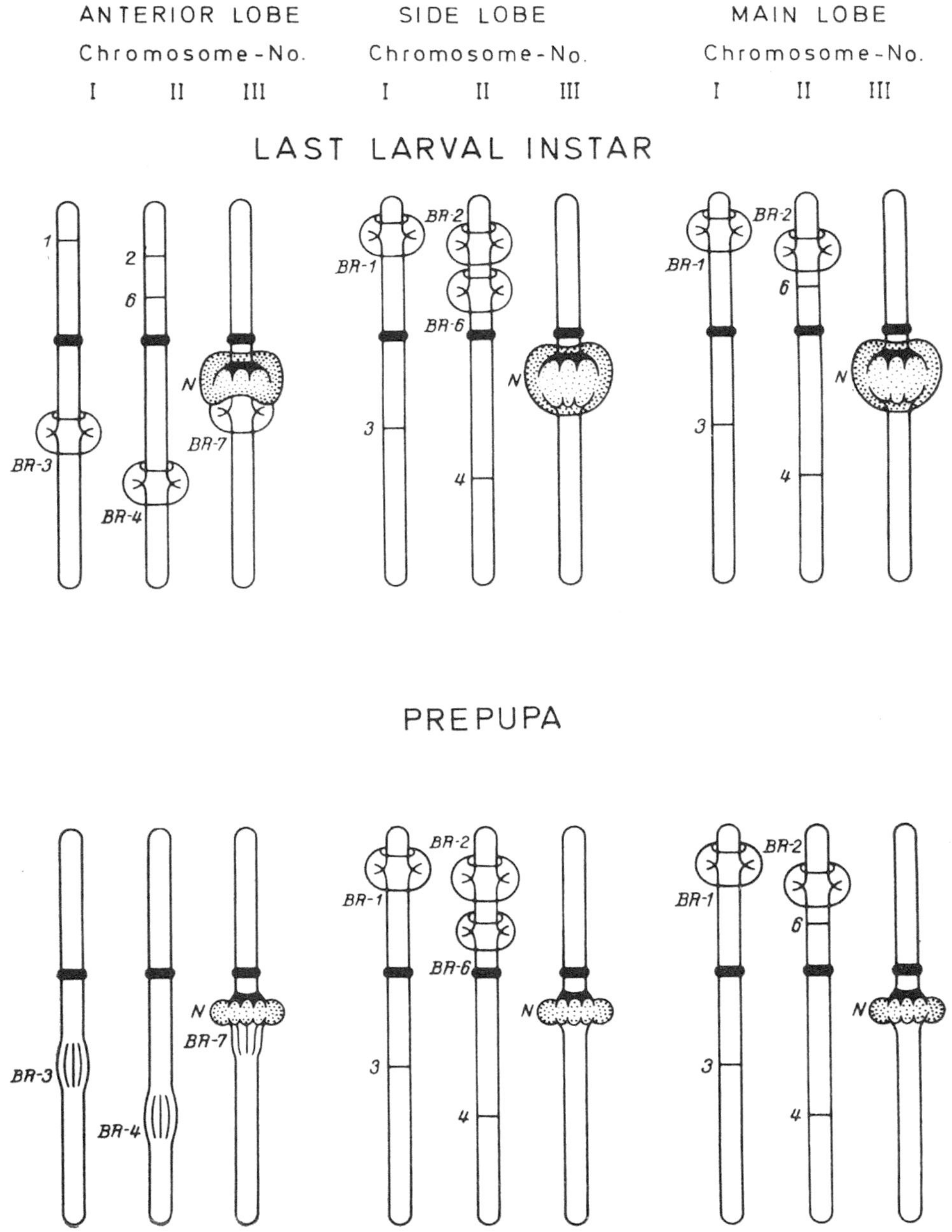


**Fig. 9.4**: Development and tissue specificity of Balbiani ring activities in the anterior, side and main lobe of larval salivary gland of *Acricotopus lucidus* (R. Panitz, in [2]).

The repression of Balbiani rings BR3 and BR4 in prepupae is due to the action of the steroid hormone ecdysone (cf. **9.2.2.2**). – Chromosomal loci of non-activated BRs are marked by *horizontal lines*. *N*, nucleolus.

BR7 that are formed on the chromosomes of the anterior lobe are not found on the chromosomes of the main and side lobes, whereas Balbiani rings BR1 and BR2 are specific to the main and side lobes. Again, Balbiani ring BR6 only occurs on the chromosomes of the side lobe. BR3 and BR4 of the anterior lobe are repressed in the prepupal stage under the influence of ecdyson (Fig. **9**.4, cf. Panitz, in Ref. [2]).

All Balbiani rings of *Chironomus* salivary glands, as far as is known, are strictly tissue-specific. This underlines their importance in the synthesis of tissue-specific products, i.e., the secretory proteins of the salivary glands (cf. also **9**.4.1.1). The DNA puffs of sciarids are also tissue-specific occurring exclusively in the salivary gland cells.

### 9.2.2.2 Hormones as Inductors of Specific Changes of Puff Patterns

Ecdyson, the moulting hormone of insects, induces significant changes in the puff pattern at specific stages in development. Detailed analyses of the action of ecdyson on the puffing activity in *Chironomus*, *Drosophila*, *Acricotopus* and *Sciara* salivary gland chromosomes yielded important information on the control of puff formation in polytene chromosomes [2].

If 0.2 μg of ecdyson is injected into larvae of *C. tentans* a puff forms after 10–15 min, followed a few minutes later by a second puff. At the same time another puff is inactivated. Induction of puffs can be inhibited by actinomycin D and is dependent on the dose of ecdyson. Threshold values of $10^{-7}$ and $10^{-6}$ μg ecdyson/mg body weight were determined, which is equivalent to 10–100 molecules of ecdyson per cell nucleus. The first hormone-specific reaction is followed a few hours later by the activation of new puffs which appear at definite times, specific to the locus involved. Since the induction of most of these secondary puffs, but not of the primary puffs, can be prevented by inhibitors of protein synthesis (such as puromycin and cycloheximide) it is assumed that secondary puff formation is not directly evoked by ecdyson but rather by protein products of the primary puffs. Hence, primary and secondary puffs are part of a determined gene expression program induced by ecdyson [2].

Instead of inducing puffs as it does in *Chironomus* (and *Drosophila*) ecdycson leads to the regression of two Balbiani rings in *Acricotopus* (cf. Fig. **9**.4). Since this takes place within 30 min and cannot be suppressed by inhibition of protein synthesis, the regression of the two Balbiani rings is in all probability the primary response to ecdyson. Interestingly, no ecdyson-specific puff reactions have been observed on chromosomes of the main and side lobes of the salivary gland [2].

Like ecdyson, juvenile hormone, which determines the type of moulting, also induces development-specific changes in the puff pattern. Although studied in far less detail than ecdyson, juvenile hormone can be assumed to behave antagonistically to ecdyson with respect to puff pattern changes. For example, whereas the BR1 of the salivary gland chromosomes of *C. tentans* is apparently stimulated by ecdyson, it is repressed by juvenile hormone; the BRc of *C. thummi* is repressed in the immature prepupa (which contains high levels of ecdyson) but stimulated to puff formation by juvenile hormone.

### 9.2.2.3 Puff Induction and Repression by Heat-Shock

Apart from the insect hormones ecdyson and juvenile hormone, a large number of substances and methods of treatment have been employed in the past to evoke specific puff pattern changes and to investigate their connections with gene expression. In this context the puff induction observed in *Drosophila* after a rise in temperature is of general interest since it has served as a model system for analysing the RNA- and protein-synthetic activities of induced genes (cf. **9**.3.2 and Ref. [4])

In *D. melanogaster*, increasing the temperature from 25 °C to 37 °C for a short period of time stimulates eight loci of the larval salivary gland chromosomes to form puffs (Fig. **9**.6). The puffs can be induced by heat treatment of the larvae in vivo as well as on isolated salivary glands in vitro. Their size increases in proportion to temperature within the range of 25–37 °C. Induction of the puffs takes place within a few minutes and they achieve their maximum size within 20–30 min. When the temperature is lowered they regress within 3–4 h. Not only does a rise in temperature induce the appearance of new puffs or of puffs normally active at other phases of development, it can also prevent regression of development-specific puffs in six loci in *D. melanogaster*. Regression of almost all other puffs is induced simultaneously, so that a rise in temperature effects a complete reprogramming of the pattern of gene activity of the *Drosophila* genome. Such drastic alterations in gene activity can be observed in all tissues in vivo as well as in tissue cultures of *Drosophila* [4].

## 9.3 Molecular Basis of Gene Expression of Puff DNA

### 9.3.1 Synthesis and Processing of High Molecular Weight Puff RNA of *Chironomus*

Fractionation of the RNA of *Chironomus* salivary gland chromosomes by means of gel electrophoresis leads to a heterogeneous pattern of distribution that, apart from 4–5 S RNA, includes mainly high molecular weight RNA of about 18–70 S (cf. Fig. **9**.5 A). Following longer labelling it is also possible to demonstrate in preparations of chromosomal RNA the presence of (pre)ribosoma. RNA and Balbiani ring RNA, which constitute the larger portion of nucleoplasmatic RNA (Fig. **9**.5 B) and appear as "impurities" in extracts of chromosomal RNA.

In situ hybridization of high molecular weight RNA (for method see Fig. **9**.10) of individual chromosomes with salivary gland chromosomes shows that the RNA of single chromosomes hybridizes with a very large number of bands of the total chromosomal set. Thus in all probability many high molecular weight RNA molecules contain similar sequences. DNA reassociation experiments suggest these sequences to be transcribed from intermediate repetitive DNA sequences which are dispersed throughout the entire genome (cf. Table **1**.1). Sequences of this kind might, for example, correspond to the transcripts of mobile dispersed-repetitive genes or the $\alpha\beta$-DNA elements of heat shock puffs (cf. **9**.3.2) that have been identified in polytene chromosomes of *Drosophila* [4].

A large number of high molecular weight chromosomal RNA molecules are apparently immediately degraded at the chromosomes and/or processed to smaller molecules. In the nucleoplasm and cytoplasm, apart from Balbiani ring RNA, molecules of 16–30 S are often found (Fig. **9**.5 B). Most of them contain 3′-terminal poly (A) sequences and some are transported more rapidly into the cytoplasm than BR-RNA [18].

### 9.3.2 The Heat-Shock Gene Expression Program of *Drosophila melanogaster*

The *Drosophila* genes inducible by heat-shock (cf. **9**.2.2.3) offer an attractive system on which to study gene expression of eukaryotic cells. The virtue of the system is that by means of a simple stimulus, namely increase of temperature from 25 to 37 °C, the transcription of a battery of genes can be switched off within a few minutes and that of a smaller number switched on. In this way it is possible to study the expression of a few genes independently of the activity of most of the other genes and to carry out a thorough analysis, thanks to a number of practical advantages such as the fact that the object can be genetically manipulated [4].

Induction of the heat-shock puffs, of which those of the loci 87A, 87C, 63DC and 95D are the most prominent, is accompanied by intensive RNA synthesis. This can be shown directly by [$^3$H]-uridine incorporation followed by autoradiography and by the accumulation of RNA polymerase II demonstrated by means of immunological

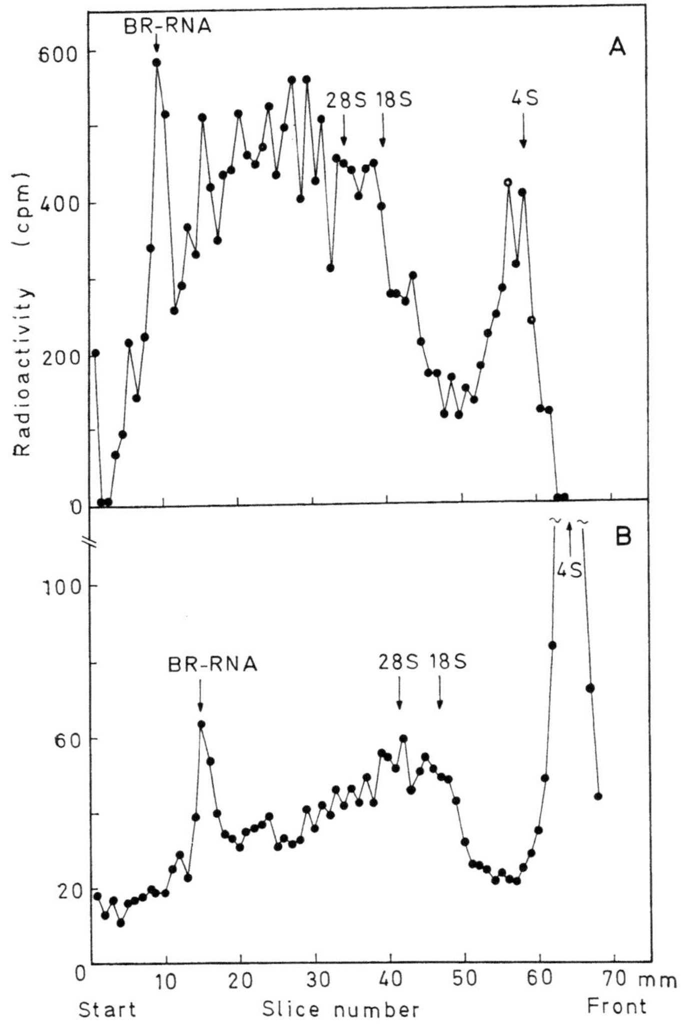


**Fig. 9.5**: Analysis of synthesis of chromosomal (**A**) and of nucleoplasmic RNA (**B**) from salivary gland cells of *Chironomus thummi* by agarose gel electrophoresis (Serfling, Orig. (A) and [17]).

After three h labelling with [$^3$H]-uridine RNA was extracted from about 100 chromosomes I–III (the "long" chromosomes in Fig. **9**.1) and of nucleoplasm from about 50 cell nuclei (for details see **9**.4.1.2). The RNA was separated electrophoretically on 1% agarose gels. Note the appearance of both heterogeneous RNA molecules and BR-RNA molecules ($M_r$ about 20 Md). The latter are accumulated in nucleoplasm (and on chromosomes) after long-term labelling. In these and the following electrophoretic profiles the electrophoretic mobility of both rRNA fractions of 28 S ($M_r$ 1.4 Md) and of 18 S ($M_r$ 0.72 Md), used as unlabelled markers in all runs, is indicated for comparison.

**Table 9.1**: The main heat-shock puffs of *Drosophila melanogaster*, their corresponding genes, messenger RNA's and polypeptides (from Ashburner & Bonner [4])

| Puff loci | Gene copies | Length of coding sequence (kb) | Length of mRNA (kb) | $M_r$ of polypeptide (kd) |
|---|---|---|---|---|
| 63DC | 1 | 2.6 | 3.05(A1) | 84 |
| 87A7 | $\geq 2$ | 2.05–2.25 | 2.55(A2) | 70 |
| 87C1 | $\geq 3$ | | | |
| 95D | 1 | 2.1 | 2.4(A3) | 68 |
| 67B | 1 | ? | (12S) (A5 + A6) | 27 |
| | 1 | | | 26 |
| | 1 | | | 23 |
| | 1 | | | 22 |

techniques (cf. **9**.1.3). Puffs synthesize disrete polysomal mRNA fractions wich can be separaetd by sucrose density gradient centrifugation into two peaks of 20S and 12S (cf. Fig. **9**.6B). Using PAA gel electrophoresis the 20S mRNA peak can be fractionated into three mRNA species (termed A1, A2, A3), which have been translated in vitro into the three heat-shock polypeptides of 84, 70 and 68 kd. The 12S mRNA peak contains at least two fractions (A4 and A5) representing the messengers for the 22, 23, 26 and 27 kd polypeptides. In situ hybridization of distinct radioactively labelled heat-shock mRNA species with the salivary gland chromosomes made possible the allocation of the individual mRNAs (and thus polypeptides) to the induced puffs. For example, the puffs 87A and 87C synthesize RNA molecules of the A2 fraction, which represent the mRNA for the 70 kd polypeptide. The results of these experiments together with analyses of cloned DNA fragments of individual heat-shock puffs are summarized in Table **9**.1.

Cloning of longer DNA segments of the heat-shock puffs in bacterial vectors (cf. Chap. **2**) and their characterization by digestion with restriction enzymes, Southern

▶

**Fig. 9.6**: Fine structure and expression of heat-shock puffs of *Drosophila melanogaster* (combined from Refs. [2, 4, 13]).

**A.** Specific induction of puffs 87A and 87C by a temperature rise from 25 °C to 37 °C within 40 min.

**B.** Characterization of heat-shock RNA. The RNA synthesized at 37 °C was passed through an oligo(dT)-cellulose column, and the mRNA bound to the column was fractionated by sucrose gradient centrifugation.

**C.** Change of polysome profile by heat-shock treatment. Thick arrow: Heat-shock polysomes corresponding in size to polysomes with about 30 ribosomes.

**D.** Electropherograms of proteins from normal and heat-shock treated *Drosophila* cells. Note the preferential synthesis of a few proteins ($M_r$ 70 kd, 84 kd, 68 kd and 22–27 kd) in heat-shock treated cells.

**E.** Restriction map of DNA from clone 132E3. Its DNA was identified to code for the heat-shock protein of 70 kd and localized in puff 87C.

$z_c$, coding sequence; $z_{nc}$, noncoding sequence; X, Y, DNA segments in front of the 5′-end of the z sequences which are either homologous *(X)* or non-homologous *(Y)* to corresponding segments in front of other z copies of puffs 87A and 87C. *Arrows* mark the direction of transcription.

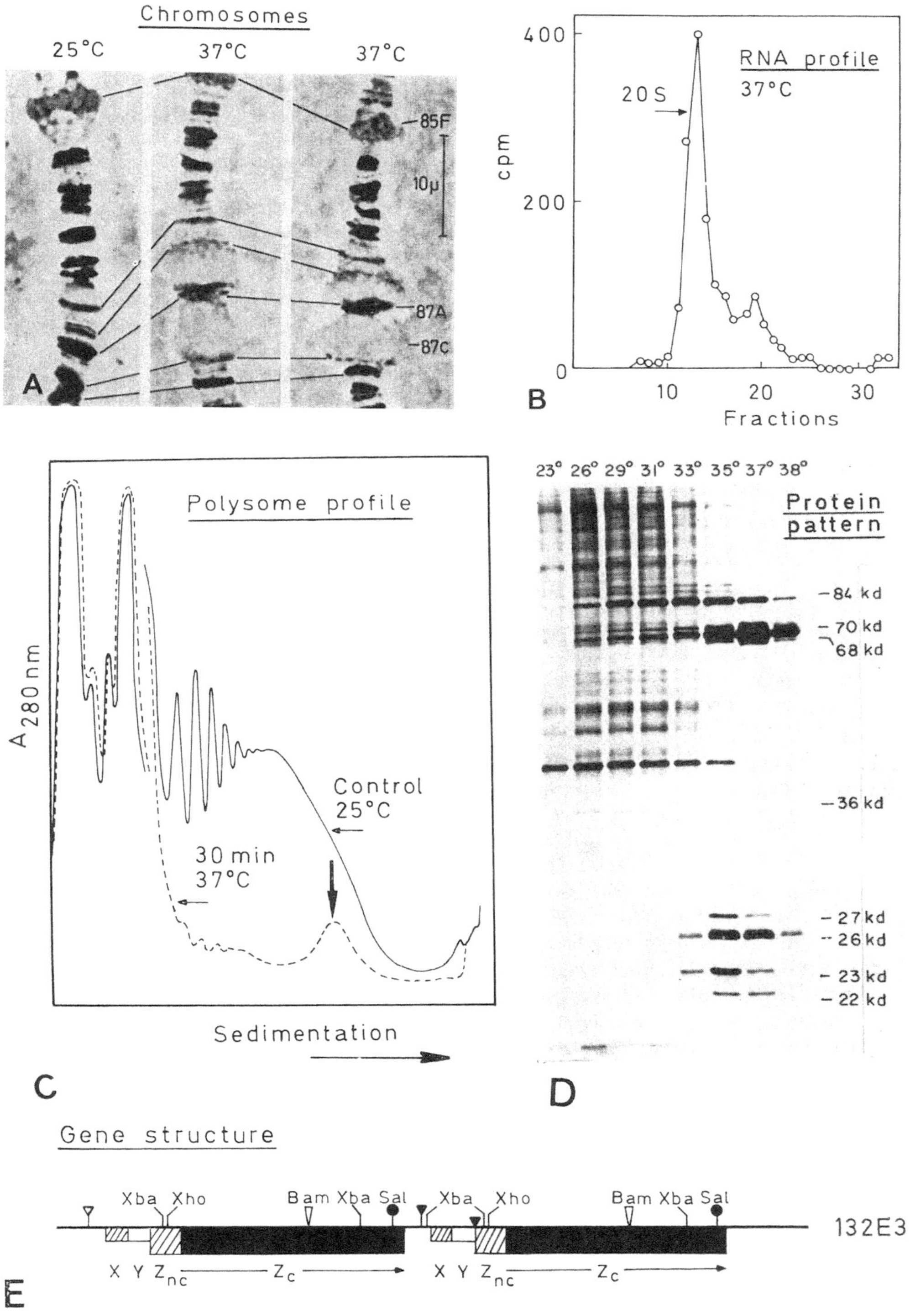
Chromosomes
25°C
37°C
37°C
85F
10µ
87A
87C
A
RNA profile
37°C
20S
cpm
400
200
0
10
20
30
Fractions
B
Polysome profile
A 280 nm
Control 25°C
30 min 37°C
Sedimentation
C
23° 26° 29° 31° 33° 35° 37° 38°
Protein pattern
84 kd
70 kd
68 kd
36 kd
27 kd
26 kd
23 kd
22 kd
D
Gene structure
Xba Xho Bam Xba Sal Xba Xho Bam Xba Sal
132E3
X Y Znc Zc X Y Znc Zc
E

blotting hybridization (cf. **2**.6) and electron microscopic techniques such as R-loop- and heteroduplex-mapping, have made it possible to draw far-reaching conclusions about the fine structure and expression of heat-shock loci. This has been particularly successful in the case of the most prominent heat shock puffs 87A and 87C [4, 9].

By incorporating *Drosophila* DNA in the plasmid pSF 2124 and selecting the resultant recombinant bacterial colonies by means of colony hybridization (cf. **2**.6) with polysomal heat shock mRNA, Gehring and Tissieres and co-workers obtained two recombinant plasmids (56H8 and 132E2) that hybridized intensively in situ with the 87A and 87C puffs. In the so-called HART assay (cf. **2**.8) the two plasmids with 8.2 kb (56H8) and 9.7 kb (132E3) *Drosophila* DNA inhibit translation of the 70 kd polypeptide, so that they definitely possess the coding sequences for this polypeptide. Figure **9**.6 E shows results of restriction analysis of one of the two plasmids. The two plasmids contain sequences in common, that can be allocated to DNA segments measuring 3 kb. These sequences are present singly in plasmid 56H8 but occur twice in plasmid 132E3.

On the basis of several characteristics the 3 kb segments can be subdivided into three sections, the x, y and z elements. The latter type of element, measuring 2.5 kb, consists of a coding portion ($z_c$) and a 5′-terminal non-coding portion ($z_{nc}$) (Fig. **9**.6E), and is of very similar structure in all 3 kb units. The same applies to the 0.18 kb x elements of the clones, whereas all y elements separating the x and z elements are of different structure. Deletions of x elements do not affect expression of $z_c$ sequences. The $z_c$ elements of all genes so far investigated lack "intron" sequences, typical for many eukaryotic genes (cf. Figs. **1**.8 and **7**.2). As shown by sequence analysis, they code for a soluble protein of a hitherto unknown sequence [9]. In front of all $z_c$ segments of three independently cloned heat-shock genes originating from puff 87A as well as 87C, there is the Hogness box 5′-TATAAATA-3′ ending at position −23 (cf. Table **1**.8) and at position −70 to −80 the sequence 5′-GTTGGTTGG-3′ which matches a consensus sequence of many eukaryotic genes at this position in only four of nine nucleotides, but matches that of the fibroin gene at six out nine nucleotides (cf. **9**.**4**.**2**). It is, therefore, tempting to suggest that this sequence motive is class-specific differing – as the structure and properties of RNA polymerase II – between such non-related species as insects and vertebrates. Moreover, it is note-worthy that the 5′-flanking regions of the three sequenced heat-shock genes are very similar until position −250 from the proposed site of transcription initiation. These regions of unknown function correspond well in length with the $z_{nc}$ elements mentioned above [9, 22].

The genes of puffs 87A and 87C coding for the 70 kd polypeptides occur in several copies. Measurements of the number of genes revealed 5 to 9 copies per genome. In an isogenic *Drosophila* strain two gene copies with identical structure were demonstrated in the 87A7 locus and three in the 87C1 locus. The two genes of locus 87A7 occur in a divergent orientation about 1.7 kb apart. The 87C1 locus, on the other hand, carries an isolated 70 kd gene separated from two tandem 70 kd genes by a region of about 38 kb containing the so-called $\alpha\beta$-sequence elements (see below) [10]. A similar number and distribution of genes could also be demonstrated in the DNA of *Drosophila* tissue culture ($K_c$) cells.

Apart from containing the genes of the 70 kd polypeptide, the DNA of the 87C puff has also been shown to contain moderately repetitive sequence elements, whose RNA occurs in polysomal heat-shock puffs but is apparently not translated. Using cloned DNA segments it was possible to identify in the 87C locus 10–15 tandemly repeated copies in the order $\alpha\beta\alpha$ as well as additional $\alpha$-segments joined to a third so-called $\gamma$-sequence element.

These 20 or more repetitive elements are organized in three clusters separated by spacer regions of varying sequences. $\gamma$-Elements (but not $\alpha$- and $\beta$-elements) also occur in the locus 87A, whereas $\alpha$- and $\beta$-elements can be demonstrated in chromocentric DNA and in other loci. In these sites, however, the $\alpha$- and $\beta$-elements are not arranged in tandem but in a dispersed repetitive order. They are thus very similar to the "copia" elements and other species of mobile dispersed repetitive genes of *Drosophila* (cf. **4**.**3**.**4**).

The function of the polysomal RNA transcribed from the $\alpha\beta$-elements is unknown. Like its coding DNA it appears to have no influence on the regulation of the induction and repression of heat-shock puffs nor on the synthesis of any heat-shock proteins, since 87C deletion mutants without $\alpha\beta$-elements are capable of normal synthesis of all heat-shock proteins [4].

In contrast to the 87A and 87C genes, those of the heat-shock puffs 63DC and 95D are non-repetitive. Only the 63DC gene contains an intron of 0.8 kb. In the HART in vitro assay (cf. **2**.8) the cloned DNA sequences of the 63DC puffs selectively inhibit synthesis of the 84 kd protein and those of puff 85D inhibit that of the 68 kd proteins. This permits the unequivocal allocation of these polypeptides to the loci coding for them (cf. Table **9**.1). The locus 67B contains the four unique genes for the heat-shock proteins of $\overrightarrow{27\ \text{kd}}$, $\overleftarrow{26\ \text{kd}}$, $\overrightarrow{23\ \text{kd}}$ and $\overrightarrow{22\ \text{kd}}$ in the indicated orientation (arrows). They are separated by spacers of 1.2 to 4.5 kb DNA [23, 24].

## 9.4 High Molecular Weight Structural Proteins of Insects are Coded by Internal Repetitive Genes: A Comparison of Balbiani Ring and Fibroin Genes

The salivary glands of chironomids and the silk glands of the silk moth larvae *(Bombyx mori)* are labial glands of homologous origin [3]. Their highly specialized cells produce large quantities of a few structural proteins. In the case of *Chironomus* they are required for cylindrical tubes in which the larvae live. The fibroin of the silk glands of *Bombyx*, on the other hand, is needed for spinning the pupal silk worm cocoons. These proteins represent mainly high molecular weight proteins with a repetitive amino acid sequence coded by genes with an internally repetitive structure. In the case of *Chironomus* salivary glands these genes are represented by Balbiani ring DNA, consisting of multitudinous short repetition units (cf. **9**.4.1.3). Similarly, the fibroin gene ($\sim$ 16 kb of DNA) contains, besides an "intron" segment, mainly short repetitive DNA segments that code for the amino acid sequence -Ser-Gly-Ala-Gly-Ala-Gly-, repeated thousands of times in fibroin (cf. **9**.4.2). Both gene systems give rise to high molecular weight pre-mRNA species which are transported into the cytoplasm without any appreciable shortening and are there translated by very large polysomes.

These examples serve to illustrate the common characteristics of the synthesis of *Chironomus* salivary gland secretory proteins and the synthesis of fibroin in the silk moth larvae. Both systems, which will be discussed in more detail, can be regarded as examples of genes coding for internally repeated high molecular weight structural proteins of insects.

### 9.4.1 Balbiani Ring Genes of *Chironomus*

#### 9.4.1.1 Balbiani Rings and the Synthesis of Secretory Proteins

Balbiani rings are the most active puffs of the *Chironomus* salivary gland chromosomes and their occurrence is strictly tissue-specific (cf. **9**.2). It is therefore tempting to assume that they code the mRNA necessary for the synthesis of the secretory proteins. The first and so far only firm evidence for a direct connection between the occurrence of a Balbiani ring and the synthesis of a protein was obtained by cytogenetic analysis of puffing and electrophoretic investigations on the proteins of larval salivary glands of *C. tentans* and *C. pallidivittatus* [3, 7].

The salivary glands of *Chironomus* consist of a main lobe comprising 30–35 cells and an anterior lobe with 3–5 cells near the duct leading out of the gland (Fig. **9**.7). The so-called special cells

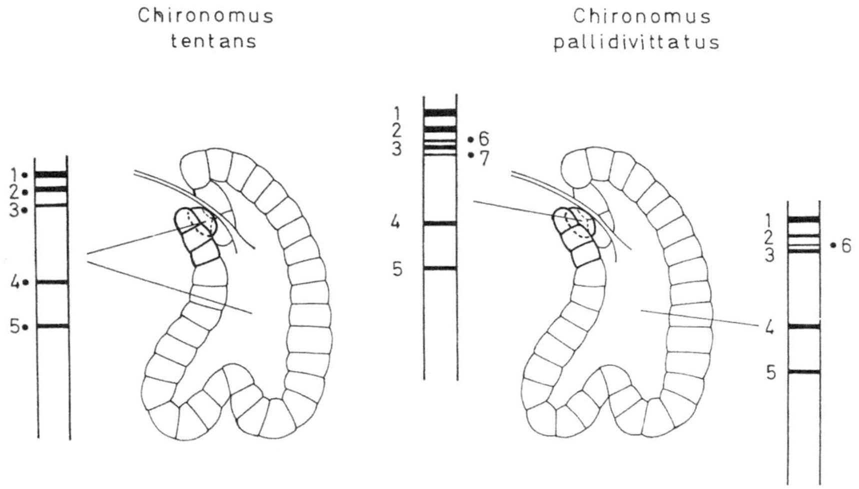


**Fig. 9.7**: Gel electrophoretic analysis of secretory proteins from main and anterior (special) lobe of larval salivary glands of *Chironomus tentans* and *Chironomus pallidivittatus* (after Grossbach [8]).

Secretory proteins were isolated from the lumen of salivary gland lobes and fractionated on 7 % PAA gels after cleavage of their disulphide bridges. Note the appearance of two additional bands in the protein pattern of salivary glands from *C. pallidivittatus* which is correlated with the activity of two Balbiani rings BR3 (band no. 6) and BR4 (band no. 7).

of the anterior lobe of the salivary gland of *C. pallidivittatus* synthesize a granular secretion that differs distinctly from that produced by the anterior lobe of the salivary gland of *C. tentans*. Cytogenetical analysis revealed that the information for the synthesis of the granular secretion is coded in the DNA of the Balbiani ring BR4 that occurs on the chromosomes of the special cells of *C. pallidivittatus* but is missing in *C. tentans*. Since the polypeptide pattern of the special cell secretion of *C. pallidivittatus* exhibits an additional polypeptide whose appearance correlates with that of BR4 it can be concluded that the RNA synthesized in the BR4 (or at least in part of it) is responsible for the synthesis of this secretion protein component [7]. In *C. thummi*, the extra polypeptide of special cells ($M_r$ 160 kd) is highly glycosylated and represents the main synthesis product of these cells [8].

In the secretion of salivary glands of *C. tentans* and *C. thummi* two very large proteins with an $M_r$ in the range of 700–800 kd were detected corresponding in length to about 55 % of 75S BR-RNA. It has been shown for *C. tentans* that the fingerprint peptide patterns of these proteins resemble those of in vitro translation products of cytoplasmic 75S BR-RNA. These proteins might, therefore, correspond to translation products of BR-RNA [16]. Up to now, however, the synthesis of 700–800 kd proteins could not unequivocally be demonstrated by in vitro or in vivo translation of 75S BR-RNA. Moreover, the fingerprint patterns of shorter secretory polypeptides of *C. thummi* secretion resemble very much those of large proteins. It is, therefore, an open question, whether the Balbiani ring DNA codes – as formerly expected [3, 5] – for only one huge polypeptide, or for several polypeptides of different lengths, but similar properties and functions (cf., e.g., proteins coded by the heat-shock puff 67B, Table **9**.1).

### 9.4.1.2 Synthesis of High Molecular Weight Balbiani Ring RNA

Balbiani rings are large enough to be isolated under the stereomicroscope at a 300-fold magnification, using the glass needles of a micromanipulator. After extraction and radioactive labelling of the BR-RNA it can be analysed by means of electrophoretic separation and hybridization experiments, and then characterized with respect to important parameters such as their synthesis, size and repetition degree of their DNA sequences [3, 5, 6, 17].

To obtain labelled BR-RNA salivary glands are first incubated for 30–180 min in an insect medium containing radioactively labelled precursors. The glands are then fixed and the components of the cells (cytoplasm, nucleoplasm, chromosomes, nucleoli and Balbiani rings) are isolated in a non-aqueous medium (ethanol: glycerol 1: 1). For a preparation of BR-RNA 50–150 BRs are required, dependent on the length of incubation. The BRs are collected in a drop of mineral oil, lysed in 10 $\mu$l of pronase-SDS solution and the RNA obtained is separated electrophoretically with unlabelled RNA on agarose gels. The distribution of the radioactively labelled BR-RNA in the agarose gel is measured in a liquid scintillation counter after slicing of the gels or by fluorography. These methods permit studying the RNA synthesis of single nuclear components, e.g., of the nucleolus and individual Balbiani rings, in more detail than was possible with the cell nucleus fractionation techniques employed until now. A great advantage is that the fixation of the salivary glands prevents mixing of the RNA of different nuclear compartments.

The electrophoretic separation of microisolated RNA of the *Chironomus* Balbiani rings yields, in contrast to the preparations of chromosomal RNA, remarkably homogeneous radioactivity profiles of asymmetrical form (cf. Fig. **9**.5 A with Fig. **9**.8 B). Only in the case of the RNA of the BRb of *C. thummi* two distinct RNA peaks were obtained in the radioactivity profile (see Fig. **9**.8 A), either representing the synthesis of two (different) RNA fractions or resulting from RNA processing of one fraction already within BRb. Comparative separations of the RNA of Balbiani rings BRc and BRb with pre-ribosomal and ribosomal RNA (see Fig. **9**.8) on 1% agarose gel gave a $M_r$ of about 20–35 Md. However, a thorough electron microscopical analysis of isolated *C. tentans* BR-RNAs suggest that these figures are apparently too high and that values of about 12 Md reflect the actual molecular weight of the longest Balbiani ring RNA molecules [5].

As far as the rates of BR-RNA initiation and elongation are concerned, transcription of the BR genes does not differ appreciably from that of bacterial genes (cf. also Table **4**.6). Employing the initiation inhibitor 5,6-dichloro-1$\beta$-ribofuranosylbenzimidazol (DRB) it could be shown that the RNA of BR2 of *C. tentans* is synthesized in about 20–30 min. Using this value together with a gene size of $\sim$ 37 kb and an initiation rate of 2–3 RNA chains per min (deduced from electron micrographs), an elongation rate of 25 nucleotides/s was calculated. More recent analyses of electron micrographs have yielded rates of 6 RNA molecules/min and elongation rates of 31 nucleotides/s. For *E. coli* initiation rates of 1–2 RNA chains/min and operon and elongation rates of 37–45 nucleotides per s (at 37 °C) or 16–17 nucleotides per s (at 15 °C) have been calculated (cf. [11] and Table **4**.6).

Spreading the DNA of isolated chromosomes IV of *C. tentans* according to the technique developed by Miller for demonstrating transcription complexes gives an idea of the transcription of the BR-DNA (Fig. **9**.9). The active BR transcription units are about 7.7 $\mu$m long and have approx. 125 growing RNP chains. The number of these growing chains varies considerably from gene to gene showing that transcription of the individual genes of the polytene BR is individually regulated. The chromatin fibres inside the transcription complex, in contrast to inactive fibres, possess only few nucleosomes and are apparently stretched to a greater extent than the latter. Every transcription complex is preceded by a fibrillar chromatin segment measuring about 0.18 $\mu$m, that might possibly represent the region of initiation of RNA synthesis (see arrow in Fig. **9**.9).

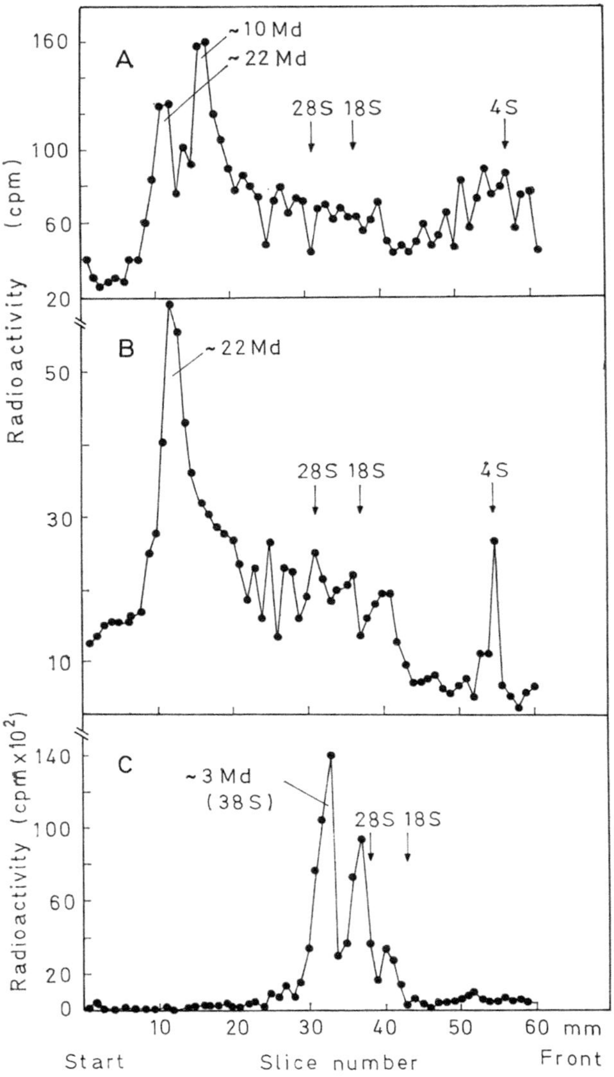

A
~10 Md
~22 Md
28S 18S
4S
B
~22 Md
28S 18S
4S
C
~3 Md
(38S)
28S 18S
Radioactivity (cpm)
Radioactivity (cpm x10²)
160
100
60
20
50
30
10
140
100
60
20
0
10
20
30
40
50
60 mm
Start
Slice number
Front


Fig. 9.8

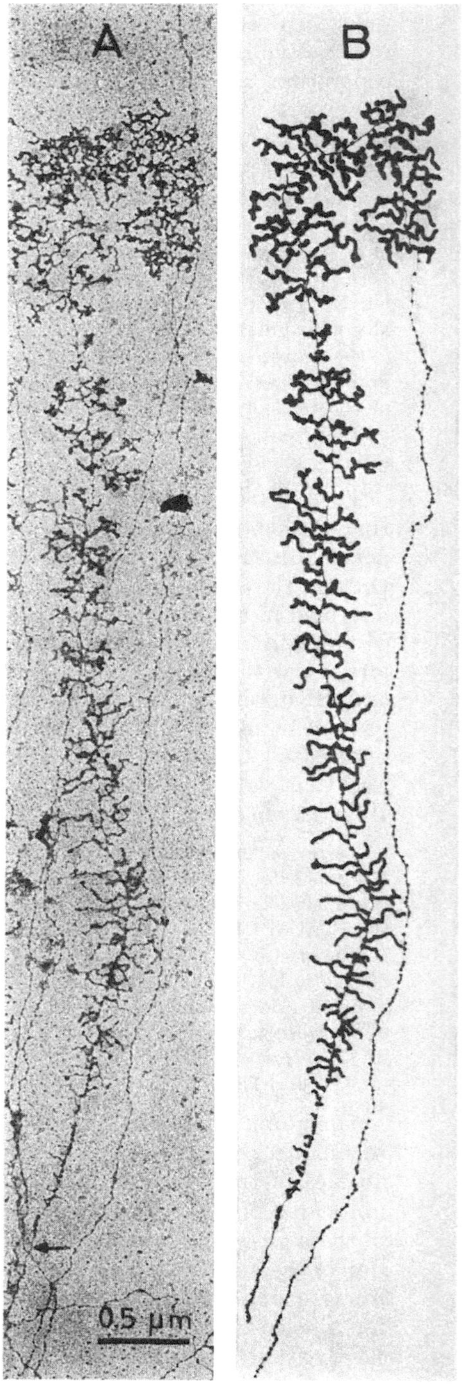

A
B
0.5 µm


Fig. 9.9

### 9.4.1.3 Molecular Characterization of BR-RNA and BR-DNA

Apart from experimental data concerning the synthesis and molecular weight of BR-RNA little is known on its molecular characteristics. Hybridization of micro-isolated BRc-RNA from *C. thummi* with DNA in excess showed that the BR-RNA contains repetitive sequences transcribed from internally repetitive gene sequences. From the specific binding of BR-RNA on poly(U)-sepharose or oligo(dT)-cellulose columns it seems likely that the BR-RNA is polyadenylated shortly after transcription and bears a 3′-terminal poly(A) sequence comparable to that of other eukaryotic mRNAs (cf. **1**.3.3.1).

The quantities of purified BR-RNA available are too small for experiments necessary for detailed analyses of their sequence organization. Following the development of methods for the generation of recombinant DNA (cf. **2**.3) this problem can nowadays be circumvented by cloning BR-DNA.

Since its base composition and sequence organization differ from those of genomic DNA, BR-DNA can be separated from the former by CsCl-actinomycin D equilibrium density centrifugation. After separation, in order to clone BR-DNA, poly(C)-homonucleotides were added by treatment with terminal transferase to the 3′-end of the BR-DNA. Under suitable conditions this homonucleotide stretch hybridizes with the poly(G)-homonucleotide ends of an appropriately treated plasmid vector pBR322, cleaved at the Pst I site of the ampicillin resistance gene (cf. Fig. **2**.3). After ligation and transformation of *E. coli* the hybrid plasmids can be multiplied. The recombinant clones containing BR-DNA were selected with [$^{125}$I]-labelled BR-RNA by colony hybridization (cf. **2**.6). In situ, the cRNA of two of the selected recombinant clones hybridized either specifically with the BRc (clone pCthBRc-19) or with the BRb (clone pCthBRb-16) of the salivary gland chromosomes (Fig. **9**.10) and hence represent purified BRc- and BRb-DNA respectively [21].

---

◀

**Fig. 9.8**: Gel electrophoretic characterization of RNA synthesis products of two Balbiani rings BRc **(A)** and BRb **(B)** and of the nucleolus **(C)** from *Chironomus thummi*, respectively (from Serfling [17]).

The RNA of larval salivary glands was labelled for 2 h **(A, B)** or 3 h **(C)** by incubation of glands in an isotope-containing insect medium, and about 100 chromosomes IV were microisolated and microdissected into BRs and nucleoli. The BR-RNA and nucleolar RNA were extracted by SDS-pronase lysis and fractionated on 1 % agarose gels along with reference RNAs. A comparison of the radioactivity profiles A–C shows that in Balbiani rings probably only one (as for BRb: see B) or a few main fractions (such as in BRc: see A) are synthesized which are significantly larger than nucleolar pre-rRNA.

◀

**Fig. 9.9**: Electron micrograph of a spread transcription unit of Balbiani rings from *Chironomus tentans* (from Lamb & Daneholt [11], with permission).

Larval salivary glands of *C. tentans* were lysed in a salt-detergent medium (containing, e.g., 0.2 % triton X 100 and 1 % nonidet P.42) by several-fold pipetting of glands. Afterwards, chromosomes IV carrying the three Balbiani rings of salivary gland chromosomes were transferred into an 100 μl droplet of spreading solution (distilled water, adjusted by borate buffer to *p*H 9 and containing 200 μg/ml tRNA), maintained for 60 min at 20 °C and then centrifuged (10 min, 1,100 ×g) through a 0.1 M sucrose – 10 % formaline cushion onto an electron microscope grid. After washing the preparations the spread material is stained by 1 % phosphotungstic acid (dissolved in 70 % ethanol), air-dried and photographed under the electron microscope. *Arrow* labels the region of the apparent transcription origin. **A** electron micrograph; **B** schematic drawing of the micrograph.

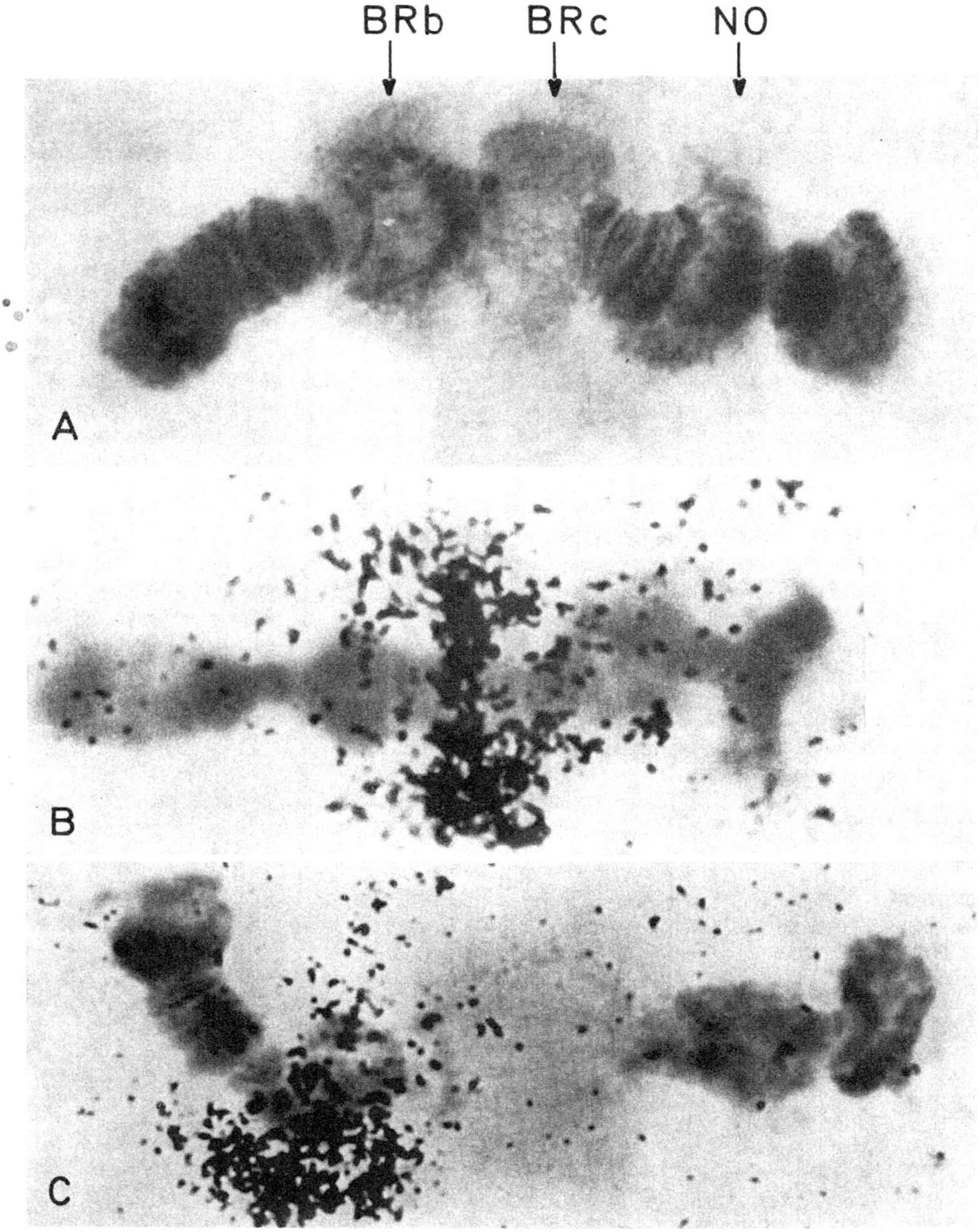


**Fig. 9.10:** In situ hybridization of Balbiani ring RNA of *Chironomus thummi* with chromosomes IV of salivary gland nuclei (from Serfling [17]).

Larval salivary glands are fixed in ethanol-acetic acid (3:1) for 45 s, transferred into 45% acetic acid and squashed. The squash preparation is frozen on dry ice and, after removing the cover glass, it is treated with RNase. After severalfold washes in 2×SSC (1×SSC: 0.15 M NaCl, 0.015 M Na-citrate, *p*H 7.0) the DNA of chromosomes is denatured (in 0.2 N HCl) for 20 min at 37 °C. Hybridization with radioactively labelled BR-RNA (cf. legend to Fig. 9.8 for preparation) dissolved in 3×SSC, 50% formamide proceeds at 40 °C overnight. Then, the preparations are washed, again treated by RNase and covered with an appropriate fotoemulsion. After exposition for several weeks the staining (silver grains) of the film demonstrates the hybridization which

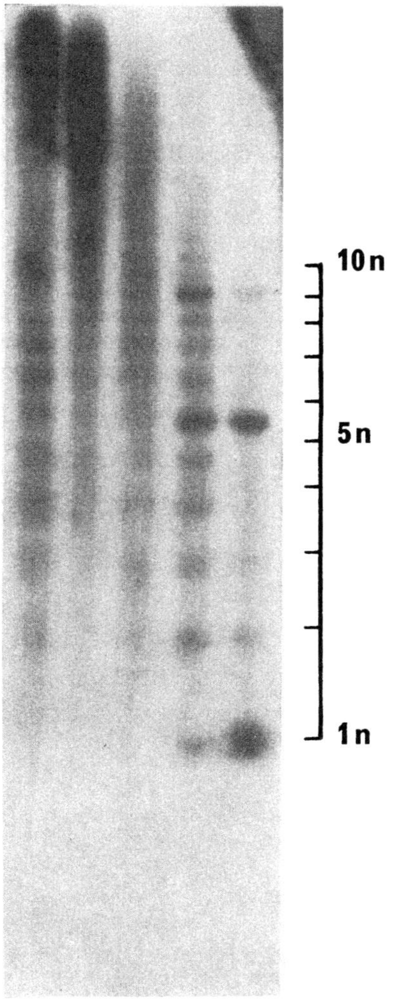


**Fig. 9.11**: Detection of repetitive structure of BRc DNA from *Chironomus thummi* (from Wobus et al. [21], with permission).

20 $\mu$g DNA of *C. thummi* larvae was fractionated after restriction digestion on a 1.5 % agarose gel, transferred by Southern blotting onto nitrocellulose filters (cf. **2.6**) and hybridized with [$^{32}$P]-labelled DNA of BRc clone pCth BRc-19. Larval DNA was digested by the following restriction enzymes: *Lane 1*, Ava II; *lanes 2 + 3*, Alu I (two different digestion conditions); *lanes 4 + 5*, Hinf I (two different digestion conditions). Note that the appearance of DNA ladders is due to a regular arrangement of restriction recognition sites within BR-DNA and is, therefore, an indication for the internal repetition of BR-DNA. Depending on the restriction enzyme used *n* corresponds to about 250 bp (cf. **9.4.1.3**).

◀

took place in situ between the BR-RNA (or in other cases between a cloned BR-DNA fragment, cf. **9.4.1.3**) and the chromosomal DNA sequences. *NO*, nucleolus organizer.

**A**, Chromosome IV, stained by carmine/orceine; **B**, same, hybridized to BRc RNA; **C**, same, hybridized to BRb RNA.

The BRc-DNA cloned in the recombinant plasmid has a highly repetitive structure similar to that of complex satellite DNA. Digestion of the approximately 1.3 kb DNA sequences with restriction enzymes Alu I, Ava II and Hinf I reveals the existence of multiple DNA segments approximately 250 bp in length. These can be especially well demonstrated using Southern blot hybridization in partially digested and electrophoretically fractionated genomic DNA (Fig. **9**.11). The BRc-DNA segments are identified in the genome DNA by hybridization with labelled BRc clone DNA or BR-RNA. Since both in situ hybridization of cloned DNA and hybridization of microisolated BR-RNA with the giant chromosomes revealed only the Balbiani rings to be labelled (cf. Fig. **9**.10), the fragments identified within the genomic DNA definitely represent the BR-DNA.

Detailed restriction mapping and sequence analyses of the cloned BRC-DNA fragment led to the identification of four tandemly repeated DNA units of 54 to 58 bp in length and, in 3′ direction, of four repeat units of 252 bp in length. The first share homologies between 60 % and 98 % to a prototype sequence which can be derived from the sequence data of the four repeats. The four

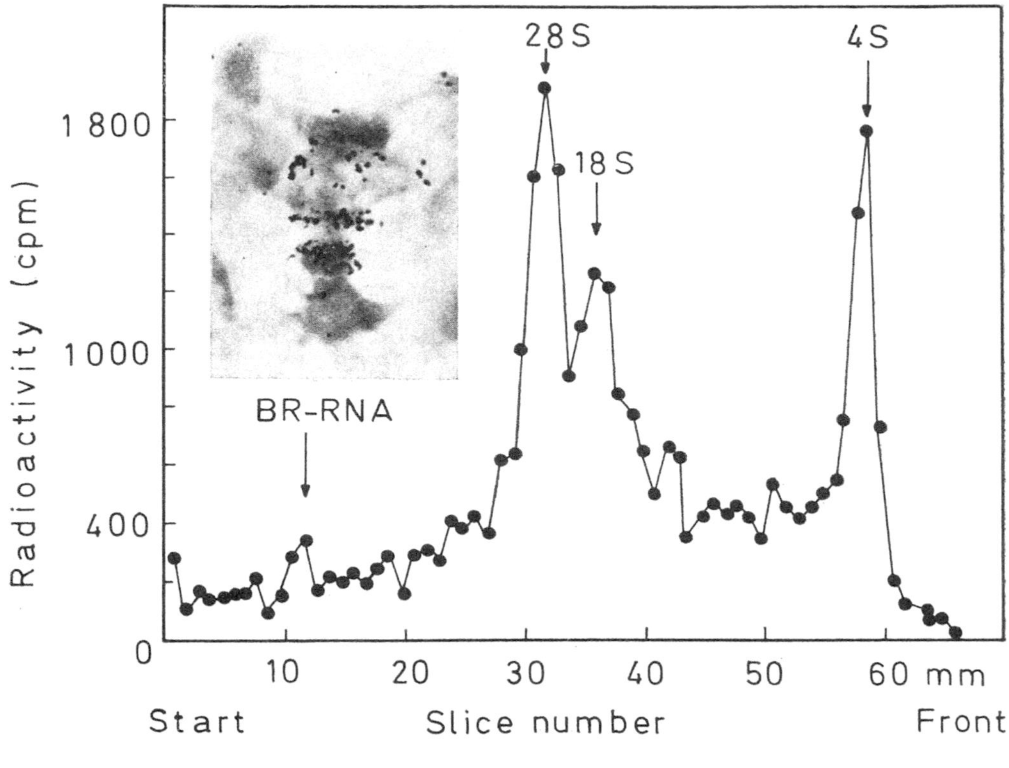


**Fig. 9.12:** Analysis of cytoplasmic RNA of larval salivary gland cells of *Chironomus thummi* (from Serfling & Huth [18]).

Salivary gland RNA was labelled in vivo for 7d and extracted after microdissection of cytoplasm by pronase-SDS lysis (see **9.4.1.2** for details). The RNA was fractionated on 1 % agarose gels which have been sliced for measuring the radioactive profile or for in situ hybridization experiments. Note the appearance of BR–RNA within cytoplasmic RNA that could unequivocally be demonstrated in gel slices 11 and 12 by melting the agarose slices on squash preparations of chromosome IV followed by in situ hybridization. The result of in situ hybridization is shown in the inset.

252 bp are (almost) completely identical in primary structure and represent, in all likelihood, the major repeat unit of whole BRc gene of 37 kb. This is shown in Fig. 9.11 where partial restriction enzyme digestion patterns of chromosomal DNA are presented showing fragment ladders of 250, 500, 750 bp etc. after Southern blotting and hybridization with labelled DNA of the BRc clone pCth BRc-19. Careful consideration of all data suggest a plausible scenario of BRc gene evolution by unequal crossing over which bears some similarity to the postulated process of satellite DNA evolution [21].

#### 9.4.1.4 Nucleo-Cytoplasmic Transport of BR-RNA

Electron micrographs have revealed RNP granules of 40–50 nm diameter in the nucleoplasm and in the cytoplasmic zones adjacent to the nucleus. They resemble those of the BR's in size and form. This finding, which does not support the idea of intranuclear reduction in size of BR-RNA, has been confirmed by biochemical data. It has been shown that not only nucleoplasmic but also cytoplasmic RNA preparations contain BR-RNA, its size resembling that of the longest of the RNA molecules isolated from the Balbiani rings (compare Figs. **9**.5, **9**.8 and **9**.12). Electrophoretic analyses thus revealed no intranuclear nucleolytic processing of the BR-RNA. However, this does not exclude the possibility that BR-RNA undergoes a variety of alterations during nucleo-cytoplasmic transport (e.g., methylation, changes in conformation etc). In addition to electron microscopical analyses, in situ hybridization experiments (Fig. **9**.12) and kinetic analyses also indicate unequivocally that a large part of the high molecular weight RNA in the nucleoplasm and cytoplasm is BR-RNA.

### 9.4.2 Structure and Expression of the Fibroin Gene in Silk Glands of *Bombyx mori* Larvae

The silk glands of *Bombyx mori* are paired organs consisting of anterior, central and posterior segments. They comprise 250–500 highly polyploid cells, of which those of the posterior segment synthesize almost exclusively fibroin in the fifth larval stage. Because of the high level of polyploidy (about $2^{17}$ N) up to 60 ng of fibroin can be produced per day by a single silk gland cell. Silk gland cells thus represent one of the most efficient protein synthesis systems, even surpassing, per unit of tissue weight, the serum albumin synthesizing system of avian liver cells by a factor of 60 (Suzuki, in Ref. [3]).

Fibroin consists of 60 % of the so-called crystalline and of 40 % of the amorphous structural elements. The former component consists entirely of the sequence Gly-Ala-Gly-Ala-Gly-[Ser-Gly-(Ala-Gly)$_n$]$_8$-Ser-Gly-Ala-Ala-Gly-Tyr, where n is normally 2. The amorphous elements are composed of several shorter sequences containing almost the same hydrophobic amino acids as the crystalline part. All these sequence elements together form a highly repetitive polypeptide of 350–420 kd, whose gene is present in various alleles in a single population [3, 12].

The fibroin gene is present in the genome in a single copy and accounts for about 0.004 % of the genomic DNA. Since its base composition (about 60 % G + C) differs distinctly from that of genomic DNA (G + C content 39 %) and because of its unusual sequence organization due to internal repetition, fibroin DNA can be readily separated from genomic DNA by means of CsCl gradient centrifugation, and thus enriched. This works especially well in CsCl-actinomycin gradients since actinomycin binds preferentially to DNA segments rich in G + C, and thus appreciably alters their buoyant density. Ohshima and Suzuki [14] used this technique to accumulate fibroin DNA for cloning in the plasmid pMB 9.

The fibroin DNA clone of Ohshima and Suzuki [14] was obtained by centrifugation (twice or more) of carefully isolated *Bombyx* DNA in CsCl-actinomycin gradients. Fibroin DNA forms two bands in CsCl-actinomycin gradients, corresponding to fragmented and unfragmented fibroin DNA. The former is concentrated about 1000 times by centrifuging twice, so that even in a few recombinant transformands constructed with this DNA several plasmids appear that contain fibroin gene material. However, the clone obtained in this manner corresponds only to those portions of the gene with fibroin-specific segments of a maximum length of 4 Md. Longer fibroin DNA segments, which make up a very large part of the fibroin gene, were only obtained by using unfragmented fibroin DNA, which is not so well concentrated by the centrifugation method. Following Eco RI digestion of this fraction and incorporation into Eco RI-treated pMB9-DNA a clone was obtained containing about 14 Md (= 21 kb) fibroin DNA, i.e., nearly all of the coding- and 5'-flanking-sequences of the fibroin gene. This clone, termed pFb29, has been used in nearly all subsequent analyses of fibroin gene structure [14, 19, 20].

Digestion with restriction enzymes and sequence analyses of the DNA of clone pFb29 brought interesting results concerning the structure of the segments flanking the 5'-end as well as of longer portions of the coding sequences of the gene. The fibroin gene has an "intron" segment of 970 bp, situated 64–66 bp downstream from the 5'-end of the gene, i.e., from its cap position. The intron was sequenced in its entire length. Similar to other introns of protein-coding genes the termini correspond to the "GT-AT rule" and to the "consensus"-sequences of exon-intron boundaries that are apparently involved in pre-mRNA processing (cf. Fig. **1**.10). Situated in the neigbourhood of both intron termini there is a stretch of about 20 bp representing inverted repeats to each other.

The "cap" sequence 5'-ATCAGCATCAG-3' of the fibroin gene differs from analogous sequences in globin- and other mammalian-genes, whereas the sequence 5'-TATAAAA-3', the so-called "Hogness box", typical for protein-coding genes of eukaryotes at position −30 to −24 (cf. Table **1**.8), is also present in front of the 5'-end of the fibroin gene. A second sequence, 5'-GTTGTTGACA TTTTT-3', situated at position −85 to −71 upstream of the "cap"-site, can also be important in initiation of transcription (cf. legend to Fig. **4**.14 and **9**.3.2). Hogness box and the −77 region may together constitute essential elements of the promoter region of the fibroin gene [19, 20].

The structure of the fibroin gene is shown diagrammatically in Fig. **9**.13. The major part of the gene is composed of highly repetitive sequences coding Gly-Ala peptides, beginning at nucleotide position +1448 and apparently extending up to the 3'-end of the gene. The 5'-part of the gene differs distinctly from this sequence and is interrupted by the intron which spans nucleotides +64–66 to +1034–1036. Owing to lack of protein-sequence data it is uncertain where the protein coding sequences of the gene begin and where the intron is exactly situated within the gene [19].

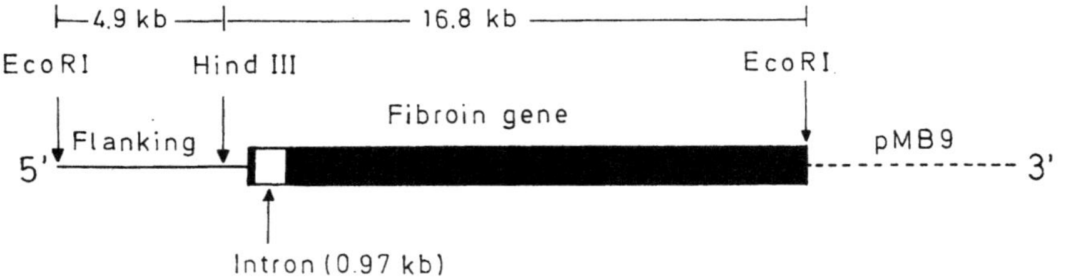


**Fig. 9.13**: Structure of the fibroin gene of the silk moth *Bombyx mori* (after Tsujimoto & Suzuki [19]).

The structure of the gene was determined by restriction mapping and partial sequence analyses after cloning of the Eco RI fragment in plasmid vector pMB9. *Full boxes* correspond to exon sequences, the *open box* denotes the intron near the 5'-terminus of the gene. For further details cf. **9**.4.2.

Some pulse-labelled, nuclear fibroin-RNA molecules have a larger apparent molecular weight than polysomal fibroin mRNA. This, in addition to other experimental results, suggests the existence of high molecular weight RNA precursors that represent continuous transcripts of both intron and protein-coding exon sequences. These molecules, of about 17 kb, are then transformed by intranuclear processing into cytoplasmic fibroin mRNA molecules of about 16 kb (5.8 Md), possessing all the typical properties of eukaryotic mRNAs (cf. **1**.3.3.1), i.e., (1) a 3′-terminal poly(A) sequence of 100–150 nucleotides, (2) the sequence $A_2UA_3$ in the non-coding 3′-segment and (3) a 5′-terminal "cap" with the sequence $m^7G(5')ppp(5')A^mpU^mpCp$.

The appearance of recombinant fibroin plasmids with deletions [14] and of partial fibroin translation products [12] can be regarded as an expression of the internal repetitive structure of the fibroin gene and of fibroin mRNA. During the translation of fibroin mRNA in vitro or in organ culture of isolated silk glands the partial fibroin translation products lead to the formation of a large number of distinct short fibroin polypeptides. These can only be transformed into complete fibroin polypeptides of 365–415 kd by improvement of the protein synthesis system, e.g., by supplementing with organspecific tRNA's (cf. **4**.4.4.6 and Ref. [12]). Both phenomena, the formation of deleted recombinant plasmids and of partial protein products are also observed in the cloning of Balbiani ring DNA [21] and in the translation of BR-mRNA and provide further evidence that the structure and expression of both internal repetitive gene systems have much in common.

## 9.5 Conclusions

The experimental results on the structure and activity of polytene chromosomes of *Diptera* presented here provide new data on the structure and function of eukaryotic genes and chromosomes in general. Due to the non- and under-replication of certain DNA types (such as α-heterochromatin and ribosomal DNA) during polytenization, meiotic and polytenic chromosomes differ not only – as hitherto accepted – by an enormous rise of DNA content, but also by the different relative amount of their DNA classes. Along with experimental data on other systems [such as the immunoglobulin genes of mammals (cf. Chap. **13**)] these findings suggest that specific changes in DNA amount and organization are not an exception, but rather a typical feature of cellular differentiation (cf. also **4**.3). Analyses of the fine structure and informational content of chromomeres provide a first insight into the molecular nature of these cytologically defined structures. Along with data on the molecular linkage of members of several gene families (e.g., the ovalbumin, globin and histone genes) within DNA stretches of almost 100 kb and more, these results lend support to the assumption that one chromomere normally harbours a number of genes with identical or very similar structure and function. Examples for this conclusion are also the heat-shock chromomeres 87A3 and 87C1 of *Drosophila melanogaster* salivary gland chromosomes which comprise up to five and more copies of the 70 kd heat-shock protein gene, or the heat-shock puff locus 67B which codes for the four small heat-shock polypeptides of 22, 23, 26 and 27 kd (see Table **9**.1). Apart from the protein-coding genes, however, moderately repeated DNA sequences resembling in many respect to the mobile-dispersed repeated elements of *Drosophila* and *Saccharomyces* (cf. **4**.3.4) have been detected in heat-shock chromomeres, too. Sequence families of this type seem to be a typical component of many chromomeres of *Drosophila* chromosomes.

The fine structure of 5′-"prelude"-segments of *Drosophila* heat-shock genes in loci 87A and 87C demonstrates the ubiquitious distribution of functionally important sequence motives, such as the "Hogness box" 5′-TATAAATA-3′ and the sequence 5′-GTTGGTTGG-3′ at positions around —30 and —75, respectively, in front of each gene copy. The primary structure of the latter sequence differs somewhat from those of vertebrate genes, but matches well with that of the fibroin gene and might, therefore, correspond to a recognition signal for the RNA polymerase II of insects which differs in several properties from that of higher eukaryotes.

The fibroin gene of silkmoths and the Balbiani ring genes of *Chironomus* represent a new type of so-called internal-repeated genes coding for structural proteins of a very high $M_r$. Expression of both genes shares a number of common pecularities, such as their transcription into huge RNA molecules of 37 kb (BR genes) or 17 kb fibroin gene) which are transported from chromosomes into cytoplasm whithout a significant reduction in size. The core structure of fibroin as well as Balbiani ring genes is composed of mainly one (or at most a very few) prototype sequence(s) of 50–25 bp which is repeated several hundredfold. It remains to be elucidated whether these particular properties of fibroin and Balbiani ring genes are restricted to these genes or hold true for many structural protein-coding genes, including those of vertebrates and plants.

## 9.6 Selected References

*Books and Reviews*

1a. Cold Spring Harbor Symp. Quant. Biology Vol. *38*: Chromosome Structure and Function. Cold Spring Harbor Lab., New York 1974.

1b. Cold Spring Harbor Symp. Quant. Biol., Vol. 42: Chromatin. Cold Spring Harbor Lab., New York 1978.

2. Beermann, W. (ed.): Developmental Studies in Giant Chromosomes. Res. Probl. Cell Diff., Vol. 4, Springer Verlag, Berlin 1972.
3. Beermann, W. (ed.): Biochemical Differentiation in Insect Glands. Res. Probl. Cell Diff., Vol. 8, Springer Verlag, Berlin 1977.
4. Ashburner, M. & J. J. Bonner: The induction of gene activity in *Drosophila* by heat shock. Cell *17* (1979): 241–254.
5. Case, S. T. & B. Daneholt: Cellular and molecular aspects of genetic expression in *Chironomus* salivary glands. Biochemistry of Cell Diff. II, Vol. 15. University Park Press, Baltimore 1977, pp. 45–77.
6. Edström, J. E.: Polytene chromosomes in studies of gene expression. In: Busch, H. (ed.), The Cell Nucleus, Vol. 2, Academic Press, New York 1974, pp. 293–332.

*Original papers*

7. Grossbach, U.: Chromosomen-Aktivität und biochemische Zelldifferenzierung in den Speicheldrüsen von *Camptochironomus*. Chromosoma *26* (1969): 136–187.
8. Kolesnikov, N. N., E. I. Karakin, T. E. Sebeleva, L. Meyer & E. Serfling: Cell specific synthesis and glycosylation of secretion proteins in the salivary glands of *Chironomus thummi* larvae. Chromosoma *83* (1981): 661–677.
9. Ingolia, T. D., E. A. Craig & B. J. McCarthy: Sequence of three copies of the gene for the major *Drosophila* heat shock induced protein and their flanking regions. Cell *21* (1980): 669–679.
10. Ish-Horowicz, D. & S. M. Pinchin: Genomic organization of the 87A3 and 87C1 heat-induced loci of *Drosophila melanogaster*. J. Mol. Biol. *142* (1980): 231–245.
11. Lamb, M. M., and B. Daneholt: Characterization of active transcription units in Balbiani rings of *Chironomus tentans*. Cell *17* (1979): 835–848.
12. Lizardi, P. M.: Genetic polymorphism of silk fibroin studied by two-dimensional translation pause fingersprints. Cell *18* (1979): 581–589.

13. McKenzie, S. L., S. Henikoff & M. Meselson: Localization of RNA from heat-induced polysomes at puff sites in *Drosophila melanogaster*. Proc. Natl. Acad. Sci. USA *72* (1975): 1117–1121.
14. Ohshima, Y. & Y. Suzuki: Cloning of the silk fibroin gene and its flanking sequences. Proc. Natl. Acad. Sci. USA *74* (1977): 5363–5367.
15. Pelling, C.: Ribonukleinsäure-Synthese der Riesenchromosomen. Autoradiographische Untersuchungen an *Chironomus tentans*. Chromosoma *15* (1964): 71–122.
16. Rydlander, L., A. Pigon & J.-E. Edström: Sequences translated by Balbiani ring 75S RNA in vitro are present in giant secretory protein from *Chironomus tentans*. Chromosoma *81* (1981): 101–113.
17. Serfling, E.: The transcripts of Balbiani rings from *Chironomus thummi*: Giant RNA molecules with messenger characteristics. Chromosoma *57* (1976): 261–284.
18. Serfling, E. & A. Huth: Balbiani ring RNA in the cytoplasm of *Chironomus thummi* salivary gland cells. Chromosoma *66* (1978): 205–223.
19. Tsujimoto, Y. & Y. Suzuki: Structural analysis of the fibroin gene at the 5′end and is surrounding regions. Cell *16* (1979): 425–436.
20. Tsujimoto, Y. & Y. Suzuki: The DNA sequence of *Bombyx mori* fibroin gene including the 5′flanking, mRNA coding, entire intervening and fibroin protein coding regions. Cell *18* (1979): 591–600.
21. Wobus, U., H. Bäumlein, R. Panitz, E. Serfling & F. C. Kafatos: Periodicities and tandem repeats in a Balbiani ring gene. Cell *22* (1980): 127–135.
22. Holmgren, R., V. Corces, R. Morimoto, R. Blackman & M. Meselson: Sequence homologies in the 5′ regions of four *Drosophila* heat-shock genes. Proc. Natl. Acad. Sci. USA *78* (1981): 3775–3778.
23. Corces, V., R. Holmgren, R. Freund, R. Morimoto & M. Meselson: Four heat shock proteins of *Drosophila melanogaster* coded within a 12-kilobase region in chromosome subdivision 67B. Proc. Natl. Acad. Sci. USA *77* (1980): 5390–5393.
24. Craig, E. A. & B. J. McCarthy: Four *Drosophila* heat shock genes at 67B: characterization of recombinant plasmids. Nucleic Acids Res. *8* (1980): 4441–4457.

# 10. Expression of Isozymes and Their Function in Differentiation

K. P. Rindt

## 10.1 Introduction

With the introduction of efficient methods for protein separation, e.g., gel electrophoresis and affinity chromatography, and for their specific histochemical or immunochemical characterization, it became evident that within a single organism or even a single cell there are often ,,multiple enzymes'' catalyzing the same reaction. They may differ with respect to their occurrence in different cells or subcellular compartments, their time of synthesis and degradation during ontogeny and/or their properties, such as affinity to substrates or interaction with allosteric effectors. In many instances these "isozymes" were shown to be synthesized in response to hormonal treatment or to changes of external conditions. We want to give some examples of how the pattern of isozymes can vary in prokaryotic and eukaryotic cells, how their formation is controlled, to which extent their use as indicators reflects differentiation processes, and how some isozyme systems themselves might be crucial factors for the control of cell differentiation.

Originally the term "isozyme" was applied pragmatically for those enzyme species catalyzing the same reaction but being separable by any physicochemical method. But this separability might stem from quite different reasons:

1. The iso-forms observed are encoded by different genes, i.e., they represent primarily different polypeptide chains;
2. the iso-forms are the result of post-translational modifications or they represent allosteric modifications (e.g., by ligand binding).

The primarily different forms often consist of more than one subunit. If they are built up of the same type of subunit, homopolymeric parental complexes as well

---

Abbreviations: AC, adenylate cyclase; $AHS_{V1}$, $AHS_{V2}$, $AHS_R$, valine-sensitive (V1, V2) and valine resistant (R) acetohydroxyacid synthase isozymes, resp.; $AK_T$, $AK_M$, $AK_L$, aspartate kinase isozymes related to threonine, methionine, and lysine biosynthesis; cAMP, cGMP, cyclic AMP and cyclic GMP; BamHI, restriction endonuclease from *Bacillus amyloliquefaciens;* C, catalytic subunit of cAMP-PK; calmodulin or CDR, calcium-dependent regulatory proteine D2 protein, a protein with T protein properties encoded by a SV40-adenovirus hybrid; DHD, dihydroxhacid dehydrase; eIF-2, eukaryotic initiation factor 2; Gpp(NH)p, guanyl-5-hl imidodiphosphate (analogue of GTP); $HSDH_T$, $HSDH_M$, homoserine dehydrogenase isozymes related to threonine and methionine biosynthesis; *ilv*, refers to genetic elements involved in the regulation of isoleucine, leucine and valine biosynthesis; IR: acetohydroxyacid isomeroreductase; kd: kilodalton; Md, megadalton; $M_r$, apparent molecular weight as determined, e.g., by gel electrophoresis; PK, protein kinase (cAMP-PK, cGMP-PK, Ca-PK: cAMP-, cGMP-, and calcium--dependent PK isozymes); protein I, a characteristic nerve membrane protein; PyrK-$A_4$, PyrK-$B_4$, PyrK-$C_2$, muscle-, liver-, and spleen-type isozymes of pyruvate kinase (indices refer to the number of subunits); R, regulatory subunit of cAMP-PK; rDNA, DNA coding for ribosomal RNA (rRNA); RNAP, RNA polymerase; RNP, ribonucleoprotein; SV40, simian virus 40; T protein, regulatory protein with PK activity encoded by SV40; $TD_s$, $TD_d$, biosynthetic and biodegradative isozymes of threonine deaminase.

as heteropolymeric or hybrid complexes can be observed. For example, the parental polypeptide chains A and B of vertebrate lactate dehydrogenase form a set of tetramers: two homotetramers ($A_4$ and $B_4$) and three heterotetramers ($A_3B$, $A_2B_2$ and $AB_3$). But in many cases the subunits are functionally different, only one of them being catalytically active. Its affinity to substrate can be modified by "regulatory" subunits which may exhibit affinity to metabolites with control function. Adenylate cyclase, some protein kinases and RNA polymerases belong to this type of isozymes and will be dealt with in some detail (cf. **10**.3.3 and **4**.4.2.6). It is often difficult to draw a line between polypeptide chains that must be considered genuine subunits of such complex isozymes and those that are less tightly bound or even represent regulatory proteins with multiple functions, as it is true for the calcium--dependent regulatory protein (cf. **10**.3.3.5) and for the components of hormone-triggered adenylate cyclase systems (cf. **10**.3.3.2), possibly also for the cAMP-binding protein of protein kinases (cf. **10**.3.3.3).

As to postsynthetical modifications, irreversible modifications of protein structure, e.g., removal of peptide sequences by limited proteolysis, or reversible modifications, e.g., phosphorylation and ADP ribosylation of proteins, are observed. When dealing with this type of isoforms we should, according to the recommendations of the International Union of Biochemistry, speak of multiple enzymes, whereas the term isozymes should be restricted to those species shown to be encoded by different genes. It seems necessary, however, to point out that this differentiation, though important for the purpose of nomenclature, may be confusing with respect to enzyme function. It is obvious that the same type of functional response to a metabolic signal can be achieved by both synthesis of specialized isozymes or modification of preexisting enzymes.

This may be briefly illustrated by the phage-induced reprogramming of the host's apparatus for RNA synthesis. Reprogramming may be brought about:

1. by synthesis of a new, phage-encoded RNA polymerase (e.g., phage T7, cf. **10.3.3.7.3**),
2. by addition of a new subunit with regulatory function (e.g., phage $\lambda$, cf. **14.2.3.4**),
3. by exchange of one subunit by a new one (e.g., phage SPO1, specifying "$\sigma$-like" proteins that replaces the $\sigma$-factor thereby altering the RNA polymerase specificity for promoter recognition, cf. Fig. **4**.18) or
4. by modification of host RNA polymerase (e.g., phosphorylation of its $\beta'$-subunit by a phage T7-specified protein kinase or its ADP ribosylation by a phage T4-encoded enzyme, cf. **10.3.3.7.3**).

Though we will mainly deal with isozymes sensu stricto, we have to speak about a very prominent example of protein modification, i.e., protein phosphorylation, if dealing with signal transformation and protein kinase isozymes (cf. **10**.3.2 and **10**.3.3).

## 10.2 Isozymes in Amino Acid Metabolism of Enterobacteria

Many steps in amino acid metabolism of enterobacteria can be catalyzed by two or more isozymes. Since these isozymes differ not only in their catalytic properties, but also in the way their synthesis is regulated, they represent examples for which a biological function appears particularly transparent. In the following section we want to concentrate on some selected isozymes involved in the biosynthesis or breakdown of amino acids belonging to the aspartate and pyruvate families. Figure **10.1** demonstrates the main routes of carbon flow in the biosynthesis of these amino acids and indicates steps for which isozymes exist in *Escherichia coli* and *Salmonella typhimurium.*

The carbon skeletons of threonine and methionine, as well as the main parts of the lysine and isoleucine skeletons, are derived from aspartate. Together with asparagine they are referred to as aspartate family. However, isoleucine might as well be considered a member of the pyruvate

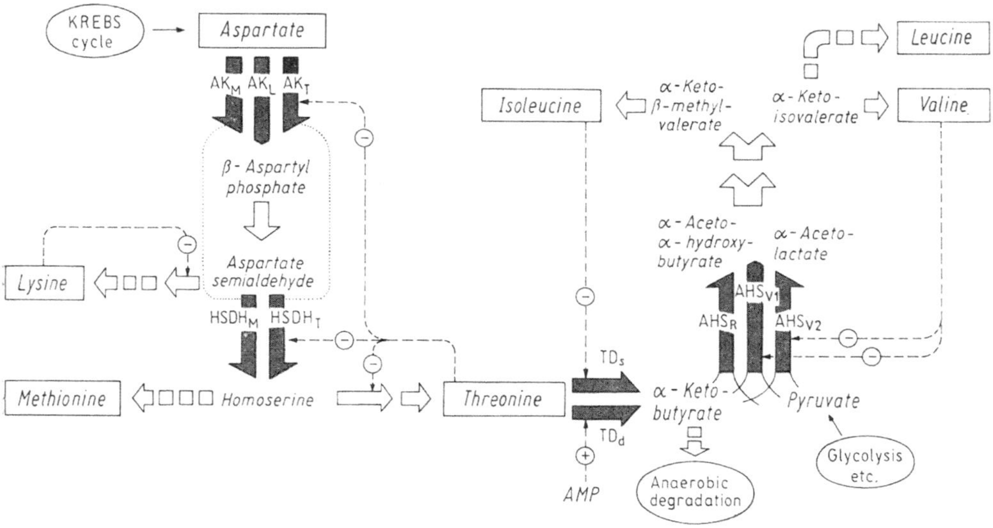


**Fig. 10.1:** Biosynthetic sequences leading from aspartate and pyruvate to the branched-chain and aspartate family amino acids in enterobacteria.

Steps catalyzed by isozymes are indicated by *hatched arrows*, whereas *double-headed arrows* refer to enzymes involved in the conversion of two substrates differing in length by one carbon atom. These sequences start with the connection of either one molecule each of α-ketoglutarate and pyruvate for isoleucine biosynthesis or of two pyruvate molecules for valine and leucine biosynthesis. Thus also acetohydroxyacid synthase (AHS) isozymes belong to this type of enzyme. $AHS_R$ and $AHS_V$ refer to valine-resistant and valine-sensitive isozymes, respectively. *Dotted lines* indicate that the two connected enzyme activities belong to one single polypeptide (multienzyme conjugate, cf. footnote 7 in Sect. **10.2.1**), *arrows with broken lines* point at sites of feedback inhibition by the corresponding amino acid (−) or, in one case, on a site of allosteric activation by AMP (+).

Control of enzyme synthesis is not indicated in this figure. $AK_M$, $AK_L$, and $AK_T$ refer to aspartate kinase isozymes repressed by methionine, lysine, and threonine, respectively. Only the gene coding for $AK_T$ is part of an operon (*thr*ABC, located at min 0 of the *E. coli* linkage map). *thr*A codes for the bifunctional $AK_T \cdot HSDH_T$ where $HSDH_T$ stands for a threonine-sensitive homoserine dehydrogenase. $AK_L$ is encoded by the *lys*C gene located at min 90 of the *E. coli* linkage map [1]. The genes coding for lysine *(lys)* and methionine *(met)* biosynthesis are scattered over the bacterial chromosome. Nevertheless, all enzymes are derepressed with limiting amounts of the corresponding amino acid. Though some *met* genes are concentrated at min 87 of the *E. coli* chromosome – amongst them *met*L and *met*M, coding for $AK_M$ and $HSDH_M$, respectively – the existence of a *met* operon coding for part of methionine biosynthetic enzymes is also unlikely. For a more detailed discussion the reader is referred to the review of Umbarger [1].

$TD_s$ and $TD_d$ refer to the biosynthetic and biodegradative isozymes of threonine deaminase (also "threonine dehydratase"). Control of $TD_s$ formation is described in connection with regulation of activity of the other *ilv* genes (cf. **10.3.2**). Data on control of $TD_d$ formation are presented in Sect. **10.2.4**. For $TD_d$ no mapping data are available because no strategy for isolation of mutants lacking this isozyme has been developed up to now.

family since two of its six carbon atoms are derived from pyruvate, as are the whole skeletons of alanine, leucine, and valine. The initial part of the pathway leading to valine and leucine formation is catalyzed by the same set of enzymes required for isoleucine biosynthesis. Therefore it seems more reasonable in this context to consider these branched-chain amino acids together, not so much because of their similar structure but because their biosynthetic pathways are so interrelated.

### 10.2.1 Bacterial Isozymes Involved in Metabolic Channeling

There are two common steps in the biosynthetic paths leading from aspartate to lysine, methionine and threonine, and one additional step common only to threonine and methionine biosynthesis (cf. Fig. **10**.1). There is only one aspartate dehydrogenase, the enzyme catalyzing the second step, whereas isozymes have been found for aspartate kinase (AK) and homoserine dehydrogenase catalyzing the first and third step in the sequence, respectively. Some data on isozyme properties and on the way their synthesis is regulated are briefly summarized in Table **10**.1.

Isozyme properties in connection with the pattern of repression control suggest:

1. that the $AK_T \cdot HSDH_T$ conjugate plays a role in threonine and isoleucine biosynthesis, whereas $AK_M \cdot HSDH_M$ is involved in methionine formation and, finally, $AK_L$ catalyzes the first step of lysine biosynthesis,
2. that there is some kind of metabolic channeling, i.e., independent carbon flow from aspartate to the corresponding amino acid via three independent pathways. Thus isozymes might provide a means for independent control of the biosynthesis of amino acids derived from aspartate. This assumption implies (a) that there might exist three independent polyisozymic complexes, (b) that intermediates would not escape from the complex until the final product has been formed and (c) that the activity of the complex as a whole should be regulated.

**Table 10.1**: Properties of aspartate kinase (AK) and homoserine dehydrogenase (HSDH) isozymes from *Escherichia coli*

| Enzymes | Regulation of enzyme activity | Regulation of enzyme biosynthesis | Remarks[a] |
|---|---|---|---|
| $AK_L$ | Allosterically inhibited by lysine | Repressed by lysine | |
| $AK_T$, ($HSDH_T$) | Allosterically inhibited by threonine | Multivalently repressed by threonine plus isoleucine | Product of the *thr*A gene which specifies a multienzyme conjugate that normally remains as intact polypeptide in cells of enterobacteria[b] |
| $AK_M$, ($HSDH_M$) | No feed-back inhibition by methionine or any other related amino acid | Repressed by methionine | In *E. coli* K-12 and *S. typhimurium* as conjugate but in *E. coli* B as separate polypeptides |

[a] For genetic organization of structural genes cf. legend to Fig. **10**.2.

[b] "Multienzyme conjugate" or "multienzyme polypeptide" should be preferred to "multifunctional enzyme" which term is often used simply to indicate an enzyme with broad substrate specificity, e.g., AHS or IR in Fig. **10**.2. At present there is no final agreement which term shall be generally used.

To our knowledge there is no experimental evidence that such tightly complexed enzymes catalyzing, e.g., the synthesis of a single amino acid, exist in enterobacteria. But the end product threonine has been shown to control the activity of an artificial system combined from all purified threonine biosynthetic enzymes, provided a threonine-sensitive $AK_T$ is used. As proposed, no intermediates, e.g., aspartate semialdehyde or homoserine, accumulate, at least not at low substrate and high enzyme concentrations, i.e., conditions prevailing in the living cell [1].

The concept of metabolic channeling might also imply that loss of one of the AK isozymes would lead to auxotrophy for the corresponding amino acid. What was observed, however, was that loss of one or even two of the three AK isozymes, e.g., by mutation, can be compensated for by the remaining one(s). This need not be contradictory to the concept of metabolic channeling because possession of different isozymes and formation of polyisozymic complexes might be of evolutionary advantage indeed, though under laboratory conditions mutant cells might well survive without such complexes. It is only this adaptive value that provides a reasonable explanation for the evolution and maintenance of isozymes each seemingly related to the biosynthesis of a single amino acid.

### 10.2.2 Biosynthetic and Degradative Processes Initiated by Alternative Isozymes

Threonine is not only incorporated into proteins but is also substrate for isoleucine biosynthesis and can be used under certain circumstances as energy source by enterobacteria. At any rate, the first metabolic step is the non-oxidative deamination to $\alpha$-ketobutyrate catalyzed by threonine deaminase (TD). In enterobacteria there are two TD isozymes. Regulation of their activity as well of their expression suggest one isozyme ($TD_s$) to be especially adapted to isoleucine biosynthesis, whereas the other ($TD_d$) appears to be involved in threonine degradation.

### 10.2.3 Biosynthetic Threonine Deaminase and the *ilv* Regulon

$TD_s$ is the product of the A gene of the *ilv*EDA operon which itself is part of the *ilv* regulon comprising at least three additional transcription units, *ilv*B, *ilv*C, and *ilv*HI. Figure **10.2** shows, in a simplified version, the enzymatic steps leading to isoleucine formation. The AHS isozymes and IR catalyze not only three steps of isoleucine formation but also three steps of valine and leucine biosynthesis. Moreover, the synthesis of these enzymes is under the control of the three branched-chain amino acids, and therefore the regulation of $TD_s$ formation is only understandable by considering the biosynthesis of *ilv* enzymes[1] as a whole.

The dual function of *ilv* enzymes makes clear that, if there were control of enzyme synthesis, e.g., by isoleucine alone, isoleucine would not only prevent its own biosynthesis but that of valine and leucine, too. Thus cells would appear auxotrophic for valine and leucine when growing in the presence of isoleucine. Therefore the pattern of multivalent repression realized in enterobacteria is one way to circumvent such metabolic complications.

For comprehension of the control of formation of $TD_s$ and the other *ilv* enzymes it is necessary to consider the isoleucine sensitivity of $TD_s$ itself as well as the expression of valine-sensitive and valine-insensitive acetohydroxyacid synthase isozymes.

---

[1] *ilv* enzymes: used here for all enzymes specified by the *ilv* regulon and involved in the biosynthesis of *i*soleucine, *l*eucine, and *v*aline.

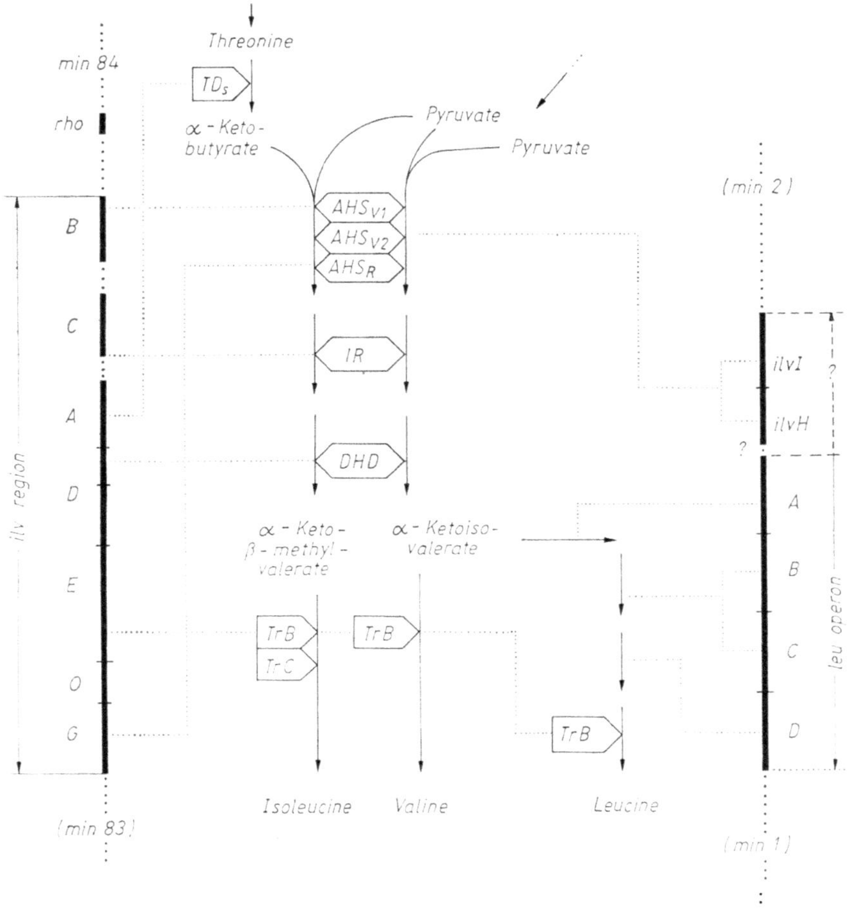


**Fig. 10.2:** Scheme of branched-chain amino acid biosynthesis *(middle)* and structural genes coding for the corresponding enzymes (*left* and *right*) (for explanations see text).

The gene order in the *ilv* region indicated here is now widely accepted, but some authors stick to a *ilv*EGDAOCB arrangement [5] which would further complicate any explanation of how *ilv* gene activity is controlled. The *leu* operon which codes for three leucine biosynthetic enzymes (names not indicated here) is under the exclusive control of leucine (for an overview see [1]). $TD_S$, AHS, cf. legend to Fig. 10.1; *IR*, acetohydroxyacid isomeroreductase; DHD, dihydroxy-acid dehydratase; *TrB, TrC*, transaminase B and C; *rho*, structural gene for the termination factor Rho. min 1, 2, 83, 84 refer to positions on the circular map of *E. coli* (cf. Fig. 5.1).

Evidently these two enzymes control the carbon flow to the branched-chain amino acids.

The $TD_S$ is a tetrameric enzyme consisting of four identical subunits [1]. Nevertheless it bears only two binding sites for the cofactor pyridoxal phosphate, two inhibitor sites with affinity for isoleucine and one activator site which binds valine. In the absence of isoleucine, $TD_S$ exhibits a hyperbolic threonine saturation curve that is shifted to sigmoidal shape in the presence of isoleucine.[2] Thus isoleucine controls the flow of carbon for its own biosynthesis. Another important aspect

[2] Valine, antagonizing isoleucine inhibition, restores the hyperbolic substrate saturation curve.

is a two-step maturation process of $TD_S$. In the first step pyridoxal phosphate is bound, in the second step enzymatically active $TD_S$ is formed in the presence of either valine or isoleucine (4b]. Simultaneous presence of both amino acids, however, prevents maturation. The fact that immature but not mature $TD_S$ binds to charged branched-chain aminoacyl tRNA species may form the molecular basis for the regulation of *ilv* gene activity by branched-chain amino acids (cf. **10**.2.3.2). As to AHS, the genome of *E. coli* K-12 contains three unlinked genes coding for two valine-sensitive ($AHS_{V1}$ and $AHS_{V2}$) and one valine-resistent ($AHS_R$) isozymes (cf. Fig. **10**.2). Contrasting to the situation in *S. typhimurium* [3], only the two Val-sensitive AHS isozymes are formed in *E. coli* under derepressing conditions, whereas *ilv*G, coding for $AHS_R$, remains cryptic. Hence, exogenous valine inhibits the total AHS activity, leading to a starvation for the other end products of the biosynthetic pathway, especially for isoleucine. The resulting growth inhibition can only be overcome by addition of Ile or by cis-dominant mutations (*ilv*$O^-$), which allow expression of the *ilv*G gene product (cf. Fig. **10**.2 and Refs. [1, 6, 9, 10]).

Although all genes of the *ilv* regulon are multivalently repressed by the branched--chain amino acids, with the exception of isomeroreductase which is inducible by its substrate, and although all *ilv* genes appear to be subject to "autogenous regulation" by $TD_S$ (cf. **10**.2.3.2), we want to concentrate on the regulation of *ilv*EDA gene product formation, especially because the *ilv*EDA operon contains the structural gene for $TD_S$ itself.

### 10.2.3.1 Multivalent Repression of Biosynthetic Threonine Deaminase

Mutations in *ilv*O do not only allow expression of *ilv*G gene product but also cause increased expression of the *ilv*EDA operon without apparently qualitatively affecting repressibility or derepressibility. Though they are mostly described as operator--constitutive mutants they much more resemble promoter mutants in other systems. According to original mapping data *ilv*O was placed at the *ilv*C end of *ilv*A, whereas *ilv*G was placed between *ilv*E and *ilv*D. But recent data are in favour of a gene order *ilv*GOEDA with transcription proceeding from E to A (Fig. **10.2**):

1. Phage Mu-1 (cf. Table **4.3**) can be inserted into any *E. coli* gene. As a consequence no active gene product can be formed. If in an operon a proximal gene is affected, transcription of more distal genes is prevented (polar effect). Insertion of Mu-1 into *ilv*A abolishes $TD_S$ activity without affecting expression of *ilv*E and *ilv*D gene products [7b]. Mu-1 insertion into *ilv*D abolishes formation of *ilv*A and *ilv*D but not *ilv*E gene products. Finally, Mu-1 insertion into *ilv*E in a strain carrying an *ilv*O mutation prevents expression of *ilv*EDA gene products, but allows formation of $AHS_R$ (*ilv*G) which is characteristic of *ilv*$O^-$ strains. Together with data obtained recently by careful gene mapping [8, 9], this can be explained only on a basis of the gene order being *ilv*GOEDA, with a direction of transcription from E to A.
2. On minimal medium there is a coordinate reduction of the *ilv*EDA gene products. However, if derepression occurs by limiting one of the branched-chain amino acids it is observed that $TD_S$, the product of the third gene on the operon, is most derepressed, whereas least derepression was found for the product of *ilv*E, the gene nearest the presumed site of transcription initiation. This polar effect suggests the existence of several termination sequences in the *ilv*EDA operon [7a, 7b].
3. There is a frame-shift site in the *E. coli* wild-type *ilv*O locus, which is displaced by mutation *ilv* 2096 [2]. As revealed by DNA sequence analysis this mutation originates from a two-base insertion permitting, at the same time, the expression of *ilv*G and an increased expression of the more distant *ilv*EDA genes. In wild-type cells, though *ilv*G is actually transcribed, there is a stop codon in the middle of the gene. As a consequence, a truncated, inactive gene product is formed [2]. According to these data *ilv*O is itself part of *ilv*G, but, in addition, has some regulatory functions for the expression of the genes downstream of *ilv*G.

Analogous with other bacterial systems, *ilv*(G)EDA expression appears to be controlled by an attenuation mechanism [11a, 11b]:

1. Mutants with altered Rho protein (cf. **1**.3.2.2) are characterized by elevated formation of *ilv*(G)EDA products under derepressing conditions. Further derepression occurs upon amino acid limitation (cf. lines 1 and 2 in Fig. **10**.3).

2. Pvu II restriction endonuclease cleavage of DNA obtained from plasmids carrying the *ilv* region lead to the isolation of a DNA fragment with a strong promoter site. This fragment very likely comprises the multivalent regulatory region of the *ilv*(G)EDA operon. The DNA sequence revealed that this fragment in fact contains at least one promoter-type sequence (cf. Table **1**.8). It is located at the beginning of a region coding for a 32-amino-acid polypeptide. The coding sequence is followed by a transcription termination site separated by about 80 bp from the beginning of the first structural gene assumed to be *ilv*G.

3. The secondary structure of the proposed *ilv* leader RNA is basically similar to that of other enterobacterial systems controlling formation of amino acid biosynthetic enzymes. The RNA contains sequences able to form two mutually exclusive stem-loop structures: (1) terminator-protector conformation and (2) preemptor conformation. Terminator conformation would lead to attenuation (termination of transcription), whereas the preemptor precludes terminator formation and thus results in deattenuation (cf. details presented for the *trp* operon, Fig. **4**.19).

4. The amino acid sequence of the putative leader peptide, translated from a region of leader RNA forming the protector stem-loop structure, is Met-Thr-Ala-Leu-Leu-Arg-Val-Ile-Ser-Leu-Val-Val-Ile-Ser-Val-Val-Val-Ile-Ile-Ile-Pro-Pro-Cys-Gly-Ala-Ala-Leu-Gly-Arg-GlyLys-Ala-(stop). The N-terminal region, which contains 14 branched-chain amino acid residues out of 20, is crucial for attenuation. The occurrence of tandem repeats of branched-chain amino acids is thought to ensure *ilv*

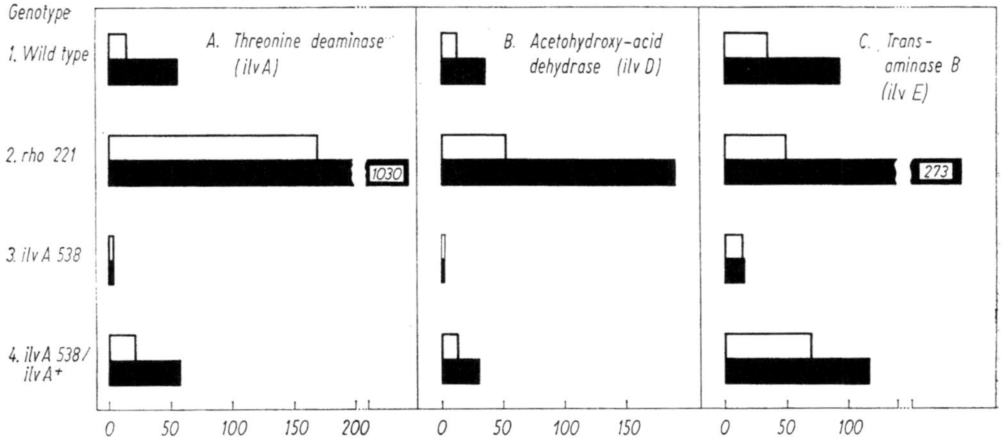


**Fig. 10.3:** Effect of mutations in the *rho* and *ilvA* genes on the expression of *ilv*EDA enzymes in *Escherichia coli* K-12 and evidence for trans-dominance of *ilv*A$^+$ over *ilv*A 538 (Combined from Smith et al. [7a] and Levinthal et al. [7b]).

*Open boxes* represent specific activities (nmol product formed per min per mg protein) under repressing conditions with 0.6 mM each of isoleucine, leucine and valine, whereas *full boxes* refer to specific activities with limiting isoleucine (derepression). Isoleucine limitation in *E. coli* K-12 is achieved by addition of only valine to the growth medium. Since under these conditions only valine-sensitive AHS is formed (cf. **10.2.3**), the biosynthesis of all branched-chain amino acids is prevented by valine. Leucine can be formed from valine, therefore valine supplementation of minimal medium leads exclusively to lack of intracellular isoleucine. The partial diploid strain **(4)** contains mutation *ilv*A 538 in the genome plus a wild type *ilv*A$^+$ allele as part of the transducing phage ∅80d *ilv*A$^+$.

gene expression to be under the exclusive control of these amino acids. Steric considerations, as well as calculation of free energies of suggested base-paired regions in the *ilv* leader RNA are in support of a model in which ribosome stalling in the protector region, brought about by lack of either of the branched-chain amino acids, would favour the preemptor conformation and thus prevent terminator formation (cf. **4**.4.2.7). The predicted, and in fact observed, result of branched-chain amino acid deprivation would be deattenuation or transcriptional read-through. On the other hand, abundance of charged tRNAs would lead to formation of the terminator stem-loop structure and result in attenuation [11a, 11b].

### 10.2.3.2 Biosynthetic Threonine Deaminase as Part of a Complex Gene Control System

Investigations into *E. coli* strains carrying a mutation in *ilv*A indicate that the *ilv*A product itself is somehow involved in the control of the *ilv*(G)EDA operon (autoregulation) [4a, 4b, 5]. Indeed, many such strains are characterized by (1) expression of a $TD_S$ with unusual regulatory properties, and (2) alterations in the expression of *ilv*(G)EDA and other gene products. As one example, in strains carrying the *ilv*A 466 allele, $TD_S$ is no longer sensitive to isoleucine inhibition. At the same time the *ilv*(G)EDA operon becomes hyperderepressible, i.e., there is elevated formation of *ilv*(G)EDA products under derepressing conditions.

The allele *ilv*A 538, on the other hand, codes for a isoleucine-hypersensitive $TD_S$. Strains carrying this allele exhibit pleiotropic regulatory effects, e.g., low and invariant expression of *ilv* genes including the *ilv*EDA operon, and loss of repression control of branched-chain aminoacyl tRNA synthetases [7a]. When strains of *E. coli* K-12 bearing the *ilv*A 538 allele on the chromosome were lysogenized with defective phages carrying the *ilv*$A^+$ allele (e.g., phage $\Phi$80d *ilv*$A^+$), derepressibility of the *ilv*(G)EDA operon is restored (cf. lines 3 and 4 in Fig. **10**.3 with line 1).

Since *ilv*$A^+$ is trans-dominant over *ilv*A 538, it can be concluded that the product of *ilv*$A^+$, the wild-type $TD_S$, acts as a positive control element in regulation of *ilv* gene activity, and that this positive control function is lost by mutation to *ilv*A 538.

It is impossible at present to unify within one model the often contradictory results reported in the literature on the regulation of *ilv*(G)EDA gene activity in *E. coli* K-12. There are indications that there are two control circuits, one of them represented by the attenuation mechanism. $TD_S$ as regulatory molecule may be somehow involved in the second control circuit. Action of both circuits involves charged or uncharged branched-chain aminoayl tRNA species. In the second circuit, $TD_S$ has been suggested to act as repressor in its immature form when it is able for binding charged branched-chain aminoacyl tRNA species [4b]. But evidence given in Sec. **10**.2.3.2 suggests that $TD_S$ acts as positive control element. This may mean an antiterminator function but other types of action, e.g., on initiation of mRNA formation, may be equally well envisaged. Loss of $TD_S$ by deletion of the *ilv*DAC genes does not result in loss of multivalent repression [6]. This is indicated by derepression of the remaining gene of the *ilv*(G)EDA operon, *ilv*E, in the deletion strain. Thus one circuit, i.e., control by attenuation, is still functioning if the other, $TD_S$-dependent, circuit is defective.

Any detailed discussion of how *ilv*(G)EDA gene activity is regulated at a molecular basis seems unwise in this context. Certainly, experiments with pure *ilv*DNA fragments, in a combined transcription-translation system, will soon reveal the nature of any positive and negative control element and definitely answer the questions of the position of *ilv*G, of its transcription relatively to the other genes of the *ilv*(G)EDA operon. Elucidation of the DNA sequence surrounding the *ilv* regulatory region should explain the nature of *ilv*O mutations. At present we are left with the fact that *ilv*O$^+$ and *ilv*O$^-$ strains exhibit essentially no differences in the nucleotide sequence within an about 240 bp-long DNA piece beginning about 80 bp in front of the start codon but ending about 80 bp behind the stop codon of the leader peptide and comprising the whole leader RNA sequence [11a, 11b].

### 10.2.4 Biodegradative Threonine Deaminase: Multivalent Induction and Role of cAMP

The tetrameric biodegradative threonine deaminase ($TD_d$) is also an allosteric enzyme. Its activation by AMP is a consequence of ligand-induced oligomerization and is accompanied by a marked reduction of its $K_m$ for threonine. Since the enzyme is formed only under very limited growth conditions its metabolic function has long remained obscure. In *S. typhimurium* and *E. coli*, $TD_d$ is formed only when growing

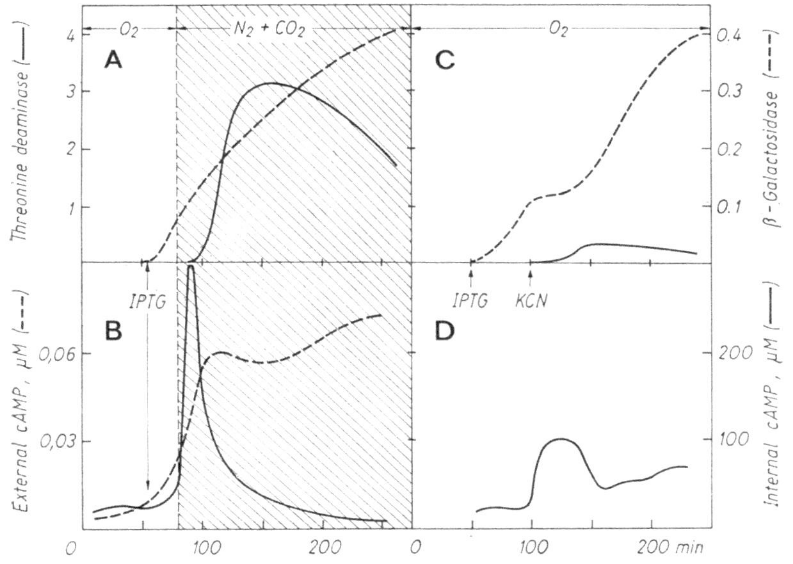


**Fig. 10.4**: Effects of transition from aerobic to anaerobic growth *(left)* and administration of cyanide under aerobic conditions *(right)* on the inducibility of degradative threonine deaminase *($TD_d$)* and β-galactosidase *(top)* and the levels of cyclic AMP *(bottom)* (Phillips et al. [12]).

Cells of *Escherichia coli* (Crookes strain) were grown aerobically on a medium containing glycerol, fumarate, and 10 g per 1 of 19 amino acids plus 4 mM threonine, i.e., conditions adequate to provide for maximum TD induction after transition to anaerobic conditions. Anaerobic conditions were established by replacing air by a mixture of $N_2$ and $CO_2$ (95 : 5). For induction of β-galactosidase, cultures were made 1 mM of isopropyl-β, D-thiogalactoside *(IPTG)*. When added, the final concentration of potassium cyanide was 2.5 mM.

In order to prevent interference of measurement of $TD_d$ activity with trace amounts of its biosynthetic counterpart *($TD_s$)*, which is repressed during growth on the medium used, the assay mixture was made 1 mM of isoleucine, the allosteric inhibitor of $TD_s$ (cf. **10.2.3**). Measurements of cAMP, $TD_d$, and β-galactosidase were made at intervals of 10 min. $TD_d$ and β-galactosidase values represent specific activities, given as 1 μmol of α-ketobutyrate produced and 1 μmol of o-nitrophenyl-β-D-galactoside hydrolyzed, respectively, per min per mg protein.

anaerobically on a glucose-free, amino acid-containing medium. Detailed investigations on $TD_d$ induction in enterobacteria revealed

1. that presence of threonine is indispensible for $TD_d$ formation, but that for maximum induction there is an additional requirement for leucine, valine, and possibly aspartate (multivalent induction), and
2. that cyclic AMP (cAMP) can overcome glucose inhibition of $TD_d$ formation in strains defective in adenylate cyclase (*cya*$^-$ strains). Thus catabolite repression via cAMP is suggested to be involved in regulation of $TD_d$ synthesis.[3]

In Fig. **10**.4A formation of $TD_d$ and $\beta$-galactosidase is compared with intracellular and extracellular cAMP levels during a shift from aerobic to anaerobic conditions. $\beta$-Galactosidase was used for comparison because it is also subject to feed-back inhibition but apparently not dependent on anaerobic conditions for relief from catabolite repression. Though the levels of cAMP were similar during exponential growth either aerobically or anaerobically (about 5–15 $\mu$M), there is a sharp increase in internal cAMP reaching 300 $\mu$M about 10 min after an abrupt transition from aerobic to anaerobic conditions. Whereas $\beta$-galactosidase can be induced by isopropyl-$\beta$,D-thiogalactoside (ITPG) at comparatively low levels of cAMP during both aerobiosis and anaerobiosis, exceedingly high levels of internal cAMP appear to be essential for $TD_d$ induction by threonine. From measurement of external cAMP during the shift it has been concluded that the achievement of the observed peak of internal cAMP is largely the result of a decreased cAMP excretion rate during the anaerobic shock (cf. Fig. **10**.4B).

Though $TD_d$ formation is normally observed under anaerobic conditions only, formation can be evoked to a limited extent also under aerobic conditions if respiration is inhibited, e.g., by cyanide. Also in this case, $TD_d$ formation is preceded by the achievement of transient peak levels of internal cAMP which, however, are less marked than during anaerobic shock (cf. Fig. **10**.4C, D).

## 10.3 Isozymes of Eukaryotic Cells

Extensive studies on isozymes in eukaryotic organisms have revealed many differences with respect to tissue or cell specificity, compartmentation within cells, and expression during ontogeny. Many of these isozymes differ in their substrate affinity or specificity, but others apparently do not. Moreover, isozymes do not necessarily exert regulatory functions since they often catalyze steps that are not rate--limiting. Therefore, after more than twenty years of research in isozymology, some investigators feel that the initial hope that investigation of isozymes would be of much help in elucidating regulation of metabolism and differentiation processes has not been fulfilled. But there should be agreement that isozymes and their highly specific expression, as for some examples outlined below, would have been abolished during evolution if they had not been of any value for the corresponding organism. Hence we might expect that differences present in vivo are often no longer detectable after cell disruption. Kinetic parameters of enzymes are supposed to be altered by extraction, dilution and purification.

---

[3] At present there is little information on the molecular mechanism of the control of $TD_d$ synthesis, mainly because genetic analysis has been hampered due to lack of a strategy for obtaining mutants in the structural or regulatory genes. As indicated by its sensitivity to rifampin and chloramphenicol, $TD_d$ appearance requires de novo protein synthesis via mRNA formation [13].

### 10.3.1 Isozymes Are Involved in Most Steps of Sugar Metabolism in Vertebrates

Pathways of sugar metabolism belong to the most thoroughly investigated of all. Sugar metabolism plays a central role in that it connects several important anabolic and catabolic processes. Isozymes have been detected for almost all of its steps. Their common presence in a single cell, or preferential expression of certain isozymes in specialized cells, may reflect their integration at distinct places into the metabolic network of these cells. This can be outlined for only a few selected examples. Figure **10**.5 illustrates the variety of isozymes of sugar metabolism in rat tissues.

Although there are some isozymes for which no kinetic differences have been known so far, we must envisage the possibility that there are specific sets of isozymes forming more or less stable polyisozymic complexes, which, through metabolic channelling, are supposed to be of advantage for the regulation of metabolic sequences. If isozymes occur in only one compartment of a cell, it seems probable that they form such polyisozymic complexes with distinct functions. This was suggested for the cytoplasm-restricted branches of sugar metabolism, e.g., gluconeogenesis, glycolysis, and a separate path for glycerol-phosphoenolpyruvate transformation. For aggregation to take place in the proper way, differences in the affinity of one distinct isozyme to an enzyme (isozyme) catalyzing a subsequent step in a reaction are most important. Differences in catalytic or regulatory properties may develop as a result of aggregation.

In support of this hypothesis are the following obervations:

- formation of enzyme complexes is favoured at high protein (and medium salt) concentrations, i.e., conditions approximating those really existing within the cell;
- the activity of enzymes in complexes is regulated in a coordinate manner;
- certain isozymes (e.g., lactate dehydrogenase $A_4$ but not $B_4$) bind to actin. Together with other isozymes they evidently form a glycolytic complex [16].

Actins are important components of eukaryotic cytoskeletons (cf. **4**.1.4). One or more types of at least six different isoactins are present in any vertebrate cell [17, 30]. Thus investigations of isozyme patterns must be extended to the possible formation of polyisozymic complexes with specific components of the cytoskeleton as well as to the questions of coordinated expression of distinct isozymes and specific isoforms of structural proteins and of how the activity of specialized polyisozymic complexes is regulated [16].

### 10.3.2 Three Functionally Distinct Pyruvate Kinase Isozymes in Mammals

One example only, the tissue distribution and developmental pattern of pyruvate kinase isozymes, will be outlined in more detail. Pyruvate kinase (PyrK) occupies a key position in both sugar metabolism and ATP formation. Regulation of its activity represents a major mechanism enabling the cell to switch from glycolysis to gluconeogenesis and vice versa. In vertebrates there are at least three individual isozymes with different regulatory properties that appear to be synthesized according to the needs of the corresponding cell type. These parental forms are:

1. The "muscle"-type pyruvate kinase which is a Michaelian enzyme shown to be a homotetramer ($M_r$, 230 kd) and therefore denoted as PyrK-$A_4$.
2. PyrK-$A_4$ is immunologically related to the "spleen"-type pyruvate kinase. This dimeric enzyme with $M_r = 110$ kd is the first to appear during ontogeny. Originally denoted as "$M_2$" it is now best referred to as PyrK-$C_2$. It is an allosteric enzyme.

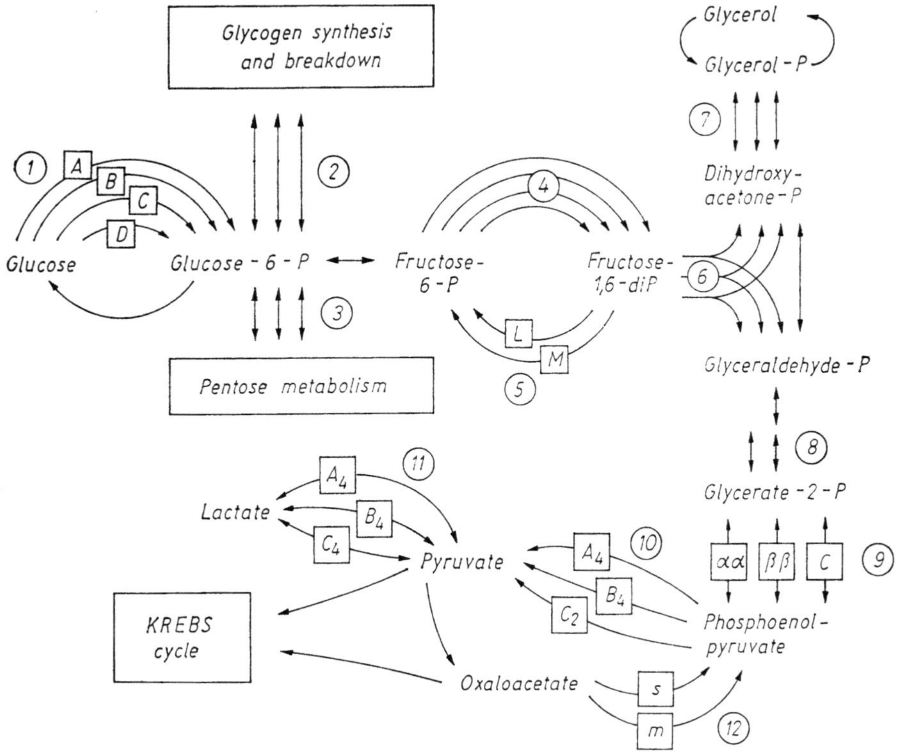


**Fig. 10.5**: Schematic representation of some known isozymes involved in hexose metabolism in rat tissues (after Ureta [15]).

For the sake of clearness only some of the intermediates are indicated. Each *arrow* represents one enzyme or isozyme of different genetic origin. *Numbers in circles* refer to remarks given below. *Letters in boxes* refer to common designations of isozymes. P (e.g., glucose-6-P) refers always to the phosphate ester of the corresponding sugar or related compound.

At least four independent hexokinase isozymes (reaction *1*) are known. They differ in their developmental pattern, tissue and subcellular distribution (cytoplasmic and membrane-bound forms) and they respond differently to hormones and allosteric effectors, respectively. Thus a prerequisite for a specific role of hexokinase isozymes seems to be fulfilled. However, no satisfactory explanation can be given for a role of isozymes of phosphoglucomutase *(2)*, which catalyzes the first step of the glycogenic pathway. The molecular basis of glucose-6-phosphate dehydrogenase isozymes *(3)* is not very clear so far. Some may represent interconvertible forms. Differences in subcellular distribution have been shown. For both, phosphofructokinase *(4)* and fructose-1,6-bisphosphatase *(5)*, tissue-specific isozymes have been detected, e.g., characteristic liver and muscle forms of the latter (L and M in Fig. **10.4**). Similarly, the existence of muscle-, liver-, and spleen-specific aldolase isozymes *(6)* has been reported (aldolase A, B, and C, resp.). These forms represent tetramers of one and the same subunit type, but hybrid tetramers with a varying proportion of parental subunits also exist. In this respect they resemble lactate dehydrogenase *(11)* and enolase *(9)*. The latter exists as three type of dimers ($\alpha\alpha$, $\beta\beta$, and the hybrid $\alpha\beta$). In addition, an independent brain isozyme (B) has been detected. In other mammalian cells several steps in the path between glycerol phosphate and glycerate-2-phosphate can be catalyzed by isozymes, whereas for rat tissues up to now only phosphoglycerate kinase *(8)* isozymes have been detected. As discussed with some more details in the text, pyruvate kinase isozymes *(10)* exhibit differences in their kinetics, tissue distribution and developmental pattern, as do lactate dehydrogenase isozymes *(11)*. The soluble and mitochondrial *(s, m)* phosphoenolpyruvate carboxykinase *(12)* isozymes represent a typical example for isozymes differing in kinetics and localization in subcellular compartments. We should emphasize that several enzymes of the

3. The "liver"-type isozyme, also characteristic of kidney, is an allosteric enzyme, too. This tetrameric enzyme is referred to as PyrK-$B_4$ ($M_r$, 210 kd). It can be allosterically activated by the "feed-forward" activator fructose bisphosphate, whereas it is inhibited by alanine and ATP [23].

**Table 10.2**: Amount of pyruvate kinase isozymes in extracts from various rat tissues (from Muroya et al.: J. Biochem. *79* (1976): 203–215).

PyrK isozymes of crude extracts were transformed to the three major forms by ammonium sulfate precipitation (see text) and separated by isoelectrofocusing. The specific activity is given in units per g protein. pI, isoelectric point

| Tissue | Specific activity of isozymes | | |
|---|---|---|---|
| | Liver type ($B_4$, pI = 6.2) | Muscle type ($A_4$, pI = 7.4) | Spleen type ($C_2$, pI = 7.8) |
| Brain | 0 | 1,900 | 0 |
| Heart muscle | 0 | 950 | 0 |
| Skeletal muscle | 0 | 8,900 | 0 |
| Sarcoma tissue[a] | 0 | 140 | 1,260 |
| Liver (normal) | 190 | 14 | 36 |
| Liver (from sarcoma-bearing rats) | 164 | 20 | 196 |
| Liver (after high carbohydrate diet) | 657 | 14 | 42 |
| Spleen | 0 | 73 | 537 |
| Kidney | 50 | 67 | 443 |
| Erythrocytes | 3 | 1 | 17 |

[a] A sarcoma is a cancer of muscle tissue s.

The distribution of pyruvate kinases in rat tissues was particularly carefully investigated (cf. Table **10.2**). It has been shown that there are at least seven forms separable by isoelectrofocusing. By ammonium sulphate precipation, however, they can be converted to the three major forms mentioned above. Enzyme precipitation results in removal of fructose bisphosphate (FBP) bound to PyrK-$B_4$ and PyrK-$C_2$. This can be shown by incubation of purified, FBP-depleted PyrK-$B_4$ and PyrK-$C_2$ in the presence of FBP. Four additional "isozymes" appear, that upon isoelectrofocusing comigrate with forms detected in crude extracts. Thus these forms apparently represent more or less stable allosteric modifications and not hybrid forms between parental subunits, as suggested previously.

Table **10.2** presents an overview about the distribution of the three PyrK isozymes in rat tissues. The results are further illustrated by the following conclusions and remarks:

1. The predominant fetal enzyme in all organs is PyrK-$C_2$. The shift to the tissue-specific pattern shown in Table **10.2** is a mainly postnatal differentiation process.

◀

pathways leading to pentose sugars and encompassing glycogen synthesis and breakdown *(boxes)* are represented by isozymes (for reviews see [15, 22, and 23]). The latter are subject to extensive regulation by phosphorylation/dephosphorylation, similar to phosphofructokinase *(4)*, fructose-1,6-bisphosphatase *(5)*, and pyruvate kinase *(10)* isozymes (cf. **10.3.2**). Thus, besides true isozymes, additional multiple enzyme species can be detected after protein separation by electrophoresis or other methods.

2. In some adult tissues only one gene for pyruvate kinase is expressed (cf. muscles and brain), contrasting to other tissues where products of all three genes are to be found. Their amount and proportion, however, varies considerably.
3. In liver, only the predominant PyrK-$B_4$ is subject to dietary control and responds to hormonal treatment.
4. The influence of the presence of sarcoma on the expression of liver PyrK isozymes is remarkable: It is not only the sarcoma tissue itself that shifts to the fetal isozyme pattern but there is an as yet unexplained influence on the expression of liver PyrK-$C_2$, too. This expression of typical "fetal" isozymes, often referred to as "oncofetal enzymes", seems to be characteristic, and therefore can be used as indicator of neoplastic transformation (cf. **19**.3.1). Similarly, expression of characteristic isozyme patterns indicates the development and/or manintenance of a distinct specialization state of cells in a given tissue or cell culture. Loss of specialization leads to the disappearance of tissue-specific isozymes with concomitant expression of the fetal types.

In the light of a more detailed treatment of protein kinases in the last section of this chapter (cf. **10**.3.3) the selective phosphorylation of the PyrK-$B_4$ isozyme appears worth mentioning. Phosphorylation of this isozyme increases two to three fold after treatment of liver cells with glucagon or epinephrine. The effect ist mediated by cAMP (cf. **10**.3.3.3) and correlates with the inhibition of enzyme activity. Thus, there is a multiple control of the flux of phosphoenolpyruvate in liver (1) by changing the level of pyruvate kinase $B_4$, (2) by allosteric activation (by FBP) and inhibition (by ATP and alanine), and (3) by regulating PyrK-$B_4$ activity via hormone-controlled phosphorylation. The liver isozyme obviously has a dual role in that it can switch the flux of carbon from glycolysis to gluconeogenesis and vice versa, whereas the muscle isozyme, which is much more active and unregulated[4], appears to be particularly adapted to facilitate extensive glycolysis in muscle and brain.

## 10.3.3 Mammalian Signal Transformation: Involvement of Protein Kinases and Other Isozyme Systems

### 10.3.3.1 The Concept of Signal Transformation

Amongst the multifold reactions leading to protein modification (cf. Table **1**.11) protein phosphorylation evidently plays a major role. It appears to be particularly suited to exert a regulatory function in metabolism and differentiation since (1) the process is reversible and (2) functional properties of enzymes and other proteins can be largely modified by phosphorylation. This implies that the degree of protein phosphorylation has to be precisely controlled. In eukaryotes, phosphorylated proteins are only one link in a central, multicomponent signal transformation system (cf. **4**.2) which involves at least three subsequent steps (cf. Fig. **10**.6):

1. In the first step a hormone and other extracellular signals, respectively, bind to their specific receptor proteins in the outer cell membrane. The signal-receptor complex thereby activates membrane-bound adenylate cyclase. The result is a rise in the intracellular level of cyclic AMP (cAMP) which is the intracellular equivalent ("second messenger" according to Sutherland [18]) of the triggering signal.

---

[4] Reports on inhibition of PK-$A_4$ by creatine phosphate are based on a hitherto unknown contaminant in commerical preparations (Fitch et al.: J. Biol. Chem. *254* (1979): 11357).

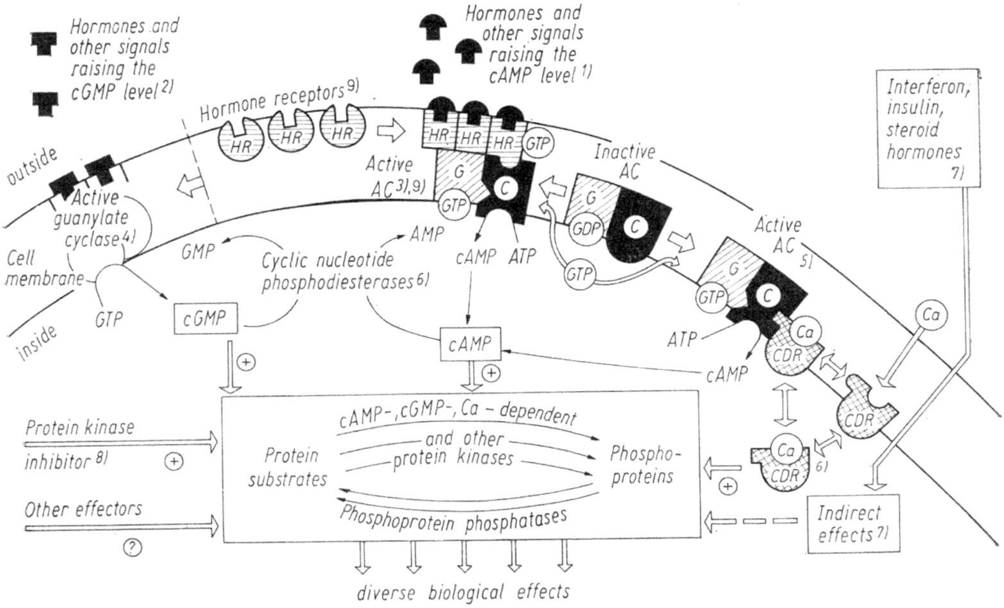


**Fig. 10.6:** Schematic presentation of the signal transformation system involving protein phosphorylation and its control elements in vertebrate cells.

*Full arrows* represent metabolic processes; *open arrows* indicate transport across membranes, diffusion, or translocation and regulatory influences; *AC*, adenylate cyclase with its catalytic *(C)* and guanyl nucleotide-binding *(G)* subunits; *HR*, hormone receptor; *CDR*, calcium-dependent regulatory protein (also "calmodulin").

Remarks referring to numbers given in the figure:

*1* Many peptide hormones, e.g., glucagon, prostaglandin $E_1$, epinephrine ($\beta$), serotonin, isoproterenol, insulin (?); mitogenic factors (?);

*2* Epinephrine ($\alpha$), histamine, acetylcholine;

*3* The active transmembrane complex consists of hormone receptor clusters + hormone + AC subunits (C and G). Subunit *G* bears a site for GTP binding and cleavage. The hormone receptor bears an additional GTP-binding site,

*4* Details of the structure of active guanylate cyclase are practically unknown.

*5* For $Ca^{2+}$-calmodulin activation of AC and the role of GTP compare text (**10.3.3.2** and **10.3.3.5**).

*6* Calmodulin has been shown to occur in plasma membranes, nuclei and cytoplasm of different cells. The calmodulin level of adrenal cortex can be largely increased by corticotropin injection. This may suggest a role of calmodulin in the regulation of hormone effects on the nucleus, possibly mediated by $Ca^{2+}$-dependent protein kinases (cf. Harper et al., Proc. Natl. Acad. Sci. USA *77* (1980): 366–370). $Ca^{2+}$-calmodulin activation of a cyclic nucleotide phosphodiesterase is not shown in **Fig. 10.6.**

*7* Steroid hormones enter the cell and affect protein phosphorylation indirectly, e.g., via gene activation (de novo synthesis of protein substrates, protein kinases etc.). Similarly, insulin and interferon affect protein phosphorylation indirectly, the latter by induction of a double-stranded RNA-dependent protein kinase ([23, 28], cf. **4.4.4.7**).

*8* cf. **10.3.3.3** and **10.3.3.6**

*9* It is well known that some hormones enhance while others inhibit cAMP formation in the presence of GTP. Therefore it has been postulated that there might be two different G-binding subunits, one interacting with receptors that bind stimulatory ligands (e.g., $\beta$-adrenergic and glucagon-binding receptors), the other with receptors that bind inhibitory ligands (e.g., $\alpha$-adrenergic receptors). The two G-binding proteins are able to interact with the catalytic subunit of AC, the former thereby enhancing, the latter inhibiting cAMP formation. Thus the picture is certainly more complex than indicated. According to this hypothesis [19] activation of hormones is followed by disaggregation of inactive receptor complex rather than by aggregation to clusters. This, however, appears contradictory to the well-documented formation of hormone-receptor clusters which precede cap formation and internalization (cf. **10.3.3.3.2** and **10.3.3.4**)

2. In the second step cAMP activates a characteristic protein kinase which as a consequence may be translocated to a specific compartment to phosphorylate a distinct protein.
3. After its phosphorylation the target protein exerts its action on other cell components. If it is an enzyme involved in metabolism, the result normally will be activation or inhibition of a metabolic chain. However, if it is a protein involved in gene expression the phosphorylated protein may represent only another link in a cascade-like system mediating the transfer of regulatory information to an ultimate target molecule.

In view of their importance it is not surprising that these signal transformation systems are highly complex, that several sets of isozymes are involved, and that they interact with other cellular control systems [22–26].

### 10.3.3.2 Role of Adenylate Cyclase

Receptors able for binding non-steroid hormones are randomly distributed and diffuse freely in the phospholipid layer of the cytomembrane. Through their specificities they determine the ability of cells for responding to distinct hormonal signals. After hormone binding they aggregate to clusters of different size and interact with inactive, membrane-bound adenylate cyclase (hereafter briefly referred to as AC). AC is composed of two dissimilar subunit types. The C subunit bears the catalytic site for the formation of cAMP from ATP. The C subunit bears one guanylate-binding site ("G site"). Binding of GTP to this G site is required for AC to be maximally activated by the hormone-receptor complex. The latter bears an additional guanylate-binding site, but only the G site associated with AC is characterized by GTPase activity [32]. Since the GDP formed during GTPase reactions renders AC in a state of low, "basal" activity the system requires a constant supply of GTP for maximum activity. Self--inactivation of AC by its GTPase activity can be artificially overcome by binding the non-hydrolyzable GTP analogue, guanyl-5-yl imidodiphosphate [Gpp(NH)p]. Similarly, inactivation can be prevented by inhibition of GTPase activity, e.g., by NAD-dependent ADP ribosylation of the G subunit, a reaction catalyzed by an endogenous enzyme or by cholera toxin [20]. In some tissues, e.g., brain, AC activity can be stimulated by $Ca^{2+}$-calmodulin (cf. **10**.3.3.5).

Upon affinity chromatography of solubilized AC on CDR-Sepharose two fractions were obtained. One of them was no longer sensitive to $Ca^{2+}$-calmodulin stimulation. But after incubation with a protein fraction obtained from detergent-solubilized cytomembranes, $Ca^{2+}$-calmodulin stimulation was restored. As reconstitution was dependent on the presence of Gpp(NH)p it was concluded that at least one guanylate-binding protein, lost during affinity chromatography, is required for stimulation of AC activity by $Ca^{2+}$-calmodulin, too ([36], cf. also Fig. **10**.6).

Thus both external signals (hormones, calcium) and internal effectors (GTP) regulate intracellular cAMP levels. GTP is an important measure of the metabolic state of the cell. Binding of GTP alone activates AC to a certain degree and appears to be functioning in the regulation of cAMP levels during the cell cycle. Moreover, GTP binding is a prerequisite for maximum response of AC to hormones or calcium.

Although hormone-specific isozymes of adenylate cyclase, e.g. ,"glucagon-adenylate cyclase", are frequently mentioned in the literature, one should keep in mind that up to now there is no evidence for differences in the C subunit. Differences observed are based either on secondary modifications, e.g., ADP ribosylation of the G subunit, or on the presence of other polypeptides (hormone receptors, calmodulin) that are normally not considered true subunits of AC since they associate with it only after binding of hormone and calcium, respectively (for reviews cf. Ref. [19, 20, 35]).

### 10.3.3.3 Protein Kinase Isozymes

Cyclic AMP as second messenger activates specific cAMP-dependent protein kinases. Most protein kinases (PK) catalyze the transfer of $\gamma$-phosphate from ATP to serine and threonine but very rarely also to histidine and lysine residues of several protein substrates, thereby altering their functional properties. But there are many reports on protein kinases which are not cAMP-dependent, e.g., cyclic GMP-dependent or calcium-dependent forms, an interferon-inducible, double-stranded RNA-specific protein kinase (cf. **4**.4.4.7 and Ref. [28]), some endogenous and some virus-induced, tumor-specific PK's phosphorylating tyrosine instead of threonine and serine residues (cf. **10**.3.3.7.2) and other isozymes for which so far no trigger has been found. These "non-specified" protein kinases are particularly abundant in the nucleus. Part of them may represent catalytic subunits of a cAMP-type isozyme.

In the following we will concentrate on cAMP-, cGMP-, and calcium-dependent protein kinases (cAMP-PK, cGMP-PK, and Ca-PK, respectively). At least two cAMP-PK isozymes are present in most animal cells. They are involved in the transmission of the "cAMP signal" to target proteins. The cGMP-PK isozyme is restricted to a few cell types (e.g., lung, cerebellum, and heart) that simultaneously exhibit guanylate cyclase activity[5] [24].

cAMP-PK and cGMP-PK reveal some remarkable similarities. At least two types of cAMP-PK are present in all tissues examined. They elute at different sodium chloride concentrations from DEAE-cellulose columns, "type I" at about 0.15 M, "type II" at about 0.25 M. In response to low cAMP concentrations ($10^{-6}$ M) both isozymes dissociate in vitro, releasing the catalytically active subunit (C) and the complex of the regulatory subunit (R) with cAMP. The isozymes differ mainly with respect to their regulatory subunits. However, recent results indicate differences in C, too.[6] Their significance for PK action has to be elucidated, but it seems most probable that some differences in substrate specificity, reported for type I and type II isozymes, are due to differences in C rather than in R. These cAMP-dependent protein kinases are the only so far identified eukaryotic proteins capable of allosterically interacting with cAMP.

Activation of cAMP-PK's proceeds according to the equation

| | $R_2C_2 + 4\ cAMP \rightleftharpoons$ | $R_2(cAMP)_4 +$ | $2C$ |
|---|---|---|---|
| | Holo-enzyme (inactive) | Regulatory subunits | Catalytic subunits (active) |
| Type I ($M_r$): | 186 kd | 2×54 kd | 38 kd |
| Type II ($M_r$): | 170 kd | 2×47 kd | 38 kd |

In contrast to cAMP-PK's cGMP-PK does not dissociate as a consequence of effector binding. Yet there is some evidence for a common evolutionary origin of both: (1) the molecular weights are similar (cGMP-PK: $M_r = 165$ kd) and (2) under certain conditions, which may involve limited proteolysis, cGMP-PK releases a free, "putative" subunit that is still catalytically active but no longer responds to cGMP. However, it can be inhibited by complexing with the free regulatory subunit R of a cAMP-PK, indicating a relationship between the isozymes. There are speculations that the cAMP-PK might have evolved via gene splitting from an ancestral cGMP-PK-like enzyme or, vice versa, that cGMP-PK arose by gene fusion from an ancestral cAMP-PK-like

[5] It can be activated by hormones in a way comparable to adenylate cyclase (cf. Fig. **10**.6).

[6] Finger-print analysis of tryptic peptides derived from catalytic subunits of type I and type II porcine muscle cAMP-PK clearly indicate some differences, though most peptides appear to be identical (Zoller et al.: J. Biol. Chem *254* (1979): 2408; cf. also **10.3.3.5**).

enzyme [25]. These interesting hypotheses have to be tested for their validity by immunochemical methods and sequence analysis.

Phosphorylase b kinase, myosin light chain kinase, and a brain membrane protein kinase represent calcium-dependent PK isozymes that depend for activity on the formation of a complex with Ca-calmodulin (cf. **10**.3.3.5 and **10**.3.3.8; [22b, 23, 26b].

The latter enzymes, however, do not phosphorylate only one single protein as their name might suggest, and it should be emphasized in this context that in several cases designation of protein kinases according to their first identified substrate protein does not seem justified. Though frequently differing in substrate specificity, only few of them appear to use a single substrate, most of them being "multifunctional" protein kinases. In other cases the PK might act specifically, but the target protein has not been identified as yet. Substrates used to measure protein kinase activities (histones, phosphovitin, protamin etc.) are in most cases not the natural substrates. As a rule, many more proteins become phosphorylated in vitro than in vivo. These differences might be due to compartmentation of either substrates or protein kinases in vivo. At any rate, phosphorylation of proteins in vitro has to be considered with some caution when drawing conclusions about the role of distinct phosphorylated proteins in control of differentiation or metabolism.

#### 10.3.3.4 All Steps Involved in Signal Transformation are Reversible

It appears to be important for the functioning of the overall signal transformation chain that for any of its steps there are synthesizing and degrading capacities in the cell:

- Shortly after hormone binding followed by receptor aggregation the clusters of hormone-receptor complexes are internalized by endocytosis (cf. Fig. **8**.12). The cell regains its full ability to respond to the same hormonal signal only after sufficient new hormone receptor molecules have been incorporated into the cytomembrane (cf. **4**.2.4.2 and Ref. [21a, 21b]).
- At least under some metabolic conditions the functioning of the adenylate cyclase depends on the constant supply of GTP and its breakdown by the GTPase activity of the cyclase itself.
- The second messengers themselves are degraded by different cyclic nucleotide phosphodiesterase isozymes, some of which act more specifically on cAMP, others on cGMP. Since the levels and ratios of these isozymes vary and since at least one species is activated by calmodulin, the cyclic nucleotide levels appear to be controlled by both synthesis and degradation.
- Dissociation and directed transport of cAMP-PK (cf. **10**.3.3.6) are reversible as soon as the cAMP level drops. It should be mentioned that the mechanism of directed transport is not known at this time. Some data suggest microtubules as being involved in the process.
- The phosphate group transferred to substrate proteins can be removed by phosphoprotein phosphatases. There is certainly one major, "multifunctional" phosphoprotein phosphatase, but there are reports on some minor, more specific forms as well.

Phosphorylation of certain enzymes can be regulated by ligands that obviously alter the conformation of substrate proteins. To give an example, binding of glucose to phosphorylated glycogen phosphorylase enhances, whereas that of AMP inhibits its dephosphorylation by multifunctional phosphoprotein phosphatase [23]. In this respect there are striking similarities with protein degradation (cf. **4**.4.3.3).

### 10.3.3.5 Isozymes Involved in Signal Transformation are Modulated by Protein Factors and Enzymes

Several protein factors selectively modulate distinct isozymes of the signal transformation chain. Calmodulin, the best known of them, is an ubiquitous, highly conserved protein (MW 17 kd) that forms a complex with calcium [27]. The four calcium binding domains give rise, at least theoretically, to 16 different conformational states, dependent on calcium concentration. Calmodulin mediates the control by calcium of a number of enzymes including AC (cf. **10**.3.3.8 and Fig. **10**.6), Ca-dependent PK isozymes (cf. **10**.3.3.3 and **10**.3.3.8), and a cyclic nucleotide phosphodiesterase (cf. **10**.3.3.4).

Calcium is another important second messenger, known to influence intracellular cAMP levels. AC as cAMP forming enzyme and phosphodiesterase as cAMP degrading enzyme appear to be activated by calcium, albeit at different concentrations. The molecular basis for this difference may be selective affinity of the two enzymes for one of the proposed calmodulin conformations. The direct influence on the system controlling the cAMP level is certainly one important aspect of the observed functional coupling of the two second messenger systems.

While cGMP-PK requires a stimulatory modulator for activity, the free catalytic subunit (C) of cAMP-PK is strongly and selectively inhibited by a specific inhibitor often used to differentiate between cAMP-independent PK's and C (cf. **10**.3.3.6). As in different cells the proportion can vary from almost 100% cAMP-PK inhibitor (e.g., muscle) to 100% cGMP-PK stimulators (intestine), the modulators are supposed to play an important role in regulating the in vivo activity of distinct PK isozymes [24].

Another modulator activates a cAMP-PK that may represent an isozyme of the catalytic subunit of cAMP-PK and is contained in cAMP-PK preparations purified from rat muscle. Activation, which is only observed with enzyme bound to gels, is not reversed after the modulator has been removed and is due to a stable conformational change rather than to proteolytic cleavage. Modulators are suggested to cause not only selective activation but also to confer substrate specificity on protein kinases [29, 40]. Similarly, occurrence of phosphodiesterase and phosphoprotein phosphatase modulators has been reported. The inhibitor of the major phosphoprotein phosphatase is itself subject to control by protein kinases since it is an inhibitor only in its phosphorylated form. In rabbit skeletal muscle the phosphorylation state of the inhibitor is increased by adrenalin and decreased by insulin [44].

A general feature of protein kinases is their capacity to autophosphorylate internal serine and threonine residues and, in rare but important cases, also tyrosine residues (cf. **10**.3.3.7.2). Autophosphorylation of the regulatory subunit of cAMP-PK results in enhanced dissociation of holoenzyme in the presence of cAMP and may be physiologically relevant. Other reported enzymatic modifications of enzymes involved in signal transformation are: activation by proteolytic cleavage (AC and a phosphodiesterase isozyme) and poly (ADP) ribosylation of an AC subunit. While the latter is physiologically relevant this has still to be shown for protein cleavage.

Summarizing, as a regulatory system of central importance the signal transformation chain is not simply regulated by hormone-controlled cAMP formation. Much more there is a fine tuning at several levels through the action of various metabolites, protein factors and protein-modifying enzymes. Protein phosphorylation is dynamic and transient in nature since the overall system of signal transformation is characterized by turnover processes. The ability of the cell to respond to a signal in the proper way depends on a constant flux of energy through the system. In view of the evidence for protein kinases that are triggered by substances other than cAMP, we should realize that the signal transformation outlined here represents only one aspect of how phosphorylation of important regulatory proteins is controlled.

### 10.3.3.6 Recompartmentation of Protein Kinase Isozymes

A central part of the idea of signal transformation ist the cAMP-induced translocation (or "recompartmentation") of protein kinases to target compartments where they evoke a specific biological response. So we have to look for the original location of protein kinases prior to a hormone stimulus and for potential differences between type I and type II cAMP-PK's.

Mature erythrocytes appear to be particularly suited for investigations of cAMP--PK compartmentation as they contain only two well-defined compartments, cytoplasm and cytoplasmic membranes (cf. **4.5.4.4**). In these cells type II cAMP-PK is restricted to the cytoplasm, whereas the type I isozyme was shown to be associated with the membrane. The type I isozyme can be released intact from the particulate fraction by Triton X-100 treatment. If in a reconstitution experiment the protein kinase-depleted erythrocyte membrane was incubated with either type I or type II isozyme, only the former became (re)associated. Therefore it must be concluded that the regulatory subunit $R_1$ has not only specific affinities for cAMP and for its catalytic subunit but possesses an additional binding site for an intrinsic membrane protein of the erythrocyte. This might give an explanation of how the regulatory subunit

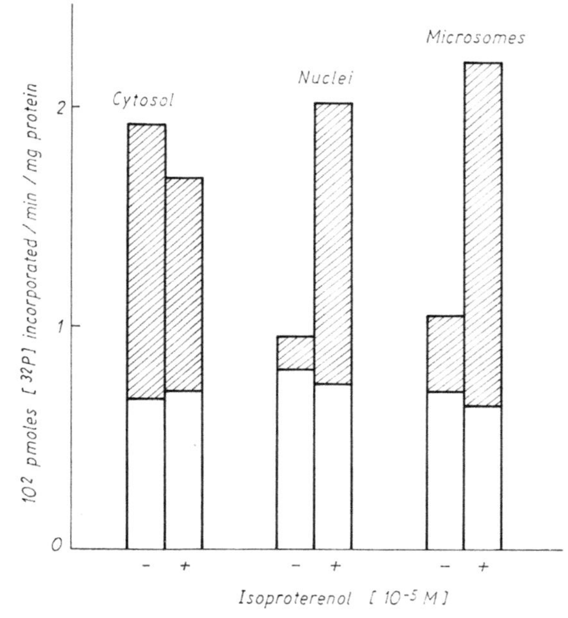


**Fig. 10.7**: Effect of cAMP-PK-specific, heat-stable inhibitor on protein kinase activities of subcellular fractions from rat glioma cells before and after isoproterenol treatment (according to Jungmann et al. [33]).

Subcellular fractions were prepared from control cells and from cells treated for 15 min with isoproterenol. Transfer of phosphate from [$\gamma$-$^{32}$P] ATP to protamine (catalyzed by endogenous protein kinases) was determined in the absence and presence of saturating amounts of heat-stable protein kinase inhibitor and $10^{-6}$ M cAMP.

*Total bars* represent protein kinase activities in the absence, *open bars* the remaining activities in the presence of inhibitor (= cAMP-independent protein kinase activities). *Crossed bars* therefore represent cAMP-dependent protein kinase(s). It should be emphasized that *bars* represent specific activities of PK. Hence figures of the total amount of PK translocated from cytoplasm to other compartments cannot be deduced from the data shown.

determines substrate specificity: it serves the "positioning" of the catalytic subunit in the proximity of an substrate protein located in the membrane [41a, 42].

It should be mentioned, however, that type I activity is not necessarily membrane-associated (e.g., cytoplasmic cAMP-PK in rat glioma cells, cf. below). There are also reports that in other cell types (e.g., human lymphocytes, brain, and heart) it is the type II cAMP-PK that is associated with the membrane [26a, 41b]. So it must be concluded that the nature of the membrane proteins determines which of the isozymes becomes bound to the membrane (cf. **10**.3.3.8).

Evidence for protein kinase translocation stems from experiments with rat C6 glioma cell cultures. In these cells isoproterenol, a catecholamine that mimicks effects of $\beta$-adrenergic substances such as epinephrine, stimulates lactate dehydrogenase formation via synthesis of specific messenger RNA for this enzyme. Coinciding with the elevation of intracellular cAMP levels (from less than 0.3 to about 4.0 nmol per milligram protein within 10 min after isoproterenol stimulation) there is a reduction of type I cAMP-PK in the cytosol fraction. The type II isozyme level remains unchanged. Examination of the nuclear fraction revealed an increase in specific protein kinase activity. Moreover, when the type of PK was analysed by means of the cAMP--PK-specific inhibitor (cf. **10**.3.3.5) a pronounced increase in the nuclear level of inhibitor-sensitive protein kinase was shown (cf. Fig. **10**.7), whereas the level of inhibitor-insensitive PK was unaffected. Sephadex G-200 chromatography resolved two nuclear protein kinase fractions not detectable prior to isoproterenol treatment. The estimated molecular weights (150 and 44 kd, resp.) and the cAMP-binding capacity of the larger species suggest the smaller as representing the subunit only and the larger as representing the holoenzyme of a type I cAMP-PK. Further evidence for the identity of the 150 kd protein kinase with type I cAMP-PK were obtained by means of immunochemical and chromatographical methods [33]. Hence experimental evidence points to distinct biological roles of type I and type II cAMP-PK: the isozymes might respond differentially to stimuli as a consequence of compartmentation.

### 10.3.3.7 Phosphorylation of Nuclear Proteins

At least two important criteria have to be fulfilled to demonstrate that protein phosphorylation is involved in the control of metabolism or gene activity.

1. It has to be demonstrated that the degree of phosphorylation of *distinct* protein species (enzymes, membrane and other structural as well as regulatory proteins) is correlated with biologically meaningful changes of their functional properties. At the moment that can be satisfactorily done only with certain enzymes where functions and properties are known (an exception is presented in Sect. **10**.3.3.7.3).
2. Phosphorylation/dephosphorylation of distinct proteins has to be shown in vitro (by using protein kinases and phosphoprotein phosphatases of the same tissue or cell type) and in vivo or in an intact cell system as a response to a stimulus (hormones etc.).

Though many details have been worked out concerning the hormonal and dietary control of glycogen synthesis and breakdown in liver cells by phosphorylation [22a, 22b], we want to restrict ourselves to the phosphorylation of proteins possibly involved in regulation of gene activity.

#### **10.3.3.7.1** Phosphorylation of Nuclear Proteins Triggered by cAMP

The major class of nuclear proteins are the histones (cf. Table **1**.2). They not only serve as good substrates for various protein kinases in vitro but are also reported to be phosphorylated to a different degree in vivo. Especially phosphorylation of histone H1, the linker between adjacent nucleosomes, is correlated with overall gene activity. Phosphorylation of histones decreases their basicity and, consequently, their general affinity to the phosphate backbone of DNA (cf. **1**.2.3.4 and **4**.4.2.5.3).

The nucleosome model proves histones to function as *the* structural proteins of chromatin. Yet numerous findings point out that isoforms of histones may be associated with chromatin of cells in different developmental stages. Moreover, in addition to phosphorylation, other types of modification (cf. Table **1**.2) contribute considerably to the multiplicity of histone forms. Hence any model of differential gene function must include histone variability as an essential aspect. Histones, i.e., especially H1, H2a and H2b, which contain highly modified and variable N-terminal polypeptide domains (cf. Fig. **1**.3) might be considered as "super repressors". Their phosphorylation is a necessary but not sufficient prerequisite for differential gene activation. The more acidic nuclear non-histone proteins and nuclear enzymes also undergo phosphorylation/dephosphorylation cycles and appear to be exchanged during development. They are suggested as playing a more specific role as (mainly positive) regulators of gene activity when complexed with specific DNA sequences, with RNA polymerase or enzymes involved in RNA processing etc.

Although there are many reports on phosphorylation of specific nucleus-associated proteins, there are only few examples unequivocally proving phosphorylation of such proteins to be triggered by cAMP. One stems from the investigation of protein phosphorylation in cells and cell-free systems from certain mammalian tumors. Some mammary tumors appear to be steroid hormone-dependent. Their growth can be arrested by ovariectomy (cf. **19**.4.1). Tumor regression can also be achieved by treatment of tumor cells with dibutyryl-cAMP, an analogue of cAMP. There is one major nuclear protein species that becomes phosphorylated during tumor growth ("tumor growth-associated protein"), whereas another basic non-histone protein becomes selectively phosphorylated after tumor growth has been arrested ("regression-associated protein", RAP, $M_r = 76$ kd).

Recently it has been shown [31] that RAP can also be selectively phosphorylated in a cell-free system with slices of tumor tissue incubated in the presence of $10^{-7}$ M cAMP. Under these conditions a marked increase of a cytosolic cAMP receptor protein ($M_r = 56$ kd) was observed. Incubation of nuclei from untreated tumor nuclei with cytosol from cells incubated with cAMP resulted in (1) uptake of the cAMP receptor and (2) phosphorylation with [$\gamma$-$^{32}$P] ATP of nuclear RAP.

Phosphorylation of RAP was prevented if the tumor slices were incubated simultaneously with 17-$\beta$-estradiol and cAMP. Hence, in this system, the steroid hormone can reverse the effect of cAMP. The cAMP receptor taken up by nuclei obviously represents a regulatory subunit, whereas the phosphorylating protein represents the catalytic subunit of a cAMP-PK. This is supported by two lines of evidence [31]:

- Binding of a 56 kd protein to the nucleus and the phosphorylation of RAP also occur when purified cAMP-PK from bovine heart tissue is used;
- Phosphorylation of RAP can be prevented by the heat-stable protein kinase inhibitor (cf. **10**.3.3.5).

It should be emphasized that also in this case both the catalytic and the regulatory subunits of a cAMP-PK are translocated to the nucleus (cf. **10**.3.3.6). Hence activation by cAMP, but not necessarily dissociation into its subunits, seems to be a prerequisite for translocation of cAMP-PK. It can even be speculated that the regulatory subunit plays an essential role in achieving specificity of nuclear protein phosphorylation. However, in some other systems (e.g., rat heart) only the catalytic subunit is translocated while the regulatory subunit remains bound to the membrane [41b].

### 10.3.3.7.2 Specialized Protein Kinases Encoded by Vertebrate Tumor Viruses

Some mammalian and avian oncogenic viruses, e.g., Rous sarcoma virus and simian virus 40 (SV40), have been in the centre of experimental investigations for many years. The genome of SV40 (cf. Fig. **10**.8) consists of a known sequence of only 5,226 base pairs with the capacity to code for about 170 kd of protein. Since a good deal of the DNA codes for the three late (mainly capsid) proteins, with molecular weights ranging between 10 and 50 kd, there is not much coding capacity left for virus-specified functional proteins. In fact, only two early proteins, encoded by partially overlapping sequences of the SV40 A gene, have been recognized and are referred to as "T antigens". These are the "small T protein" ($M_r$ 17 kd) and the "large T protein" ($M_r$ 96 kd). So

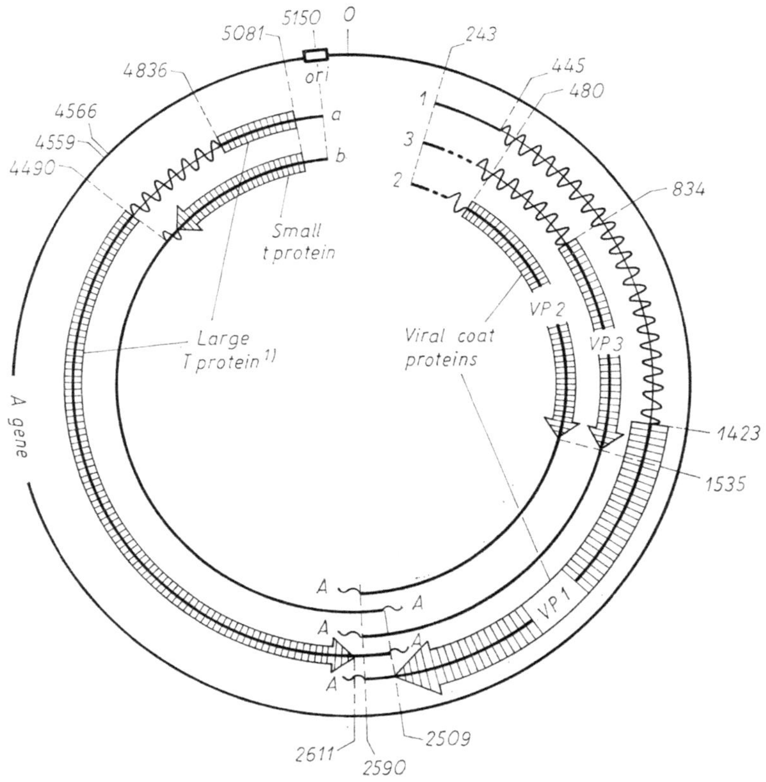


**Fig. 10.8:** Schematic presentation of the SV 40 genome and the gene products coded by it. (Compiled from Ready et al.: Science *200* (1978): 494, and Fiers et al.: Nature *273* (1978): 113).

The *outer circle* represents the circular DNA. The reference point (0) for counting the 5224 bp of DNA is a base located in the centre of a remarkable, 27-base-pair long palindrome sequence located near the origin of DNA replication (*ori*).
*a*, *b* refer to messengers for large and small T antigen (T protein and t protein, respectively).
*1*, *2*, and *3* refer to messengers for the viral capsid proteins VP 1, VP 2, and VP 3, respectively.
Messenger RNA transcribed from SV 40 DNA is indicated by *solid lines*. *Zigzag lines* indicate segments that are spliced out. A *wavy line* with an *A* illustrates the 3′-terminal poly A tail. *Blocked, hatched arrows* indicate regions on mRNA that code for the five virus-specific proteins. The *width of the arrows* indicates amount of gene product. It should be emphasized that VP 1, the major capsid protein, is read in another frame than VP 2 and VP 3, and that large T protein is read from two non-contiguous DNA segments (bp 5081–4836 and 4490–2611).

it is highly probable that at least one of these proteins is able to disturb thoroughly balanced cellular processes, with the result that the cell produces virus particles or becomes transformed to a neoplastic state (cf. Fig. **19**.5). It has been shown, indeed, that the T proteins affect several viral and cellular processes:

1. they bind to a specific region of viral DNA and initiate replication;
2. they stimulate host protein synthesis;
3. they stimulate RNA synthesis in isolated nuclei and nucleoli;
4. they are involved in the induction and maintenance of the neoplastic state.

A potential explanation of these multiple and seemingly unrelated effects became available when a protein kinase activity associated with both T antigen and a related protein called D2 was discovered.[7] A highly purified D2 protein was checked for the substrate specificity of the protein kinase associated with it. To this end D2 was incubated in the presence of [$\gamma$-$^{32}$p] ATP with nuclei from uninfected monkey cells. After gel electrophoretic separation and radioautography of nuclear proteins it was shown that, besides T protein itself, only four as yet uncharacterized species became selectively labelled. These proteins are already present in uninfected cells and therefore cannot be specified by the viral genome. One of them, the "53K protein" ($M_r = 53$ kd), the nature of which is unknown, appears to form a tight complex with large T protein. It can be detected as phosphoprotein also in cells transformed by other viruses or, even more interestingly, by chemical carcinogens. Thus phosphorylation of this protein may be critical for transformation [37a, 37b].

In support of such a central role of nuclear protein phosphorylation in cell transformation are two facts:

1. The loss of transformation activity beyond a permissive temperature by a mutation in the SV40 A gene which specifies a temperature-sensitive T protein.
2. The many reports on the appearance of tumor-induced protein kinase activities not detectable in normal tissues. Such newly acquired protein kinases have been demonstrated in nuclei of Ehrlich ascites tumors, Novikoff ascites hepatoma as well as in Rous or avian sarcoma and polyoma virus-infected cells. In the latter two cases the protein kinases, i.e. the *src* protein and middle T antigen respectively, were shown to be specified by viral genes. All or most of them are unusual in that they phosphorylate proteins on tyrosine rather than on serine or threonine residues.

Like many other PK's, the *src* protein undergoes autophosphorylation. Virtually all of this protein ($M_r$ 60 kd) is bound to the the cytoskeleton which is known to make intimate contact with both the cytomembrane and the nuclear lamina. It phosporylates a cytoskeletal protein ($M_r$ 50 kd), an event that is essential for transformation by Rous sarcoma virus. DNA sequences closely related to the *src* gene and a 60 kd protein structurally related to the *src* protein are present in uninfected cells ranging from birds to man, yet in amounts 30–50fold lower than *src* protein in Rous sarcoma virus-transformed cells. Thus the *src* protein may be related to a normal cell protein involved in some universal cellular function such as maintenance of control over cell division, and malignancy may result from an uncontrolled increase in the phosphorylation of a cytoskeletal element by the viral protein kinase ([39a, 39b]; cf. **19**.2.2.1 and **19**.2.2.2).

It is obvious that phosphorylation of nuclear and/or cytoskeletal proteins is an essential step in transformation. The molecular events in the path between phosphorylation and transformation, however, remain obscure (for reviews see [38a, 38b]).

---

[7] D2 is one of several proteins with T antigen properties produced in monkey cells after infection with an adenovirus 2-simian virus 40 hybrid. It appears to be a hybrid protein ($M_r$ 107 kd), since the larger part is encoded by the SV40 gene, whereas a small part (about 10 kd) is encoded by an as yet unspecified adenovirus gene [37a, 37b, 38a, 38b].

### 10.3.3.7.3 Phosphorylation of Certain Non-Histone Proteins Directly Affects Transcription

Though it is evident that cAMP-dependent phosphorylation of certain nuclear proteins is involved in regulation of gene activity in general, there are, at present, no indications of how these proteins act on a molecular basis. However, two examples may show that other protein kinases phosphorylate proteins clearly involved in regulation of transcription.

First a prokaryotic system will be mentioned to demonstrate that potential regulation of gene activity by protein phosphorylation is not restricted to eukaryotes.

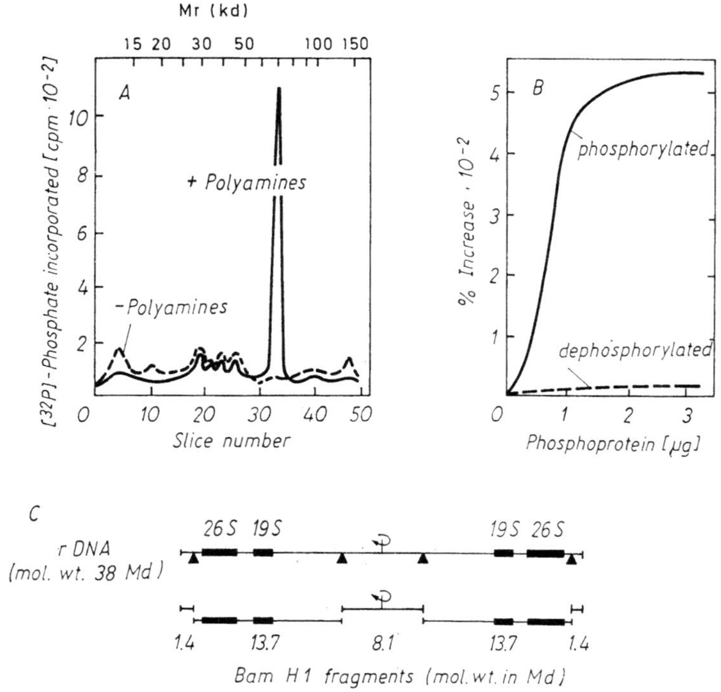


**Fig. 10.9:** Stimulation of ribosomal RNA synthesis by phosphorylation of a distinct nucleolar non-histone protein in a cell-free system from *Physarum polycephalum* (Kuehn et al. [34]).

**A** Phosphoproteins associated with the rDNP complex. Isolated nucleoli were phosphorylated with [$\gamma$-$^{32}$P]-ATP in the absence (---) or presence (—) of 0.33 mM each of putrescine, spermine, and spermidine. The rDNP was solubilized from chromatin with buffer containing EDTA. The soluble proteins were separated by rod gel electrophoresis. Incorporation of [$^{32}$P]-phosphate into proteins was measured after slicing. The main fraction phosphorylated corresponds to a protein with $M_r$ of 70 kd that exerts its biological effect as a dimer of $M_r$ 140 kd.

**B** In vitro transcription of the purified rDNP complex containing RNA polymerase I and rDNA template.

Incorporation of [$^{3}$H]-UMP into RNA was detected in the presence of increasing amounts of purified 140 kd nuclear protein, either phosphorylated (—) or dephosphorylated by treatment with alkaline phosphatase-agarose (---). Values represent percent increase over control values (without 140 kd protein).

**C** Schematic presentation of the structure of rDNA from *P. polycephalum* with regions coding for the 26S and 19S rRNAs and cleavage sites for restriction endonuclease BamH1 which generates fragments of three different sizes. One of them (8.1 Md) includes the axis of symmetry of the palindromic rDNA. It was shown to associate specifically with the phosphorylated 140 kd protein.

A unique prokaryotic protein kinase, encoded by an early phage T7 gene, is transcribed in *Escherichia coli* by host RNA polymerase (RNAP). It phosphorylates the $\beta'$-subunit of RNAP and thus interferes with transcription starting from all host-specific and early phage-promoter sequences. At the same time termination at phage-internal termination signals in the early gene region is enforced. The shut-down of host RNAP does not disturb transcription of late genes of phage T7 because, together with the protein kinase, a phage-specific RNAP is formed. Thus T7 protein kinase is involved in a kind of timing mechanism which, 7–10 min after infection, switches off synthesis of host gene and early T7 gene transcripts and allows formation of late T7 gene products to proceed preferentially (for more details cf. **14**.2.2).

An example for an eukaryotic system throws some light on the molecular basis of how gene activity might be regulated by protein phosphorylation [34]. In the polynuclear syncytium of the slime mold *Physarum polycephalum* the polyamines putrescine, spermine, and spermidine stimulate synthesis of ribosomal RNA (rRNA), apparently via phosphorylation of a distinct nuclear non-histone protein ($M_r$ = 140 kd). The mechanism of this phosphorylation and the protein kinase involved have not yet been elucidated. The nucleolar DNA coding for rRNA (rDNA) in *P. polycephalum* consists of multiple copies. Each rDNA copy (molecular weight 38 Md) is a palindrome with two coding sequences for 5.8S, 19S, and 26S rRNA. The rDNA has been recently isolated as a transcriptionally active deoxyribonucleoprotein (rDNP) complex containing nucleolar RNA polymerase I, rDNA, and several nucleolar non-histone proteins, including the 140 kd protein mentioned above. Fig. **10**.9 shows the specific phosphorylation of this protein (A) and its stimulatory effect on cell-free rRNA synthesis (B).

Only the phosphorylated form of the 140 kd protein shows high affinity binding for the central part of labelled rDNA and its Bam H1 restriction fragment, respectively:

1. The affinity of the 140 kd protein is at least 17 times higher in its phosphorylated than in its unphosphorylated state.
2. The phosphorylated protein binds to the axis of symmetry in the rDNA, i.e., at a considerable distance from the coding regions. This was evidenced by showing that the 8.1 Md fragment bound to the protein as strongly as uncleaved rDNA, whereas there was almost no binding of the 13.7 Md fragment.

By hybridization experiments it was further shown that the products of in vitro transcription in the presence of phosphorylated 140 kd protein represented, in fact, 26S and 19S rRNA. So this is a good example that the activity of distinct genes can be regulated by phosphorylation of distinct nuclear non-histone proteins.

### 10.3.3.8 Competence and Compartmentation of Cells Provide the Basis for Specific Action of Multifunctional Protein Kinases

It seems that in most cases phosphorylation is not part of an all-or-non control mechanism, but rather a component of the "homoeostatic mechanism by which multicellular organisms regulate and coordinate metabolic (and developmental) processes in the face of a constantly changing external and internal environment" (Greengard [26b]). But on the other hand there are observations which suggest that an alteration in the phosphorylation pattern can represent a crucial step. This is documented by the role of virus-encoded protein kinases for cell transformation (cf. **10**.3.3.7.2) or by cAMP-dependent protein kinase-induced regression of steroid hormone-dependent mammary tumors (cf. **10**.3.3.6.1). The multiple role of protein phosphorylation, as well as the multiplicity and ubiquity of proteins affected by it, may explain why eukaryotes have developed this system to the complexity observed. In view of the existence of

- a great number of protein substrates subject to phosphorylation,
- a limited number of, mostly multifunctional, protein kinases and
- only a few phosphoprotein phosphatases

the question must be raised how the phosphorylation state of a protein can be controlled, and how specificity of phosphorylation, i.e., in some cases modification of only a single functional protein in a cell, can be achieved. It has been mentioned that the conformation of the substrate protein itself determines the rate of protein dephosphorylation by a phosphoprotein phosphatase (cf. **10**.3.3.4). In a comparable way effector-induced alterations in the conformation of a protein may alter its susceptibility to a protein kinase. Hence, the degree of phosphorylation of a particular protein, which represents a balance between the actions of protein kinases and phosphoprotein phosphatase, might be determined (1) by directly affecting activities of one or both of the counteracting enzymes, e.g. by modulator proteins, and (2) by changes in substrate protein conformation, induced, e.g., by low molecular weight effectors.

The answer to the question of specificity of phosphorylation does not seem difficult, at least at first glance, if a protein kinase of high specificity is involved.[8] But, on the one hand, there is not, and cannot be, a specialized protein kinase/phosphatase pair for any single protein whose functional properties are altered by phosphorylation. On the other hand, if we consider a protein kinase of high specificity, it very often represents only the last link in a signal transformation chain beginning with activation of a multifunctional protein kinase.

Hence the main reason for specificity of protein phosphorylation and its highly dynamic changes must be looked for in the competence of cells, which has to be understood in a most comprehensive sense:

1. *Presence of Distinct Hormone Receptors:* A cell will respond to an extracellular signal only if its membrane contains the corresponding receptor. During ontogeny, cells become highly specialized in this respect.
2. *Type of Signal Transformation Chain:* As detailed in Fig. **10**.6, signal binding to its receptor may result in formation of different second messengers within (cAMP, cGMP) or their transport into ($Ca^{2+}$) target cells.
3. *Type of Protein Kinase:* In correspondence with different second messengers there are cAMP-, cGMP-, and Ca-calmodulin-dependent protein kinases with differing specificity. The appearance in a cell of one of the second messengers often results in phosphorylation of distinct proteins. But Ca-calmodulin and cAMP may also trigger the phosphorylation of the same protein in vivo, as it was shown for a specific brain membrane protein [26a]. On the other hand there is an increasing number of specialized PK isozymes, phosphorylating only one or a very few proteins and being activated by phosphorylation, in some cases by autophosphorylation, or by effectors other than cAMP, cGMP, or Ca-calmodulin (cf. **4**.4.4.5, **4**.4.4.7, **10**.3.3.7.2 and **10**.3.3.7.3).
4. *Compartmentation of Protein Kinases:* The positioning and possibly redistribution of protein kinases were discussed in Sect. **10**.3.3.6. It is worth mentioning that in some cells protein kinases are evidently constituents of the outer cell membrane by which they are enabled to interact directly with their membrane protein substrate.

[8] Examples are an epidermal growth factor-dependent PK (Cohen et al.: J. Biol. Chem, *255* (1980): 4834), a corticotropin-stimulated PK (McPherson & Rama chandran: Biochem. Biophys. Res. Comman. *94* (1980): 1057), virus-encoded PK's (cf. **10**.3.3.7.2), and a polyamine-triggered PK (cf. **10**.3.3.7.3).

it is highly probable that at least one of these proteins is able to disturb thoroughly balanced cellular processes, with the result that the cell produces virus particles or becomes transformed to a neoplastic state (cf. Fig. **19**.5). It has been shown, indeed, that the T proteins affect several viral and cellular processes:

1. they bind to a specific region of viral DNA and initiate replication;
2. they stimulate host protein synthesis;
3. they stimulate RNA synthesis in isolated nuclei and nucleoli;
4. they are involved in the induction and maintenance of the neoplastic state.

A potential explanation of these multiple and seemingly unrelated effects became available when a protein kinase activity associated with both T antigen and a related protein called D2 was discovered.[7] A highly purified D2 protein was checked for the substrate specificity of the protein kinase associated with it. To this end D2 was incubated in the presence of [$\gamma$-$^{32}$p] ATP with nuclei from uninfected monkey cells. After gel electrophoretic separation and radioautography of nuclear proteins it was shown that, besides T protein itself, only four as yet uncharacterized species became selectively labelled. These proteins are already present in uninfected cells and therefore cannot be specified by the viral genome. One of them, the "53K protein" ($M_r$ = 53 kd), the nature of which is unknown, appears to form a tight complex with large T protein. It can be detected as phosphoprotein also in cells transformed by other viruses or, even more interestingly, by chemical carcinogens. Thus phosphorylation of this protein may be critical for transformation [37a, 37b].

In support of such a central role of nuclear protein phosphorylation in cell transformation are two facts:

1. The loss of transformation activity beyond a permissive temperature by a mutation in the SV40 A gene which specifies a temperature-sensitive T protein.
2. The many reports on the appearance of tumor-induced protein kinase activities not detectable in normal tissues. Such newly acquired protein kinases have been demonstrated in nuclei of Ehrlich ascites tumors, Novikoff ascites hepatoma as well as in Rous or avian sarcoma and polyoma virus-infected cells. In the latter two cases the protein kinases, i.e. the *src* protein and middle T antigen respectively, were shown to be specified by viral genes. All or most of them are unusual in that they phosphorylate proteins on tyrosine rather than on serine or threonine residues.

Like many other PK's, the *src* protein undergoes autophosphorylation. Virtually all of this protein ($M_r$ 60 kd) is bound to the the cytoskeleton which is known to make intimate contact with both the cytomembrane and the nuclear lamina. It phosporylates a cytoskeletal protein ($M_r$ 50 kd), an event that is essential for transformation by Rous sarcoma virus. DNA sequences closely related to the *src* gene and a 60 kd protein structurally related to the *src* protein are present in uninfected cells ranging from birds to man, yet in amounts 30–50fold lower than *src* protein in Rous sarcoma virus-transformed cells. Thus the *src* protein may be related to a normal cell protein involved in some universal cellular function such as maintenance of control over cell division, and malignancy may result from an uncontrolled increase in the phosphorylation of a cytoskeletal element by the viral protein kinase ([39a, 39b]; cf. **19**.2.2.1 and **19**.2.2.2).

It is obvious that phosphorylation of nuclear and/or cytoskeletal proteins is an essential step in transformation. The molecular events in the path between phosphorylation and transformation, however, remain obscure (for reviews see [38a, 38b]).

---

[7] D2 is one of several proteins with T antigen properties produced in monkey cells after infection with an adenovirus 2-simian virus 40 hybrid. It appears to be a hybrid protein ($M_r$ 107 kd), since the larger part is encoded by the SV40 gene, whereas a small part (about 10 kd) is encoded by an as yet unspecified adenovirus gene [37a, 37b, 38a, 38b].

gene for $TD_S$ into the *ilv* regulon, which specifies all enzymes involved in the biosynthetic pathway common to isoleucine, leucine and valine (cf. **10**.2.3). What appears particular, however, is its role in the control of its own formation and that of both the other *ilv* enzymes and the branched-chain aminoacyl tRNA synthetases (cf. **10**.2.3.2). With respect to this multiple control function $TD_S$, similar to bacterial glutamine synthetase (cf. Fig. **4**.3), has a central position in the regulatory networks of bacterial amino acid metabolism.

3. Apart from its regulatory properties, the involvement of an isozyme in metabolic channeling, e.g., through formation of complexes with other specialized isozymes or by association with subcellular structures, deserves particular interest. However, evidence for the participation of enterobacterial aspartokinase and acetohydroxyacid synthase isozymes and of eukaryotic isozymes involved in sugar metabolism in polyisozymic complexes is mainly indirect at present (cf. **10**.2.1, **10**.2.3, **10**.3.1 and **10**.3.2).
4. Isozymes play an important role in signal transformation and regulation of differential gene expression in eukaryotes. Adenylate cyclase, protein kinase, cyclic nucleotide phosphodiesterase, phosphoprotein phosphatase, and RNA polymerase isozymes are characteristic examples (cf. **10**.3.3 and **4**.4.2.6). With the exception of the latter, they function as components of signal transformation systems mediating the information carried by external signals in a tissue- and cell-specific way to distinct sites within a cell. Thus, a combination of distinct isozymes not only characterizes metabolic properties of a given cell type but also provides the necessary basis for a specific response of cells on external signals (cf. **10**.3.3.8), i.e., for the tight integration of cell differentiation into the network of developmental control elements of multicellular organisms.

## 10.5 Selected References

*Bacterial Isozymes and Related Subjects*


1. Umbarger, H. E.: Amino acid biosynthesis and its regulation. Ann. Rev. Biochem. *47* (1978): 533–606.
2. Lawther, R. P., D. H. Calhoun, C. W. Adams, C. A. Hauser, J. Gray & G. W. Hatfield: Molecular basis of valine resistance in *Escherichia coli* K-12. Proc. Natl. Acad. Sci. USA *78* (1981): 922–925.
3. Berg. C. M. & K. J. Shaw: Organization and regulation of the ilv GEDA operon in *Salmonella typhimurium* LT2. J. Bacteriol. *145* (1981): 984–989.

4a. Calhoun, D. H. & G. W. Hatfield: Autoregulation of gene expression. Ann. Rev. Microbiol. *29* (1975): 275–299.

4b. Calhoun, D. H. & G. W. Hatfield: Autoregulation: A role for a biosynthetic enzyme in the control of gene expression. Proc. Natl. Acad. Sci. USA *70* (1973): 2757–2761.

5. Abrescia, P., J. Guardiola, M. Foresti, A. Lamberti & M. Iaccarino: Threonine deaminase: Autogenous regulator of the ilv genes in *Escherichia coli* K-12. Mol. Gen. Genet. *171* (1979): 261–275.
6. Kline, E. L., C. S. Brown, W. G. Coleman & H. E. Umbarger: Regulation of isoleucine-valine biosynthesis in an ilvDAC deletion strain of *Escherichia coli* K-12. Biochem. Biophys. Res. Commun. *57* (1974): 1144–1151.

7a. Levinthal, M., M. Levinthal & L. S. Williams: The regulation of the ilvADGE operon: Evidence for positive control by threonine deaminase. J. Mol. Biol. *102* (1976): 453–465.

7b. Smith, J. M., D. E. Smolin & H. E. Umbarger: Polarity and the regulation of the ilv gene cluster in *Escherichia coli* strain K-12. Mol. Gen. Genetics *148* (1976): 111–124.

8. Baez, M., D. W. Patin & D. H. Calhoun: Deletion mapping of the ilvGOEDAC genes of *Escherichia coli* K-12. Mol. Gen. Genet. *169* (1979): 289–297.

9. Patin, D. W. & D. H. Calhoun: Mapping of *Escherichia coli* K-12 with bacteriophage λ dilv. J. Bacteriol. *137* (1979): 1234–1242.
10. Smith, J. M., F. J. Smolin & H. E. Umbarger: Mutations affecting the formation of acetohydroxy acid synthase II in *Escherichia coli* K-12. Mol. Gen. Genet. *169* (1979): 299–314.
11a. Lawther, R. P. & G. W. Hatfield: Multivalent translational control of transcription termination at attenuator of ilvGEDA operon of *Escherichia coli* K-12. Proc. Natl. Acad. Sci. USA *77* (1980): 1862–1866.
11b. Nargang, F. E., C. S. Subrahmanyam & H. E. Umbarger: Nucleotide sequence of ilvGEDA operon attenuator region of *Escherichia coli*. Proc. Natl. Acad. Sci. USA *77* (1980): 1823–1827.
12. Phillips, A. T., R. M. Egan & N. Lewis: Control of biodegradative threonine dehydratase inducibility by cyclic AMP in energy-restricted *Escherichia coli*. J. Bacteriol. *135* (1978): 828–840.
13. Yui, Y., Y. Watanabe, S. Ito, Y. Shizuta & O. Hayaishi: Multivalent induction of biodegradative threonine deaminase. J. Bacteriol. *132* (1977): 363–369.

*General Literature on Eukaryotic Isozymes*

14. Markert, C. L. (ed.): Isozymes. Academic Press, New York 1975.
15. Ureta, T.: The role of isozymes in metabolism: A model of metabolic pathways as the basis for the biological role of isozymes. Curr. Top. Cell. Regul. *13* (1978): 233–258.
16. Masters, C. J. & R. S. Holmes: Haemoglobin, Isoenzymes and Tissue Differentiation. North Holland Publishing Company, Amsterdam 1975.
17. Korn, E. D.: Biochemistry of actomyosin-dependent cell motility (A review). Proc. Natl. Acad. Sci. USA *75* (1978): 588–599.
18. Sutherland, E. W.: Studies on the mechanism of hormone action. Science *177* (1972), 401–408.

*Signal transformation and Protein Kinase Isozymes*

19. Robdell, M.: The role of hormone receptors and GTP-regulatory proteins in membrane transduction. Nature *284* (1980): 17–22.
20. Ross, E. M. & A. G. Gilman: Biochemical properties of hormone-sensitive adenylate cyclase. Ann. Rev. Biochem. *49* (1980): 533–564.
21a. Goldstein, J. G., R. G. W. Anderson & M. S. Brown: Coated pits, coated vesicles, and receptor-mediated endocytosis. Nature *279* (1979): 679–685.
21b. Pearse, B.: Coated vesicles. Trends Biochem. Sci. *5* (1980): 131–134.
22a. Chock, P. B., S. G. Rhee & E. R. Stadtman: Interconvertible enzyme cascades in cellular regulation. Ann. Rev. Biochem. *49* (1980): 813–843.
22b. Cohen, P.: The role of cyclic AMP-dependent protein kinase in the regulation of glycogen metabolism in mammalian skeletal muscle. Curr. Top. Cell. Regul. *14* (1979): 117–196.
23. Krebs, E. G. & J. A. Beavo: Phosphorylation-dephosphorylation of enzymes. Ann. Rev. Biochem. *48* (1979): 923–959.
24. Kuo, J. F., M. Shoji & W.-N. Kuo: Molecular and physiopathologic aspects of mammalian cyclic GMP-dependent protein kinase. Ann. Rev. Pharmacol. Toxicol. *18* (1978): 341–355.
25. Lincoln, T. M. & J. D. Corbin: On the role of the cAMP and cGMP-dependent protein kinases in cell function. J. Cycl. Nucleotide Res. *4* (1978): 3–14.
26a. Greengard, P.: Cyclic nucleotides, phosphorylated proteins, and the nervous system. Federation Proc. *38* (1980): 2208–2217.
26b. Greengard, P.: Phosphorylated proteins as physiological effectors. Science *199* (1978): 146–152.
27. Klee, C. B., T. H. Crouch & P. G. Richman: Calmodulin. Ann. Rev. Biochem. *49* (1980): 489–515.
28. Ochoa, S. & C. de Haro: Regulation of protein synthesis in eukaryotes. Ann. Rev. Biochem. *48* (1979): 549–580.
29. Ogawa, H., H. Shiraki, Y. Matsuda & H. Nakagawa: Interaction of adenylosuccinate synthetase with F-actin. Eur. J. Biochem. *85* (1978): 331–337.
30a. Vandekerckhove, J. & K. Weber: The complete amino acid sequence of actins from bovine aorta, bovine heart, bovinefast skeletal muscle, and rabbit slow skeletal muscle. Differentiation *14* (1979): 123–133.
30b. Firtel, R. A.: Multigene families encoding actin and tubulin. Cell *23* (1981): 6–7.

31. Cho-Chung, Y.-S., D. Archibald & T. Clair: Cyclic AMP receptor triggers nuclear protein phosphorylation in a hormone-dependent mammary tumor cell-free system. Science *205* (1979): 1390–1392.
32. Iyengar, R. & L. Birnbaumer: Coupling of the glucagon receptor to adenylyl cyclase by GDP: Evidence for two levels of regulation of adenylyl cyclase. Proc. Natl. Acad. Sci. USA *76* (1979): 3189–3193.
33. Jungmann, R. A., M. I. Mednieks & J. S. Schweppe: Nuclear compartmentation of protein kinase and the regulation of gene expression. In: Nover, L., F. Lynen & K. Mothes (eds.): Cell Compartmentation and Metabolic Channeling. Gustav Fischer Verlag, Jena and Elsevier/North-Holland Biomedical Press, Amsterdam 1980, pp. 415–426.
34. Kuehn, G. D., H.-U. Affolter, U. J. Atmar, T. Seebeck, U. Gubler & R. Brown: Polyamine-mediated phosphorylation of a nuclear protein from *Physarum polycephalum* that stimulates rRNA synthesis. Proc. Natl. Acad. Sci. USA *76* (1979): 2541–2545.
35. Strosberg, A. D., G. Vauquelin, O. Durieu-Trautmann, C. Delavier-Klutchko, S. Bottari & C. Andre: Towards the chemical and functional characterization of the $\beta$-adrenergic receptor. Trends Biochem. Sci. *5* (1980): 11–14.
36. Toscano, W. A., K. R. Westcott, D. C. LaPorte & D. R. Storm: Evidence for a dissociable protein subunit required for calmodulin stimulation of brain adenylate cyclase. Proc. Natl. Acad. Sci. USA *76* (1979): 5582–5586.
37a. Tjian, R. & A. Robbins: Enzymatic activities associated with a purified simian virus 40 T antigen-related protein. Proc. Natl. Acad. Sci. USA *76* (1979): 610–614.
37b. Baumann, E. A. & R. Hand: Protein kinase activity associated with the D2 hybrid protein related to simian virus 40 T antigen: Some characteristics of the reaction products. Proc. Natl. Acad. Sci. USA *76* (1979): 3688–3692.
38a. Crawford, C. V.: The T antigen of simian virus 40 and polyoma virus: Their role in transformation. Trends Biochem. Sci. *5* (1980): 39–42.
38b. Langan, T.: Malignant transformation and protein phosphorylation. Nature (London) *286* (1980): 329–330.
39a. Selfton, B. M., T. Hunter, K. Beemon & W. Eckhart: Evidence that the phosphorylation of tyrosine is essential for cellular transformation by Rous sarcoma virus. Cell *20* (1980): 807–816.
39b. Burr, J. G., G. Dreyfuss, S. Penman & J. M. Buchanan: Association of the src product of Rous sarcoma virus with skeletal structures of chicken fibroblasts. Proc. Natl. Acad. Sci. USA *77* (1980): 807–816.
40. Gagelman, M., J. Reed, D. Kübler, W. Pyerin & V. Kinzel: Evidence for a "mute" catalytic subunit of cyclic AMP-dependent protein kinase from rat muscle and its mode of activation. Proc. Natl. Acad. Sci. USA *77* (1980): 2492–2496.
41a. Dreyfuss, G., K. J. Schwartz & E. K. Blout: Compartmentalization of cyclic AMP-dependent protein kinases in human erythrocytes. Proc. Natl. Acad. Sci. USA *75* (1978), 5926–5930.
41b. Hayes, J. S., L. L. Brunton & S. E. Mayer: Selective activation of particulate cAMP-dependent protein kinase by isoproterenol and prostaglandin. J. Biol. Chem. *255* (1980): 5113–5119.
42. Rubin, C. S.: Characterization and comparison of membrane-associated and cytosolic cAMP-dependent protein kinases. Studies on human erythrocyte protein kinases. J. Biol Chem. *254* (1979): 12439–12449.
43. Lohmann, S. M., U. Walter, P. E. Miller, P. Greengard & P. Camilli: Immunohistochemical localization of cGMP-dependent protein kinase in mammalian brain. Proc. Natl. Acad. Sci. USA *78* (1981): 653–657.
44. Foulkes, J. G., L. S. Jefferson & P. Cohen: The hormone control of glycogen metabolism: Dephosphorylation of protein phosphatase inhibitor-1 in vivo in response to insulin. FEBS letters *112* (1980): 21–24.
45. Lohmann, S. M., U. Walter & P. Greengard: Identification of endogenous substrate proteins for cAMP-dependent protein kinase in bovine brain. J. Biol. Chem. *255* (1980): 9985–9992.
46. Fuller, D. J. M., C. U. Byus & D. H. Russel: Specific regulation by steroid hormones of the amount of type I cyclic AMP-dependent protein kinase holoenzyme. Proc. Natl. Acad. Sci. USA *75* (1978): 223–227.

# 11. The Expression of Secondary Metabolism – an Aspect of Cell Specialization

M. Luckner

## 11.1 What is Secondary Metabolism?

Many of the events and compounds involved in metabolism are almost identical in all living organisms. They are of such importance that they are collectively known as "primary metabolism" and "primary metabolic products" (cf. Table **4**.1). Examples are seen in the synthesis and degradation of nucleotides and nucleic acids, many amino acids and proteins and the most important sugars and carboxylic acids. In contrast, certain metabolic pathways and substances occur only in a few species of organisms, in single "chemical races" or only during certain stages of the differentiation of specialized cells. Such pathways are collectively referred to as "secondary metabolism" and the compounds formed are called "secondary natural products".

The compounds encountered in secondary metabolism exhibit an almost infinite variability of structures. They include nitrogen-, oxygen- and sulphur-containing aliphatic, carbocyclic and heterocyclic compounds. Saturated and unsaturated substances, glycosides, peptides, hydroxamic acids and azomethines have been isolated, and widely differing functional groups such as hydroxy, epoxi, ester, ether, amino, nitro and carboxylic groups have been identified. Tens of thousands of secondary natural products are known to-day, and with improvement in analytical methods the number of known compounds is increasing at considerable speed.

Secondary natural products are produced in microorganisms, higher plants and animals. They are usually only formed during certain phases of development of the producing organism and only in certain cells, tissues or organs. This phase dependence in synthesis is due in turn to the phase-dependent production of the enzymes involved. The expression of secondary metabolism is thus an aspect of cell differentiation and secondary metabolism can be defined as the *biosynthesis, transformation and degradation of endogenous compounds by specialization proteins* (cf. Table **4**.1).

Like many morphological traits is the formation of secondary natural products of no significance to the individual cells producing them. But their role in the development and functioning of the organism as a whole may be decisive. This is well illustrated by the participation of secondary natural products in ecological relationships (natural repellents, attractants, phytoalexins), or in their function as signal substances within the programs of gene expression and development (most animal and plant hormones).

In order to illustrate some aspects of the expression of secondary metabolism two examples will be treated in the following sections: the flavonoid metabolism of parsley cell culture and the alkaloid metabolism of the mould *Penicillium cyclopium*. The first of these two systems is of relative simplicity and will serve here for a description

---

Abbreviations: AA, anthranilate adenylyltransferase; CD, cyclopeptine dehydrogenase; CH, cycloheximide; CmH, cyclopenin m-hydroxylase; DE, dehydrocyclopeptine epoxidase; FS, flavanone synthase (FS was recently identified as chalcone synthase, the flavanone being formed only by the following enzymatic isomerization); p.i., after inoculation; Pfr, physiologically active form of phytochrome; Pr, physiologically inactive form of phytochrome; PAL, phenylalanine ammonia-lyase.

of molecular biological aspects of the induction of a group of enzymes involved in secondary metabolism. The second system will serve to illustrate the integration of secondary metabolism in the developmental program of a simple eukaryotic organism.

## 11.2 Phase-Dependent Formation of Enzymes of Secondary Metabolism

### 11.2.1 The Role of Light in Triggering Formation of Enzymes Involved in Flavonoid Metabolism in Parsley Cell Cultures

At a certain phase in leaf development the shoots of garden parsley (*Petroselinum hortense*) synthesize large quantities of flavonoid glycosides. Formation of these substances is part of the complex program of leaf development, which includes the formation of a large number of specialized cells, e.g., epidermal, pallisade, parenchyma, tracheidal, fibre and phloem cells, etc. Regulation of the synthesis of flavonoids is extremely difficult to investigate in leaves on account of the interplay between different processes of cell specialization. A much more convenient test material is provided by parsley cells cultivated in vitro. These are characterized by rapid growth and an almost complete absence of any demonstrable morphological cell specialization. Nevertheless, they are still capable of forming flavonoid glycosides, which makes it possible to study the regulation of flavonoid metabolism more or less uncomplicated by other processes of cell specialization.

Cell cultures are initiated by placing sterilized petioles of parsley plants in a nutrient medium containing relatively large amounts of substances with auxin activity, e.g., 2,4-dichlorophenoxy acetic acid, to induce and promote cell division. After some time, a callus, i.e., an aggregate of unorganized cells, begins to grow, usually starting from meristematic cells. If fragments of this callus are shaken in nutrient medium a suspension of small cell aggregates is formed, each consisting of about 20–100 cells.

For the experiments to be described here suspension cultures were cultivated in the dark at 27 °C on rotary shakers (400 ml nutrient medium in 2l-Erlenmeyer flasks). The flasks were inoculated with 40 ml of a 7-day-old cell culture and after the cultures had achieved the desired stage of development (cf. Fig. **11.4**) 40 ml portions were removed for further cultivation (zero time). In order to induce flavonoid metabolism (see below) cultures were irradiated as described for each experiment. At the time points given cells were filtered off, frozen in liquid nitrogen, and stored at −20 °C until required. For enzyme determination cells were mechanically ruptured, and larger cell debris were removed by centrifugation. The supernatant was crudely purified by addition of a small quantity of Dowex 1×2 (equilibrated with 0.1 M Tris-HCl, *p*H 7.5) and afterwards used for the determination of the enzymes. Flavones and flavone glycosides were determined by means of their absorption at 340 nm.

Figure **11.1** shows the enzyme reactions involved in the formation of malonylapiin, one of the major flavonoid glycosides in parsley cultures. The enzymes can be divided into two groups:

a) enzymes of cinnamic acid metabolism (group I with three enzymes) and
b) enzymes of flavonoid metabolism (group II with 13–14 enzymes).

Group I enzymes provide compounds that are not only precursors of flavonoids but of a large number of other secondary products as well, whereas group II enzymes are involved solely in the production of flavonoid glycosides.

Flavonoid biosynthesis in parsley cell cultures is triggered by UV light. UV irradiation forms or activates phytochrome, a signal transformation protein which is fre-

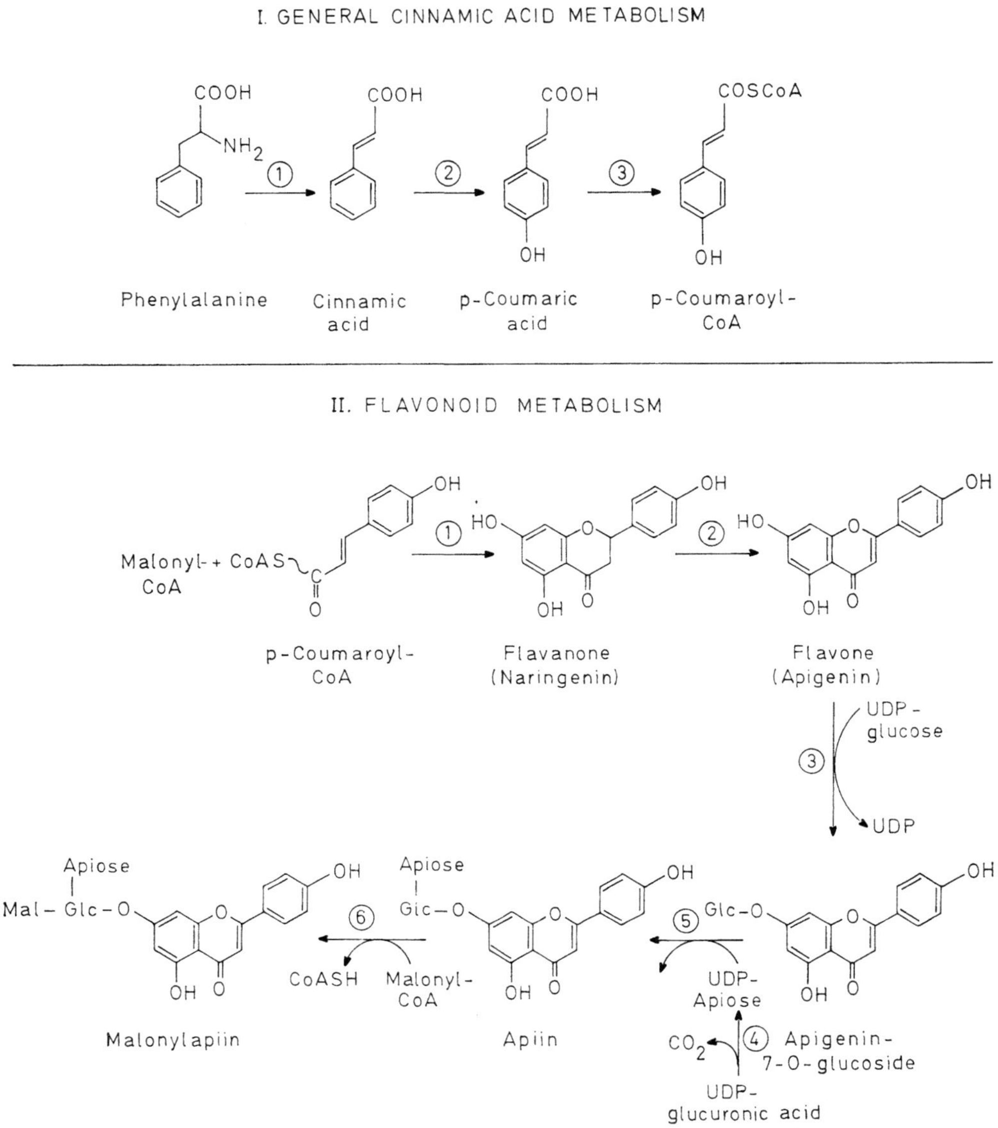


**Fig. 11.1:** Formation of the flavone glycoside malonylapiin in parsley cell cultures (Hahlbrock et al. [7]).

Enzyme group I: *(1)*, phenylalanine ammonia-lyase (PAL); *(2)*, cinnamic acid 4-hydroxylase; *(3)*, p-coumaric acid: CoA ligase.

Enzyme group II: *(1)*, flavanone synthase (FS); *(2)*, flavanone oxidase; *(3)*, UDP-glucose: flavone/flavonol-7-O-glucosyl transferase; *(4)*, UDP-apiose synthetase; *(5)*, UDP-apiose: flavone-7-O-glucosyl-2″-O-apiosyl transferase; *(6)*, malonyl-CoA: flavonoid glycoside malonyl-transferase and other enzymes not given in this figure.

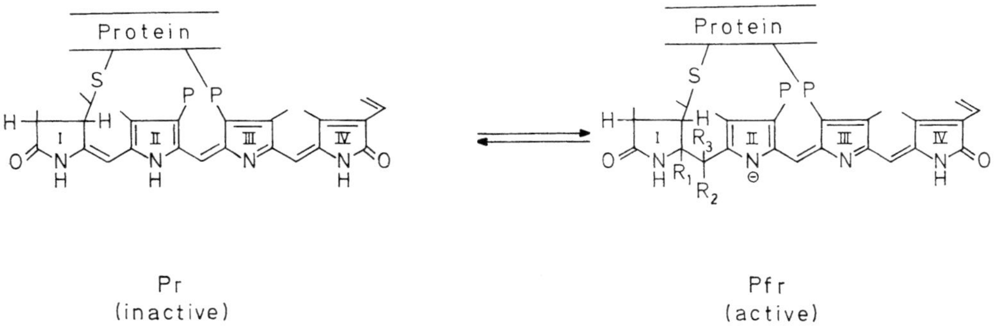


**Fig. 11.2**: Photointerconversion of phytochrome (from Mohr, H. & P. Schopfer, Lehrbuch der Pflanzenphysiologie, 3. Aufl., Springer Verlag, Berlin 1978, p. 315, with permission).

Phytochrome is a chromoprotein with $M_r \sim 120$ kd for the monomer. The covalent linkages between the apoprotein and the tetrapyrrol chromophore involve a thioether bond *(S)* and a bond to one of the propionic acid side chains *(P)*. The nature of the three substituents *($R_1$, $R_2$, $R_3$)* in Pfr is not yet clarified. Spectroscopic evidence point to a photoaddition to this methine-bridge upon Pr → Pfr transition.

quently encountered in light reception in higher plants. Phytochrome possesses a protein-bound chromophore structurally related to bile pigments (Fig. **11**.2).

Phytochrome exists in two spectral forms, which are interconvertible by means of irradiation. Irradiation with far red light (maximum at 730 nm) leads to formation of an inactive form (Pr) whereas a physiologically active form (Pfr) is produced by the action of red light (maximum 660 nm) (see Fig. **11**.2). Biophysical experiments suggest that this photoconversion proceeds via a series of short-lived intermediates whose life time at 0 °C is in the region of milliseconds to seconds. In the dark, phytochrome exists only in the form of Pr, but under irradiation at whatever wavelength, an equilibrium is established between Pr and Pfr. Under the influence of white daylight the physiologically active form predominates [4a].

15–30 s after the commencement of irradiation and the resultant formation of Pfr permeability of the protoplasmic membrane is greatly enhanced. A subsequent effect is an alteration in the localization of phytochrome within the cell. Whereas Pr is evenly distributed throughout the cytoplasm, within a few minutes after the onset of irradiation Pfr accumulates within certain cell compartments, possibly identical with the physiologically active receptor regions.

The effect of phytochrome on secondary metabolism, however, requires a longer period of time for its expression. In parsley cell cultures activity of the enzymes of group I increases for about 20 h and then begins to drop with a half-life of about 7–10 h (Fig. **11**.3). Similar changes in enzyme activity have also been demonstrated for enzymes of groups II, although their maximum activity is not achieved until 37–69 h after the beginning of irradiation (Fig. **11**.3). Thus, there is a distinct difference between light-induced alterations in activity of the enzymes of group I and of those of group II.

Initially, the amount of enzyme synthesized is directly dependent on the duration of UV irradiation (Table **11**.1), but after 20 min a saturation level is reached. If this stage is followed by red/far-red irradiation it can be shown that under UV irradiation active phytochrome has been formed or released and that it is involved in the regulation of enzyme induction (Table **11**.1). Brief irradiation with far-red light transforms the active phytochrome formed under UV irradiation into its inactive form, thus reducing the UV effect. This far-red effect can be reversed by brief red-light treatment. If not preceded by UV irradiation red and far-red are just as ineffective as white light.

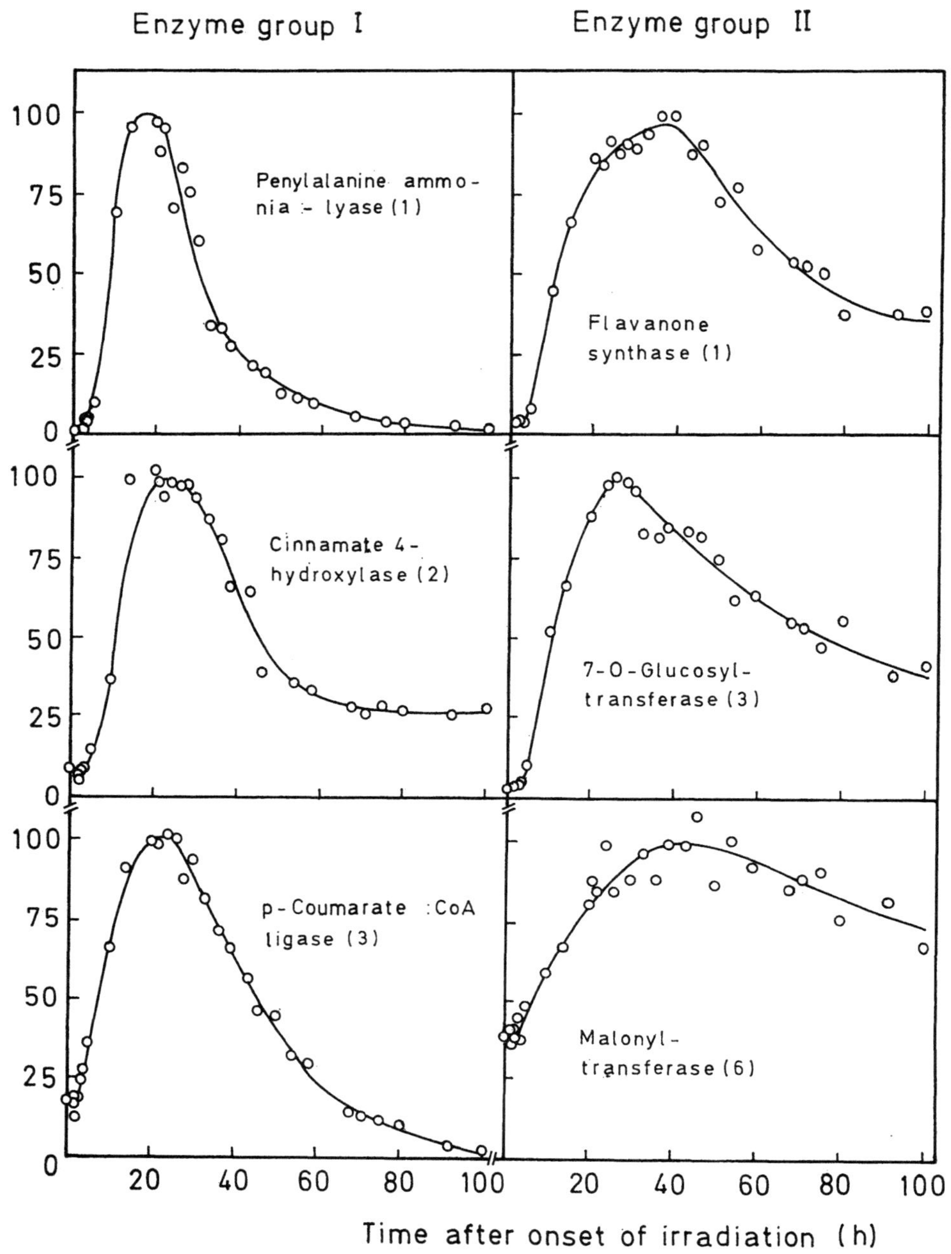


**Fig. 11.3:** Changes in the in vitro activity of enzymes of groups I and II (cf. **Fig. 11.1**) in cell cultures of *Petroselinum hortense* (Hahlbrock et al. [7]).

The cultures were continuously irradiated starting at zero hours. Cells used for experiments were harvested 240 h after inoculation (cf. **Fig. 11.4**).

*Ordinate:* All values are given in pkat/mg protein: phenylalanine ammonia-lyase: 100 = 100; cinnamic acid 4-hydroxylase: 100 = 3.5; p-coumaric acid: CoA ligase: 100 = 70; flavanone synthase: 100 = 1.0; 7-O-glucosyl transferase: 100 = 50; malonyl-transferase: 100 = 80.

**Table 11.1**: Activity of PAL and accumulation of flavonoids after irradiation of parsley cell cultures (Wellmann [8])

| Irradiation program[a] | Flavonoids[b] | PAL[b] |
|---|---|---|
| 20 min UV | 100 | 100 |
| 10 min UV | 49 | 50 |
| 5 min UV | 26 | 24 |
| 20 min UV + 10 min red | 98 | 102 |
| 20 min UV + 10 min far-red | 72 | 71 |
| 20 min UV + 10 min far-red + 10 min red | 101 | 101 |
| 10 min red | 1 | 0 |
| 60 min white | 4 | 3 |

[a] UV light, 320 to 380 nm, 1.2 W/m²; far-red light, 718 nm, 3.5 W/m²; red light, 660 nm, 0.67 W/m²; white light, > 400 nm, ca. 50 W/m². Radiation was followed by a dark period lasting till maximum accumulation of flavonoids and of PAL activity, respectively.

[b] Relative values; values after 20 min UV radiation = 100%.

### 11.2.2 Induction of Phenylalanine Ammonia-Lyase and Flavanone Synthase in Parsley Cell Cultures

Investigations on the mechanism of enzyme induction have so far been concentrated on phenylalanine ammonia-lyase (PAL, group I) and flavanone synthase (FS, group II). Using the following techniques it could be shown that for example PAL is formed de novo as a result of irradiation.

1. In vivo inhibition of the increase in enzyme activity using inhibitors of protein synthesis, e.g., cycloheximide.
2. Incorporation into the enzyme of [$^{35}$S]-methionine or of amino acids labelled with [$^{15}$N] during the rise in enzyme activity. By means of gel electrophoresis and density gradient centrifugation respectively the isotope label can be shown to reside in the newly formed enzyme molecules.
3. Precipitation of the enzyme with a specific antibody showed that irradiation leads to a parallel rise in enzyme activity and quantity.

The quantitative aspects of induction, even in the simple systems investigated, are influenced by a large number of factors. For example, the rate of induction depends on the stage of development of the cell culture at the time of irradiation. If the latter takes place during the logarithmic growth phase the induction rate is low, but irradiation immediately following inoculation and shortly before the beginning of the stationary phase leads to the highest induction rates (Fig. **11**.4). Cells cultivated in vitro thus exhibit changes in their competence to light that are typical of the intact plant. Also dependent on the stage of development is the time elapsing between the beginning of irradiation and the onset of maximum enzyme activity. Further, differences have been observed in the length of time over which a prolongation of irradiation elicits a further rise in enzyme activity. Quantitative studies on induction should therefore be performed under meticulously standardized conditions.

Experiments with actinomycin, a transcription inhibitor, suggest that the synthesis of PAL-mRNA starts immediately after the beginning of irradiation. In experiments in which isolated mRNA was translated in vitro or in vivo it could be confirmed that formation of PAL-mRNA and FS-mRNA does in fact commence at this point.

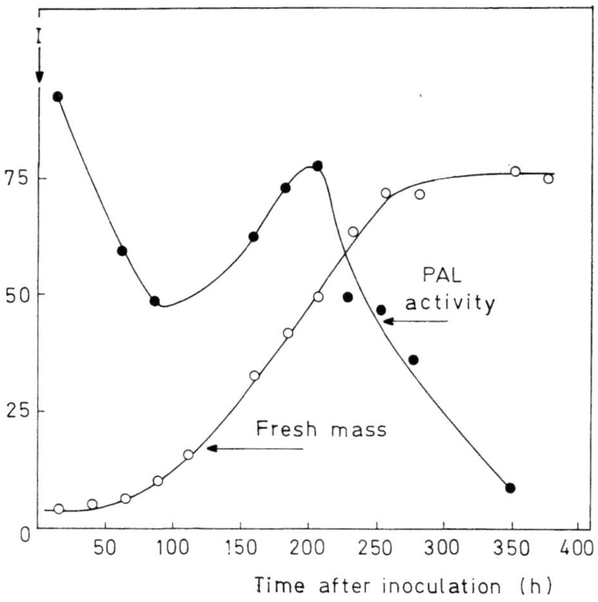


**Fig. 11.4**: Changes of fresh weight and inducibility of phenylalanine ammonia-lyase after irradiation during the growth cycle of parsley cell cultures (Hahlbrock [9]).

Fresh weight (100 = 20 g/40 ml culture); PAL activity after induction by irradiation (100 = 200 pkat/mg protein).

For the in vitro translation, polysome preparations were incubated with a soluble fraction of wheat germ or reticulocyte lysates, ATP, GTP and creatine phosphate, creatine phosphokinase, amino acids essential for protein synthesis and [$^{35}$S]-methionine. For in vivo translation, polysomes or mRNA preparations were injected into living oocytes of *Xenopus laevis*, after which the oocytes were incubated in a medium containing [$^{35}$S]-methionine (cf. **2**.8).

The PAL and FS formed in vitro or in vivo was precipitated by means of a specific immune serum following the addition of non-labelled enzyme. The precipitate was purified by polyacrylamide gel electrophoresis in the presence of SDS, and the radioactivity at the site of the PAL- and FS-subunits was determined. Under both in vitro and in vivo translation conditions the quantity of enzyme formed was directly proportional to the quantity of PAL-mRNA and FS--mRNA in the polysome or mRNA preparations analysed. The quantities of both mRNA species begin to rise exponentially immediately after commencement of irradiation, whereby the increase in FS-mRNA is at first slower, since the concentration of FS-mRNA in the non-irradiated cells is lower than that of PAL-mRNA.

Figure **11**.5A shows the rate of synthesis of PAL and FS as measured by means of the in vivo translation test. The total activities (Fig. **11**.5B) calculated from these rates, allowing for the continuous degradation of the enzyme proteins, are in good agreement with the experimental values for the activities of PAL and FS. This demonstrates that the enzyme synthesis depends on the quantity of specific mRNA and that the formation of both enzymes is regulated at the transcription level.

The elevated levels of PAL- and FS-mRNA following irradiation are returned to their original values after 20–25 h, even under long-term irradiation. This effect is partly due to the relatively short half-life of both species of mRNA (3.5–5 h) and partly to the as yet unclarified reduction in their rates of synthesis which have returned to their initial values during this time.

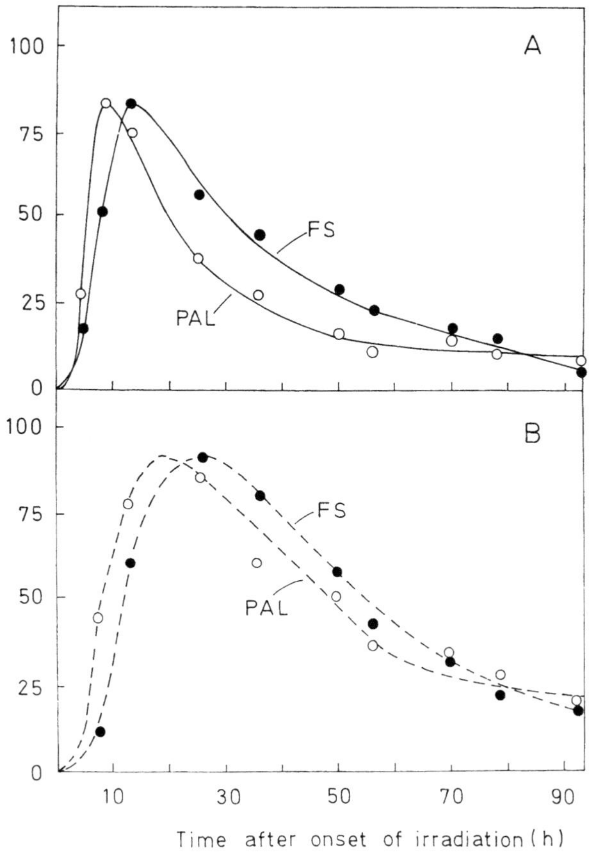


**Fig. 11.5**: Changes in synthesis of phenylalanine ammonia-lyase and flavanone synthetase in continuously irradiated parsley cell cultures (Hahlbrock et al. [13]).

**A**: Amount of PAL-mRNA and FS-mRNA determined by synthesis of PAL and FS in vitro (100 = 0.4 (PAL) and 1.3 (FS) % of total synthesis).

**B**: Enzyme activity (100 = 150 (PAL) and 1.0 (FS) pkat/mg protein).

- - -: Enzyme activity calculated from the values determined for the rate of synthesis and the rates of enzyme degradation by the following equation:

$$\frac{dE(t)}{dt} = k_s(t) - k_a \cdot E(t)$$

*E*, enzyme amount; *t*, time; $k_s$, rate of enzyme synthesis (cf. A); $k_a$, rate of enzyme degradation, calculated from the half-life of enzyme activity after termination of induction (PAL: 7–10 h, FS: 5–6 h).

In order to investigate the question of whether the coordinated formation of the enzymes of groups I and II is a consequence of the formation of polycistronic mRNA, the sizes of PAL- and FS-mRNA were determined by means of density gradient centrifugation and polyacrylamide gel electrophoresis. In each case the size proved to be about twice that required for coding the enzyme subunits. Both mRNA species contain poly(A) sequences and probably also the cap structure typical of eukaryotic mRNA (cf. Fig. 1.9). The coding capacity of FS-mRNA is therefore far too low for the 13–14 enzymes belonging to group II, thus excluding the possibility of a polycistronic mRNA for the enzymes of this group. The size of the PAL-mRNA would only be sufficient for the enzymes of group I if there were no larger non-coding regions. Hence here again, the results do not suggest a polycistronic mRNA.

In summary, the results demonstrate that the synthesis of the enzymes PAL and FS are regulated at the level of transcription, but that the enzyme amounts present depend on the rate of synthesis as well as on the rate of enzyme degradation.

## 11.3 Integration of Secondary Metabolism in the Program of Cell Specialization and Development

### 11.3.1 Expression of Alkaloid Synthesis during Chemical Specialization of Hyphal Cells of the Mould *Penicillium cyclopium*

Integration of secondary metabolism into differentiation programs, already mentioned in the preceding section, has been studied in detail in the mould *Penicillium cyclopium.*

In these experiments *P. cyclopium* was cultivated using chemically defined nutrient media. These include a carbohydrate source (usually glucose), a nitrogen soure (usually $NH_4^+$), $PO_4^{3-}$ and certain other ions. The mould develops on liquid nutrient media as well as on media solidified with agar. If left undisturbed (surface cultures) the mould forms a firm covering on the surface of the nutrient solution, if shaken it grows under submerged conditions in the form of larger or smaller aggregates or as a fine suspension. The nutrient medium is usually inoculated with conidiospores, although fragments of hyphae can also be used.

Figure **11**.6 illustrates the development of the mould. At first the spores swell and form rapidly growing and branching hyphae. It is at the end of this phase of vegetative growth that the penicilli develop. They are made up of three types of cell: the conidiophores, the metulae and the phialides (sterigmata), all of which are quite distinct in form from the cells of the hyphae. At intervals of about 90–120 min a unicellular, mononuclear conidiospore is detached at the tip of the phialides in a quantal cell division cycle (cf. **15**.2). The conidia mature within 12–24 h.

Conidiospores are distinguishable from phialides and other cells of the mould by their spherical form, a relatively thick cell wall containing a high proportion of melanin and by their green pigment. They are loosely connected with one another and form cell chains, with the mature conidia at the distal end and the newly formed cells at the base. Under favourable conditions the cell clone formed by a phialide consists of more than 100 spores. The conidia can easily be separated from one another and from the mycelium by mechanical means.

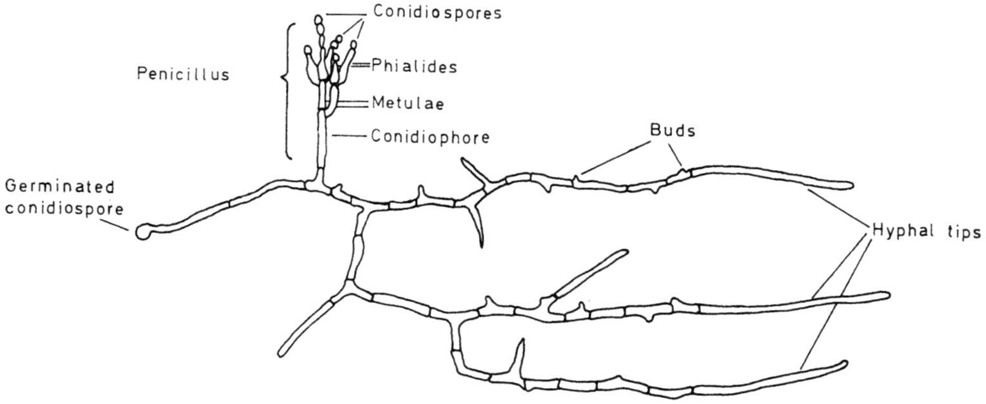


**Fig. 11.6:** Types of cells formed during the development of *Penicillium cyclopium* (Luckner [18]).

Thus the cycle of vegetative development in *P. cyclopium* comprises four distinct cell types arising from each other in the following sequence:

→ hyphal cells → conidiophores, metulae → phialides —
— conidiospores ← —

The expression of alkaloid metabolism is intimately connected with the synthesis of these cell types.

In surface cultures of *P. cyclopium* on a liquid nutrient a layer of hyphal cells is formed on the surface of the medium. Decreasing the nutrient supply 48 h after innoculation results in more or less synchronous formation of penicilli and conidiospores on the side of the mycelium exposed to air. Totally three stages of development can be distinguished in the cultures (Fig. **11**.7): the germination phase, the phase

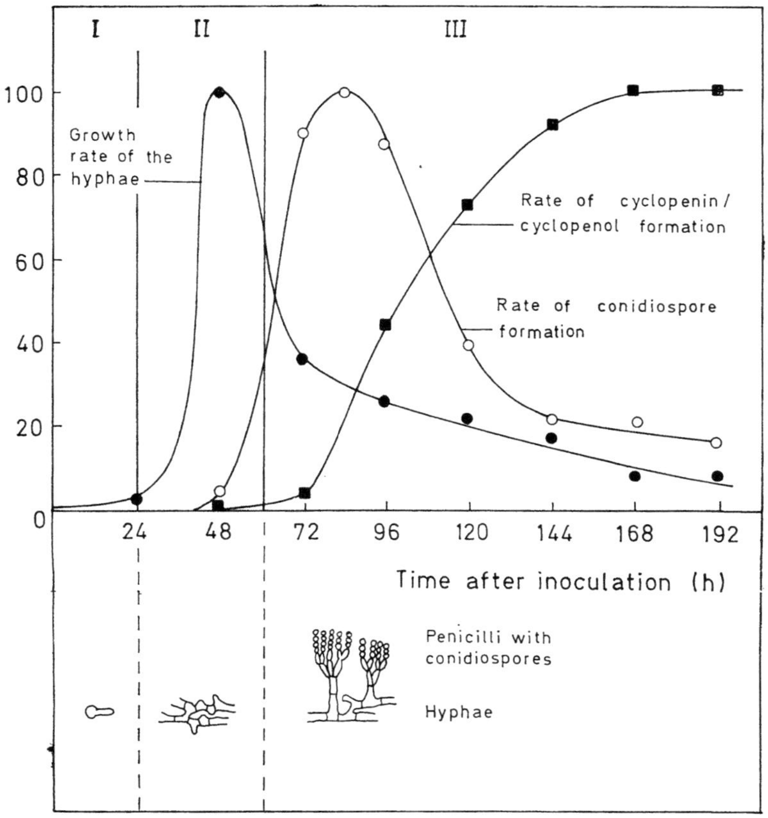


**Fig. 11.7:** Growth and cell specialization in surface cultures of *P. cyclopium* (Framm et al. [19]).

Cultures were grown on a nutrient solution containing 5 % glucose, 0.12 % [$NH_4^+$] and 0.025 % phosphate (NLI). Synchronization of development with respect to the transition from trophophase to idiophase was brought about by shift-down in the nutrient conditions 48 h p.i. (cf. [16]). All values are given in units/cm² culture area · sec.

Growth rate of the hyphae (100 = 22 ng); rate of conidiospore formation (100 = 105 conidia); rate of cyclopenin-cyclopenol formation by the hyphae (100 = 12 pmol). I, germination phase; II, trophophase; III, idiophase.

of hyphal growth (trophophase), and the phase of cell specialization (idiophase). The latter phase is characterized by the development of penicilli and conidiospores, and by the beginning synthesis of the alkaloids cyclopenin and cyclopenol (Fig. **11**.7).

The enzymes of cyclopenin/cyclopenol biosynthesis (Fig. **11**.8) cannot be detected during the trophophase, but their activities exhibit a coordinated rise at the beginning of the idiophase (Fig. **11**.9). This rise in activity can be suppressed by inhibitors of gene expression (except in the case of anthranilate adenylyl-transferase), e.g., by 5-fluorouracil and cycloheximide. Comparative studies on other enzymes have revealed that the inhibitors have no special effects on alkaloid metabolism. The inhibitor experiments thus suggest that in *P. cyclopium* biosynthesis of the enzymes of alkaloid metabolism depends on a continuous synthesis of RNA and proteins. The main site of regulation is at the level of transcription.

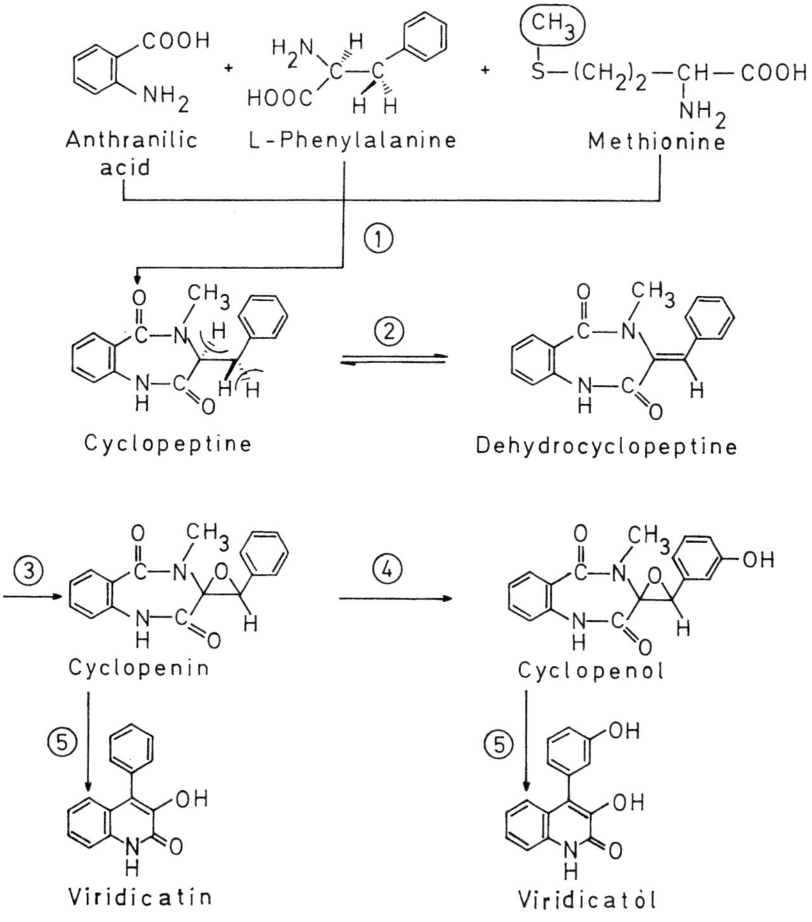


**Fig. 11.8:** Formation of the alkaloids of the cyclopenin-virdicatin group in *P. cyclopium* (Luckner [22]).

*(1)* Cyclopeptine synthetase complex: hypothetic; presumably catalyzing activation of anthranilic acid and phenylalanine, formation of peptide bonds, and methylation; intermediates are evidently covalently bound to the enzyme complex; *(2)* cyclopeptine dehydrogenase; *(3)* dehydrocyclopeptine epoxidase; *(4)* cyclopenin m-hydroxylase; *(5)* cyclopenase.

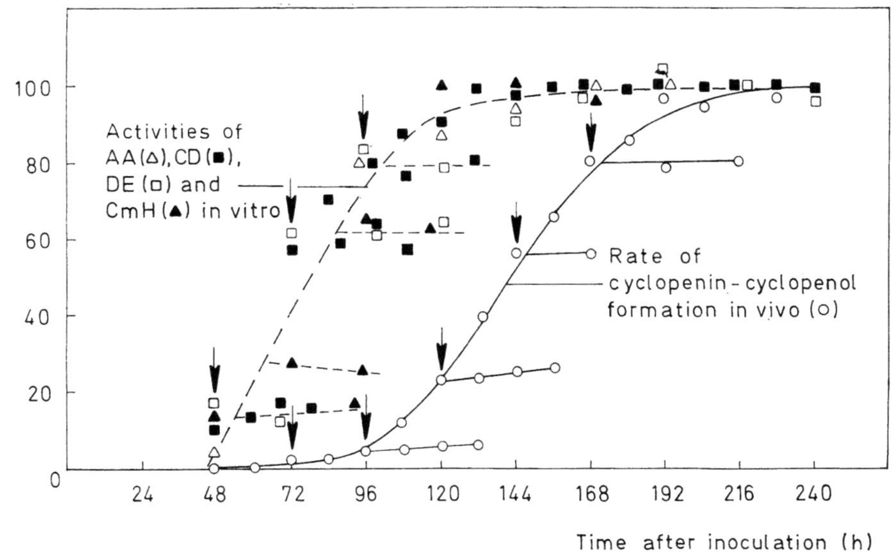


**Fig. 11.9:** In vitro activities of the enzymes of alkaloid biosynthesis and rates of cyclopenin-cyclopenol formation in vivo in hyphae of *P. cyclopium* (Voigt et al. [21]).

*P. cyclopium* was cultivated in surface cultures by discontinuous replacement of the nutrient solution (cf. Fig. 11.7). At time points indicated by arrows cycloheximide (100 μg/ml) was added to the culture medium. All values are given in units/cm² culture area.
Anthranilate adenylyltransferase activity *(AA)* (100 = 5.6 pkat); cyclopeptine dehydrogenase activity *(CD)* (100 = 40 pkat); dehydrocyclopeptine epoxidase activity *(DE)* (100 = 42 pkat); cyclopenin m-hydroxylase activity *(CmH)* (100 = 12 pkat); cyclopenin-cyclopenol formation 100 = 9 pmol/sec).

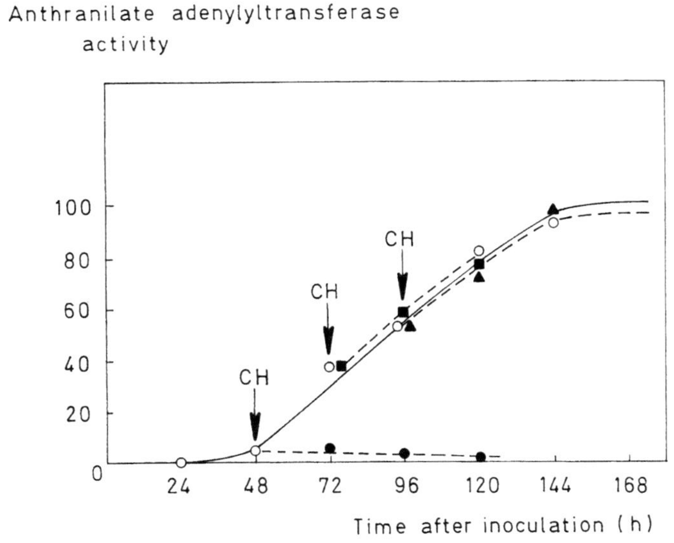


**Fig. 11.10:** In vitro measurable activity of anthranilate adenylyl-transferase in hyphae of *P. cyclopium* after addition of cycloheximide (100 μg/ml) (Luckner [22]).

Experimental details cf. Fig. 11.7. ○—○, control (without CH); ●—●, CH added 48 h p.i., ■—■, CH added 72 h p.i., ▲—▲, CH added 96 h p.i.; 100 = 6.0 pkat/cm² culture area.

In some instances, however, transcriptional control is evidently superimposed by post-translational regulation. Examples for this are the appearance and increase of anthranilate adenylyl-transferase (Fig. **11**.10) and cyclopenase activities (cf. Sect. **11**.3.2), which cannot be prevented by cycloheximide in concentrations suppressing protein biosyntnesis. The mechanism by which the increase of these two enzyme activities is regulated remains unclarified.

Under certain experimental conditions the rise in enzyme activities measurable in vitro and the alkaloid formation in vivo at the beginning of the idiophase do not run parallel (Fig. **11**.9). The rate of alkaloid synthesis rises about fivefold after the enzymes of cyclopenin/cyclopenol biosynthesis have already attained their maximum activity. Use of inhibitors of gene expression has shown that even after maximum activity of the enzymes has been attained the rise in the in vivo rate of synthesis is dependent on the continuous formation of RNA and protein. The in vivo rate of alkaloid formation must therefore be limited by the synthesis of a protein that is formed considerably later than the enzymes responsible for cyclopenin/cyclopenol synthesis. The biological function of this putative protein, limiting the in vivo rate of synthesis is as yet unknown. It is possible that it limits the synthesis or availability of precursors or cosubstrates.

In summary, these results demonstrate that during the transition from trophophase to idiophase at least four processes are necessary before alkaloid biosynthesis can take place in the hyphae of *P. cyclopium:*

a) formation of a precursor of the enzyme anthranilate adenylyl-transferase (terminated about 48 h p.i.),
b) activation of anthranilate adenylyl-transferase (48–144 h p.i.),
c) synthesis of cyclopeptin dehydrogenase, dehydrocyclopeptin epoxidase and cyclopenin m-hydroxylase (48–144 h p.i.) and
d) formation of the protein limiting the in vivo rate of alkaloid biosynthesis (72–216 h p.i.).

These processes cannot be blocked by inhibitors of DNA biosynthesis, thus indicating that in *Penicillium cyclopium* cells become specialized without concomitant cell division. This is in contrast to frequent observations in animal systems.

### 11.3.2 Alkaloid Metabolism During the Formation of Conidiospores

Unlike hyphal cells that only specialize during the idiophase, conidiospores are already specialized upon detachment. This is shown, for example, by the fact that, in contrast to the hyphae, they contain the enzymes of cyclopenin/cyclopenol biosynthesis as constitutional proteins. Activities of these enzymes are already demonstrable in very young spores and remain almost constant during spore maturation (Fig. **11**.11). Their formation is thus part of the cyclic gene expression program "conidia detachment" of the phialides (cf. **15**.2). Formation of the conidiospores is one of the rare cases where enzymes of secondary metabolism are directly formed de novo in dividing cells.

There is yet another difference between the alkaloid metabolism of hyphae and spores. Independently of the expression of the cyclopenin/cyclopenol-synthesizing enzymes, the enzyme cyclopenase is formed in the spores. This enzyme converts the benzodiazepine alkaloids cyclopenin and cyclopenol into viridicatin and viridicatol (Fig. **11**.8 no. 5). Cyclopenase cannot be demonstrated in hyphal cells of different ages or in young conidiospores, but only appears during conidiospore maturation. Since its activity increase is not arrested by inhibitors of gene expression such as cycloheximide, it cannot be directly coupled with the de novo formation of protein.

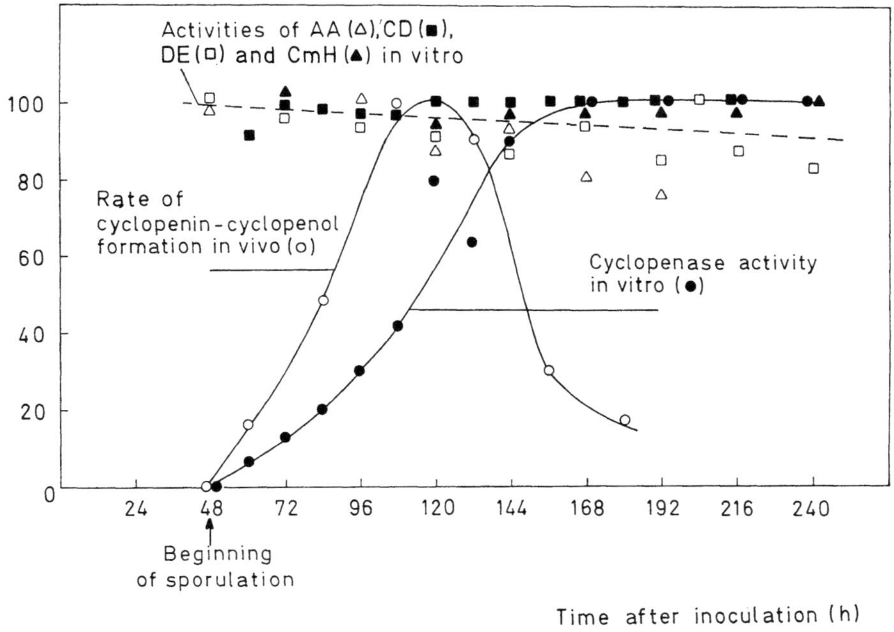


**Fig. 11.11:** In vitro activities of the enzymes of alkaloid biosynthesis and rates of cyclopenin-cyclopenol formation in vivo during maturation of conidiospores of *P. cyclopium* (Voigt et al. [21]).

*P. cyclopium* was cultivated in surface cultures by discontinuous replacement of the nutrient solution (cf. Fig. 11.7). At the time points indicated conidiospores were brushed off. Enzyme values refer to one mg dry weight.
Anthranilate adenylyltransferase activity *(AA)* (100 = 1.15 pkat); cyclopeptine dehydrogenase activity *(CD)* (100 = 16 pkat); dehydrocyclopeptine epoxidase activity *(DE)* (100 = 0.007 pkat); cyclopenin m-hydroxylase activity *(CmH)* (100 = 0.17 pkat); cyclopenase activity (100 = 250 pkat (!)); rates of cyclopenin-cyclopenol formation (100 = 0.5 pmol/sec).

A proenzyme of cyclopenase might possibly be synthesized during detachment of conidiospores, concurrently with the other enzymes of alkaloid metabolism, and then converted to the active state during maturation of conidiospores.

In summary, the activities of enzymes of alkaloid metabolism become manifest at strictly determined stages in the gene expression program of spore formation and maturation. Figure **11**.12 shows this in connection with other parameters. At present, three groups of events can be distinguished in this program:

a) Detachment of the conidiospores: this process includes the formation of enzymes of cyclopenin/cyclopenol synthesis, of invertase and of phenol oxidase. The latter evidently is localized in the penicilli and probably participates in the synthesis of the spore pigment (see below).
b) Synthesis of the characteristic green spore pigment.
c) Appearance of the enzyme cyclopenase on the inner side of the cell membrane of spores and accumulation of the alkaloids cyclopenin and cyclopenol in the spore wall, i.e., on the outer side of the cell membrane.

Separation by the cell membrane of cyclopenase from its substrates cyclopenin and cyclopenol in the cell wall (or in the culture filtrate) is a characteristic of this particular strain of *P. cyclopium*.

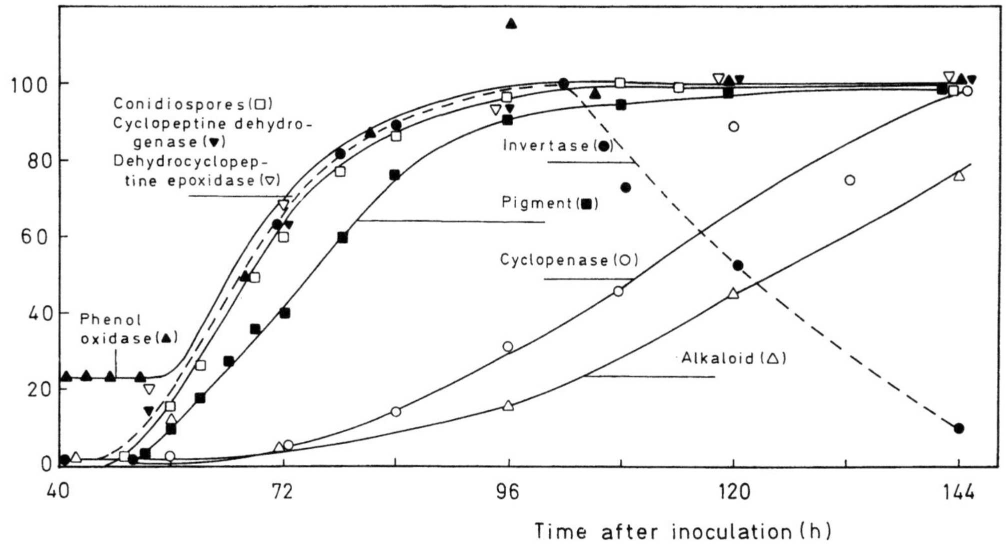


**Fig. 11.12:** Biochemical characteristics connected with detachment and maturation of conidiospores of *P. cyclopium* (Nover and Luckner [16]).

*P. cyclopium* was cultivated as given in Fig. 11.7. All values are given in units/cm$^2$ culture area.

Phenol oxidase activity (relative units); number of conidiospores (100 = $1.7 \cdot 10^8$); cyclopeptine dehydrogenase activity (100 = 5.0 mU); dehydrocyclopeptine epoxidase activity (100 = 1.8 $\mu$U) invertase activity (100 = 54 mU); content of pigments (relative units); cyclopenase activity (100 = 100 mU); accumulation of alkaloids (100 = 27 $\mu$g).

If, due to lowering of the ATP level in spores and the subsequent alterations in phospholipid turnover, permeability of the cell membrane is increased, cyclopenin and cyclopenol are rapidly converted into viridicatin and viridicatol. On account of their low solubility in aqueous media these quinoline alkaloids are stored largely in or on the mycelium, i.e., they disappear from the culture filtrate. The metabolic inactivity of cyclopenase observed in native mature spores clearly indicates the role of compartmentalization and permeability in the regulation of secondary metabolism.

## 11.4 Influence of Signals on the Expression of Alkaloid Metabolism

An important piece of evidence indicating the integration of alkaloid metabolism in the developmental program of *P. cyclopium* is the existence of a period during the early trophophase in which expression of the alkaloid metabolism and other characteristics of the idiophase (e.g., sporulation) can be determined. Although the expression of the idiophase in cultures of *P. cyclopium* grown in surface cultures is subject to endogenous control (cf. **11**.3.1), two types of compounds have been found that are capable of enhancing expression of idiophase characteristics:

a) phenylalanine and certain phenylalanine analogues and
b) a group of low molecular weight substances that occur in the mycelium of *P. cyclopium* but have no direct connection with the alkaloid precursors or with intermediates and cosubstrates of alkaloid biosynthesis (P-factor).

At the beginning of the growth phase there is a brief period of competence for both groups of signals. Since they have no effect if added later, they can exert no direct influence on the alkaloid metabolism in the idiophase. Following addition of the P--factor the rates of protein and nucleic acid synthesis and the protein content at the beginning of the idiophase are the same as in untreated controls. The effect of the signals on alkaloid metabolism and sporulation can therefore not be due to a general acceleration of primary metabolism during the idiophase. It is possible that the factors influence some early steps in the differentiation program of hyphae. These steps might govern the degree of expression of cell specialization during the idiophase by enhancing formation of regulatory proteins or endogenous factors. However, as yet no detailed knowledge is available.

Similar developmental phases during which the formation of secondary substances can be determined are known in other microorganisms as well. Examples are the stimulation of ergoline biosynthesis by tryptophan in *Claviceps* spec. or the increase of streptomycin formation in *Actinomyces streptomycini* by a natural factor produced by the bacterium itself (A-factor). In both cases the "signals" must be added immediately after inoculation of cultures. Here, too, the competence period occurs long before the period during which the secondary natural products are in fact formed (for references cf. [3]).

In all of these cases the effectors only have an indirect effect on the expression of secondary product formation via further steps. The existence in the developmental program of such a phase of determination is an important indication that secondary metabolism must be considered within the context of metabolism as a whole and that its regulation is integrated in the entire program of development.

## 11.5 Alkaloid Metabolism in Mutants with Modified Developmental Programs

The position of alkaloid metabolism in the developmental program of *P. cyclopium* has also been studied in mutants. As a result of the action of mutagenic agents on *P. cyclopium* two types of mutants were found that were suitable for investigations of this kind (Table **11**.2):

1. Of relatively frequent occurrence are strains in which none of the idiophase characteristics, including alkaloid metabolism, are developed. Strain *rev-met* 83a is of particular interest. It relatively frequently splits off revertants that are similar to the wild-type strain.
   Normalization of this kind of mutants indicates that not only the triggering, but also the release of idiophase events from the blockage in the *rev-met* 83a strain can be achieved in a single step and that expression of idiophase is controlled by a central program.
2. Careful biochemical analysis of other mutants showed that strains exist in which only one or a few of the typical idiophase events are blocked as a result of the mutation, whereas the rest are practically unchanged. For example, in strains *rev-met* 280, *rev-eth* 1 and *dev* 63 either only the alkaloid formation in the hyphae or only sporulation is suppressed. In addition, in strains *dev* 63, *dev* 109 and *SM* 72b there is no correlation between the changes in alkaloid content of the spores and the activity of cyclopenase (Tab. **11.2**). These results suggest that different parts of the total program "idiophase", although triggered at one and the same time, can be regulated individually and thus possess a certain degree of independence.

**Table 11.2**: Expression of idiophase events in mutants of *Penicillium cyclopium* [20]

| Strains | Idiophase events | | | | |
|---|---|---|---|---|---|
| | Alkaloid production by the hyphae | Sporulation | Spore maturation | | |
| | | | Alkaloid production | Pigmen-tation | Cyclopenase activity |
| *SM* 72 (wild-type strain) | ++++ | ++++ | ++++ | ++++ | ++++ |
| *rev-meth* 83a | + | + | | + | ++++ |
| Revertants of *rev-meth* 83a to strains of the *SM* 72 type | +++ to ++++ | ++ to +++ | | +++ | +++ to ++++ |
| *dev* 63 | +++ | ± | ++ | + | ± |
| *dev* 109 | ++ | ++ | ++ | +++ | ++++ |
| *SM* 72b | + | ++++ | ± | +++ | +++ |
| *res-eth* 1 | ++ | +++ | +++ | ± | +++ |
| *rev-meth* 280 | + | +++ | ++ | ++++ | +++ |

Crosses allow comparison with the wild-type strain: ++++, 75–100%; +++, 50–75%; ++, 25–50%; +, 10–25%; ±, less than 10%.

## 11.6 Conclusions

In summing up it can be said that the most important principle in the regulation of the expression of secondary metabolism is de novo synthesis of the enzymes involved. In most cases formation of these enzymes is regulated at the transcription level (cf. example in **11**.2.2). However, further mechanisms may be superimposed on the regulation of transcription. The following have been discussed:

- activation of proenzymes, e.g., the rise in activity of anthranilate adenylyl-transferase and cyclopenase not influenced by cycloheximide (cf. **11**.3.1),
- factors influencing the in vivo activity of enzymes (cf. **11**.3.1) and
- degradation of enzymes (cf. **11**.2.1).

Expression of secondary metabolism involves many steps which are temporarily fixed with respect to other characteristics of cell specialization (cf. **11**.3.1 and **11**.3.2). A fixation of this kind combined with the frequent occurrence of determinated sequences of events (cf. **11**.4) indicates that secondary metabolism is integrated into the program of cell specialization and development. This integration explains the long-known dependence of secondary natural product formation on the state of development and specialization of the cells, organs and organisms, a finding which is of economic significance in the production of antibiotics and alkaloids by microorganisms on an industrial scale. The trophophase-idiophase behaviour of bacteria and fungi (cf. **11**.3.1) reduces the yields in time and space, and thus the economic efficiency of the process. Moreover, the probable reason, why it is usually difficult to trigger formation of secondary products in cell cultures of higher plants, is the regulatory integration of their biosynthesis, storage and possible degradation into the networks of developmental control.

In addition, the investigations described show that an experimental approach to the induction of secondary product synthesis is possible. Since their formation is not an indispensible part of the programs of gene expression and development, it can be influenced by alterations in the genetic material (cf. **11**.5) as well as by signals (**11**.4) or by changes in the environment without heavily decreasing the organism's chances of survival. Data so far available on this subject, although fragmentary, give reason to hope that advances in the understanding of the molecular mechanisms involved in the regulation of secondary metabolism will eventually lead to better means of influencing the production of secondary natural products.

## 11.7 Selected References

*Secondary Metabolism – General Aspects*

1. Luckner, M.: Was ist Sekundärstoffwechsel? Pharmazie *26* (1971): 717–724.
2. Luckner, M.: Secondary Metabolism in Plants and Animals. Chapman and Hall Ltd., London 1972.
3. Luckner, M., L. Nover & H. Böhm: Secondary Metabolism and Cell Differentiation (Molecular Biology, Biochemistry and Biophysics, Vol. 23). Springer Verlag, Berlin 1977.
4. Luckner, M.: Expression and control of secondary metabolism. In: Bell, E. A. & B. V. Charlwood (eds.): Encyclopaedia of Plant Physiology – New Series, Vol. 8: Secondary Plant Products. Springer-Verlag, Berlin 1980, pp. 23–63.

*Parsley Cell Cultures*

4a. Mohr, H.: Phytochrome and gene expression. In: Nover, L., F. Lynen & K. Mothes (eds.), Cell Compartmentation and Metabolic Channeling. G. Fischer Verlag and Elsevier/North-Holland Biomedical Press, Jena and Amsterdam 1980, pp. 399–414.

5. Hahlbrock, K. & H. Ragg: Light-induced changes of enzyme activities in parsley cell suspension cultures. Effects of inhibitors of RNA and protein synthesis. Arch. Biochem. Biophys. *166* (1975): 41–46.
6. Hahlbrock, K.: Regulation of phenylalanine ammonia-lyase activity in cell-suspension cultures of *Petroselinum hortense*. Apparant rates of enzyme synthesis and degradation. Eur. J. Biochem. *63* (1976): 137–145.
7. Hahlbrock, K., K. H. Knobloch, F. Kreuzaler, J. R. M. Potts & E. Wellmann: Coordinated induction and subsequent activity changes of two groups of metabolically interrelated enzymes. Light-induced synthesis of flavonoid glycosides in cell suspension cultures of *Petroselinum hortense*. Eur. J. Biochem. *61* (1976): 199–206.
8. Wellmann, E.: Regulation of flavonoid biosynthesis by ultraviolet radiation and phytochrome in cell suspension cultures of parsley. In: Luckner, M., K. Mothes & L. Nover (eds.): Secondary Metabolism and Coevolution. Nova Acta Leopoldina, Suppl. 7, Joh. Ambrosius Barth Verlag, Leipzig 1976, pp. 319–326.
9. Hahlbrock, K.: Coordinated regulation of the enzymes of flavonoid biosynthesis. In: Nover, L. & K. Mothes (eds.), Cell Differentiation in Microorganisms, Plants and Animals. G. Fischer and North-Holland Publ. Comp., Jena and Amsterdam 1977, pp. 524–537.
10. Ragg, H., J. Schröder & K. Hahlbrock: Translation of poly(A)-containing and poly(A)-free messenger RNA for phenylalanine ammonia-lyase, a plant-specific protein, in a reticulocyte lysate. Biochim. Biophys. Acta *474* (1977): 226–233.
11. Schröder, J., F. Kreuzaler & J. Schmock: Translation of plant-specific messenger RNAs in living animal cells. FEBS Letters *81* (1977): 10–12.
12. Schröder, J.: Light-induced increase of messenger RNA for phenylalanine ammonia-lyase in cell suspension cultures of *Petroselinum hortense*. Arch. Biochem. Biophys. *182* (1977): 488–496.
13. Hahlbrock, K., B. Betz, S. E. Gardiner, F. Kreuzaler, U. Matern, H. Ragg, E. Schäfer & J. Schröder: Enzyme induction in cultured cells. In: Thorpe, T. A. (ed.), Frontiers of Plant Tissue Culture. University of Calgary Offset Printing Services 1978, pp. 317–324.

14. Schröder, J., F. Kreuzaler, E. Schäfer & K. Hahlbrock: Concomitant induction of phenylalanine ammonia-lyase and flavanone synthase mRNAs in irradiated plant cells. J. Biol. Chem. *254* (1979): 57–65.

15. Kreuzaler, F., H. Ragg, W. Heller, R. Tesch, I. Witt & D. Hammer: Flavanone synthase from *Petroselinum hortense*. Eur. J. Biochem. *99* (1979): 89–96.

15a. Hahlbrock, K., J. Schröder & J. Vieregge: Enzyme regulation in parsley and soybean cell cultures. Adv. Biochem. Engineering *18* (1980): 39–60.

*Penicillium cyclopium*

16. Nover, L. & M. Luckner: Expression of secondary metabolism as part of the differentiation processes during the idiophase development of *Penicillium cyclopium* Westling. Biochem. Physiol. Pflanzen *166* (1974): 293–305.

17. El Kousy, S., E. Pfeiffer, G. Ininger, W. Roos, L. Nover & M. Luckner: Influence of inhibitors of gene expression on processes of cell specialization during the idiophase development of *Penicillium cyclopium* Westling. Biochem. Physiol. Pflanzen *168* (1975): 79–85.

18. Luckner, M.: The integration of benzodiazepine and quinoline alkaloid formation into the developmental program of *Penicillium cyclopium*. In: Nover, L. & K. Mothes (eds.), Cell Differentiation in Microorganisms, Plants and Animals. G. Fischer Verlag and North-Holland Publ. Comp., Jena and Amsterdam 1977, pp. 538–558.

19. Framm, J., L. Nover, A. El Azzouny, H. Richter, K. Winter, S. Werner & M. Luckner: Cyclopeptin und Dehydrocyclopeptin, Zwischenprodukte der Biosynthese von Alkaloiden der Cyclopenin-Viridicatin-Gruppe bei *Penicillium cyclopium* Westling. Eur. J. Biochem. *37* (1973): 78–85.

20. Schmidt, I., L. Nover, G. Ininger & M. Luckner: Gewinnung und Charakterisierung von Mutanten des Pilzes *Penicillium cyclopium* mit verändertem Entwicklungsprogramm. Z. Allg. Mikrobiol. *18* (1978): 219–224.

21. Voigt, S., S. El Kousy, N. Schwelle, L. Nover & M. Luckner: Activities in vitro and in vivo of enzymes of benzodiazepine alkaloid biosynthesis during development of *Penicillium cyclopium*. Phytochemistry *17* (1978): 1705–1709.

22. Luckner, M.: Alkaloid biosynthesis in *Penicillium cyclopium* – does it reflect general features of secondary metabolism? J. Natural Products (Lloydia) *43* (1980): 21–40.

# 12. Biosynthesis and Accumulation of Storage Proteins in Developing Plant Seeds

K. Müntz

The seeds of flowering plants represent the highest form to which an organ of propagation and survival has developed during the course of plant evolution. In most cases seeds separate from the mother plant after achieving the required state of maturity. They have now completed the first part of their embryonic development and enter a temporary resting stage. At this stage seeds consist of at least two organs: the seed coat (testa) and the embryo. Many seeds also contain a nutrient tissue, the endosperm, in which macromolecules (proteins, carbohydrates, lipids) are stored. Their degradation products serve as precursors and energy reserves for further embryonic development after the resting period has terminated, until autotrophic metabolism takes over in the young plants. Typical representatives of seeds with an endosperm of this kind are the caryopses of cereals. Even seeds that possess no endosperm in the mature state contain, over a long period of their development, this type of a nutrient tissue, but this is gradually broken down during seed formation and the products of degradation are translocated to the developing embryo, e.g., in seeds of cotton and legumes. Biosynthesis and storage of reserve products take place in the large cotyledons of the embryo of such seeds.

Irrespective of whether biosynthesis and storage of reserve products take place in the endosperm or in the cotyledons of the embryo itself, the formation and accumulation of such materials result from extensive cell specialization, including also the provision of the conditions necessary for mobilization of the storage material at the time of germination. The molecular basis for this cell specialization consists of complex programs of differential gene expression, involving the temporal and spatial coordination of the realization of certain parts of the genetic information. Details of the regulation of these programs are for the most parts unknown. The greatest advances in this direction have been made over the past ten years in studies on developing cereal caryopses and legume seeds, of which the latter are particularly rich in storage proteins. Their biosynthesis and storage will therefore be discussed in this chapter, as examples of analogous processes in the development of the seeds of other higher plants.

## 12.1 Stages of Seed Development

### 12.1.1 Growth Characters of Seeds

Seed development in legumes begins with the double fertilization typical of angiosperms. In the example chosen here (the garden pea, *Pisum sativum*) it takes place during the 24 h before the flower fully opens. Transverse division of the zygote leads to the development of a two-cell proembryo: its apical cell divides again transversely

---

Abbreviations: d, dictyosomes; e, protein lumps in vacuoles; ER, endoplasmic reticulum; rER, rough endoplasmic reticulum; kd, kilodaltons; $M_r$, relative molecular weight; n, cell nucleus; p, protein bodies; s, starch grains; SDS, sodium dodecylsulfate; w, cell wall; βME, β-mercaptoethanol; f, free (polysomes); mb, membrane-bound (polysomes).

into an apical embryo mother cell and the basal cell situated between the former and the suspensor cell. Further mitoses of the embryo mother cell lead to the formation of a radially symmetrical globular cell aggregate which is pushed forward to the highest point of the embryo sack (Fig. **12.1**) by the growth of the multinuclear suspensor cells. At first the cell aggregate is only 0.2 mm in diameter (stage 2) but enlarges to 0.6 mm (stage 3) before stage 4, i.e., organogenesis of the cotyledons, commences. Stage 4 begins with periclinal cell divisions on both sides of the apex and results in two bulges consisting of arched layers of cells. The embryo thus exhibits bilateral symmetry and becomes heart-chaped (stage 4). At this point the multinuclear syncytium of the endosperm attains its maximum extent, and its breakdown in the ensuing stages of development apparently serves to provide a source of nutrient for the rapidly growing cotyledons during stages 5 and 6 [23]. During stage 6 the embryo

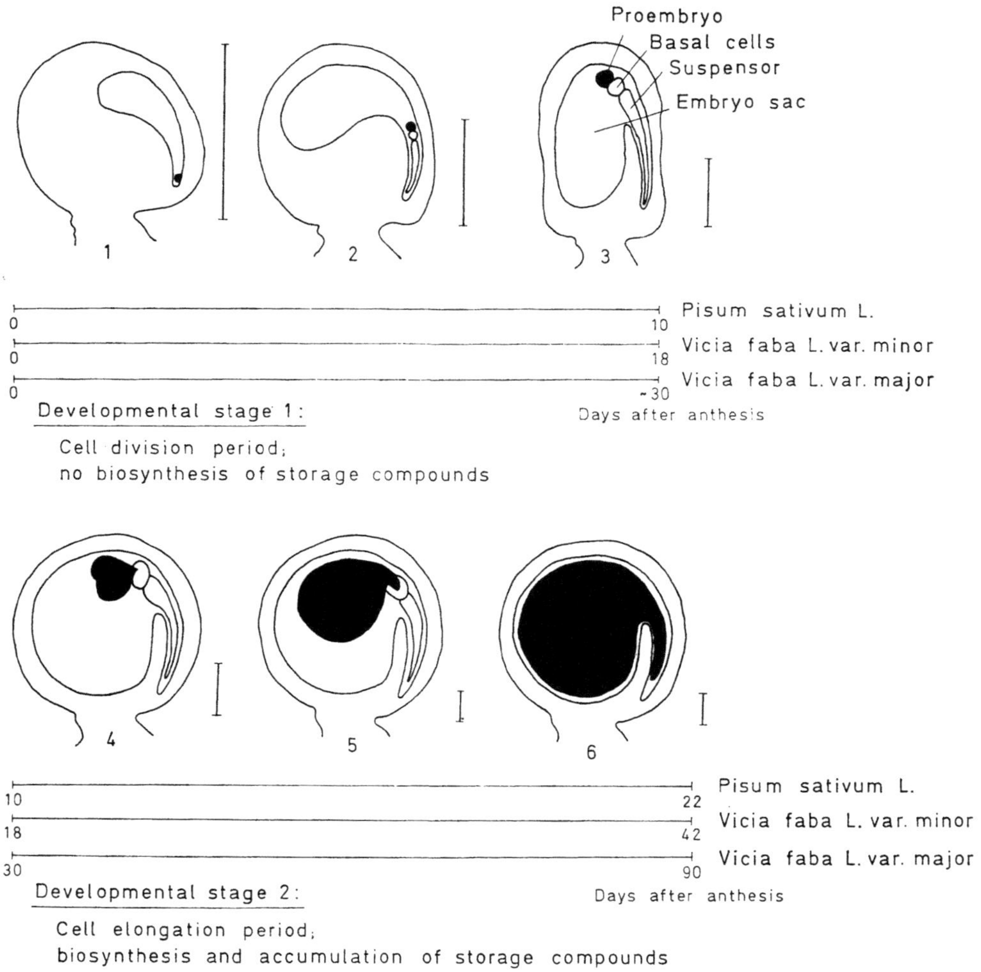


**Fig. 12.1:** Morphological development of legume seeds.

*Bars* indicate 1 mm in nature (from [23]). *Time scales* are approximations since the time required for seed development varies in dependence of environmental and genetic factors. Date taken from Ref. [33] for *Pisum sativum* L., from Ref. [30] for *Vicia faba* L. var. *minor* and from Ref. [22] for *Vicia faba* L. var. *major*.

fills the entire embryo sack. Finally, endosperm and suspensor are broken down and the seed undergoes desiccation and enters a resting stage. Seed development of the *Vicia* species and in many other legumes proceeds in a similar manner [8].

### 12.1.2 Cytological Characters of Seed Development

Two techniques have been applied to analyse the basis of cell growth during seed development: (1) Cytophotometry of the cell nucleus of cotyledon tissues at various stages of development and (2) techniques in which the stained cell nuclei can be isolated and counted using the same principle as applied for counting bacteria under the microscope. It has been shown that during a first developmental period the growth of legume cotyledons is mainly the result of cell divisions. Mitosis stops before the large increase in the size of the cotyledons starts (Fig. **12**.1). Up to the end of this stage the embryos grow relatively little. In the storage organs or cotyledons, however, all of the cells that will later be capable of protein storage are formed in this period [22]. In the embryo axis, radicles and shoot apex cell division continues longer than in the cotyledons. Since cell specialization in the cotyledons does not take place synchronously there is a more or less clearly recognizable period of transition from growth by cell division to growth by cell elongation. Finally, the increase in size and dry weight of the seeds comes to an end, and, during desiccation and the onset of the resting stage, some signs of shrinkage and a slight decline in dry weight can be detected.

As regards development of the cotyledon tissue in which storage and synthesis of reserves take place, seed ontogenesis can be divided into three stages [8]: (1) the phase of cell division within the cotyledons with no noticeable synthesis or storage of reserve material in the mesophyll; (2) the phase of cell elongation in the cotyledon mesophyll and the period of formation and storage of reserve material (cf. Fig. **12**.4 A); (3) the phase of desiccation and the onset of the resting stage. Analogously, the endosperm cells of cereal caryopses exhibit three phases of development with synthesis and accumulation of reserves also coinciding with the period of pronounced cell elongation in the endosperm.

In this phase all the synthetic and degradative processes in legumes and cereals result in a strong fresh and dry matter increase (cf. Fig. **12**.4 A), which proceeds almost linearly in the major part of the second phase of seed ontogenesis. In legumes this increase is linearly correlated with the growth of the cotyledon length [22]. The transition from cell division to the phase of cell elongation and synthesis as well as storage of reserves (Fig. **12**.1, phases 3–4) is accompanied by marked alterations in subcellular structures (Fig. **12**.2): (1) a substantial increase in the rough endoplasmic reticulum (rER) and its membrane-bound polysomes (mb-polysomes), (2) considerable alterations in the vacuolization of the storage cells, e.g., disappearance of the large central vacuoles and the appearance (usually only transitory) of several to many small vacuoles containing protein lumps (Fig. **12**.2 A), (3) the formation of large starch grains and (4) the formation of protein bodies in which storage proteins accumulate (Fig. **12**.2 B) [33]. In lipid-storing seeds lipid vacuoles are also formed. In cells of mature storage tissue of dicotyledonous plants and cereals the cell nucleus, some mitochondria, occasionally proplastids, and small amounts of cytoplasm without polysomes but with ribosomes are visible. Vacuoles containing cell sap are missing and the cells are for the most part filled with reserve products deposited in starch grains, lipid vacuoles or protein bodies (Fig. **12**.2 B) whose numbers vary considerably according to species and the quantities of the different macromolecular storage products present [33].

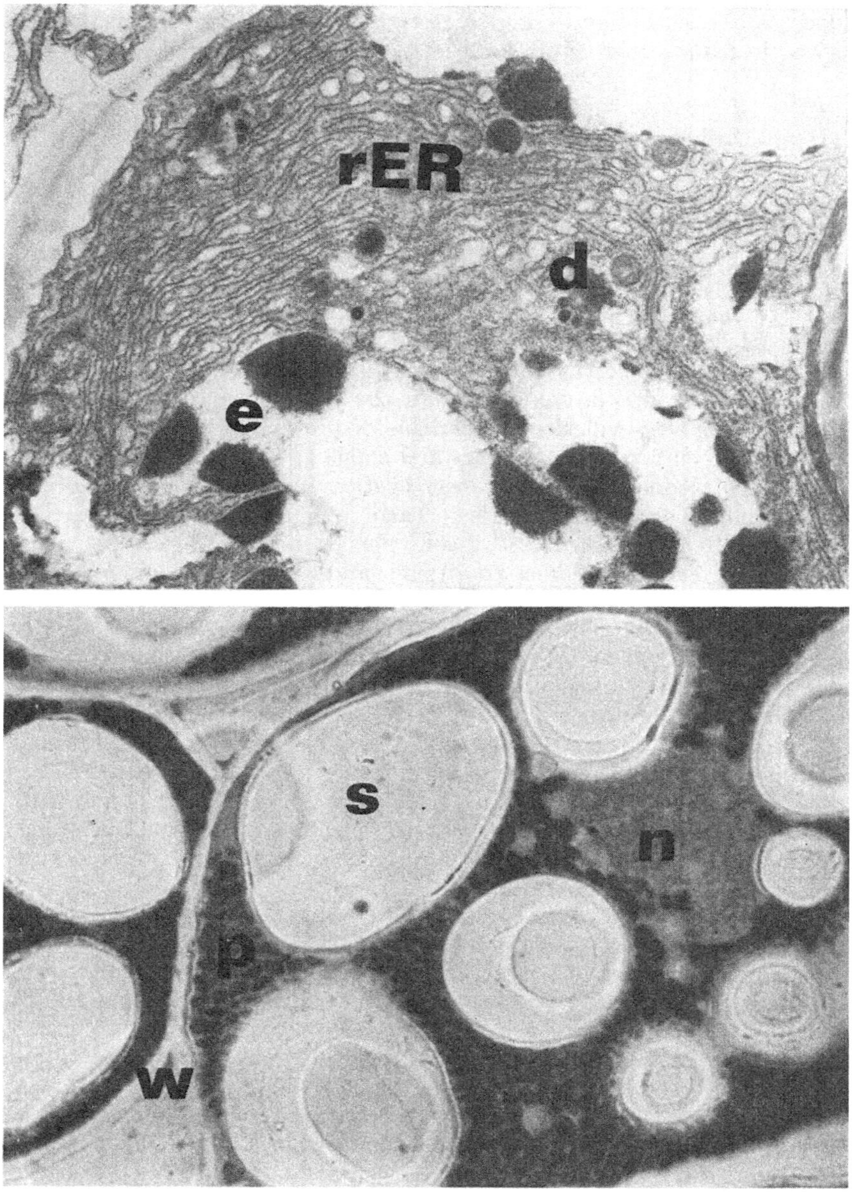


A

B

**Fig. 12.2**: Ultrastructure of cells from developing (A) and mature (B) field bean seeds (*Vicia faba* L. var. *minor*) (from Weber & Neumann [33]).

**A**: Cotyledon cells 31 days after anthesis *(stage 2)* with highly developed rER membranes, many dictyosomes *(d)* accompanied by osmiophilic granules; vacuoles with protein lumps *(e)* attached to the inner surface of the tonoplast (electron micrograph, *mangific.* 16,000 fold);
**B**: Cross-section of mature cotyledons from *Vicia faba* L. var. *minor (stage 3)*; protein bodies *(p)* were stained with Hg-Bromophenolblue, cell nucleus *(n)*, starch grains *(s)*, and a small amount of cytoplasm attached to the cell wall *(w)* are also visible; (light micrograph, *magnific.* 900 fold).

### 12.1.3 DNA Accumulation

All quantitative analyses of nucleic acid accumulation in the cotyledons of developing peas and field beans have shown that the nucleic acid content of the cotyledons increases in parallel with the cell number during the 1st phase (mitotic phase) whilst the amount of DNA per cotyledon cell remains constant [24, 25].

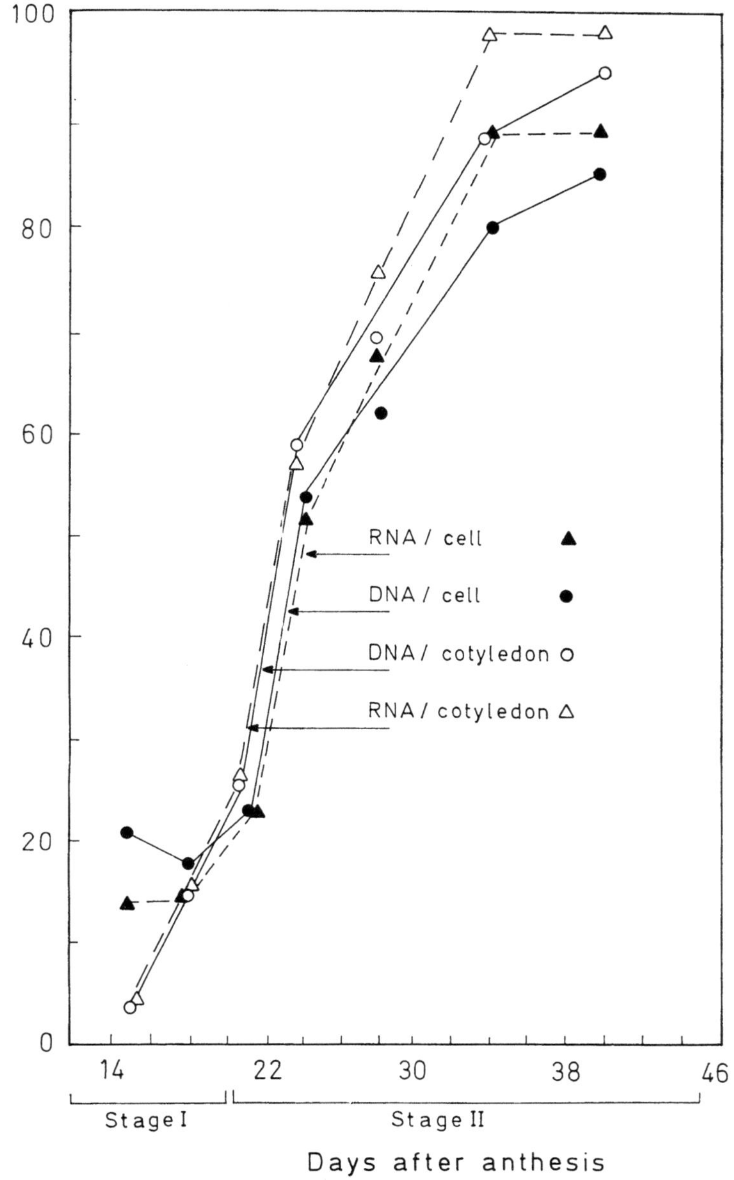


**Fig. 12.3:** DNA and RNA accumulation during developmental stages 1 and 2 of field bean cotyledons (*Vicia faba* L. var. *minor*); transition from stage 1 to 2 at days 18–20 (from Manteuffel et al. [22]).
RNA/cell: 100 = 1.34 ng; DNA/cell: 100 = 0.27 ng. RNA/cotyledon: 100 = 0.9 mg; DNA/cotyledon: 100 = 0.18 mg.

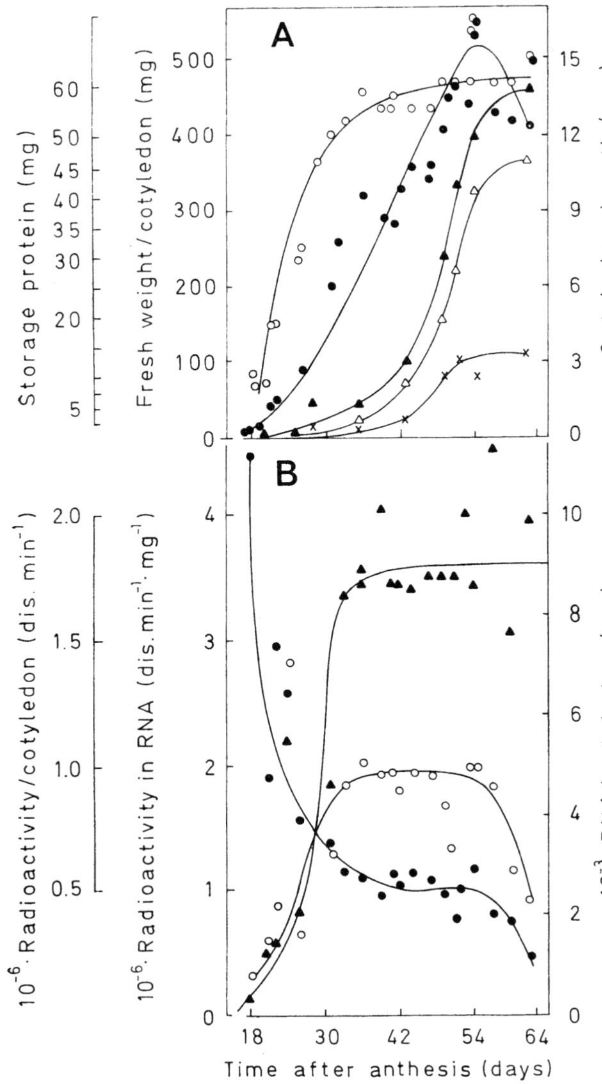


**Fig. 12.4**: Cotyledon development and RNA synthesis in seeds of *Vicia faba* L. var. *minor* (from Püchel et al. [28]).

**A**: Growth: increase of fresh weight (●) and length of the cotyledons (○), and accumulation of storage proteins: globulins (▲), legumin (△), and vicilin (×).

**B**: RNA synthesis: RNA content per cotyledon (▲), incorporation of radioactive precursors by cotyledons (○), and specific activity of radioactively labelled RNA (●). The latter decreases during seed development, since the later the harvest the more non-labelled RNA has already been accumulated before incubation with [$^{3}$H]-uridine was started. Between day 30 and 55 after anthesis (stage 2) the plateau phase of precursor incorporation parallels the constant levels of specific radioactivity of RNA and of RNA concentration per cotyledon, thus indicating that RNA synthesis and breakdown are in equilibrium.

Although the number of cells in the cotyledons does not increase after the end of phase 1, replication of DNA continues (endoreduplication, Fig. **12**.3). Quantitative cytophotometry and biochemical analyses have shown that this increase in DNA repeatedly leads to doublings of the amount of DNA in initially diploid cells [22, 24, 25]. Such findings led to the idea of reduplication of the entire genome [15], which, for *Pisum sativum* cotyledons, involves polyploidization as well as polytenization [36].

The question of endoreduplication of the entire genome has been carefully examined using DNA-DNA hybridization [24]. The reassociation kinetics of DNA from diploid *Vicia faba* cells and of DNA from the cotyledons with an elevated cellular content of DNA did not show any differences expected if the increase in DNA content in the nuclei of the cotyledon cells were the result of selective gene amplification (cf. **4**.4.1). Although these results support the idea of genome multiplication, they do not exclude the possibility that one or other of the individual genes might in fact undergo selective amplification since the method does not permit the discrimination of such fine differences.

### 12.1.4 Accumulation of RNA and Ribosomes

During the phase of cell elongation the RNA content per cotyledon cell of *Vicia faba* and *Pisum sativum* multiplies by the same factor as the DNA content [22]. Changes in the rate of accumulation of the two nucleic acids follow the same time course and the rise in RNA content per cell does not lag behind that of DNA (Fig. **12**.3).

Several days before the end of protein biosynthesis in developing legume seeds the RNA content per cotyledon has already reached a constant level (Fig. **12**.4B) that is mainteined during the entire resting stage of the seeds. It only begins to change with the onset of germination [28].

Studies on *Phaseolus* seeds had suggested that changes in the rate of RNA accumulation are only due to changes in synthesis and not to changes of RNA degradation. In contrast, it could be shown in the field bean that, when a constant RNA content per cotyledon has been reached, a state of equilibrium exists between synthesis and breakdown [28]. This means that transcription of all RNA fractions must still be taking place (Fig. **12**.4B) until storage protein synthesis ceases during stage **3**.

Density gradient fractionation of ribosomes from *Vicia* cotyledons and electron microscope studies revealed the occurrence of large quantities of mb-polysomes parallel to the increasing rate of synthesis of storage proteins [14, 26, 27]. This polysome fraction disappears again when reserve protein synthesis stops (Fig. **12**.5). Experiments on the incorporation of radioactive precursors of rRNA synthesis ([$^{32}P_i$] or [$^{3}$H]-uridine) revealed that the membrane-associated polysomes of the second phase of development are synthesized de novo and not recruited from preformed ribosomal subunits of the cells [26].

### 12.1.5 Protein Accumulation

#### 12.1.5.1 Biochemical Characters of Storage Proteins

According to the classical fractionation scheme of Osborne [10] seed proteins of plants can be classified as albumins, globulins, prolamines and glutelines (Table **12**.1). The enzyme proteins of seeds are albumins and globulins. The typical storage

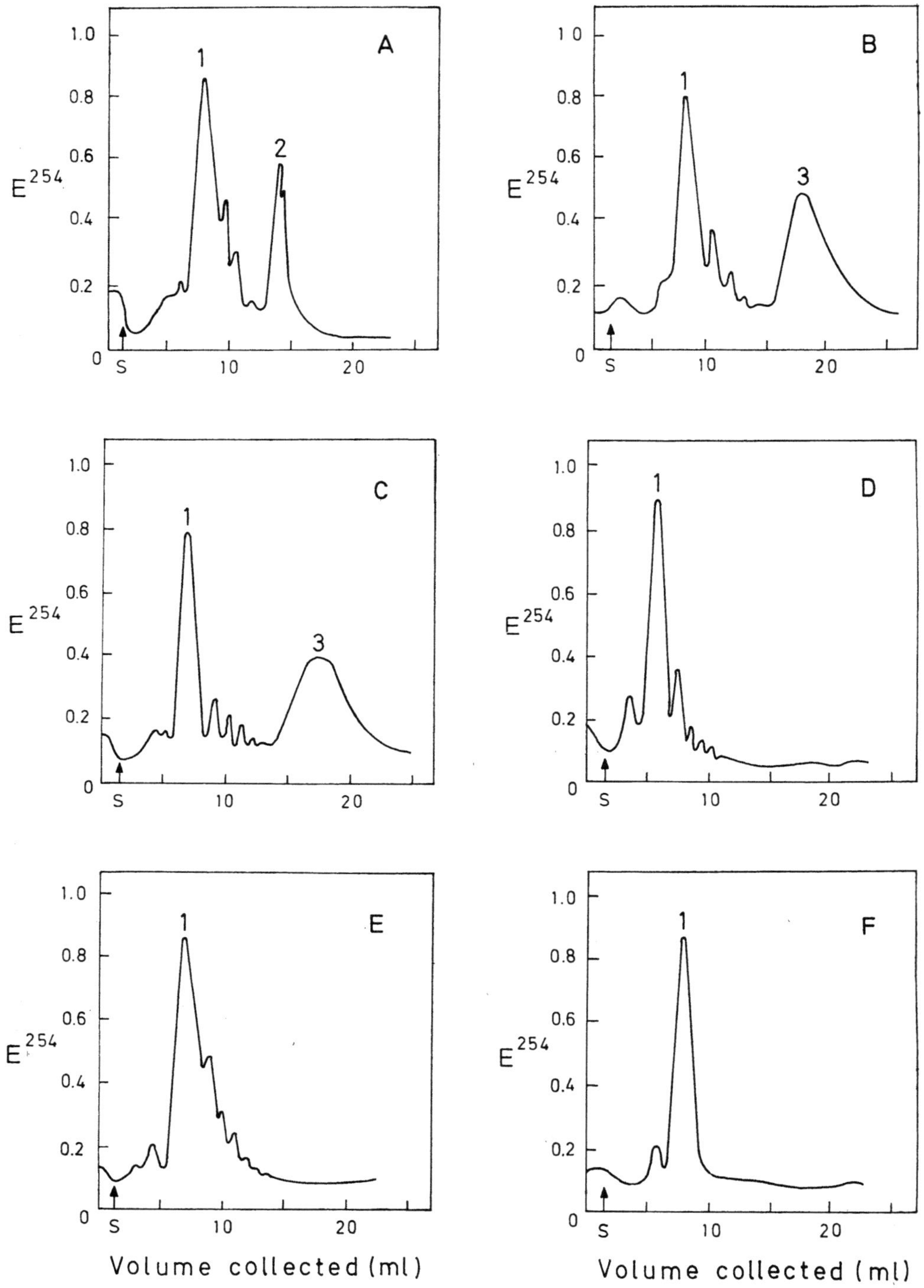


Fig. 12.5

**Table 12.1:** Modified scheme of seed protein fractionation according to Osborne [10], starting from defatted seed flour

| Extraction | Protein fractions | |
|---|---|---|
| 1. Buffered solution of neutral salt (*p*H 7.0) | Albumins and globulins | |
| | Soluble supernatant ← ALBUMINS | Dialysis against running destilled water |
| | ↓ Insoluble precipitate GLOBULINS | |
| | *7S Globulins* of legumes | *11–13S Globulins* of legumes |
| | Vicilin *(Vicia, Pisum)* | Legumin *(Vicia, Pisum)* |
| | β-Conglycinin *(Glycine)* | Glycinin *(Glycine)* |
| | β-Conglutin *(Lupinus)* | α-Conglutin *(Lupinus)* |
| | Phaseolin *(Phaseolus)* | G 2 protein *(Phaseolus)* |
| 2. 50–70% ethanol, n-propanol or isopropanol solution with β-mercaptoethanol | PROLAMINS | |
| | Zein from *Zea mays* L. | |
| | Hordein from *Hordeum vulgare* L. | |
| | Gliadin from *Triticum sativum* L. | |
| | Oryzenin from *Oryza sativa* L. | |
| 3. Low concentration base or acid solutions | GLUTELINS | |

proteins of cereals are prolamines and glutelines, whereas in dicotyledonous plants they are in most cases globulins [1]. During seed germination all these proteins act as reserves whose proteolytic degradation yields amino acids for protein synthesis, growth and development of the seedlings. In a narrow sense storage proteins are proteins with no enzyme activity; they accumulate in protein bodies of the storage tissue of seeds, are characterized by a relatively high N/C ratio and represent exclusively macromolecular storage compounds of the amino acids necessary for seedling development. Their synthesis and accumulation is limited to the second phase of cell specialization in the storage tissue. As examples, the globulins of legume cotyledons and the prolamines of cereal endosperms will now be characterized in more detail.

The globulin fraction accounts for about 70–85% of the protein content of the protein bodies of legume seeds, i.e., about 15–25% of the dry matter of the cotyledons. Sucrose density gradient

**Fig. 12.5:** Fractionation of microsomes from developing cotyledons of the broad bean (*Vicia faba* L. var. *major*) (from Payne et al. [27]).

*Peak 1* – monoribosomes, *peak 2* – f-polysomes, and *peak 3* – mb-polysomes; cotyledons were harvested **(A)** 40 d, **(B)** 50 d, **(C)** 60 d, **(D)** 70 d, **(E)** 80 d, and **(F)** 100 days after anthesis. Centrifugation of polysome containing cotyledon extracts through 15–45 percent isokinetic sucrose density gradient. The appearance of a large peak of mb-polysomes coincides with the period of maximum storage protein synthesis (Fig. **12.1**, stage 2).

centrifugation of the globulins usually yields four fractions, the proportions being species-dependent and influenced by environmental factors: 2S, 6–8S, 11–13S and 15–18S. Whereas the first three fractions are in the native form, the fraction with the highest sedimentation coefficient usually results from association of globulins. In most legumes globulins of the 6–8S and the 11–13S fractions are usually predominant (cf. Table **12.**1). Their molecular weights are 150–200 kd and 350 kd respectively, their isoelectric points are more or less acidic, in the region of *p*H 4–6. Nearly all legume globulins are glycoproteins. The 11–13S globulins in most of the species so far investigated consist of two types of subunits, the one acidic ($M_r$ = 35–40 kd) and the other basic ($M_r$ 20–25 kd). Probably six molecules of each subunit form the high molecular weight oligomeric globulin. In many legumes the 6–8S globulin also appears to consist of several different subunits (e.g., soybean $\beta$-conglycinin), but their structural analysis has not progressed as far as that of the larger globulins [1]. Genetic analyses in *Pisum* and *Phaseolus* have shown that the globulin polypeptide chains exhibit monogenic intermediary inheritance. The amino acid composition of the globulins is characterized on the one side by 20–40 % glutamic acid plus aspartic acid residues, mostly in the amide form, and on the other side by a small or even very low content of tryptophane and SH-containing amino acids (methionine, cysteine). This is why the biological values are as much as 50 % lower than those of plant leaf proteins or animal proteins.

In cereals, prolamines make up about 30–50 % of the grain proteins in corn and wheat but only 10% in rice. The prolamine contents of the other cereals are somewhere between these two extremes. Cereal prolamines are usually covalently bound to the glutelines and can only be extracted with alcoholic solvents in the presence of reducing agents [5] that cleave disulphide bonds (Table **12.**1). Therefore reliable data are only available for the $M_r$ of the prolamine subunits (e.g., zein 1, $M_r$ 22–23 kd and zein 2, $M_r$ 19–21 kd). Like the globulins of the legumes, these are acidic proteins with a high content of glutamic acid plus aspartic acid residues and their corresponding amides. Prolamines differ from legume globulins in having a high proline content and a very low proportion of basic amino acid residues. The relatively low biological value of prolamines is due to their extremely small lysine content.

### 12.1.5.2 Accumulation of Storage Proteins During Seed Development

Biosynthesis and accumulation of storage proteins start at the transition from the first to the second phase of development of the cotyledons (cf. Fig. **12.**1). The globulins vicilin and legumin from the field bean (*Vicia faba* L.) and the garden pea (*Pisum sativum* L.) have been chosen as examples by which to describe the sequence of events involved in the accumulation of storage proteins.

The rate of synthesis of the globulins increases rapidly at the beginning of the second phase of development and then attains a constant value maintained over several days and resulting in a linear increase in the globulin content of the cotyledons [22]. When the seeds begin to desiccate during the transition to stage 3, the rate of synthesis is lowered and finally falls down to zero (Fig. **12.**4A). The maximum rate of synthesis is reached sooner for vicilin (V) than for legumin (L). On the other hand the rate of synthesis of vicilin drops earlier and its synthesis stops before that of legumin [28]. This is why biosynthesis of vicilin predominates at the begining of the second stage of development, whereas later chiefly legumin is formed, and in *Vicia faba* the ratio of V : L is displaced during this phase from $> 1$ at the beginning to $< 1$ at the end. Furthermore, the occurrence and ratios of the minor globulin components also change [22].

Under normal conditions of development mature seeds of legumes exhibit a species-specific ratio for the components of the globulin fraction, although environmental conditions can bring about considerable modifications. In the absence of sulphur, for example, the formation of S-containing $\alpha$-conglycinins in lupins is almost completely replaced by increased synthesis of S-free $\beta$-conglutinins; or, if embroygenesis is prematurely terminated by drought, the composition of the proteins remains fixed at the stage typical of immature seeds.

## 12.2 Regulation of Storage Protein Biosynthesis

### 12.2.1 Endoreduplication of DNA

The parallel increases in RNA synthesis and DNA content per cell, connected with the transition from the phase of cell division to that of cell elongation in combination with the delayed increase in the rate of globulin synthesis, have led to the speculation that endoreduplication of DNA in the mesophyll cells of cotyledons and biosynthesis of storage proteins may be causally related in the legumes under consideration [15]. This could be mediated via RNA synthesis in the form of a gene-dosage effect comparable to that influencing, in some cases, the proteins in the cereal endosperm, where a double set of maternal chromosomes is present.

Provided RNA polymerase activity is not the rate-limiting factor, increased amount of templates for the transcription of mRNA, tRNA and rRNA could represent a basis for the increasing rates of RNA- and storage protein accumulation at the time of transition from phase 1 to phase 2 of cotyledon development (cf. Figs. **12**.1 and **12**.4). Such a gene-dosage effect was clearly demonstrated in the suspensor cells of *Phaseolus coccineus* which, like the cells of the embryo, derive from the zygote. These suspensor cells contain polytene giant chromosomes formed by endoreduplication. Few experimental attempts have so far been made to reveal the significance of DNA endoreduplication in legume storage tissue for the regulation of storage protein biosynthesis. No clear picture has yet emerged, although to some extent the results contradict the idea of a gene-dosage effect.

For example, experiments with isolated nuclei from pea cotyledons have indicated that during the period of DNA reduplication in non-dividing cells the endogenous RNA polymerase activity is low and the transcription activity measured per $\mu$g DNA substantially decreases concomitant to the increasing amount of DNA per nucleus [25]. On the other hand, it is obvious that the temperature prevailing during seed development greatly influences the results of this type of experiment. In pea seeds grown at 15 °C the transcription activity was limited by the availability of template activity of DNA; in seeds grown at 30 °C, however, the limiting factor was the endogenous polymerase activity. It remains therefore unresolved whether endoreduplication of the DNA in cotyledon cells of legume seeds directly influences the mass synthesis of storage proteins or not.

### 12.2.2 Evidence for Transcriptional Control of the Formation of Storage Proteins

Biosynthesis of rRNA, tRNA and poly(A)-containing RNA, i.e., mRNA, continues throughout the entire period of storage protein biosynthesis in *Vicia faba* [28]. Although there is no noticeable change in the ratios of the various RNA clases, biosynthesis of storage proteins in mature cotyledons of field beans is paralleled by a large increase in the incorporation of labelled uridine into the poly(A)-containing RNA (12–13 times greater incorporation than in the poly(A)-free RNA, i.e., rRNA + tRNA). In vitro translation of mb-polysomes and poly(A)-containing RNA from cotyledons of the phases of development connected with intensive globulin biosynthesis revealed that, of the radioactivity incorporated into newly synthesized polypeptide chains and measured in the trichloroacetic acid precipitate, 30%–50% are precipitable with monospecific antibodies for vicilin and legumin [28]. The 3′-terminal poly(A) sequence of the corresponding mRNA fraction contains approximately 160 AMP residues. Experiments using m7-GMP to inhibit the in vitro translation of poly(A)-containing RNA in *Vicia faba*, *Pisum sativum* and *Glycine max* showed [7, 31] that a cap-like structure must be present at the 5′-terminal end of the mRNA (cf. Fig. 1.9). The poly(A) sequence and the cap structure, in analogy to the mRNA biosynthesis of other eukaryotic organisms, are probably added during the post-synthetic processing of the primary transcript in the nucleus (cf. Fig. 1.8). With a mean coding length of about 1800 nucleotides the poly(A)-containing RNA of field beans is large enough to serve as template for the largest known subunits of the storage globulins of this plant [7].

**Table 12.2:** Incorporation of $[^3H]$-uridine into poly(A)-containing RNA of developing cotyledons of the field bean during embryogenesis.

Cotyledons of the respective developmental stage were isolated aseptically and incubated with liquid nutrient medium containing the labelled uridine; poly(A)-containing RNA was separated from poly(A)-free RNA by affinity chromatography of total cellular RNA on poly(U)-Sepharose columns [28]

| Days after flowering | Poly(A)-containing RNA | |
|---|---|---|
| | % of total RNA | % of $[^3H]$-uridine incorporated into total RNA |
| 18 | 0.71 | 7.08 |
| 26 | 1.41 | 9.95 |
| 31 | 1.22 | 7.00 |
| 36 | 1.30 | 9.74 |
| 50 | 1.17 | 21.40 |
| 53 | 1.16 | 15.77 |

A considerable body of experimental evidence supports the assumption that in legumes and in cereals transcription is the main level of control of storage protein biosynthesis:

1. There is a rapid turnover of poly(A)-containing RNA during biosynthesis of storage protein in the field bean (Table 12.2) [28].
2. $\alpha$-Amanitin (20$\mu$g/ml), already proved to be a selective inhibitor of the biosynthesis of poly(A)-containing RNA in the field bean, was shown to inhibit simultaneously the incorporation of $[^3H]$-uridine into poly(A)-containing RNA (79% inhibition) and the labelling of vicilin and legumin with $[^{14}C]$-leucine (50% inhibition) [28].
3. Equal quantities of poly(A)-containing RNA isolated from different stages of development of seeds of *Vicia faba* were placed in the cell-free translation system of wheat germ, and the proportion of globulin polypeptide chains among the translation products was quantitatively determined [7, 8]. Globulins only appear among the products of in vitro biosynthesis from a certain stage onwards. The later quantitative changes in the proportion of 6–8S and 11–13S globulins reflect the in vivo situation parallelly analysed by serological methods using the same couple of cotyledons from which the poly(A)-containing RNA was isolated (Fig. 12.6).
4. After in vitro translation of zein mRNA of corn the zein polypeptide chains exhibit approximately the same kind of heterogeneity as found in vivo and already shown to be genetically determined [29].

Qualitative und quantitative analyses of developmentally specific mRNA populations heavily depend on the application of cell-free translation systems known from prokaryotes as well as from animal and plant eukaryotes, e.g., the reticulocyte and the wheat germ systems. Prerequisites or an efficient cell-free translation system are low endogenous levels of RNase activity and of mRNA. The endogenous mRNA level can be further lowered by short incubation of the tissue extract with *Staphylococcus aureus* nuclease, which has subsequently to be inactivated.

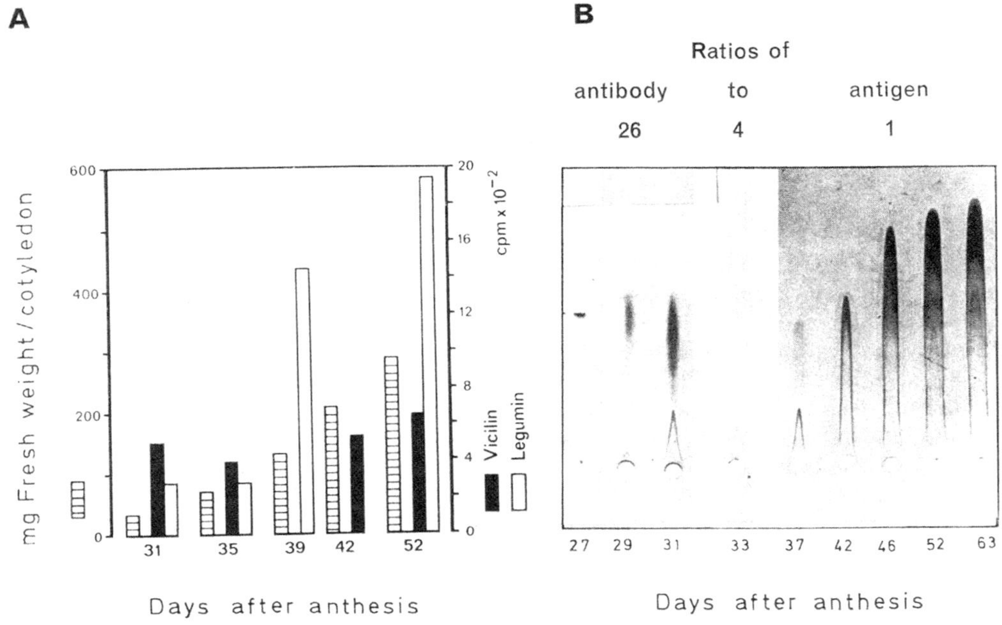


**Fig. 12.6:** In vivo- and in vitro-biosynthesis of globulins of developing field bean seeds (*Vicia faba* L. var. *minor*).

**A:** In vitro-biosynthesis of vicilin and legumin polypeptides in the standardized cell-free wheat germ system was programmed by equal amounts of poly(A)-containing RNA that was isolated from field bean cotyledons of different harvests during developmental stage 2 (compare Fig. **12.1**) as indicated on the abscissa. Vicilin and legumin polypeptides were immunoprecipitated by specific antibodies; radioactivity of the solubilized immunoprecipitates was counted in a liquid szintillation counter. For comparison cotyledon growth is indicated in terms of the fresh weight increase.

**B:** Rocket immunoelectrophoresis of globulin extracts from developing field bean cotyledons of stage 2 (compare **Fig. 12.1**) on agarose slabs containing legumin antibodies. In some cases samples were taken from the same seeds which were used for RNA extraction (see A), extracting always 1 cotyledon for RNA and the other for protein analysis.

The ratio of legumin mRNA to other mRNA in the poly(A)-containing RNA increases in a similar manner as does the in vivo legumin accumulation indicating a regulation of globulin biosynthesis at the transcription level.

In order to prepare a cell-free protein biosynthesis system cells must be lysed and a cytoplasmic fraction containing ribosomal subunits and all the necessary soluble translation factors has to be isolated. For the wheat-germ system the supernatant of a $23{,}000 \times g$ centrifugation of a defatted extract from homogenized hard wheat embryos is used. The wheat germ extract represents the translation machinery which has to be completed by the addition of an "energy-generating" system, the appropriate mRNA and radioactive amino acids for the synthesis of labelled polypeptide chains. The system must be run at a constant *p*H 7.4–7.6 and at definite concentrations of mono- and divalent cations. Compounds for the reduction of disulphide bridges and maintenance of SH groups, and for the stabilization of the mRNA (e.g., spermine or spermidine) have to be included. The energy-generating systems consist of ATP and GTP, and either phosphoenol--pyruvate plus pyruvate kinase or creatine phosphate plus creatine phosphate kinase. The system is working well if the incorporation of radioactive amino acids into the polypeptides formed exhibits linear dependence on the amount of added mRNA and on the incubation time, if high degrees of labelling are reached in comparison to the control containing only endogenous mRNA, if the products show a high specific radioactivity, which in most cases is a prerequisite for further analysis of the translation products, and if complete polypeptides are formed.

Specific analysis of the translation products is necessary if, as in most cases, pure mRNA for only one type of polypeptide cannot be used to direct in vitro biosynthesis. The protein in question may be isolated by specific immune reaction with the respective antibodies, and the resulting precipitates then have to be further analysed by polyacrylamide gel electrophoresis under denaturing conditions.

### 12.2.3 Molecular Cloning of cDNA Copies of Storage Protein mRNA and Gene Analysis

Single and double-stranded cDNA has been prepared by reverse transcription (cf. **7**.4.5 and **19**.2.2.2) of purified zein mRNA (*Zea mays* storage protein) [4], of G 1 globulin mRNA (7S storage protein of *Phaseolus vulgaris*) [2], of total poly(A)-containing RNA from mb-polysomes of barlex endosperm [3] and of developing cotyledons of the field bean [7] and pea [35]. After "tailing" the double-stranded cDNA with oligonucleotides that permit insertion into correspondingly "tailed" linearized plasmid DNA (cf. e.g., Fig. **2**.3), the plasmids (mainly pBR 322) were used to transform plasmid-free strains of *E. coli*.

Selection of transformed *E. coli* strains was based on the following two antibiotic resistance markers: Plasmid pBR 322 contains a gene for tetracyclin and one for ampillicin resistance. Insertion of the tailed cDNA takes place into the ampillicin site split by Pst I restriction endonuclease and subsequently "tailed". As a concequence the ampillicin resistance is lost. Transformed *E. coli* strains can be selected on the basis of tetracyclin resistance, whereas in a second selection step bacterial strains wearing plasmids with eukaryote cDNA inserts can be isolated by means of ampillicin sensitivity (for details of the method cf. Chap **2**, especially Fig. **2**.3).

Strains of transformed *E. coli* with cDNA inserts corresponding to storage protein mRNA have subsequently to be isolated by hybridization with [$^{32}$P]-labelled storage protein mRNA or poly(A)-containing RNA (cf. **2**.6). The specificity of the cDNA sequences has been tested by the hybrid arrested in vitro translation (HART assay, cf. **2**.8) or in vitro translation of storage protein mRNA which can be selected via hybridization with corresponding cDNA with subsequent isolation of the mRNA for translation analysis.

Parallel, the cloning of genomic DNA from *Zea mays* and *Phaseolus vulgaris* in lambda-phage has been performed (genomic DNA library, cf. **2**.3). Comparison of partial sequences of genomic DNA with cDNA clones coding for G 1 protein of *Phaseolus vulgaris* resulted in the discovery of at least three intervening sequences (introns) in the corresponding gene. Splicing should therefore form part of the mRNA formation for this legume storage protein [20, 37].

Detailed DNA/mRNA hybridization studies as well as the analysis of the polymorphism of storage protein polypeptides of maize and soybean led to the conclusion that multigene families code for storage proteins [9, 11].

### 12.2.4 Control of Translation and Processing of Polypeptide Precursors of Storage Proteins

In the biosynthesis of storage proteins and its regulation there are three main aspects of the control of translation and post-translation processes:

1. The appearance of large amounts of *membrane-bound* (mb) *polysomes* composed of newly synthesized ribosomal subunits (cf. Fig. **12**.2 A) coincides with the period

of most active storage protein formation. The time course of increase and decrease of mb-polysomes suggests a coordinated membrane- and acceptor site formation for polysomes, which can be looked upon as the sites of storage protein biosynthesis. There is experimental evidence that in legumes and cereals the biosynthesis of reserve proteins takes place on mb-polysomes.

(a) Following pulse-chase labelling of sections through *Vicia faba* cotyledons with [$^{14}$C]-leucine by counting the silver grains in autoradiograms indicated a transfer of radioactivity from rER to the protein bodies [12]. (b) The membranes of protein bodies of corn bear mb-polysomes from which practically pure zein mRNA can be isolated [21]. (c) A comparison of the in vitro translation products from cell-free systems into which free (f) or mb-polysomes had been introduced showed that in corn, barley, *Phaseolus vulgaris* and field beans biosynthesis of storage protein proceeds practically exclusively on mb-polysomes [13, 17, 21, 28]. (d) Antibody-binding experiments on free and mb-polysomes isolated from cytoledons of field beans in an active state of storage protein biosynthesis and largely freed of non-specifically bound globulins showed that the quantity of labelled vicilin antibodies reacting with mb-polysomes was twice that bound to free polysomes [28].

2. Storage proteins are very likely synthesized as *precursor polypeptides* that are processed by cleavage of signal peptides (cf. Table **4**.10) and subsequent glycosylation, since many of the storage proteins have been shown to be glycoproteins [1].

In vitro translation of poly(A)-containing RNA from storage tissue in an active state of storage protein biosynthesis results in products shown to have higher molecular weights (SDS gel electrophoresis) than the reference samples obtained from mature cereal caryopses [16, 21]. The differences in $M_r$ are in the range of 2–4 kd (corn: zein, barley: hordein). Employing membranes of the ER it was possible to simulate processing of the in vitro translation products (Fig. **12**.7), i.e., the putative signal Apeptide of the precursors war removed [17, 21]. Synthesis of the polypeptide chains on mb-polysomes as well as the in vitro synthesis of precursor molecules with subsequent processing by limited proteolysis suggest that, analogously to the secretory proteines of animals and in agreement with the signal hypothesis of Blobel and Dobberstein (cf. **4**.4.5.1), the polypeptide chains of plant storage proteins originally contain N-terminal extensions which are usually removed co-translationally.

The 11–13S globulins of grain legumes are generally composed of several copies of two types of polypeptide chains, the larger one with $M_r$ approximately 40,000 and the smaller one with $M_r$ approximately 20,000. In the native hetero-oligomeric globulin disulphide bridges form covalent linkages between the two types of polypeptides. Therefore, denaturation without reduction of the disulphide bridges leads to the formation of subunits with $M_r$ approximately 60,000 which afterwards may be split into single polypeptide chains by reduction and subsequent alkylation of the sulfhydryl groups.

In vitro translation of poly(A)-containing RNA from peas [18], soybeans [9] and field beans [8] always resulted in the formation of polypeptides with $M_r$ approximately 60,000, whether the translation products have been analysed under reducing or non-reducing conditions (Fig. **12**.7). Immunologic investigation and peptide mapping have proven the identity of the in vitro products with the polypeptide chains from native 11–13S globulins from the respective seeds (Fig. **12**.7). These findings suggest that both polypeptides of this type of storage globulins should be translated from a common mRNA leading to the formation of a precursor polypeptide chain which is composed of both subunits covalently linked. Taking into account (1) that the common precursor of the large and small polypeptide of 11–13S globulins should be synthesized by mb-polysomes and (2) that some indications are available for the presence of a signal peptide-like N-terminal extension of the precursor, the hypothesis of an at least two-step processing has been developed: (a) cleavage of the signal peptide by limited proteolysis should represent the first step, whereas (b) the separation of the two different polypeptides should follow as the second step. Biosynthesis

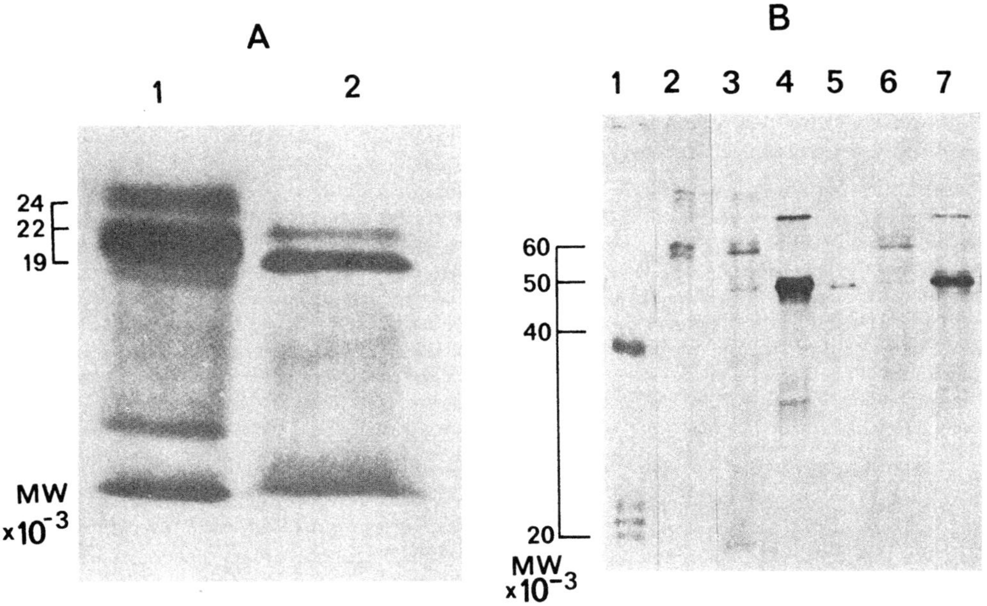


**Fig. 12.7**: In vitro-synthesis of cereal and legume storage proteins.

**A**: Autoradiographs of electrophoretically fractionated zein polypeptides after in vitro translation of detached mb-polysomes **(1)** and microsomes **(2)** of maize endosperm, indicating the precursor molecules of zein (1) with $M_r$ 24,000 and 21,000, and the mature zein polypeptides (2) with $M_r$ 22,000 and 19,000, respecitvely [21].

**B**: Autoradiographs of globulin polypeptides from field beans after SDS-gel electrophoresis. In vivo labelled legumin from developing cotyledons, denatured under reducing **(1)** or non-reducing conditions **(2)**. Products of in vitro translation of detached mb-polysomes in the cell-free wheat germ system: Legumin after denaturation under reducing **(6)** or non-reducing conditions **(3)**; vicilin after denaturation under reducing **(7)** or non-reducing conditions **(4)**. Lane **(5)** represents the control with non-specific antibodies. Isolation of the different globulin polypeptides was performed by immunologic methods using Protein A-Sepharose.

and processing of the polypeptides constituting the 11–13S globulins of these legumes seems to proceed similar to the formation of some animal secretory polypeptides, e.g. of insulin (cf. Fig. **8**.6).

Until now, no evidence has been presented concerning co-translational glycosylation of the nascent polypeptide chains of storage proteins that are vectorially sequestered from the mb-polysomes into the rER cisternae. Nevertheless, the participation of the ER in protein glycosylation has been demonstrated in cotyledons of *Pisum sativum* and in storage proteins of *Phaseolus vulgaris* L. 7S globulin and phytohemagglutinine have been specifically glycosylated in vitro using extracts of bean cotyledons [19]. Experiments with castor bean endosperm indicated that protein glycosylation proceeds, at least in part, via lipid polysaccharide intermediates in the ER. Glycosyl transferases mediating the transfer of glucose and mannose from UDP-[$^{14}$C]-glucose or GDP-[$^{14}$C]-mannose, respectively to lipid acceptors have been localized in the ER of etiolated pea stems. These results suggest the presence of glycosylation mechanisms similar to those described for mammalian systems (cf. Fig. **13**.10).

3. *Changes of* the *tRNA pattern* seem to reflect the increased incorporation of glutamic acid, aspartic acid and leucine that occurs when the synthesis of storage proteins, like the prolamines of cereals and the 7S and 11–13S globulins of legumes, represents a predominating metabolic event during developmental stage 2 of the seeds.

After the initial failure to demonstrate, in cotton and wheat, an adaptation of the quantitative composition of the different tRNA or isoacceptor tRNA species for certain amino acid residues characteristic of the storage proteins, this was in fact finally demonstrated in the case of zein synthesis in corn. Paralleling intensive zein synthesis 30 days after pollination, the tRNA shows a higher acceptor capacitiy for glutamic acid, leucine and alanine than the embryo tissue. At the same time the proportions of isoacceptor tRNA species for these amino acids change [32]. Although proline is also one of the quantitatively predominant amino acids in zein, no correspondigly high acceptor activity of the endosperm tRNA could be detected. Furthermore the very small proportion of lysine in zein is not reflected in a correspondingly low tRNA level. It therefore remains uncertain whether the observed adaptation of some of the tRNA of the total tRNA population of corn endosperm plays a regulatory role at the translational level in the biosynthesis of reserve protein or not.

## 12.3 Biogenesis of Protein Bodies and the Storage of Protein Reserves

Since the biosynthesis of storage protein polypeptides takes place on mb-polysomes and the polypeptides are usually vectorially sequestered into the rER cisternae as precursors and then undergo maturation processes, the question arises of how the protein bodies, containing the storage proteins of mature seeds, are formed.

The protein bodies of dicotyledonous and many monocotyledonous plants are vacuole-like organelles confined by a simple membrane and containing a protein

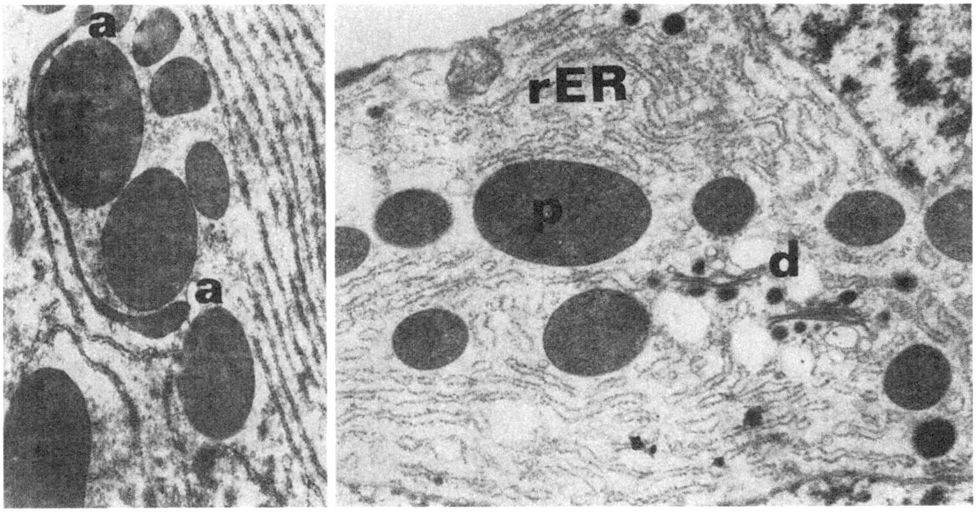


A B

**Fig. 12.8:** Formation of protein bodies in developing cotyledons of field bean seeds from stage 2 (*Vicia faba* L. var. *minor* (from Weber & Neumann [33]).

**A:** Electron micrographs (*magnific.* 16,000 fold) of cotyledon cells 35 days after anthesis: Enlarged cisternae of the ER *(a)* with segregating osmiophilic vesicles (protein bodies) are visible; **B:** Electron micrograph (*magnific.* 24,000 fold) of cotyledon cells 36 days after anthesis: Cells are rich in rER *(a)*, contain protein bodies *(p)* and dictyosomes *(d)* which are accompanied by strongly osmiophilic granules. Preparation technique: fixation with glutaraldehyde and osmium tetroxide was followed by block staining with uranyl acetate and embedding into Araldit. Ultrathin sections were cut and afterwards stained with lead acetate.

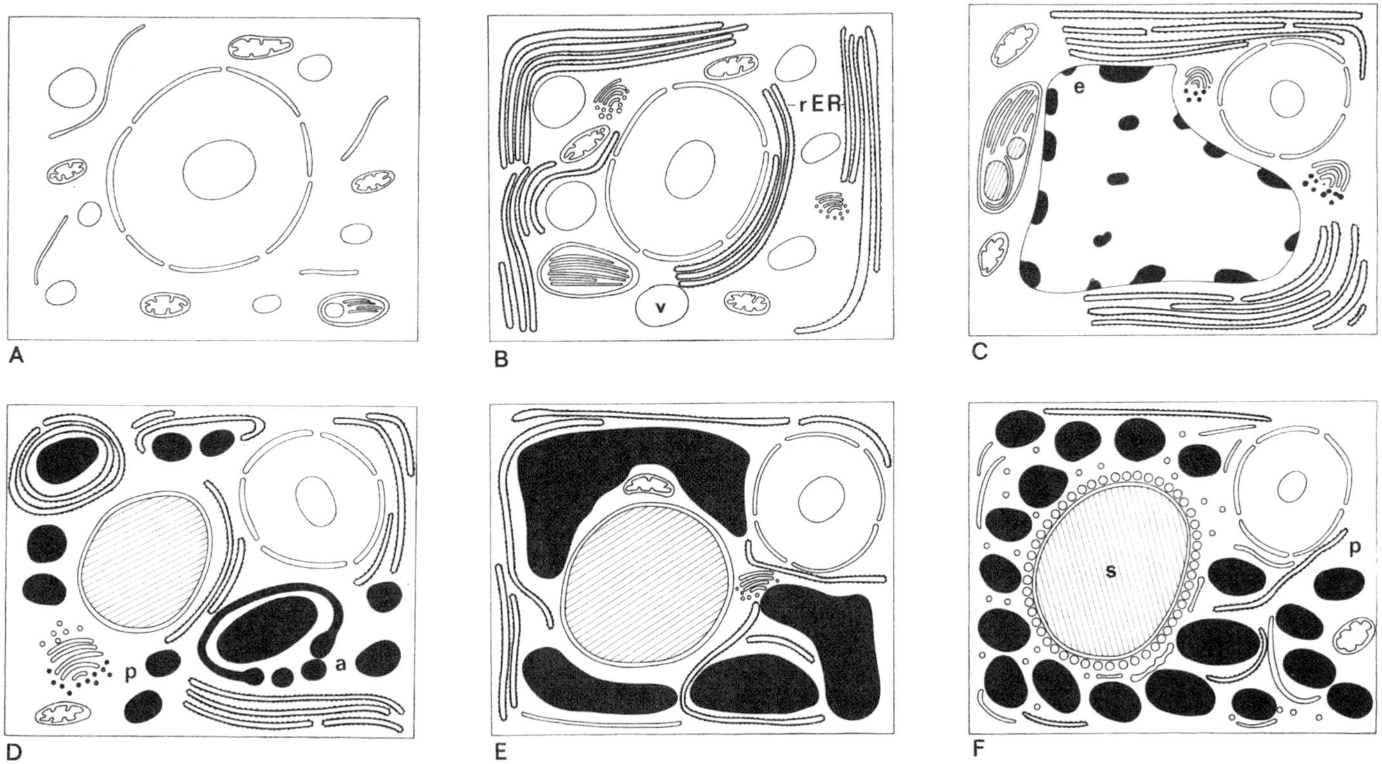


**Fig. 12.9:** Hypothetic scheme of protein body formation in cotyledon cells of developing field bean seeds (*Vicia faba* **L.** var *minor*) (from Weber & Neumann [33]).

**(A)** Embryonic cell from stage 1 (0–24 days after anthesis); **(B)** development of the rER and vacuoles *(v)* (25–28 d); **(C)** protein lumps *(e)* attached to the inner surface of the vacuolar membrane (29–32 d); **(D)** Formation of enlarged cisternae of the ER *(a)* and of protein bodies *(p)*, (33–38 d); **(E)** appearance of large protein containing vesicles of the ER (39–43 d); **(F)** Degradation of the rER; cells contain large amounts of starch grains *(s)* and protein bodies (> 44 d).

matrix in which in some plant species phytin globoids and/or protein crystalloids are deposited [33]. Proteins, chiefly storage proteins in the narrower sense, make up 80–95% of the protein body substance. There is no unequivocal evidence that the protein bodies of barley are surrounded by a biomembrane, although vesicles of the ER are frequently observed to be in immediate contact with the protein bodies. Protein bodies in corn are surrounded by a membrane covered with polysomes which, with the aid of marker enzymes, was shown to resemble membranes of the rER. In electron micrographs continuities between the membranes of protein bodies and those of the rER have been observed. Translation of poly(A)-containing RNA from polysomes of the protein body membrane and from the rER led in both cases to approximately equal percentages of zein in the total protein formed with the cell-free wheat germ system. This suggests that the protein bodies of corn arise directly from the rER and represent prolamine deposites within the rER [21].

Apart from the unexplained situation in barley, evidence for mb polysomes on protein bodies is still so far lacking for other plants. In legumes the sites of synthesis and storage of globulins are clearly not identical [33]. Since most legume globulins are glycosylated [1] it is frequently assumed that dictyosomes are involved in the biosynthesis of storage proteins. In electron micrographs the osmiophilic granules visible on the periphery of the dictyosomes are interpreted in this way (Fig. **12**.8B). In *Pisum* and *Ricinus* it has been shown that the vesicles of the ER are also equipped for glycosylation and hence there is no reason for doubting the assumption that protein bodies can directly arise from the smooth ER (sER).

In electron micrographs of the cotyledons of many dicotyledonous plants vacuoles containing plaques of protein can be observed, at least at the beginning of the period of formation of reserve products (Fig. **12**.2A). This is the basis of the hypothesis that protein bodies might arise from vacuoles and gradually fill up with storage protein. The latter is then transported from the site of synthesis to the site of storage in protein granulae of dictyosomal origin (Fig. **12**.8B). Electron microscopic studies on mustard have yielded evidence that both, direct formation from the ER and filling of the vacuoles, may be involved in protein body formation. But without confirmation in the form of specific immunohistochemical evidence of the identity of the proteins in the different vesicles that may be related with protein body synthesis (Fig. **12**.9) the biogenesis of the protein-storing organelles of dicotyledons remains unexplained, though recently the presence of bean globulin in the ER was demonstrated [34].

## 12.4 Conclusions

The events of cell specialization related to the biosynthesis of storage proteins and the accumulation of these reserves in vacuole-like organelles, are very similar to the differentiation processes preceding formation and accumulation of plant secondary products (cf. Chap. **11**). One very fundamental difference, however, is that the stored substances themselves carry genetically determined information in the amino acid sequence and are destined for reactivation. Thus there is a large degree of agreement between the cell specialization in biosynthesis and accumulation of storage proteins on the one hand, and similar processes in the differentiation and formation of specialized proteins in animal cells on the other, e.g., hemoglobin in erythrocytes (cf. **4**.5.4), salivary gland proteins of *Diptera* (cf. **9**.4.1), silk fibroin biosynthesis (cf. **9**.4.2). The initial phase involving cell divisions and possible endoreduplication of the genetic material is followed by a second phase of development involving the specialization of the cells responsible for biosynthesis, storage or secretion of the proteins concerned. Storage of proteins in vacuole-like organelles separate from the cytoplasm itself can be regarded as a variant of protein secretion. The results obtained from studies involving various seed plants, and in particular cereals and legumes, can best be summarized as follows:

Formation and accumulation of storage proteins starts with the production of the appropriate mRNA at the beginning of the phase of cell specialization. Up to the point where a constant rate of protein accumulation has been attained a quantitative increase of the translation apparatus takes place, coordinated with a rise in the percentage of the storage protein-specific mRNA in the total poly(A)-containing RNA of the cells. Since the mRNA for the storage proteins contains the information for a signal sequence at the 5′-end, the biosynthesis of the storage proteins must take place on mb-polysomes, the nascent polypeptide chains being secreted cotranslationally into the rER vesicles where the signal sequence is removed. Formation of the vacuole-like storage organelles begins at more or less the same time as the biosynthesis of storage proteins and is almost certainly coordinated with it. In the simplest cases, e.g., corn, these protein bodies are swellings of the rER. In most other plants, especially dicotyledons, the biogenesis of the protein bodies is more complicated. Glycosylation of storage proteins appears to take place before they are transferred to the storage site. Deposition of the storage proteins in protein bodies, apart from rendering the amino acid reserve osmotically harmless, also protects from proteolysis. Evidence for the validity of the hypothesis concerning differentiation processes in the formation and storage of reserve proteins in plants has been obtained from investigations on corn, barley, soybeans, kidney beans, peas and field beans. Isolated results obtained using other cereals, legumes, crucifers and cotton also support the assumption that a model of the kind presented above may be generalized for all spermatophytes.

## 12.5 Selected References

*Reviews*


1. Derbyshire, E., D. J. Wright & D. Boulter: Legumin and vicilin, storage proteins of legume seeds. Phytochemistry *15* (1976): 3–24.
2. Hall, T. C., S. M. Sun, B. U. Buchbinder, J. W. Payne, F. A. Bliss & J. D. Kemp: Bean seed globulin mRNA: Translation, characterization, and its use as a probe towards genetic engineering of crop plants. In: Leaver, C. J. (ed.), Genome Organization and Expression in Plants. Plenum Press, New York 1980, pp. 259–272.

2a. Müntz, K., R. Manteuffel & G. Scholz (eds.): The Regulation of Protein Biosynthesis and Degradation During Embryogenesis and Germination of Plant Seeds. Proc. 2nd Seed Protein Symp., Abh. Akad. Wiss. DDR, Abt. Math., Naturwiss., Techn., No N5, Berlin 1981.

3. Ingversen, J., A. Brandt & V. Cameron-Mills: The structure, biosynthesis and intracellular transport of barley reserve proteins. In: [2a], pp. 41–56.
4. Larkins, B. A., K. Pedersen, W. J. Hurkman, A. K. Handa, A. C. Mason, C. Y. Tsai & M. A. Hermodson: Maize storage proteins. Characterization and biosynthesis. In: Leaver, C. J. (ed.), Genome Organization and Expression in Plants. Plenum Press, New York 1980, pp. 203–218.
5. Miflin, B. J. & P. R. Shewry: The biology and biochemistry of cereal seed prolamins. In: Seed Protein Improvement in Cereals and Grain Legumes. Proc. Symp. IAEA/FAO, Neuherberg, September 4–8, 1978, IAEA, Wien 1979, pp. 137–158.
6. Millerd, A.: Biochemistry of legume seed proteins. Ann. Rev. Plant Physiol. *26* (1975): 53–72.
7. Müntz, K., H. Bäumlein, R. Bassünder, R. Manteuffel, M. Püchel, P. Schmidt, and U. Wobus: The biosynthesis of storage proteins in developing field bean seeds. In: [2a], pp. 57–72.

8. Müntz, K., H. Bäumlein, R. Bassüner, R. Manteuffel, M. Püchel, P. Schmidt & U. Wobus: Regulation von Biosynthese und Akkumulation der Reserveproteine während der Entwicklung pflanzlicher Samen. Biochem. Physiol. Pflanzen *176* (1981): 401–422.
9. Nielsen, N. C., M. Moreira, P. Staswick, M. A. Hermodson, N. Tumer & V. H. Than: The structure of glycinin from soybeans. In: [2a], pp. 73–92.
10. Osborne, T. B.: The vegetable proteins. Monographs in Biochemistry. Longmans, Green and Co., London 1924.
11. Pedersen, K. & B. A. Larkins: Structure and expression of zein genes in maize endosperm. In: [2a], pp. 31–40.

*Original Papers*

12. Bailey, C. J., A. Cobb & D. Boulter: A cytoledon slice system for the electron autoradiographic study of the synthesis and intracellular transport of the seed storage protein of *Vicia faba*. Planta *95* (1970): 103–118.
13. Bollini, R. & M. J. Chrispeels: The rough endoplasmic reticulum is the site of reserve protein synthesis in developing *Phaseolus vulgaris* cotyledons. Planta *146* (1979): 487–501.
14. Briarty, L. G., D. A. Coult & D. Boulter: Protein bodies of developing seeds of *Vicia faba*. J. Exp. Bot. *20* (1969): 358–372.
15. Broekaert, D. & R. van Parijs: The relationship between the endomitotic cell cycle and the enhanced capacity for protein synthesis in *Leguminosae* embryogeny. Z. Pflanzenphysiol. *86* (1978): 165–175.
16. Brandt, A. & J. Ingversen: Isolation and translation of hordein messenger RNA from wild type and mutant endosperms in barley. Carlsberg Res. Commun. *43* (1978): 451–469.
17. Cameron-Mills, V., J. Ingversen & A. Brandt: Transfer of in vitro synthesized barley endosperm proteins into the lumen of the endoplasmic reticulum. Carlsberg Res. Commun. *43* (1978): 91–102.
18. Croy, R. R. D., J. A. Gatehouse, J. M. Evans & D. Boulter: Characterization of the storage protein subunits synthesized in vitro by polyribosomes and RNA from developing pea (*Pisum sativum* L.). I. Legumin. Planta *148* (1980): 49–56.
19. Davies, H. M. & D. P. Delmer: Seed reserve-protein glycosylation in an in vitro preparation from developing cotyledons of *Phaseolus vulgaris*. Planta *146* (1979): 513–520.
20. Hall, T. C. & F. A. Bliss: Bean globulin: The structure and biogenesis of phaseolin (G1--globulin) from french bean and the isolation of a phaseolin gene. Abstr. 6th Annual EMBO--Meeting, September 29 – October 1, Heidelberg 1980, pp. 18–19.
21. Larkins, B. A. & W. J. Hurkman: Synthesis and deposition of zein in protein bodies of maize endosperm. Plant. Physiol. *62* (1978): 256–263.
22. Manteuffel, R., K. Müntz, M. Püchel & G. Scholz: Phase-dependent changes of DNA, RNA, and protein accumulation during ontogenesis of broad bean seeds (*Vicia faba* L. var. *minor*). Biochem. Physiol. Pflanzen *168* (1976): 595–605.
23. Marinos, N. G.: Embryogenese der Erbse *(Pisum sativum)*. 1. Mitt. Die zytologische Umgebung des sich entwickelnden Embryos. Protoplasma (Wien) *70* (1970): 261–279.
24. Millerd, A.: DNA and RNA synthesis during growth by cell expansion in *Vicia faba* cotyledons. In: Pollack, J. K. & J. W. Lee (eds.), The Biochemistry of Gene Expression in Higher Organisms. Reidee Publ. Comp., Dordrecht-Boston 1973, pp. 357–366.
25. Millerd, A. & D. Spencer: Changes in RNA-synthesizing activity and template activity in nuclei from cotyledons of developing pea seeds. Austral. J. Plant Physiol. *1* (1974): 331–341.
26. Payne, P. I. & D. Boulter: Free and membrane bound ribosomes of the cotyledons of *Vicia faba* (L.). I. Seed development. Planta *84* (1969): 263–271.
27. Payne, E. S., A. Brownrigg, A. Yarwood & D. Boulter: Changing protein synthetic machinery during development of seed of *Vicia faba*. Phytochemistry *10* (1971): 2299–2303.
28. Püchel, M., K. Müntz, B. Parthier, O. Aurich, R. Bassüner, R. Manteuffel & P. Schmidt: RNA metabolism and membrane bound polysomes in relation to globulin biosynthesis in cotyledons of developing field beans. Europ. J. Biochem. *96* (1979): 321–329.
29. Soave, C., A. Viotti, N. di Fonzo & F. Salamini: Maize prolamin: Synthesis and genetic regulation. In: Seed Protein Improvement in Cereal and Grain Legumes. Proc. Symp. IAEA/FAO, Neuherberg, September 4–8, 1978, IAEA, Wien 1979, pp. 165–174.

30. Sun, S. M., M. Mutschler, F. A. Bliss & T. C. Hall: Protein synthesis and accumulation in bean cotyledons during growth. Plant Physiol. *61* (1978): 918–923.
31. Takagi, S. & T. Mori: The 5′-terminal structure of poly(A)-containing RNA of soybean seeds. J. Biochem. *86* (1979): 231–238.
32. Viotti, A., C. Balducci & J. H. Weil: Adaptation of the tRNA population of maize endosperm for zein synthesis. Biochim. Biophys. Acta *517* (1978): 125–132.
33. Weber, E. & D. Neumann: Protein bodies as storage organelles in plant seeds. Biochem. Physiol. Pflanzen *175* (1980): 279–306.
34. Baumgartner, B., K. T. Tokuyasu & M. J. Chrispeels: Immunocytochemical localization of reserve protein in the endoplasmic reticulum of developing bean *(Phaseolus vulgaris)* cotyledons. Planta *150* (1980): 419–425.
35. Evans, J. M., R. R. D. Croy, P. A. Brown & D. Boulter: Synthesis of complementary DNA to partially purified mRNAs coding for storage proteins of *Pisum sativum* L. Biochim. Biophys. Acta *610* (1980): 81–95.
36. Marks, G. E. & D. R. Davies: The cytology of cotyledon cells and the induction of giant polytene chromosomes in *Pisum sativum*. Protoplasma *101* (1979): 73–80.
37. Sun, S. M., J. L. Slightom & T. C. Hall: Intervening sequences in a plant gene – comparison of the partial sequence of cDNA and genomic DNA of French bean phaseolin. Nature *289* (1981): 37–41.

# 13. Determination and Function of the Immune System

R. F. Richter & L. Nover

## 13.1 Introduction

### 13.1.1 The Immune System

The existence and highly efficient function of multicellular organisms depend on the cooperative interaction of their cells mediated by direct cell-cell contact via components of the cell membranes as well as by long range effects involving electric or chemical signals, e.g., hormones. Proteins and other high molecular weight compounds play prominent roles as cell membrane constituents as well as intraorganismic chemical signals. Evolution of the highest developed animals not only led to a constant improvement of these means for intercellular communication but simultaneously resulted in the acquisition of an elaborate defense system (immune system) directed against the same type of compounds if produced by other organisms. This discrimination between "self" and "not-self" is of vital importance for the maintenance of the integrity and the protection of multicellular animals against foreign invading organisms.

An immune reaction can be roughly characterized by the following process: If a foreign protein or polymeric carbohydrate happens to enter an animal, specialized immunoglobulins *(antibodies)* are formed and secreted to the body fluids. They have a specific affinity for binding to the invading not-self material *(antigens)*. Antigen and antibody form a complex that can be taken up and digested by special cells (phagocytes). The specificity of the antigen-antibody reaction depends on the flexibility of the immunoglobulin structure and on the affinity of its antigen binding sites for determinant groups *(haptens)* exposed at the surface of the antigen molecule.

Whole cells and parts of them as well as proteins, carbohydrates, lipids, or even low molecular weight compounds, when bound to high molecular weight carriers, can be effective as antigens. Besides the secretion of antibodies into the body fluids (humoral immune response) there are special immune cells with proteins on their surface which react with antigens as specifically as antibodies and are responsible particularly for the defense against foreign cells. In the following we will neglect this cell-mediated immune response and only deal with some remarkable properties of the *humoral immune system.*

---

Abbreviations: Immunoglobulins (Ig) are classified (cf. Table **13.1**) according to the type of heavy chains involved (H-chains of the $\alpha$-, $\delta$-, $\varepsilon$-, $\gamma$-, or $\mu$-type): IgA, IgD, IgE, IgG and IgM. Ig light chains (L-chains) are of the $\varkappa$ or $\lambda$-type (cf. Table **13.1**). Segments of the Ig genes are designated with V for the variable region, D for the diversity region, J for the joining region and C for the constant region (cf. Figs. **13.4** and **13.6**). BALB/c, NZB denote inbred strains of mice. MHC, major histocompatibility complex (cf. **13.3.1**); bp, kb, base pairs and kilobase pairs of DNA; kd, kilodaltons; MW, molecular weight; $M_r$, relative molecular weight (determined, e.g., by electrophoretic methods).

### 13.1.2 Structure and Heterogeneity of Immunoglobulins

Immunoglobulins (Ig) are glykoproteins. The quarternary structure of the monomer looks like a Y (Fig. **13**.1). It consists of two light chains (L-chains, MW 22.5 kd) and two heavy chains (H-chains, MW 50–75 kd) connected with each other by disulphide bridges and non-covalent bonds. For the function of the molecule the two antigen-binding domains are of special interest. They are formed by interaction of the so-called V-regions of the H- and L-chains. If one compares the amino acid sequences of different immunoglobulins of the same organism these V-regions exhibit a great variability particularly in their so-called hypervariable parts (Fig. **13**.1). This

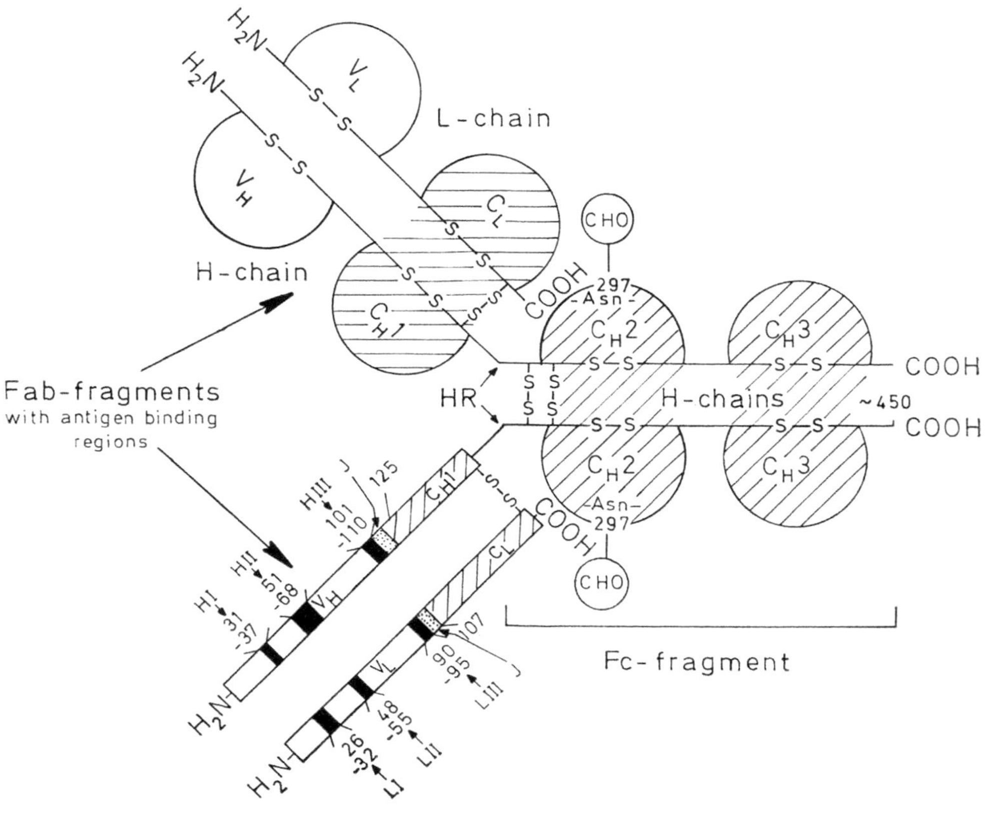


**Fig. 13.1:** Basic structure of immunoglobulins exemplified for IgG.

Two heavy *(H)* and two light *(L)* chains form a tetrameric molecule *($H_2L_2$)* stabilized by 4 interchain disulphide bonds (—S—S—). The total Ig molecule has a Y-shape with two protein domains in the C-terminal stem *($C_H3$, $C_H2$)* and each two domains in the N-terminal arms *($C_H1/C_L$* and *$V_H/V_L$)*. The two antigen binding regions in the V-domains result from the interaction of the hypervariable sequences *(HI–HIII* and *LI–LIII)*. Their precise position in the V-domain is detailed in the lower arm of the IgG molecule. The tetrameric molecule can be split by limited proteolysis at the hinge region *(HR)* to give the stem *(Fc)* and two *a*ntigen *b*inding arms *(Fab)*. All immunoglobulins correspond to this basic structure. The heavy chains contain one or more branched carbohydrate chains *(CHO)*. Details of the stepwise synthesis of the carbohydrate chain during processing of Ig precursor molecules are given in Fig. **13**.10. IgM molecules contain an additional constant domain ($C_H4$) with alternative C-terminal ends (cf. **13**.2.2.3).

contrasts to the barely changing amino acid sequences in the constant domains of the Ig molecules ($C_L/C_H1$, $C_H2$, $C_H3$).

With the electron microscope and with X-ray diffraction methods the functional domains of the immunoglobulins can be differentiated from each other as distinct, more or less globular regions. Each of the domains is stabilized in itself by a disulphide bridge. This structure confers a great conformational flexibility to the whole molecule and is a prerequisite for its multiple functions. Particularly the short linker sequence of 12–13 amino acid residues between the constant part (Fc) and the antigen-binding part (Fab), the so-called hinge region (HR), plays an essential role in this respect.

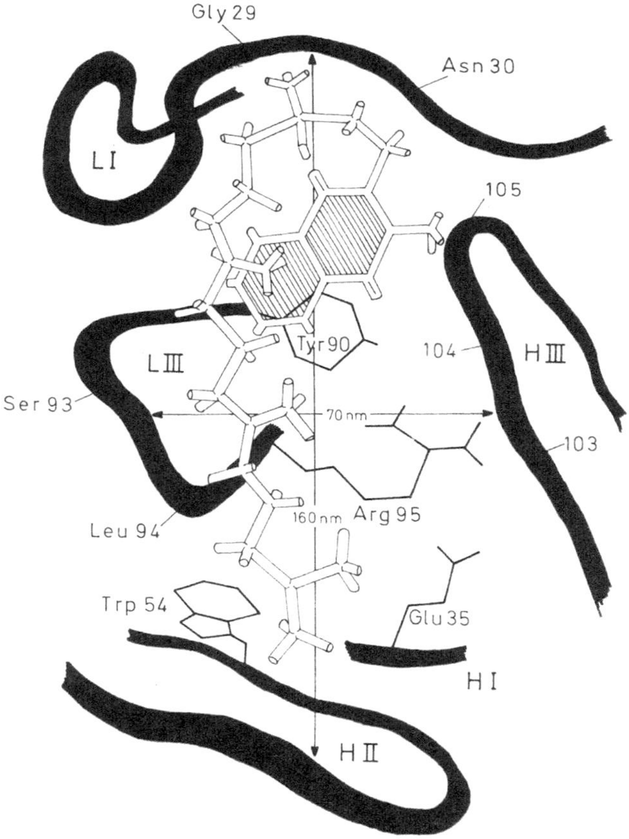


**Fig. 13.2:** Model of the antigen binding region of the Fab fragment "New" from human myeloma tumor cells (from Richards et al. [12], modified).

The model results from X-ray diffraction analyses of crystallized Fab fragments, which are specific for $\gamma$-hydroxy vitamin $K_1$. For the isolation of sufficient amounts of pure Fab fragment in vitro cultivated myeloma tumor cells are used because they result from the transformation of a single immune competent cell and hence produce a single type of antibody.
The antigen binding pocket of about $160 \times 70 \times 60$ nm$^3$ is formed by hypervariable regions HI, HII, HIII, LI and LIII. The amino acid residues are indicated by numbers and in some prominent cases also by their type of side chains. The hapten (vitamin $K_1$) fits into the pocket with the phytyl side chain in front and the naphthoquinone ring in the rear.

The predominant secondary structure of the individual domains is very similar [6, 7]. Each subunit contains two slightly twisted β-pleated sheets of three (3β1–3β3) and four (4β1–4β4) antiparallel strands. The conformation is stabilized by numerous interstrand hydrogen bonds. About 50% of the amino acid residues are part of these β-sheets. They are highly conserved and exhibit an alternating pattern of polar and apolar side chains. In the variable domains the three-stranded β-sheet is extended by an extra loop to a five-stranded β-sheet. Interestingly, the three hypervariable regions (cf. Figs. **13.1** and **13.2**) form the connecting loops between strands of the β-sheet on one side of the domain. The close structural relationships between the different domains of an Ig molecule suggest that it arose, during evolution, by repeated gene duplication. This hypothesis is also supported by the fact that still nowadays all domains are coded by distinct gene segments separated by intervening sequences (cf. Figs. **13.4** and **13.6**).

Details of the inner structure of the variable domains in the Fab part with the antigen binding region are of considerable interest. The model constructed from results of X-ray diffraction patterns (Fig. **13.2**) demonstrates a pocket of about $160 \times 70 \times 60$ $nm^3$ to be formed by the hypervariable sequences HI, HII, HIII, LI and LIII, in which the determinant groups of the antigen are stereospecifically fixed by numerous non-covalent bonds.

The totality of antibodies of one specificity, i.e., of antibodies recognizing the same antigen, is highly heterogeneous. In the first line, this results from the extraordinary multiplicity of variable regions and their possible combinations. This will be discussed

**Table 13.1**: Properties and composition of human immunoglobulins

| Ig class | IgM[2] | IgD | IgG | IgE | IgA[2] |
|---|---|---|---|---|---|
| Subunit composition[1] | $H_2L_2$ and $(H_2L_2)_5$ | $H_2L_2$ | $H_2L_2$ | $H_2L_2$ | $H_2L_2$ and $H_2L_2)_{2,3}$ |
| Types of H-chains | $\mu_1, \mu_2$ | $\delta$ | $\gamma_1$–$\gamma_4$ | $\varepsilon$ | $\alpha_1, \alpha_2$ |
| Types of L-chains | $\varkappa$, $\lambda_1$–$\lambda_3$ | $\varkappa$, $\lambda_1$–$\lambda_3$ | $\varkappa$, $\lambda_1$–$\lambda_3$ | $\varkappa$, $\lambda_1$–$\lambda_3$ | $\varkappa$, $\lambda_1$–$\lambda_3$ |
| Mol. weight (kd) | 185 and 900 | 180 | 150 | 195 | 160 (for monomer) |
| Carbohydrate content (%) | 12 | 14 | 3 | 11 | 8 |
| Normal adult serum concn. (g/100 ml) | 0.04–0.15 | 0.003 | 1.0–1.4 | $10^{-3}$ to $10^{-5}$ | 0.2–0.3 |
| Rate of synthesis (g/day/70 kg body wt.) | 0.1 | 0.03 | 2.3 | $2 \cdot 10^{-4}$ | 1.7 |
| Half-life (days) | 5 | 3 | 23 | 3 | 6 |
| Distribution | Mainly intra-vascular | Mainly intra-vascular | Intravascular and extra-vascular fluid | Intravascular and skin, respiratory and gastrointestinal tracts | Intravascular and internal secretions |

[1] Immunoglobulin molecules are always built of identical H and L chains.

[2] In addition to the corresponding L and H chains the IgM pentamer and the IgA oligomers contain a so called J-chain joining two H chains by disulphide bridges (Koshland, M. E.: Advanc. Immunol. *20* (1975): 41–69). J chain synthesis triggers excretion of IgM pentamers and IgA oligomers.

in detail in the following sections (**13**.2 and **13**.3). To a minor degree the heterogeneity derives from differences of the constant regions as outlined in Table **13**.1. With respect to their heavy chains immunoglobulins can be grouped in classes (isotypes) and subclasses, respectively. The most frequent antibodies are of the IgG-, IgA- and IgM-types. For the IgG-type there is a total of sixteen possible combinations of the four different H-chain types ($\gamma_1$–$\gamma_4$ in man, $\gamma_1$, $\gamma_{2a}$, $\gamma_{2b}$, $\gamma_3$ in mice) with the four different L-chain types ($\varkappa$, $\lambda_1$–$\lambda_3$).

## 13.2 Determination and Maturation of Immune Competent Cells

### 13.2.1 Phases of Cell Specialization and the Origin of Antibody Diversity

In the adult organism immune cells derive from the pluripotential hematopoietic stem cells in the bone marrow (cf. Fig. **4**.39). Two stages of cell specialization can be distinguished:

Hematopoietic stem cell → Lymphoid stem cell → Immune competent cell → Antibody producing B cell or T-effector cell

Phase I (determination): Hematopoietic stem cell → Lymphoid stem cell → Immune competent cell

Phase II (maturation): Immune competent cell → Antibody producing B cell or T-effector cell

In the first step of phase I the pluripotential hematopoietic stem cells give rise to the lymphoid stem cells. Like other stem cells, they are capable of permanently producing, by "quantal" mitoses (cf. **4**.5.4.1 and **15**.2), cells of the next level of specialization without diminishing their own stock. Such mitoses are influenced by unknown extracellular signals.

As possible signals for cell specialization in phase I two polypeptide hormones have been detected in chicken: thymopoietin (MW 5562, 49 amino acid residues) for the T cell line and bursopoietin for the B cell line. Similar functions are discussed for thymosin and ubiquitin in mammals.

Two lines of immune competent cells originate from the lymphoid stem cells:

1. Precursors of *T-cells* migrate to the thymus, where they are imprinted to different types of immune competent T lymphocytes. They play a central role for the triggering and control of the humoral immune response (T helper and suppressor cells respectively, cf. **13**.3.1).

Special large epithelial cells (thymic nurse cells) evidently play an essential role for T cell education and multiplication [45]. In the thymus glands of rat and mouse these epithelial cells were found to engulf lymphocytes without degrading them but rather providing the microenvironment for their proliferation and differentiation. Thymic nurse cells were reported to release up to 50 educated lymphocytes.

2. In the center of this chapter are the immunoglobulin producing *B cells* for which the total phase of imprinting proceeds in the bone marrow while maturation to the antibody producing plasma cells (phase II) takes place in the so-called secondary lymphoid organs, i.e., in the spleen and lymph nodes. Both organs contain a number of subcompartments whose cellular microenvironments are crucial for different parts of the humoral immune response [43].

As an antigen circulates it first encounters the immune system either in the blood by entering the central part of the spleen (red pulp) or in the lymph by entering the medulla of the lymph nodes. From the central part of the secondary lymphoid organs the antigen travels through the marginal zone to the mantle layer in spleen and through the corticomedullary border to the cortex in the lymph nodes. The central parts in both types of organs are the sites of antibody production by plasma cells, whereas the marginal zones between central part (medulla) and peripheral part (cortex) are the sites of plasmablast proliferation (cf. **13.3.2**) and B cell/T cell/macrophage interactions (cf. Fig. **13.7**). Finally, the cortical layers serve as microenvironment for B memory cell generation (cf. **13.3.2**) and storage of T and B cells.

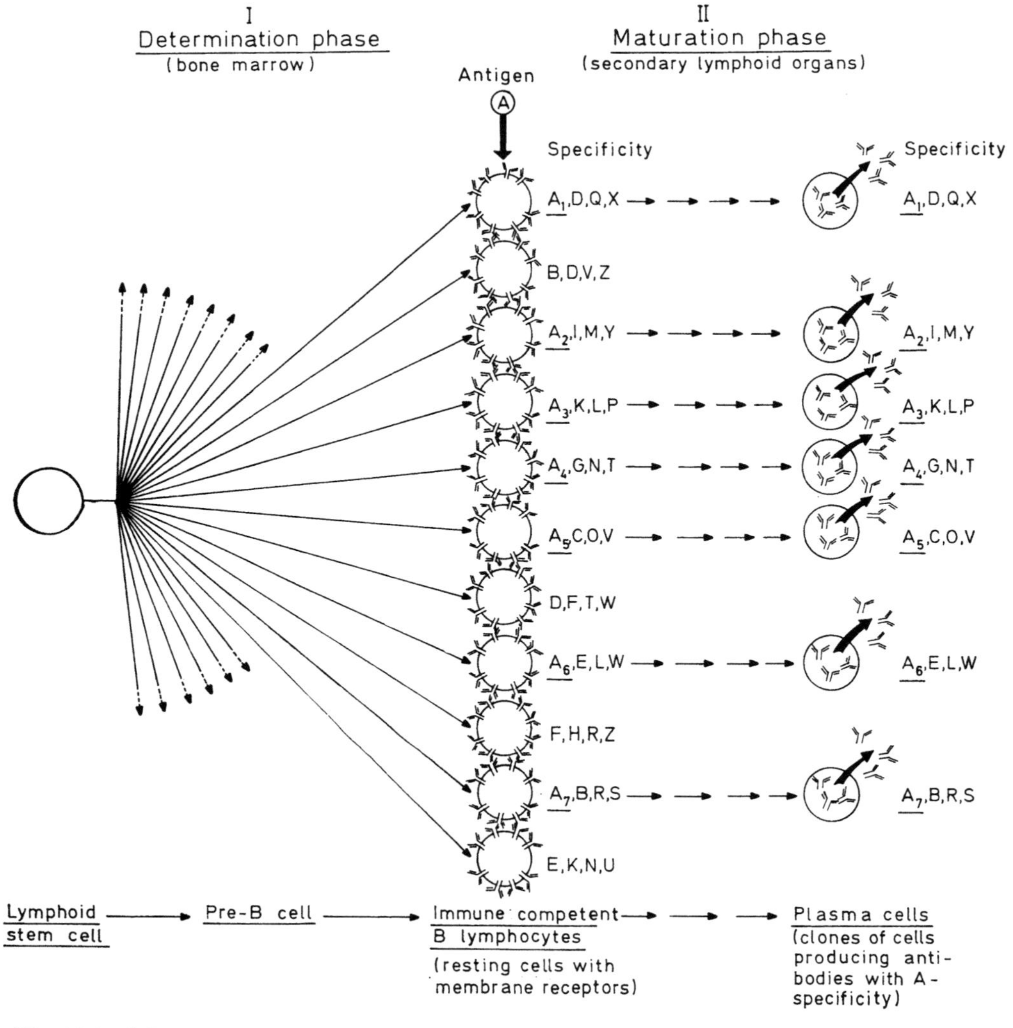


**Fig. 13.3**: Cell determination and maturation in the B-cell lineage.

In the absence of antigen contact the lymphoid stem cells of the bone marrow give rise to a great number of different miniclones of immune competent cells (**I**, determination phase). By interaction with a given antigen (antigen A) all cells of clones with A-specific membrane receptors proliferate and mature to plasma cells excreting large amounts of A-specific antibodies (**II**, maturation phase). About $10^3$–$10^4$ different clones are normally involved in an immune response. All are specific against A but the antibodies of each cell clone react furthermore with numerous structurally non-related haptens bound by functional groups of the same antigen binding pocket. This polyfunctionality, in Fig. **13.3** assumed to involve 4 different haptens, in reality extends to about 100. It is an essential means for an "economic" use of the limited genetic material and does not contrast to the high specificity of the overall antibody population (cf. Fig. **13.8**).

The generation of immune competent B lymphocytes in the bone marrow (Fig. **13**.3, determinative phase) proceeds in the absence of any antigen. The most important events are rearrangements within the genetic material coding for the immunoglobulins (cf. the following Section). They lead to the fusion and activation of distinct gene segments and provide the bases for the creation of a seemingly unlimited number of cell clones (B resting cells). Each of them is characterized by its own type of specific immunoglobulin in the outer cell membrane (Fig. **13**.3). The immunoglobulins in the cell membrane (about $10^4$–$10^5$ molecules/cell) are mainly of the IgM-type (cf. **13**.2.4). They function as antigen receptors. In phase II (Fig. **13**.3) antigen binding to B resting cells selectively triggers proliferation and maturation of those cell clones that contain the corresponding antigen-specific receptors in their cell membranes (clone selection). The multistage maturation to antibody-secreting plasma cells involves many cell divisions and finally leads to a cell that is specialized to the synthesis and secretion of huge amounts of a single type of antibody whose combination of variable regions is identical with that in the membrane receptors of the B resting cell. In many cases B cell maturation is connected with a second type of gene rearrangement (heavy chain switch, cf. the following section).

### 13.2.2 Structure and Rearrangements of Immunoglobulin Genes

Our present-day understanding of the function of the immune system is decisively influenced by the rapidly advancing knowledge about the organization of Ig gene systems. The progress is mainly based on the application of methods briefly outlined in Chap. **2**, i.e. (1) cutting by restriction endonucleases of genome DNA from embryonic (undetermined) and plasmacytoma (immunologically determined) cells, (2) cloning and selection of the Ig gene containing DNA fragments and (3) characterization and eventually sequencing of selected fragments. Comparison of the restriction pattern for both cell types gave direct evidence for genome rearrangements accompanying the two phases of cell specialization in the immune system (Fig. **13**.3).

The scientific and practical value of another method can barely be overestimated. It is the *hybridoma technique* described 1975 by Köhler and Milstein [15]. Individual lymphoid cells of a population are fused with special mutant myeloma cells, growing permanently in vitro, but being themselves unable to form Ig molecules. By the following selection and characterization of the hybridoma cell clones it is possible to dissect the highly complex immune response into its individual components (monoclonal antibodies). With the same technique intermdiates of the immune cell lineages can be cloned and characterized as well.

Figure **13**.4 summarizes basic features of the gene structure in embryonic and immune competent cells and presents some details of Ig pre-mRNA processing. Though the concrete data refer to the mouse $\varkappa$-L-chain system, the scheme is principally valid for the corresponding system of other Ig chains, too. Peculiarities of the gene system coding for the heavy chains will be mentioned at several points. They will be discussed in more detail in connection with Fig. **13**.6. But we will first turn our attention to the general principles of Ig gene expression outlined in Fig. **13**.4.

#### 13.2.2.1 Fusion of V- and C-Gene Segments

The genetic material for immunoglobulins is contained in separate gene segments that code for the variable and constant parts of the molecule. Independent sets of V-gene segments with $10^2$–$10^3$ different copies each map on the chromosomes at a certain distance of the C-gene segments for the $\varkappa$-L-chain and for the heavy chains respectively.

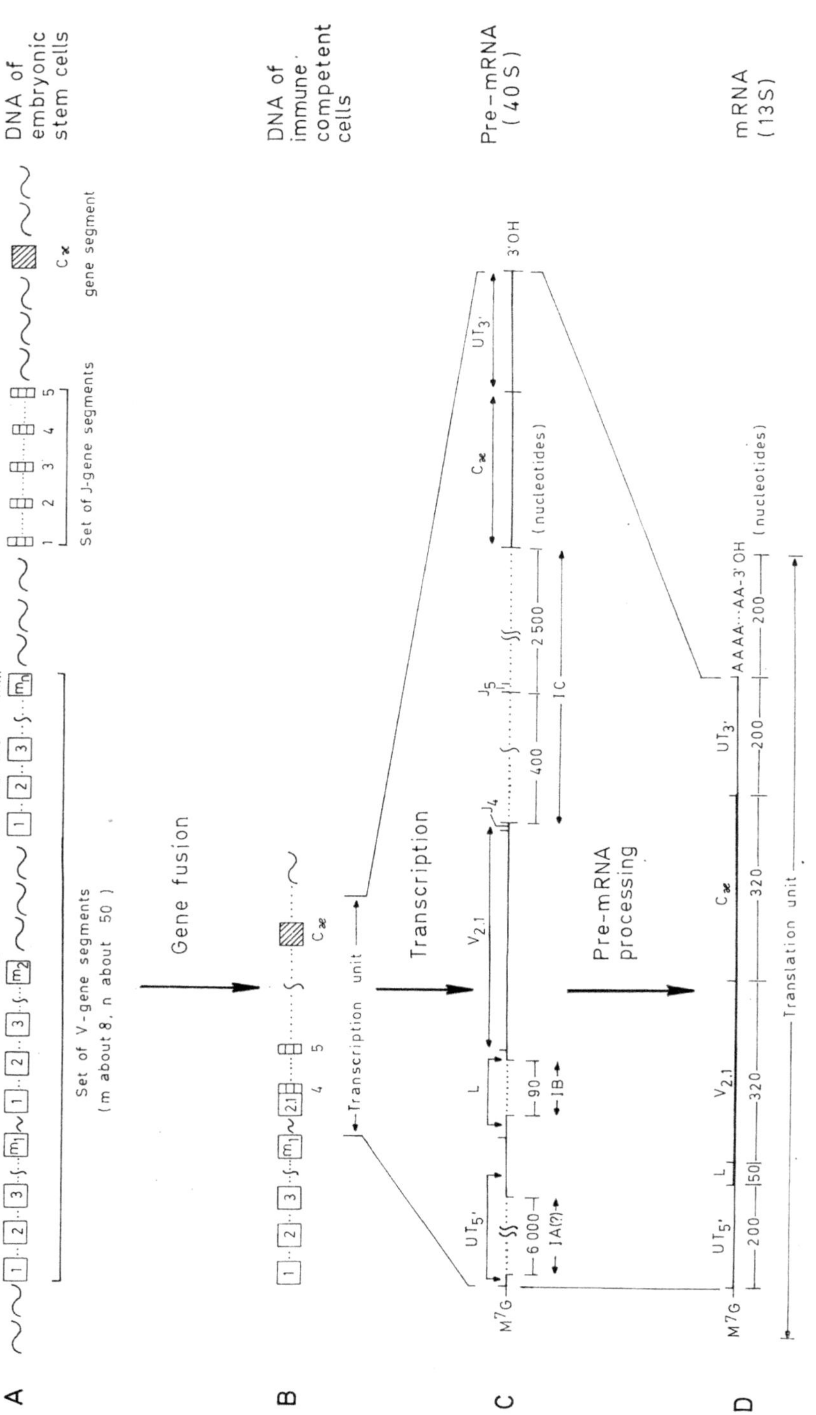
A
B
C
D
DNA of embryonic stem cells
DNA of immune competent cells
Pre-mRNA (40 S)
mRNA (13S)
Set of V-gene segments (m about 8, n about 50)
Set of J-gene segments
Cϰ gene segment
Gene fusion
Transcription
Pre-mRNA processing
Transcription unit
Translation unit
(nucleotides)
M7G
3'OH
AAAA···AA-3'OH
UT5'
L
V2.1
Cϰ
UT3'
IA(?)
IB
IC
6 000
90
400
2 500
200
50
320
320
200
200

Fig. 13.4

For the mouse genome the genetic material of the $\varkappa$-L-chain was mapped on chromosome 6, that of the heavy chains on chromosome 12 [16]. It was an important finding that the order of the $C_H$-gene segments corresponds to the sequence of their possible expression during B cell maturation, i.e., $5'$-$\mu$-2.5-$\delta$-?-$\gamma_3$-34-$\gamma_1$-21-${}_{2b}\gamma$-15-$\gamma_{2a}$-15-$\varepsilon$-12-$\alpha$-$3'$ (spacers given in kb [20, 21]). The special implications of this arrangement for the heavy chain switch will be discussed in connection with Fig. **13**.6 and Table **13**.2. Contrasting to the small $C_L$-gene segments, $C_H$-segments comprise ~2 kb for the $\gamma$-type (cf. Fig. **13**.6) and 7–10 kb for the $\mu$- and $\delta$-types [19, 20]. They contain 3 to 7 exons and corresponding introns of variable length (cf. summaries [13, 15a].

It is a peculiarity of the mouse immune system that L-chains of the $\lambda$-type make only 5 % of total L-chains with 80–90 % being of the $\lambda_1$-type. The genes are localized on chromosome 16, arranged in two groups whose relative position with respect to each other is unknown [25] (spacers given in kb):

$5'$-$V_1$-?-$J_3$-1.26-$C_3$-1.95-$J_1$-1.3-$C_1$-$3'$ and $5'$-$V_2$-?-$J_2$-1.37-$C_2$-1.7-$J_4$-1.33-$C_4$-$3'$.

Synthesis of the continuous information from the Ig gene segments proceeds at two consecutive levels: (1) Somatic gene rearrangements, causing deletion of considerable parts of the genetic material between the gene segments to be fused, produce the *transcription units* (Fig. **13**.4, A → B) and (2) processing of the primary transcript gives rise to the *translation unit* (Fig. **13**.4, C → D). Only after gene fusion are parts of the genetic material fit for productive expression (cf. details given in Sect. **13**.2.4). The important result of gene fusion is the activation of any one of the L-chain V-segments and H-chain V-segments respectively whose combination in the immunoglobulin determine the future antigen response of the immune competent cell clone.

V-gene segments occur in more or less numerous different copies [6, 14] that can be arranged in different families of closely related sequences (cf. Fig. **13**.4). The number of families for the $\varkappa$-L-chain of mouse is estimated to be about 50 (n ~ 50) with about 8 segments in each family (m ~ 8). That means a sum of $n \times m \sim 400$ different V-gene segments contained in the germ-line. Moreover the number of V-gene segments for the mouse H-chains is estimated to be about 250. Consequently a total of $250 \times 400 \sim 10^5$ Ig variants results by free combination of H- and L-chains.

This germ-line coded antibody diversity is strongly increased by *somatic mutation* [14, 21a, 24, 28] including the flexibility of the V/C recombination frames (cf. Fig. **13**.5) but above all numerous single base exchanges in the V-gene segments. Up to 90 % of antibody diversity evi-

---

◀

**Fig. 13.4**: Generation of the transcriptional and translational units of Ig light chains by gene fusion, transcription and pre-mRNA processing (compiled from data cited in Refs. [15a, 22–24, 26, 37 and 38]). For detailed explanations see text.

**A**: V-gene segments of the unrearranged, embryonic chromosome are indicated by a two-numbered code referring to the family and to the individual gene segment (e.g., 1.1. · · · n.m.). Besides this, five gene segments for J and one for the $\varkappa$-constant region are shown. The spacer DNA between individual V-gene segments can be as large as 10–15 kb. The size of the spacer between the last V- and the first J-gene segment is unknown.

**B**: Gene fusion is assumed to take place between $J_4$ and the V-gene segment 2.1. The DNA between the two segments is deleted (cf. Fig. **13**.6).

**C**: Transcription yields a high mol. weight mRNA precursor with a $5'$-cap ($M^7G$) and with two long (IA ~ 6,000 nucleotides, IC ~ 2,900 nucleotides) and one short intervening sequences (IB ~ 90 nucleotides). The existence of the two intervening sequences IB and IC is experimentally well established. In contrast, existence of IA is hypothetic and only deduced from the length of the primary transcript.

**D**: During processing of pre-mRNA the intervening sequences are removed and a poly(A) sequence is added to the $3'$-untranslated part *($UT_{3'}$)*. Translation starts at the initiation triplett at the beginning of the leader sequence *(L)* coding for the signal peptide (cf. Fig. **13**.90).

dently arise by these mechanisms [24, 28]. This is especially prominent for some Ig L-chains coded by chromosomes with a single (mouse λ-L-chains) or with only few V-gene segments (human ϰ-L-chains based on ~20 germ-line V-segments [21a]). Furthermore, analysis, by the hybridoma technique, of the total complexity of the mouse immune response to phosphorylcholine [28] revealed single base exchanges scattered over the whole $V_H$-gene segments with increased frequency when comparing late (IgG, IgA) with early Ig isotypes (IgM, cf. **13**.2.2.2).

A short sequence of about 39 base pairs (J-gene segment) maps about 4 kb upstream of the C-gene segments and serves as joining site between C- and V-gene segments. In fact, as indicated in Fig. **13**.4 for the ϰ-L-chain, there are four to five different J-segments only one of which is selected by chance for fusion. Thus, the total number of Ig variants is increased by the factor 16 if 4 J-segments are assumed to exist for each type of Ig chain. The 13 amino acid residues coded by the J-segment have

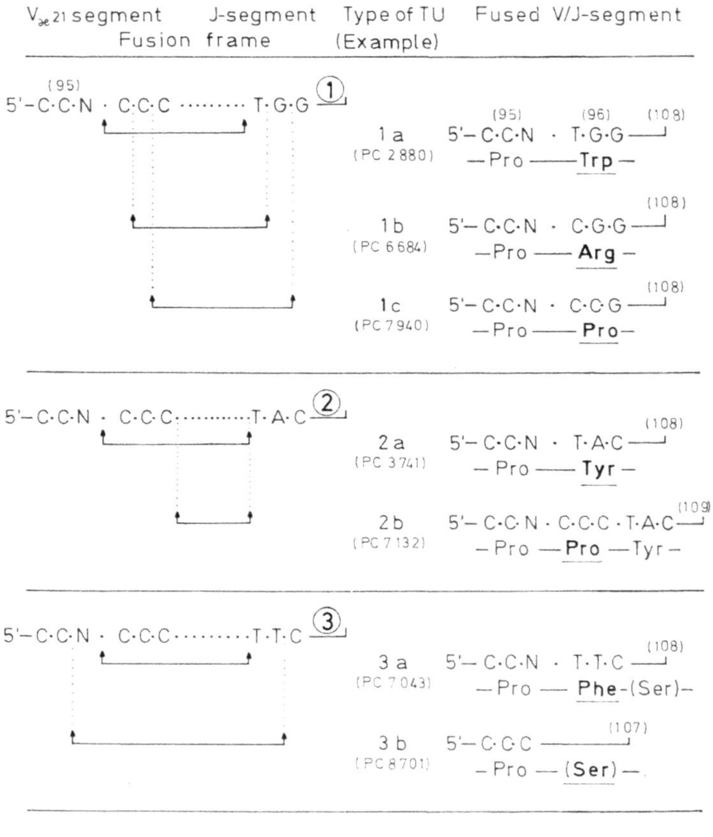


**Fig. 13.5:** Generation of antibody diversity by fusing different J-segments with one $V_ϰ$-segment and by shifting the fusion frame (from Weigert, M. et al., Nature *283* (1980): 497).

In the example 7 known amino acid sequences at the V/J joining site from NZB mice myeloma antibodies were correlated with the nucleotide sequences of the $V_ϰ$21 segment and the known types of J-segments from BALB/c mice. Amino acid residue 96 in the fused V/J-segment is highly variable due to shifts in the joining frame (arrows). Moreover, examples of V/J-segments extended and shortened by one amino acid residue were observed (e.g., types 2b and 3b). The types of transcription units *(TU)* are denoted by *small case letters* and the number of J-segment they are derived from. Examples for plasmacytoma cell lines *(PC)* producing the particular antibody type are given in *brackets*.

essential influence on the antigen binding properties of the Ig molecule because they become part of the hypervariable regions LIII and HIII (cf. Fig. **13**.4). Moreover, it is evident from a thorough comparison of the sequence data of the positions around the joining site (amino acid residues 95/96) that the fusion frame is not rigidly determined (Fig. **13**.5) allowing for some additional variability in this important position. Even deletion or addition of one amino acid residue can result from the ambiguity of the joining site (examples 2b and 3b in Fig. **13**.5).

In contrast to Ig L-chain genes the functional V-gene segments coding for the V-domains of H-chains must be composed by fusion of three separated parts. In the germ-line state the bodies of these V-gene segments lack, in addition to the J-segment, part of the sequence coding for the hypervariable region HIII. About 10 separate gene segments (D-segments), mapping 5′ upstream of the four $J_H$-segments, encode the lacking amino acid residues. D-segments are flanked on both sides by the recombination sequence indicated below [27, 27a, 34a], Thus, only a dual recombinatorial joining of independent gene segments on the H-chain chromosome generates the complete V-gene segment present in an immune competent cell [27a, 34a].

Sequence analyses in the 3′-flanking regions of V-gene segments and the 5′-flanking regions of J-gene segments have revealed two conserved blocks of a heptamer plus a nanomer at each site. The two blocks are separated either by one or two turns of the DNA double helix, i.e., by 11 or 22 bp. The consensus sequences for the 3′- and 5′-flanking regions are mutually complementary [27a, 34a]:

5′-V-gene segment-CACAGTG ... 11 or 22 bp ... ACAAAAACC ⌉
~
3′-J-gene segment-GTGTCAC ... 22 or 11 bp ... TGTTTTTGG ⌋

Interestingly, always one partner in the recombination displays the two recognition sequences one turn of the helix apart, and the other displays them two turns apart. Correspondingly, it is assumed that the recombination enzyme complex contains two DNA-binding proteins specific for the same recognition sequences but distinguished by the distance of the two binding sites. The two DNA strands are cut and rejoined in vicinity of the heptamers. Intrastrand base-pairing may facilitate the ligation reaction.

### 13.2.2.2 Heavy Chain Switch

The deletion mechanism underlying V-C gene fusion in the determinative phase [26, 33–36] also applies to a second type of somatic gene rearrangement that proceeds during B cell maturation (phase II of Fig. **13**.3) and leads to the well-known heavy chain switch. As already mentioned, the overwhelming number of B-lymphocytes exhibit membrane receptors of the IgM or IgM plus IgD isotypes, whereas during B cell maturation a stepwise transition from "early" (IgM, IgD) to "late" isotypes (IgG, IgA) is frequently observed. The switch from one isotype to another retains the original combination of V-regions and hence the antigen specificity. It concerns only the types of H-chain constant regions (cf. Table **13**.1) expressed in the particular cell clone.

In analogy to the V-, D- and J-gene segments, sites for intrachromosomal recombination exist in the 5′ flanking regions of all C-gene segments, too [switch regions (S), cf. Fig. **13**.6]. An important implication of the model shown in Fig. **13**.6 is the sequential deletion of increasing parts of the genetic material coding for the constant parts of the Ig H-chains. It is a remarkable fact that the gene segment order on mouse chromosome 12 in fact corresponds to the order of Ig isotypes expressed during B cell maturation. Direct proof of the deletion mechanism was obtained when analyzing DNA restriction fragments of selected plasmacytoma cell lines of BALB/c mice by Southern blotting hybridization (cf. **2**.6). A successive loss of gene fragments was observed if the cell types were aligned in the order of isotype expression (Table **13**.2).

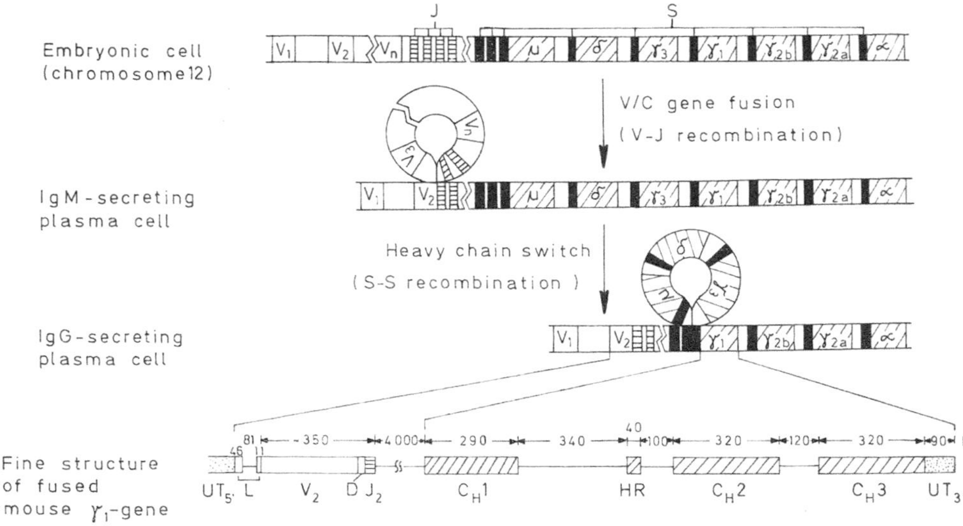


**Fig. 13.6**: Expression of Ig heavy chain genes is accompanied by two types of intrachromatid recombinations with deletions of the DNA in-between (compiled from Refs. [18, 37]).

The embryonic mouse chromosome 12 shows the set of V-genes *($V_1 \cdots V_n$)*, four J-gene segments separated by $\sim$ 7.5 kb from the group of constant region gene segments aligned in the order of their expression during development ($\mu$, $\delta$, $\gamma_3$, $\gamma_1$, $\gamma_{2b}$, $\gamma_{2a}$, $\alpha$). Each of them is flanked at the 5′-end by one or several ($\mu$-gene segment) DNA regions for recombination *(S)*. The *V/C gene fusion* involves recombination at the J-segments, one of which becomes part of the productive V-gene segment (cf. Figs. **13.4**, **13.5**). In fact, recombinatorial joining of one of the V-gene segments with one of the D-segments must precede V/C gene fusion via J-segments [27a, 34a]. These D-segments were not included into the figure. They map 5′ upstream and in close vicinity of the four J-segments. The *H-chain switch* proceeds by recombination between two S-regions. Though there are clear differences between the S-regions flanking $C_\mu$-, $C_\gamma$- and $C_\alpha$-genes, they are all composed of manyfold tandemly repeated units of 30–50 kb [31, 31a]. Expression of Ig classes may be developmentally regulated by S-region-specific switch proteins [31].
The fine structure of the $\gamma_1$-gene generated by the two types of fusion exhibits six exons *(boxes)*, i.e., (from *left* to *right*): (1) the sequence coding for the 5′ untranslated *($UT_{5'}$)* plus most of the leader part *(L)* of the mRNA, (2) the remainder of the leader part plus the sequence for the V-domain composed of the V-, D- and J-segments and (3–6) sequences for the $C_H1$, $C_H2$ and $C_H3$ domains and the hinge region (HR) in between. They are separated by five introns *(lines)* of varying lenghth (cf. the base-pairs indicated on top of the gene structure). Nothing is known so far about a putative further intron in the 5′ untranslated part as discussed in the legend to Fig. **13.4**. For a survey of the fine structure of other H-chain genes cf. Ref. [13].

Two types of results cannot be explained by this simple deletion model of H-chain switch:

1. Some cell clones express both IgM and IgD because a given V-domain can be combined simultaneously with two types of C-domains. This results from a joint transcription unit: V-D-J-7.5 kb-$C_\mu$-2.5 kb-$C_\delta$ (totally $\sim$30 kb), generating different primary transcripts [19, 20], e.g., by selective stop of transcription at internal poly (A) addition sites. By subsequent differential pre-mRNA processing the corresponding translation units are formed. This process is remarkable because both $\mu$- and $\delta$-H-chains themselves occur with several different forms of C-terminal amino acid sequences produced from the same transcription unit (cf. **13**.2.2.3).
2. Other results in conflict with the sequential deletion of $C_H$-segments (Table **13**.2) were reported [36a]. Thus, the mouse plasmacytoma MC101, though expressing IgG1, contains parts of the $S_\alpha$-region at the recombination site 5′ upstream of its $C_\gamma$-segments. Hence a twofold switch must be assumed in the history of this clone, one normal (IgM → IgA) and one reversed (IgA → IgG1).

**Table 13.2**: Evidence for sequential deletion of $C_H$-gene segments upon switching between Ig-isotypes (from Rabbitts et al. [33] and Cory et al.: Nature *285* (1980): 450–456).

DNA restriction fragments from plasmacytoma cells were analyzed for $C_H$ gene segments by hybridizing with the corresponding radioactively labelled cDNA (cf. **2**.6)

| Cell type (BALB/c mice) | Ig-type produced | $C_H$-gene segment expressed | Gene segments detected by hybridization assay[a] | | | | |
|---|---|---|---|---|---|---|---|
| | | | $\mu$ | $\gamma_1$ | $\gamma_{2b}$ | $\gamma_{2a}$ | $\alpha$ |
| HPC 76 | IgM | $\mu$ | + | + | + | + | + |
| MOPC 21 | IgG1 | $\gamma_1$ | – | + | + | + | + |
| MPC 11 | IgG2b | $\gamma_{2b}$ | – | – | + | + | + |
| MOPC 173 | IgG2a | $\gamma_{2a}$ | – | – | – | + | + |
| MOPC 315 | IgA | $\alpha$ | – | – | – | – | + |

[a] cf. the order of gene segments given in Fig. **13**.6; +, gene segment present, i.e., extent of hybridization roughly corresponds to that in embryonal cells if calculated per haploid genome; –, gene segment absent or detected at a much lower level than in embryonal DNA.

### 13.2.2.3 Ig Transcription Unit and Pre-mRNA Processing

In the transcription unit (Fig. **13**.4B) the genetic information coding for the Ig molecule is still interrupted by several non-coding sequences (introns) that are eliminated by splicing during pre-mRNA processing (steps C → D). For the Ig $\varkappa$-L-chain two introns have been shown to separate the activated V-segment from the C-segment (IC, 3–4 kb) and to interrupt the signal sequence (IB, ~100 bp) that codes for the N-terminal extension of the Ig molecule (cf. Fig. **13**.9). A third intron (IA) of about 6 kb must be assumed to occur in the 5′-untranslated sequence of the pre-mRNA if one considers the sizes of the primary transcript (about 10 kb) and of its processing intermediates.

The organization of the genetic material coding for immunoglobulins can serve as an illustrative example that functional parts (domains) of a protein may exist as preformed coding units (exons) at the DNA level separated from each other by intervening sequences (introns). This is especially prominent when considering the heavy chains of Ig molecules (compare Figs. **13**.1 and **13**.6). In fact, each domain of the H-chains, i.e., $V_H$, $C_H1$, HR, $C_H2$, $C_H3$ and eventually $C_H4$ is present also as distinct coding part in the transcription unit [18–20] and pre-mRNA. This type of sequence organization was found for all Ig genes so far studied in detail. It points to the possible evolution of Ig molecules by duplication of short gene pieces that contained the coding capacity of one functional domain of the recent Ig molecule. However, this seemingly clear picture of the gene organization reflecting the basic structural elements of the gene product and its origin during evolution cannot be applied equally well to other examples of split genes (cf. Fig. **7**.2).

Processing of $\varkappa$-pre-mRNP proceedes through the following RNA intermediates (Fig. **13**.4, transition C → D):

| 40S → | 24S → | 13S → | 13S |
|---|---|---|---|
| (~ 10 kb) | (~ 4.8 kb) | (1.3 kb) | (1.3 kb) |
| nucleus | | | cytoplasm |

From experiences with other systems (cf. Fig. **1**.8) it can be assumed that modification of the 5′-end of pre-mRNA by synthesis of the cap takes place already during transcription. Transition from the 40S to the 24S RNA intermediate means removal of the putative intron IA but also includes modification reactions, e.g., eventual methylations, addition of poly(A) to the 3′-end. By removal of the remaining two introns the mature translation unit (13S mRNA) is produced and enters the cytoplasm. It was estimated [38] that in plasma cells about 100 molecules pre-mRNA per gene are transcribed simultaneously generating pools of 540 molecules of 40S pre-

-mRNA (half-life ~10 min), 900 molecules of 24S pre-mRNA (half-life ~18 min), 4,200 molecules of nuclear 13S mRNA and 54,000 molecules of cytoplasmic 13S mRNA (half-life 20 hours).

Recent analyses of heavy chain-specific mRNA species in a murine pre-B cell clone (70 Z/3), which can be stimulated by mitogen (lipopolysaccharide) to express surface IgM, revealed the presence of four different types of mature $\mu$-chain mRNAs in stimulated as well as unstimulated cells [39]. They differ in size (2.1, 2.4, 2.7, and 3.0 kb respectively) as well as in their cytoplasmic localization. The 2.4 and 2.7 kb species are mainly associated with membrane-bound polysomes, whereas the other two are found in the soluble fraction only. There is experimental evidence that the mRNAs code for at least three different forms of $\mu$-chains: (1) cytoplasmic $\mu$-chains (2.1 kb mRNA), (2) $\mu$-chains that become part of secreted IgM (2.4 kb mRNA) and (3) $\mu$-chains destined to be incorporated into surface membrane IgM (2.7 kb mRNA).

This interpretation is supported by sequence analysis of the membrane-bound ($\mu_m$) and secreted ($\mu_s$) $\mu$-chains and of the pertinent segments of the Ig $\mu$-gene. In addition to their four constant domains, the $\mu_s$-chains contain a short C-terminal sequence of 20 amino acid residues including a cysteine, which serves the linkage of $\mu_s$-chains with the J chain in secreted IgM, and an asparagine site for attachment of a carbohydrate side chain (CHO):

—GlyLysProThrLeuTyrAsnValSerLeuIleMetSerAspThrGlyGlyThrCysTyrCOOH
(CHO attached at Asn; S—J chain attached at Cys)

The sequence is coded by a direct extension of the fourth constant domain. In contrast, $\mu_m$-chains have a different and much longer C-terminal extension of the fourth domain. It is composed of a short extracellular hydrophilic linker of 12 amino acid residues (with 6 glutamic acid residues), a very hydrophobic intramembrane part with 26 amino acid residues and a short tail —LysValLysCOOH on the cytoplasmic site. The coding sequence for this C-terminal part is contained in two additional exons separated by a large intron of 1,850 bp from the main coding part of the $\mu$-gene. Thus, biosynthesis of $\mu$-chains starts with a single primary transcript. The subsequent pre-mRNA processing either removes all coding extensions of the fourth domain ($\rightarrow$ 2.1 kb mRNA for cytoplasmic $\mu$-chians) or leaves the extension for the 20 amino acid residues of secreted IgM ($\rightarrow$ 2.4 kb mRNA) or removes this extension and instead adds, by two splicing events, the two extra exons for 41 amino acid residues of the membrane-bound IgM ($\rightarrow$ 2.7 kb mRNA).

An alternative explanation for the origin of the three types of $\mu$-specific mRNA is a control of transcription termination at different sites downstream of the coding segment for the fourth constant domain. At any rate, the sequential appearance of cytoplasmic, membrane-bound and secreted $\mu$-chains in the life span of a B cell clone must include corresponding changes in the gene expression program. Cell hybridization experiments point to a positive regulatory factor involved in these $\mu$-chain shifts.

### 13.2.3 Generation of Antibody Diversity

For many years research on the immune system was dominated by the central question of how the formation of a seemingly unlimited number of different antibodies can be explained. For mammals the total of antibody specificities was estimated to amount to $10^6$–$10^8$. Two hypotheses have been proposed and frequently modified. According to the "germline hypothesis" diversity is based on a corresponding multiplicity of genes already present in the germ cells[2], while the hypothesis of "somatic

[2] Hood, L., & D. W. Talmage: Science *168* (1970): 325–334.

mutation" supposed only a few antibody coding genes of the germ cells to mutate during individual development.[3]

The detailed analyses of the genetic systems coding for immunoglobulins, as well as of somatic gene rearrangements preceding gene expression, have clearly shown that both hypotheses reflect distinct parts of the mechanisms that contribute to the generation of the antibody diversity:

1. There are three sets with many different V-gene segments ($10^2$–$10^3$ copies each) that code for the large N-terminal part of the Ig V domains. In addition, small but functionally important parts at the C-terminal ends of V-domains are encoded by short extrasegments existing in multiple copies. These are the J-segments of L-chains and the J- and D-segments of H-chains (cf. **13**.2.2).
2. Several somatic mechanisms increase the genetic diversity encoded in the germ-line by a factor of $10^3$–$10^5$. These are:
   - free combination of V-, D- and J-gene segments during gene fusion,
   - increase of V-segment multiplicity by somatic mutation, intrafamily recombination and flexibility of fusion frames between V-, D- and J-segments,
   - random activation of a V/J–C pair of gene segments for each chain and free combination of H- and L-chains to the functional immunoglobulin.

The existence of each partial mechanism is more or less well supported by experimental results. However, their relative contributions to the total diversity of a given antibody population are obscure. Doubtless a practically unlimited multiplicity of antibody specificities is realized with relatively little genetic material. This principle is further emphasized by an unexpected peculiarity of antibodies: they are polyfunctional (cf. Figs. **13**.3 and **13**.8), i.e., due to the spatial arrangement of many functional groups in the antigen binding pocket (Fig. **13**.2) several non-related determinants can be fixed alternatively. Hence, the high selectivity of the overall antigen response results from the sum of specificities of an antibody population. The principle is illustrated in Fig. **13**.3 on the basis of four different antigens, but it is estimated that up to 100 structurally unrelated compounds are bound by each antibody. Experimental evidence for this polyfunctionality and its influence on the humoral antigen response will be discussed in connection with Fig. **13**.8.

### 13.2.4 Control of Gene Fusion

The decisive role of two types of gene rearrangements for expression of immunoglobulin genes is experimentally well grounded. These are the V/C-gene fusions and the H-chain switch (cf. **13**.2.3). It is very likely that both lead to irreversible changes of the genetic material because distinct parts between the fused gene segments are deleted (Fig. **13**.6). This mechanism contrasts to other gene rearrangements, e.g., during mating type switch in yeast (cf. **4**.3.3) or flagellar phase variation in *Salmonella* (cf. **4**.3.2).

Though chance plays a crucial role for the generation of antibody diversity by V/C gene fusion, it is evident from several observations that the overall process is tightly controlled within the genetic and developmental frame of the immune cell system.

1. Comparison of antibody diversity of two well characterized inbred strains of mouse (BALB/c and NZB) demonstrates distinct differences [29]. That means, different parts of the total repertory of possible V/C-combinations are expressed dependent on ill-defined genetic and epigenetic factors influencing either the determination phase (V/C-gene fusions) or the maturation and propagation of distinct cell clones

[3] Lederberg, J.: Science *129* (1959): 1649–1653; Smithies, O.: Science *157* (1967): 267–273.

after antigen contact. Genes of the major histocompatibility complex (MHC-complex on mouse chromosome 17) may be involved (cf. **13**.3.1).

2. The gene rearrangements are limited to distinct phases of B cell development. V/C-gene fusion proceeds in phase I whereas the heavy chain switch is part of the B cell maturation (phase II, Fig. **13**.3), i.e., it requires activation of the B resting cells by antigen.
3. From genetic analyses it has been well known for many years that the information of only one of the two homologous chromosomes coding for immunoglobulins is expressed at the protein level *(allelic exclusion)*. A straight forward explanation would be that gene fusion somehow is restricted to one of the two chromosomes. Though investigations on the genomic DNA and composition of Ig-specific pre-mRNA in nuclei of various plasmacytoma cell lines essentially supported this explanation [30, 30a], they simultaneously revealed a much more complicated picture.

Three different types of gene rearrangements were detected for the mouse chromosome 6 coding for the $\varkappa$-L-chain [30]: (1) $\varkappa^0$: the germline configuration with V- and J/C-gene segments apart. Surprisingly, the J/C-gene segment is transcribed into an 8.4 kb unstabile pre-mRNA fragment that is degraded in the nucleus. (2) $\varkappa^+$: the productive rearrangement as outlined in Fig. **13.4**. The primary transcript is rapidly processed to functional $\varkappa$-L-chain mRNA. (3) $\varkappa^-$: a defective rearrangement may either lead to an erroneous fusion between V- and J/C-gene segments or to deletion of the J/C-gene segment. In the first case the transcription product cannot be processed properly in the second case no transcription was observed. Hence, we are left with totally six different genotypes:

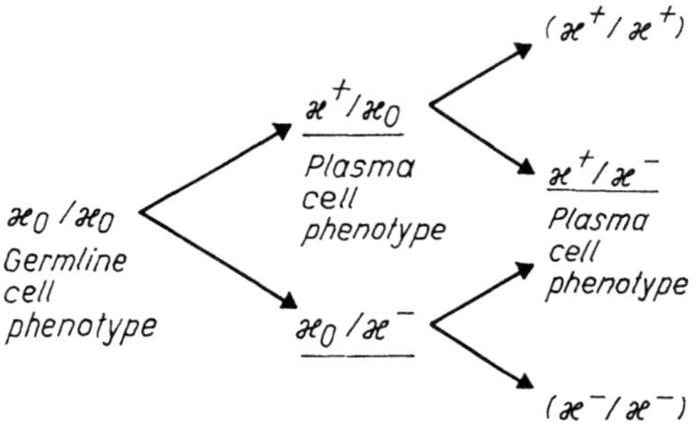


A total of 30 plasmacytoma cell lines were analyzed [30]. They belonged to one family ($V_{K21}$) with respect to the type of $\varkappa$-L-chain produced. 15 were found to be of the $\varkappa^+/\varkappa_0$ type (allelic exclusion due to restriction of gene fusion to one chromosome) whereas 15 were of the $\varkappa^+/\varkappa^-$ type (allelic exclusion due to defective rearrangement on one of the two chromosomes). Existence of the $\varkappa^0/\varkappa^-$ and $\varkappa^-/\varkappa^-$ type can of course not be detected by these methods based on the analysis of Ig-producing plasmacytoma cell lines. Further, it can not be ruled out that some cell lines of the $\varkappa^+/\varkappa^-$ type in fact are of the rare $\varkappa^+/\varkappa^+$ type expressing a second Ig molecule unrelated to $V_{K21}$.

4. V/C gene fusion evidently proceeds sequentially in the determinative phase during which immune competent B cell clones are generated. The first stage of immunoglobulin gene activation is V/C gene fusion on the chromosome carrying the heavy chain gene segments. The pre-B lymphocytes produce short-lived $\mu$-chains, which reside in the cytoplasm, but no light chains (cf. **13**.3.2). Very soon after the V/C fusion on one of the light chain coding chromosomes synthesis of Ig light chains is initiated and, consequently, complete IgM appear in the outer membranes (immune competent B lymphocytes, cf. Fig. **13**.3). A recent model [30a] assumes that, in fact, rearrangement of the light chain genes again proceeds sequentially. In mouse it starts with chromosome No. 6 containing the genetic material for the

ϰ-L-chains. If the V/C fusion is productive, i.e., one of the possible fusion frames shown in Fig. **13**.5 have been used, integration of intact IgM into the outer membranes stops gene rearrangements on other chromosomes. If, however, the first V/C gene fusion is defective, the second chromosome 6 and afterwards, the λ-L--chain coding chromosomes 16 are rearranged as well until a productive combination of heavy and light chains is achieved.

The evidence for this model is deduced from the analysis of λ-L-chain expressing myeloma cell lines. It was found that ϰ-genes are frequently rearranged as well in such λ producers and that these cell lines may contain abundant and defective ϰ-mRNA encoding aberrant ϰ-related polypeptides which are rapidly degraded.

In summary, the results discussed in points (3) and (4) show that rearrangements of the Ig genes frequently fail. Genes defective in transcription, pre-mRNA processing or translation may arise. Thus, the term "*allelic exclusion*" stands for a complex phenomenon. Though only a single type of functional L-chain coded by one chromosome is produced under normal conditions Ig genes on other chromosomes may be rearranged and even expressed. But due to mistakes in the fusion frames gene expression is defective [28a, 30, 30a].

## 13.3 The Humoral Immune Response

### 13.3.1 Triggering of the Immune Response – Signal Transformation

As discussed in the preceding sections, B lymphocytes arise from stem cells in the course of quantal mitoses. Precursors of the immune competent B cells shown in Fig. **13**.3 are "pre-B cells" with cytoplasmic μ-chains [51]. Afterwards cells appear with a special type of short-lived ($t_{1/2} = 45$ min), monomeric IgM molecules in the outer cell membrane. In the course of numerous proliferative mitoses they develop to immune competent B cells that remain in the $G_0$ phase (resting cells). Cells of this type are devoid of DNA synthesis and have long-lived ($t_{1/2} = 20$ h) monomeric IgM antibodies in their outer cell membranes. After contact with antigens and/or factors inducing cell division (mitogens) they are able to go on proliferating and to mature to plasma cells that have no or only few membrane-bound antigen receptors and are specialized for mass secretion of one antibody type.

The first part of the reaction sequence in phase II is a complex signal transformation that, with many antigens,[4] demands interaction between three cell types: B cells, special T lymphocytes (helper cells) and macrophages (Fig. **13**.7). Besides antigen receptors (e.g., of the IgM-type) and mitogen receptors in the outer cell membrane of B cells, a further membrane glycoprotein plays an important role for signal transformation, the so-called Ia-protein [42, 44]. It is evidently an integral part of the membrane complexes formed after antigen contact by both B cells and macrophages. In Fig. **13**.7 two current models of cell-cell interaction for thymus dependent antigens are shown: (1) The antigen binds first to T helper cell receptors and thus is presented in an immunogenic form to the B resting cell. Simultaneously mitogenic signals, formed by the stimulated T helper cell act on the activated mitogen receptors at the

---

[4] In contrast to these so-called thymus-dependent antigens (e.g., proteins) there are thymus--independent ones (e.g., polysaccharides, lipopolysaccharides) capable of stimulating B resting cells directly, because they exhibit simultaneously antigen and mitogen activities.

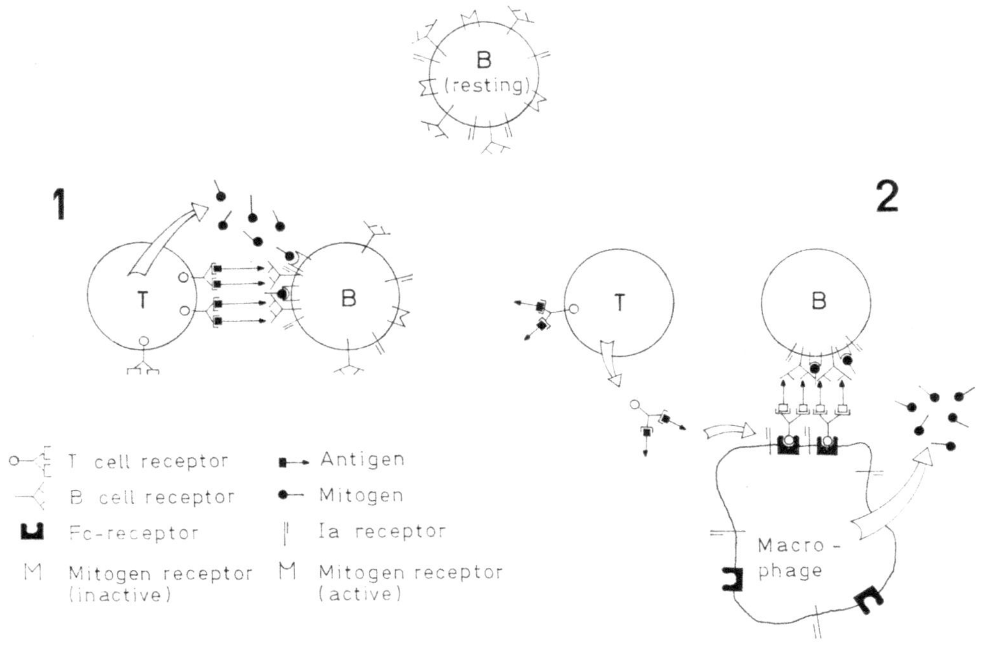


**Fig. 13.7:** Models of cell-cell interactions that promote the triggering of the immune response.
By antigen stimulation the B resting cell is stimulated to proliferate and mature to the antibody-secreting plasma cell. For stimulation most antigens require interaction with T cells and/or macrophages. Two types of cell-cell interactions are discussed.
**1.** The antigen binds primarily to T helper cells that present the antigen to the B cell receptors and simultaneously produce a mitogenic signal. Essential for B cell stimulation is the formation fo microclusters in the membrane containing antigen receptors, activated mitogen receptors and Ia receptors.
**2.** The primary contact of the antigen with the T helper cell results in secretion of the antigen--receptor complex that is presented to the B cell by a macrophage. Aggregation of membrane receptors and stimulation of B cell proliferation by a mitogen is basically similar to the situation in part 1.

B cell surface and trigger proliferation. In Fig. **13**.7/2 the primary antigen-receptor complex is released from the T cell membrane. It binds subsequently to special membrane proteins (Fc-receptors) of macrophages [47] and thus initiates a similar sequence of cell-cell interaction as shown in (1), but this time involving a macrophage.

Ia-proteins are coded by genes that are part of a complex locus called major histocompatibility complex (H-2 locus on mouse chromosome 17, HL-A locus on human chromosome 6). Besides the antibody coding gene systems the MHC locus was shown to be of crucial importance for all aspects of the immune response (cf. the summaries [8–11, 44, 46]). Among numerous other genes at least five loci for Ia-glycoproteins have been mapped in the MHC-region of mouse. Ia-protein consists of two subunits ($\alpha$, $M_r$ 35 kd and $\beta$, $M_r$ 28 kd). The well-known immune response genes (Ir-genes) located in the same gene complex are probably identical with the Ia-genes.

A striking and fast partial reaction of the signal transformation chain is a recompartmentation of proteins involved in the recognition complex of B cell membranes (antigen receptors, Ia-proteins, mitogen receptors). At first these move freely in the membrane fluid. But very soon after antigen contact they aggregate to microclusters and patches [48, 49] fixed in the membrane by interaction with components of the cytoskeleton (actin, α-actinin). It is assumed that, similar to the signal transformation reactions observed for many hormones (cf. Fig. **10**.6), clustering of proteins in the B cell membrane triggers formation of an intracellular "second messenger" (e.g., cAMP or cGMP) which might activate corresponding protein kinases.

The cooperation of the three cell types in triggering the humoral immune response serves as another interesting example for intercellular interactions during cell differentiation (cf. **4**.6). It includes not only the mutual activation in production and processing of mitogenic factors, that promote cell division in distinct clones of B resting cells, but also leads to an increased proliferation of macrophages and T helper cells distinguished by the corresponding Ia-proteins (macrophages) and other membrane glycoproteins (T helper cells) that may be structurally related to the immunoglobulins. Moreover, signals from the T helper cells trigger the proliferation of different T regulatory cells, which, e.g., by inhibitory signals, stop further proliferation of the cell clones (B cells, T helper cells) involved in the immune response. Hence, the whole system is autoregulating. The immune reaction proceeds in a network of cells connected with each other by surface proteins involved in direct cell-cell contact, but also in the reception and transformation of intercellular signals (antigens, antibodies, mitogens, hormone-like factors). Though many genetic and functional details of this cell interaction are still obscure most studies on autoimmune diseases indicate that the normal failure of the immune system to respond to self-antigens in fact is due to the strict regulation of immune homeostasis by the cell-cell interactions within the T cell compartment (T helper and suppressor cells, cf. the summaries [44 and 46]).

### 13.3.2 Formation of Clones of Antibody Secreting Plasma Cells

At the end of the determinative phase (cf. Fig. **13**.3) we find, e.g., in the human body, about $10^{11}$ immune competent B lymphocytes being able, by virtue of their particular cell surface receptors, to react on $10^6$–$10^8$ different antigens with cell proliferation and production of specific antibodies. By antigen contact distinct cell clones are selected and triggered to proliferate (clone selection theory of Burnet [1]). Every antigen is confronted with a multiplicity of such clones. That is illustrated schematically in Fig. **13**.3 with a few examples. Thus, contact of B resting cell clones with a single antigen (A) leads to the secretion of a great number of different antibodies that only share one property: they are all A-specific. But, due to the polyfunctionality (cf. **13**.2.3), each antibody is capable of combining with many structurally divergent determinants. However, the titre of these non-A related specificities is usually too low to be detected, except with special analytical procedures. An illustrative example is given in Fig. **13**.8.

By isoelectrofocusing of rabbit antibodies against antigen A up to 100 A-specific protein bands are detected. In fact the heterogeneity is much greater since each band contains several proteins. If the same electropherogram is treated with a structurally non-related antigen B (cf. legend to Fig. **13**.8), a few bands are found reacting with antigen B as well. By the following immunization of the same animal with antigen B it can be demonstrated that the latter result is not an artefact but a direct consequence of the polyfunctionality of antibodies. A two-phase reaction is observed in this case. In the beginning great quantities of antibodies appear with A/B-specificity (up two 100-fold more than during the first immune response, cf. lane **3**). Only in the late phase of the second immunization the usual pattern of antibody heterogeneity is again observed (lane **4**).

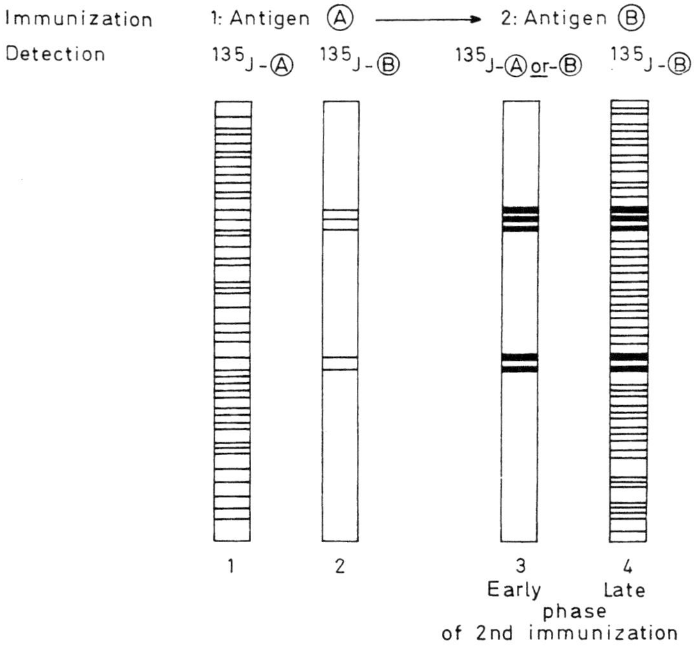


**Fig. 13.8:** Analysis of antibody heterogeneity by isoelectric focusing and demonstration of polyfunctionality (from Richards et al. [12], modified).

Rabbits were first immunized with antigen A and after some time with antigen B. Antibodies obtained from blood serum were separated on polyacrylamide gels by isoelectric focusing and detected with iodine-labelled antigen A or B. Antigen A contained dinitrophenol as determinant group, whereas either ribonuclease or vitamin $K_1$ coupled to a carrier protein were used as antigen B.

*Lanes 1, 2:* After the first immunization about 100 bands show A-specificity and five of these have B-specificity, i.e., five of the numerous cell clones with A-specificity produce antibodies recognizing the structurally unrelated B-antigen.

*Lanes 3, 4:* The second immunization results (1) in a very rapid production of the five types of antibodies already found during the first immunization (early phase of B-response) and (2) only in the later phase the total heterogeneity of B-specific antibodies can be observed.

The experiment in Fig. **13**.8 shows a further characteristic phenomenon of the immune reaction frequently referred to as "immunological memory". After first contact of an organism with an antigen, a longer interval passes before the antibodies in the body fluids have reached their maximal concentrations. The response then rapidly declines. If after a certain time immunization is repeated with the same antigen, antibody synthesis is stimulated distinctly faster: the system "remembers" the first antigen contact. Basis of this remembrance is a considerable number of special B-cells with augmented antigen receptors in their outer membranes (memory cells) that are produced, together with plasma cells, during the first antigen response. Hence, the increased production of A/B-specific antibodies in the second immune response (Fig. **13**.8B) is based on selective proliferation and specialization of those clones of memory cells which possess A/B-specific receptors and developed in the course of the first immune response.

Maturation of B cell clones is a multistep process with numerous cell divisions [51]. An early intermediate are fastly proliferating plasmablasts. They contain the usual monomeric IgM receptors in their outer membranes but already start to secrete pentameric IgM. This requires three types of changes in the gene expression program:

1. Synthesis of the joining chain is triggered. This J-chain can be considered as a structural backbone of the IgM pentamers (cf. Table **13**.1). Whereas production of $\mu$-chains starts very early in the development of immune competent cells, secretion of IgM pentamers is a much later event limited to the maturation phase and bound to the synthesis of J-chains.

2. There is now clear evidence for $\mu$-chains, that are part of the membrane IgM and those of the secreted IgM, to be different with respect to their molecular weights and to result from two different splicing programs during pre-mRNP processing (cf. **13**.2.2). $\mu$-chains of membrane-bound IgM contain an additional hydrophobic C-terminal extension of their $C_H4$-domain. The exon coding for this extension is eliminated during pre-mRNP processing that leads to formation of mature mRNA for $\mu$-chains to be secreted in the IgM pentamers. Hence evidently one transcription unit for $\mu$-chains, that is generated during V/C-gene fusion, codes for two or even more types of mature mRNA, whose translation leads to the synthesis of $\mu$-chains with distinct differences in their cellular compartmentation (secreted vs, membrane--bound vs. cytoplasmic). The latter type is the earliest form of $\mu$-chains to be observed during the determination phase in cells of the bone marrow that precede the B resting cells in the B cell lineage (cf. **13**.2.2).

3. The translation from two different types of mRNA causes not only differences in compartmentation but also in postsynthetic modification of both $\mu$-chain types by glycosylation [51]. Membrane-bound IgM monomers are only modified by the core building blocks of the typical oligosaccharide moiety (cf. Fig. **13**.10), i.e., by N-acetyl glucosamin and mannose. In contrast, the carbohydrate chains of secreted IgM pentamers are extended shortly before secretion by addition of galactose, neuraminic acid and fucose.

Already with the proliferation of plasmablasts a gradual increase begins of the components of the gene expression chain that are necessary for mass Ig-synthesis (membranes of rough and smooth ER, carbohydrate precursors, glykosyl transferases etc.). Besides, due to the heavy chain switch (cf. **13**.2.2), there is a transition to other Ig-types of the same specificity. Finally, the plasma cells lose most of their membrane receptors. Maturation of B cell clones therefore essentially means functional adaptation of the protein synthetic apparatus and specialization for mass secretion of a single immunoglobulin species.

### 13.3.3 Synthesis of Antibodies by Membrane-Bound Ribosomes

The last stage of the humoral immune response is the mass synthesis and secretion of antibodies by distinct plasma cell clones produced in phase II. Insight into some details of this process could only be gained after the corresponding Ig-specific mRNA had been isolated from myeloma tumor cells. In fact, in vitro translation experiments with $\varkappa$-L-chain mRNA gave the first convincing evidence for the mechanism of sorting out distinct mRNP species to be translated by membrane-bound polysomes. As outlined in Chap. **4** (cf. Table **4**.10) a hydrophobic, N-terminal extension (signal peptide) of the nascent polypeptide chain mediates the association of the ribosome--mRNP complex with the rER membranes (Fig. **13**.9). The $\varkappa$-L-chains segregate co-translationally into the cisternes of the rough ER, where they undergo first steps of an extended proteinogen processing. This includes the endoproteolytic removal of the N-terminal signal peptide, formation of disulphide bonds, assembly of two

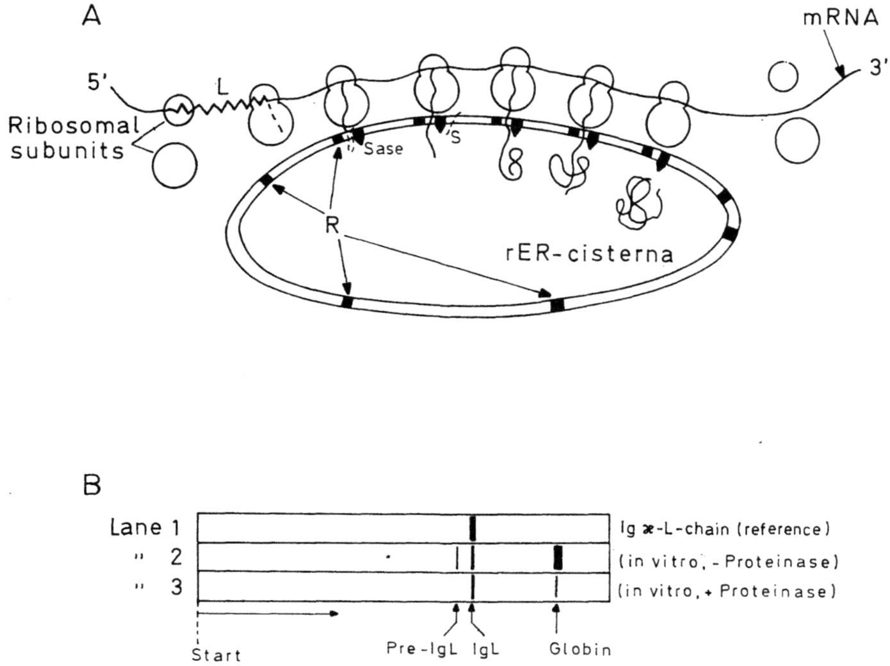


**Fig. 13.9**: Synthesis of immunoglobulins by membrane-bound ribosomes (from Blobel & Dobberstein [50], modified).

**A**: Initiation and first steps of Ig-synthesis proceed on free ribosomes. After a considerable part of the mRNA leader sequence (L coding for about 20 amino acids, cf. Table **4**.10) has been translated the translation complex, by means of the emerging signal peptide *(S)*, attaches to receptor proteins *(R)* in the rER membrane. Hence, the newly formed Ig molecule is injected co-translationally into the ER cisterna, where the signal peptide is soon removed by a membrane-bound, rER-specific endoprotease (signalase, *Sase*).

**B**: Comparison of in vitro translation of purified mRNA of the mouse ϰ-L-chain and globin mRNA in the presence of rER membrane vesicles from dog pancreas. A reticulocyte lysate was used as translation system. Translation products were separated by acrylamide gel electrophoresis. *Lane 1,* reference protein Ig ϰ-L-chain. *Lane 2,* products of in vitro translation without proteinase degradation. Besides the fast moving globin there is a strong band of Ig-L-chain molecules and a faint band representing unprocessed pre-Ig-L-chain molecules containing the N-terminal signal peptide. *Lane 3,* same as lane 2 but with proteinase degradation. Only the molecules of Ig-L-chain injected cotranslationally into the ER-vesicles are resistant to proteinase degradation.

L- and two H-chains to the $H_2L_2$-immunoglobulin and stepwise modification of distinct amino acid residues by glycosylation (Fig. **13**.10).

Synthesis of an L-chain takes about 40 s, that of an H-chain about 60–90 s. A small portion of L-chains always occurs in free form. This intracisternal L-chain pool evidently serves the assembly of Ig precursor tetramers by combining with the nascent H-chains and forming disulphide bonds [54].

Though total synthesis of the polypeptide chains takes only 90 s, mature Ig-molecules can be detected not earlier than after 30 min. Following the biosynthetic pathway of radioactively labelled Ig-precursors they remain about 5 min at the site of synthesis (rER). Then they pass to the smooth ER, to the Golgi apparatus (GA) and finally to the outer cell membrane. Transport vesicles between the different compartments are presumably specialized forms of coated vesicles that also serve

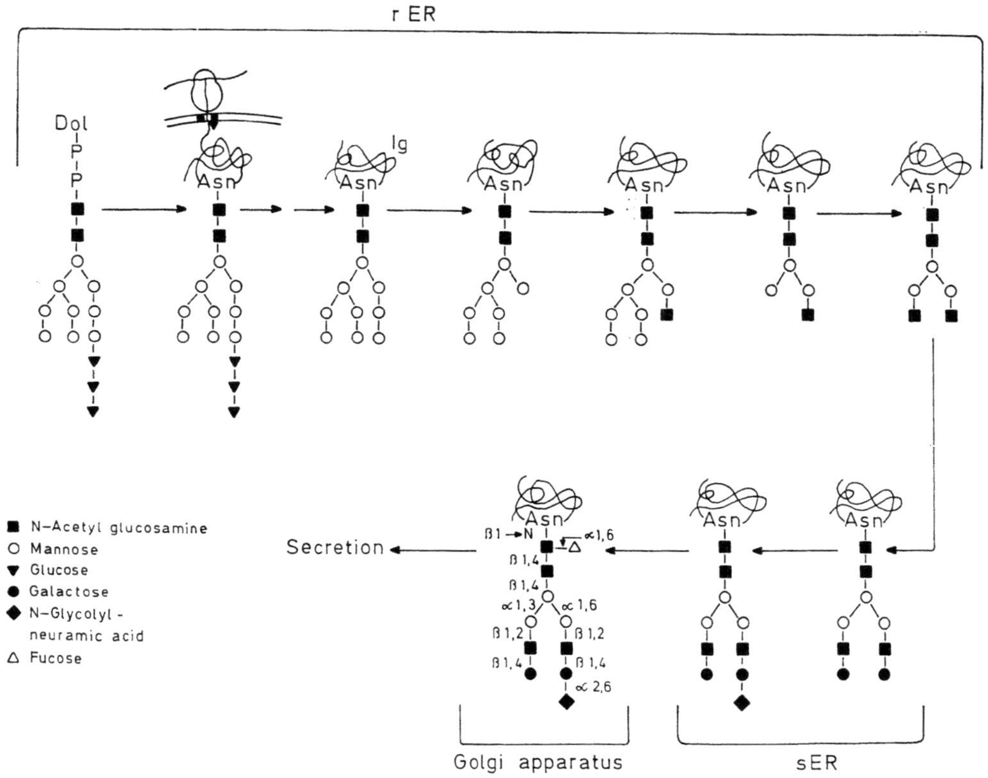


**Fig. 13.10**: Processing of high mannose oligosaccharides to form the complex carbohydrate side chains of immunoglobulins (from Tabas et al. [52] and Kornfeld et al. [53], modified).

Transfer of a preformed oligosaccharide chain from the universal donor molecule (dolichol pyrophosphate, Dol-P-P) proceeds in statu nascendi of the Ig heavy chain. Acceptor amino acid residue of the Ig $\gamma$-chain is the asparagine residue in position 297 (cf. Fig. **13**.1). Processing of the oligosaccharide chain by degradation and then stepwise addition of new sugar residues accompanies the sequential transition of the maturing Ig molecule from the rER to the sER, to the Golgi apparatus and finally to the secretory vesicles. The type of glycosidic bonds within the completed carbohydrate moiety are specified in the last part of the figure.

the internalization of extracellular proteins (cf. Fig. **4**.1). Already during translation the nascent polypeptide chains are modified by transfer of the preformed basic element of the carbohydrate chain (Fig. **13**.10). Mannose-N-acetylglucosaminoligosaccharide-dolichol phosphate functions as co-substrate. Generation of the final carbohydrate moiety by partial degradation and subsequent attachment of the branch carbohydrates reflects a general mechanism of glycoprotein synthesis in fungi, plants and animals. Numerous glycosyl transferases in the different membrane types (rER, sER, GA) may be essential control elements of the directed intracellular transport of Ig-chains from their site of translation to secretion. But details of this control are largely unknown.

## 13.4 Conclusions

During evolution vertebrates acquired a polycellular system (immune system) that is able, by means of specialized proteins (immunoglobulins, Figs. **13.1** and **13.2**) integrated into the cell surface membrane or secreted into the body fluids, to recognize, to inactivate and, finally, to degrade not-self high molecular weight materials, e.g., proteins, polysaccharides or whole cells, having invaded the organism. As far as the humoral immune response is concerned two phases of cell specialization in the immune system are observed (Fig. **13.3**):

1. In the determination phase stem cells of the bone marrow generate and maintain a population of about $10^{11}$ immune competent B lymphocytes (B resting cells) that belong to $10^6$–$10^8$ different clones if classified with respect to their specific membrane-bound antigen receptors. The great multiplicity of receptor types and, correspondingly, of antibodies results from free combination of gene segments by gene fusion (Figs. **13.4**–**13.6**) and from somatic alterations of the gene pool coded by the germline cells.
2. Whereas the determination phase requires no antigen for clone generation, the second phase (B cell maturation) is based on clone selection after antigen contact. After a complex signal transformation that frequently involves the interaction of immune competent B resting cells with T helper cells and eventually macrophages (Fig. **13.7**) the selected B cells undergo many cell divisions and specialize to Ig-secreting plasma cells. In most cases maturation is characterized by another gene rearrangement, the heavy chain switch (Fig. **13.6** and Table **13.2**).

Analysis of the programmed expression of the immunoglobulin genes accompanying cell specialization in the immune system has provided important new insights into cell differentiation mechanisms in general. Three points deserve special notice: (1) Irreversible chromosome rearrangements with deletion of a considerable part of the genetic material generate the Ig transcription units (cf. Figs. **13.4** and **13.6** and Table **13.2**). (2) The translation unit derives from an elaborate pre-mRNA processing (cf. **13.2.2.3**). In two instances there is good experimental evidence for the regulated production of two or three different translation units (mature mRNA's) out of a single primary transcript (pre-mRNA). Thus, differential processing may influence not only the Ig isotype expressed (cf. **13.2.2.2**) but also the intracellular localization of the Ig molecules (cf. **13.2.2.3**). (3) A multistep co- and post-translational processing of the Ig precursor polypeptides accompanies the maturation and correct positioning of the Ig molecules (cf. Figs. **13.9** and **13.10**).

Though many fascinating details of the immune system especially concerning the peculiarities of Ig gene structure and generation of antibody diversity have recently evolved, many important problems remain unsolved. These are the mechanism of the intrachromosomal rearrangements, factors and regulatory mechanisms that govern the generation and maturation of immune competent cell clones or the detailed analysis of the structure and function of other gene systems involved, e.g., the MHC complex. But above all it is the problem of intercellular interactions required for T cell education (cf. Sec. **13.2.1**), for signal transformation after antigen contact (Fig. **13.7**), for maintenance of immune homeostasis (cf. Sect. **13.3.1**) as well as for all forms of cell-mediated immune response that were not discussed in this chapter.

## 13.5 Selected References


*General References*

1. Burnet, F. M.: Self and Not-Self. Melbourne University Press 1969.
2. Roitt, J.: Essential Immunology. 2nd edn., Blackwell Scientific Publications. Oxford 1974.
3. Nisonoff, A., J. E. Hopper & S. B. Spring: The Antibody Molecule. Academic Press, New York 1975.
4. Cold Spring Harbor Symposia on Quantitative Biology, Vol. XLI: Origin of Lymphocyte Diversity. Cold Spring Harbor Lab., Cold Spring Harbor, New York 1977.
5. Immunological Reviews (Transplantation Reviews). Editor: Möller, G. (ed.), Munksgaard. Copenhagen.
6. Hilschmann, N., W. Palm, H. U. Barnikol, S. Barnikol-Watanabe, J. Bertram, J. Horn, M. Engelhard, M. Schneider & L. Dreker: Myeloma proteins and cell differentiation. Proc. XI Intern. Cancer Congress, Vol. 1: Cell Biology and Tumor Immunology. Elsevier, Amsterdam 1975.
7. Amsel, L. M., & R. J. Poljak: Three-dimensional structure of immunoglobulins. Ann. Rev. Biochem. *48* (1979): 961–997.
8. Klein, J., A. Juretic, C. N. Baxevanis & Z. A. Nagy: The traditional and a new version of the mouse H-2 complex. Nature *291* (1981): 455–460.
9. Klein, J.: The major histocompatibility complex of the mouse. Science *203* (1979): 516–521.
10. Uhr, J. W., J. D. Capra, E. S. Vitetta & R. G. Cook: Organization of the immune response genes. Science *206* (1979): 292–297.
11. Snell, G. D.: Recent advances in histocompatibility immunogenetics. Adv. Genetics *20* (1979): 291–355.
12. Richards, F. F., W. H. Konigsberg, R. W. Rosenstein & J. M. Varga: On the specificity of antibodies. Science *187* (1975): 130–137.
13. Nover, L., R. F. Richter & E. Serfling: Structure and expression of immunoglobulin genes. Biol. Zentralbl. *101* (1982): in press.
14. Seidman, J. G., A. Leder, M. Nau, B. Norman & P. Leder: Antibody diversity. Science *202* (1978): 11–17.
15. Köhler, G. & C. Milstein: Continuous culture of fused cells secreting antibody of predefined specificity. Nature *256* (1975): 495–497; Milstein, C.: Monoclonal antibodies from hybrid myelomas. Proc. R. Soc. Lnd. B *211* (1981): 393–412.

15a. Adams, J. M.: The organization and expression of immunoglobulin genes. Immunol. Today *1* (1980): 10–17.

*Genetics, Gene Structure, Gene Rearrangements*

16. Hengartner, H., T. Meo & E. Müller: Assignment of genes for immunoglobulin *K* and heavy chains to chromosomes 6 and 12 in mouse. Proc. Natl. Acad. Sci. USA *75* (1978): 4494–4498.
17. Croce, C. M., M. Shander, J. Martinis, L. Cicurel, G. G. D'Ancona, T. W. Dolby & H. Koprowski: Chromosomal location of genes for human immunoglobulin heavy chains. Proc. Natl. Acad. Sci. USA *76* (1979): 3416–3419.
18. Honjo, T., M. Obata, Y. Yamawaki-Kataoka, T. Kataoka, T. Kawakami, N. Takahashi & Y. Mano: Cloning and complete nucleotide sequence of mouse immunoglobulin $\gamma 1$ chain gene. Cell *18* (1979): 559–568.
19. Maki, R., W. Roeder, A. Traunecker, C. Sidman, M. Wabl, W. Raschke & S. Tonegawa: The role of DNA rearrangement and alternative RNA processing in the expression of immunoglobulin delta genes. Cell *24* (1981): 353–365.
20. Moore, K. W., J. Rogers, T. Hunkapiller, P. Early, C. Nottenburg, I. Weissmann, H. Bazin, R. Wall & L. E. Hood: Expression of IgD may use both DNA rearrangement and RNA splicing mechanisms. Proc. Natl. Acad. Sci. USA *78* (1981): 1800–1804.
21. Roeder, W., R. Maki, A. Traunecker & S. Tonegawa: Linkage of the four $\gamma$ subclass heavy chain genes. Proc. Natl. Acad. Sci. USA *78* (1981): 474–478.

21a. Bentley, D. L., T. H. Rabbitts: Human $V_\varkappa$ immunoglobulin gene number: Implications for the origin of antibody diversity. Cell *24* (1981): 613–623.

22. Sakano, H., K. Hüppi, G. Heinrich & S. Tonegawa: Sequences at the somatic recombination sites of immunoglobulin light-chain genes. Nature *280* (1979): 288–294.
23. Seidman, J. G., E. E. Max & Ph. Leder: A *K*-immunoglobulin gene is formed by site-specific recombination without further somatic mutation. Nature *280* (1979): 370–375.
24. Valbuena, O., K. B. Marcu, M. Weigert & R. P. Pery: Multiplicity of germline genes specifying a group of related mouse K chains with implications for the generation of immunoglobulin diversity. Nature *276* (1978): 780–784.
25. Blomberg, B., A. Traunecker, H. Eisen & S. Tonegawa: Organization of four mouse $\lambda$ light chain immunoglobulin genes. Proc. Natl. Acad. Sci. USA *78* (1981): 3765–3769.
26. Max, É. E., J. G. Seidman & Ph. Leder: Sequences of five potential recombination sites encoded close to an immunoglobulin *K* constant region gene. Proc. Natl. Acad. Sci. USA 76 (1979): 3450–3454.
27. Gough, N. M. & O. Bernard: Sequences of the joining region genes of immunoglobulin heavy chains and their role in generation of antibody diversity. Proc. Natl. Acad. Sci. USA *78* (1981): 509–513.
27a. Early, P., H. Huang, M. Davis, K. Galame & L. Hood: An immunoglobulin heavy chain variable region gene is generated from three segments of DNA: $V_H$, D and $J_H$. Cell *19* (1980): 981–992.
28. Crews, St., J. Griffin, H. Huang, K. Calame & L. Hood: A single $V_H$ gene segment encodes the immune response to phosphorylcholine: Somatic mutation is correlated with the class of the antibody. Cell *25* (1981): 59–66.
28a. Early, Ph. & L. Hood: Allelic exclusion and nonproductive immunoglobulin gene rearrangements. Cell *24* (1981): 1–3.
29. Loh, E., B. Black, R. Riblet, M. Weigert, J. M. Hood & L. Hood. Myeloma proteins from NZB and BALB/c mice: Structural and functional differences. Proc. Natl. Acad. Sci. USA *76* (1979): 1395–1399.
29a. Joho, R., & I. L. Weissman: V-J joining of immunoglobulin *K* genes only occurs on one homologous chromosome. Nature *284* (1980): 179–181.
30. Perry, R. P., D. E. Kelley, C. Coleclough, J. G. Seidman, Ph. Leder, S. Tonegawa, G. Mathyssens & M. Weigert: Transcription of mouse *K* chain genes: Implication for allelic exclusion. Proc. Natl. Acad. Sci. USA *77* (1980): 1937–1941.
30a. Alt, F. W., V. Enea, A. L. M. Bothwell & D. Baltimore: Activity of multiple light chain genes in murine myeloma cells producing a single, functional light chain. Cell *21* (1980): 1–12.

*Heavy Chain Switch*

31. Davis, M. M., S. K. Kim & L. E. Hood: DNA sequences mediating class switching in $\alpha$-immunoglobulins. Science *209* (1980): 1360–1365.
31a. Kataoka, T., T. Miyata & T. Honjo: Repetitive sequences in class-switch recombination of immunoglobulin heavy chain genes. Cell *23* (1981): 357–368.
32. Kataoka, T., T. Kawakami, N. Takahashi & T. Honjo: Rearrangement of immunoglobulin $\gamma$1-chain gene and mechanism for heavy-chain class switch. Proc. Natl. Acad. Sci. USA *77* (1980): 919–923.
33. Rabbitts, T. H., A. Forster, W. Dunnick, D. L. Bentley: The role of gene deletion in the immunoglobulin heavy chain switch. Nature *283* (1980): 351–356.
34. Cory, S., & J. M. Adams: Deletions are associated with somatic rearrangement of immunoglobulin heavy chain genes. Cell *19* (1980): 37–51.
34a. Sakano, H., R. Maki, Y. Kurosawa, W. Roeder & S. Tonegawa: Two types of somatic recombination are necessary for the generation of complete immunoglobulin heavy-chain genes. Nature *286* (1980): 676–683.
35. Davis, M. M., K. Calame, P. W. Early, D. L. Livant, R. Joho, I. L. Weissman & L. Hood: An immunoglobulin heavy-chain gene is formed by at least two recombinational events. Nature *283* (1980): 733–738.
36. Maki, R., A. Traunecker, H. Sakano, W. Roeder & S. Tonegawa: Exon shuffling generates an immunoglobulin heavy chain gene. Proc. Natl. Acad. Sci. USA *77* (1980): 2138–2142.
36a. Obata, M., T. Kataoka, S. Nakai, H. Yamagishi, N. Takahashi, Y. Yamawaki-Kataoka, T. Nikaido, A. Shimizu & T. Honjo: Structure of a rearranged $\gamma$1 chain gene and its implication to immunoglobulin class-switch mechanism. Proc. Natl. Acad. Sci. USA *78* (1981): 2437–2441.

*Pre-mRNP processing*

37. Rabbitts, T. H.: Evidence for splicing of interrupted immunoglobulin variable and constant region sequences in nuclear RNA. Nature *275* (1978): 291–296.
38. Gilmore-Herbert, M. & R. Wall: Nuclear RNA precursor in the processing pathway to MOPC 21 *K* light chain messenger RNA. J. Mol. Biol. *135* (1979): 879–891.
38a. Rogers, J., P. Early, C. Carter, K. Calame, M. Bond, L. Hood & R. Wall: Two mRNAs with different 3′ ends encode membrane-bound and secreted forms of immunoglobulin $\mu$ chain. Cell *20* (1980): 303–312; Early, P., J. Rogers, M. Davis, K. Calame, M. Bond, R. Wall & L. Hood: Two mRNAs can be produced from a single immunoglobulin $\mu$ gene by alternative RNA processing pathways. Cell *20* (1980): 313–319.
39. Perry, R. P. & D. E. Kelley: Immunoglobulin messenger RNAs in murine cell lines that have characteristics of immature B lymphocytes. Cell *18* (1979): 1333–1339.
40. Singer, P. A., H. H. Singer & A. R. Wiliamson: Different species of messenger RNA encode receptor and secretory IgM $\mu$ chains differing at their carboxy termini. Nature *285* (1980): 294–300.

*B Cell Interaction with Macrophages and T Lymphocytes, Signal Transformation*

41. Paul, W. E., & B. Benacerraf: Functional specificity of thymus-dependent lymphocytes. Science *195* (1977): 1293–1300.
42. Berzofsky, J. A., L. K. Richman & D. J. Killion: Distinct H-2-linked Ir genes control both antibody and T cell responses to different determinants on the same antigen, myoglobin. Proc. Natl. Acad. Sci. USA *76* (1979): 4046–4050.
43. Lumb, J. R.: Perspectives on the in vivo location of cellular interactions in the humoral immune response. Ann. Rev. Microbiol. *33* (1979): 439–457.
44. Williamson, A. R.: Three-receptor, clonal expansion model for selection of self-recognition in the thymus. Nature *283* (1980): 527–532.
45. Wekerle, H., & U.-P. Ketelsen: Thymic nurse cells – Ia-bearing epithelium involved in T-lymphocyte differentiation? Nature *283* (1980): 402–404.
46. Reinherz, E. L., & St. F. Schlossman: The differentiation and function of human T lymphocytes. Cell *19* (1980): 821–827.
47. Zuckerman, St. H., & St. D. Douglas: Dynamics of the macrophage plasma membrane. Ann. Rev. Microbiol. *33* (1979): 267–307.
48. Flanagan, J., & G. L. E. Koch: Cross-linked surface Ig attaches to actin. Nature *273* (1978): 278–281.
49. Geiger, B., & S. J. Singer: The participation of $\alpha$-actinin in the capping of cell membrane components. Cell *16* (1979): 213–222.

*Proteinogen processing*

50. Blobel, G., & B. Dobberstein: Transfer of protein across membranes I. and II. J. Cell. Biol. *67* (1975): 835–851 and 852–862.
51. Melchers, F.: Organization of immunoglobulin synthesis. In: Nover, L., F. Lynen & K. Mothes (eds.), Cell Compartmentation and Metabolic Channeling. G. Fischer Verlag and Elsevier/North-Holland Biomedical Press, Jena/Amsterdam 1979, pp. 547–452.
52. Tabas, I., S. Schlesinger & S. Kornfeld: Processing of high mannose oligosaccharides to form complex type oligosaccharides on the newly synthesized polypeptides of the vesicular stomatitis virus G protein and the IgG heavy chain. J. Biol. Chem. *253* (1978): 716–722.
53. Kornfeld, S., E. Li & I. Tabas: The synthesis of complex-type oligosaccharides II. Characterization of the processing intermediates in the synthesis of the complex oligosaccharide units of the vesicular stomatitis virus G protein. J. Biol. Chem. *253* (1978): 7771–7778.
54. Tartakoff, A., & P. Vassalli: Plasma cell immunoglobulin M molecules. Their biosynthesis, assembly, and intracellular transport. J. Cell Biol. *83* (1979): 284–299.

# 14. Phage Coded Programs of Gene Expression in *Escherichia coli*

L. Nover and R. Piechocki

## 14.1 Introduction

### 14.1.1 Phage-Infected Cells as Model Systems for Cell Differentiation

Viruses infecting bacteria are called bacteriophages. First detected by medicinial microbiologists at the beginning of this century, they evoked hopes to become powerful agents against bacterial infections. This hope, however, proved unrealistic. But due to the efforts of Delbrück in the early 1940's, phages gained a prominent position as objects to study DNA replication and the nature and function of genese

**Table 14.1**: Bacteriophages

| Phage[a] | Genome | | | Virion | | Related phages |
|---|---|---|---|---|---|---|
| | Type[b] | MW (Md) | kb | Morphology | Dimension nm | |
| T4 | dsDNA, linear (CPS, TR) | 130 | 166 | Icosahedral head, contractile tail | 65×95 95×20 | T2, T6 |
| T7 | dsDNA, linear (U, TR) | 26 | 38 | Octahedral head, non-contractile tail | 47×47 15×15 | T3 |
| $\lambda$ | dsDNA, linear (U, CE) | 32 | 48 | Icosahedral head non-contractile tail | 54×54 150×15 | $\Phi$80, $\Phi$434 |
| SP01 (cf. **4.4.2.5**) | dsDNA, linear | 100 | 40 | Icosahedral head, contractile tail | 100 165×20 | SP8, SP82 |
| $\Phi$X174 | ssDNA, circular (U) | 1.7 | 5.5 | Icosahedral, tail-less | 25 | G4, M13 |
| MS2 (cf. **4.4.4.3**) | ssRNA, linear (U) | 1.1 | 3.3 | Icosahedral, tail-less | 25 | R17 |
| Q$\beta$ (cf. **4.4.4.3**) | ssRNA, linear (U) | 1.3 | 4.5 | Icosahedral, tail-less | 25 | |

[a] All phages, except $\lambda$, are virulent. The common host for all phages, except SP01, is *Escherichia coli*. Host of SP01 is *Bacillus subtilis*.

[b] ds, double-stranded; ss, single-stranded; U, unique base sequence (non-permuted); CPS, cyclically permuted base sequence; TR, terminal redundant; CE, cohesive ends.

Abbreviations: For a survey of phages (e.g., $\lambda$, T4 and T7) cf. Table **14.1**; phage encoded genes are detailed in Fig. **14.3** (T7), Figs. **14.6** and **14.12** ($\lambda$) and Fig. **14.13** (T4): gene products (proteins) are denoted by gp and the name of their structural genes; bp, kb, base pairs and kilobase pairs; kd, kilodaltons; $M_r$, relative molecular weight (determined, e.g., by electrophoretic methods); MW, molecular weight; 0, operator; p, promoter; t, terminator; L1, L2, L*(int)*, L*(cI)*, R1, R2, transcripts coded by the $\lambda$-phage genome (cf. Figs. **14.6** and **14.11**); RNAP, RNA polymerase.

In fact, experiments with phages and their bacterial host cells have made decisive contributions to the fundaments of modern molecular biology.

A great number of bacteriophages are presently known but only few have been studied thoroughly (cf. the selection of phages treated in different chapters of this book, Table **14.1**). Within this chapter we would like to discuss typical aspects of phage development using experimental data obtained with three phages of *Escherichia coli*, i.e., with T4, T7 and $\lambda$. All three are rather large and complex phages which, despite many important differences, possess basic similarities in their structure and in the gene expression programs involved in their biogenesis.

### 14.1.2 Survey of Structure and Developmental Cycles of Bacteriophages

As is true for all plant and animal viruses, bacteriophages are built up of nucleic acid (DNA or RNA) and a protein shell surrounding it (Table **14.1**). For simple spherical phages, e.g., Q$\beta$ or $\Phi$X174, the protein shell consists of one major protein plus a few minor ones. In contrast, 20 to 50 different protein species contribute to the elaborate structures of the large phages, e.g., of T4, T7, $\lambda$ and SP01. As example the basic structure of phage T4 is shown schematically (Fig. **14.1**).

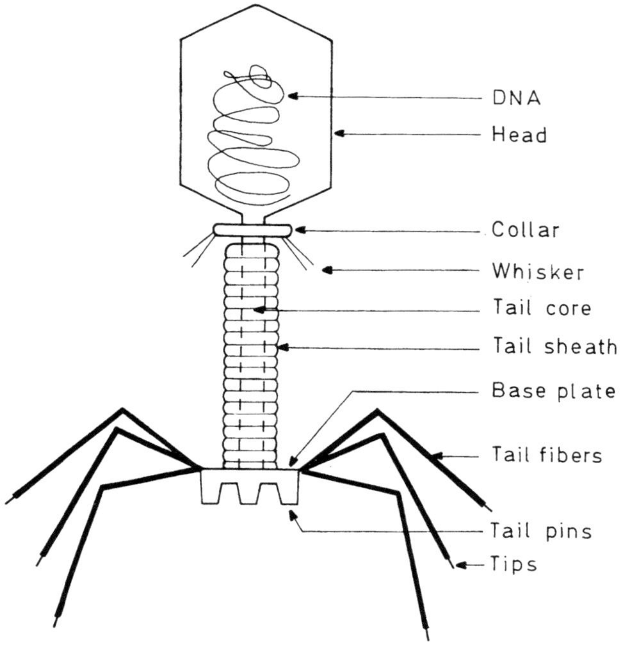


**Fig. 14.1:** Scheme of phage T4 structure.

When infecting their host cells phages recognize special structures of the bacterial cell wall and outer membrane respectively (phage receptors), e.g., parts of the surface lipopolysaccharides (phages T4 and T7) or the *lam B* gene product, a protein involved into maltose transport (phage $\lambda$). After injection of their genetic material and eventually of some minor proteins development of phages may proceed on two different pathways (Fig. **14.2**):

1. Virulent phages (e.g., T4, T7) embark on the lytic pathway, i.e., the host functions of gene expression and DNA replication are shut down, the phage genome is replicated and, finally, up to 100 new phage particles are formed and released from the cell.

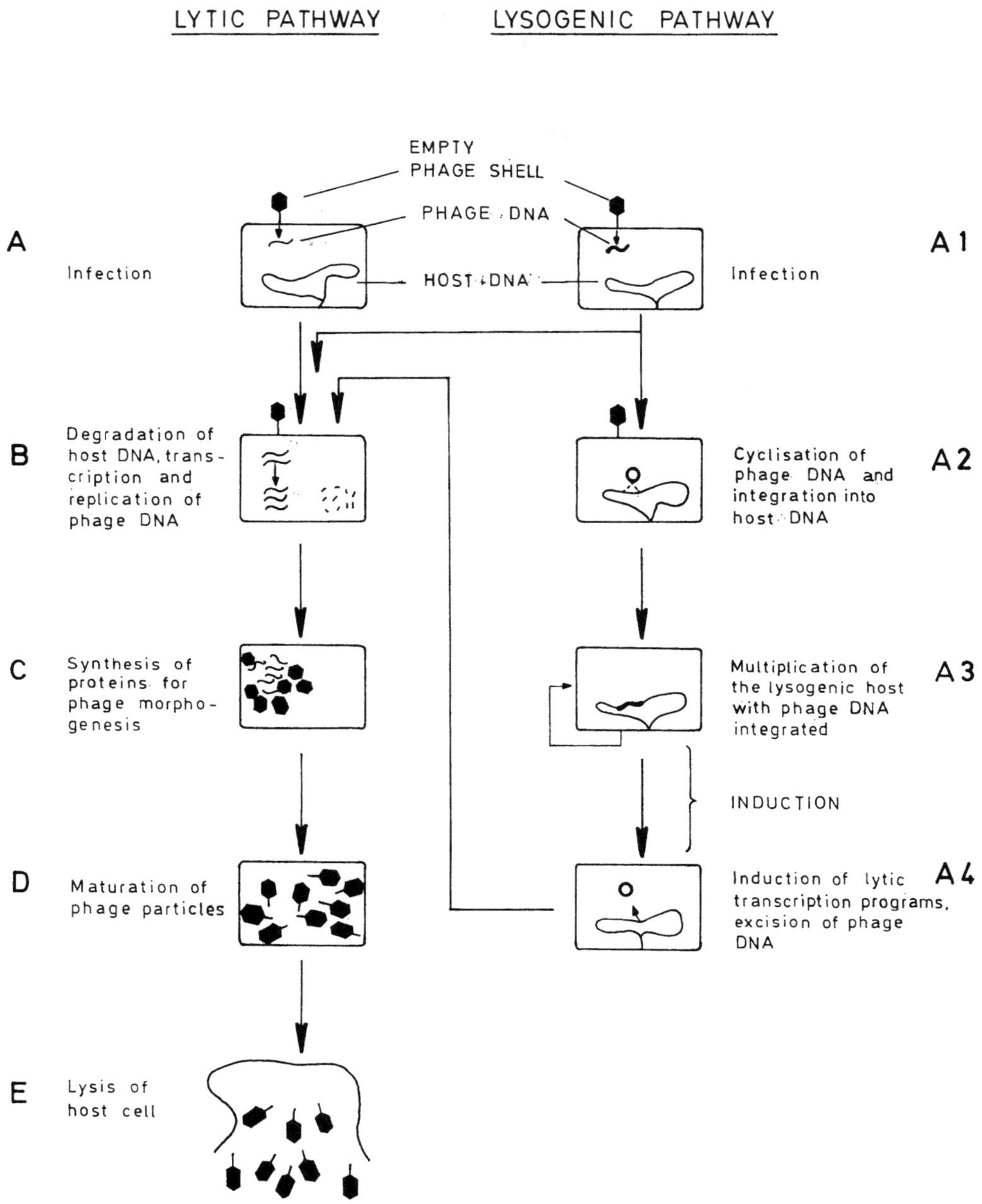


**Fig. 14.2:** Scheme of developmental cycles of virulent and temperate phages.

*Lytic Pathway* **(A–E)**: Immediately after absorption of the phage particle and injection of its DNA into the host cell, the "immediate early" genes are transcribed by the host RNAP. The early proteins contribute to the modification of the host gene expression machinery, stop replication of the host genome and possibly trigger its degradation **(B)**. Replication of the phage genome and synthesis of the "late" (morphogenetic) proteins are the basis for assembly and maturation of new phage particles **(B–D)**. After lysis of the host cell wall a burst of new phages is set free. This scheme is valid for the three phages discussed in this chapter, i.e., for phages T4, T7 and $\lambda$.

2. Temperate phages (e.g., λ) either follow the lytic pathway or become integrated into the host chromosome (lysogenic pathway). Details of the regulatory decision between both pathways are discussed elsewhere in this chapter (cf. **14.2.3.9**). The integrated phage genome (prophage) is multiplied with the host genome and remains in the cryptic state unless external or internal factors interfer with the regulatory equilibrium sustaining the lysogenic state (prophage induction, cf. **14.2.3.8**).

In the following we will discuss four main aspects of the interaction between phages and their host cells (1) DNA restriction and modification, (2) shut down and reprogramming of the host gene expression system (exemplified by phage T7), (3) programmed gene expression as basis for phage development (exemplified by phage λ) and (4) post-translational control of the morphogenetic program (exemplified by phage T4).

## 14.2 Selected Aspects of Phage Infection and Development

### 14.2.1 DNA Restriction and Modification

As detailed elsewhere in this book (cf. **2.2**) bacterial cells possess enzymic systems to specifically degrade foreign genetic material invading the cell. These restriction--modification systems (RM-systems) comprise two types of enzymes: (1) a sequence-specific endonuclease and (2) an enzyme modifying by methylation the DNA sequence recognized by the corresponding endonuclease, so as to protect the host genome against degradation. It is evident from this that the "positive" interaction of phages with their host cells also depends on effective mechanisms to overcome these restriction systems and finally even to degrade host DNA itself.

The three *E. coli* phages treated in this chapter are restricted by RM-systems of type I [6, 7]. In contrast to the type II systems (cf. Table **2.1**), these are rather complicated multimeric enzymes of the type $\alpha_2\beta_2\gamma$ ($M_r$: $\alpha$ = 135 kd, $\beta$ = 60 kd, $\gamma$ = 55 kd). They require ATP, $Mg^{2+}$ and S-adenosyl methionine (SAM) for their catalytic activity. There is evidence that the endonucleolytic cleavage requires ATP and that the enzyme must be activated by SAM to bind to the unmodified DNA recognition sequence. The cleavage sites of type I restriction systems are not enzyme-specific. They are localized at a certain distance from the recognition sites.

Phage T4 DNA is cleaved by two very special endonucleases, Eco $r_{2,4}$ and Eco $r_6$. Both are found in *E. coli* strains B and K and are specific for the hydroxymethyl cytosine-containing DNA of T-even phages. In contrast, phages T7 and λ are restricted by RM-systems Eco B and Eco K found in *E. coli* strains B and K respectively. A detailed knowledge about the recognition sites of these type I RM-systems is lacking except for one site of Eco B recently sequenced[1] (asterisks indicate the bases modified by methylation):

$$5' - \text{TG}\overset{*}{\text{A}} - 8\ \text{bases} - \text{TGCT} - 3'$$
$$3' - \text{ACT} - 8\ \text{bases} - \overset{*}{\text{A}}\text{CGA} - 5'$$

[1] Sommer, R., & H. Schaller: Mol. Gen. Genetics *168* (1979): 331.

◀

*Lysogenic Pathway* **(A1–A3)**: Infection of *E. coli* with a temperate phage (e.g., phage λ) can either lead to the lytic pathway **(A–E)** or result in the metastable integration of the phage DNA into the host genome (for details cf. Fig. **14.5**). The lysogenic bacterium with the prophage can multiply indefinitely. Under certain conditions the prophage can be "induced" to desintegrate from the host genome and embark on the lytic pathway **(A4, B–E)**.

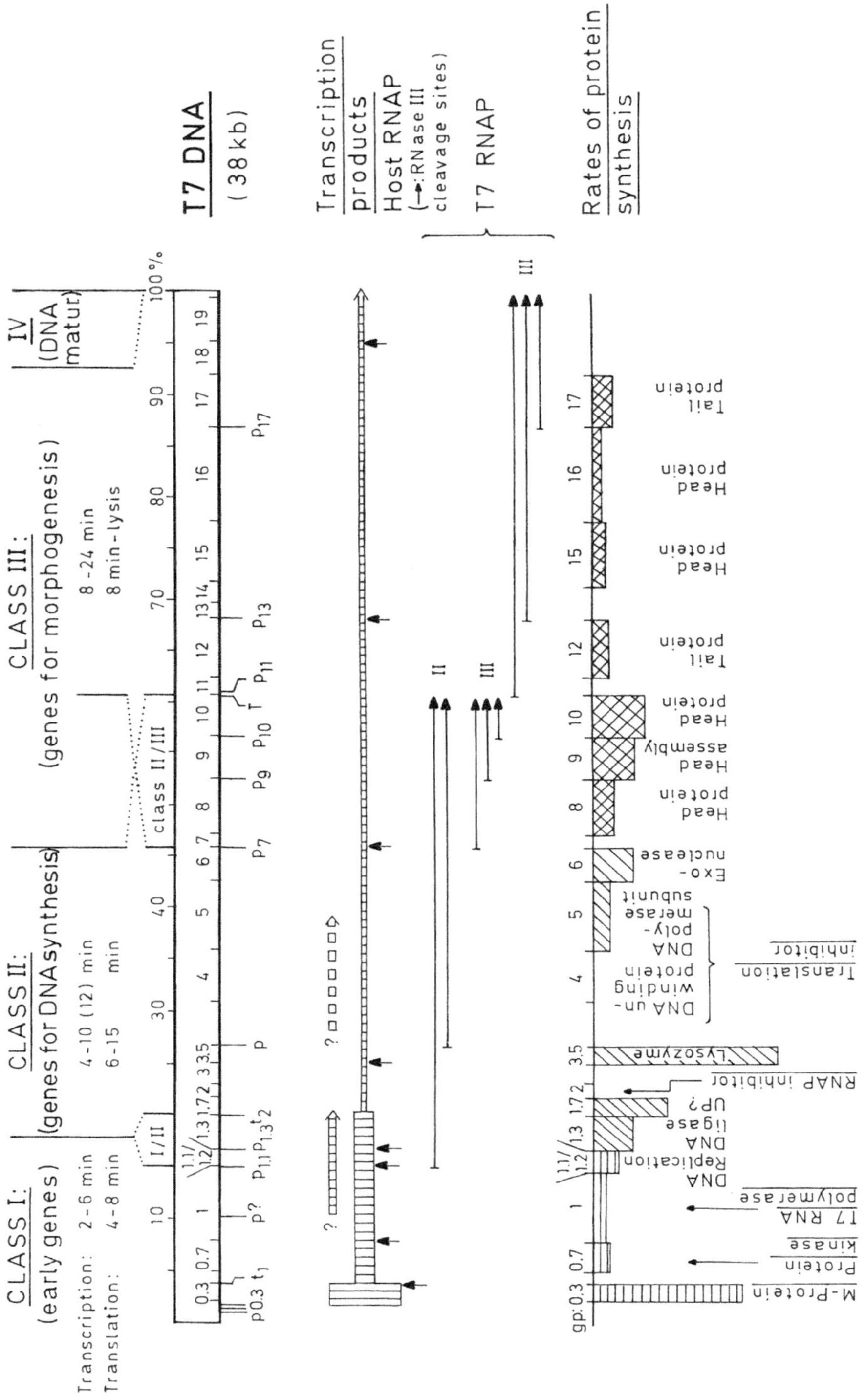
CLASS I:
(early genes)
Transcription: 2-6 min
Translation: 4-8 min
CLASS II:
(genes for DNA synthesis)
4-10 (12) min
6-15 min
CLASS III:
(genes for morphogenesis)
8-24 min
8 min-lysis
IV
(DNA matur)
T7 DNA
(38 kb)
class II/III
I/II
100 %
Transcription products
Host RNAP
(→: RNase III cleavage sites)
T7 RNAP
Rates of protein synthesis
M-Protein
Protein kinase
T7 RNA polymerase
DNA Replication
DNA ligase
UP?
RNAP inhibitor
Lysozyme
DNA un-winding protein
DNA poly-merase subunit
Translation inhibitor
Exo-nuclease
Head protein
Head assembly
Tail protein
gp:

Fig. 14.3

Three types of phage-specific weapons are involved to overcome the host genetic system and its protection by RM-systems: (1) Phage-DNA modification, (2) antirestriction proteins and (3) endonucleases to degrade host DNA.

1. *Phage-DNA Modification:* In phage T4 cytosine residues of DNA are replaced by hydroxymethyl cytosine residues which to a certain extent are modified by glucosylation [5, 9] thus preventing attack by practically all host restriction endonucleases. In case of insufficient or lacking glucosylation, which requires the joint action of phage and host genes during phage DNA replication, T4 DNA is sensitive to Eco $r_{2,4}$ and Eco $r_6$ restriction.

The DNA of phages T7 and $\lambda$ is protected against restriction by site-specific methylation by the host's own RM-system, e.g. by RM-system Eco B in *E. coli* strain B and RM-system Eco K in strain K. Hence resistance is limited to the situation where $\lambda$ or T7 infect the same type of host cell they originated from.

2. *Antirestriction:* Two immediate early proteins of T4 (gp *arn* and gp *arx*) are inhibitors of Eco $r_{2,4}$ (gp *arn*) and of the exonuclease V (gp *arx*), which is responsible for further degradation of the DNA fragments generated by Eco $r_{2,4}$. Similar antirestriction activities were found for the $\lambda$ gene product gp *ral* (cf. gene *ral* mapping between genes *cIII* and *N*, Fig. **14.4**). For T7 gp 0.3 (cf. Fig. **14.3**) was found to interact directly with Eco K endonuclease.

3. *Host DNA Degradation:* Despite the modification of host DNA against its own restriction endonucleases it is not resistant against phage DNases. Host DNA degradation is characteristic for all lytic cycles, the building blocks being reused for phage DNA synthesis.

## 14.2.2 Reprogramming of the Host's Synthetic Machinery by Phage Proteins Exemplified by the Phage T7 Development

### 14.2.2.1 The Genome of Phage T7

The phage T7 genome is a linear double-stranded DNA with $\sim$ 38 kb (cf. Table **14.1**). It codes for about 30 proteins, whose genes were localized by mapping numerous nonsense and deletion mutations and by determining the size of the gene products

---

◀

**Fig. 14.3**: Phage T7: Genetic map, phases of transcription and rates of protein synthesis (compiled from ref. [10–15]).

The genetic map is characterized by the sequence and relative size of 25 genes belonging to three functional and transcriptional classes. The genome is transcribed from *left* to *right*. Host RNAP starts at one of the three leftmost promoters ($A_1$, $A_2$ and $A_3$) collectively referred to as $p_{0\cdot3}$ [17] and terminates at $t_1$ and $t_2$ (class I transcripts), with a low rate of transcription into the late genes. Viral RNAP starts for class II transcripts at promoters $P_{1\cdot1}$ and $P_{1\cdot3}$ and for class III transcripts at promoter $P_8$, $P_9$, $P_{10}$, $P_{11}$, $P_{13}$ and $P_{17}$ (cf. Table **14.3**). The polycistronic mRNAs of all three classes are processed by RNase III at the sites indicated by the *arrows* (cf. also Fig. **14.4**). In the *lower part* of the figure the rates of protein synthesis are indicated as found after feeding radioactively labelled amino acids [15]. They are dependent on a multiplicity of regulatory effects (cf. text). Hence, the figure does not show the actual amount of the individual proteins present or active at different time points of the infection cycle.

Many more details of the class I/class II region evolved by restriction analysis and sequencing of the first 12,100 bp till the beginning of gene 4 [10, 11]. Totally 21 proteins are encoded in this region, ranging in size from 29 to 883 amino acid residues. Except those coded by the DNA synthesis region the role of most other proteins is unclear. Products of genes 1.1 (42 aa residues) and 1.2 (85 aa residues) are presumably involved in T7 DNA replication (cf. Saito, H.: Proc. Natl. Acad. Sci. USA 77 (1980): 3917). Three genes between 2 and 3.5 code for a singlestrand DNA binding protein (232 aa residues, gene 2.5), an unspecified protein (139 aa residues, gene 2.8) and for an endonuclease (149 aa residues, gene 3). Contrasting to the genetic map in the figure, the lysozyme gene (3.5) is much closer to gene 4, beginning at 26.8 % of T7 DNA.

by SDS gel electrophoresis. All genes are transcribed on the same strand of the DNA. This contrasts to the transcription of the T4 and $\lambda$ genomes [cf. Figs. **14.6** and **14.13**]. Three functionally different regions in the genetic map can be discerned (Fig. **14.3**): (I) the early genes in the so-called control region, (II) genes for DNA synthesis and (III) genes for morphogenesis plus two genes for DNA maturation (genes 18, 19). They correspond largely to the three classes of transcripts (I, II, III) defined by the sequence of their formation and by the RNAP involved. In the discussion of T7 development we will restrict ourselves to some important aspects of host-virus interaction and gene expression regulation mostly connected with gene products of the control region (cf. Table **14.2**).

**Table 14.2**: Regulatory proteins involved in the reprogramming of the host cell to T7 multiplication

| | Code site[a] (gene) | Code region[a] | $M_r$ (kd) | Effects |
|---|---|---|---|---|
| M-protein | 0.3 | I | ~ 20 | Membrane alteration, $K^+$-ionophore (?), antirestriction |
| Protein kinase | 0.7 | I | 40 | Phosphorylates $\beta$'-subunit of host RNAP |
| T7 RNA polymerase | 1 | I | 107 | RNAP specific for late phage promoters |
| Transcription inhibitor | 2 | II | 14 | Inhibits host RNA polymerase |
| Translation inhibitor | ?[b] | II | 80 | Inhibits translation of host mRNA |

[a] cf. Fig. **14.3**
[b] Coded somewhere in the region of genes 4 and 5.

### 14.2.2.2 Two Types of RNA Polymerase

After the linear T7 DNA has been introduced into the host cell, transcription starts by host RNAP at three adjacent strong promoters located 445, 560 and 670 base pairs from the left end (p 0.3). It proceeds through the whole genome to the right end. But soon after infection synthesis of gp 0.7, the protein kinase, leads to a phosphorylation of the $\beta'$-subunit of RNAP. The modified enzyme now recognizes two termination signals in the early region, $t_1$ and $t_2$, each one reducing the number of transcribing RNAP molecules by 75%. Thus transcription of the late genes practically ceases. However, by this time synthesis of virus-specific RNAP (gp 1) has been completed. In contrast to the complex structure of host RNAP (cf. **1.3.2.1**) T7 RNAP is a monomeric protein which recognizes several internal promoters ($P_{1.1}$, $P_{1.3}$, $P_7$, $P_9$, $P_{10}$, $P_{11}$, $P_{13}$, $P_{17}$) in regions II and III of the viral DNA (Fig. **14.3**). It reads about twice as fast as the host enzyme and is highly specific for T7 DNA. Even the DNA of the closely related phage T3 is transcribed with very low efficiency.

The transcription products of T7 DNA (cf. Fig. **14.3**) are mostly polycistronic mRNA containing the information of two or more than ten cistrons. It is well known that at least the polycistronic mRNA of the early region genes must be cut into monocistronic pieces before translation can start. This is accomplished by the sequence specific RNase III, which is also required for the host pre-rRNA processing (cf. **1.3.3.2**). The late transcripts are likewise processed by RNase III and by one further RNase hitherto not classified. Analysis of the RNase III recognition sites, which, in contrast to the rRNA precursor, mostly undergo single-strand cleavage, revealed a typical dsRNA hairpin with a central loop (Fig. **14.4**). Most bases at the border and within this loop are identical in three different cleavage sites sequenced [20].

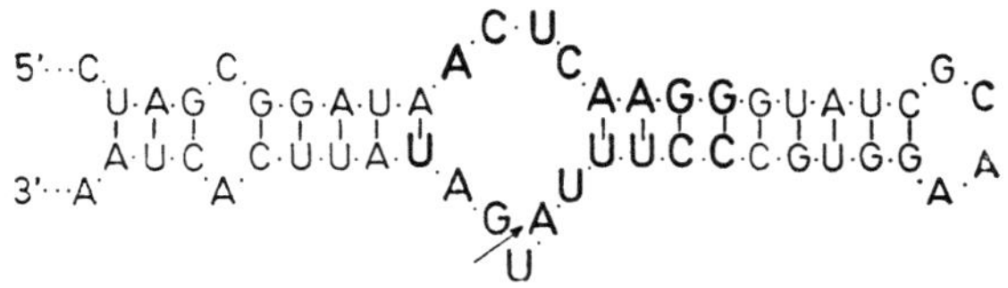

**Fig. 14.4**: Sequence and RNA secondary structure surrounding an RNAse III processing site of T7 early region pre-mRNA (from Robertson & Barany [20], modified).

The sequence corresponds to the processing site at map position 2.19% between the 5′-leader and the body of gene 0.3 mRNA (following at the 3′-end). RNase III cuts in the large unpaired bubble between the two double-helical stems *(arrow)*. *Bold-faced bases* are conserved in all three RNAse III recognition sites which undergo single-strand cleavage.

### 14.2.2.3 Two Types of Promoters

The number, distribution and binding-specificity of promoters as well as terminators have decisive influence on the differential transcription of T7 DNA and, hence, on the temporal program and rates of protein synthesis (cf. Fig. **14**.3). Besides the very strong promoters at the leftward end, host RNAP may recognize only two weak internal promoters, one in the coding region of gene 1 and another possibly in the region of gene 4, to transcribe the information of the translation inhibitor. In contrast, virus RNAP only starts from internal promoters, especially from the strong promoters of group III.

In the medium phase of infection (4–10 min) T7 RNAP has a strong preference for the group II promoter(s). The later shift to group III promoters is somehow controlled by lysozyme. It is interesting to notice that the genes of three major head proteins (gp 8–10) are transcribed in both

**Table 14.3**: Phage T7 promoters.

Promoters of class I, II and III are compared (non-coding DNA strands). The bases denoted by bold-faced letters indicate an almost identical 17-base pair sequence preceding the initiation point at +1. This structure of class II and III promoters contrasts to that of class I promoters, which closely resembles the *E. coli* consensus promoter (RS, primary recognition site; PB, Pribnow box).

| Promoter type (cf. Fig. **14**.3) | Map pos. (%) | Sequence (positions −30, −20, −10, +1 marked) | Ref. |
|---|---|---|---|
| $P_{17}$ (class III) | 87 | 5′ . . . AACTTAGGGAGCGTAGGAAATAATACGACTCACTATAGGGAGA . . . 3′ | [16] |
| $P_8$ (class III) | 46.5 | 5′ . . . TAAACTCAAGGTCCCTAAATTAATACGACTCACTATAGGGAGA . . . 3′ | [16] |
| $P_{1.3}$ (class II) | 15.9 | 5′ . . . . . . . . . . AAACTATGGTTGACACATACGACTCAGTATAGGGACA . . . 3′ | [18] |
| $P_{1.1}$ (class II) | 14.6 | 5′ . . . GCGTTCGCGTAACGCCAAATCAATACGACTCACTATAGAGGGA . . . 3′ | [16] |
| $P_{0.3}$ (class I) | 1.25 | 5′ . . . ATTGACTTAAAGTCTAACCTATAGGATACTTACAGCCATCGAGA . . 3′ | [17] |
| *E. coli* consensus promoter (cf. Table **1**.8) | | 5′ . . . . TTGACA (RS) . . . . . . . 12–14 bp . . . . . . TATAAT (PB) . . 5–6 bp . Pu . . . . . . . . 3′ | |

phases. The six group III transcripts are formed in equimolar amounts. They are separated into two groups by the T7 RNAP terminator (T) at the end of gene 10.

Comparison of DNA sequences of four group III and one group II promoters (specific for T7 RNAP) with the sequence of one of the left-most promoters ($p_{0.3}$) reveals decisive differences (Table **14**.3). Only $p_{0.3}$ shows the basic structure of *E. coli* promoters though somewhat distorted. In contrast, in all T7 RNAP-specific promoters, which have a 27 base pair region of nearly absolute homology around and preceding the transcription initiation site, the typical A=T-rich region required for RNAP binding is centered at base pair —17 instead of —32 as usually found in *E. coli* (cf. Table **1**.8).

### 14.2.2.4 Patterns of Protein Synthesis

The pattern of synthesis of early T7 proteins (cf. Fig. **14**.3) reflects four regulatory effects: (1) differential transcription controlled by terminators $t_1$, $t_2$ and T as well as by the initiation frequency and specificity of the promoters involved, (2) rate of processing to monocistronic mRNA by RNase III plus an unspecified RNase for the late transcripts, (3) mRNA stability and (4) rate of translation initiation [15]. Functional half-lives of the mRNA are on the average 3.5–4 min, whereas protein kinase mRNA has a half-life of only 2 min. The relative translation efficiencies are 1.7 (M-protein), 2.2 (protein kinase), 0.8 (RNAP), 0.7 (gp 1.1) and 1.8 (DNA ligase). Thus, kinase mRNA is translated 2–3 times more efficiently than RNAP mRNA, a fact compensating for the short half-life. As a consequence, protein kinase is synthesized fast at a time when its activity is needed, but, due to the high turnover of its message, it is particulary sensitive to regulatory influences. Moreover, an autocatalytic inactivation of protein kinase by phosphorylation is observed.

### 14.2.2.5 Inhibitors of Host Transcription, Translation and Membrane Function

About 6–8 min after infection two further alterations of the host gene expression machinery finish the reprogramming for phage protein synthesis (cf. Table **14**.2):

1. Transcription from the early phage promoter and all host promoters is stopped by a transcription inhibitor binding to the holoenzyme of host RNAP. The inhibitor is a product of gene 2, which evidently has some additional functions in DNA packaging.
2. Another, "late" protein, whose coding site is not yet known, modifies the host ribosomal machinery and interferes with translation initiation of host mRNA. The mechanism is unclear. So far no modification of the ribosomal RNA or of initiation factors were found.

When discussing the regulatory circuits used by phage T7 to abuse the host cell synthetic machinery for its own purpose we must, finally, mention the membrane-modifying protein coded by gene 0.3. It is synthesized in large amounts early during infection but is evidently also part of the phage particle and injected during infection. Its incorporation into the host cell membrane results in a very rapid increase of membrane fluidity and sudden efflux of $K^+$-ions. It can be speculated that the large amount of energy set free by the $K^+$-efflux might be used for virus synthesis. Furthermore, the deenergetization of the host membrane and the changes in the cellular ion milieu surely contribute to the reprogramming and, finally, to the death of the host cell. Recently gp 0.3 was also shown to interact directly with host restriction endonuclease, thus preventing unmodified, e.g., newly synthesized, phage DNA from degradation.

## 14.2.3 Transcription Programs of Phage Lambda

### 14.2.3.1 Circularization, Integration and Excision of the Lambda Genome

The genome of the temperate phage $\lambda$ comprises a double-stranded DNA with MW $32\times10^6$ corresponding to about 48 kb. It undergoes characteristic changes upon interaction with the host cell (Fig. **14.5**). The linear infectious form of the molecule,

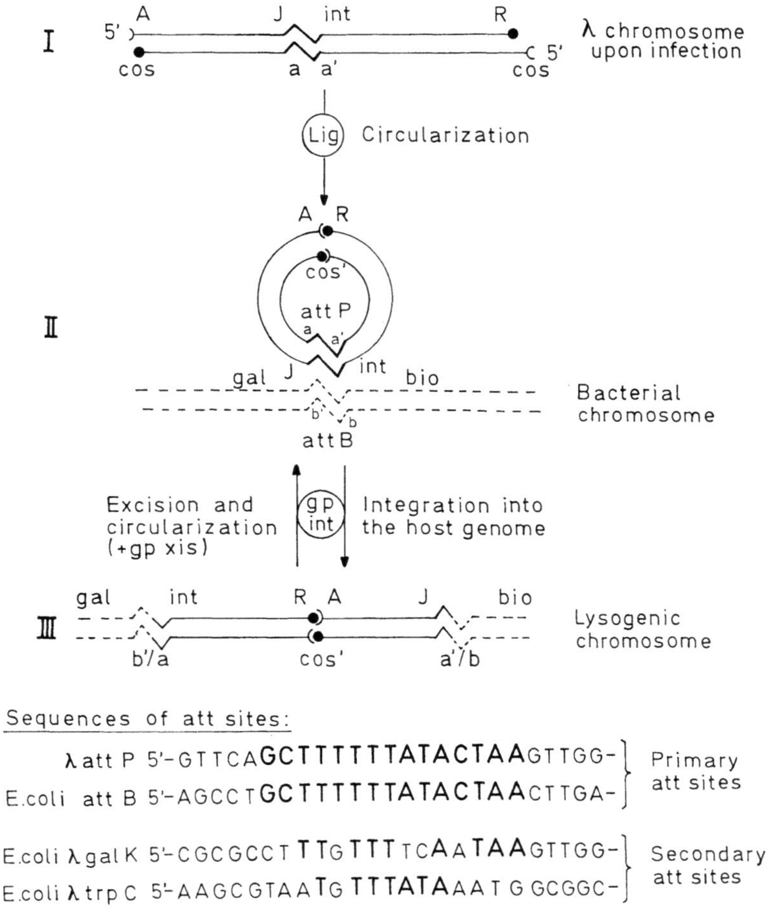


**Fig. 14.5**: Circularization, integration and excision of $\lambda$-DNA (from Campbell [25] and Szybalski [23]).

Circularization of the infectious $\lambda$-genome (I $\rightarrow$ II) by host polynucleotide ligase *(Lig)* catalyzed fusion of the 5'-cohesive ends *(cos)* brings the morphogenetic genes *(A–J)* on the left arm in continuity with their promoter *($p_Q$)* localized on the right arm in the *S*, *R* gene region (cf. Fig. 14.6). Integration (II $\rightarrow$ III) or excision (III $\rightarrow$ II) of the $\lambda$-genome requires the action of gp *int*. Under normal conditions the $\lambda$-genome is integrated at a specific receptor site *(attB)* in the *E. coli* chromosome, between genes for galactose utilization *(gal)* and biotin biosynthesis *(bio)*. As shown in the *lower part* of the figure *attB* shares a sequence of 15 nucleotides with the corresponding site *(attP)* in the phage genome. Besides the primary attachment site *(attB)* the g*E. coli* genome contains several secondary *att* sites with much lower effectivity for $\lambda$-genome integration. They contain also sequence elements homologous to *attP* [28, 29].

possessing single-stranded cohesive ends of 12 nucleotides, is stabilized by circularization soon after injection into the bacterium. The circular form of the genome is the basis for the transcription programs and for replication (lytic pathway) or for integration as prophage into the host genome (lysogenic pathway).

The integration and excision reaction of bacteriophage $\lambda$ is the most thoroughly characterized example of site-specific recombination. It depends on the interaction between two DNA recognition sites, *attP* on the phage genome and *attB* on the bacterial chromosome, mapping at 17.5 min of the circular *E. coli* chromosome between the genes for galactose utilization *(gal)* and biotin biosynthesis *(bio)*. Although this attachment sites are distinct from each other, there is a common 15 bp sequence (Fig. **14**.5).

Integration requires the action of gp *int*, whereas excision proceeds only in the presence of gp *int* and gp *xis* (cf. also **14**.2.3.9). Furthermore, at least two host-specified proteins, encoded by the genes *him A*, *him B*, *him C* and *hip*, are also involved. They play major roles not only in $\lambda$-recombination but also in genetic transposition, plasmid stability and chromosome duplication [27].

In vitro and in vivo studies revealed the following features of the integration event:

1. DNA supercoiling is required for recombination promoted by gp *int*. This is strictly true for the *attP* site, whereas the *attB* site can be contained either in supercoiled or in linear DNA.
2. Integration in vivo requires a higher level of gp *int* than excision. The simplest model is that integrative recombination is promoted by a higher order multimeric complex of gp *int* than excisive recombination. 10–20 gp *int* molecules per attachment site are required to promote recombination.
3. Breakage and reunion seems to take place without degradation and resynthesis of DNA and without an obvious requirement of chemical energy. Therefore the free energy of the broken phosphodiester bond must be stored and reused in the rejoining step. Gp *int* itself is assumed to have DNA topoisomerase activity (cf. **1**.2.1) and to form the multimeric protein core for the phage DNA to be wrapped around. Thus, the spatial organization of this phage *att* site-nucleoprotein complex provides most of the structural and enzymatic information necessary for the site-specific integration by interaction with the bacterial recipient site *(attB)* [30].

With the exception of the *cI-rex* operon, all other phage scriptons must be repressed in the stabile lysogenic state. This maintenance mode of transcription (cf. Fig. **14**.6 A) allows for the autoregulated synthesis of *cI* repressor (about 16–20 molecules/lysogenic cell) which makes the lysogenic host cell immune against superinfection by phages of the same or a related type. During prophage induction (cf. **14**.2.3.8) the integration sequence is run in the reverse direction (excision).

Errors during the excision process can result in phage genomes carrying small parts of the host chromosome, e.g., genes for galactose utilization or biotin synthesis. Such phage hybrids frequently lack part of their own genome. They are defective in certain essential functions and can be multiplied only in the presence of helper phages supporting for this function. However, they have proved of extreme experimental value as means for controlled transfer of genes from one bacterial strain to another *(transduction)*. Meanwhile the restriction of transducible genes to those adjacent to the *attB* receptor region was eliminated by deleting this region from the host chromosome. Prophage integration in such hosts proceeds at so-called secondary integration sites of the chromosome (cf. Fig. **14**.5), though with much lower frequency. Hence, much more bacterial genes are amenable to transduction provided suitable selection procedures for the desired hybrid phages are designed (cf. Chap. **5**). Recently, this important technique was improved by in vitro construction of hybrid phages carying even eukaryotic genes (cf. **2**.4).

### 14.2.3.2 The Genetic Map of Phage Lambda

Details of the genetic map are shown at the top of Fig. **14**.6 by using the prophage state of the DNA integrated into the host chromosome between genes for galactose utilization and biotin biosynthesis. About half of the 48 kb coding capacity is required for the morphogenetic proteins (*A–J* region), which are further detailed in Fig. **14**.12. Besides this, there are genes for host cell lysis (*R*, *S*), DNA synthesis (*O*, *P*), recombination (*red*, *gam*) and the integration-excision process (*xis*, *int*). However, special interest deserves a set of regulatory genes (*cIII*, *N*, *rex*, *cI*, *cro*, *cII* and *Q*), controlling the transcriptional program at different developmental stages of the phage. Their function and mutual interaction will be in the center of the following sections (**14**.2.3.3–**14**.2.3.9).

It is essential to recall that the far advanced knowledge of the $\lambda$ map is the key to the sophisticated analysis of the regulatory circuits to be discussed. It stems from the generation and thorough biochemical as well as genetic characterization of a large number of mutants. Classical genetic methods (crossing mutant phages by co-infection of host cells and analysing the complementation capability and recombination frequency) were more recently complemented by physical and chemical methods for gene fine structure analysis, e.g., electron microscopic deletion mapping, restriction mapping (cf. **2**.9) and direct DNA sequencing (cf. **2**.10).

### 14.2.3.3 Four Transcription Phases and the Role of Regulatory Proteins

Both strands of the $\lambda$-DNA serve as template for the formation of leftward and rightward transcripts (L1, L2, R1 and R2) which partially overlap in the regulatory region and in the *int-a-b*$_2$-region. Several transcription units (scriptons) are defined by their promoters ($p_{rm}$, $p_L$, $p_R$ and $p_Q$) and further characterized by operators ($O_R$, $O_L$) and/or by terminators interrupting them (Fig. **14**.6). In contrast to the results reported for phage T7 development, all $\lambda$-transcripts are produced by the host RNA polymerase, albeit modified by a number of phage regulatory proteins to overcome the diverse termination signals. These scripton-specific antitermination factors (gp *N* and gp *Q*) and the crucial role of other regulatory proteins (gp *cI*, gp *cro*, gp *cII*, gp *cIII*) will be described in the following sections.

Four transcriptional phases can be discerned, one for the lysogenic state (phase A) and three consecutive phases (B–D) for the lytic cycle:

*Prophage Maintenance Transcription (Phase A)*: Besides some minor short transcripts from promoters $p_Q$ and $p_0$ the only essential mRNA produced starts from $p_{rm}$ and extends leftward through genes *cI* and *rex*. The product of *cI* (*cI* repressor) interfers directly or indirectly with the expression of all other scriptons.

*Immediate Eearly Transcription (Phase B)*: After inactivation of the *cI* repressor (phage induction, cf. **14**.2.3) host RNAP is now able to transcribe leftward from $p_L$ and rightward from $p_R$, but in both gases stops at the next termination signal ($t_{L1}$ and $t_{R1}$ respectively). mRNA for two regulatory proteins is synthesized. The product of gene *N* (gp *N*) serves as antiterminator in the following transcription phase, whereas the product of gene *cro* (gp *cro*) acts at $O_R$ as antirepressor of *cI* repressor synthesis, which would interfere with the further progress of the lytic cycle.

*Delayed Eearly Transcription (Phase C)*: Host RNAP, now modified by gp *N* (cf. **14**.2.3.4), is able to pass the termination signals $t_{L1-3}$, $t_{R1}$ and $t_{R2}$. Thus genes *O*, *P* for DNA synthesis (rightward transcription, R1) and for recombination and DNA-excision (leftward transcription, L2) are expressed. The rightward transcript also codes for a further antitermination factor (gp *Q*) required for the next phase.

*Late Transcription (Phase D)*: The accumulation of gp *cro* results in a decrease of transcription from promoters $p_L$ and $p_R$. The product of gene *Q* enables the RNAP to continue the short leader RNA synthesized from the very strong promoter $p_Q$ by passing the two terminators $t_{Q1}$ and

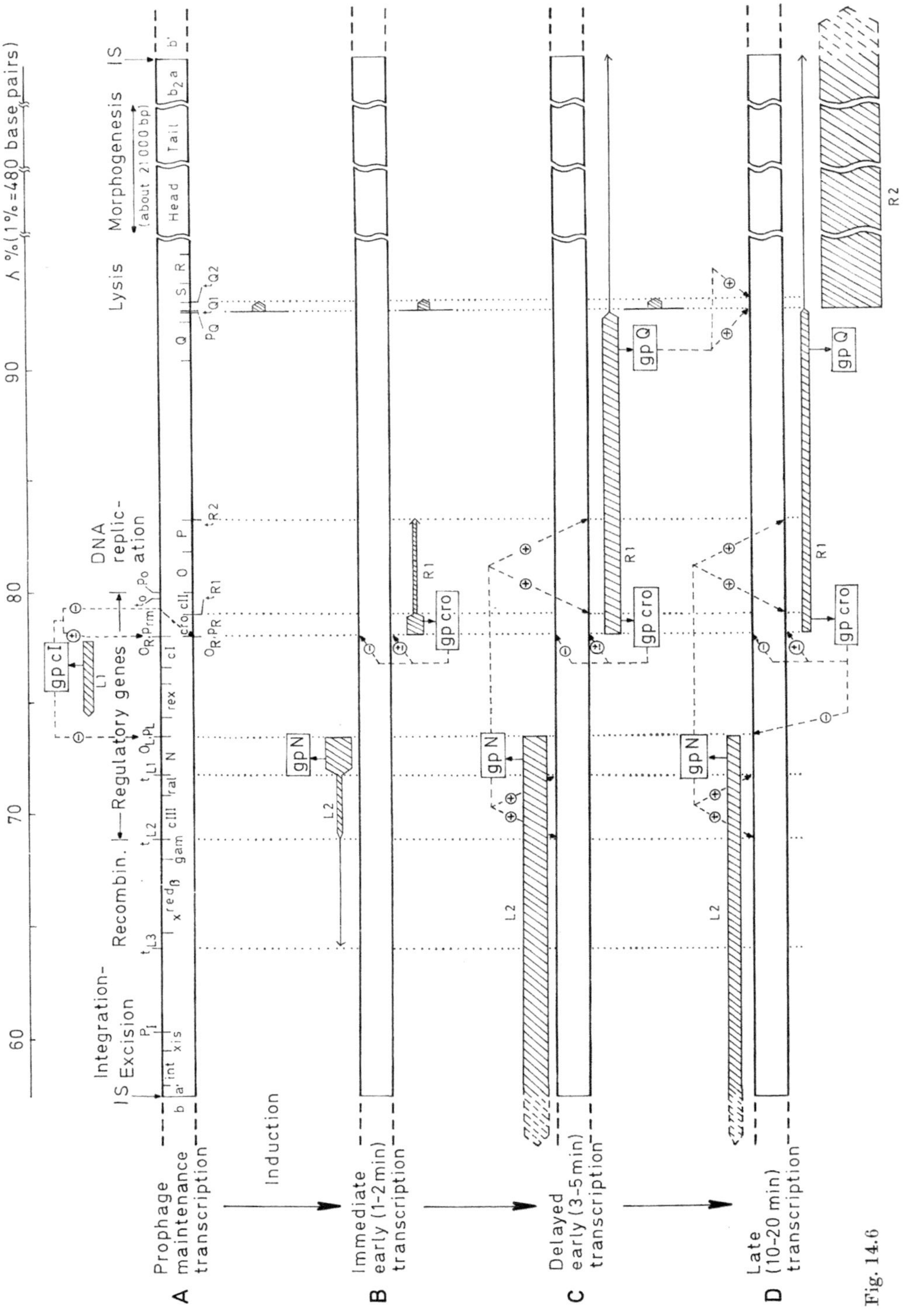
λ % (1% = 480 base pairs)
60
70
80
90
IS
Integration-Excision
Recombin.
Regulatory genes
DNA replication
Lysis
Morphogenesis
(about 21000 bp)
int
xis
gam
cIII
N
rex
cI
O
P
Q
S
R
Head
Tail
gp cI
gp N
gp cro
gp Q
L1
L2
R1
R2
A Prophage maintenance transcription
Induction
B Immediate early (1-2 min) transcription
C Delayed early (3-5 min) transcription
D Late (10-20 min) transcription

Fig. 14.6

$t_{Q2}$. The large amount of a giant polycistronic mRNA (transcript R2) contains the information for all morphogenetic proteins and for the lytic proteins. It is worthwhile to recall that the continuity of the late gene region *A–J* with its promoter $p_Q$ originates only post infection by circularization of the $\lambda$ genome (cf. Fig. **14.5**).

Many details of these transcription patterns resulted from RNA-DNA hybridization experiments with separated strands of $\lambda$-DNA. These are obtained after heat denaturation of $\lambda$-DNA and subsequent centrifugation in poly(U,G)-containing CsCl gradients. Because of differences in the overall base composition, the DNA strand for the leftward transcripts (L-strand) bands at a lower density ($\zeta = 1.738$ g/cm$^3$) than that for the rightward transcripts (R-strand, $\zeta = 1.759$ g/cm$^3$). If after labelling with radioactive isotopes RNA of phage-infected cells is isolated and hybridized to the L- or R-strand of $\lambda$-DNA the DNA-RNA hybrid is resistent to pacreatic RNase, whereas non-bound RNA is degraded. Combination of this basic procedure with different phage mutants as sources for the DNA as well as for the infection of bacterial cells made major contributions to our present picture of the structure and function of the $\lambda$-genome. Even many genes of the host chromosome were amenable to this technique by using transducing phages (cf. **5.5**).

### 14.2.3.4 Antitermination by gp *N*

A characteristic product of the immediate early gene expression after prophage induction (phase B) is the antitermination protein gp *N*. It is a short-lived, highly basic protein ($M_r$ 13.5 kd) that binds tightly to the host RNAP. For its antitermination function gp *N* requires interaction with the host termination subunit gp *nus*A, which, shortly after initiation of transcription, replaces the initiation subunit $\sigma$ of the RNAP holoenzyme [35]. The direct binding of gp *N* to RNAP is documented (1) by certain RNAP mutants unable to respond to gp *N*, (2) by the isolation of gp *N* from RNAP preparations and (3) by the finding that RNAP (gp *N*) shows increased initiation rates from promoters $p_L$ and $p_R$.

Because all terminators of phage $\lambda$ counteracted by gp *N* are $\varrho$-dependent it is useful to recall some facts connected with termination (cf. **1.3.2.2**): (1) Due to the structural peculiarities of terminator signals (formation of a stem-loop structure by intrastrand base-pairing, oligo (dT) stretch at

---

◀

**Fig. 14.6**: The genetic map of phage $\lambda$ and the four transcription phases (from Szybalski [23], modified).

The $\lambda$-genome codes for four major transcripts *(L1, L2, R1, R2)*, whose time point and extent of formation is influenced by phage-specified regulatory proteins binding to the operator-promoter regions *(O, p)* or enabling RNAP to overcome distinct termination signals *(t)*. Transcript L1 from the leftward scripton is essentially the only transcript in the lysogenic state *(maintenance transcription)*. After prophage induction L1 formation stops and the lack of *cI* repressor allows transcription to proceed from $p_R$ (begin of R1 with gene *cro*) and $p_L$ (begin of L2 with gene *N*) respectively *(immediate early transcription)*. The antitermination function of gp *N* promotes the read-through at terminators $t_{L1}$, $t_{L2}$ and $t_{L3}$ (L2) and at terminators $t_{R1}$ and $t_{R2}$ (R1). Thus, in the *delayed early phase* transcript L2 is extended just to the left end of the phage genome containing the genes for prophage excision and DNA recombination. The extension of R1 includes genes for DNA synthesis and a further gene with antitermination function *(Q)*. Gp *Q* is required to overcome the two strong terminators $t_{Q1}$ and $t_{Q2}$ which so far restrict the transcript R2 to a very short leader RNA. The giant, polycistronic transcript R2 dominates in the *late transcription* phase. It is processed and gives rise to all morphogenetic phage proteins (cf. **14.2.3.10**). $-\oplus\rightarrow$, $-\ominus\rightarrow$, $-\oplus/\ominus\rightarrow$ denote the influence of regulatory proteins on transcription that can be positive, negative and, dependent on the concentration, positive and negative respectively (cf. details in Sect. **14.2.3.4–14.2.3.7**). Gene products (proteins) are denoted by gp and boxed. The *width* of the *arrows* gives an estimate of the rate of transcription as deduced from hybridization experiments with the separated L and R strands of the $\lambda$-genome.

On the possible role of two further rightward transcripts as well as the promoters ($p_I$ and $p_0$) and regulatory proteins (gp *cII*/gp *cIII*) involved cf. **14.2.3.9**.

**Table 14.4**: In vitro transcription upstream and downstream of the termination signal $t_{R1}$ in the presence and absence of termination factor rho (from Rosenberg et al. [34])
The radioactively labeled in vitro transcription products starting from $p_R$ (cf. Fig. **14.**6) were digested by $T_1$ RNase and the oligonucleotides separated by two dimensional chromatography. The spots corresponding to the indicated oligonucleotides upstream (T2, T4) and downstream (Ta, Tc) of $t_{R1}$ were cut out and their radioactivity was determined. The numbers in the table indicate molar ratios of the individual oligonucleotides. The genetic map represents the regulatory region between the coding sequences for genes *cro* and *cII*, showing the localization of the oligonucleotides tested. For further details and explanations of mutations *cin* and *cnc* cf. legend to Fig. **14.7**.

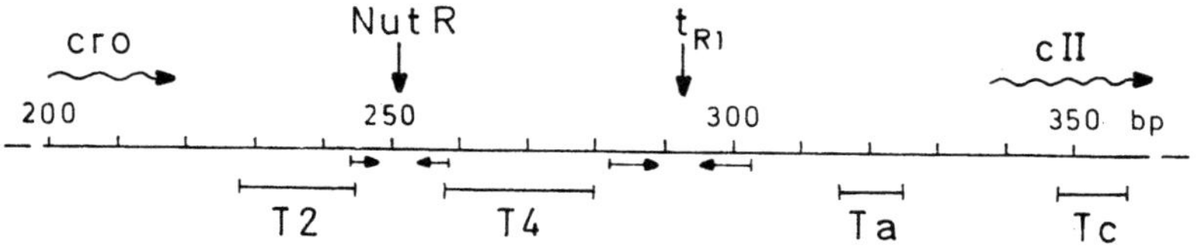


| | DNA | rho factor | Incub. time (s) | Oligonucleotides (molar ratios) T2 (upstream) | T4 | $t_{R1}$ | Ta (downstream) | Tc |
|---|---|---|---|---|---|---|---|---|
| A | λ + | – | 25 | 0.95 | 0.80 | | 0.00 | 0.00 |
| | | | 40 | 1.01 | 0.90 | | 0.05 | 0.05 |
| | | | 300 | 0.98 | 0.93 | | 0.94 | 0.95 |
| B | λ + | + | 300 | 0.95 | 1.01 | | 0.16 | 0.19 |
| | λ *cin* | – | 300 | 0.91 | 0.92 | | 0.92 | – |
| | λ *cin* | + | 300 | 0.97 | 0.96 | | 0.07 | 0.05 |
| | λ *cnc* | – | 300 | 1.01 | 0.96 | | 0.93 | 0.90 |
| | λ *cnc* | + | 300 | 0.94 | 0.92 | | 0.89 | 0.90 |

the non-coding DNA strand) the propagation rate of RNAP is slowed down (cf. data given in Table **14.**4). (2) The hexameric termination factor $\varrho$ attaches to distinct primary binding sites of the nascent RNA chain (sites for rho utilization = *rut*) and thus acquires ATPase activity. (3) With concomitant ATP hydrolysis $\varrho$ somehow "moves" along the RNA strand until it catches up with RNAP. The movement may in fact be due to a progessive wrapping of the nascent RNA strand around the $\varrho$-factor. (4) The mechanism by which the contact of RNAP with $\varrho$ causes termination is unknown.

When looking for mutations preventing gp *N* antitermination, sequences in both the rightward and leftward transcripts were characterized which are evidently required for gp *N* utilization (*Nut R* and *Nut L*). *Nut R* is situated in direct neighbourhood of $t_{R1}$ (cf. Fig. **14.**7) whereas *Nut L* maps at the beginning of gene *N*, only 55 bp downstream from the initiation point of promoter $p_L$. Sequencing of the intracistronic] regulatory region between genes *cro* and *cII* (Fig. **14.**7) revealed the structures for the termination as well as for the gp *N*-utilization site (*Nut R*) (Fig. **14.**7B). Both are characterized by DNA-sequences with inversed rotational symmetry, which enable the nascent RNA, or, alternatively, the non-coding DNA strand, to form stem-loop structures at positions 244–258 (*Nut R*) and 280–303 ($t_{R1}$) respectively. It is interesting to notice that mutations increasing the symmetry of the terminator region (*cin*) improve its effectivity in $\varrho$-dependent termination (80% → 90%). In contrast, mutations which disturb the symmetry (*cnc*) destroy the terminator function (Table **14.**4). Analysis of the in vitro transcription products starting from $p_R$ and proceeding into the intracistronic region between genes *cro* and *cII* also showed that there is indeed a slow-down of RNAP progression in the termination region (Table **14.**3, part A), which, however, results in termination only in the presence of $\varrho$-factor.

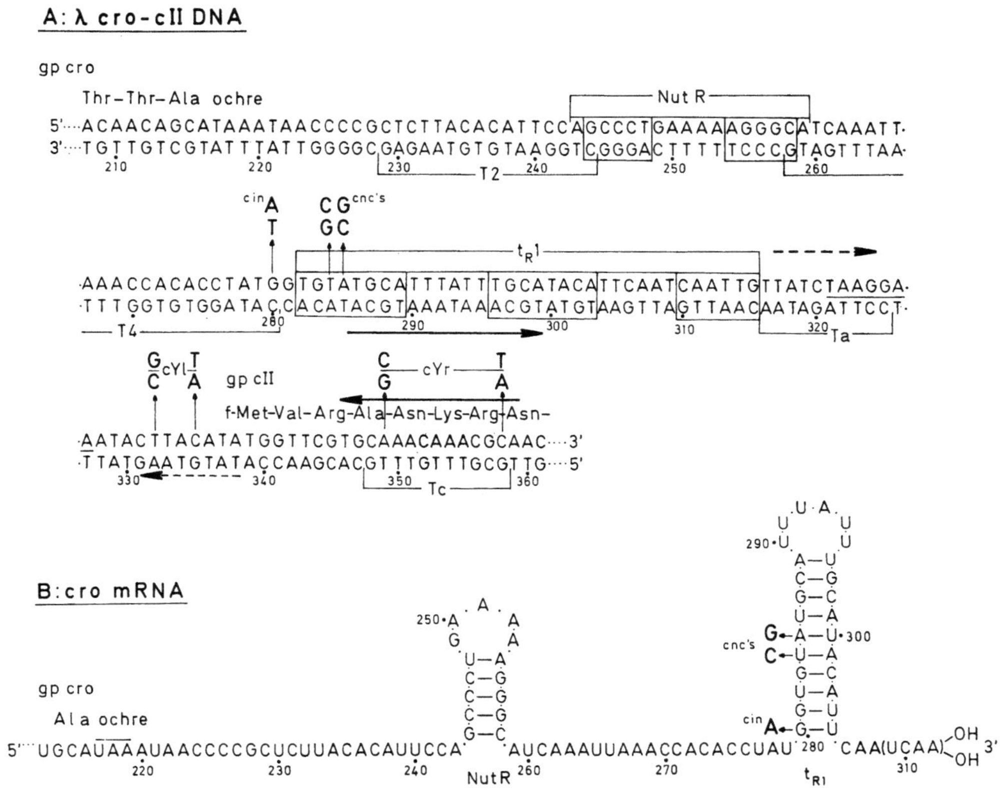


**Fig. 14.7**: Nucleotide sequence of the *cro-cII* intercistronic region with *Nut* R and $t_{R1}$ (from Schwarz et al. [33] and Rosenberg et al. [34]).

The DNA sequence is shown in part **A**, starting at the end of gene *cro* through the intercistronic region to the beginning of gene *cII*. The gp *N* utilization site *(Nut R)* and the terminator of the *cro* transcript *($t_{R1}$)*, abolished by gp *N*, are sequences with high degree of rotational symmetry (marked by *boxes*). These enable the RNA or the non-coding DNA strand to form stem loop structures (shown in part **B** of the figure for the RNA) which presumably serve as recognition sites for gp *N* and the ϱ-factor. Two types of mutations in the $t_{R1}$ region enhance *(cin)* or abolish *(cnc)* ϱ-mediated termination (cf. results in Table **14.4**). Mutation *cin* increases, mutations *cnc* decreases the stability of the $t_{R1}$ stem **(B)**. For the meaning of the four oligonucleotides T2 and T4 proximal and Ta and Tc distal to the termination signal cf. legend to Table **14.4**. It is a remarkable fact that in addition to *Nut* R and $t_{R_1}$ the intercistronic region between genes *cro* and *cII* contains a further cis-acting regulatory region (*cYl* and *cYr*, bp 330–360) which presumably serves as promoter ($p_{re}$) for the leftward establishment transcription of the *cI* gene (cf. **14.2.3.9**). This region is characterized by the indicated point mutations interfering with the establishment transcription. Furthermore, both *cYr* and *cYl* are inverted repeats of DNA sequences located at bp 285–300 and 317–324 respectively (cf. the *full* and *dashed arrows*).

To summarize, both termination factor ϱ and antitermination factor gp *N* need specific recognition sequences (*rut* and *Nut*). They are part of the scripton to be regulated but in some cases are localized at a considerable distance from the termination signal itself. It is worthwile mentioning in this respect that *Nut L* situated immediately downstream from promoter $p_L$, and *rut L*, located about 250 bp upstream of $t_{L1}$, both influence the termination events not only at $t_{L1}$ but also at $t_{L2}$ which is still about 1300 bp downstream from $t_{L1}$. As assumed for *rut* (cf. Fig. **14.7**) *Nut*

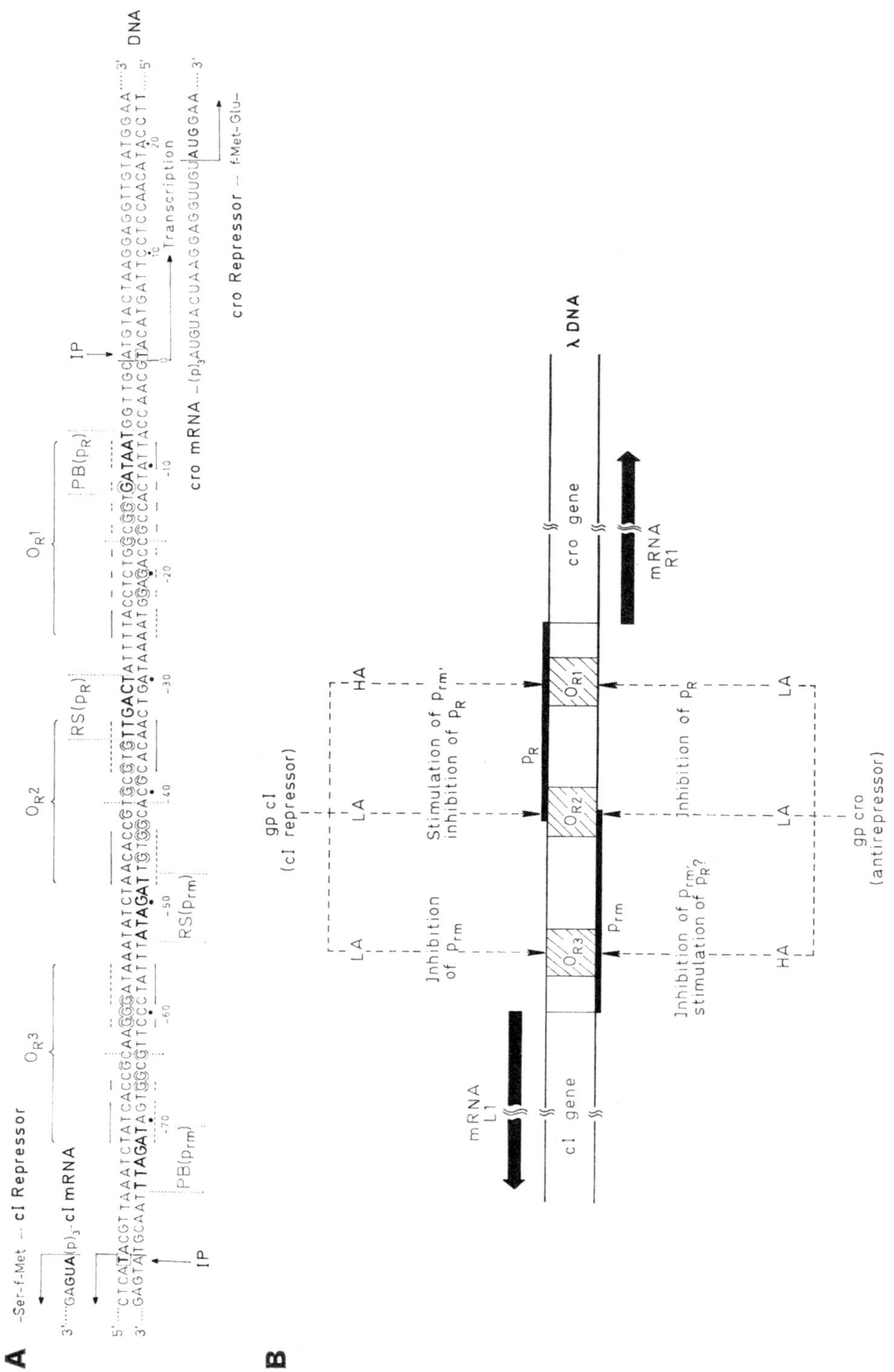
A
cI Repressor
cI mRNA
cro mRNA
cro Repressor
Transcription
DNA
B
gp cI
(cI repressor)
gp cro
(antirepressor)
Inhibition of $p_{rm}$
Stimulation of $p_{rm}$, inhibition of $p_R$
Inhibition of $p_R$
Inhibition of $p_{rm}$, stimulation of $p_R$?
cI gene
cro gene
mRNA L1
mRNA R1
λ DNA

Fig. 14.8

must not be recognized on the DNA but as well on the nascent RNA chain. It is evident that RNAP · gp *N* passing the *Nut* sequence is somehow stabilized against the termination effect of rho but the molecular details are totally unclear. In contrast to the antitermination effect in the attenuation model (cf. Fig. **4**.19) there is no evidence for a role of ribosomes in λ-phage termination.

### 14.2.3.5 Antitermination by gp *Q*

The *Q*-gene product (gp *Q*) is far less well investigated than gp *N*. Like its prominent brother it modifies RNAP to overcome two strong gp *Q*-specific termination signals, $t_{Q1}$ and $t_{Q2}$ situated 15 bp and 198 bp downstream from the initiation point of promoter $p_Q$. The short leader RNA sequences present in all transcription phases (cf. Fig. **14**.6) are thus extended into the morphogenesis region. The termination signals $t_{Q1}$ and $t_{Q2}$ are *ϱ-in*dependent. So far there is no experimental evidence for the localization of a hypothetical *Q*-utilizing sequence *(Qut)* which should exist because of the scription-specific action of gp *Q*.

### 14.2.3.6 The Two Repressors gp *cro* and gp *cI*

The *cro*-gene product (gp *cro*) is the product of the immediate early transcript (R1) synthesized rightward from promoter $p_R$. It is a protein of MW 7360 with 66 amino acid residues. It binds as a dimer with its C-terminal domains to operators $O_L$ and $O_R$, thus counteracting the effect of *cI* repressor [40, 40a].

The *cI*-gene product (gp *cI*, *cI* repressor) is the only prominent product of phage gene expression in the lysogenic state. It consists of 236 amino acid residues (MW 30 kd) binding as a dimer to the same operator sites as gp *cro*. The interaction of both regulatory molecules will be described in detail. The *cI* repressor is built of two domains that are decisive for its different functions in the life cycle of the λ-phage. The amino-terminal domain (amino acid residues 1–92) is joined by a connector of about 40 amino acids with the carboxyl-terminal domain (amino acid residues 132 to 236). The former domain interacts with the operator DNA, whereas the latter serves

◀

**Fig. 14.8**: Sequence of the *cI-cro* intercistronic region and the regulatory influence of gp *cI* and gp *cro* on the leftward and rightward transcription (from Szybalski [23]).

**A**: The nucleotide sequence contains (from *left* to *right*) the start of the *cI* gene (transcript L1), the *cI* gene promoter ($p_{rm}$, −40 to −80), three binding sites for the two regulatory proteins *($O_{R1}$, $O_{R2}$* and *$O_{R3}$)*, the *cro* gene promoter ($p_R$, −40 to +1) and the begin of the *cro* gene (transcript R1). $p_{rm}$, $p_R$ and the three operator sequences physically overlap indicating the close interaction of RNAP and the two antagonistic λ repressors (gp *cI* and gp *cro*) in this region. The promoter sequences are characterized by the usual RNAP recognition sites *(RS)* and Pribnow box *(PB)*. $O_{R1}$–$O_{R2}$ exhibit sequence elements with inverse rotational symmetry *(arrows)*. Guanin residues protected against chemical methylation by binding of gp *cI* and gp *cro* are marked by ⌒ and ⌣ respectively.

**B**: The multifold positive and negative influence of the two antagonistic repressors gp *cI* and gp *cro* are summarized in the scheme. They stimulate or inhibit not only their own expression but, by inhibiting the expression of the antagonistic repressor, also tend to keep the phage in distinct transcriptional states (lysogenic state in the presence of gp *cI*, lytic state in the presence of gp *cro*; cf. also Fig. **14**.9).

for dimerization. High affinity binding to DNA is only observed with *cI* dimers. Phage induction involves the proteolytic cleavage of *cI* repressor between amino acid residues 111 and 112 (cf. Fig. **14**.10) and hence a drastic reduction of operator binding affinity.

The experimental work with the λ-phage system and particularly the investigations on the gene expression programs after prophage induction were decisively facilitated by using mutants with temperature sensitive (ts) *cI* repressor. In these mutants alterations in the amino acid sequence of the repressor led to a molecule which operates normally at about 30 °C (permissive temperature) but is rapidly inactivated, e.g., by changing its conformation, if the lysogenic bacteria are transferred to 41 °C (non-permissive temperature). Hence such lysogenic strains can be induced coordinately at a high rate by shift-up of the temperature.

### 14.2.3.7 Regulatory Interaction of gp *cro* and gp *cI* in the $O_R$ Region

The intercistronic region between genes *cro* and *cI* comprises about 80 bp with three binding sites ($O_{R1}$, $O_{R2}$ and $O_{R3}$) for the two repressor proteins, gp *cro* and gp *cI*. The complete nucleotide sequence of this region is shown on top of Fig. **14**.8. The three operator sites are formed by 17 bp each and show a high degree of rotational symmetry. The two A=T-rich spacers in between belong to two promoter sequences ($p_R$ and $p_{rm}$) for the rightward (*cro* gene) and leftward (*cI* gene) transcription respectively. The high density of regulatory information within these 80 bp DNA sequence reflects its central role in the decision between the lysogenic and lytic pathways of phage λ development (cf. **14**.2.3.8 and **14**.2.3.9).

Detailed analyses of the complex interactions of the two regulatory proteins and RNAP, starting transcription from $p_R$ and $p_{rm}$, revealed the following facts [40]:

1. Both gp *cI* and gp *cro* bind as dimers to the minor groove of the DNA double helix.
2. Gp *cI* has highest affinity for $O_{R1}$ and much lower for $O_{R2}$ and $O_{R3}$. However, because of a cooperative effect between two gp *cI* dimers in adjacent operators, $O_{R1}$ and $O_{R2}$ are practically always occupied simultaneously. In this configuration transcription from $p_{rm}$ is stimulated and that from $p_R$ is repressed (Fig. **14**.8, bottom). Only very rarely does the concentration of gp *cI* increase to a level where $O_{R3}$ is occupied and hence transcription from $p_{rm}$ is also repressed. This positive and negative autogenous control system for the leftward transcript starting at $p_{rm}$ (L1) is typical for the lysogenic state (cf. Fig. **14**.6, phase A).
3. In contrast to gp *cI*, gp *cro* binds most firmly to $O_{R3}$ and less to $O_{R2}$ and $O_{R1}$ without cooperativity between gp *cro* dimers. A faint autogenous positive control by gp *cro* is somehow connected with its high-affinity binding to $O_{R3}$, whereas the autorepression of gp *cro* reflects binding to $O_{R1}$ and $O_{R2}$ at higher concentrations (Fig. **14**.8, bottom). Association of gp *cro* with all three operators interferes with transcription from $p_{rm}$ (*cI* transcription). The described regulatory effects of gp *cro* characterize the transcription phases B–D in the lytic cycle (cf. Fig. **14**.6).

Since both repressor proteins have been isolated and purified to homogeneity it was possible to demonstrate their multifold regulatory effects in the *cI*-*cro* intercistronic region in an in vitro assay system (Fig. **14**.9).

Transcription under the control of the two proteins was measured with a Hae III restriction DNA fragment of 790 bp (cf. top of Fig. **14**.9). The two RNA fragments synthesized by the in vitro system were separated by gel electrophoresis. In part A of Fig. **14**.9 the amount of gp *cro* in the incubation mixture was increased and that of gp *cI* remained constant, in part B the opposite holds true.

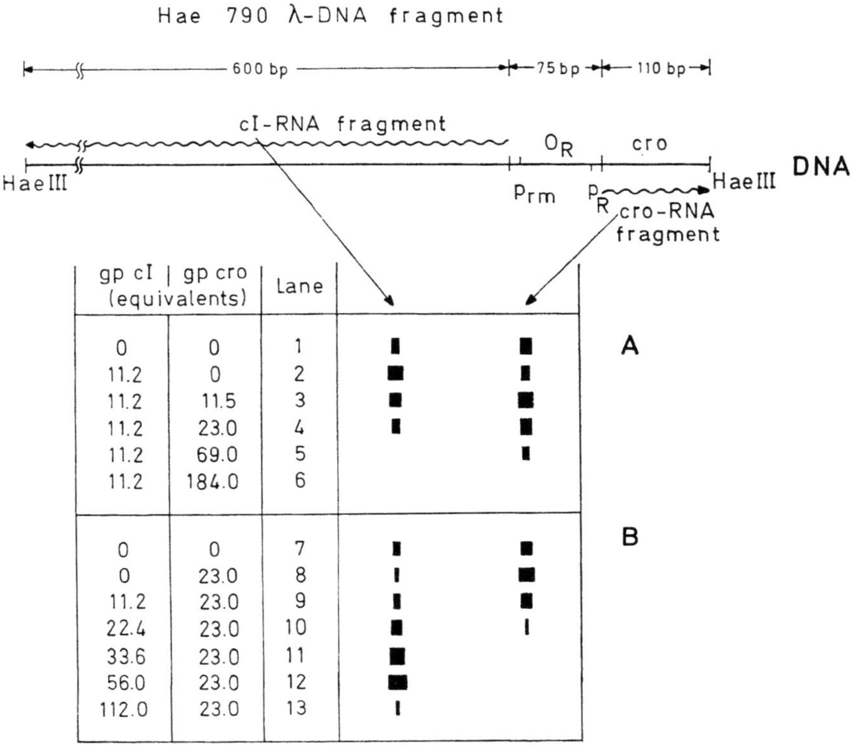


| gp cI (equivalents) | gp cro (equivalents) | Lane |
|---|---|---|
| 0 | 0 | 1 |
| 11.2 | 0 | 2 |
| 11.2 | 11.5 | 3 |
| 11.2 | 23.0 | 4 |
| 11.2 | 69.0 | 5 |
| 11.2 | 184.0 | 6 |
| 0 | 0 | 7 |
| 0 | 23.0 | 8 |
| 11.2 | 23.0 | 9 |
| 22.4 | 23.0 | 10 |
| 33.6 | 23.0 | 11 |
| 56.0 | 23.0 | 12 |
| 112.0 | 23.0 | 13 |

**Fig. 14.9:** In vitro transcription products from a Hae III λ-DNA fragment containing parts of the genes *cI* and *cro* and the $O_R$ region. Influence of *cro* and *cI* repressors (after Takeda [38]).

A restriction fragment of 790 bp was obtained from λ-DNA by cutting with endonuclease Hae III (cf. Table **2**.1). Besides the whole operator/promoter region ($O_R$, $p_R$, $p_{rm}$) it carries a small adjacent fragment of gene *cro* and a larger one of gene *cI* (cf. the map at the *top* of the figure). The two in vitro transcription products can be easily distinguished by their size after electrophoretic separation on polyacrylamide gels (A, B). The mixture for in vitro transcription contained the necessary substrates and cofactors, 0.08 μg/50 μl of the Hae III fragment and the indicated amounts of gp *cro* and/or gp *cI*. The original concentrations for the repressor proteins were recalculated to give molar equivalents with respect to the operator regions present in the incubation mixture (= 1). Both repressors were assumed to bind as dimers.

The following effects are observed (cf. also Fig. **14**.8, bottom):

gp *cro:*
a) inhibits transcription from $p_{rm}$ (lanes 2–6),
b) at lower concentrations faintly stimulates transcription from $p_R$ (autogenous positive control, lanes 1, 4 and 7, 8) and
c) at higher concentrations turns off transcription from $p_R$ (autogenous negative control, lanes 3–6).

gp *cI:*
a) inhibits transcription from $p_R$ (lanes 8–13),
b) at lower to medium concentrations stimulates transcription from $p_{rm}$ (autogenous positive control, lanes 8–13) and
c) at higher concentrations turns off transcription from $p_{rm}$ (autogenous negative control, lanes 12, 13).

A close inspection of Fig. **14**.9 also gives an impression of what is going on shortly after phage $\lambda$ infected a host cell, i.e., when both repressors are formed simultaneously. The binary decision between lysogenic or lytic pathway, in fact, depends on whether gp *cI* or gp *cro* become dominant over its counterpart. Thus, in this situation the operator region $O_R$ functions as a bi-stable switch with $O_{R2}$ as a kind of transition element: the repressor which, after binding to its high affinity site ($O_{R1}$ and $O_{R3}$ respectively), first occupies $O_{R2}$ determines the regulatory response and hence the developmental pathway to be taken [38].

Many other details of the interactions between DNA, RNAP and the two regulatory proteins are far from clear. Whereas the repressor effects may simply be due to steric hindrance of the binding and/or propagation of RNAP, the activator effects may involve subtle changes in the DNA conformation and/or direct contacts between the polymerase molecule and gp *cI* or gp *cro* (cf. the discussion on the interaction of the catabolite activating protein and the araC protein with RNAP, Sects. **4**.4.2.2 and **5**.5.6). Special problems are also posed by the mutual interaction between the two opposing repressor proteins and the alterations in the regulatory effects of gp *cro* observed when RNAP is modified by gp *N*, e.g., in transcription phases C and D (cf. Fig. **14**.6).

### 14.2.3.8 Prophage Induction – An SOS Phenomenon

Prophage induction requires decrease of *cI* repressor concentration in the cell to a level where transcription from $p_L$ and $p_R$ can proceed and hence gp *N* and gp *cro* are formed. Such a situation might occur if during bacterial conjugation the prophage is transferred from the lysogen to a repressor-free receptor cell (zygotic induction). On the other hand transition from the lysogenic to the lytic pathway is triggered by interference with host DNA synthesis or structure, e.g., by UV or X-ray irradiation, by treatment with base analogues or mitomycin C, by thymin deprivation etc. Generally speaking, all cancerogenous agents may lead to prophage induction and hence the lysogenic *E. coli* cell is a convenient test system for cancerogens. This "natural" induction permits the prophage to escape from a sickly or dying cell. It takes more than half of an hour and involves complex regulatory circuits connected with the so-called SOS functions [**41**, **42**].

Briefly, damages in the DNA (presumably pieces of single-stranded DNA) act as signals triggering the coordinate expression of SOS functions which include (a) induction of prophage $\lambda$, (b) expression of colicinogenic plasmids, (c) an error-prone DNA repair activity connected with mutagenesis (SOS repair), (d) inhibition of cell division that leads to filamentation, (e) inhibition of postirradiation DNA degradation and (f) "aberrant" reinitiation of DNA synthesis at the chromosomal origin. Induction of SOS functions requires the activity of two host regulatory genes *recA* and *lexA*. A current hypothesis of the SOS phenomena and the interaction of gp *recA* and gp *lexA* is presented in Fig. **14**.10.

The dicisive event for $\lambda$ phage induction is the proteolytic degradation of the *lex A* gene product that functions as repressor of the *recA* gene. The result is an overproduction of gp *recA*. The free *recA* protein has activity of an ATP-dependent protease, cleaving the *cI* repressor in its hinge region between amino acids 111 and 112 and hence drastically decreasing its operator binding affinity (cf. **14**.2.3.6). The protease function of highly purified gp *recA* was shown to depend on the association with single-stranded DNA pieces (optimum concentration about 6 nucleotides per 40 kd monomer of gp *recA*). This proprotease complex is activated by binding ATP as effector, but ATP hydrolysis is not required for the reaction [43, 45]. The *recA* gene product is a bifunctional enzyme. Besides its protease activity it catalyzes DNA strand exchange leading to recombinational DNA repair. In the latter case, however, ATP hydrolysis is a necessary step.

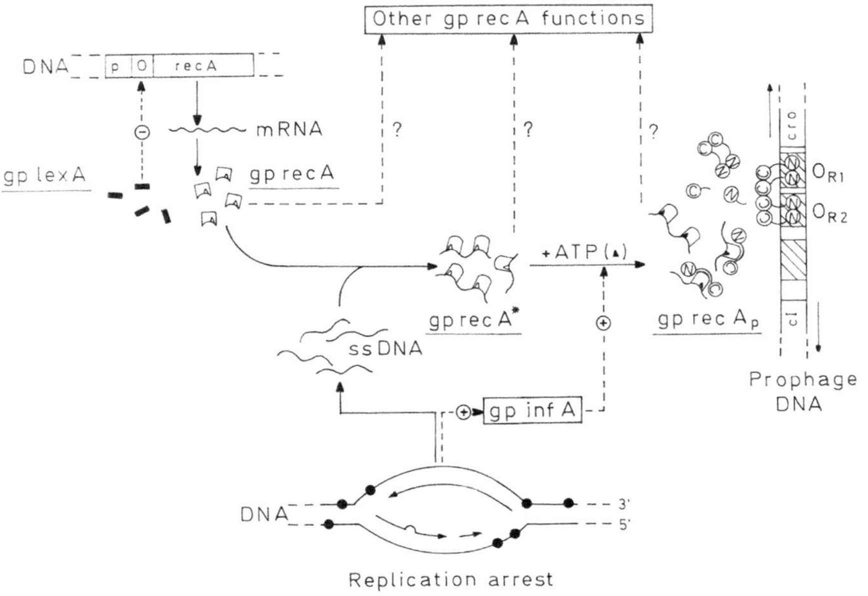


**Fig. 14.10**: Role of the *recA* gene product for prophage induction (from Bailone et al. [44], Craig & Roberts [45] and Little et al. [46]).

The *recA* gene product occurs in three functional states: (1) the primary translation product (gp *recA*) with binding sites for ATP (▲) and ssDNA; (2) the DNA-activated form (gp *recA**) which, by combining with ATP, is transformed to (3), the protease state (gp $recA_{\mathrm{p}}$). The latter transformation is somehow dependent on the interaction with the *infA* gene product. Both the single-strand DNA pieces *(ssDNA)* required for state 2 and the *infA* gene product are formed under conditions of arrested replication, e.g., because of DNA lesions (●), thymine starvation or drug treatment. The gp *recA* proteases cleaves the *cI* repressor molecules ((N)⌒(C)) in the hinge region between the N- and C-terminal domains, thus decreasing the repressor concentration under the threshold required for prophage maintenance. Expression of the *recA* gene is repressed by the *lexA* gene product. Under conditions of gp *recA* protease formation the *lexA* gene product is proteolytically inactivated, too, resulting in an increased expression of gp *recA* [46].

### 14.2.3.9 Lysis Versus Lysogeny – The Regulatory Role of gp *cII*

The lytic and lysogenic responses are mutually exclusive – one leads to cell death, the other to survival of a host cell with modified genetic material. A striking feature in the choice between the lytic and lysogenic response is that the phage embarks a considerable distance on both before it is committed to one or the other. The decision between lysogenic and lytic response is thoroughly influenced by the physiological state of the host cell and the multiplicity of the phage infection. In rich media lysogenization is obtained only at a multiplicity of more than 5 phages/cell, whereas in minimal media already one phage/cell is sufficient. The overall biological effect is that phage $\lambda$ multiplies if the relative number of phages in a host cell population is low and if its physiological state is favorable. On the other hand, $\lambda$ takes a "waiting position", i.e., it lysogenizes its host cell, if the two prerequisites are not fulfilled.

Following infection of a nonlysogenic cell, transcription initiated at the promoters $p_L$, $p_R$ and $p_{rm}$ results in the formation of some proteins (gp *cro*, gp *O*, gp *P*, gp *xis* gp *Q*) that tend to push the cell toward lytic development and others (gp *cII*, gp

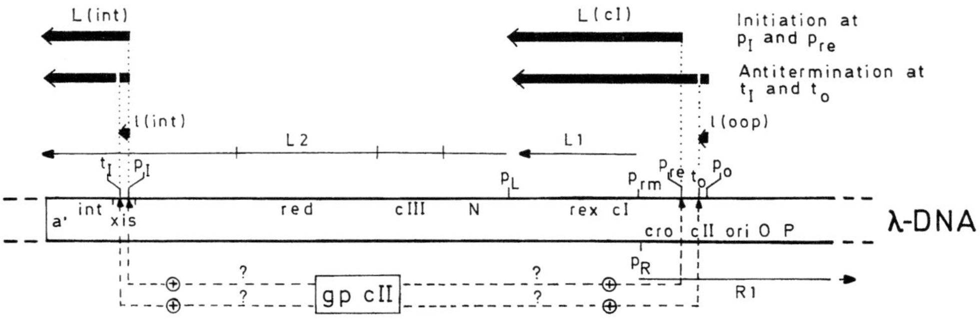


**Fig. 14.11:** Establishment of the lysogenic state is under the control of gp *cII*. Initiation vs. antitermination hypothesis.

The partial map of λ-DNA shows the regulatory region *P–int* (cf. the details in Fig. **14**.6). Some of the transcripts discussed in Sect. **14.2.3.3** are indicated by *thin arrows*, including the $p_{rm}$ directed maintenance transcript (L1) characteristic for the lysogenic state. In addition, *full arrows* at the *top* of the figure represent the establishment transcripts (L*(int)* and L*(cI)*) read under the influence of the regulatory protein gp *cII*. The two controversial hypotheses about gp *cII* action (initiation [49–51] vs. antitermination [23]) are discussed in the text. *l(oop)* and *l(int)*, putative leader RNAs initiated at promoters $p_0$ and $p_I$ and ended at terminators $t_0$ and $t_I$ respectively. As outlined in the text the recent sequencing data [50, 51] strongly favour the initiation hypothesis.

*cIII*, gp *cI*, gp *int*) that promote the lysogenic response. As described for phage induction (cf. **14**.2.3.3) an ordered sequence of transcription phases is observed. In the immediate early phase, 1–2 min after infection, the newly formed transcripts correspond roughly to those found in phases A plus B (cf. Fig. **14**.6), followed by a transition to a mixture of phases A plus C about 5 min after infection. The regulatory details involved are practically the same as discussed for the lysogenic state and the subsequent prophage induction, with the exception that the newly synthesized *cI* repressor is not destroyed, i.e., both antagonistic repressors (gp *cI* and gp *cro*) act on the intercistronic regulatory region $O_R$ (cf. **14**.2.3.7).

The late transcriptional phase after phage infection either corresponds to phase D (Fig. **14**.6) if the phage embarks on the lytic pathway or there is a stepwise repression of gp *cro* synthesis and transit to the prophage mode of transcription (phase A in Fig. **14**.6).

Two further regulatory proteins translated from phase C transcripts play a prominent role in pushing the developmental pathway to lysogeny. These are gp *cII* and a gp *cII*-stabilizing protein (gp *cIII*). Action of gp *cII* leads to the formation of two new leftward transcripts (Fig. **14**.11).

One suggestive explanation for the positive regulatory influence of gp *cII* on transcription of *cI* and *int* genes comes from the observation that, in all phases of λ-DNA transcription, two short leader RNAs are initiated at some distance from genes *cI* and *int* respectively (Fig. **14**.11). These might be elongated by an antitermination effect of gp *cII* (antitermination hypothesis). The presumptive leader RNA for *cI* starts from $p_0$ and ends at $t_0$ (*opp* RNA, 81 nucleotides) that for gene *int* starts from $p_I$ and ends at $t_I$ (*int* leader about 60 nucleotides) [23].

In contrast to the antitermination hypothesis, gp *cII* is assumed to act as initiation factor helping host RNAP to recognize internal phage promoters otherwise not used, i.e., $p_I$ and a presumptive promoter $p_{re}$ for establishment of *cI* repressor dominance (initiation hypothesis). The $p_{re}$ promoter was assumed to map within the intercistronic region between genes *cro* and *cII* (cf. the DNA sequence in Fig. **14**.7) because a number of closely related, cis-dominant mutations in this region ("y"-region, bp 364 to 319 of Fig. **14**.7) interfere with gp *cII* stimulation of *cI* transcription.

The initiation hypothesis is strongly supported by sequence analyses of the two putative promoter regions [50, 51], which, at least partly, overlap with protein coding regions. The regulatory impact of this particular arrangement will be discussed below:

```
 - - - Asn Arg Lys Asn Ala Arg Val Metf
              gp cII  <-----------          A    C    C
                                            ↑    ↑    ↑
5'-CTCGTTGCGTTTGTTTGCACGAACCATATGTAAGTATTTCCTTAGA-3'   pre
      -40         -30         -20         -10         1
5'-TTTCTTGCGTGTAATTGCGGAGACTTTGCGATGTACTTGACACTTC-3'   pI
                                        ↓
                                        A
                              fMet Tyr Leu Thr Leu Gln - - -
                              ------> gp xis
```

Both promoters contain highly A=T-rich regions with somewhat distorted Pribnow boxes at —10 (compare the bold-faced residues with the sequence of the *E. coli* consensus promoter in Table 1.8: TATAAT). In addition, a large stretch with almost identical base pairs is centered around —32 (underlined). It is suggested that this sequence element is responsible for the binding of gp *cII* which, in turn, facilitates the interaction of RNAP with these unusual promoter types.

The two Pribnow boxes are characterized by mutations altering promoter functions. Two mutations in the $p_{re}$ region (cf. cY1 in Fig. **14**.7), interfer with the gp *cII*-dependent transcription initiation because they lead to a further deviation from the *E. coli* consensus sequence. In contrast, the mutation in $p_I$ results in an "improvement" of this promoter and thus in a gp *cII*-*in*dependent int$^c$ phenotype.

The regulatory phenomena connected with the peculiarities of gene structure and hence with transcription and translation of the two establishment transcripts (L*(cI)* and L*(int)*, cf. Fig. **14**.11) enlarge the fascinating repertoire of control mechanisms contributing to the highly efficient developmental program of phage $\lambda$:

1. The $p_{re}$ transcript (L*(cI)*) starts in opposite direction to the $p_R$ transcript (R1) which includes genes *cII* and *Q*. The convergent transcription from $p_{re}$ and $p_R$ is mutually exclusive. Thus, presence of sufficient levels of gp *cII* turns off R1 transcription, i.e., gp *cII* formation is self-limiting and, because gp*Q* is lacking, transcription of the "late" $\lambda$-genes (R2) is blocked (cf. Fig. **14**.6).

2. Comparison of the two transcripts used for gp *cI* translation shows another remarkable feature. Both differ largely in their translation efficiency, which is very bad for the $p_{rm}$ directed transcript (L1) starting immediately in vicinity to the AUG translation initiation codon (cf. the sequence Fig. **14**.8 top). In contrast, the establishment transcript (L*(cI)*) contains a long leader sequence with a ribosome binding site. Thus, the low and high levels of gp *cI* required for maintenance and establishment of the lysogenic state respectively are achieved, inter alia, by using two transcripts with different translation efficiency for the same gene product.

3. Two different transcripts are also used to adjust the appropriate levels of gp *int* (MW 40.3 kd) and gp *xis* (MW 8630 d) either to promote phage DNA integration (high levels of gp *int*, no gp *xis*) or excision (low levels of gp *int* plus gp *xis*, cf. Fig. **14**.5). The *xis*-*int* region is transcribed in phase C of the lytic development as part of the large leftward transcript (L2) starting from $p_L$ (Fig. **14**.6). If, however, in the presence of gp *cII* promoter $p_I$ is open, the same region is transcribed as a short

leftward transcript (L(*int*)) that starts ~ 15 bases downstream of the AUG initiator triplet of gp *xis*. Thus, this transcript, formed e.g., under conditions of establishment of the lysogenic state, contains a large fragment of the gp *xis* coding sequence as leader sequence and serves as template for translation of gp *int* only. Finally, the coding parts of both genes overlap by 20 bp with gene *xis* upstream of gene *int*. It is evident that translation of the *xis* message blocks ribosome binding upstream of the *int* message, i.e., this part of the L2 transcript gives high levels of gp *xis* and low levels of gp *int* as is required for excision. These interesting features of a combined transcription and translation control are illustrated in the following scheme showing the basic structural elements of the two transcripts [50, 51]:

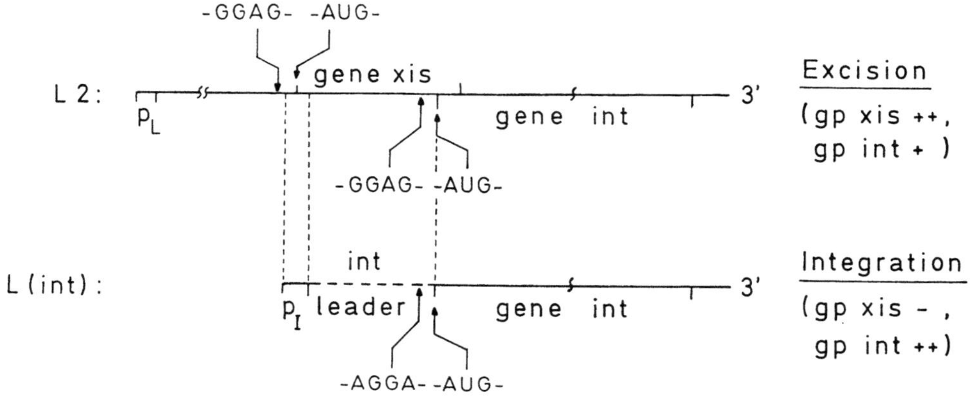


In summary, we can emphasize the following facts:

1. In the late phase after phage infection two additional leftward transcripts starting at $p_{re}$ and $p_I$ respectively augment the production of gp *cI* and gp *int*.
2. Formation of the two new transcripts is totally dependent on the preceding synthesis of the regulatory protein gp *cII* which has an extremely short half-life and is somehow stabilized by gp *cIII*.
3. Establishment of the appropriate levels of *cI* repressor as well as of the integration-excision proteins (gp *int* and gp *xis*) is thoroughly influenced by the coding capacity and translation efficiency of the four types of transcripts involved.

### 14.2.3.10 Post-transcriptional Control of Synthesis of Morphogenetic Proteins

As described in Sect. **14.2.3.3** transcription of the more than 20 genes of the λ morphogenesis region starts from promoter $p_Q$. It is catalyzed by the gp *Q*-modified RNAP. The primary transcription product is a very large polycistronic mRNA, which later on is processed into five or more pieces. The mRNA concentration of different parts of the *A–J* region was tested by hybridizing in vivo 5-[$^3$H]-uridine labelled RNA with different deletion mutants containing increasing length of the *A–J* DNA. It turned out that nearly equimolar amounts were present irrespective whether short- or long-term labelled RNA was used (Fig. **14.12**). In contrast, the relative amounts of proteins synthesized from the individual genes vary about 870-fold with gp *A* about 350 and gp *D* about 280,000 copies per cell. Hence the concentration of the individual proteins required for morphogenesis is exclusively regulated at the posttranscriptional levels. Basically similar results were reported also for the two other phages (T4 and T7) discussed in this chapter. The regulatory principles involved are still totally unknown. All mechanisms found to participate

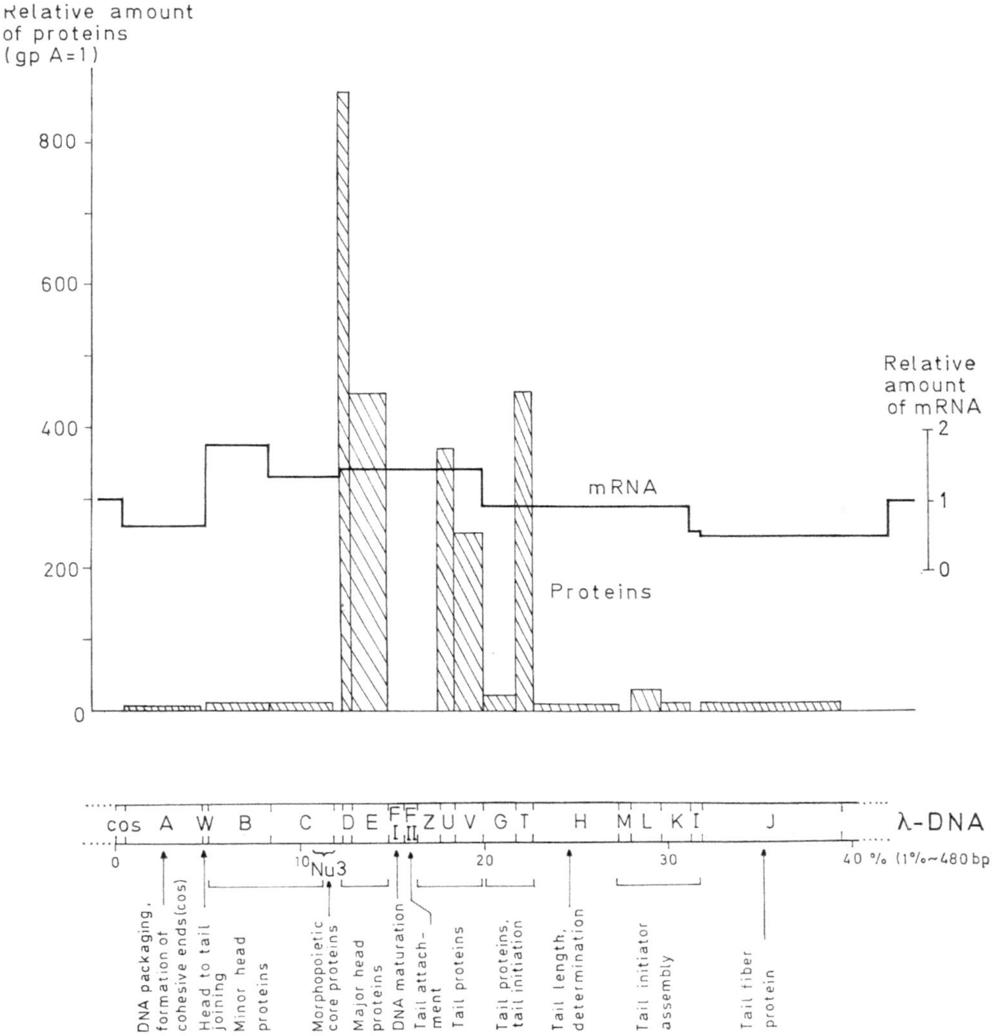


**Fig. 14.12:** Relative amounts of mRNA and number of protein copies synthesized from the late genes (*A–J* region) of phage *λ* (after Ray & Pearson [52]).

The *horizontal line* in the *upper part* indicates the relative concentrations of mRNA as analysed by DNA/RNA hybridization with 5-[$^3$H]-uridine labelled RNA. The *height* of the *bars* represents the relative number of protein copies per gene per cell normalized to gp A. In the *lower part* a detailed map of the left arm of the *λ*-chromosome shows the *A–J* region. The map has been corrected according to the data given by Echols & Murialdo [30], Szybalski & Szybalski [31] and the recent finding that gene *Nu 3* is part of gene *C* (Shaw & Murialdo [53]).

in posttranscriptional control of protein synthesis (cf. Chap. **4**) might contribute to the overall effect, e.g., processing of the polycistronic mRNA, biological but not chemical inactivation of parts of the mRNA, preferred translation of individual mRNA pieces because of differences in the ribosome binding sites and/or interaction with regulatory proteins etc.

Finally it should be mentioned that morphogenesis of phage $\lambda$ proceeds essentially in the same manner as will be described for T4 in the following sections. Phage heads and tails are assembled from the preformed proteins with key steps controlled by morphopoietic proteins and by proteinogen processing. $\lambda$-DNA synthesized in the medium phase of the lytic pathway is packaged into the empty head shells. After joining heads and tails the mature phage particles are released into the medium [**55**].

Recently some interesting details appeared of the organization and expression of four genes involved into $\lambda$ prohead assembly [53], i.e., of genes *B*, *C*, *Nu3* and *E*. gp *Nu3* acts as a scaffolding protein in prohead construction, directing the association of the major capsid protein gp *E*. It is subsequently degraded, whereas the other minor protein gp *C* is later on fused to a minor fraction of gp *E*, and the protein-hybrid is cleaved proteolytically to generate two proteins X1 and X2, which constitute minor components of the mature prohead.

Exact mapping of the morphopoietic gene *Nu3* revealed that it represents an example of an overlapping gene, its coding sequence being part of gene *C*, i.e., gp *Nu3* ($M_r$ 19 kd) represents the C-terminal part of gp *C* ($M_r$ 58 kd). It is a remarkable aspect of the control of gene expression in the *A–J* region that two- to threefold more gp *Nu3* is synthesized than gp *C*, suggesting a strong internal ribosome binding site for *Nu3* translation within gene *C*. Furthermore translation of the *C* gene message interferes with translation of the *Nu3* gene message. The very simple explanation that gp *Nu3* is in fact a proteolytic processing product of gp *C* was ruled out.

### 14.2.4 Morphogenesis of Phage T4

The virulent bacteriophage T4 belongs to the most complicated and fascinating viruses of *Escherichia coli*. This refers to the size and organization of its genome as well as to the structure and mechanisms of assembly of the phage particles. Morphogenesis of bacteriophages is a characteristic part of late gene expression programs in the infected bacteria. Unlike the early stages of development, they are determined by controls on the post-transcriptional level. Here we will deal only with one well--investigated aspect of phage T4 morphogenesis: the ordered assembly of the phage particles out of preformed elements. The morphogenetic program, i.e., the coordination of the single steps to regular sequences involving helper (morphogenetic) proteins and precursors of some structural proteins, may serve as a model for morphogenesis of other viruses (cf. Table **14**.1) and moreover of organelles in eukaryotic cells (cf. e.g., the transformation of the prolamellar body of proplastids during the light-induced biogenesis of chloroplasts, Chap. **6**).

#### 14.2.4.1 The Genetic Map of T4 and Characterization of Genes Involved in Morphogenesis

Besides numerous genes for DNA synthesis and for realization of the genetic information of phage T4 there are at least 45 genes for its morphogenesis. They map in several clusters scattered on the whole genome. Most of them are transcribed as late genes about 10–20 min after infection.

Effective transcription of late genes is evidently dependent on T4-DNA replication, i.e., on the presence of replicative intermediates in the cell (DNA with short single-strand gaps), which later on are joined. Late gene transcription without replication is only possible in phage mutants with defects in gene 30 (DNA ligase) and 46 (T4 exonuclease) [57].

Morphogenesis genes and the proteins coded by them were characterized by isolation and genetic analysis of numerous conditional mutants (nonsense or ts-mutants) as well as by gel electrophoretic separation of the proteins. Especially experiments with nonsense mutations in one of the morphogenetic genes helped to work out the fundaments of phage T4 morphogenesis. If such nonsense mutants are used to infect a non-permissive wild-type host cell intermediates of the phage assembly line accumulate as shown in the squares around the genetic map (Fig. 14.13). Apparently,

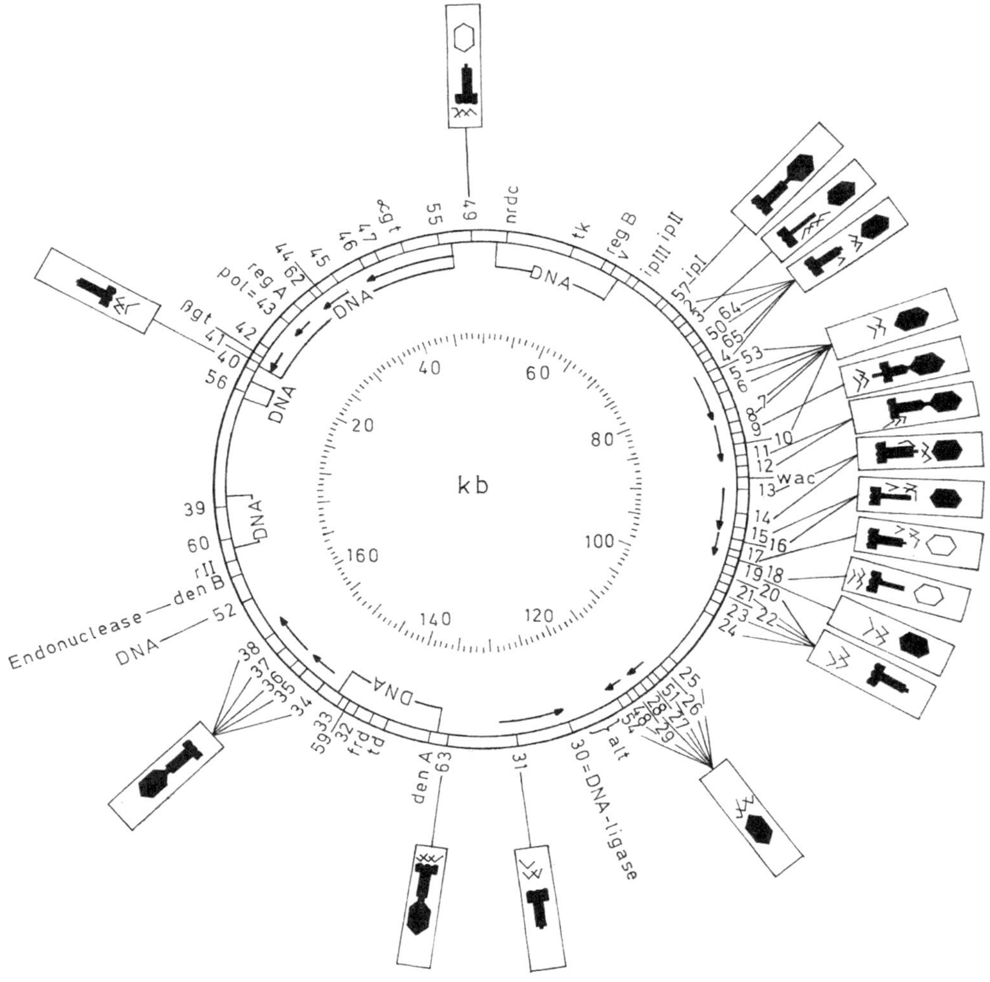


**Fig. 14.13:** Genetic map of phage T4 with particular emphasise on the morphogenesis genes (compiled from Wood et al. [54] and Mosig, G., In: CRC Handbook of Biochemistry and Mol. Biol., Nucleic Acids, Vol. II, CRC Press 1976, pp. 664–675).

The about 45 T4 morphogenesis genes map in several clusters at diverse sites of the phage genome. They belong to several transcription units denoted by *arrows*. The structural symbols in *boxes* arranged around the T4 genome indicate the components accumulating in cells that were infected with mutant phages defective in the particular morphogenesis gene. Genes participating in T4 DNA synthesis are marked on the inner circle. For details cf. Ref. [54, 57].

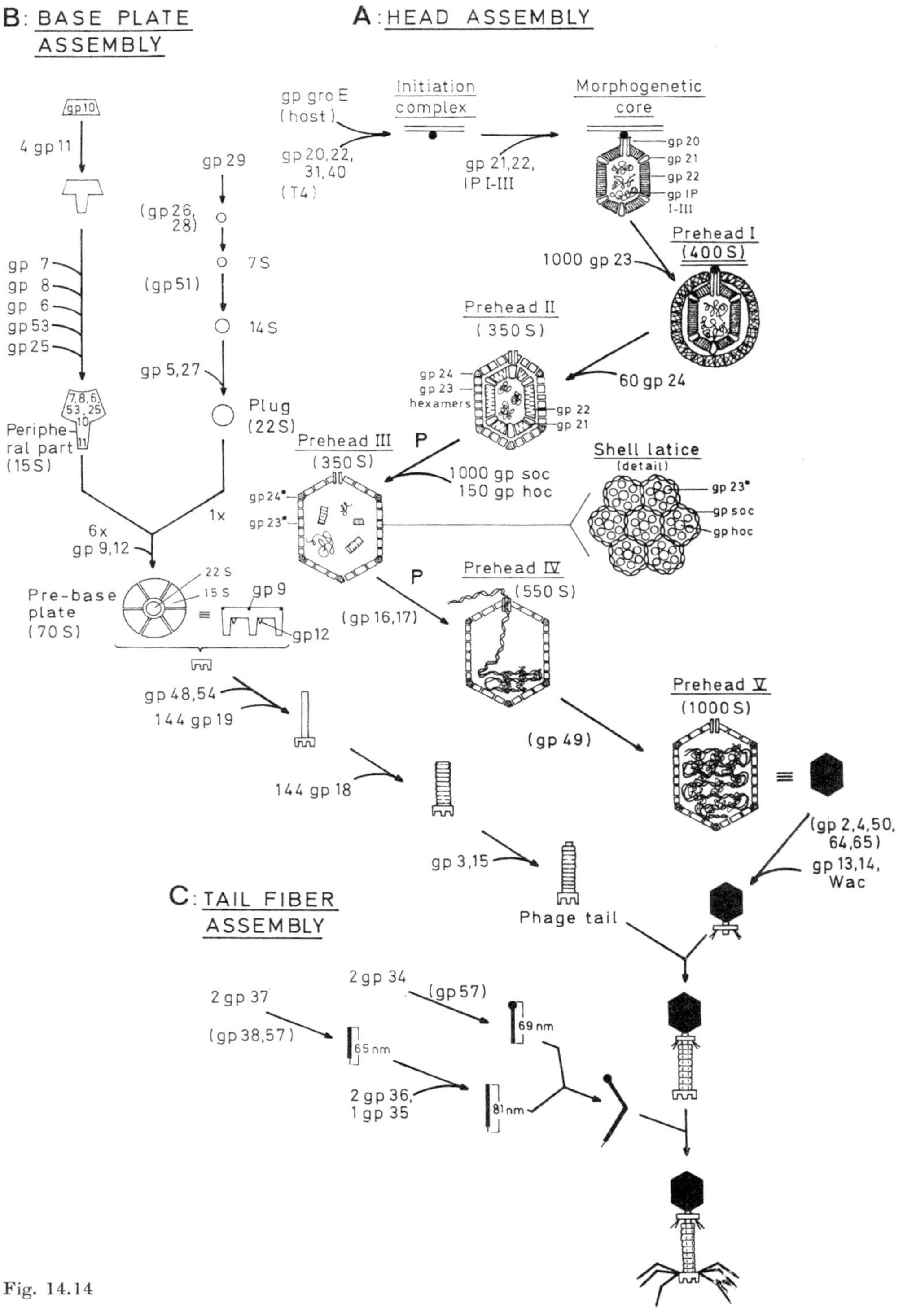
B: BASE PLATE ASSEMBLY
A: HEAD ASSEMBLY
C: TAIL FIBER ASSEMBLY
gp 10
4 gp 11
gp 7
gp 8
gp 6
gp 53
gp 25
7,8,6 53,25 10 11
Peripheral part (15S)
gp 29
(gp 26, 28)
7S
(gp 51)
14S
gp 5,27
Plug (22S)
6x gp 9,12
1x
Pre-base plate (70S)
22 S
15 S
gp 9
gp12
gp 48,54
144 gp 19
144 gp 18
gp 3,15
Phage tail
gp groE (host)
gp 20,22, 31,40 (T4)
Initiation complex
gp 21,22, IP I-III
Morphogenetic core
gp 20
gp 21
gp 22
gp IP I-III
Prehead I (400S)
1000 gp 23
Prehead II (350S)
60 gp 24
gp 24
gp 23 hexamers
gp 22
gp 21
P
Prehead III (350S)
1000 gp soc
150 gp hoc
gp 24*
gp 23*
Shell latice (detail)
gp 23*
gp soc
gp hoc
P
Prehead IV (550S)
(gp 16,17)
Prehead V (1000S)
(gp 49)
(gp 2,4,50, 64,65)
gp 13,14, Wac
2 gp 37
(gp 38,57)
65 nm
2 gp 34
(gp 57)
69 nm
2 gp 36, 1 gp 35
81 nm

Fig. 14.14

certain mutants can only make heads and tail-fibres, others only tail-fibres and tails but no heads, and still others only defective tail structures and empty heads. The analysis of these cell extracts from non-permissive hosts already gave first hints at morphogenesis principles of phage T4: at the beginning heads, tails and tail-fibres of the phage are made independently of each other. Particles without tail-fibres can develop out of heads and tails, however, tail-fibres and tails never join without an intact phage head. Further details of the morphogenesis were elucidated with in vitro complementation experiments. Extracts from suppressor negative cells infected with different nonsense mutants of T4 were mixed. The mixture may produce intact, infectious phage particles, if appropriate mutants are combined.

The following example demonstrates the procedure: In extracts from cells, infected with a phage defective in genes 13 and 14, in which for that reason the joining of head and tail structures did not take place, the formation of infectious phage particles in vitro could be obtained if extract of a "head-gene mutant", but not if extract of a "tail-gene mutant" was added. The function of genes 13 and 14 consequently must be connected with the completion of heads as a prerequisite to the association with tail structures. This procedure with the corresponding mutants of phage $\lambda$ recently gained great importance as method for in vitro encapsidation of large, semisynthetic DNA molecules for cloning experiments (cf. **2**.5).

### 14.2.4.2 Morphogenesis Proceeds by Three Independent Patways

In Fig. **14**.14 the multistage morphogenesis process of phage T4 is outlined. Besides the essential structural proteins, the deficiency of which leads to more or less abnormal phage particles, numerous morphopoietic proteins and, particularly in the heads, several accessory protein species are involved. Part A of the scheme shows the formation of the head shell, initiated by a membrane-bound morphogenesis scaffold, and the later encapsidation of the phage DNA. It is dominated by a sequential proteinogen processing catalyzed by a special phage-coded proteinase (Table **14**.5). Part B leads in three, at first independent sequences to the development of the complicated phage tail. Part C describes the generation of the tail fibres which are attached to the last intermediate of morphogenesis after the mature head and tail have been joined.

Upon infection of a host cell this tail serves as organell for recognition and attachment (base plate with tail fibres) and for injection of phage DNA and some helper proteins (tail tube and sheath). Shortly after phage attachment to the host cell the hexagonal base plate rearranges into a star-shaped configuration, releasing the tip of the tail tube and triggering the contraction of the sheath, which forces the tail tube through the bacterial cell envelope [65].

---

◀

**Fig. 14.14**: Morphogenesis program of phage T4 (compiled from Wood et al. [54], Casjens & King [55], Kellenberger [56] and Showe & Onorato [59]).

Morphogenesis starts from the preformed proteins on three independent pathways: **(A)** *head assembly* including DNA incorporation, **(B)** *assembly of* the *base plate* and the *tail* and **(C)** *assembly of* the *tail fibres*. For a detailed explanation of the individual steps cf. text. Gene products which become structural components of the phage particle or of its morphogenetic intermediates are indicated at the arrows with *gp* and the *gene numbers* adjoined. Morphopoietic proteins are put in *brackets*. An *asterisk* marks proteins that are proteolytically processed (cf. Table **14**.5).

### 14.2.4.3 The Four Stages of Head Assembly are Controlled by Proteinogen Processing and DNA Packaging

The phage head is a prolate icosahedron of triangulation number T = 13 elongated on its fivefold axis of symmetry with a length-to-width ratio of 1.37. More than 20 proteins are involved in its biogenesis only part of them being essential structural elements (gp 20, 23, 24, 13, 14), others are dispensable gp *ipI*, *ipII*, *ipIII*, *pip*, *wac*, *alt*, *hoc*, *soc*) and a third group has morphopoietic functions (e.g., gp 21, 22, 31, 40 etc.). Morphogenesis of the head proceeds in four steps: (1) prehead formation, (2) proteinogen processing and core removal, (3) DNA packaging and (4) head completion (Fig. **14**.14 A).

1. *Prehead Formation:* In the first step a morphogenesis scaffold (prehead core) is made from gp 21, 22, *ipI–III* and *pip* which, together with a gp 20 protein complex (15 molecules), form a polymerization complex at the bacterial cell membrane. The host protein gp *groE* takes part in this process, too. Gp *groE* is also necessary for the morphogenesis of other T-phages and phage λ. In the cell it is present as disc-like oligomeric complex with 7-fold rotational symmetry. Its function is not known in detail [62].

It is interesting to notice in this connection that an unknown early protein of T4 (but also of T7) functions as inhibitor of the host membrane-associated protease system which recognizes and degrades abnormal proteins in an ATP-requiring reaction (cf. **1**.3.6). This might be a prerequisite for the accumulation of phage proteins required for morphogenesis and for the stability of the head initiation complex [63].

Around the prehead core 1,000 molecules of the major head shell protein gp 23 polymerize to a hexagonal network (prehead I) stabilized at its icosaheder vertices by deposition of 60 molecules gp 24 (prehead II). For the formation of normal preheads the correct quantitative proportion of gp 22, 23 and 24 is evidently very important, i.e., it depends on the strict control of the late gene expression programs during phage development. If these programs are disturbed or if certain mutants are used for infection abnormal structures (e.g., miniheads, giant heads, polyheads, preheads) accumulate [59].

2. *Proteinogen Processing:* Incorporation of gp 24 starts an extensive morphogenetic proteinogen processing (Table **14**.5), which is in its essential steps prerequisite for incorporation of further structural proteins and for phage DNA encapsidation. Proteolytic enzyme is a phage-specific endoprotease being as a proenzyme (gp 21) part of the prehead core [58, 59]. After autocatalytic activation the enzyme cleaves gp 23 and 24 of the phage shell (gp 23, 24 → gp 23*, 24*), thus generating the conditions for complementation of the head shell network. One thousand molecules of gp *soc* and about 150 molecules of gp *hoc* are incorporated finally, resulting in a solid lattice of hexagonal units of the structure 6 gp *soc* + 6 gp 23* + 1 gp *hoc* (prehead III). In this respect, it is surprising that functional and almost normal phage shells are also made without addition of gp *soc* and *hoc*, i.e., by the corresponding nonsense mutants. Gp *soc* and gp *hoc* are dispensible proteins.

The proteinogen processing on its further course involves gp *ipI–ipIII*, gp *pip* and last not least gp 22, the major constituent (500–600 molecules) of the prehead core. The latter is degraded to oligopeptides. The proteolysis program, still going on during DNA encapsidation, ends apparently with the autocatalytic inactivation of the protease itself.

3. *DNA Packaging* [64] into the preformed head shells is ATP-dependent. Two special proteins localized at the neck opening (gp 20 and gp 17) may function as a kind of DNA pumping and sizing device. These assumptions are supported by the strong binding affinity of gp 20 for DNA and by the evidence for endonuclease activity of gp 17. As intermediate of the DNA encapsidation a complex sedimenting at 550 S can be isolated with about half of the total DNA. Further packaging is assisted by an endonuclease (gp 49) and finally leads to a head precursor with 1,100 S. The DNA encapsidated in T4 heads has the lowest specific volume so far known for DNA. It is condensed to a ball of garn which is stabilized by association with the processing products of the internal proteins (gp *ipI–ipIII*), with the polypeptide cleavage products of gp 22 and with spermidin and putrescin. To estimate the overall condensation it should be recalled that a DNA double-strand of about 80 μm is packaged in a head shell of 120 nm length and 80 nm width.

**Table 14.5:** Properties and processing of T4 head proteins

| | Prehead | | Mature head | | Function |
|---|---|---|---|---|---|
| | Copies | $M_r$ (kd) | Copies | $M_r$ (kd)[a] | |
| *Head shell* | | | | | |
| gp 20 | 30 | 65 | 15 | 65 | Prehead initiation, DNA encapsidation |
| gp 23 | 1,000 | 56 | 1,000 | 46* | Head shell lattice |
| gp *soc* | 1,000 | 10 | 1,000 | 10 | Head shell lattice |
| gp *hoc* | ~150 | 40 | ~150 | 40 | Head shell lattice |
| gp 24 | 60 | 48 | 60 | 45* | Head shell vertices |
| *Head core* | | | | | |
| gp 21 | ~100 | 27.5 (→ 18.5*) | 0 | (degraded) | Proteinase |
| gp 22 | ~600 | 27–32 | 0 | (degraded to peptides*) | Core proteins, DNA stabilization |
| gp *ipIII* | ~400 | 23.5 | ~400 | 21* | |
| gp *ipI + II* | 750 | 10, 11.7 | ? | 8.9*, 9.7* | |
| gp *pip* | 200 | 13 | 0 | (degraded to peptides) | |
| gp *alt* | ~40 | 79 | ~40 | 61* | ADP ribosyltransferase[b] |

[a] Gene products proteolytically modified by action of gp 21* are denoted by *.
[b] Upon infection gp *alt** modifies the α-subunit of host RNAP.

4. After DNA packaging the *head maturation* continues with participation of gp **2, 4, 50, 64** and **65**, their function being little unterstood. By attachment of the neck piece (gp **13, 14**) and the neck whiskers (gp *wac*) the head is prepared for connection with the preformed tail.

### 14.2.4.4 Tail Assembly

Assembly of the phage tail starts at three points independently (the two pieces of the base plate and the tail fibres). The gradual self-assembly of the six peripheral pieces of the base plate with the wedges is of special interest. It begins with a complex of two gp 10 to which four molecules of the wedge protein (gp 11) are added. It is followed by gradually increasing the core of the base plate by sequential addition of gp 7, 8, 6, 53 and 25. Evidently, this part of the assembly line proceeds without any helper proteins but only in the order shown in Fig. **14.14**B. The pre-base plate formed out of a plug and six peripheral pieces is made competent for the following steps of morphogenesis by addition of gp 12 next to gp 11 and of three further proteins: gp 9 marks the attachment sites of the tail fibres, gp 48 and 54 initiate the development of tail tube and sheath. The latter comprises first the addition of 144 molecules of gp 19 to give the base plate-tube intermediate which serves as a nucleation core for the rapid and cooperative association of 144 molecules of gp 18 (tail sheath [67]). The labile structure is stabilized by addition of gp 3 (tube tip) and gp 15 for sheath stabilization and head attachment. It is a deep mystery how the exact length of the tube-sheath entity (100 nm) is determined. The morphogenesis ends with addition of the mature head and the six tail fibres to generate the infective phage particle.

The tail fibres of phage T4 are about 150 nm long and 3 to 4 nm thick structures with a kink in the middle dividing each fibre into a proximal and distal half. The tip and the major part of the distal half fibre consists of two copies of gp 37 ($M_r$ = 120 kd). They serve to attach the phage to the bacterial outer membrane by interaction of the tip with the lipopolysaccharide receptor, whereas the proximal half fibres play a part in triggering the transformation of the base plate, which is the initial step in tail contraction and subsequent injection of the phage DNA [68].

## 14.3 Conclusions

The interactions of bacteriophages with their host cells provide an exciting example for the reprogramming of gene expression that results in the specialization of the cell to a task defined by the invading genetic material of the phage. The far advanced genetic and molecular biological techniques have facilitated deep insights into the regulatory mechanisms involved. They concern (1) the direct interaction of the two genetic systems by restriction, modification and antirestriction proteins, (2) the shut--down of host-specific gene expression functions at all levels (transcription, translation, protein degradation) and, last but not least, (3) the gene expression programs destined to ensure effective phage multiplication under appropriate conditions. The latter exhibit a multiplicity of regulatory effects which are realized by cooperation between phage-encoded proteins and components of the host's gene expression machinery. They sum up to a highly ordered sequence of determinative events finally leading to cell death and release of mature phage particles.

It is an interesting peculiarity of temperate phages, e.g., phage λ, to be able to decide between two types of the host cell's response (lysis vs. lysogeny) by means of a metastabile switch in the intercistronic region of genes *cro* and *cI*. This mechanism represents an illustrative example of how determinative decisions between two alternative pathways can be brought about by the interplay of two antagonistic regulatory proteins binding to the same DNA sequence elements. The tendency to embark on the lysogenic pathway is enhanced by a special regulatory protein (gp *cII*) that allows host RNAP to initiate transcription at two otherwise inaccessible promoters ($p_I$ and $p_{re}$). The two additional transcripts code for the phage DNA integration protein (gp *int*) and the characteristic repressor of the lysogenic state: gp *cI*.

A specially intriguing aspect of the phage developmental cycle is the morphogenesis of the elaborate phage particles from a pool of preformed proteins. The overall irreversibility of the assembly line results (1) from the sequential self-assembly ordered by conformation changes of the interacting proteins, (2) from the catalytic action of numerous morphopoietic proteins, (3) from the participation of a morphogenetic scaffold for head assembly and (4) from a complex program of proteinogen processing which, inter alia, removes most components of the scaffold.

## 14.4 Selected References

### 14.4.1 Reviews and Books

1. Kaper, J. M.: The Chemical Basis of Virus Structure, Dissociation and Reassembly. North-Holland/American Elsevier, Amsterdam-New York 1975.
2. Mathews, Ch. K.: Bacteriophage Biochemistry. Van Nostrand and Reinhold, New York 1972.
3. Winkler, U., W. Rüger & W. Wackernagel: Bakterien-, Phagen- und Molekulargenetik. Springer Verlag, Berlin 1972.
4. Grossman, L. & K. Moldave (eds.): Methods in Enzymology, Vol. XXI, Section VI. Gene Localization Techniques. Academic Press, New York 1971, pp. 341–480.

### 14.4.2 DNA Restriction and Modification


5. Revel, H., & S. E. Luria: DNA glucosylation in T-even phage: Genetic determination and role in phage – host interactions. Ann. Rev. Genetics *4* (1970): 177–192.
6. Arber, W.: DNA modification and restriction. Progr. Nucleic Acid Res. Mol. Biol. *14* (1974): 1–37.
7. Arber, W.: Promotion and limitation of genetic exchange. Science *205* (1979): 361–365.
8. Smith, H. O.: Nucleotide sequence specificity of restriction endonucleases. Science *205* (1979): 455–462.
9. Dharmalingam, K., & E. B. Goldberg: Restriction in vivo III. General effects of glucosylation and restriction on phage T4 gene expression and replication. Virology *96* (1979): 393–403.


### 14.4.3 Transcription Programs of Phage T7


10. Dunn, J. J. & F. W. Studier: Nucleotide sequence from the genetic left end of bacteriophage T7 DNA to the beginning of gene 4. J. Mol. Biol. *148* (1981): 303–330.
11. Studier, F. W., A. H. Rosenberg, M. N. Simon & J. J. Dunn: Genetic and physical mapping in the early region of bacteriophage T7 DNA. J. Mol. Biol. *135* (1979): 917–937.
12. McAllister, W. T., & H.-L. Wu: Regulation of transcription of the late genes of bacteriophage T7. Proc. Natl. Acad. Sci. USA *75* (1978): 804–808.
13. Pachl, C. A., & E. T. Young: The size and messenger RNA activity of bacteriophage T7 late transcripts synthesized in vivo. J. Mol. Biol. *122* (1978): 69–101.
14. Pfennig-Yeh, M., H. Ponta, M. Hirsch-Kauffmann, H. J. Rahmsdorf, P. Herrlich & M. Schweiger: Early T7 gene expression. Rates of RNA synthesis and degradation, protein kinase dependent termination of transcription and efficiency of translation. Mol. Gen. Genet. *166* (1978): 127–140.
15. Herrlich, P., M. Schweiger, H. J. Rahmsdorf, H. Ponta, N. Spoerel, M. Pfennig-Yeh, P. Wolfram, U. Rahmsdorf, B. Auer & E. F. Wagner: Virus T7: A probe of cellular control mechanisms Proc. FEBS Meeting Dresden, Vol. 51. Pergamon Press, Oxford 1979, pp. 337–346.
16. Rosa, M. D.: Four T7 RNA polymerase promoters contain an identical 23 bp sequence. Cell *16* (1979): 815–825.
17. Siebenlist, U.: Nucleotide sequence of the three major early promoters of bacteriophage T7. Nucleic Acids Res. *6* (1979): 1895–1907.
18. Panayotatos, N., & R. D. Wells: Recognition and initiation site for four late promoters of phage T7 is a 22-base pair DNA sequence. Nature *280* (1979): 35–39.
19. Datta, A. K., & S. K. Niyogi: Biochemistry and physiology of bacterial ribonucleases. Progr. Nucleic Acid Res. Mol. Biol. *17* (1976): 271–308.
20. Robertson, H. D., & F. Barany: Enzymes and mechanisms in RNA processing. Proc. FEBS Meeting Dresden, Vol. 51. Pergamon Press, Oxford 1979, pp. 285–295.
21. De Wyngaert, M. A., & D. C. Hinkle: Bacterial mutants affecting phage T7 DNA replication produce RNA polymerase resistant to inhibition by the T7 gene 2 protein. J. Biol. Chem. *254* (1979): 11247–11253.


### 14.4.4 Development of Phage λ

*Reviews*


22. Hershey, A. A. (ed.): The Bacteriophage Lambda. Cold Spring Harbor Laboratory, Cold Spring Harbor, New York 1971.
23. Szybalski, W.: A network of developmental controls in coliphage Lambda. In: Nover, L., & K. Mothes (eds.), Cell Differentiation in Microorganisms, Plants and Animals. G. Fischer Verlag, Jena and North-Holland Publ. Comp., Amsterdam 1977, pp. 262–316.
24. Echols, H.: Bacteriophage development. In: Leighton, T., & W. F. Loomis (eds.), The Molecular Genetics of Development, Academic Press, New York 1980, pp. 1–16.


*Integration and Excision*


25. Campbell, A. M.: Episomes. Adv. Genetics *11* (1962): 101–145.
26. Nash, H. A.: Integration and excision of bacteriophage λ. Curr. Topics Microbiol. Immunol. *78* (1977): 171–199.

27. Cold Spring Harbor Symp. Quant. Biol. *43* (1979), Cold Spring Harbor, New York, pp. 1089–1146, several papers.
28. Bidwell, K., & A. Landy: Structural feature of λ site-specific recombination at a secondary *att* site in *gal T*. Cell *16* (1979): 397–406.
29. Christie, G. E., & T. Platt: A secondary attachment site for bacteriophage λ in *trpC* of *E. coli*. Cell *16* (1979): 407–413.
30. Hsu, P.-L., W. Ross & A. Landy: The λ phage *att* site: Functional limits and interaction with Int protein. Nature *285* (1980): 85–91.

*Genetic Map*

31. Echols, H. & H. Murialdo: Genetic map of bacteriophage Lambda. Microbiol. Rev. *42* (1978): 577–591.
32. Szybalski, E. H. & W. Szybalski: A comprehensive molecular map of bacteriophage lambda. Gene *7* (1979): 217–270.

*Antitermination*

33. Schwarz, E., G. Scherer, G. Hobom & H. Kössel: Nucleotide sequence of *cro*, *cII* and part of the *O* gene in phage λ DNA. Nature *272* (1978): 410–414.
34. Rosenberg, M., D. Court, H. Shimatake, C. Brady & D. L. Wulff: The relation between function and DNA sequence in an intercistronic regulatory region in phage λ. Nature *272* (1978): 414–423.
35. Greenblatt, J.: Regulation of transcription termination by the N gene protein of bacteriophage lambda. Cell *24* (1981): 8–9.
36. Hu, S. L., J. S. Salstrom & W. Szybalski. Stimulation of Tof function and $p_L$-proximal transcription by the *N*-gene product of coliphage Lambda. Virology *98* (1979): 411–423.
37. Hu, S. L., & W. Szybalski: Control of rightward transcription in coliphage Lambda by the regulatory functions of phage genes *N* and *cro*. Virology *98* (1979): 424–432.

*Repressor and Antirepressor*

38. Takeda, Y.: Specific repression of in vitro transcription by the *cro* repressor of bacteriophage lambda. J. Mol. Biol. *127* (1979): 177–190.
39. Johnson, A. D., B. J. Meyer & M. Ptashne: Interactions between DNA bound repressors govern regulation by the λ phage repressor. Proc. Natl. Acad. Sci. USA *76* (1979): 5061–5065.
40. Ptashne, M., A. Jeffrey, A. D. Johnson, R. Maurer, B. J. Meyer, C. O. Pabo, T. M. Roberts & R. T. Sauer: How the λ repressor and *cro* work. Cell *19* (1980): 1–11.
40a. Anderson, W. F., D. H. Ohlendorf, Y. Takeda & B. W. Mathews: Structure of the cro repressor from bacteriophage λ and its interaction with DNA. Nature *290* (1981): 754–758.

*Prophage Induction*

41. Witkin, E. M.: Ultraviolet mutagenesis and inducible DNA repair in *Escherichia coli*. Bact. Rev. *40* (1976): 869–907.
42. Hanawalt, Ph. C., P. K. Cooper, A. K. Genesan & Ch. A. Smith: DNA repair in bacteria and mammalian cells. Ann. Rev. Biochem. *48* (1979): 783–836.
43. Phizicky, E. M., & J. W. Roberts: Induction of SOS functions: Regulation of proteolytic activity of *E. coli* RecA protein by interaction with DNA and nucleoside triphosphate. Cell *25* (1981): 259–267.
44. Bailone, A., A. Levine & R. Devoret: Inactivation of prophage repressor in vivo. J. Mol. Biol. *131* (1979): 553–572.
45. Craig, N. L., & J. W. Roberts: *E. coli* recA protein-directed cleavage of phage λ repressor requires polynucleotide. Nature *283* (1980): 26–30.
46. Little, J. W., S. H. Edmiston, L. Z. Pacelli & D. W. Mount: Cleavage of the *Escherichia coli lexA* protein by the *recA* protease. Proc. Natl. Acad. Sci. USA *77* (1980): 3225–3229.

*Establishment of Lysogeny*

47. Hayes, S., & C. Hayes: Control of bacteriophage λ repressor establishment transcription from the *y-cII-oop-O-P* region. Mol. Gen. Genetics *170* (1979): 75–88.
48. Knoll, B. J.: An analysis of repressor overproduction by the λ *cIII* mutants. J. Mol. Biol. *132* (1979): 551–555.

49. Jones, M. O., R. Fischer, I. Herskowitz & H. Echols: Location of the regulatory site for establishment of repression by bacteriophage λ. Proc. Natl. Acad. Sci. USA *76* (1979): 150–154.
50. Hoess, R. H., C. Foeller, K. Bidwell & A. Landy: Site-specific recombination functions of bacteriophage λ: DNA sequence of regulatory regions and overlapping structural genes for Int and Xis. Proc. Natl. Acad. Sci. USA *77* (1980): 2482–2486.
51. Abraham, J., D. Mascarenhas, R. Fischer, M. Benedik, A. Campbell & H. Echols: DNA sequence of regulatory region for integration gene of bacteriophage λ. Proc. Natl. Acad. Sci. USA *77* (1980): 2477–2481.

*Post-transcriptional Control*

52. Ray, P. N., & M. L. Pearson: Evidence for post-transcriptional control of the morphogenetic genes of bacteriophage lambda. J. Mol. Biol. *85* (1974): 163–175.
53. Shaw, J. E., & H. Murialdo: Morphogenetic genes *C* and *Nu3* overlap in bacteriophage λ. Nature *283* (1980): 30–35.

### 14.4.5 Phage T4 Morphogenesis

*Reviews*

54. Wood, W., R. S. Edgar, J. King, I. Lielausis & M. Henninger: Bacteriophage assembly. Federat. Proc. *27* (1968): 1160–1166.
55. Casjens, S. & J. King: Virus assembly. Ann. Rev. Biochem. *44* (1975): 555–611.
56. Kellenberger, E.: Differential gene product activation and differential rates of gene product synthesis as regulating principles in virus assembly. In: Nover, L., & K. Mothes (eds.): Cell Differentiation in Microorganisms, Plants and Animals. G. Fischer Verlag, Jena, and North Holland Publ. Comp., Amsterdam 1977, pp. 317–339.
57. Witmer, H. J.: Regulation of bacteriophage T4 gene expression. Progr. Mol. Subcell. Biol. *4* (1976): 17–51.

*Head Assembly*

58. Showe, M. K., E. Isobe & L. Onorato: Bacteriophage T4 prehead proteinase. II. Its cleavage from the product of gene 21 and regulation in phage infected cells. J. Mol. Biol. *107* (1976): 55–69.
59. Showe, M. K., & L. Onorato: Kinetic factors and form determination of the head of bacteriophage T4. Proc. Natl. Acad. Sci. USA *75* (1978): 4165–4169.
60. Yanagida, M.: Molecular organization of the shell of T-even bacteriophage head. II. Arrangement of subunits in the head shells of giant phages. J. Mol. Biol. *109* (1977): 515–537.
61. Leibo, S. P., E. Kellenberger, C. Kellenberger van der Kamp, T. G. Frey & C. M. Steinberg: Gene 24-controlled osmotic shock resistance in bacteriophage T4: Probable multiple gene functions. J. Virol. *30* (1979): 327–338.
62. Holm, Th., B. Holm, A. Engel, M. Wurtz & Ph. R. Smith: Isolation and characterization of the host protein groE involved in bacteriophage lambda assembly. J. Mol. Biol. *129* (1979): 359–373.
63. Simon, L. D., K. Tomczak & A. C. St. John: Bacteriophages inhibit degradation of abnormal proteins in *Escherichia coli*. Nature *275* (1978): 424–428.
64. Earnshaw, W. C., & S. R. Casjens: DNA packaging by the double-stranded DNA bacteriophages. Cell *21* (1980): 319–331.

*Tail Assembly*

65. Crowther, R. A., E. V. Lenk, Y. Kikuchi & J. King: Molecular reorganization in the hexagon to star transition of the baseplate of bacteriophage T4. J. Mol. Biol. *116* (1977): 489–523.
66. Wood, W. B., & M. P. Conley: Attachment of tail fibres in bacteriophage T4 assembly: Role of the phage whiskers. J. Mol. Biol. *127* (1979): 15–29.
67. Tschopp, J., F. Arisaka, R. van Driel & J. Engel: Purification, characterization and reassembly of the bacteriophage T4 D sheath protein P18. J. Mol. Biol. *128* (1979): 247–258.
68. Earnshaw, W. C., E. B. Goldberg & R. A. Crowther: The distal half of the fibre of bacteriophage T4. Rigidly linked domains and cross-β structure. J. Mol. Biol. *132* (1979): 101–131.
69. Crowther, R. A.: Mutants of bacteriophage T4 that produce infective fibreless particles. J. Mol. Biol. *137* (1980): 159–174.

# 15. Programmed Gene Expression During Cell Division Cycles

M. Luckner

Growth and development of all types of organisms depend on cell division. During this process DNA, proteins and the other cell constituents are duplicated and two daughter cells are formed from the mother cell. The reactions involved are part of a cyclic gene expression program. This is shown by their exact timing, i.e., their temporal determination relative to other events (phase dependence) and their dependence on earlier events (determination) (cf. **4**.5.1). In the following two of the most important types of cell division, the symmetrical "proliferative" cell cycle and the asymmetrical "quantal" cell cycle, will be discussed.

## 15.1 The Proliferative Cell Cycle

### 15.1.1 Phases of the Proliferative Cycle

The progress of a eukaryotic cell through the proliferative cell cycle is marked by the following periods: (1) the phase of DNA replication (S phase), (b) mitosis (M), i.e., the distribution of the duplicated chromosomes between two daughter nuclei and formation of a primary cell wall and (c) phases of apparent non-activity so-called gaps ($G_1$ and $G_2$ phases) (Fig. **15**.1).

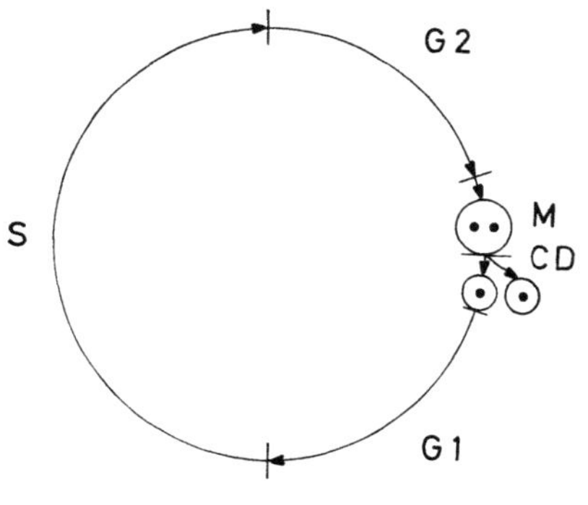


**Fig. 15.1**: The mitotic cell division cycle (from Mitchison [1], modified).
*S* phase, phase of DNA synthesis; *M*, mitosis; *CD*, cell division; *G1* and *G2* phases, "gaps" between M/CD and S phase; **Interphase**: period between cell division and the following mitosis.

It is generally recognized that the period of DNA synthesis (S phase) and the post--DNA synthetic gap (G2 phase), as well as mitosis and cell division periods are relatively invariable, in contrast to the G1 phase. This implies that variations in the length of the cycle are primarily regulated by the length of the G1 period. In fast-growing cells the G1 phase may be lacking completely. Moreover in rapidly dividing prokaryotic

Abbreviations: BrdU, bromodeoxyuridine; CD, cell division; EGF, epidermal growth factor; M, S, G1 and G2, phases of cell cycle (cf. Fig. **15**.1); *cdc* denotes genes involved in the cell division cycle of yeast.

cells two cell division cycles can be superimposed, i.e., these cells contain two nucleoids, which are relatively independent of each other, the first being replicated in the period when the second already divides.

Cells with long turnover times spend a large proportion of their life in a resting state with respect to proliferation, the G0 state, which is not part of the G1-S-G2-M-CD proliferation cycle. In the G0 state the cells can perform their specific biochemical functions, and from this state they enter into the cell cycle again if the appropriate extracellular signal is supplied (cf. **4**.5.4.1 and **13**.3.1).

### 15.1.2 Cell Cycle Synchronization

The results describing the pattern of gene expression during the cell cycle rely on two main bodies of evidence. On the one hand they base on the work done with single cell techniques which rely on microscopic measurements. These have the advantage that they can reveal finer details of growth pattern and the variation between individual cells. The measurement of isolated events, however, is very limited for technical reasons.

The second body of evidence comes from synchronous cultures, where the cycles of individual cells have been brought into phase. The advantage of such synchronous cultures is that they provide enough material to allow the usual bulk biochemical measurements to be made; but they have attendant faults. Usually the degree of synchrony is not complete, so that cell divisions are spread over a considerable period of time and are not precisely simultaneous. Therefore, finer details of synthetic patterns are obscured. Furthermore, cultures lose their synchrony after a few cycles if they are left to run free, because of the natural variation of cycle time from cell to cell. The most important drawback, however, is that many methods used for synchronization may influence the behaviour of the cells and thus lead to artefacts.

There are two groups of synchronizing procedures: (a) selection synchrony in which cells at one particular stage of the cycle are selected out of a normal asynchronous culture and grown as a separate culture and (b) induction synchrony in which treatment with light, temperature, inhibitors etc. is applied to asynchronous cultures of wild-type strains or appropriate mutants. Especially when applying the latter methods it must be proven carefully whether they generate normal cultures in which the cell division cycle proceeds in a way identical with that in the cells of a random asynchronous culture under the same experimental conditions.

Most normal in this respect are obviously cells which show synchrony also in their natural environment. One example is the slime mould *Physarum polycephalum*, which in the plasmodial stage forms macroscopic syncytial cells containing $10^8$ to $10^9$ nuclei. These nuclei divide synchronously within a short period (early prophase 15–20 min, prophase 5 min, metaphase 7 min, anaphase 3 min, telophase 5 min) every 8 h. *Physarum* is therefore an extremely attractive organism for studying molecular events in the cell cycle.

### 15.1.3 Periodic and Continuous Events During the Proliferative Cell Cycle

#### 15.1.3.1 Replication of DNA

Synthesis of DNA during the mitotic cycle has been investigated in detail with *Physarum polycephalum*. Nuclear DNA starts to replicate immediately following mitosis and does so for about one third of the intermitotic time, i.e., *Physarum* lacks the G1 phase. Replication is most active during the first 2 h of the mitotic cycle so that at the end of this period approximately 80% of the DNA is replicated. It takes place in a well-ordered temporal sequence: DNA replicated in a distinct part of the S period will be replicated in the same part of the following S period. Addition of cycloheximide to S phase cultures allows only a fraction of the unreplicated DNA to be replicated further. A more detailed analysis has led to the proposal that the genome of *Physarum* is made up of at least ten families of replicons replicating their DNA in sequence, whereby each family has to receive a start signal in form of one or several newly made proteins.

Nuclei isolated from the S phase (but not from G2 phase) carry out in vitro DNA synthesis. The same holds true for homogenates devoid of intact nuclei. However, only about 15% of the genome may undergo replication in these preparations. This result indicates that replicons initiated in vivo continue to synthesize DNA in vitro. But no in vitro re-initiation occurs, due to the fact that in these in vitro systems there is no protein synthesis, and, consequently, the regulator proteins necessary for re-initiation of DNA synthesis are not formed.

In contrast to the normal nuclear DNA, the nuclear satellite DNA, which contains most of the organism's ribosomal DNA and is found in the nucleolus, replicates throughout the whole intermitotic period with the exception of the first hour of the S period. By density labelling it was shown that the rDNA replicates "anarchically", i.e., rDNA molecules are selected at random for replication, so that molecules can replicate more than once per cycle or not at all, though the population as a whole usually doubles precisely. The probability that a given molecule will not replicate during a particular cycle is nearly 0.5.

### 15.1.3.2 Synthesis of RNA

Pulse-labelling experiments with [$^3$H]-uridine and analysis of incorporation of radioactivity in RNA by autoradiography have shown that RNA is synthesized continuously throughout the whole interphase, but that synthesis is markedly depressed

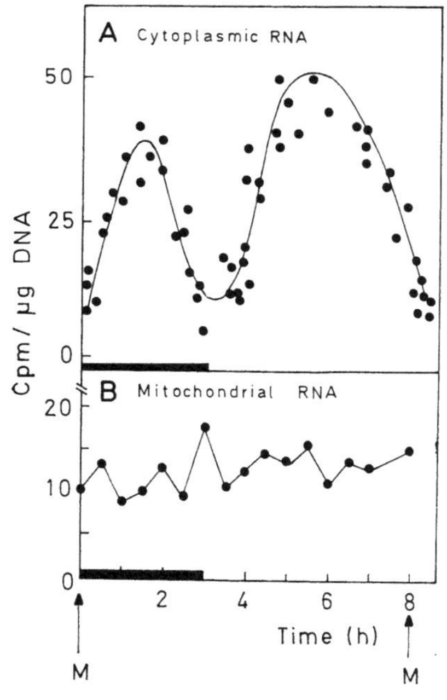


**Fig. 15.2**: Synthesis of cytoplasmic and mitochondrial RNA in *Physarum* (from Grant [7], modified).

Radioactivity in RNA 10 min (A) and 15 min (B) after administration of [$^3$H]-uridine.

The experiment was started at the second mitosis after fusion of the microplasmodia to the macroplasmodium. *M*, mitosis; *black bar*, period of DNA synthesis (S phase), followed by G2 phase and mitosis.

between S phase and G2 phase and during nuclear division (Fig. **15**.2). RNA synthesis is readily observed during early prophase, but decreases to an undetectable level during metaphase and anaphase. Like DNA synthesis it is resumed immediately after anaphase. A similar biphasic pattern of RNA synthesis, although less pronounced, has been demonstrated in nuclei isolated at different stages during the mitotic cycle.

The rate of mitochondrial RNA synthesis remains constant throughout the mitotic cycle. That of ribosomal RNA shows a continuous increase, the change in rate between the beginning and the end of interphase being five- to six-fold. The ratio of tRNA to total nucleic acid does not alter and no significant quantitative or qualitative changes occur in tRNA composition.

### 15.1.3.3 Formation of Enzymes Related to Nucleic Acid Metabolism

Because of the phase-dependent formation of DNA and total RNA it might be expected that also the activity of enzymes involved in the biosynthesis of these macromolecules might fluctuate during the mitotic cycle. Indeed it has been found that thymidine kinase and deoxycytidine kinase exhibit a peak of activity during the S phase, whereas histone H1 phosphokinase activity shows maximum activity during G2. The phosphate content of histone H1 increases threefold during late G2 and starts to decrease at the onset of mitosis. It has been suggested that phosphorylation of the serine-rich ends of histone H1 (cf. Fig. **1**.3) might lead to a lowered affinity for DNA, thereby enabling H1 to control chromosome condensation-decondensation cycles [6].

In contrast to these proteins, the activities of thymidine monophosphate kinase, thymidine diphosphate kinase, uridine kinase and deoxyadenosine kinase do not exhibit any marked changes during the mitotic cycle. Also the DNA polymerase activity so far solubilized from *Physarum* shows a constant specific activity which does not reflect the in vivo rates of DNA synthesis. The phase-dependent formation of DNA and RNA therefore is not dependent on the phase-dependent activity of all enzymes of nucleic acid metabolism. As shown below, it may rather be regarded as an aspect of a general gene expression program that includes the periodic activity of certain key enzymes of nucleic acid biosynthesis and other fields of metabolism, as well as the constant synthesis of many proteins.

### 15.1.3.4 Synthesis of Enzymes in Other Fields of Metabolism

Either periodic or continuous synthesis has been shown also with other enzymes. On the one hand there are "step enzymes" whose activity doubles during a short period within the cycle. By means of single cell experiments it has been shown for instance that in *Saccharomyces cerevisiae* there is an increase of $\beta$-galactosidase activity at a relatively short period immediately before the end of the cell division cycle, i.e., at a stage when cells have nearly full-sized buds (Fig. **15**.3). Rapid doubling of the level of $\beta$-galactosidase activity has been demonstrated also with synchronous cultures, or cultures separated into fractions by zonal centrifugation representing different stages of the cell cycle. Similar results were obtained, for example, with ornithine transaminase (Fig. **15**.4). The period of doubling of enzyme activity, however, in this case is somewhat earlier in the cycle.

Nevertheless, most enzymes show a continuous increase in activity through the cell cycle (Fig. **15**.5; note that, in contract to Figs. **15**.3 and **15**.4, enzyme activity is given per ml culture in this case). The rate of the activity increase doubles once per cycle within a very short period, a pattern which has been found also for dry

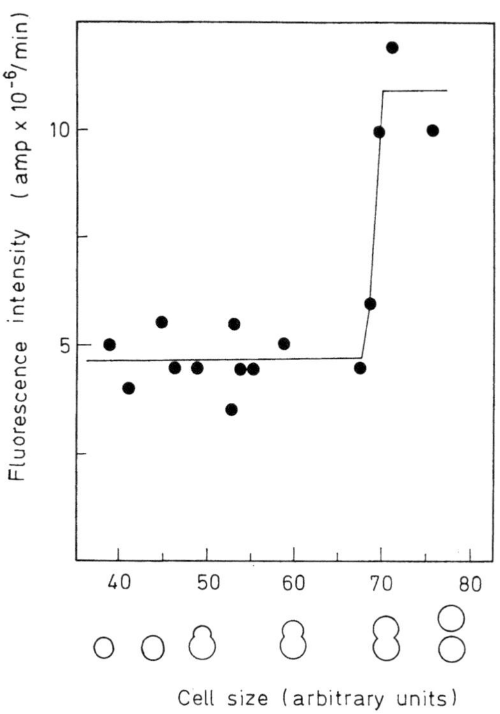


Fig. 15.3

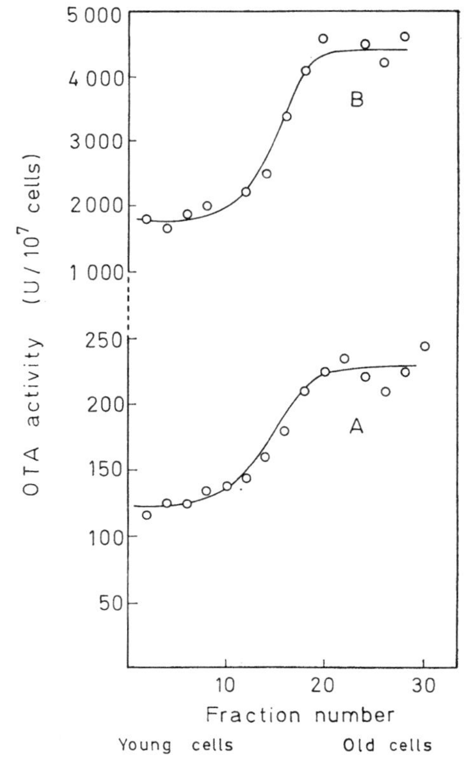


Fig. 15.4

**Fig. 15.3:** $\beta$-Galactosidase activity in single cells of *Saccharomyces lactis* during cell division cycle (from Halvorson [11], modified).

Random suspensions of yeast cells were treated with diluted isoamyl alcohol at 0 °C for 10 min with occasional shaking to increase membrane permeability. Afterwards they were diluted with phosphate buffer, and sprayed in a silicon oil chamber. After introduction of the substrate fluorescein-di-$\beta$-galactoside the liberation of fluorescein was measured with a microphotometer. The volume of cells was calculated from microphotographs. It was used for determination of the stage of the individual cell in the division cycle. The shape of mother and daughter cells is drawn schematically below the ordinate.

**Fig. 15.4:** Induction capacity of ornithine transaminase (OTA) during the yeast cell cycle (from Halvorson [11], medified).

Cells of a non-synchronized culture of *Saccharomyces cerevisiae* were grown on a synthetic medium containing 0.2% succinate and 0.1% $(NH_4)_2SO_4$ **(A)** as well as on this medium supplemented with 0.3% arginine to induce the synthesis of ornithine transaminase **(B)**. The two cultures were allowed to grow for 40 min. After that time cycloheximide (100 $\mu$g/ml) was added to stop protein biosynthesis. The cultures were chilled and the cells were harvested by filtration. The cells were resuspended and centrifuged in a zonal rotor. Thirty fractions corresponding to different stages of the cell cycle were collected. These samples were filtered, lyophilized and analyzed for ornithine transaminase activity.

Age during the cell cycle increases with fraction number.

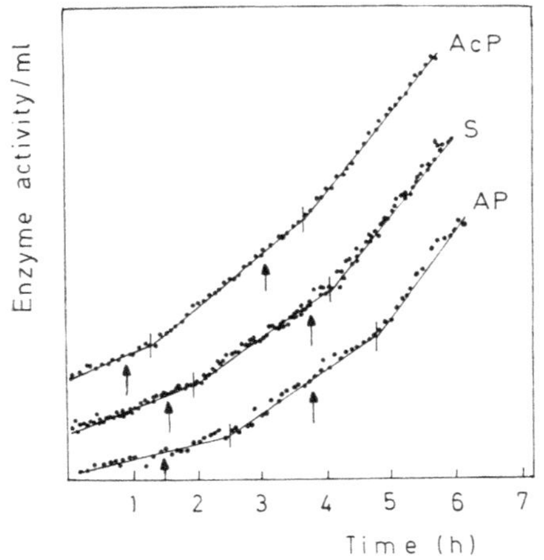


**Fig. 15.5:** Continuous-linear synthesis of enzymes in synchronized cultures of *Schizosaccharomyces pombe* (from Mitchison [12], modified).

Cells of a non-synchronized culture in the logarithmic growth phase were fractionated according to cell size by sucrose gradient centrifugation. A fraction of cells at the same developmental stage was used for starting the synchronized culture. At the time indicated samples were taken and enzyme activities were determined. *Arrows* indicate the peaks in cell divison. The lines and the rate transition points *(vertical lines)* were calculated by statistical methods. *AcP*, acid phosphatase; *S*, sucrase; *AP*, alkaline phosphatase.

mass accumulation, rRNA, ribosomal protein, mRNA and $CO_2$ production. However, the rate changes of these events differ in their timing in the cycle and in their dependence on DNA synthesis and cell division.

### 15.1.3.5 Early Models to Explain the Phase Dependence of Events

In order to explain the exact timing of events within the proliferative cell cycle the following ideas have been proposed:

a) The periodic expression of certain parts of the genome is the result of *oscillating induction* and *repression*.

This model implicates that the genome is continuously available for transcription and that periodic synthesis of special proteins is the result of either an oscillatory induction or repression. In this model enzyme synthesis results in an increased rate of end product formation and, when a critical concentration of the end product is reached, enzyme synthesis is repressed. During subsequent growth the concentration of this end product falls until it is below a critical concentration and enzyme synthesis starts again. Similarly, oscillating induction causes periodic enzyme synthesis in an analogous manner because enzyme synthesis ceases as a result of the utilization of inducer and begins again when the inducer concentration of the cell reaches a critical concentration. This model does not explain, however, by which way the individual processes are coupled with each other to build the gene expression programs.

Experimental attempts to prove this model with *Saccharomyces cerevisiae* have not given any evidence for the connection between induction or repression of certain proteins and the cell division cycle. Two enzymes of L-arginine degradation, arginase and ornithine transaminase, which are both induced by arginine, have been investigated. These enzymes are synthesized once

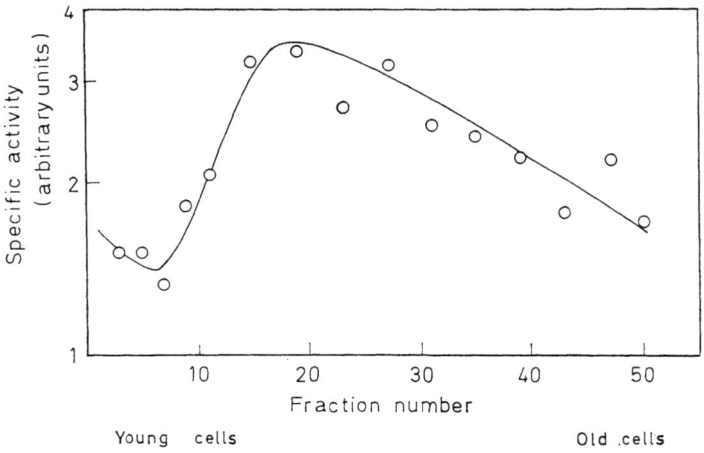


**Fig. 15.6:** Specific activity of UDP galactose-4-epimerase in cells of an operator-constitutive mutant of *Saccharomyces cerevisiae* in dependence on the position in the cell division cycle (from Halvorson [11], modified).

Cells of a culture in the exponential growth phase were harvested, chilled to stop protein biosynthesis and separated according to size and thus stage within the cell division cycle by zonal centrifugation (cf. Fig. 15.4). After centrifugation fractions were collected and assayed for UDP galactose-4-epimerase activity.

per cell cycle. However, in spite of the fact that they have the same inducer, synthesis of arginase occurs earlier in the cycle than that of ornithine transaminase. Furthermore, it was shown that the arginine pool does not show any significant fluctuation during the cell cycle. Thus, although arginine regulates the amount of enzyme synthesized at every stage of the cycle (Fig. 15.4), the timing of arginase and ornithine transaminase synthesis appears to be independent of the internal concentration of arginine.

A further critical test of this model has been afforded by operator constitutive mutants. In these mutants the rate of synthesis of the enzyme involved is no more controlled by the level of low molecular inducers. According to the discussed model the mutation should lead to continuous enzyme synthesis throughout the cell cycle. However, mutants constitutive for UDP-galactose-4--epimerase still show the usual step-wise synthesis of this enzyme, whether they are measured in synchronously dividing cells or in cells separated by centrifugation according to their state in the cell cycle (Fig. 15.6). These results make it highly unlikely that periodic protein synthesis in *Saccharomyces cerevisiae* is caused by periodic fluctuations of inducer or repressor levels.

b) Periodic protein synthesis is the result of an ordered transcription of the genome (*"linear reading"*).

In this model genes coding constitutive gene products are available for transcription only during a specific period of the cell cycle. Hence the chromosomal position of the structural gene determines the time of expression during the cell cycle.

In agreement with this idea it has been demonstrated that the temporal sequence of expression of four different genes on chromosome 2 of *Saccharomyces cerevisiae* is the same as the order of these genes on this chromosome. Figure **15**.7 demonstrates that transposition of a gene alters the time of expression during the cell cycle. It contains a genetic map of chromosome 2 of two strains of *Saccharomyces cerevisiae*. In strain S 288C the genes $gal_1$ and $lys_2$ encoding the enzymes galactokinase and $\alpha$-aminoadipic acid reductase, are closely linked, whereas in strain 61009 they are separated by a greater distance. As to be expected this transposition increases the

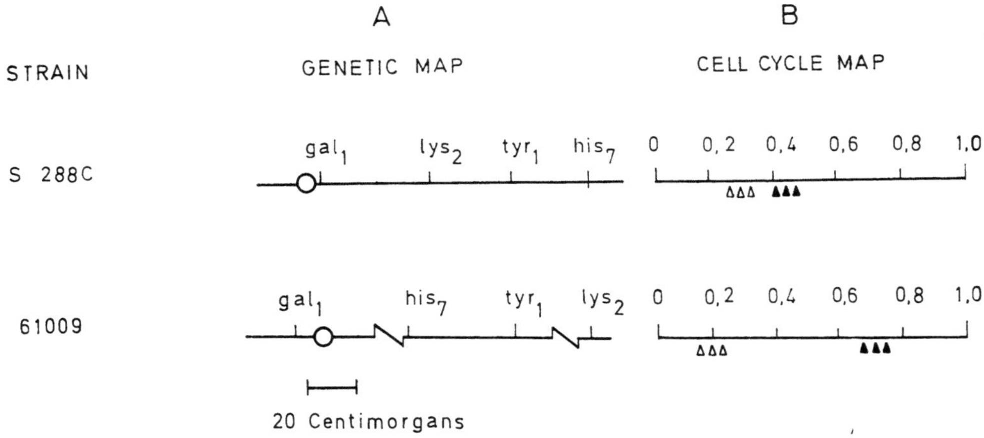


**Fig. 15.7:** Timing of synthesis of two step enzymes during the cell cycle of two strains of *Saccharomyces cerevisiae* (from Halvorson [11], modified).
**A:** Position of genes on the genetic map (chromosome No. 2, right arm)
*gal*$_1$, structural gene for galactokinase; *lys*$_2$, structural gene for α-amino adipic acid reductase; ○, centromer; *tyr*$_1$, *his*$_7$, genes causing auxotrophy with respect to tyrosine and histidine, respectively.
**B:** Time of synthesis during the cell cycle ▲, galactokinase; △, α-amino adipic acid reductase.

temporal difference in expression in the cell cycle from 0.15 to 0.50 units. The "linear reading" model therefore is consistent with certain features of protein synthesis during the cell cycle.

### 15.1.4 The Programmed Expression of Events Throughout the Proliferative Cell Cycle

As indicated in the introduction the cell cycle may be regarded as a program whose progression depends on consecutive and/or simultaneous action of a series of genes. Most of the recent progress in elucidation of such gene expression programs bases on the investigation of appropriate mutants. Malfunction or a quantitative change in the action of any one of the genes involved in the proliferative cell cycle owing to mutation might lead to alteration or even arrest of the program. Since the proliferative cell cycle is a vital process, many of the cell division cycle mutations (*cdc*-mutations) are expected to be lethal, unless they are of conditional nature and enable a more or less normal cell reproduction to occur under permissive conditions.

Many cell cycle mutants were obtained from *Saccharomyces cerevisiae*. Most of the mutants are heat-sensitive. For these the low temperature is permissive and the high temperature is non-permissive. As yet in *Saccharomyces cerevisiae* mutations in at least 32 different complementation groups could be demonstrated, suggesting that a minimum of 32 gene products is required for the specific steps of the cell division cycle.

When arrested at the restrictive temperature, a cell cycle mutant exhibits a complex terminal phenotype consisting of effects which may be described morphologically, physiologically and biochemically. The properties of cells arrested at the restrictive temperature can be like those of cells in any one of the stages of the cell cycle. Some of the mutants arrest as unbudded cells, while others arrest with a terminal phenotype with a large bud, but an undivided nucleus, etc.

However, there are also cells with a morphology not encountered among non-mutant strains. Many cell cycle mutants continue to grow at the restrictive temperature, in spite of the cessation of cell division, thus producing unusually large cells. Cytokinesis mutants continue to bud and go through nuclear divisions at the non-permissive temperature and form large, multinucleate and branched cells, and so on.

The different types of mutants obtained have been used to order a number of stages of the proliferative cell cycle relative to their dependence and determination. In these experiments most important results were obtained by using the following experimental procedures [13]:

a) Determination of the State of Arrest of Double Mutants

When two mutations cause different terminal phenotypes, one can determine the order and dependence of their respective gene functions by examining the behaviour of a double mutant obtained by mating. The double mutant with arrest at the restrictive temperature will show a terminal phenotype of either one of the two mutations, if both mutations are at the same branch of the cycle program, or with a new phenotype combining the characteristics of the two terminal phenotypes, if the two mutations affect genes that act during independent (parallel) stages of the cycle. If the two mutations affect consecutive gene functions in the same dependent pathway, most cells will arrest at the restrictive temperature with the phenotype of the first of the two mutants. Depending on the difference between the points of expression within the cell cycle of the two mutations, a fraction of the cells, however, might be caught between the two points at the time of transfer to the restrictive temperature, and therefore will arrest with the terminal phenotype of the second mutation.

b) Reciprocal Shift Experiments

In these experiments single mutants were used which may be arrested at another point of the cycle by use of stage-specific inhibitors, e.g., hydroxyurea, which arrests the cells by blocking DNA synthesis.

A reciprocal shift experiment consists of two parallel cultures of a cell cycle mutant growing under permissive conditions. One culture is first arrested at the non-permissive temperature and then returned to the permissive temperature, but in the presence of the inhibitor. The second culture is first arrested by being exposed to the inhibitor at the permissive temperature and then transferred to the restrictive temperature with the inhibitor removed. Three types of relations between the mutation and the inhibitor may be seen in these experiments:

1. Indepedence: When the temperature-sensitive function and the inhibitor-sensitive function act independently in the cycle, the cells in the two main cultures of the experiment will be able to complete ongoing cycles.
2. Interdependence: When the two functions are required together for the same event in the cycle, both reciprocal shifts will not enable the cells to traverse the cycle.
3. Sequential dependence: When one function is dependent on the other in a sequential manner, the culture in which the order of the restrictive conditions is the same as the order of dependence will not complete the ongoing cycle, while the reciprocal shift will enable the cells to traverse the cycle.

As a result of these and similar experiments it has been shown that a large part of the cell division cycle of *Saccharomyces cerevisiae* obviously is formed by two parallel programs of gene expression: the mitotic division cycle and the budding cycle (Fig. **15**.8). One example may demonstrate this [15]:

In certain mutants DNA synthesis is blocked in the prereplicative phase of the cell cycle (G1 phase) in the sequence *cdc* 25 $\rightarrow$ *cdc* 4 $\rightarrow$ *cdc* 7 $\rightarrow$ initiation of DNA synthesis. Electron micros-

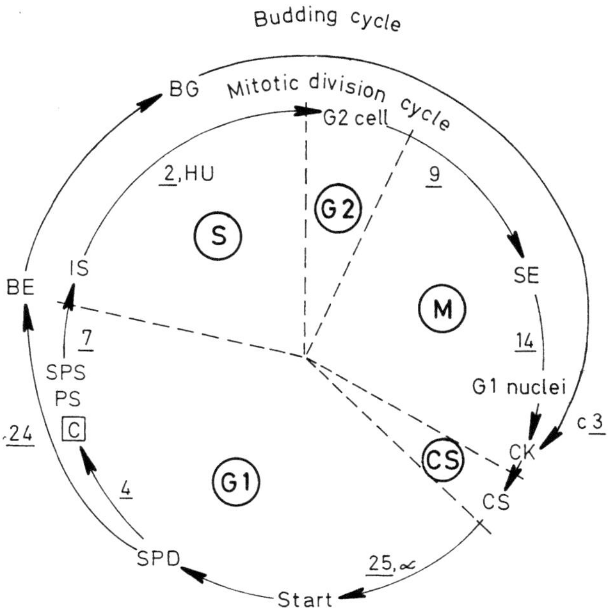


**Fig. 15.8:** The mitotic cell cycle of yeast (from Simchen [13], modified).
The scheme is based on terminal phenotypes of mutants and double mutants, on reciprocal shift experiments with α-mating hormone and hydroxyurea, on cyclo-heximide inhibition of DNA synthesis and on commitment experiments.
*Underlined numbers* refer to *cdc*-genes. *SPD*, duplication of spindle plaques (the spindle plaque is a microtubule-organizing center located in the nuclear envelope); *SPS*, spindle plaque separation to the two poles; *PS*, protein synthesis required for DNA replication; *C*, point of commitment to the mitotic cell cycle, beyond which the cells were committed to the ongoing cycle, even though challenged with conditions that favour meiotic cell division, i.e., formation of haploid vegetative cells; *IS*, initiation of S phase; *BE*, bud emergence; *BG*, bud growth; *NM*, migration of the nucleus in the developing bud; *SE*, spindle elongation; *CK*, cytokinesis (cell division by formation of the wall between mother and daughter cell, cf. Fig. 4.30). *CS*, separation of mother and daughter cells; α, *HU*, stages sensitive to α-mating hormone and hydroxy urea respectively.

copy revealed that *cdc* 25 at non-permissive temperature is arrested at the stage of formation of a single spindle plaque, i.e., exactly the same stage in which cells are arrested by the α-factor (cf. 4.3.3).[1] In arrested *cdc* 4 cells duplication of the spindle plaque had occurred, but their subsequent separation in the nuclear envelope had failed. In *cdc* 7, at the non-permissive temperature, nuclear growth continues, yielding a terminal phenotype in which the dumbbell-shaped nucleus occupies a medial position between the mother cell and a large bud. The mitotic spindle is complete but has not reached its full length. A point of major interest is that *cdc* 7 mutants, on release from the non-permissive temperature at 38 °C, are able to complete a round of DNA synthesis even in the presence of cycloheximide. This is unlike the situation with *cdc* 4 mutants which fail to initiate DNA synthesis on return to 23 °C in the presence of cycloheximide. This fact and other considerations indicate that an essential protein-synthesis program occurs between *cdc* 4 and *cdc* 7, but that the initiation of DNA synthesis occurs without protein synthesis.

In summary these results point out that the events typical for different stages of the cell division cycle show the following characteristics: (a) exact timing relative to

[1] The α-factor, a linear tridecapeptide, is a mating specific pheromone, which in addition to other actions synchronizes the cell division cycles of the mating partners in yeast. This synchrony is a prerequisite for mating between a and α cells.

their position within the cycle, (b) dependence on processes preceding their expression and (c) determination of their expression if certain preceding stages of the cell cycle have been passed. These features indicate that the processes are integrated in a general gene expression program whose molecular organization, however, is still unknown.

### 15.1.5 Mitogenic Signals in Multicellular Organisms

One of the prominent features of multicellular organisms is the strict regulation of growth of their different specialized tissues. This coordination is brought about by intercellular signals regulating mitosis. Many signals influencing distinct phases of the mitotic cell cycle have been described in the literature. Most, however, are only insufficiently characterized.

First let us again consider the relatively simple system of *Physarum*. If sections of the plasmodia, due to undergo mitosis at different times, are allowed to fuse, the newly formed plasmodium undergoes a synchronous mitosis at a time intermediate to the unfused controls. If the plasmodia are of different size the product plasmodium synchronizes closer to the timing expected for the larger of the two plasmodia. Thus nuclei within the syncytium of *Physarum* seem to respond to accumulation of a common division signal. This signal behaves as if it is mediated by a concentration or ratio of concentrations of one or more molecular species which can be transmitted through the cytoplasm. If, however, one of the plasmodia is within the last quarter of the cycle, then the fusion product will undergo an asynchronous mitosis, with each half of the nuclei entering mitosis as though it were still part of their parent plasmodia. This result agrees with the idea that the postulated mitogenic signal reaches a critical state just before mitosis by which this process is irreversibly determined. The chemical nature of this signal, however, as yet is still unknown.

Most complicated is the regulation of cell division in animals. Many different signals may be involved. There is a number of trophic hormones required for the proliferation and maintainance of certain types of cells. In fact most hormones exhibit an influence on cell division and specialization at least under certain conditions (e.g., for luteinizing hormone and follicle stimulating hormone cf. **4**.2.4.3 and for estrogens cf. **7**.2). So-called mitogenic factors trigger cell proliferation in the immune system (cf. Fig. **13**.7) and in other types of cells (epidermal growth factor, nerve growth factor).

Mouse epidermal growth factor (Mouse EGF), an extensively investigated representative, is a single polypeptide chain of 53 amino acid residues, devoid of alanine, phenylalanine and lysine. It increases the rate of cell division in epidermal and epithelial tissues in animals, but cell culture studies suggest that additional cell types might be responsive in vitro [18]. EGF is bound to specific receptors (glycoproteins) of the cell membrane (40,000 to 100,000 binding sites per human fibroplast cell, dissociation constant $2\text{–}4 \cdot 10^{-10}$ M). The EGF-receptor complex is internalized into the cells and ultimately degraded in lysosomes (cf. Fig. **4**.1).

The processes of binding and internalization have been directly visualized by use of a biologically active EGF-ferritin conjugate. At 4 °C the EGF-ferritin binds to the plasma membrane in a random distribution. Within 1 min at 37 °C the EGF-ferritin complex redistributes on the surface of the membranes to form a number of small clusters that were subsequently internalized into pinocytotic vesicles by receptor-mediated endocytosis (cf. Fig. **4**.1). Within 30 min approximately 90 % of the ferritin was seen in lysosomes.

EGF has been demonstrated to elicit significant biological responses in intact animals, organ cultures and cell culture systems which ultimately result in cell proliferation. The molecular basis of these biochemical and morphological events, however, is still unknown. The EGF receptor complex influences phosphorylation of certain membrane proteins in cell-free systems. Hence the complex might act as a protein kinase or as effector or regulatory subunit of such a kinase (cf. Chap. **10**). But since neither the indentity nor function of EGF-stimulated phosphorylated membrane components have been determined, it is currently not possible to relate these findings to distinct changes of metabolism and gene expression induced in intact cells by EGF.

For many years is has been speculated that endogenous, tissue-specific but species--non-specific proteinaceous inhibitors (chalones) are involved in the regulation of cell proliferation in animals. Chalones are thought to be produced in every cell and to be released to the surrounding milieu. A chalone concentration higher than a critical value should inhibit proliferation of the neighboured cells. Cells located in tissues in the stationary phase of growth, e.g., animal epidermis cells, undergo divisions in such a way that about 50% cells divide (stem cells), whereas the other 50% follow a program of cell specialization which ultimately leads to cell death. It has been assumed that the chalone concentration in the surrounding of an individual cell is used as a signal to maintain this equilibrium. If cells are removed, e.g., by wounding, the chalone level decreases and the number of cells proliferating will increase until equilibrium is reached again.

In the last few years a vast body of evidence has accumulated for the existence of endogenous effectors of cell proliferation from work on a wide variety of tissue extracts and fractions. But in nearly 20 years of intensive research after the first statement of the chalone hypothesis, no reproducible and well-authenticated homogenous chalone has been prepared. Evidence for the existence of chalones in the classical sense is thus still sparse [17].

## 15.2 Quantal Cell Division Cycle

Contrasting to proliferative mitoses quantal mitoses generate daughter cells that are different from each other and/or from the mother cell (cf. cell divisions and maturation in the erythroid lineage of mammals, Fig. **4**.39; the determination and maturation of immune competent cells, Chap. **13**; spermatogenesis in mammals, Fig. **4**.46; and the formation of the conidiospores of the mould *Penicillium cyclopium*, **11**.3.1). The concept of quantal mitoses was developed by Holtzer [20]. It is based on the observation that processes of cell differentiation often are intimately connected with cell division and implies: (1) that only during DNA replication it is possible to create a new state of potential gene activity enabling the doughter cell to undergo further stages in an determinated sequence of differentiation steps and (2) that each quantal cell cycle offers the possibility for a binary decision only. According to this concept, cell specialization is visualized as proceeding by a fixed number of such deterministic quantal cell cycles proceeding under the control of instructive signals.

Illustrative examples showing the main features of quantal cell cycles are the generation of specialized cell compartments during early *Drosophila* development (for details see **4**.6.3) and mammalian myogenesis.

A proposed lineage for myoblasts, fibroblasts and chondroblasts starting with cells in the "primitive mesochyme" is shown in the following scheme[2]:

---

[2] Dienstman, S. R., & H. Holtzer, in: Cell Cycle and Cell Differentiation. In: Reinert, J., & H. Holtzer (eds.), Springer, Berlin 1975, pp. 1–25; PPMbFb compartment: cells ancestral to myogenic and fibrogenic lineages; PPCbFb compartment: cells ancestral to fibrogenic and chondrogenic lineages.

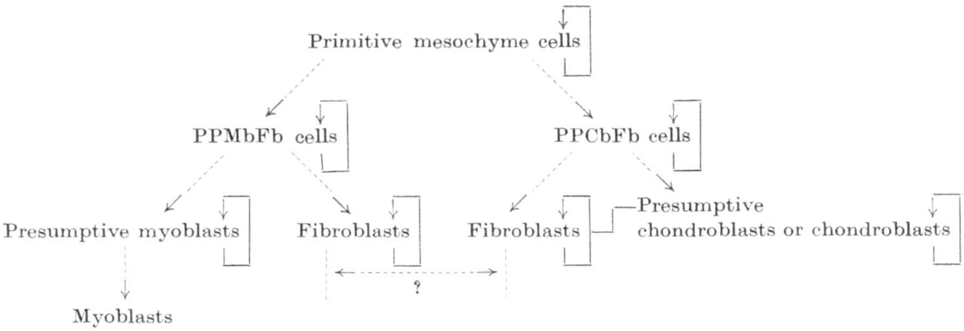


*Dashed arrows* indicate quantal mitoses generating daughter cells with new properties in a fixed series of steps. *Full arrows* point out that cells within the compartment may replicate and are capable of increasing the population within the compartment. The presumptive myoblast compartment for example consists of cells yielding either postmitotic myoblasts or more replicating presumptive myoblast cells. It is not clear whether a given PPMbFb cell actually divides asymmetrically yielding one presumptive myoblast and one presumptive fibroblast cell or whether such cells only divide symmetrically yielding two presumptive myoblast and two presumptive fibroblast cells, respectively. As terminally differentiated fibroblast cells may emerge very early in development and readily replicate, there may be no self-renewing presumptive fibroblast compartment.

Currently there is no way of identifying presumptive chondroblasts. The double-headed arrow and the question mark indicate uncertainty about the possible relationships of fibroblasts that may emerge via different lineages.

Two kinds of arguments have been brought forward against the concept of quantal mitoses:

1. Experimental evidence favours the existence of pluripotential stem cells in animals (cf., e.g., Fig. **4**.39), i.e., stem cells that are capable to produce directly daughter cells of more than two lineages.
2. Stem cell populations themselves are evidently not homogenous. There is evidence for the continuous generation in a probabilistic manner of cells belonging to different subpopulations of the stem cell compartment [23, 24]. These are characterized by the capability, e.g., by means of the corresponding receptor proteins, to respond selectively to distinct types of extracellular signals with proliferation and further specialization. Hence the putative signals are in fact not instructive and quantal cells cycles are not deterministic. There is rather a principle of clone selection and amplification as encountered, e.g., in the immune system (cf. Chapt. **13**).

A decision between the two divergent views of the quantal cell cycle (deterministic vs. probabilistic mechanisms of cell lineage generation) is difficult at present. This is due to our very restricted knowledge about stem cells on the whole, about signals influencing their differentiative capacity and about the early steps of stem cell differentiation on the molecular level.

### 15.2.1 Inhibition of Quantal Cell Division by Bromodeoxyuridine

Specialization of animal cells, e.g., during formation of the retina or during myogenesis, is inhibited by suitable concentrations of 5-bromo-2′-deoxyuridine (BrdU), an antimetabolite of deoxythymidine. Cell proliferation and other essential processes of basic metabolism, however, are not influenced by these concentrations. Hence BrdU depresses specifically the formation of specialization proteins.

Inhibition of synthesis of specialized proteins is dependent on the amount of BrdU incorporated into DNA. BrdU incorporation during a single S phase, i.e., into one DNA strand only, is sufficient to depress, for instance, formation of special surface proteins of myoblasts necessary for the fusion of these cells during myogenesis. Restoration of the ability to fuse proceeds step by step during several cell cycles in absence of BrdU and corresponds to the extent of diluting BrdU out of the DNA.

By means of synchronized cultures of myogenic cells it was demonstrated that BrdU influences DNA replication during the early S phase. If DNA synthesis is inhibited by hydroxyurea, cytosine arabinoside, cycloheximide, or simply by high cell density, BrdU becomes ineffective. There is also no influence on cell specialization if deoxythymidine is administered together with BrdU. These results indicate that BrdU is incorporated into DNA instead of deoxythymidine. Probably this substitution causes stronger binding of proteins to DNA due to the stronger electron negativity of the bromine atom.

### 15.2.2 Conservative Segregation of Parental Histones During DNA Replication – a Rationale of Quantal Cell Cycles?

In addition to results indicating that the deviating states of differentiation of mother and daughter cells after quantal cell cycles may depend on the unequal distribution of maternal stable mRNP present in the cytoplasm, as is evident in the early cleavage stages of embryo development (cf., e.g., **4**.6.3), there are experiments pointing out that also the unequal distribution of maternal histones may be involved. Histone proteins segregate during chromosome replication in a non-random fashion to the two daughter strands. One of the daughter strands is supplied with the histones of the mother chromatin, i.e., in the corresponding daughter cell the structural information for a given region of the mother cell chromatin is retained. On the contrary, the other strand will form nucleosomes with a set of core histones synthesized de novo during replication. Thus a new pattern of histone modifications might be imprinted during chromatin replication and passed to one of the two daughter cells only.

The asymmetric segregation of parental proteins during chromosome replication has been demonstrated by use of density-labelled amino acids. In these experiments parental histone octamers, as well as long nucleosomal sequences formed of them, were shown to segregate from the unreplicated DNA to progeny DNA molecules as intact units, free of newly synthesized histones. Similarly newly assembled histone octamers on the progeny DNA were essentially free

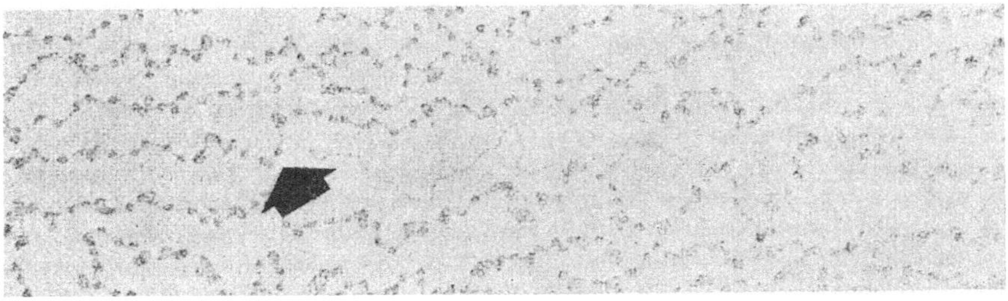

**Fig. 15.9**: Presumptive asymmetric replication forks generated by DNA-replication in the presence of cycloheximide (from Riley and Weintraub [22], with permission).

Nuclear spreads of the chromatin of chicken leukaemia cells replicated in the presence of 100 $\mu$M cycloheximide. Notice the long stretches of newly synthesized non-beaded (protein-free) DNA which is sometimes as long as 100 kb and the less visible single-stranded DNA near the fork which has half the diameter of the double-stranded DNA present in some distance from the fork.

of parental histone contributions. Furthermore, it was demonstrated that nucleosomes retain their integrity for at least 3–4 rounds of DNA replication.

When DNA is replicated in the absence of protein biosynthesis, the parental nucleosomes segregate to only one of the two daughter strands, leaving the other strand free of nucleosomes. Long stretches of protein-free, non-beaded DNA may be observed in these experiments (Fig. 15.9). The non-beaded DNA has the same diameter as naked DNA and is preferentially sensitive to DNase I digestion.

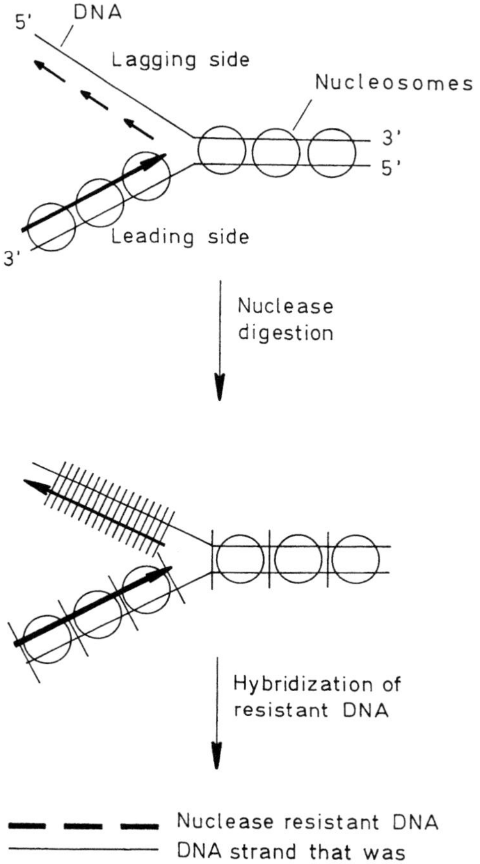


**Fig. 15.10:** Mode of segregation of nucleosomes after DNA replication in presence of cycloheximide (from Seidman et al. [21], modified).

During DNA replication the newly formed strand grows from the 5'- to the 3'-end. Thus starting from the replication origin only one of the two new DNA strands can be synthesized continuously (leading side of the replication fork). In contrast at the "lagging" side short pieces of about 1,000 nucleotides are formed initially and lateron are connected by DNA ligase. If replication proceeds in the presence of cycloheximide de novo formation of nucleosomes is prohibited due to the lack of histones. The parental histones seggregate conservatively with the DNA double helix at the leading side only. Hence after digestion with staphylococcal nuclease part of the newly synthesized DNA is completely hydrolyzed (lagging side), whereas, as the result of shielding by nucleosomes at the leading side, another part forms fragments of about 200 bp (cf. Fig. 1.4). These fragments were shown to hybridize to the DNA-strand that served as template for the leading side of the replication fork.

Sites of hydrolysis are indicated by *vertical lines.*

There is biochemical evidence that the leading side of the DNA replication fork is covered by the parental histones, while the lagging side of the fork receives the newly synthesized nucleosomes. This was shown with the virus SV40 chromosome as a model system. It is well known that the SV40 genome after entering the host cells associates with the histones forming a circular "minichromosome" with ~20 nucleosomes. It replicates bidirectionally from a single point of origin abusing the host cell replicative machinery. Therefore it appears very likely that SV40 minichromosome replication is very similar to the replication of cellular chromatin.

It was demonstrated that the nuclease-resistant SV40 DNA daughter strands formed in presence of cycloheximide, i.e., in the absence of protein synthesis, hybridize predominantly to the template strand of the leading side of the replication fork (Fig. **15**.10). This result is consistent with the idea that the parental histones preferentially segregate to and protect the leading and not the lagging DNA strand. Furthermore it has been shown that the same parental DNA strand serves as template for viral mRNA synthesis. If this observation can be extended to eukaryotic chromatin in general, it would imply that an essential means for the mitotic inheritance of stabile transcription patterns within stable cell compartments might be the conservative propagation of structural information stored in the nucleosomal core histones that are associated with the coding strand within a chromatin domain.

## 15.3 Conclusions

Though there is good evidence now that different types of cell division cycles are directed by programs of gene expression, most details of the molecular structure of these programs, i.e., processes involved in the regulation of the different steps, the regulative connection of the steps, the mode of action of mitogenic signals etc., have hitherto remained unknown. It is one of the challenging tasks of modern biology to get an insight into the construction of such gene expression programs. May be that, due to the relative simplicity and the ease in obtaining and analysing suitable mutants, investigations on the organization of cell division can help in a break-through in this important field.

## 15.4 Selected References

*Proliferative Cell Division – General Aspects*

1. Mitchison, J. M.: The Biology of the Cell Cycle. Cambridge University Press, London 1971.
2. Balls, M., & F. S. Billett (eds.): The Cell Cycle in Development and Differentiation. Cambridge University Press, London 1973.
3. Edmunds, L. N., K. J. Adams: Clocked cell cycle clocks. Science *211* (1981): 1002–1013.
4. Pardee, A. B., R. Dubraw, J. L. Hamlin & R. Kletzien: Animal cell cycle. Annu. Rev. Biochem. *47* (1978): 715–750.
5. Cooper, S.: A unifying model for the G1 period in prokaryotes and eukaryotes. Nature (London) *280* (1979): 17–19.
6. Matsumoto, Y., H. Yasuda, S. Mita, T. Marunouchi & M. Yamada: Evidence for the involvement of H1 histone phosphorylation in chromosome condensation. Nature (London) *284* (1980): 181–183.

*Mitotic Cycle of Physarum polycephalum*

7. Grant, W. D.: RNA synthesis during the cell cycle of Physarum polycephalum. In: Balls, M., & F. S. Billett (eds.), The Cell Cycle in Development and Differentiation. Cambridge University Press, London 1973, pp. 77–109.

8. Braun, R., L. Hall, M. Schwärzler & S. S. Smith: The mitotic cycle of *Physarum polycephalum*. In: Nover, L., & K. Mothes (eds.), Cell Differentiation in Microorganisms, Plants and Animals. G. Fischer Verlag, Jena, and Elsevier/North Holland, Amsterdam 1977, pp. 402–423.
9. Turnock, G.: Patterns of nucleic acid synthesis in *Physarum polycephalum*. Progr. Nucleic Acid Res. Mol. Biol. *23* (1979): 53–104.

*Proliferative Cell Division Cycle in Yeasts*

10. Hartwell, L. H.: *Saccharomyces cerevisiae* cell cycle. Bacteriol. Rev. *38* (1974): 164–198.
11. Halvorson, H. O.: A review of current models on temporal expression in *Saccharomyces cerevisiae*. In: Nover, L., & K. Mothes (eds.), Cell Differentiation in Microorganisms, Plants and Animals, G. Fischer Verlag, Jena, and Elsevier/North Holland, Amsterdam 1977, pp. 361–376.
12. Mitchison, J. M.: Enzyme synthesis during the cell cycle. In: Nover, L., & K. Mothes (eds.), Cell Differentiation in Microorganisms, Plants and Animals. G. Fischer Verlag, Jena, and Elsevier/North Holland, Amsterdam 1977, pp. 377–401.
13. Simchen, G.: Cell cycle mutants. Annu. Rev. Genet. *12* (1978): 161–191.
14. Fraser, R. S. S., & P. Nurse: Novel cell cycle control of RNA synthesis in yeast. Nature (London) *271* (1978): 726–730.
15. Edwards, D. R. W., J. B. Taylor, W. F. Wakeling, F. Z. Watts & I. R. Johnston: Studies on the prereplicative phase of the cell cycle in *Saccharomyces cerevisiae*. Cold Spring Harbor Symp. Quant. Biol. *43*, Part I (1979): 577–586.
16. Thorner, J.: Intercellular interactions of the yeast *Saccharomyces cerevisiae*. In: Leighton, T. J., & W. F. Loomis (eds.), The Molecular Genetics of Development. Academic Press, New York 1980, pp. 119–178.

*Mitotic Signals*

17. Allen, J. C., & C. J. Smith: Chalones: A reappraisal. Biochem. Soc. Trans. *7* (1979): 584–592.
18. Carpenter, G., & S. Cohen: Epidermal growth factor. Annu. Rev. Biochem. *48* (1979): 193–216.

*Quantal Cell Division*

19. Lough, J., & R. Bischoff: Differential sensitivity to 5-bromodeoxiuridine during the S-phase of synchronized myogenic cells. Dev. Biol. *50* (1976): 457–475.
20. Holtzer, H., & N. Rubinstein: Binary decisions, quantal cell cycles and cell diversification. In: Nover, L., & K. Mothes (eds.), Cell Differentiation in Microorganisms, Plants and Animals. G. Fischer Verlag, Jena, and Elsevier/North Holland, Amsterdam 1977, pp. 424–437.
21. Seidman, M. M., A. J. Levine & H. Weintraub: The asymmetric segregation of parental nucleosomes during chromosome replication, Cell *18* (1979): 439–449.
22. Riley, D., & H. Weintraub: Conservative segregation of parental histones during replication in the presence of cycloheximide. Proc. Natl. Acad. Sci. USA *76* (1979): 328–332.
23. Lajtha, L. G.: Stem cell concepts. Differentiation *14* (1979): 23–34.
24. Pearson, M. L.: Muscle differentiation in cell culture: A problem in somatic cell and molecular genetics. In: Leighton, T., & F. W. Loomis (eds.), The Molecular Genetics of Development. Academic Press, New York 1980, p. 360.

# 16. Control of Morphogenesis in *Acetabularia*

R. Wollgiehn

Since the classical investigations of Hämmerling 50 years ago the unicellular alga *Acetabularia* has proved to be an excellent object for the study of many problems concerning cell differentiation, the interaction of cell components, as well as intracellular regulation. What makes it so suitable are its unusual size for a unicellular organism and its structure, which at first sight are suggestive of a multicellular rather than a unicellular organism. A particularly interesting property of *Acetabularia* is its ability to survive longer periods of time without a cell nucleus and even to carry out complicated morphogenetical processes.

## 16.1 The Object

Algae of the *Acetabularia* species are natives of the Mediterranean and other warm southern waters but can also be very successfully cultivated in the laboratory. The adult plant resembles an open umbrella. Most species attain a height of up to 6 cm, and *Acetabularia major* of the South Seas even grows to a height of 20 cm. The gametangium known as the cup, is borne on the tip of a thread-like stalk and varies in structure from species to species (Fig. **16**.1 and **16**.2). At the posterior end of the stalk, a rhizoid is formed, which during the vegetative phase contains the single cell nucleus. Thus *Acetabularia* is a strictly polarized structure with the nucleus-containing rhizoid at one end, and the stalk and cap at the other, the growing apical end. The fine structure of *Acetabularia* cells differs scarcely from that of other plant cells. Electron micrographs reveal a large number of chloropasts besides the usual structures such as mitochondria, Golgi apparatus, ribosomes and endoplasmatic reticulum.

*Acetabularia* is characterized by a developmental cycle in which the individual phases are well distinguished (Fig. **16**.3).

The zygote (a) elongates at one end into a cylindrical stalk (b), and the rhizoid, containing the nucleus, begins to form at the other end. A number of whorls of hairs now forms successively at the apex of the stalk (c). A large part of the stalk is occupied by a vacuole, and the surrounding protoplasm contains numerous chloroplasts. When the stalk has attained its full length a cap consisting of about 75 radial segments (rays) is formed (d). The diameter of the cap is about 10 mm. The cell grows very slowly, and even under favourable conditions does not achieve its maximum size for about 4 months. During this time the "primary nucleus" in the rhizoid undergoes a considerable increase in size. Whereas the nucleus of the zygote has a volume of only about $4 \times 10^{-9}$ $mm^3$, the primary nucleus, previous to cap formation, has a volume of about $1 \times 10^{-3}$ $mm^3$.

This increase in nuclear volume is connected with characteristic structural alterations. For example, the number of pores in the nuclear membrane is greatly increased, both per nucleus and per unit of area. A striking change is seen in the structure of the nucleoli: Instead of the numerous elongated, loosely structured nucleoli of young cells, older cells contain a single, round, condensed vesicular nucleolus. The total nucleolar mass per nucleus rises 100 fold during the course of development, with an accompanying large increase in nucleolar DNA. This increase in DNA reflects a selective amplification of rRNA genes, resulting in an intensive synthesis of

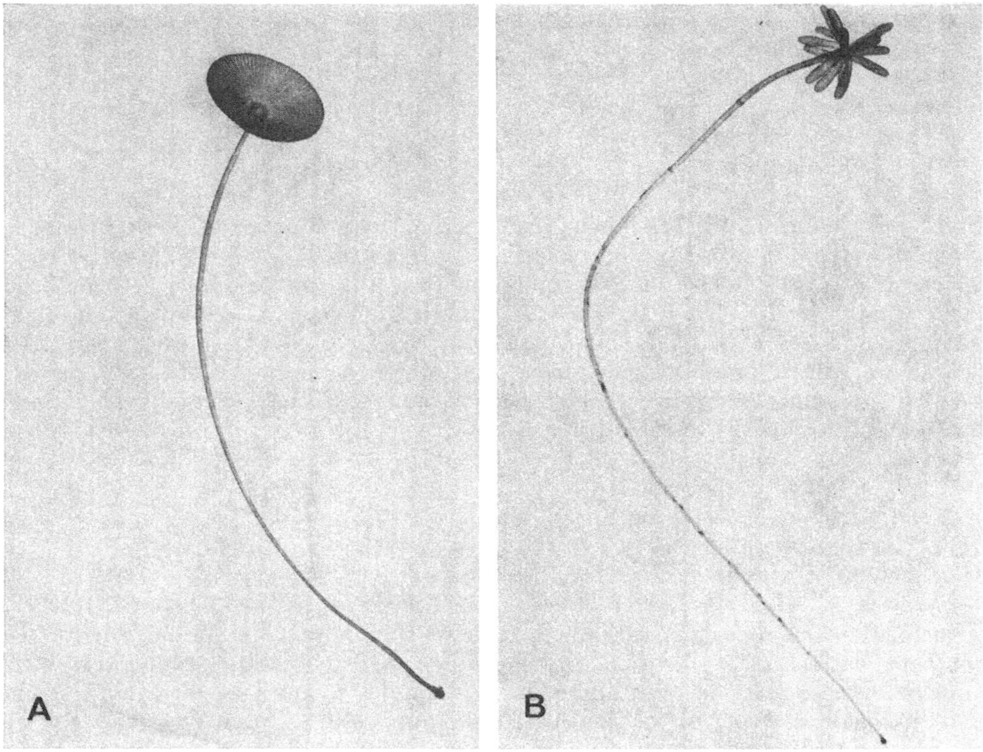


**Fig. 16.1:** *Acetabularia mediterranea* (A) and *Acetabularia cliftonii* (B).
Cells are about 5 cm in length. Photographs were kindly supplied by Dr. S. Berger, Wilhelmshaven.

rRNA in the growing cell nuclei. No dramatic changes appear to take place in the other parts of the nucleus. The DNA content of the nucleoplasm changes only very little during nuclear growth and mRNA is formed continuously.

When the cap has reached its full size, changes take place in the cell nucleus. It becomes smaller, the number of nucleoli is reduced to one, clearly visible chromosomes and a large spindle can now be observed. Finally, via a considerable number of steps the nucleus divides and forms between 7,000 and 15,000 diploid secondary nuclei. The secondary nuclei are then translocated by a strong, directed plasma flow into the cap, where each nucleus and its surrounding protoplasm become enclosed within a thick membrane to form a cyst. Repeated mitoses now take place within these cysts that each becomes multinuclear. Meiotic division leads finally to about $10^8$ haploid gametes. For the formation of the secondary nuclei considerable synthesis of DNA is obviously necessary. Assuming a DNA content of $10^{-13}$ g per gamete, it can be calculated that during the phase of secondary nuclei formation ($\sim$ 200 h) $10^{-6}$ g DNA are synthesized per cell.

Upon disintegration of the caps the cysts are set free (e, f). After a resting period the cysts open and the gametes emerge (g), fusion takes place between plus-gametes and minus-gametes and zygotes are formed (h), which in turn grow into plants.

The eminent suitability of *Acetabularia* for studies of a cell-biological nature is chiefly due to its developmental cycle and morphogenesis. From the germination of the zygote up to the formation of the stalk, whorls and the specific cap there is a strictly temporal organization of differentiation, which includes the synthesis of specific cell components and enzymes at definite phases of development. A further

**Fig. 16.2**: Cap structures of different *Acetabularia* species.
**A**: Mature cap of *A. wettsteinii*; **B**: Underside of cap of *A. crenulata*; **C**: cap of *A. mediterranea*. The scanning electron micrographs were kindly supplied by Dr. S. Berger, Wilhelmshaven.

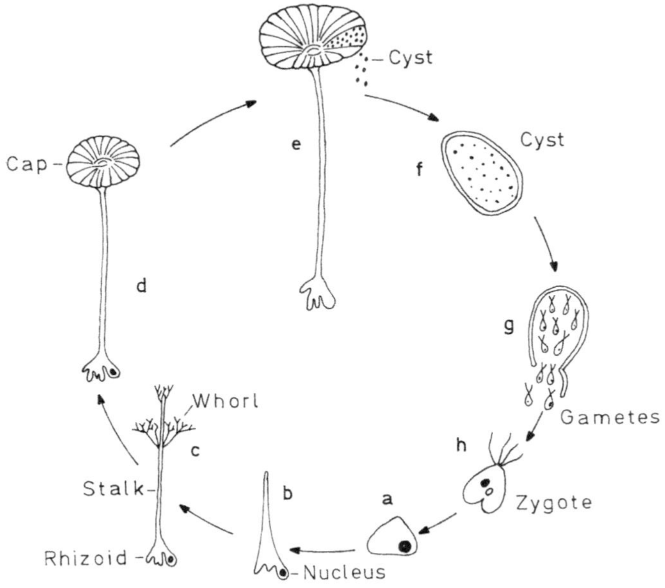


**Fig. 16.3:** Scheme of the developmental cycle of *Acetabularia mediterranea* (from Zetsche [15]). For a–h and further explanations see text.

asset of these cells is the localization of the primary nucleus in the rhizoid, which makes it possible to divide the cell into nucleate and enucleate portions, or to transplant the nucleus in order to study its effect on the growth and developmental capacity of the cell. Such investigations have revealed that not only the nucleate but also the enucleate portions of the cell are capable of growth and regeneration. Nucleate rhizoids develop into complete plants and enucleate immature stalks continue to grow for several weeks and develop normal caps. Thus it appears that differentiation in *Acetabularia* is at certain phases largely independent of direct control by the cell nucleus. On the other hand, transplanted nuclei from one species of *Acetabularia* can radically change the recipient cell and imprint on it the properties of the donor cell. This again demonstrates that the cell nucleus plays a fundamental role as carrier of morphogenetic information. Details of experimental procedures and the interpretation of results will be discussed in the following sections. They give insight into some basic mechanisms of cell differentiation, particularly the interactions between nucleus and cytoplasm, an important aspect of gene expression control.

## 16.2 Nuclear-Cytoplasmic Interactions as Demonstrated by Surgical Experiments

### 16.2.1 Influence of the Cytoplasm of the Nucleus

In the normal developmental cycle of *Acetabularia* formation of the cap is followed by the beginning of disintegration of the primary nucleus and formation of secondary nuclei. It is possible to bring about a temporal displacement of this process by surgical means. If the cap of a mature cell is cut off, the primary nucleus shrinks and no

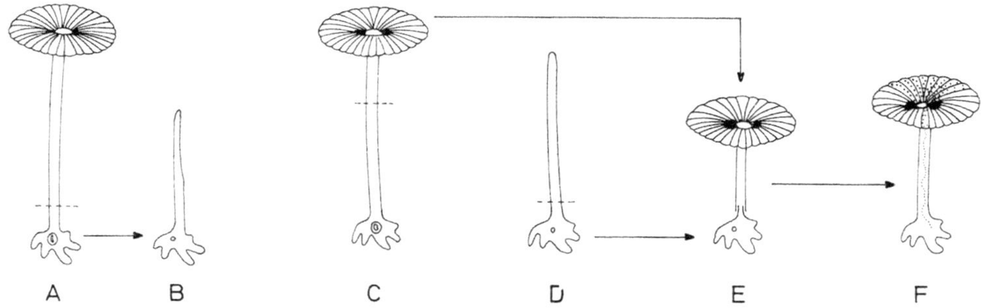


**Fig. 16.4**: Experiments with *Acetabularia mediterranea* to demonstrate the influence of cytoplasmic factors on the primary nucleus (from Mohr & Sitte [9], modified).

*Left:* After removing the cap of a mature cell the primary nucleus shrinks without desintegration (**A, B**).

*Right:* Transplantation of a mature cap onto an immature rhizoid (**C–E**) leads to precocious nuclear desintegration with formation of secondary nuclei (**F**).

disintegration occurs (Fig. **16**.4). This can be repeated over longer periods of time simply by removing the regenerating caps. However, as soon as a regenerated cap is left on the cell, normal formation of secondary nuclei takes place with subsequent development of cysts and gametes. If the reverse experiment is made by grafting a mature cap on to a young rhizoid (Fig. **16**.4), disintegration of the primary nucleus sets in prematurely. These experiments show that the course of development in the primary nucleus is regulated from the cap, or in other words that the time at which the primary nucleus divides depends on the attainment of a certain state of differentiation in the cytoplasm. Thus the cytoplasm exerts a direct influence on the nucleus.

The same effect can also be clearly demonstrated by means of nuclear transplantation experiments. If the primary nucleus of an immature cell is transplanted into the cytoplasm of a mature cell, it very rapidly assumes the ultrastructure characteristic of older nuclei, with typical nucleoli and large numbers of pores in the nuclear membrane. In the reverse experiment correspondingly reversed results are obtained, the mature nucleus taking on the structure of a young nucleus when implanted into the cytoplasm of an immature cell. The signals triggering these alterations have not yet been identified, but it can be assumed that proteins from the cytoplasm enter the nucleus and induce the characteristic structural and resultant functional changes.

### 16.2.2 Influence of the Nucleus on the Cytoplasm

The most striking morphological trait of *Acetabularia* is its cap, which exhibits a different form and structure in each species. That it is the nucleus and not the cytoplasm that is responsible for this specific cap structure can be demonstrated unequivocally in two types of experiments described below:

1. If part of the stalk of *A. mediterranea* is transplanted to the nucleate rhizoid of an *A. wettsteinii* individual the composite organism produces an *A. wettsteinii* cap, i.e., one corresponding to the nucleated cell fragment (Fig. **16**.5).
2. The influence of the cell nucleus on the formation of a species-specific cap can be demonstrated still more clearly by means of heterologous exchange of nuclei. Although isolated pieces from the middle of the stalk stay alive for a considerable

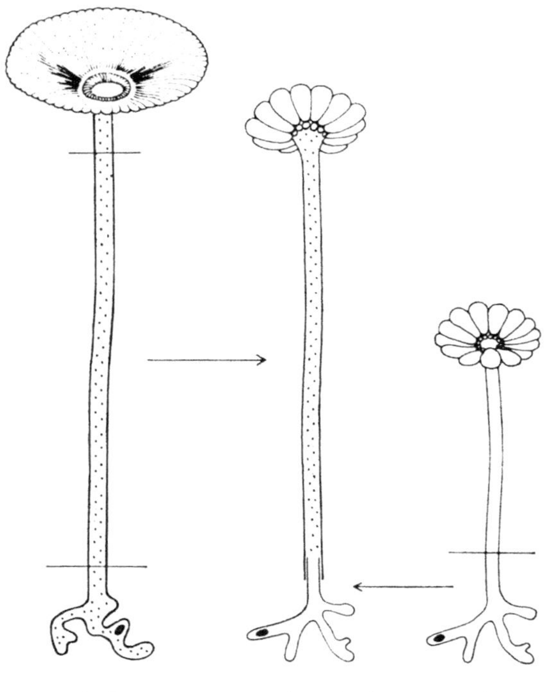


**Fig. 16.5**: Role of the cell nucleus of *Acetabularia* on the formation of a species-specific cap (from Mohr & Sitte [9], modified).

An enucleate part of an *A. mediterranea* stalk *(med)* is transplanted onto the rhizoid of *A. wettsteini (wettst)*. The graft *(med-wettst)* forms a cap that is typical for *A. wettsteinii*.

time, they are not able to develop a cap, in contrast to the apical part of the cell and the rhizoid with its nucleus (Fig. **16**.6 A). But if a primary nucleus, that has been freed of all cytoplasm, is implanted into such a piece of stalk, regeneration begins after a short time and a normal cap is formed, the structure of which is determined solely by the cell nucleus. A nucleus from *A. mediterranea* gives rise to a cap of the species *A. mediterranea* (Fig. **16**.6 B) and a nucleus from *A. crenulata* gives rise to a cap typical of the species *A. crenulata* (Fig. **16**.6 C).

The influence of the cell nucleus on the cytoplasm has, however, also to be considered from another aspect. If the rhizoid (i.e., the cell nucleus) is removed from immature algae or if RNA synthesis in the immature plant is suppressed by means of actinomycin treatment, cap formation is precocious, i.e., the cap develops before the stalk has reached its normal length. It therefore appears, that by some unknown mechanism the cell nucleus controls the temporal pattern of the differentiation program of the cytoplasm. A possible explanation might be that the cell nucleus produces repressor compounds temporarily suppressing the realization of several steps of differentiation.

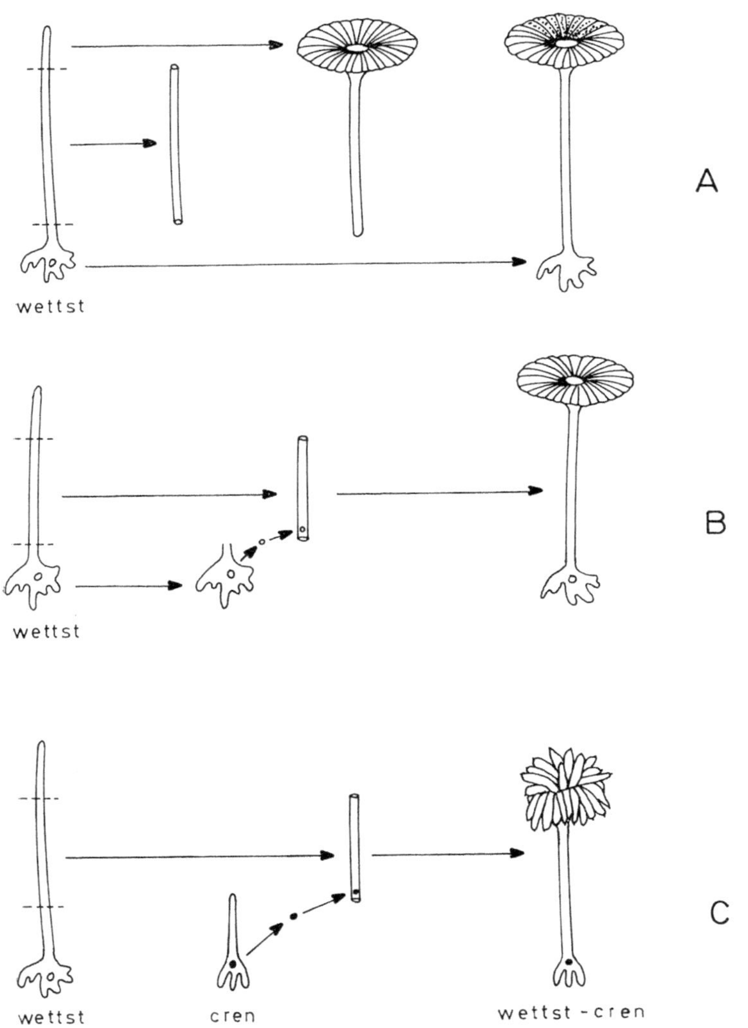


**Fig. 16.6:** Experiments with *Acetabularia* to demonstrate the influence of the primary nucleus on the cytoplasm (from Mohr & Sitte [9], modified).

**A.** Enucleate posterior stalk sections normally exhibit no morphogenetic capacity. Anterior cell parts cut from capless plants form a stalk and a normal cap. The nucleus-containing rhizoid regenerates to a normal cell.

**B.** and **C.** Implantation of a nucleus into the cytoplasm of a stalk section results in formatio of a complete cell, the morphological properties being determined by the nucleus. Implantation of an *A. wettsteinii* nucleus into a *wettst.* stalk results in formation of a *wettst.* cap. In contrast, implantation of an *A. crenulata* nucleus into a *wettst.* stalk causes formation of a *crenulata* cap.

### 16.2.3 Regulatory Processes Within the Cytoplasm

Despite the obvious role of the nucleus in the generation of a species-specific cap, it can be shown experimentally that the cells are also capable of forming a cap in the absence of a nucleus, although only the immature stalks and the apical parts of older cells are able to regenerate (Fig. **16**,4). Such observations are in fact not contradictory. The young stalks and apical regions already contain information from the nucleus determining formation of the stalk, the whorls and a species-specific cap, so that removal of the nucleus does not affect the developmental program of the plant. This experiment, too, shows that important regulatory mechanisms in the differentiation of *Acetabularia* are not directly dependent upon the nucleus. At a very early stage in the development of *Acetabularia* cells the information carriers for stalk, whorl and cap formation are synthesized by the nucleus and exported to the cytoplasm. Although some informational macromolecules immediately become active, e.g., for stalk formation, those concerned with further steps in differentiation are first stored in an inactive form and are only realized at a later stage. Details of these processes will be discussed later.

## 16.3 Molecular Basis of Morphogenesis in Acetabularia

### 16.3.1 Messenger RNA as "Morphogenetic Substances"?

Based upon his conclusions from a large number of experiments Hämmerling put forward the hypothesis in 1963 that there must be substances in the cytoplasm that originate in the cell nucleus and serve as mediators in the realization of morphogenesis. He gave these hypothetical substances the name "morphogenetic substances". His experiments showed that these substances must be nucleus-specific, i.e., species-specific, and have a relatively long life time. They are not randomly distributed throughout the cytoplasm but concentrated in the apical part of the stalk (cf. Fig. **16**.4).

Although direct evidence is still lacking there is good reason for assuming that these morphogenetic substances, produced by the primary nucleus, are in fact populations of mRNA molecules, whose function is to code for the many proteins necessary for the generation of species-specific morphological traits.

The first hints as to the RNA nature of these substances were based on the following experimental findings. Autoradiographic methods showed that, although a major part of the RNA synthesis in *Acetabularia* proceeds in the cell nucleus, it also takes place in the cytoplasmic organelles, i.e., on the DNA of chloroplasts and mitochondria. This of course makes difficult the unequivocal proof of RNA transport from the nucleus to the cytoplasm. Nevertheless, in a cell whose nucleus has been labelled with radioactive RNA precursors, it is possible to trace migration of the radioactivity from the nucleus to the tip of the cell if, following short-time labelling, the cell is transferred to a normal medium containing no labelled RNA precursors. However, evidence for the mRNA character of the morphogenetic substances does not rest solely on the existence of transport of RNA from the nucleus to the cytoplasm, but also upon experiments employing inhibitors of gene expression.

If RNA synthesis is inhibited by actinomycin no synthesis of morphogenetic substances takes place. This is seen from the fact that regeneration and therefore cap formation does not occur in the nucleated parts of the cell, whereas the enucleate apical parts of the cell are able to produce a cap even in the presence of actinomycin. They have already accumulated sufficient morphogenetic substances, the function of which is unaffected by the inhibitors. Thus the immature parts of the cell (cell tips) do not require de novo synthesis of RNA for cap formation. On the other hand, if protein synthesis is inhibited by puromycin or cycloheximide in the nucleated parts of the cell or in enucleate parts capable of regeneration, the process of differentiation and cap formation comes to an immediate halt. Morphogenetic substances can be formed in the nucleate cells under these conditions. However, their translation into proteins is inhibited.

Recently other methods have been employed in an attempt to demonstrate directly the mRNA character of the morphogenetic substances. This mRNA must obviously be very long-lived since *Acetabularia* is able to carry on protein synthesis and morphogenesis for several weeks following enucleation.

It could be argued that this alone is insufficient evidence of the longevity of mRNA itself, since protein synthesis could have taken place in cell organelles under the direction of their own DNA. Or the formation of stalk, whorl and cap might have been achieved with the help of stable proteins formed by short-lived messengers. Criticism of this kind can be refuted, however. The fact, mentioned earlier, that weeks after enucleation morphogenesis but not RNA synthesis can still be blocked by application of inhibitors of protein synthesis is convincing evidence of the RNA character and longevity of the "morphogenetic substances".

It is not yet satisfactorily known to what the stability of mRNA can be attributed, particularly since the cytoplasm contains considerable amounts of ribonuclease. It might be that protection against nuclease attacks is afforded by the fact that the mRNA is protein-bound, i.e., is present in the cytoplasm in the form of nucleoprotein particles and thus protected against nuclease attack. Particles of this kind sedimenting at about 20S, 40S and 120S have in fact been isolated from *Acetabularia*. Data so far available are somewhat contradictory as to their possible role as carriers of "morphogenetic substance". The large 120S particles can be excluded with certainty in this respect.

A very remarkable finding concerns 40S RNP particles that were isolated from *Acetabularia* cells and injected into enucleate stalk fragments, usually incapable of growth and regeneration. Morphogenetic processes now began on a limited scale, i.e., in about 10% of the cells formation of whorl-like structures and of rudimentary caps could be observed. The effects were not so well developed as those following implantation of nuclei in non-regenerative stalk fragments, where normal caps were formed in 40% of the experiments (see Fig. **16**.4). Nevertheless, the results do indicate a role of cytoplasmic RNP particles in cell morphogenesis. It becomes evident from these results that 40S RNP particles contain a large spectrum of different mRNA species, since growth and morphogenesis require a large number of different proteins.

Attempts to isolate the RNA itself and to demonstrate its role as morphogenetic substance have also been made. First of all the template activity of the total RNA of *Acetabularia* was tested in a cell-free, protein-synthesizing system, and it was found that the messenger activity of the RNA decreases with increasing age of the cells. It was not clear however, whether the mRNA isolated with the total RNA was really the product of nuclear genes or rather originated from cytoplasmic organelles. In the latter case it could not be regarded as morphogenetic substance in the true sense of the term. For some time it has been possible to obtain mRNA synthesized in the nucleus directly from the nuclei or from cytoplasmic polysomes, free from other RNA fractions: most nuclear mRNA has a poly (A) sequence at the 3′-end of the molecule (cf. Fig. **1**.8) which makes it possible to separate these molecules from other RNA types by affinity chromatography.

For this purpose the total RNA isolated from nuclei or cytoplasmic polysomes is passed through a column containing either oligo(dT) bound to cellulose or poly(U) bound to Sepharose. The poly(A)-containing RNA molecules are bound either to oligo(dT) or to poly(U). The poly(A)-free

RNA is not retained in the column. Poly(A)-RNA can then be eluted and represents a highly enriched mRNA fraction whose activity and function can be tested in a protein-synthesizing system.

Poly(A)-containing mRNA isolated from *Acetabularia* possesses some of the properties expected of a morphogenetic substance: This RNA is heterogenous with regard to molecular weight, which ranges from $0.5–3.0 \times 10^6$, and could thus code proteins with MW of 55–300 kd. It is only formed in nucleate parts of the cell. Its synthesis is not inhibited by rifampicin, which blocks transcription in the cytoplasmic organelles but not in the nucleus. This indicates quite clearly that this RNA originates from the cell nucleus. Since rifampicin does not affect growth and cap formation the possibility that the morphogenesis of enucleate parts of *Acetabularia* is controlled by transcription in chlorplasts or mitochondria can be excluded.

The transport velocity of the poly(A)-RNA from the nucleus to the cell tip amounts to about 5 mm daily and is thus considerably higher than that of ribosomes (1–2 mm daily). Transport of the two RNA species as nucleoprotein particles thus takes place independently. Three days after synthesis about 30% of the de novo synthesized poly(A)-RNA has accumulated in the apical region of the cells. If a cell is cut a few mm above the rhizoid synthesis of poly(A)-RNA is stimulated slightly during the phase of regeneration of the nucleate part, but less than the synthesis of ribosomal RNA. Apparently synthesis of mRNA takes place in the normally growing cell at almost the maximum rate, in contrast to ribosomal RNA, so that it is only slightly stimulated during regeneration.

It has already been mentioned that actinomycin blocks the synthesis of all kinds of RNA, and consequently inhibits growth and morphogenesis in immature, nucleate cells, but not in anucleate cell fragments. Addition of cordycepin (3′-deoxyadenosine), an inhibitor blocking the addition of the poly(A)-sequence at the 3′-end of the mRNA, brings about the same biological effect as actinomycin. In the presence of cordycepin no functional poly(A)-containing mRNA is formed, and hence growth and morphogenesis cannot proceed. As would be expected, however, enucleate cell tips form caps also in the presence of cordycepin since they have accumulated sufficient amounts of functional mRNA.

Finally, it is worth mentioning that all data so far indicate that the mRNA originating from the nucleus accumulates in the apical growing region, thus leading to an apical-basal gradient of mRNA throughout the cell. However, mRNA forms only a small proportion of the total RNA of the cell. If the total RNA content of various zones of the cell is determined there is also found an apical-basal gradient. The usual explanation offered is the same given also for the mRNA gradient, i.e., polar transport of the total RNA from nucleus to cytoplasm and subsequent accumulation in the growth zone. This now appears to have been an oversimplification and, in fact, other factors play an important role. A considerable part of the total rRNA and tRNA, which are main constituents of the total cellular RNA, is formed in the organelles (chloroplasts and mitochondria). It has been shown that the apical-basal gradient of RNA concentration is above all due to differences in the rate of RNA synthesis and stability in the various zones themselves, i.e., is caused by differences of metabolic activity in the chloroplasts of the various zones. The transport of rRNA from nucleus to cytoplasm does not play a decisive role in establishing the RNA gradients nor, as a consequence, in the accumulation of higher concentrations of RNA in the apical region.

### 16.3.2 Regulation of Enzyme Synthesis During Morphogenesis of *Acetabularia* and the Role of Translational Control

The fundamental role of mRNA as morphogenetic substances has been discussed at length in the preceding section. The process of morphogenesis, i.e., the formation of stalk, whorls and cap, depends on the coordinated and exactly timed formation of distinct protein patterns from preformed cytoplasmic mRNA species. The dynamics of the synthesis of some enzymes has already been well studied and a number of regulatory mechanisms were recognized. As was to be expected, regulation of the synthesis of many enzymes that are coded by nuclear genes takes place at the translational level, so that even in anucleate cells regulation of enzyme synthesis takes place in accordance with the stage of development of the regenerating cells. In addition, the significance of the extranuclear genetic systems and thus the interaction between cell nucleus and other cell organelles in the course of the differentiation processes have been demonstrated in *Acetabularia.*

#### 16.3.2.1 Role of the Nucleus in the Synthesis of Species-Specific Proteins: Malate Dehydrogenase

As we have already seen, the specificity of the cap is determined by the nucleus. If a primary nucleus of one species is transplanted into an immature enucleated cell of another species, the latter develops a cap characteristic of the species of the implanted nucleus (cf. Fig. **16**.6). The most plausible explanation for this is that the short-lived cytoplasmic proteins responsible for determination of a particular cytoplasmic trait – in this case cap formation – are replaced by other proteins whose mRNA comes from the new nucleus. The development of the trait thus probably depends on a specific spatial arrangement of the proteins concerned, which in turn might be dependent upon species-specific differences in the structural proteins or enzymes involved. However, we still know very little about the molecular mechanisms involved in the realization of morphological traits.

The basic idea of a species-specific cell nucleus providing the mRNA for species--specific proteins is undoubtedly correct. This is very clearly seen in lactate and malate dehydrogenases. Malate dehydrogenase occurs in the cytoplasm in a number of forms, so-called isoenzymes, that can be separated by means of electrophoresis. The isoenzyme patterns of *Acetabularia mediterranea* and of *A. crenulata* differ quite distinctly in the number and mobility of bands (Fig. **16**.7). If the nuclei of these two species are exchanged experimentally, the original cytoplasmic isoenzymes disappear within a few weeks and are replaced by those corresponding to the new nucleus. A similar kind of species-specificity can be assumed also for other proteins including those with regulatory functions, a fact which might be of importance for the specificity of morphogenesis. But until the significance of species-specific isoenzyme patterns, which probably all have the same or a very similar biochemical function, has been elucidated, we are far from understanding the role that such patterns might play in species-specific differentiation processes (cf. Chap **10**).

Also unexplained is how the long-lived cytoplasmic mRNAs can be replaced by other mRNAs under the influence of a new cell nucleus. Synthesis of excessive quantities of the new mRNA would certainly not alone be sufficient explanation, since the old mRNA has to disappear completely and at an even more rapid rate than that at which degradation normally proceeds. It is conceivable that, following the implantation of a different nucleus, the original mRNA is released from some protected form (protein-bound?) under the influence of the new mRNA and thus is exposed to degradation or is inactivated in some other way. The cell would then be able to synthesize proteins only from mRNA of the new nucleus. It has been shown experimentally that the stability of poly(A)-RNA is greater in enucleate than in nucleate cells.

The nuclear exchange experiment described above (Fig. **16**.5) yielded a further important result. At least part of the malate dehydrogenase is known to be localized in chloroplasts and mitochondria, which means that enzymes of the organelles are also coded by nuclear DNA and can be exchanged under the influence of the foreign

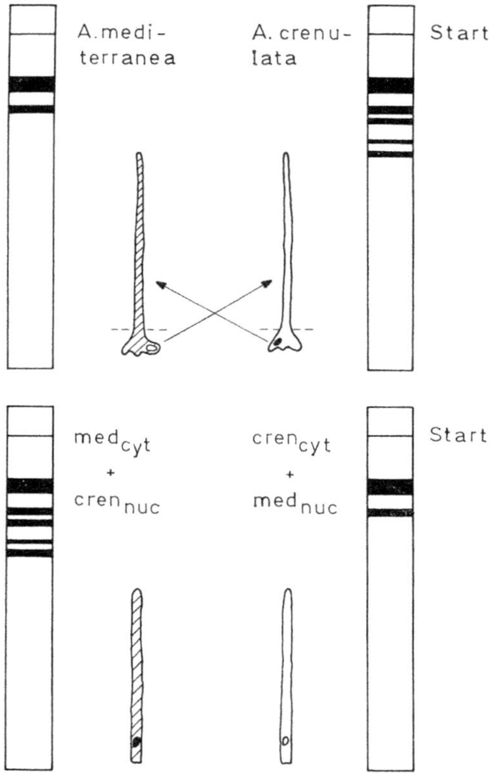


**Fig. 16.7**: Influence of the nucleus on the formation of a species-specific isoenzyme pattern of malate dehydrogenase (form Schweiger, H. G., Arzneimittelforschg. **17** [1967]: **1433**).

Cells of *A. mediterranea (med)* and *A. crenulata (cren)* differ in their isoenzyme pattern as detected by gel electrophoresis *(upper part)*. Reciprocal exchange of nuclei between young cells results in changes of the isoenzyme patterns in correspondence to the implanted nucleus *(lower part.) cyt*, cytoplasm; *nuc*, nucleus.

cell nucleus. However, details of coding sites and sites of synthesis of chloroplast proteins and the interactions of the various genetic systems in eukaryotic cells will be discussed in Chap **6**.

### 16.3.2.2 Dynamics and Control of Enzyme Synthesis as Related to Morphogenetic Events

Whereas many enzymes in both nucleate and enucleate cells are formed at constant or at slowly decreasing rates over longer periods of time, others remain at a low level of activity and only before or during certain stages of differentiation do they attain full activity, in most cases also even in the absence of a cell nucleus.

In several cases it was possible to show that the rise in activity is due to new-synthesis of enzyme protein. The dynamics of the activity and synthesis of some enzymes and the question of the control level will be dealt with below, the individual enzymes being taken as typical for an entire group.

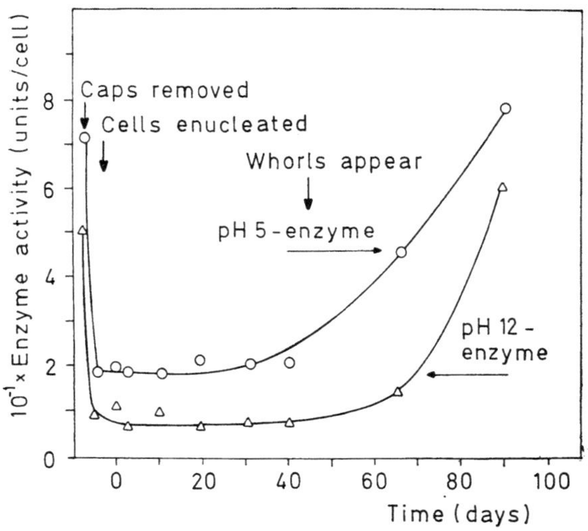


**Fig. 16.8**: Increase in the activities of two phosphatases during normal cell growth of *Acetabularia crenulata* (after Spencer & Harris [12]).

The generation time of the cells was about 8 months. Whorls appeared about 6 months after formation of the zygotes, and caps were mature about 2 months later. ○, acid phosphatase; △, alkaline phosphatase.

*Phosphatases.* In *Acetabularia crenulata* three phosphatases were found. In one enzyme an increase in activity has been shown to be associated with a particular event in morphogenesis.

Synthesis of two phosphatases during normal cell growth is shown in Fig. **16**.8. In the culture used for this experiment caps were formed when cells were 33–38 mm long. Activity of the *p*H 5 enzyme (acid phosphatase) increased continuously during the growth of the cell, but the marked increase of the alkaline phosphatase (*p*H 12 enzyme) was associated with formation of the cap. Synthesis of these two enzymes took place in the normal way when the cap was reformed, after amputation, in the regenerating nucleate cell, i.e., re-formation of the cap is again accompanied by a marked increase in the activity of the *p*H 12 enzyme.

But it is of special interest that nearly the same results were obtained in experiments with regenerating enucleate cells. If caps and nuclei containing rhizoids are removed from the cells, regeneration of the caps takes place more slowly than in cells containing nuclei. But as shown in Fig. **16**.9 there is again an increase in acid phosphatase activity during the course of the experiment, and again the marked increase in *p*H 12 enzyme activity is associated with the re-formation of the cap. The regeneration of the enucleate cell is associated with a progressive net synthesis of protein, and there are reasons to assume that the increase in the phosphatase activities do represent new synthesis of these enzymes.

These results clearly indicate that specific enzymes can be synthesized in the cytoplasm from a pool of long-lived masked mRNA in the absence of the nucleus and that the cytoplasm contains mechanisms for the continuous regulation of this synthesis over a period of many weeks.

Another interesting observation was made in connection with the acid phosphatases. They are repressible enzymes, which means that the end product, in this case inorganic phosphate, inhibits the activity of the enzyme. Removal of the phosphate from the medium ought, therefore,

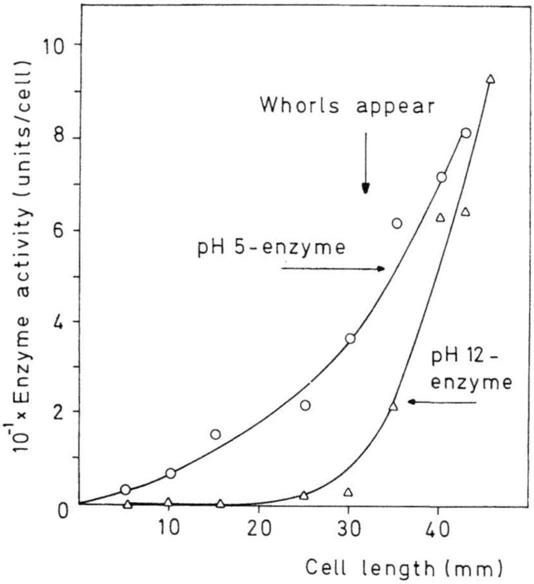


**Fig. 16.9:** Activity of two phosphatases in regenerating enucleate cells of *Acetabularia crenulata* (after Spencer & Harris [12]).

Caps were removed from cells about 35 mm in length and the cells were then kept in the dark for 8 days before removal of the rhizoids. ○, acid phosphatase; △, alkaline phosphatase.

to lead to a rise in activity. This is in fact the case in whole cells and in nucleate fragments of cells, whereas in anucleate fragments no rise in activity occurs. Whether or not enzyme activation or new-synthesis is involved has not yet been investigated, but the results indicate that nucleate cells possess regulatory mechanisms that are lost upon enucleation.

*Protein kinases.* Protein kinases are ubiquitous enzymes that transfer $\gamma$-phosphoryl groups of ATP to serine and threonine residues of proteins. Protein phosphorylation has been shown to exert a decisive influence on the activity of enzymes and regulatory proteins. Frequently, activation of a specific protein kinase and its action on distinct components of the gene expression chain is an essential part of signal transformation (for details cf. Fig. **4**.4, **4**.26, **10**.6 and Sect. **14**.2.2). Therefore it was obvious to investigate the activity of the protein kinases in the course of cell development of *Acetabularia.*

It could be shown that the activity of protein kinases does in fact increase steadily during the growth of the cell, attaining a maximum immediately before cap formation begins. It drops again during maturation and is at its lowest level in the mature cell (Fig. **16**.10). A considerable part of the enzyme activity is associated with chloroplasts.

Although the role of the protein kinases in *Acetabularia* has not yet been studied, these findings suggest that one of the last steps in gene expression, the activation of proteins by enzymic modification, plays a role in cell morphogenesis. Just as certain enzymes (e.g., thymidine kinase) are only formed immediately before a certain step of differentiation begins, the activity of others could be altered by phosphorylation. The dynamics of protein kinase activity points to a regulatory mechanism of this nature.

*Thymidine kinase.* A particularly interesting regulatory mechanism has been found for thymidine kinase of *Acetabularia.* This enzyme phosphorylates thymidine (dT),

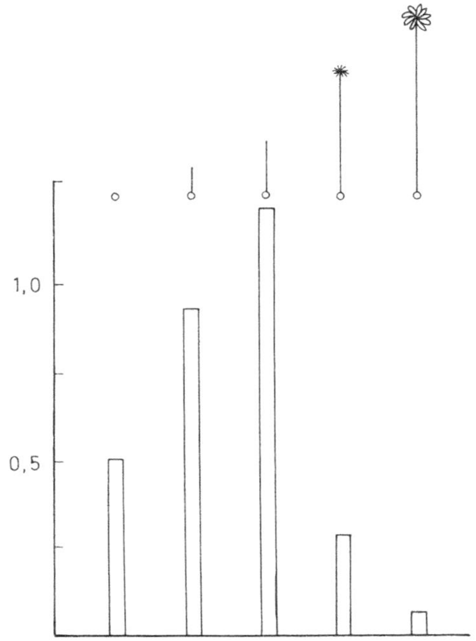


**Fig. 16.10:** Developmental changes of protein kinase activity in cells of *Acetabularia major* (from Pai et al., Protoplasma *85* [1975]: 209).

The developmental stages are indicated schematically on *top* of the figure. Protein kinase activity *(open columns)* is given as moles of [$^{32}$P] incorporated per mg protein.

to dTMP, which is then further phosphorylated by other kinases via dTDP to dTTP serving as a precursor for DNA synthesis. Synthesis of thymidine kinase is largely confined to the period of transition from the vegetative to the generative phase.

At the onset of the generative phase, when the cap has attained its maximum size and the formation of DNA and of approximately $10^8$ secondary nuclei commences, thymidine kinase activity increases sharply. Cellular organelles, and thus also their DNA are no longer formed during this phase. When cyst formation starts dT kinase disappears (Fig. **16.11**). The increase in activity of this enzyme takes place even if the cell nucleus has been removed 2–3 weeks before cap formation starts. The regulation of the enzyme is thus not dependent on the transcription of the nuclear genome.

The rise in activity can be prevented by application of antibiotics inhibiting protein synthesis, which points out that regulation is by means of new-synthesis of enzyme protein. However, only antibiotics that attack the 70S ribosomes of the cellular organelles (chloramphenicol, puromycin) are able to affect this inhibition, whereas cycloheximide, which specifically inhibits protein synthesis by the cytoplasmic 80S ribosomes, does not (cf. Table **1**.9). This implies that dT kinase is synthesized in the cell organelles, although in its active form it is largely bound to cytoplasmic membranes outside the organelles.

Employing inhibitors of transcription, it could also be demonstrated that mRNA for thymidine kinase is transcribed on extranuclear DNA (chloroplast- and/or mitochondrial-DNA), since rifampicin and ethidium bromide stop the increase in kinase activity. It should nevertheless be borne in mind that this inhibitory effect might also

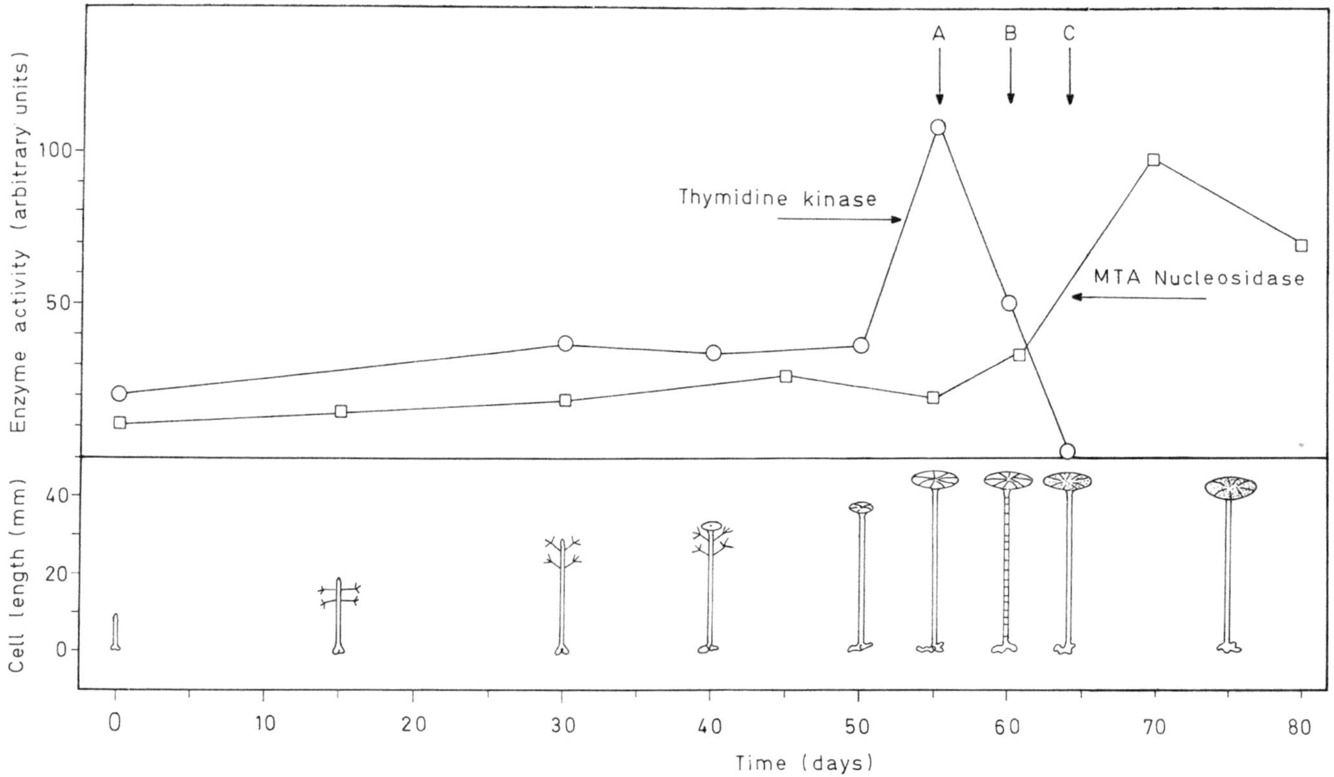


**Fig. 16.11:** Developmental changes of activity of thymidine kinase and 5′-methyladenosin nucleosidase in *Acetabularia mediterranea* (from Bannwarth et al. [1] and Yamakawa [14]).

Cells were grown under standard conditions. During the developmental stages, indicated schematically in the lower part of the figure, cells were homogenized and enzyme activities were determined. **A–C** are different developmental stages of the mature cap. Cysts are formed in stage C.

be achieved indirectly, i.e., by inhibition of the synthesis of a regulatory protein acting on protein synthesis in organelles.

At this point it should be emphasized that there are still other reasons for caution in interpreting biological experiments employing inhibitors of transcription and translation. Even if the action of an antibiotic or other inhibitor, as mentioned above, is well defined and localized on the basis of in vitro evidence, its action in the intact cell is by no means always so specific. Frequently, side effects have been observed. Erroneous interpretation can only be avoided if, apart from the inhibitor experiments, additional criteria are used in elucidating a particular question. In the case of *Acetabularia*, inhibitor experiments can never be regarded as providing more than a link in the chain of evidence.

Bearing all of these qualifications in mind, the results still warrant several important conclusions (Fig. **16**.12).

1. The mRNA for dT kinase is a transcription product of organellar DNA.
2. The enzyme is translated on 70S ribosomes of the organelles, but whether in the chloroplasts or/and mitochondria is not yet known.
3. The regulation of the dT kinase goes on even in the absence of the cell nucleus. The increase in activity can be attributed to de novo synthesis.

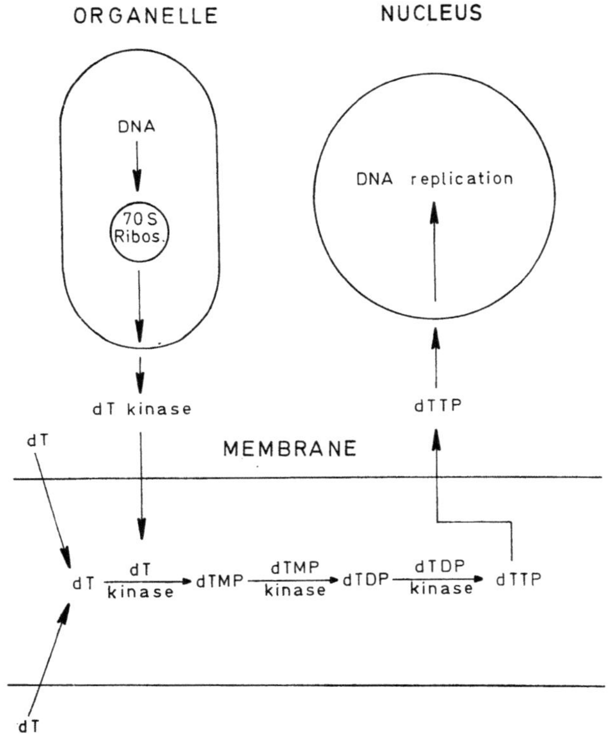


**Fig. 16.12**: Scheme of nucleo-cytoplasmic interactions controlling synthesis and activity of thymidine kinase (from Bannwarth et al. [1]).

Thymidine kinase is synthesized in cell organelles and exported into the cytoplasm. Together with two other kinases it forms a membrane-bound multienzyme chain phosphorylating deoxythymidine to deoxythymidine triphosphate that serves as precursor for DNA synthesis in the nucleus.

4. Regulation of the enzyme synthesis is independent of DNA synthesis, in spite of the temporal relationship between the increase of thymidine phosphorylation and the onset of DNA synthesis and nuclear division.
5. dT kinase, in common with dTMP kinase and dTDP kinase, is a component of cytoplasmic membranes.
6. The rise in dT kinase activity is a prerequisite for replication of the nuclear genome. Thus, the enzyme demonstrates the influence of cell organelles on the activity of the cell nucleus.
7. It is not known what signals evoke the synthesis of dT kinase at a distinct stage of cellular development. This is a point of particular interest since synthesis and activity control of the enzyme involve the coordinated interaction of nucleus and organelles at an important step of differentiation.

*Methylthioadenosine Nucleosidase (MTN).* This enzyme splits methylthioadenosine. Its function within the cell has not yet been fully elucidated, but it is known to be formed at a later stage in cell development than the enzymes described so far. In immature cells and during cap formation the activity of this enzyme is very low but rises drastically immediately after formation of the cysts and begins to decline appreciably after only a few days (Fig. **16**.11). Unlike other enzymes, e.g., alkaline phosphatase, thymidine kinase, synthesis of MTN is controlled mainly by the nucleus. Hence, there is no increase of enzyme activity in enucleated cells. The inhibition by puromycin and cycloheximide shows that the rise in activity is due to de novo synthesis of enzyme protein. Chloramphenicol, on the other hand, has no effect, so that this enzyme is obviously formed in the cytoplasm and not in cell organelles. Experiments with inhibitors of transcription have shown that the mRNA for the enzyme originates from the nucleus and is not formed until the cap has attained its maximum size. Although detectable for, about three weeks in the cell MTN mRNA is relatively short-lived as compared with other mRNA of *Acetabularia*.

## 16.4 Conclusions

If we attempt to review what is at present known about molecular biological processes in *Acetabularia* in relation to the developmental cycle as a whole we obtain a fairly clear picture of the individual regulatory processes involved in morphogenesis. Gene expression in its entirety, which determines morphogenesis, can be divided into two phases. In the first, information in the nucleus is transcribed and transported into the cytoplasm. In the second phase, the information is successively made available for individual steps of morphogenesis.

At least in the most prominent phases of the developmental cycle, such as stalk, whorl and cap formation, which can take place almost normally even after enucleation, control of morphogenesis proceeds predominantly in the cytoplasm. This means that the idea of each of the above-mentioned phases being immediately preceded by activation of the corresponding genes, whilst the genes for the ensuing phase of differentiation are still inactive, is incorrect.

What must in fact happen is that the information for the entire process up to cap formation is almost completely transcribed at a very early stage in development, and this is the only possibility of regulation at the transcriptional level. Therefore it seems that the temporal course of the differentiation process is not determined by transcription regulation.

Long before anything was known about the molecular basis, the information carriers formed by the cell nucleus (mRNA) were termed "morphogenetic substances". The mRNA is stored in the cytoplasm in a stable form (mRNA) and exhibits an apical-basal gradient, i.e., its concentration is higher in the tip region, where the morphogenetic events occur.

This information is activated in a predetermined temporal sequence for the realization of the individual steps of differentiation, independent of whether the cell contains a nucleus or not, which is manifest not only in the morphological development of the cell, but also in the temporal sequence in which individual enzymes appear. Translation regulation must therefore be regarded as an important principle in differentiation, although the role of the nucleus in the regulation of certain enzymes, e.g., methylthioadenosine nucleosidase, should not be overlooked. The interaction of nucleus, organelles and cytosol is a further important characteristic in the regulation of enzyme synthesis. In addition, there are almost certainly other regulatory mechanisms in morphogenesis acting in the cytoplasm, such as changes in activity of already existing enzymes.

Little is known as to the nature of signals that trigger the realization of the genetic information for various steps in differentiation. This is true for signals entering the nucleus from the cytoplasm, e.g., signals triggering DNA synthesis and nuclear division following cap formation. But this is also valid for signals effective within the cytoplasm and determining the sequence of morphogenetic events. In addition to such internal factors there are almost certainly external factors that play an important role. For example, it has long been recognized that differentiation is to a very large extent influenced by light intensity. But despite progress in the analysis of differentiation in *Acetabularia*, many questions remain to be answered before the molecular mechanisms, and above all the coordination of various steps in the developmental program, can be completely understood.

## 16.5 Selected references


1. Bannwarth, H., N. Ikehara & H. G. Schweiger: Nucleocytoplasmic interactions in the regulation of thymidine phosphorylation in *Acetabularia*. Proc. R. Soc. Ind. B *198* (1977): 177–190.
2. Berger, S., W. Herth, W. W. Franke, H. Falk, H. Spring & H. G. Schweiger: Morphology of the nucleo-cytoplasmic interactions during the development of *Acetabularia* cells. II. The generative phase. Protoplasma *84* (1975): 223–256.
3. Berger, S., L. Sandakhchiev & H. G. Schweiger: Fine structural and biochemical markers of *Dasycladaceae*. J. de Microscopie *19* (1974): 89–104.
4. Bonotto, S., V. Kefeli & S. Puiseux-Dao (eds.): Developmental Biology of *Acetabularia*. Elsevier/North Holland, Amsterdam 1979.
5. Brachet, J., & S. Bonotto (eds.): Biology of *Acetabularia*. Academic Press, New York 1970.
6. Franke, W. W., S. Berger, H. Falk, H. Spring, U. Scheer, W. Herth, M. F. Trendelenburg & H. G. Schweiger: Morphology of the nucleo-cytoplasmic interactions during the development of *Acetabularia* cells. I. The vegetative phase. Protoplasma *82* (1974): 249–282.
7. Hämmerling, J.: Entwicklungsphysiologische und genetische Grundlagen der Formbildung bei der Schirmalge *Acetabularia*. Naturwiss. *22* (1934): 829–836.
8. Hämmerling, J.: Nucleo-cytoplasmic interactions in *Acetabularia* and other cells. Ann. Rev. Plant. Physiol. *14* (1963): 65–92.
9. Mohr, H. & P. Sitte: Molekulare Grundlagen der Entwicklung. Akademie-Verlag, Berlin 1971, pp. 109–121.
10. Schweiger, H. G., H. Bannwarth, S. Berger & K. Kloppstech: *Acetabularia*, a cellular model for the study of nucleocytoplasmic interactions. In: Puiseux-Dao, S. (ed.), Molecular Biology of Nucleocytoplasmic Relationships. Elsevier Publ. Comp., Amsterdam 1975, pp. 203–215.

11. Schweiger, H. G.: The role of compartmentation in cell morphogenesis and development. In: Nover, L., F. Lynen & K. Mothes (eds.), Cell Compartmentation and Metabolic Channeling. Fischer-Verlag, Jena, and Elsevier/North Holland, Amsterdam 1980, pp. 455–463.
12. Spencer, T., & H. Harris: Regulation of enzyme synthesis in an enucleate cell. Biochem. J. *91* (1964): 282–286.
13. Spring, H., M. F. Trendelenburg, U. Scheer, W. W. Franke & W. Herth: Structural and biochemical studies of the primary nucleus of two green algal species, *Acetabularia mediterranea* and *Acetabularia major*. Cytobiology *10* (1974): 1–65.
14. Yamakawa, M.: 5'Methyladenosin nucleosidase in *Acetabularia:* Charakterisierung und Regulation. Dissertation Freie Universität Berlin, Berlin 1978.
15. Zetsche, K.: Steuerung der Zelldifferenzierung bei der Grünalge *Acetabularia*. Biol. Rundschau *6* (1968): 97–112.

# 17. Differentiation and Development of *Dictyostelium discoideum*

B. Parthier and W. Lerbs

## 17.1 Introduction

*Dictyostelium discoideum* is an unicellular eukaryotic organism belonging to the cellular slime moulds, an order *(Acrasiales)* in the class of mycetozoa. It is characterized by two unique developmental phases in its life cycle: a vegetative phase represented by amoeboid single cells and a generative phase showing a morphologically well-defined multicellular organism. The transition from one phase to the other is triggered by nutrient deprivation, which leads to production and emission of a diffusible signal substance and in turn to cell aggregation and morphogenetic pattern formation. Upon cell-cell contact a series of events, very rapid and slower ones, seems to be initiated and results in further differentiation and specialization of cells. Final result is a fruiting body composed of two types of cells (spores and stalk cells), whose developmental state was determined in the preaggregation phase.

The developmental sequences in the life cycle of *D. discoideum* belong to the most fascinating examples of a programmed and coordinated morphogenesis accessible to experimentalists. The organism permits to study, at a fairly lucid niveau, the organization of a complex biological system in time and space, cell-cell communication, transformation of external information, and differential activation and expression of genetic information. Thus it is often considered as a model system for embryonic development of multicellular higher metazoa, because of the obvious similarities in the steps of cell differentiation: determination, cell-cell interaction, pattern formation, and morphogenesis.

## 17.2 Biology and Life Cycle of *D. discoideum*

Biologists became interested in this organism when in 1935 Kenneth B. Raper was able to transfer it from the natural forest soil milieu to laboratory conditions. Now it is easy to cultivate *D. discoideum* strains under controlled conditions either fed with bacteria (complex medium) or axenically in a chemically defined organic nutrient broth. In addition, there are more advantages for the preferential use of this organism to study developmental biology:

- Belonging to the most simple eukaryotic cells the nucleus is haploid and contains seven chromosomes only. This facilitates the characterization and experimental handling of mutants, especially developmental mutants.
- Synchronization of mass cultures is easy; growth and developmental phases can be controlled or directed by the experimentalist.

---

Abbreviations: cAMP, cyclic adenosine-3′,5′-monophosphate; CBP, carbohydrate binding protein; $M_r$, relative molecular weight, determined, e.g., by electrophoretic methods; PDE, phosphodiesterase.

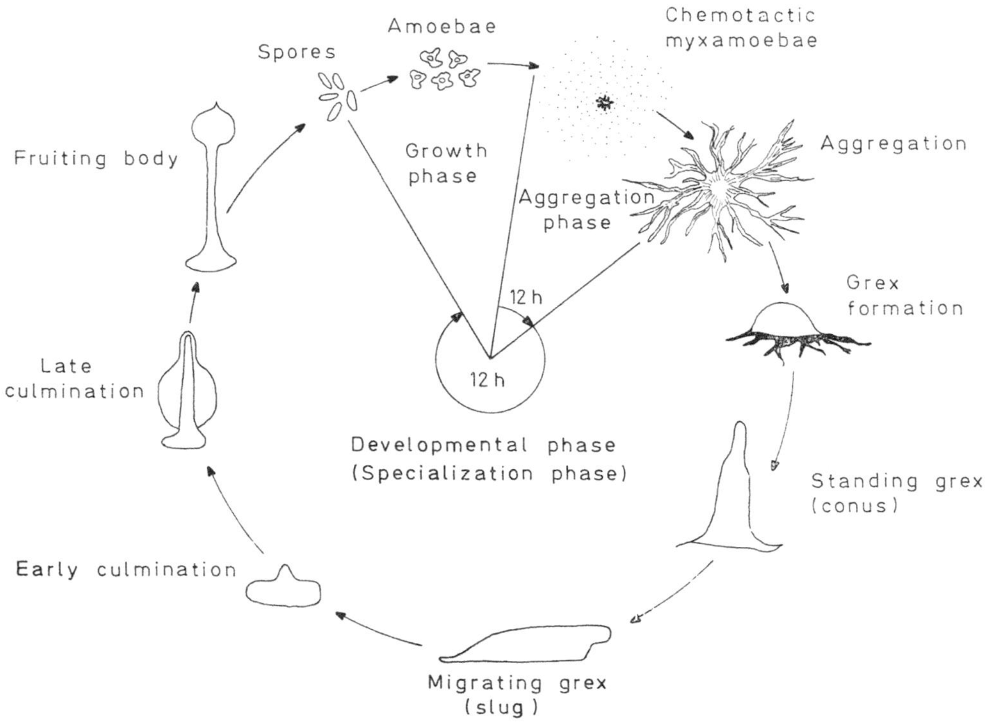


**Fig. 17.1:** Scheme of the cell cycle of *Dictyostelium disocoideum* illustrated by the various morphological patterns of the cell population.
For explanations see text.

The biphasic differentiation process is well characterized by single cells in the growth phase (trophophase) and aggregated cells in the developmental or specialization phase (idiophase). A preaggregation interphase, lasting 5–8 h after food removal, can be only discerned by biochemical and molecular biological methods. Cell-cell interaction, at first documented by signal emission and reception and thereafter by cohesive interaction between cell surfaces, results in dramatic alterations at the levels of gene expression, biochemistry and ultrastructure.

The life cycle of *D. discoideum* can be described in brief as follows (Fig. **17**.1): After ingestion of about 1,000 bacteria a single cell myxamoeba of 10 $\mu$m diameter divides by binary fission within a generation time of four h (12 h in chemically defined media). This vegetative multiplication can last infinitely provided that sufficient food and a slightly acid, moist milieu is available (in the laboratory usually on agar plates in Petri dishes or in suspensions). The amoebae form loose colonies of several million individuals.

If cells exhaust their food supply, or if they are placed on buffer-soaked millipore filters, they enter the first part of the developmental cycle (the preaggregation interphase), but after few hours they begin to migrate towards centres emitting periodic signals (pulses). The centripetal movement of the amoebae is best explained with the existence of a chemical gradient with the highest concentration of the chemotactically active substance (the signal) in the aggregation center. Discrete aggregation fields are formed (Fig. **17**.2) which show concentric wave patterns consisting of moving

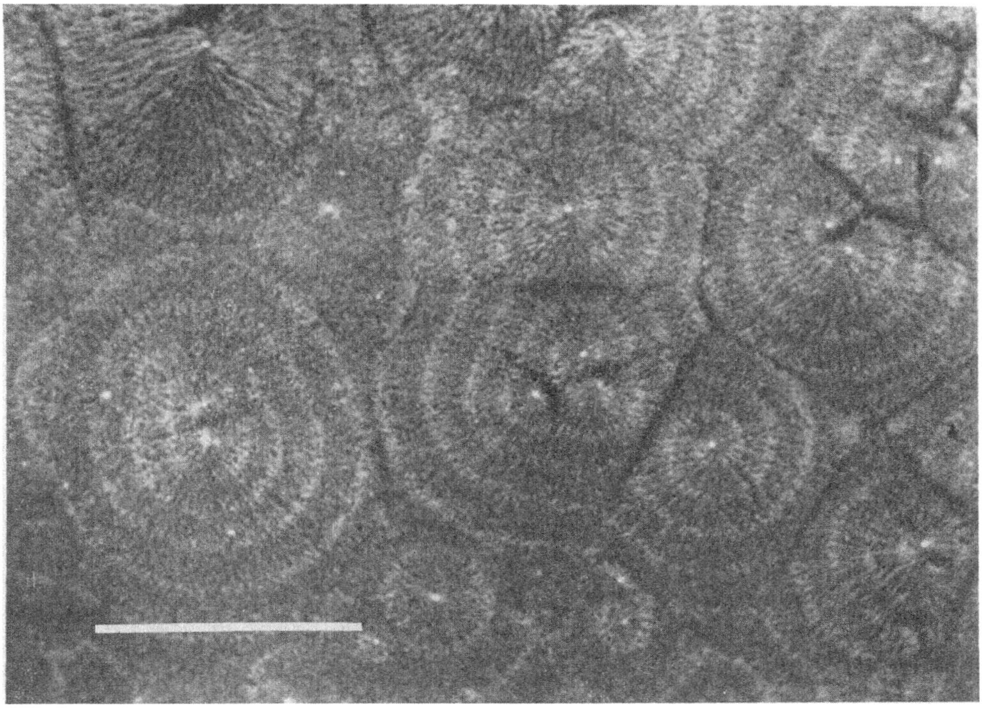

**Fig. 17.2:** Concentric wave pattern in aggregation field of *D. discoideum* (from Alcantara & Monk [12], with kind permission).
Chemotactic movement of cells after external pulse of cAMP. The light bands are bands of moving cells, the dark bands represent cells in the refractory phase. Cell density $10^5$ amoebae/cm$^2$. The *white bar* denotes 1 cm.

and non-moving cells [12]. The chemotactic signals are propagated outwards through the field by a relay mechanism, i.e., a stimulated cell induces adjacent cells to produce and emit signal pulses. The width of each band of moving cells depends on the distance the signal is propagated during the time in which, after signal reception, cells respond with chemotactic movement.

Aggregation and accumulation of cells proceed with formation of a standing grex consisting of about $10^5$ cells. This conus declines and begins to migrate as a slug of 2–3 mm length. Obviously during this period cells are determined for their fate in the resulting fruiting body. Prestalk and prespore cells can be discriminated by cytological and biochemical methods. The volume ratio of the two zones remains almost constant (anterior: posterior $\sim$1 : 2) over a wide range of slug sizes. However, determination of cells of the anterior zone to become part of the stalk and of cells of the posterior zone to develop to spores is not irreversibly fixed. If the volume ratio is disturbed, e.g., by removing parts of the slug, conversion of prestalk cells to prespore cells occurs, or vice versa, thus restoring normal proportions. It is suggested that the prespore-prestalk pattern can be modified by diffusible substances which influence the specialization program in a way allowing one cell type to be converted into the other cell type. This "activator-inhibitor model" is favoured mostly among possible models [17]. The potency of reversion is likewise demonstrated by refeeding the slug;

it soon desaggregates into single myxamoebae. This example shows that maintenance of determination depends on continued food deficiency and cell-cell contact. Determination is stabilized after 15–18 hours of development. Physical desaggregation of the slug at this period results in reaggregation and resumption of morphogenesis. However, the developmental process proceeds at a higher speed than normally, and the entire life cycle including the artificial interruption lasts likewise 24 hours.

Further development of the phototactically moving slug to fixed culmination stages is promoted under unfavourable external conditions, such as low *p*H, high ionic strength, low moisture, and particularly overhead illumination. The "point of no return" is reached after the slug has formed a cellulose envelope. The prestalk cells at the front vacuolize, form thick cellulose walls and finally die. Cells in the posterior part undergo alteration in size, changes in the internal structure of their mitochondria, alteration in chromatin structure, and are eventually converted into spore cells. Thereafter they are enclosed in extracellular layers of mucopolysaccharides and cellulose and are translocated onto the top of the stalk. The fruiting body has developed. – To close the life cycle, spores are released and germinate to haploid myxamoebae under suitable conditions. Fusion of amoebae to diploid heterozygotes does rarely occur.

In the following we will focus our contribution to (1) the prerequisites for cell aggregation (signal formation, emission and reception; chemotaxis and cohesion), and (2) the genetical and biochemical events before and after the aggregation process accompanied by cell specialization.

## 17.3 Chemotaxis and Cell Aggregation

### 17.3.1 Signal Reception and Transformation

#### 17.3.1.1 Chemotactic Movement

Chemotaxis in *Dictyostelium* has to be considered from two aspects. Firstly, vegetative amoeboid cells move toward bacterial food in response to a chemical information from these bacteria. The responsible chemoattractant seems to be folic acid and/or pterins. *Dictyostelium* cells possess receptors for both of them (cf. below).

Secondly, chemotaxis resulting in cell-cell adhesion and aggregation is a developmentally regulated process and absent in vegetative myxamoebae. This process is triggered by a chemical substance which is recognized as a hormone-like signal from outside the cell. The signal becomes transformed, can be resynthesized intracellularly and is released into the medium where it is available as signal for other cells in the population. Thus, the signal is propagated by means of cells as relay systems.

Within five h following removal of nutrients (actually, the removal of one of the essential amino acids is sufficient) no oriented movement of the cells can be observed. What happens in this latent period? What is the primary metabolic or genetic event in the causal chain between removal of essential food components to aggregation--specific signal production in few cells? One might speculate the existence of a primary signal system active prior to the aggregation-specific signal system, probably similar to that in shift-down cultures of *E. coli* or in *Myxobacteria* after nutrient starvation (cf. **4**.4.2.6 and **4**.6.1). In the latter, as in *Dictyostelium*, the resulting cell differentiation (sporulation and aggregation, respectively) seems to be triggered by ppGpp (cf. **4**.6.1).

This nucleoside tetraphosphate or related compounds have not yet been detected in *Dictyostelium*. It would be an interesting phenomenon, if this organism possesses a primary signal system of prokaryotic type and a secondary signal system of eukaryotic (animal cell) type.

After the period of latency cell behaviour changes markedly. Within a few hours single cells aggregate by moving with a rate of about 20 µm/min toward a fixed center of signal emission. Gerisch and coworkers [3, 4, 16] were able to show that the movement is a discontinuous process consisting of about two min migration phase and four min refractory break, both together mediating the impression of periodic waves in dense populations (cf. Fig. **17**.2).

### 17.3.1.2 Nature of Chemoattractant

The chemical nature of the diffusible substance with attracting properties, which causes cells to migrate, has been elucidated. First isolated in J. T. Bonner's laboratory more than 30 years ago, the original substance, named acrasin, was identified as 3′,5′-cyclic AMP. Considering a comparable role of cAMP on hormone-mediated processes of animal cells (cf. Fig. **10**.6), cAMP in its function for cell aggregation was regarded as "second messenger", although it does not match this term completely in this case.

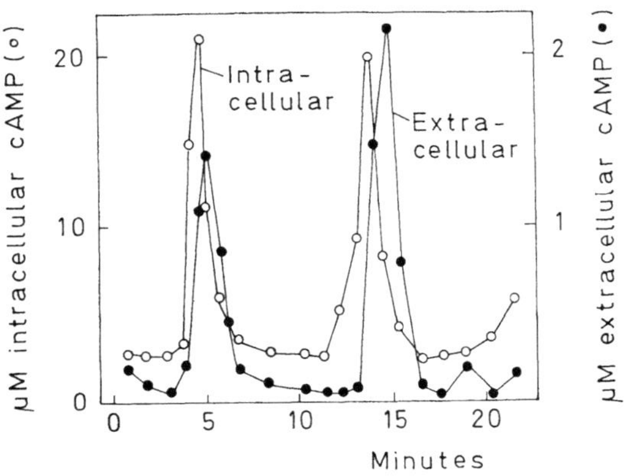


**Fig. 17.3:** Spontaneous oscillations of intra- and extracellular cAMP concentrations (from Gerisch [4], with kind permission).
cAMP is determined within the cells and in the surrounding medium. The increase in intracellular cAMP is due to increase of adenylate cyclase activity. Released (extracellular) cAMP appears with a delay of about 1 minute in the medium. Decrease of the cAMP concentrations are the results of intracellular and extracellular phosphodiesterase activities.

In the first 1–2 h upon nutrient deprivation the level of cAMP within the cells is very low. It rises 30–100-fold during the following hours before decreasing to a constant level of 0.5 µM/cell. The nucleotide is released by exocytosis into the surrounding milieu showing oscillations with pulse frequencies of about eight min ([15], cf. Fig. 17.3). Thus, it appears that the periodicity of aggregation reflects transient levels of cAMP which is synthesized and subsequently degraded by two enzyme systems with opposing functions within a coordinated regulatory network.

### 17.3.1.3 Adenylate Cyclase

Cyclic AMP synthesis from ATP is catalyzed by adenylate cyclase. Its activity is low in vegetative cells, shows maximum six h after removal of food supply, and then decreases again to a low level (Fig. **17**.4). The increase in activity is blocked by cycloheximide suggesting de novo synthesis of enzyme protein. In addition to this major increase in activity, Gerisch [3, 4] detected smaller alterations not connected with synthetic events but due to enzyme activation and deactivation processes. The periodicity of this oscillation behaviour showed intervals of minutes which correlate with the periodicity of endogenous cAMP formation measured with other methods.

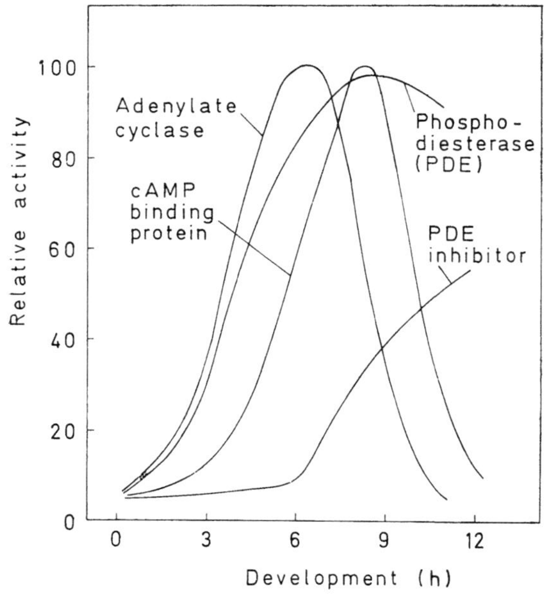


**Fig. 17.4:** Comparison of the kinetics of several enzyme activities and proteins of the signalling apparatus after induction by cAMP (compiled from Loomis [7]).

The activities are expressed in relative values. For details of the function of the proteins cf. Fig. **17**.5.

Sequence of activity increase reflects subsequent periods of de novo synthesis of enzymes of proteins involved in cAMP reception, synthesis and degradation. Obvious examples for this interdependency are the opposite behaviour of PDE activity (decrease) and PDE inhibitor (increase) in the same time period.

Since external cAMP is able to activate adenylate cyclase, which in turn increases the intracellular cAMP level, the activation (and deactivation) reactions of the enzyme must be regarded as the essential part in this whole relay system of signalling. Enzyme activation by cAMP in vitro has not yet been observed due to the extreme instability of adenylate cyclase after homogenization of cells.

The activation reaction was shown to be more complex after the involvement of 3′,5′-cyclic GMP had been observed [14, 18]. Endogenous cAMP production was measured with a one min delay upon exogenous cAMP addition, but cGMP appeared already within an interval of several seconds. Thus adenylate cyclase activation seems to be caused by cGMP which is synthesized by guanylate cyclase primarily activated as response to the addition of external cAMP [18].

### 17.3.1.4 cAMP Binding Sites

Transformation of the cAMP signal into biochemical events requires its binding to receptors located in the outer cell membrane. The number of binding sites for cAMP has been measured using radioactive cAMP. At the beginning of development $5 \times 10^3$ molecules of cAMP are bound per cell. This number increases by one or two orders of magnitude during the following 8–10 h (Fig. **17**.4). Two distinct binding proteins have been isolated showing high cAMP affinity; binding constants $K_d$ between $10^{-8}$ and $10^{-7}$ M have been determined. Short-term experiments revealed that cAMP binding likewise oscillates with a periodicity of 2–5 min.

In contrast to the developmental rise in number of cAMP binding sites, that for folic acid binding decreases from $6 \times 10^4$ sites per cell in vegetative organisms to $3 \times 10^4$ sites before aggregation [27]. A possible explanation for this phenomenon is that during cell development the folic acid receptors, apparently important for vegetative cells to find chemotactically bacterial food, are disconnected from cAMP signal processing. This demonstrates that the receptors for the different chemoattractants are highly specific, although each of them must be able to trigger the same biochemical events which result in cell movement.

### 17.3.1.5 cAMP-Phosphodiesterases and Inhibitor

Cyclic AMP which after release into the surrounding medium remains stable for a long time, would occupy all available cAMP binding sites at the cell surface. Cell migration and further development might be prevented. Therefore, to ensure that the cells detect extracellular cAMP in the form of pulses, the concentration of cAMP in the vicinity of aggregating cells must be maintained at a low level. Similar to animal cells (cf. Fig. **10**.6), *Dictyostelium* has solved this problem by synthesis and partial release into the surrounding medium of cAMP-specific phosphodiesterases (PDE)[1], which convert cAMP into 5′-AMP. Two enzyme species have been described [13, 24]. Part of both (PDE 1, $M_r$ 50 kd and PDE 2, $M_r$ 48 kd) are found in the extracellular milieu. They differ in the size of their carbohydrate moieties. While PDE 1 is synthesized during the vegetative phase, PDE 2 appears towards the end of the growth period and during early development.

The essential role of PDE for development is shown by the fact that PDE mutants do not aggregate. Addition of the enzyme or continuous removal of cAMP render these mutants competent for aggregation.

Phosphodiesterase activity is controlled by a feedback mechanism which involves a proteinaceus inhibitor specific to PDE. This inhibitor is a 50 kd sized, heat-stable glycoprotein [25]. It blocks selectively enzyme activity in vitro. Its appearance in developing cells in a complementary manner to PDE activity (Fig. **17**.4) points to its biological function: In a period, where PDE reaches its maximum level (measured by external cAMP inactivation), the inhibitor is synthesized due to an unknown derepression process. On the other hand, high cAMP concentration represses formation of the inhibitor which results in a high level of PDE activity. This balanced control system enables the cell to maintain a constant low amount of cAMP optimal for discrimination and reception of external signal pulses.

In summary, differentiation-specific chemotaxis in *Dictyostelium* is founded on the capability of cells to discriminate differences in cAMP concentration. This substance is the transmitter in a signal transfer system in which the cells amplify and propagate the signal. Chemotaxis is regulated by a sophisticated network of synthesis and

---

[1] A cGMP-specific PDE was recently observed in cAMP-specific PDE mutants. Its genetic control is exerted independently of that of cAMP-PDE [14].

activation of several enzymes in connection with feedback control mechanisms. The relay principle allows a rare number of cells in a large population to emit spontaneously cAMP. This signal stimulates surrounding cells to use the chemotactic apparatus and to bring the cells in the whole population into synchrony of aggregation competence, i.e., synchronized movements leading to aggregation.

### 17.3.2 Cell Responses after Signal Reception

The interaction between cAMP and its receptor at the membrane level triggers three types of intracellular responses discernible by their rapidity and site of action:
1. physico-chemical responses, within few seconds after external cAMP addition, at the level of membrane conformations or responses mediated by alterations of membrane function;
2. biochemical responses, within few minutes after external cAMP addition, at the level of enzyme activities;
3. responses in gene expression (RNA and protein synthesis), after several (5–10) hours, resulting in aggregation and cohesion ability.

All these responses of primary and secondary nature are coordinated events during and after signal transformation.

#### 17.3.2.1 Early Events after cAMP Addition

Searching for the most rapid responses upon addition of external cAMP the extrusion of pseudopodia was observed after few seconds. This spatially limited loosening of cell bounderies are usually directed to the sites of cAMP addition and they might mediate cell orientation towards the source of cAMP emission. It must be concluded from this observation that the signal was transmitted through the cell during this very short period. Transfer of the signal pulses via conformational alterations of constituents of the cell membrane might be suggested preferably. It is likewise possible, though not convincingly proved, that elements of the cytoskeleton are involved.

Concomitantly with this rapid response an immediate drop in extracellular *p*H was noted in connection with proton release. Likewise, a rapid ATP-dependent influx of calcium ions can be measured [26]. This cation is partially released in the following period, which appears consistent with the periodicity of cell movement and cAMP production. Thus $Ca^{2+}$ seems to be necessary for adenylate cyclase activity, since chelation of $Ca^{2+}$ with EDTA or EGTA suppresses signal emission without affecting chemotaxis, cell aggregation and sensitivity of the cells against external cAMP. Because of the extreme instability of adenylate cyclase in vitro, it remains unclear whether $Ca^{2+}$ exerts its stimulatory action on this enzyme directly or via activation of guanylate cyclase or any other still unknown reaction. A schematic survey on our present knowledge about cAMP-connected signal transformation in *D. dictyostelium* is presented in Fig. **17**.5.

To the responses in the magnitude of minutes belong the amplification of the external cAMP signal by activation of adenylate cyclase (as described above), the propagation of the signal over the aggregation field, and signal transformation leading to chemotactic movement of cells. The intracellular transformation chain should end at components of the contractile apparatus of the cell. As for other cells with amoeboid migration capability, *Dictyostelium* movement is likewise mediated by actin and myosin. Another result following cAMP addition can be observed in

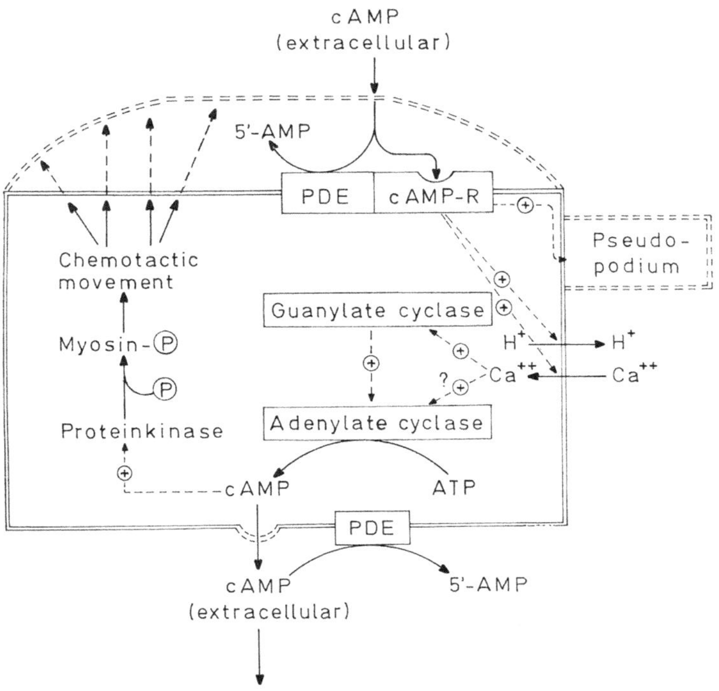


**Fig. 17.5**: Schematic survey on signal transformation and cellular responses after addition of external cAMP to *D. discoideum*.

→, fluxes of substances; – – – ⊕ – – – →, stimulatory effects on processes and on the activity of enzymes. *PDE*, phosphodiesterase(s); *cAMP-R*, binding protein for cAMP.
Upon extracellular addition to myxamoebae cAMP can be bound to cAMP-R (or partially degraded to 5′-AMP by PDE), which causes short-term effects (in few seconds) on proton export, $Ca^{2+}$-import and pseudopodia formation. Adenylate cyclase is activated via $Ca^{2+}$-activated guanylate cyclase within the following minutes, and the newly formed cAMP may give rise to chemotactic movement of the cells. But it is also excreted and serves as a signal to neighbouring cells.

cell structure modifications by methylation of proteins and phospholipids [19] and by phosphorylation of the heavy chain of myosin [21]. These modifications are involved in intracellular $Ca^{2+}$ movement eventually leading to actin-dependent ATPase stimulation and increased contractility. Movement of cells seems to be the result of two processes: general cytoplasmic contraction and local dilatations of the cell membrane.

The aspect of protein phosphorylation came into focus of interest when chemotaxis was correlated with protein modifications in the preaggregation phase of *D. discoideum* [21, 22]. At least six soluble and seven membrane-bound protein kinases have been reported. Two of the soluble enzymes phosphorylated preferentially histone H2A; myosin was the preferred substrate for one of the membrane-bound protein kinases. However, strain-specific variations make general conclusions difficult. Total protein kinase activity in strain V12/M2 increased five-fold during the first eight h of development and, in addition, two new cAMP-dependent protein kinase activities appeared, which are not present in vegetative cells [22]. These results have not been confirmed with strain AX-2 [21]; none of the protein kinases responded to cyclic nucleotides. Further studies should reveal whether or not cAMP-dependent protein kinases are involved in the chain of events between cAMP signal reception and cell differentiation. Assuming that external cAMP is unable to penetrate the cell membrane, the observation of altered myosin phosphorylation in cAMP-stimulated cells resulting in chemotactic movement [21] should be caused by intracellular rather than by external cAMP (Fig. 17.5).

#### 17.3.2.2 Cohesive Surface Binding Proteins

Parallel to the kinetics of increase of enzyme activities in the signalling apparatus as described in the preceding sections (cf. Fig. 17.4), cell aggregation shows a lag phase of 8–10 h after removal of food supply from the amoebae. Before this time occasionally contacting cells do not remain attached but move away from each other. About five h after induction the mutual affinities of cells and their cohesiveness increase, giving rise to clump and stream formations. What is the molecular background of this phenomenon?

Concomitant with the growing tendency of the cell to stick together two cell surface proteins appear, as measured by immunological techniques. One is called contact site A and the other one discoidin or synonymously CBP (carbohydrate binding protein). Contact site A is a glycoprotein ($M_r$ 80 kd) and able to immunoprecipitate with antibodies raised against proteins from developing cells. However, it is unable to precipitate those from non-developed cells [20].

CBP consists of two lectin-like polypeptides slightly different in their molecular weights (CBP-24 of $M_r$ 24 kd, CBP-26 of $M_r$ 26 kd). Both types of proteins bind specifically to galactose, a property facilitating their purification by affinity chromatography on agarose columns. The binding specificity to carbohydrate moieties suggest molecular interactions to glycoprotein receptor molecules in or attached at the outer cell membranes. It is possible, but not yet confirmed, that the receptors are identical with contact site A.

Accumulation of carbohydrate binding proteins in the period following cell starvation is due to de novo synthesis of CBP mRNA and subsequent translation into CBP polypeptides ([23], see below). These can account for more than 3% of the newly synthesized total protein fraction of cells.

Exploiting various developmental mutants (see below) provided further evidence that CBP-26 as well as adenylate cyclase, phosphodiesterase, PDE inhibitor and cAMP binding proteins are necessary for the aggregation process. All these mutants are phenotypically non-cohesive strains. In addition to a lack of one or more enzymes of the cAMP signal system in several strains, none of these strains are able to produce CBP-26 protein, while CBP-24 protein synthesis remains unaffected [20]. However, mixing wild-type and mutant strains results in aggregation competence of the mutant cells, and they proceed to further development and spore formation. This experiment demonstrates that the loss of one binding protein does not prevent attachment of mutant cells to wild-type cells, and that the binding event is essential for subsequent steps in the differentiation program.

Certain mutants provided information that phosphodiesterase and cAMP binding proteins are synthesized prior to adenylate cyclase and CBP-26, thus suggesting that coordinated regulation is not always verified in the synthesis of the molecular equipment for the aggregation process.

## 17.4 Metabolic Alterations during Cell Specialization

The fusion of hundred thousands of individual cells and their development to specialized units cooperating within a multicellular organism requires drastic changes at all levels of cellular metabolism. Starvation of nutrients brings about a high turnover of cellular components, because the energy necessary for cell specialization is now only available from endogenous sources. Compared to vegetative cells, protein, lipid, and nucleic acid contents are reduced to 40%. About 80% of the cellular proteins are turned over and the liberated amino acids are metabolized or reincorporated.

To understand the developmental consequences of all these metabolic alterations on a molecular niveau, the basic biochemical processes and their genetical background, as well as the regulatory interactions, need to be investigated. Very attractive in this regard are selected enzymes of carbohydrate metabolism, since their syntheses are closely related to developmental changes and their activities are prerequisites for the establishment of specialized morphological structures and physiological functions.

The content of carbohydrates does not change quantitatively but undergoes various qualitative modifications. Cellulose, the skeleton material of the stalk, several mucopolysaccharides forming an extracellular sheath, and trehalose as storage substance are synthesized in large amounts. The source for these products is glycogen, which has been accumulated during the vegetative phase but is now metabolized via uridine-diphosphate glucose (Fig. **17**.6). Key enzymes of these metabolic pathways are UDP-glucose pyrophosphorylase, trehalose-6-phosphate synthetase, UDP-galactose epimerase, UDP-galactose: polysaccharide galactosyl transferase and glycogen pyrophosphorylase. All are synthesized during the late aggregation phase, after about ten h of development, whereas the appropriate mRNAs seem to appear several

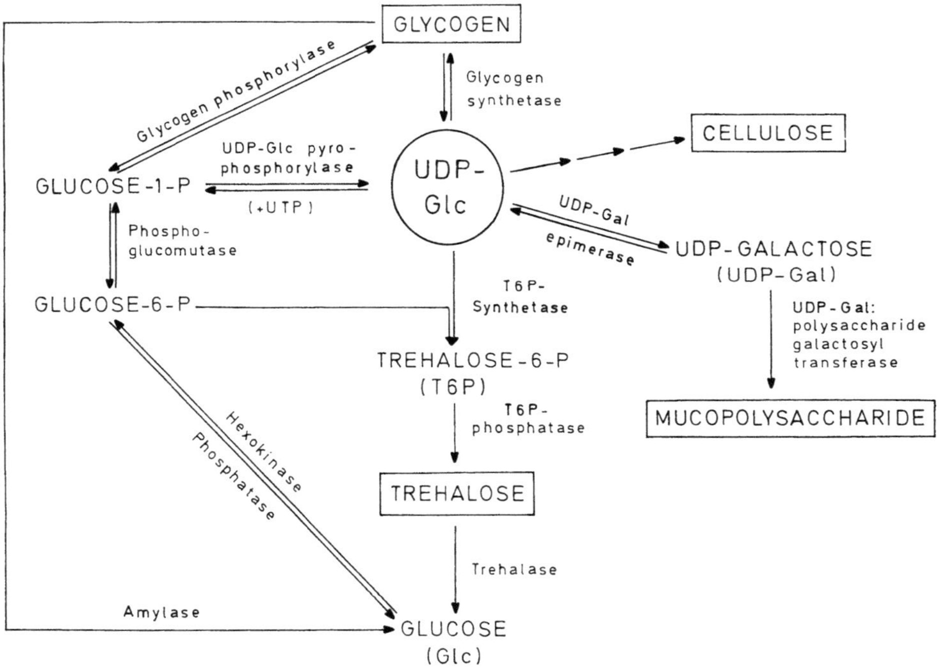


**Fig. 17.6:** Pathways and involved enzymes of carbohydrate metabolism (oligo- and polysaccharide synthesis) in *D. discoideum* (from Sussman & Newell [10], modified).

During cell specialization storage material (glycogen) is degraded to UDP-glucose, the pool from which the most important final products for spore and stalk formation, cellulose, mucopolysaccharide and trehalose *(in boxes)* are derived. Other pathways of glycogen – glucose interconversions are likewise indicated.

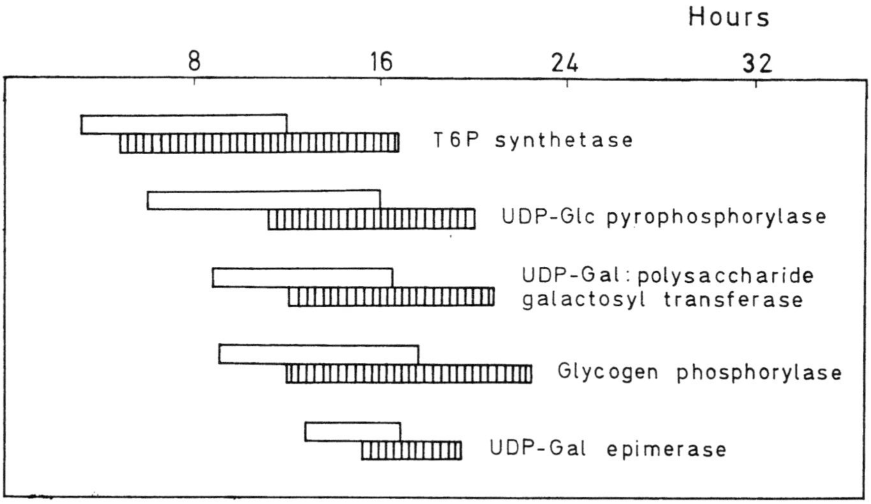


**Fig. 17.7:** Periods of transcription *(open bars)* and translation *(hatched bars)* of mRNA for some "developmentally" regulated enzymes (from Sussman & Newell [10], modified).

The figure demonstrates both the successive de novo synthesis of mRNA and certain enzymes, as revealed by the use of transcription and translation inhibitors, as well as the time delays due to mRNA processing.

hours earlier [11] (Fig. **17**.7). This time delay between mRNA synthesis and translation varies from enzyme to enzyme and, as demonstrated later (Fig. **17**.9), many other functionally unknown but developmentally regulated polypeptides show a similar delayed transcription – translation relationship, too.

Synthesis of all of these enzymes and polypeptides is an integral part of the developmental program. Even if some of the referred carbohydrate metabolizing enzymes are only representative for regulatory side branches in the network of gene expression (cf. **17**.5.1), their activity is an absolute requirement for the realization of the developmental program. Thus, "morphological" mutants may be produced which are only lacking certain enzymes of carbohydrate metabolism, e.g., mutants with a block in N-acetylglucosaminidase activity are unable to migrate [35], those lacking UDP-Glc pyrophosphorylase [32] cannot culminate and, as mentioned above, mutant strains without an intact carbohydrate binding protein (CBP-26) are unable to aggregate [40]. Similarly, inactivation of other metabolic reactions by mutagen treatment results also in an abnormal morphological development.

## 17.5 Genetical and Molecular Biological Aspects of Development

Most of our knowledge on sequences of events during programmed gene expression has been derived from studies with bacteria or viruses (cf. **4**.5). Quite different pathways of intermediary metabolism as well as their regulatory mechanisms and complex developmental programs are relatively well understood at present. The much higher complexity of an eukaryotic organism restrains many of the classic genetical or biochemical approaches, e.g., mutant and recombinant analysis as fundamental methods for the elucidation of prokaryotic operons and genomes. First attempts to bring the more or less descriptive results to a molecular level have been made during the last years.

### 17.5.1 Genome of Dictyostelium and Developmental Mutants

Balancing between a lucid developing system and eukaryotic complexity, *Dictyostelium* was chosen as a model organism for reasons of its relative simplicity in genome organization and for other reasons mentioned at the beginning of this chapter. The genome of most strains used in the laboratory is haploid. Seven chromosomes can be distinguished by their size and binding pattern when arrested at the metaphase. Nevertheless, the DNA content is eleven times greater than that of *E. coli*. A simple calculation based on some experimental results is aimed at demonstrating the consequences of this difference. The haploid genome has a molecular weight of about $3.6\times10^{10}$ dalton ($5\times10^{4}$ kb). 70% of the total DNA seems to be repetitive (cf. Table **1**.1). Assuming 1,100 bases for an average mRNA molecule, 14,000 different mRNAs might be synthesized. In accordance, an estimation of the number of genes on the basis of poly(T)-sequences present in the DNA[2] revealed a number of about 15,000 [2].

Another type of experiment allows to estimate the number of genes expressed within distinct developmental stages. If one considers the frequency of mutations observed for a well characterized protein (about $5.4\times10^{-4}$) and compares this figure with the mutation rate affecting aggregation (about $9\times10^{-2}$) and following stages of morphogenesis ($8\times10^{-2}$), one arrives at about 160 genes for aggregation and about 300 for the total developmental process [8]. These estimations may contain a high degree of inaccuracy, but they demonstrate the complexity of the system when attempting a molecular biological and genetical analysis.

According to the seven chromosomes described, seven linkage groups of ~40 established genetic marker mutants have been detected, and a number of biochemical and morphological mutants have been mapped [9]. In few cases of mutations affecting the activity of a specific enzyme, the mutant gene product or the biochemical alteration caused by the mutation has been described in detail. But no such details are known at present about mutations affecting ordered sequences of gene expression during development and their regulatory interrelations. Only few developmental mutants have been studied with respect to the biochemical blocks and their effects on following steps of differentiation (Fig. **17**.8). Mutants blocked at defined stages, e.g., between stages A and B or B and C etc., cannot express the characteristics for the following steps. Thus, a linearly dependent sequence may be described in a way that events before the mutational block are present normally, while later steps of the program fail to occur [8]. However, in no case sufficient knowledge exists about the primary mutagenic attack.

There are correlations between defined mutations in enzymes of carbohydrate metabolism and a disturbed synthesis of cytoskeleton material [32]. In developmental mutants enzymes of the carbohydrate metabolism are likewise missing, e.g., mutants WL3, WL4 and AGG206 unable to express the gene for UDP-Glc pyrophosphorylase, or mutant KY3 free of $\beta$-glucosidase activity. However, not all of the markers examined so far may define parts of the indispensable core of the developmental program; they might represent the expression of branches or "subprograms". As already mentioned, mutants lacking N-acetyl-glucosaminidase may carry out the whole developmental program but are unable to migrate. To complete the picture of the developmental program and its regulatory interaction a much more detailed analysis of developmental abnormal strains and also of the wild-type strains is required.

---

[2] Contrasting to most other eukaryotes a 25 nucleotides long poly(A) sequence at the 3′ terminal end of *Dictyostelium* mRNA is coded for by DNA [2] and not added postsynthetically (cf. Fig. **1**.8).

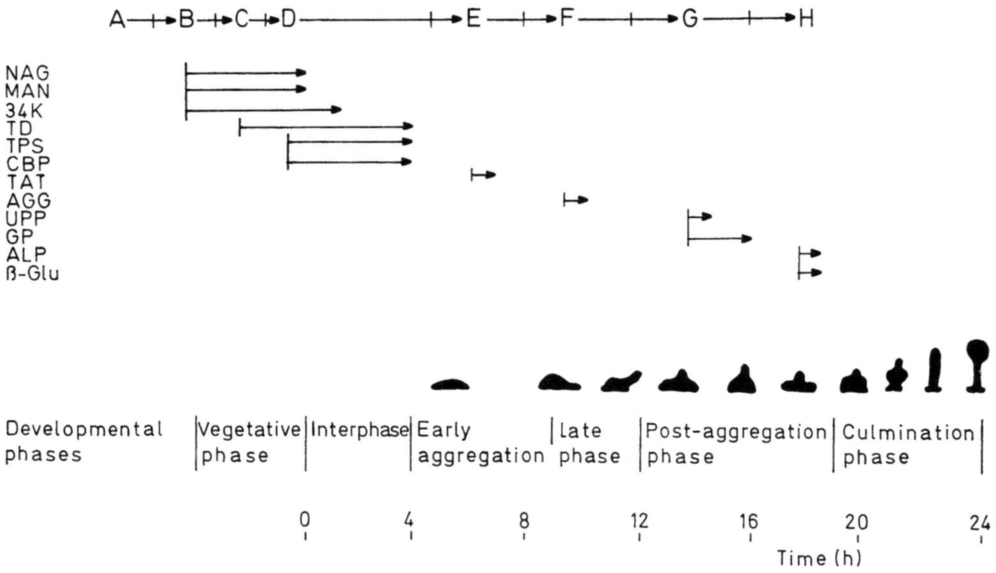


**Fig. 17.8:** Programmed sequence of events, as elucidated by the use of mutants, during the developmental phase of *Dictyostelium* (from Loomis [8], modified).

*Top:* Sequence *A* through *H* symbolizes developmental stages which are defined by the pattern of protein accumulation *(arrows)*. Mutants, denoted by *blocks* in the developmental sequence, accumulate only those enzymes expressed before the appropriate stage but fail to synthesize enzymes of the following steps. The marker enzymes are N-acetylglucosaminidase *(NAG)*, α-mannosidase *(MAN)*, threonine deaminase *(TD)*, trehalose phosphate synthetase *(TPS)*, tyrosine amino transferase *(TAT)*, UDP-Glc pyrophosphorylase *(UPP)*, glycogen phosphorylase *(GP)*, alkaline phosphatase *(ALP)*, and β-glucosidase *(β-Glu)*. 34K refers to the major developmentally regulated membrane protein and CBP refers to the carbohydrate binding protein. *Bottom:* For details of the developmental stages cf. Fig. 17.1.

### 17.5.2 Gene Expression during Development

Genetic analysis, i.e., the biochemical characterization of mutation blocks and their mapping on the genome is only one of the tools to investigate a developmental program. Another approach is to investigate mRNA and protein synthesis during proceeding development itself. The establishment of new metabolic pathways and molecular structure as characteristics of specialized developmental stages indicates an intensive de novo synthesis of mRNA and proteins. Contrary to spore germination and growth of amoebae, where relatively stable mRNAs seem to predominate, the synthesis of developmentally specific proteins is accompanied by intensive changes at the level of mRNA synthesis.

"Additive" DNA-RNA hybridization studies demonstrate great differences in the transcription pattern of non-repetitive DNA sequences. This statement was derived from studies with mRNA of distinct developmental stages [34].

If RNA sequences of two developmentally different stages are equal, an excess of RNA from one stage should not change the extent of DNA hybridization with RNA from the other stage. If the two RNAs differ from each other at least partially, the amount of hybridized RNA should increase in proportion to that difference. By this method greatest differences were observed between RNAs from vegetative cells and cells in the culmination stage. In this comparison only 40% of the newly synthesized RNA is equal, about 60% of RNA representing stage-specific molecules.

On the other hand, degradation of messenger RNA molecules occurs constantly without any drastic changes. The turnover of total translatable RNA during the life cycle varies only by ~25 %. Half-life of total RNA in cells during aggregation is about 180 min, in vegetative cells and in cells of other developmental stages it lasts 230 min. Selected mRNA species have apparent half-lives ranging from 13 min to 420 min [37, 38]. Unlike differentiating tissues of other higher organisms with extremely long half-lives of distinct mRNAs, *Dictyostelium* cells are characterized by balanced synthesis and degradation of their mRNAs. This offers a broad regulatory basis for development.

The situation is further complicated by reports on mRNAs transcribed from different DNA regions but coding for the same functional protein [36]. Thus, actin mRNA hybridizes to 17 different DNA restriction fragments, but it directs the synthesis of proteins equal in molecular weight and isoelectric point. However, due to postsynthetic modification (?), four isoforms of reactive actin are resolved, differing in their isoelectric points. At least two of them seem to be encoded in genes which are transcribed by different size classes of mRNA [8]. During development, transcription and translation of the different actin messages are selectively regulated [41]. This observation may explain possible variation in half-life of developmentally regulated mRNAs, but it opens also new aspects for not yet known control mechanisms.

Only few facts exist about detailed relationships between specific gene activation and morphogenesis. The "classical" ways of identification of in vitro or/and in vivo synthesized polypeptides and examination of regulatory mechanisms affecting their synthesis are very time-consuming and also limited with respect to the complexity of developmental changes to be analysed. Here another method, i.e., two-dimensional gel electrophoresis [39], seems to be helpful. It enables analysis in one or two gels of nearly the total cellular protein equipment. This technique separates individual proteins by combination of isoelectric focusing (protein charge) and electrophoresis (protein size). However, the biological functions of the separated polypeptides remain largely obscure.

The improved 2D-electrophoretic separation of proteins developed by O'Farrell [39] may resolve about 90 % of the total cell proteins. A crude protein solution is separated by isoelectric focusing according to the isoelectric points of the individual polypeptides. After reduction of disulfide bridges and treatment with sodium dodecylsulfate (SDS), proteins are separated in the second dimension according to their molecular weights. Techniques for the detection of radioactively labelled material or a highly sensitive protein staining method allow the discrimination between newly synthesized and already present polypeptides.

The two-dimensional analysis of the protein pattern during the developmental program of *Dictyostelium* provided data in good agreement with already known morphological and physiological markers [30]. Among three hundred protein spots identified from germinating spores, the content of only one individual protein increases; four major spore proteins are not detectable in amoebae. About 70 proteins show changes in their rate of synthesis during this time. Examination of in vitro synthesized polypeptides, whose synthesis was directed by RNA preparations from appropriate stages, e.g., from germinating spores or from amoebae, reveals that most changes observed in vivo and in vitro were attributable to the presence or absence of translatable mRNA, thus indicating that synthesis of the majority of the proteins during spore germination is transcriptionally controlled. A small number of proteins functional at later stages is translated from mRNA already present in dormant spores, i.e., representing long-lived mRNA [33]. Only few proteins were synthesized stage-specifically during spore germination suggesting that few genes are involved in controlling this process. The relatively small number of synthesized proteins (67 out

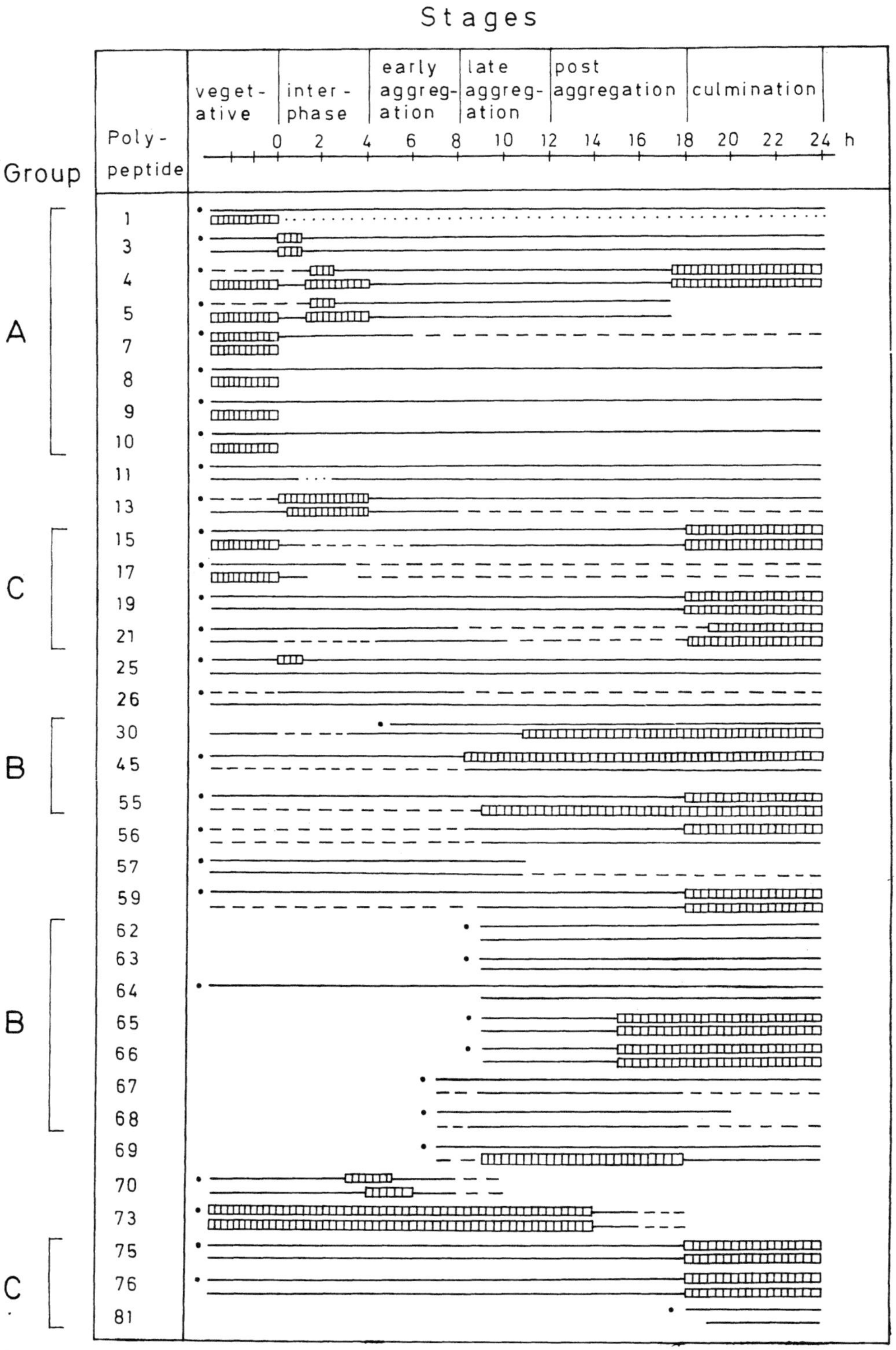
Stages
vegetative
interphase
early aggregation
late aggregation
post aggregation
culmination
Group
Polypeptide
0 2 4 6 8 10 12 14 16 18 20 22 24 h
A
1
3
4
5
7
8
9
10
11
13
C
15
17
19
21
25
26
B
30
45
55
56
57
59
B
62
63
64
65
66
67
68
69
70
73
C
75
76
81

Fig. 17.9

of total 300 made between spore and amoeba), and a qualitative change of only five polypeptides in the total protein composition may reflect the small part of the total genome contributing to spore germination.

Biochemically and genetically better characterized is the developmental program converting vegetative cells to cells of the culmination stage. Comparison of these two stages shows changes in the synthesis of about 150 polypeptides [28, 30]. This number is very close to the expected number of developmentally regulated genes derived from mutation analysis (cf. **17**.5.1). Furthermore, the pattern of protein synthesis is related to morphological alterations and allows a classification of groups of protein characteristic for distinct stages (Fig. **17**.9). The first predominant change in protein pattern is a reduction of the number of proteins synthesized in the vegetative phase (group A). A second clear-cut change takes place during the late aggregation phase. Synthesis of many new proteins is triggered (about 30, e.g. group B) and synthesis of others ceases (about ten proteins, not shown). The third major step accompanied by drastically decreased or increased de novo synthesis of proteins is the culmination stage (e.g., polypeptides of group C).

Surprisingly, pronounced quantitative and qualitative differences are evident when comparing the in vivo labelling of polypeptides with those labelled in the wheat germ system, e.g., polypeptides 1, 5, 8–10, 15, 45. But for many other polypeptides time point and rate of in vivo synthesis correlate with the availability of the corresponding mRNA as detected in the in vitro assay, e.g., polypeptides 3, 4, 13, 19, 21, 59, 65, 66, 75, 76. It is interesting in this respect that, for some proteins of vegetative cells, mRNAs apparently persist after protein synthesis has stopped. All these results point at a selective translational control, at least for distinct polypeptide species. The control point is assumed to be translation initiation [29, 31].

Several proteins of the early aggregation phase are translated even if aggregation is prevented by dilution of the cultures and by mechanical desintegration. Within 20 h in the case of suspended cells, or in plated cells up to 8 h after plating, the protein pattern is characteristic for interphase and early aggregation phase, respectively. Prolonged prevention of aggregation seems to fix the suspended cells at this stage. Getting the possibility to reaggregate, such cells become able to synthesize mRNAs and proteins typical for late aggregation stage cells. This phase-specific protein synthesis ceases and preceding developmental protein patterns are partially reestablished, if the aggregated cells are resuspended to a single cell state. It is evident that continued cell-cell interaction is a prerequisite for most changes of protein synthesis pattern observed in the late aggregation stage. Experimental disaggregation results in a rapid decline of the contents of stage-specific mRNA species. This process is prevented by addition of cAMP, which restores synthesis and prolongs the half-lives of these particular mRNA species. Interestingly, this cAMP effect is restricted to cells having previously been in contact with other cells. It is not observed in starved, aggregation--competent cells of the interphase [31]. Though a detailed explanation cannot be

---

◀

**Fig. 17.9**: Changes in protein and mRNA pattern during different phases of development of *Dictyostelium* (from Alton & Lodish [30]).

Numbers refer to polypeptides from in vivo and in vitro assays analyzed by two-dimensional electrophoresis according to the O'Farrell technique [39]. *Upper lines* (●) indicate proteins synthesized in vitro in the wheat germ system (cf. **12**.2.2), directed by *Dictyostelium* mRNA from different developmental stages. *Lower lines* represent results obtained from in vivo labelling of polypeptides. *Dotted line, broken line, full line, hatched boxes* denote increasing rates of protein synthesis estimated from the amount of label in a given protein spot.

given at present, these observations stress the profound changes of the gene expression patterns following cell-cell contacts, which, under normal conditions, initiate a determined developmental program leading to fruiting body formation. However, state-specific protein synthesis and mRNA presence allow neither the classification of biochemically related groups of polypeptides nor any suggestions about regulatory events triggering the developmental program. This has to be further elucidated in studies with developmental mutants including a detailed analysis of the physiological functions of the observed developmentally regulated proteins.

## 17.6 Conclusions

*Dictyostelium* is widely regarded as a model system for cell differentiation and development. During its life cycle the organism undergoes cell differentiation, cell aggregation, pattern formation and morphogenetic processes, which in many respects are similar to analogous events in animal development.

Cellular slime moulds, as well as embryonic vertebrate cells, show the same type of cohesion characteristics believed to be crucial in controlling morphogenesis. Both specific glycoproteins and specific carbohydrate binding proteins grant the cells cohesive properties. Similar lectin systems are implicated in cell-cell cohesion during embryogenesis of animal cells. *Dictyostelium* exhibiting molecular changes in cohesion properties during the events of chemotactic aggregation appears especially useful because of the easiness in production of aggregational and developmental mutants.

Another example of analogy represents the mechanism of signal transfer in *Dictyostelium*, which resembles signal transmission in animal neuromuscular systems. Like cAMP receptors at the slime mould cell surface, receptors for the transmitter acetylcholine are located at cell membranes of cholinergic synapses. Phosphodiesterases as signal inactivator in *Dictyostelium* plays an analogous role as cholinesterase in the nerve cell system. There is, however, a difference: The slime mould, unable to accumulate the transmitter intracellularly, activates the transmitter synthesizing enzyme (adenylate cyclase) after every pulse of excitation, while nerve cells are able to store signals. On the other hand, the refractory period for adenylate cyclase resulting in periodic aggregation is similar to habituation and adaptation system such as bacterial chemotaxis.

Thus, further studies of development of this simple eukaryotic organism should provide essential clues for causal analysis of fundamental aspects of development in general. For this aim it is essential to conceive the complex developmental phenomena on the biochemical, molecular biological, and genetical levels. In the last years tremendous progress in analysis has been made, but we are still at the beginning of understanding the intra- and intercellular correlations.

## 17.7 Selected References

*Reviews*

1. Bonner, J. T.: The Cellular Slime Moulds. Princeton University Press 1967.
2. Firtel, R. A., & A. Jacobson: Structural organization and transcription of the genome of *Dictyostelium discoideum*. Int. Rev. Biochem *15* (1977): 377–429.
3. Gerisch, G., Y. Maeda, D. Malchow, W. Roos, U. Wick & B. Wurster: cAMP signals and the control of cell aggregation in *Dictyostelium discoideum*. In: Cappucinelli, P., & J. M. Ashworth (eds.), Development and Differentiation in the Cellular Slime Moulds. Elsevier/North Holland, Amsterdam 1977.

4. Gerisch, G.: Periodische Enzymaktivierung als Kontrollfaktor multizellulärer Entwicklung. Rheinisch-westfälische Akad. Wiss. Vorträge *N 292* (1980): 7–38.
5. Jacobson, A., & H. F. Lodish: Genetic control of development of the cellular slime mould, *Dictyostelium discoideum*. Ann. Rev. Genetics *9* (1975): 145–185.
6. Loomis, W. F.: *Dictyostelium discoideum*, a developmental system. Academic Press, New York 1975.
7. Loomis, W. F.: Biochemistry of aggregation of *Dictyostelium*. Dev. Biol. *70* (1979): 1–12.
8. Loomis, W. F.: Genetic analysis of development in *Dictyostelium*. In: Leighton, T., & W. F. Loomis (eds.), The Molecular Genetics of Development. Academic Press, New York 1980, pp. 179–212.
9. Newell, P.: Genetics of the cellular slime moulds. Ann. Rev. Genetics *12* (1978): 69–93.
10. Sussman, M., & P. C. Newell: Quantal control. In: Sussman, M. (ed.), Molecular Genetics and Developmental Biology. Prentice Hall Inc., Englewood Cliffs, New Jersey 1972, pp. 275–302.
11. Sussman, M.: The genesis of multicellular organization and the control of gene expression in *Dictyostelium discoideum*. Progr. Mol. Subcell. Biol. *4* (1976): 103–131.

*Signalling and Aggregation*

12. Alcantara, R., & M. Monk: Signal propagation during aggregation in the slime mould *Dictyostelium discoideum*. J. Gen. Microbiol. *85* (1974): 321–334.
13. Brown, S. S., & C. L. Rutherford: Localization of cyclic nucleotide phosphodiesterase in the multicellular stage of *Dictyostelium discoideum*. Differentiation *16* (1980): 173–183.
14. Dicou, E. & P. Brachet: A separate phosphodiesterase for the hydrolysis of cyclic guanosine 3′,5′-monophosphate in growing *Dictyostelium discoideum* amoebae. Eur. J. Biochem *109* (1980): 507–514.
15. Gerisch, G., & B. Hess: Cyclic-AMP controlled oscillations in suspended *Dictyostelium* cells: Their relation to morphogenetic cell interactions. Proc. Natl. Acad. Sci. USA *71* (1974): 2118–2122.
16. Gerisch, G., H. Fromm, A. Huesgens & U. Wick: Control of cell contact sites by cyclic AMP in differentiating *Dictyostelium* cells. Nature (London) *255* (1975): 547–549.
17. MacWilliams, H. K., & J. T. Bonner: The prestalk-prespore pattern in cellular slime molds. Differentiation *14* (1979): 1–22.
18. Mato, J. M., & D. Malchow: Guanylate cyclase activation in response to chemotactic stimulation in *Dictyostelium discoideum*. FEBS Lett. *90* (1978): 119–122.
19. Mato, J. M., & D. Marin-Cao: Protein and phospholipid methylation during chemotaxis in *Dictyostelium discoideum* and its relationship to calcium movement. Proc. Natl. Acad. Sci. USA *76* (1979): 6106–6109.
20. Müller, K., G. Gerisch, I. Fromme, H. Mayer & A. Tsugita: A membrane glycoprotein of aggregating *Dictyostelium* cells with the properties of contact sites. Eur. J. Biochem. *99* (1979): 419–426.
21. Rahmsdorf, H. J.: Protein kinase action as a mediator of chemotaxis and differentiation in the cellular slime mould, *Dictyostelium discoideum*. Proc. 12th FEBS Meeting, Vol. 54. Pergamon Press, Oxford 1979, pp. 199–210.
22. Sampson, J.: Developmentally regulated cyclic AMP dependent protein kinases in *Dictyostelium discoideum*. Cell *11* (1977): 173–180.
23. Shinnick, T. M., & R. A. Lerner: The *cbpA* gene: Role of the 26,000 dalton carbohydrate binding protein in intercellular cohesion of developing *Dictyostelium discoideum* cells. Proc. Natl. Acad. Sci. USA *77* (1980): 4788–4792.
24. Tsang, A. S., & M. B. Coukell: Biochemical and genetic evidence for two extracellular adenosine-3′,5′-monophosphate phosphodiesterases in *Dictyostelium purpureum*. Eur. J. Biochem. *95* (1979): 407–417.
25. Tsang, A. S., & M. B. Coukell: Direct evidence for extracellular adenosine 3′,5′-monophosphate phosphodiesterase induction and phosphodiesterase inhibitor repression by exogenous AMP in *Dictyostelium purpureum*. Eur. J. Biochem. *95* (1979): 419–425.
26. Wick, U., D. Malchow & G. Gerisch: Cyclic-AMP stimulated calcium influx into aggregating cells of *Dictyostelium discoideum*. Cell Biol. Int. Reports *2* (1978): 71–79.
27. Wurster, B., & U. Butz: Reversible binding of the chemoattractant folic acid to cells of *Dictyostelium discoideum*. Eur. J. Biochem. *109* (1980): 613–618.

*Genetics and Molecular Biology*


28. Alton, T. H., & M. Brenner: Comparison of proteins synthesized by anterior and posterior regions of *Dictyostelium discoideum* pseudoplasmodia. Dev. Biol. *71* (1979): 1–7.
29. Alton, T. H., & H. F. Lodish: Translational control of protein synthesis during the early stages of differentiation of the slime mold *Dictyostelium discoideum*. Cell *12* (1977): 301–310.
30. Alton, T. H., & H. F. Lodish: Developmental changes in mRNAs and protein synthesis in *Dictyostelium discoideum*. Dev. Biol. *60* (1977): 180–206.
31. Chung, S., S. M. Landfear, D. D. Blumberg, N. S. Cohen & H. F. Lodish: Synthesis and stability of developmentally regulated *Dictyostelium* mRNAs are affected by cell-cell contact and cAMP. Cell *24* (1981): 785–797.
32. Dimond, R. L., P. A. Farnsworth & W. F. Loomis: Isolation and characterization of mutations affecting UDP glucose pyrophosphorylase activity in *Dictyostelium discoideum*. Dev. Biol. *50* (1976): 169–181.
33. Dowbenko, D. J., & H. L. Ennis: Regulation of protein synthesis during spore germination in *Dictyostelium discoideum*. Proc. Natl. Acad. Sci USA *77* (1980): 1791–1795.
34. Firtel, R. A., & J. Bonner: Characterization of the genome of the cellular slime mold *Dictyostelium discoideum*. J. Mol. Biol. *66* (1972): 339–361.
35. Freeze, H., & W. F. Loomis: The role of fibrillar component of the surface sheats in the morphogenesis of *Dictyostelium discoideum*. Dev. Biol. *56* (1977): 184–194.
36. Kindle, K., & R. Firtel: Identification of a recombinant plasmid containing *Dictyostelium* genomic sequences complementary to actin mRNA. Cell *15* (1978): 763–778.
37. Margolskee, J. P., & H. F. Lodish: Half-lives of messenger RNA species during growth and differentiation of *Dictyostelium discoideum*. Dev. Biol. *74* (1980): 37–49.
38. Margolskee, J. P., & H. F. Lodish: The regulation of the synthesis of actin and two other proteins induced early in *Dictyostelium discoideum* development. Dev. Biol. *74* (1980): 50–64.
39. O'Farrell, P. H.: High resolution two-dimentional gel analysis of proteins. J. Biol. Chem. *250* (1975): 4007–4021.
40. Ray, I., T. Shinnick & R. Lerner: A mutation altering the function of a carbohydrate binding protein blocks cell-cell cohesion in developing *Dictyostelium discoideum*. Nature *279* (1979): 215–221.
41. McKeon, M., & R. A. Firtel: Differential expression and 5′ end mapping of actin genes in *Dictyostelium*. Cell *24* (1981): 799–807.

# 18. Cell Differentiation and Tumorigenesis – Plant Crown Gall Tumors

R. Krauspe

## 18.1 Introduction

### 18.1.1 Description of the System

Crown gall tumor (CGT) is a neoplastic disease afflicting higher plants. It is induced by wound infection by a virulent (tumorigenic, oncogenic)[1] strain of the ubiquitous Gram-negative soil bacterium *Agrobacterium tumefaciens*. The bacterium has a very broad host spectrum, and crown gall tumors are widespread among gymnosperms and dicotyledonous (but not monocotyledonous) angiosperms. Three to four days after infection, cell proliferation can be observed in the wound region leading to the primary crown gall tumor (Figs. **18**.1 and **18**.2). Spontaneous secondary tumors (metastases) may also develop at a distance from the site of infection (Fig. **18**.2). Under field conditions CGT's develop in the crown region of the root. Although CGT's do not kill their host plant, they can cause severe damage and, in the case of crop plants, drastic reduction in yield.

CGT cells are characterized by three main properties which are passed unchanged from one cell generation to the next:

1. Autonomic division: the CGT cells themselves produce the phytohormones necessary for the regulation of cell growth and division (cf. **18**.3.2).
2. Synthesis of "opines". These unusual derivatives of basic amino acids are not found in normal plants, and are excreted from CGT cells into the surrounding medium (cf. **18**.3.1).
3. The tumor characteristics can be transferred to the cells of normal plants by transplantation (grafting). In some cases the CGT cells may revert to normal cells during this process (cf. **18**.4).

The rapid progress made during the past five years with regard to the causal analysis of CGT is largely due to intensified studies of the molecular genetics of the tumor-inducing bacterium, *A. tumefaciens*. The crowning feat was the identification and characterization of the "tumor-inducing principle" (TIP).

The hypothesis of a "tumor-inducing principle" had already been put forward in the 1940's by A. C. Braun and co-workers (New York) after they had shown that CGT cells are free of bacteria and that *A. tumefaciens* is only necessary for the induction but not for the maintenance of CGT's [39]. They postulated that the transformation of normal cells to tumor cells is dependent upon the transfer of products or components of the bacterial cell – the TIP – to the wound cells. Since the tumor character is stably maintained and propagated indefinitely, the transfer of genetic

---

[1] In this chapter the term "normal" refers to nontransformed, healthy plants or plant cells. The terms "virulent", "tumorigenic" and "oncogenic" are used synonymously.

Abbreviations: CGT, crown gall tumor; pTi, Ti plasmid = tumor-inducing plasmid; T-DNA, part of pTi DNA which becomes part of the CGT cell genome; $M_r$, relative molecular weight; kd, kilodalton; Md, Megadalton.

**Fig. 18.1**: Crown gall tumor at the root neck of peach cuttings naturally developed under field conditions (courtesy of Dr. I. Böttger, Department of Phytopathology of the Martin-Luther-Universität Halle).

**Fig. 18.2**: Crown gall tumor induced by wound infection of a tomato. Secondary tumors have been formed at the shoot in considerable distance from the infection site (courtesy of Dr. I. Böttger).

material from the bacteria to the plant cells seemed a likely explanation. The search for a TIP thus mainly concentrated on the nucleic acids (DNA, RNA or bacteriophages) of *A. tumefaciens*.

In the years 1974 and 1975 two groups of investigators, J. Schell's team in Ghent and that of E.W. Nester in Seattle, proved that a large plasmid in *A. tumefaciens* is responsible for tumor induction. All virulent strains possess this Ti (tumor-inducing) plasmid, whereas it is lacking in all avirulent strains. Loss or acquisition of the Ti plasmid is invariably associated with loss or acquisition of oncogenicity by the strain of *A. tumefaciens* concerned (cf. **18.2.3**). Usually only one Ti plasmid is present in each *A. tumefaciens* cell.

**Table 18.1**: Opines synthesized by crown gall tumors[a].

Generalized scheme of opine synthesis via reductive condensation:

Basic amino acid + α-keto acid ⟶ Opine

$$R_1-\underset{\displaystyle NH_2}{\underset{|}{CH}}-COOH \;+\; R_2-\underset{\displaystyle O}{\underset{\|}{C}}-COOH \;\xrightarrow[\text{NAD(P)H} \;\to\; \text{NAD(P)}]{E}\; \begin{matrix} R_1-CH-COOH \\ | \\ NH \\ | \\ R_2-CH-COOH \end{matrix} \;+\; H_2O$$

| | | |
|---|---|---|
| Classification | Octopine family [$N_2$-(D-1-carboxyethyl)--L-amino acid family] | Nopaline family [$N_2$-(D-1,3-dicarboxypropyl)--L-amino acid family] |
| Enzyme involved (E) | Octopine synthase | Nopaline synthase |
| Inducing strain of *A. tumefaciens* | Octopine strains (e.g., B6-806, Ach5) | Nopaline strains (e.g., C58, T37) |
| *Basic L-amino acid* | *α-Keto acid* | |
| | *Pyruvate* $R_2 = CH_3-$ | *α-Keto glutarate* $R_2 = HOOC-(CH_2)_2-$ |
| *L-Arginine* $R_1 = NH_2-\underset{\displaystyle NH}{\underset{\|}{C}}-NH-(CH_2)_3-$ | Octopine | Nopaline |
| *L-Ornithine* $R_1 = NH_2-(CH_2)_3-$ | Octopinic acid (Ornopine) | Nopalinic acid (Ornaline) |
| *L-Histidine* $R_1 =$ 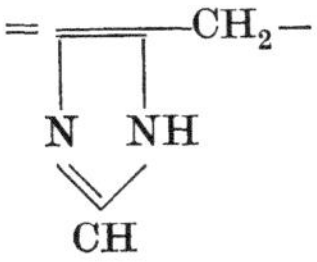 | Histopine | — |
| *L-Lysine* $R_1 = NH_2-(CH_2)_4-$ | Lysopine | Lysaline[b] |

[a] In addition to the opines given above, a special compound, agropine, has been detected in *A. tumefaciens* strains with octopine-type pTi's as well as in strains with "null-type" pTi's, now classified as agropine-type pTi's. The agropine structure ($C_{11}H_{17}NO_7$) recently turned out to be a bicyclic derivative of glutamic acid and a hexitol sugar [31].

[b] Non-amino acid part unknown, probably not α-keto glutarate.

Every CGT cell can be allocated to one of two groups according to the specific opines synthesized. Nopaline CGT's produce opines of the nopaline family, i.e., conjugates of basic amino acids with $\alpha$-keto glutarate. Octopine CGT's produce opines of the octopine family, i.e., conjugates of basic amino acids with pyruvate ([40], cf. Table **18**.1). The ability to produce opines is a tumor-specific property, completely independent of the host plant, and determined solely by the tumor--inducing strain of *A. tumefaciens* infecting the cell, i.e., nopaline-type strains as contrasted to octopine-type strains. Each tumorigenic strain of *A. tumefaciens,* and consequently each CGT, belongs to only one opine group. Tumorigenic strains of *A. tumefaciens* can utilize opines as carbon, nitrogen and energy sources, but only those opines whose synthesis they induce within the CGT. Both properties of *A. tumefaciens* – induction of CGT of a particular opine group and catabolism of the same group of opines – are determined, like oncogenicity, by the Ti plasmid. Thus, in correspondence with the bacterial strains, a distinction is made between nopaline--type and octopine-type Ti plasmids.

All of these facts fitted in well with the TIP hypothesis and it was postulated that CGT induction involves the transfer of the Ti plasmid or parts of it into the plant cell. In fact a fragment of DNA that could be isolated from CGT but not from norma

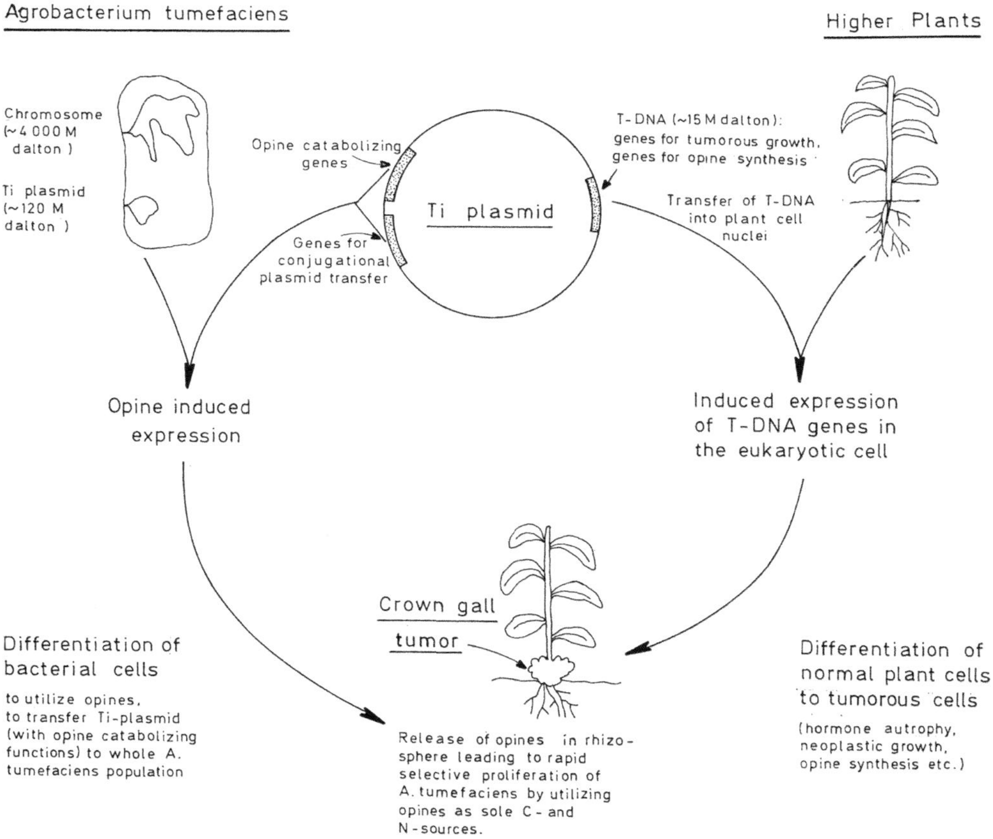


**Fig. 18.3:** Dual differentiation by genetic colonisation of plants via T-DNA transfer from Ti plasmid of *Agrobacterium tumefaciens* (after Schell [4]).

cells was found to hybridize with a particular portion of the Ti plasmid. This so-called T-DNA is a double-stranded, approximately 15 Md piece of the Ti plasmid. In CGT induction the T-DNA is selectively transferred from the Ti plasmid into the nucleus of the plant cell, is replicated and transcribed, and is responsible for the maintenance of the tumor character (cf. **18**.3.1).

Although nothing is known of the molecular mechanisms that lead to uninhibited proliferation of opine-synthesizing CGT cells after the transfer of T-DNA into the plant cell, we have a very clear overall picture of the *A. tumefaciens*-CGT system under natural conditions, with Ti plasmids and opines occupying a key position ([4, 18], Fig. **18**.3). By CGT induction, tumorigenic strain of *A. tumefaciens* forge themselves an ecological niche via gene transfer because they alone are able to benefit from the opines synthesized and secreted by the tumor cells. In this way they divert a major part of the photosynthetic capacity of the host plant to their own use. Apart from constituting the source of nutrition and energy for the tumorigenic strains of *A. tumefaciens*, opines also induce the expression of Ti plasmid genes coding for enzymes of opine degradation (cf. **18**.5). Further, they induce the expression of transfer functions (also coded by the Ti plasmid), and thus effect the rapid distribution of the Ti plasmid by conjugational transfer in a mixed population of tumorigenic and non-tumorigenic *A. tumefaciens* cells (cf. **18**.4). In this way the entire *A. tumefaciens* population is finally able to exploit opines for growth and multiplication. Hence, the opine-induced propagation of the Ti plasmid raises the degree of virulence of the total population thus increasing the chances of further CGT inductions.

The Ti plasmid can be characterized as a highly evolved catabolic plasmid which induces the synthesis of substances to be catabolized by an intriguing mechanism of substances to be catabolized by an intriguing mechanism of transfer of its own genes into higher plants, using the host's synthetic and energetic capacities. This particular kind of parasitism, of which the only example so far known is provided by *A. tumefaciens*, has been termed "genetic parasitism" or "genetic colonization", and can be regarded as a naturally occurring example of genetic engineering.

### 18.1.2 Problems Connected with Cell Differentiation in the *A. tumefaciens*-CGT System

The *A. tumefaciens*-CGT system deserves special interest for a variety of reasons: (1) it provides a model system of tumorigenesis and tumor reversion, (2) it offers the possibility of studying the regulation of cell division in plants, (3) it is of considerable interest as a, so far, unique form of parasitism, (4) it is an important topic for theoretical and practical soil microbiology and (5) the system has aroused great hopes with respect to genetic engineering in higher plants. Problems of cell differentiation are connected with each of the above-mentioned aspects of the *A. tumefaciens*-CGT system. They will be described under the following topics:

a) development of competence of the plant cell to be transformed by *A. tumefaciens* during the differentiation program termed "wound cell division";
b) development of the competence of *A. tumefaciens* to induce CGT, following reception of signals from the wound cell;
c) differential transfer and expression of Ti plasmid genes during CGT induction;
d) differential gene expression in the transformed tumor cells;
e) differentiation of *A. tumefaciens* cells following reception of signals (opines) from the CGT cells;
f) phenotypic reversion of CGT's as a differentiation process.

It should be pointed out that induction of CGT's and their genotypic reversion can be regarded as a special case of cell differentiation (cf. **4**.1), since genetic information is both gained (induction) and lost (reversion). The same applies to the induction of animal tumors by oncogenic viruses (cf. **19**.2.2). Nevertheless, we regard the transformed CGT cells as a new, distinct genetic system with its own cellular DNA plus T-DNA, which seems to be justified in view of the stability of the tumor state (cf. **18**.4). Doubtless the originally foreign T-DNA exerts a regulatory influence on the plant genome, and vice versa the latter influences expression and maintenance (cf. **18**.4) of the T-DNA. Under its original, natural conditions the *A. tumefaciens*-CGT system can be considered as an example of co-differentiation of two organisms by mutual signal exchange brought about during the course of evolution. A general analysis of the causality of the complicated interactions between the two genetic systems can be confidently expected in the near future.

## 18.2 Induction of CGT

### 18.2.1 Conditioning of the Plant Cell by Wounding

Wounding leads to reprogramming of the gene expression pattern of non-dividing highly specialized plant cells. The wound cells temporarily show mitotic activity, which is made possible by the activating and subsequently switching off the metabolic processes concerned, in a fixed temporal and spatial sequence. Thus the normal process of wound healing is governed by a gene expression program, which is only of interest (in this context) as far as it constitutes, for a brief time, a part of the program of "crown gall tumor formation". After an initial period of resistance,

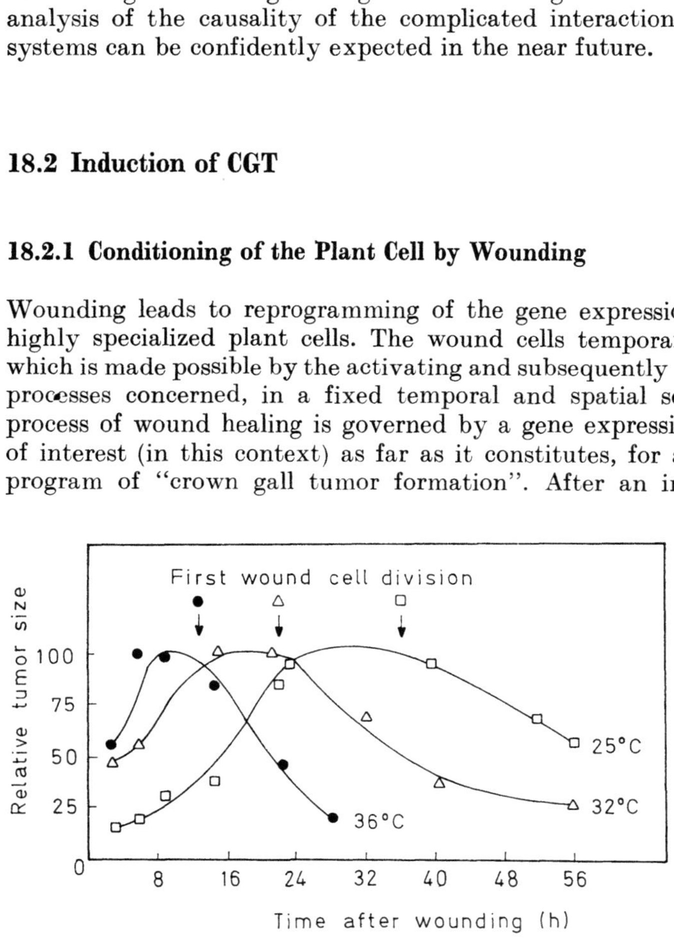


**Fig. 18.4**: Competence for tumor induction by *Agrobacterium tumefaciens* in wound cells of *Kalanchoë daigremontiana* (from Lipetz, J.: Science *149* [1965]: 805, and Cancer Res. *26* [1966]: 1597, redrawn).

After wounding plants are kept at the indicated temperatures (phase of conditioning). Afterwards they are subject to wound infection with *A. tumefaciens* for 24 h at 25 °C. The effect of tumor induction is measured by determination of the tumor size after 3 weeks of growth at 32 °C, which is non-permissive for induction. The maximum competence, in all three variants just preceding the first wound cell division, is reached at different time points after wounding in dependence of the temperature.

wound cells are increasingly susceptible to CGT induction during the early stages of wound healing (competence phase) but regain resistance to *A. tumefaciens* after the first wave of cell division. The point of maximal competence coincides in most cases with the S-phase of the cell cycle (Fig. **18**.4). In this phase it is apparently possible, by means of infection with *A. tumefaciens*, to change the gene expression program for wound healing with temporary mitotic activity to one characteristic of the tumor state with permanent mitotic activity. If young, still mitotically active cells are wounded and infected they immediately exhibit maximum competence. Apparently the stage of "actively dividing cells", whether already present or induced by injury, is essential for CGT induction. Secondary CGT's, that are induced in uninjured cells at a distance from the original site of infection (probably due to migration of *A. tumefaciens* cells through the tracheids), presumably have the same requirements for induction. Cell cycle dependency of tumor induction is likewise observed in animal systems (cf. **19**.2.3 and **19**.4.2).

### 18.2.2 Interactions between Wound Cell and Bacterium

The specific receptors to which *A. tumefaciens* attaches on the cell wall of the plant are present in normal, uninjured plants, but they are inaccessible on account of the close contact between the adjacent cells and only become exposed on injury (Figs. **18**.5 and **18**.6).

Tumorigenic strains of *A. tumefaciens* become avirulent if they are incubated with preparations of cell walls of normal dicotyledonous plants. The receptor structures of the plant, which appear to remain intact during the process of preparation, are able to block the binding sites on the surface of the bacteria [41]. Plant cell membranes, however, are ineffective in this respect. The resistance of the large majority of monocotyledonous plants to CGT induction is due to lack of receptor structures, and consequently, cell wall preparations from such plants have no effect on the virulence of *A. tumefaciens*. The same holds for cell wall preparations of CGT's a fact which can be explained by the loss or modification of receptors at or following tumor induction. This is probably why cells, once transformed, are resistant to reinfection by *A. tumefaciens*. Alteration in the cell surface structures is a widely observed phenomenon in animal tumors, too (cf. **19**.3.2), although in these cases the alterations concern the cell membrane. The plant receptors are not present at embryonal cell walls but are formed during early seedling stages [10]. Plant protoplasts can only be transformed to CGT cells after regeneration of the cell wall and thus of the receptors. The chemical and physical characterization of the receptors is still incomplete. Since about $10^5$ agrobacteria can bind to one tobacco cell, the appropriate receptors must be present in large amounts. The infecting bacteria cells may form clusters at certain sites on the cell wall [9].

The bacterial receptors (binding sites) are better characterized than those of the plant cell wall. They are formed by the polysaccharide moiety of the lipopolysaccharide of the outer bacteria wall. If plant wound cells are preincubated with this lipopolysaccharide fraction, CGT's cannot be induced by subsequent infection with *A. tumefaciens*, since their receptors are blocked. For the same reasons infection of plant wounds with a mixture of virulent and avirulent strains reduces the chances of CGT induction, since both strains possess similar binding sites and have equal chances of binding to the receptors at the plant cell wall. If formation of the cell wall in a virulent strain is disturbed, e.g., by treatment with penicillin or agrocin (cf. Table **18**.2, pt. 4) or if the cell wall is altogether absent, e.g., in L-forms, tumor induction is prohibited.

In all probability complementary surface charges and/or complementary structures of the plant and bacterial receptors are at the basis of the specificity of the recognition process. Whether the interaction between the two receptors involves the exchange or induction of synthesis of

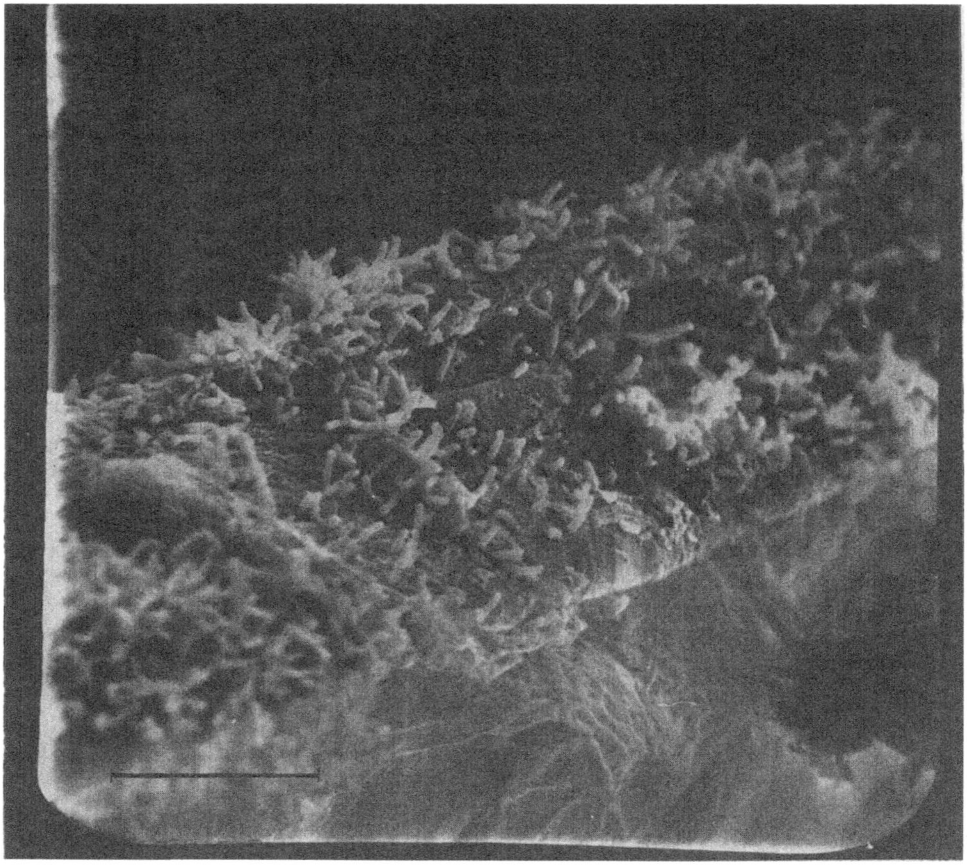

Fig. 18.5

**Fig. 18.5, Fig. 18.6:** Scanning electron micrograph of *Agrobacterium tumefaciens*-infected tobacco cells (from Smith & Hindley [9], with permission).

Figures show tobacco cell with agrobacteria attached to the cell wall. *Scale bars,* 5 mm.

chemical signals is not yet known. In some way or other the Ti plasmid of *A. tumefaciens* must be mobilized by contact with the wound cell. In addition, appropriate conditions must be established for the expulsion of T-DNA from the plasmid after its transfer into the plant cell. So far we have no exact concept of the molecular basis of this undoubtedly highly complex reaction sequence.

### 18.2.3 Ti Plasmids as Tumor-Inducing Principle

Following the discovery that virulence or avirulence of *A. tumefaciens* strains is strictly correlated with the presence or absence of a large plasmid [18, 19], the molecular analysis of its genetics became the chief goal of CGT research. Apart from the Ti plasmid, of which ususally only one copy is present in each cell, other large, non--tumorigenic plasmids may be present in *A. tumefaciens*. Ti plasmids are circular, double-stranded, supercoiled, membrane-bound DNA molecules of 95–180 Md [5].

Fig. 18.6: (Legend on p. 576)

With the loss of the Ti plasmid the virulence of the respective strain of *A. tumefaciens* is also lost. If apathogenic strains acquire a Ti plasmid by transformation or conjugation they become tumorigenic. Since loss or gain of Ti plasmids can be induced experimentally, it was possible to show that other genetic functions of *A. tumefaciens*, besides oncogenicity, are also attributable to the Ti plasmid. These are listed and explained in more detail in Table **18**.2.

The two basic types of Ti plasmids already mentioned in the introduction (octopine- and nopaline-types) differ not only in their functions, but, as is to be expected, in details of their genetic organization. Whereas the octopine Ti plasmids from different octopine strains of *A. tumefaciens* closely resemble one another or are even identical, the nopaline-type of plasmids exhibit only a limited degree of homology with one another and with the octopine-type of pTi's (20–60% homology [21]). In addition, they are as well more heterogeneous with regard to size. This information derives from a comparison of the restriction patterns of a representative number of different Ti plasmids. In the meantime detailed physical maps have become available for some of them, as a result of degradation with up to eight restriction endonucleases (Hind III, Hpa I, Sma I, Kpn I, Eco RI, Xba I, Bam HI, Bst I, cf. Table **2**.1), cloning of the restriction fragments in *E. coli*, and cross-hybridization using the Southern blotting technique (cf. **2**.6). Physical maps of this kind were a prerequisite for establishing an

**Table 18.2**: Summary of Ti plasmid encoded functions

| Genetic symbol | Characterization | Present[a] in Ti plasmids of octopine-type | nopaline-type |
|---|---|---|---|
| | *1. Tumor Inducing and Related Functions* | | |
| *onc* | **Oncogenicity**: Capacity for CGT induction, consisting of at least 4–5 different loci, spread over the entire Ti plasmid, but all of them in homology regions between octopine- and nopaline-type pTi's. One of them is the middle part of T-DNA, it is involved in maintenance of tumorous growth of CGT cells. Exact functions of the different *onc* loci are unknown. They may comprise: recognition sites on bacterial outer cell envelope, mechanisms for Ti plasmid transfer into plant cells, excising T-DNA out of the Ti plasmid, T-DNA modification, transfer of T-DNA into plant cell nucleus, integration of T-DNA into plant chromosomal DNA, initial T-DNA amplification etc. Some of the *onc* loci code for proteins. Deletion mutations or transposon insertions in anyone of the *onc* loci cause loss of oncogenicity, i.e., there is a coupled functioning of all *onc* loci. *Onc* functions are somehow dependent on *A. tumefaciens* chromosome, since they are not expressed after Ti plasmid transfer into *E. coli*; but expression is possible under the closely related chromosomal background of *Rhizobium* spec. | + | + |
| *onch* | **Host range effect on oncogenicity**. CGT formed on *Kalanchoë* and/or potatoes, but not on tobacco, pea and sunflower. Mode of action unknown. | + | + |
| *cys* | **Synthesis of cytokinin** (trans-zeatin): biosynthetic enzyme(s) or enzyme(s) of tRNA degradation, or membrane modification for release of purine derivative(s)? Left hand part of T-DNA of pTi-C58 (and other nopaline-type Ti plasmids?). | ? | + |
| *morph* | **Morphogenetic functions** for phenotypical appearence of CGT's, e.g., *Kalanchoë* CGT's with rough or smooth surface (octopine- and nopaline-type pTi's, resp.); unorganized growing CGT's or teratoma CGT's (octopine- and nopaline-type pTi's, resp.). *Morph* functions may eventually include *cys* gene. Position on pTi uncertain. | + | + |
| *ops* (*nos*, *lys*, *ocs*, *ags*) | **Genes for opine synthesis** (*ops*, general operational term), i.e., for **nopaline** *(nos)*, **lysaline** *(lys)*, **octopine** *(ocs)* and **agropine** *(ags)* **synthesis**; only active in CGT's; right-hand part of T-DNA. Genes may code for the respective opine synthase itself or for a subunit of it. Alternatively gene products may function as activator of synthesis or activity of a plant genome encoded enzyme. *Nos* controls also ornaline synthesis; *ocs* controls also ornopine, histopine and lysopine synthesis (cf. details in Table **18.1**). pTi's containing only *ags*, but not *ocs* are now classified as agropine-type pTi's [34]. | + *(ops)*<br>— *(nos, lys)*<br>+ *(ocs, ags)* | +<br>+<br>— |

Table 18.2 (Continued)

| Genetic symbol | Characterization | Present[a] in Ti plasmids of | |
|---|---|---|---|
| | | octopine--type | nopaline--type |
| | *2. Catabolic Functions* | | |
| *opc* | **Opine catabolism**; general, operational term; refers to *noc/nop*, *occ/ocp*, *arc* and *orc*. These loci are induced by the respective opines and are controlled by a common regulatory element (*ocr* or *nor*), together with *tra*. *Ops* functions are somehow dependent on *A. tumefaciens* chromosome, since they are not expressed after Ti plasmid transfer into *E. coli*; but expression is possible under the closely related chromosomal background of *Rhizobium* spec. | + | + |
| *opp* | **Opine permeases**; general operational term for *nop*, *ocp*, and *agp*, which are induced by their substrates. | + | + |
| *noc* | **Nopaline catabolism.** A complex of genes within a regulon or operon, allowing the utilization of the respective opines as sole C- and N-source. *Noc* codes for nopaline (and ornaline) oxidase, it is inducible by nopaline. *Noc* is regulatorily and/or physically coupled to *arc*, *ocr*, *nop* and *tra*. *Noc*-constitutive mutants are able to catabolize octopine, too. There are at least two *noc* loci. | – | + |
| *nop* | **Nopaline permease**, can also transport ornaline; and octopine in *noc*$^{c}$ mutants. *Nop* is inducible by nopaline. *Nop* is regulatorily and/or physically coupled to *noc*, but can be expressed independently from *noc* in *noc*$^{-}$*nop*$^{+}$ mutants (uptake but no breakdown of nopaline). | – | + |
| *occ* (= *uad*A) | **Octopine catabolism.** A complex of genes, inducible by octopine, forming a regulon or an operon allowing the utilization of the respective opines as sole C- and N-source. Codes for the oxidase(s), which cleave(s) octopine-type opines (octopine, ornopine, lysopine, histopine). *Occ* is regulatory and/or physically coupled to *arc*, *orc*, *ocp*, and *tra*. *Occ*-constitutive mutants do not catabolize nopaline. | + | – |
| *ocp* (= *uad*B) | **Octopine permease**, transports all octopine-type opines, but not nopaline and agropine. *Ocp* is inducible by octopine. *Ocp* is regulatorily and/or physically coupled to *occ*, but can be expressed independently from *occ* in *occ*$^{-}$*ocp*$^{+}$ mutants. | + | – |
| *ocr* | **Regulation of octopine catabolizing and *tra* functions.** Octopine catabolizing functions *(occ, ocp, arc, orc)* are regulatorily coupled to *tra* functions: both are octopine-inducible; both are constitutive in certain spontaneous mutants (*occ*$^{c}$ and *tra*$^{c}$: possibly *ocr* mutants). The same holds true for putative repressor mutants, in which *occ* and *tra* are inducible by nor--octopine, in contrast to induction in wild-type by octopine only. *Occ, ocp, arc, orc* and *tra* obviously form two or more operons in an *ocr*-controlled regulon (see *occ*). *Ocr* may behave as "master gene" for expression of most Ti plasmid-encoded functions, possibly also for *onc*, since regulatory mutants (phenotype *occ*$^{c}$ and *tra*$^{c}$) are more effective in inducing CGT. | + | ? |
| *nor* | A similar, or even more complex regulatory system very likely exists in nopaline-type Ti plasmids *(nor)*. | | |

Table 18.2 (Continued)

| Genetic symbol | Characterization | Present[a] in Ti plasmids of | |
|---|---|---|---|
| | | octopine--type | nopaline--type |
| *agc* | **Agropine catabolism.** Enzymes involved are unknown, possibly some sugar catabolizing functions. *Agc* is inducible by agropine, but not by octopine. *Agc* enables *A. tumefaciens* to utilize agropine as sole C- but not as N-source (cf. footnote [2] and Table **18.1**). | + | – |
| *agp* | **Agropine permease.** *Agp* is inducible by agropine, but not by octopine. Coupled expression of *agp* and *agc* is very likely. | + | – |
| *arc* *orc* | **Catabolism of arginine (arc) and ornithine (orc):** Exact reaction sequence catalysed by the gene products of *arc* and *orc* is unknown; main product is glutamic acid. Both are inducible by nopaline and octopine in the respective strains. *Arc* and *orc* are regulatorily and/or physically coupled to *noc/occ*, *nop/ocp* and *tra* and seem to be regulated by *nor/ocr*. Catabolism of both amino acids is also encoded in the *A. tumefaciens* chromosome. | + | + |
| | *3. Functions for the Conjugational Transfer of Ti Plasmids* | | |
| *tra* | **Transfer functions** are normally repressed but inducible by the respective opines and controlled, together with *opc*, *opp*, *arc* and *orc* by *ocr* (or *nor*). *Tra* functions are temperature sensitive in all *A. tumefaciens* strains ($t \geq 30$ °C). They are obviously not involved in CGT induction. *Tra* functions are encoded in regions of homology between octopine- and nopaline-type pTi's, except one of three *tra* functions in pTi-C58, which is physically linked to $agr^{s}_{84}$ and codes probably for some cell wall structure(s). *Tra* constitutive mutants have been isolated. After induction by the respective opines, rate of conjugational transfer of Ti plasmid is much higher in octopine- than in nopaline-strains (factor $10^1$–$10^5$). | + | + |
| | *4. Other Functions* | | |
| *ape* | **Exclusion of bacteriophage AP 1.** *Ape* is localized in regions of homology between octopine- and nopaline-type Ti plasmids. AP 1 cannot complete a single growth cycle; no production of viable phage progeny; killing of the bacterial host cell. | + | + |
| $agr^{s}_{84}$ | **Sensitivity to agrocin 84,** produced by *A. tumefaciens* K84. Agrocin synthesis is encoded on pTi-K84. Agrocins are intra--species bacteriocids of different agrobacteria. There are specific receptors on the cell membrane of sensitive bacteria. Agrocin 84 is a nucleotide analogue acting indirectly on bacterial cell wall synthesis, thereby reducing the affinity of *A. tumefaciens* for plant cell wall receptors. Octopine strains are usually $agr^{r}_{84}$. | – | + |
| $agr^{s}_{S1005}$ | **Sensitivity to agrocin S1005,** which is produced by *A. radiobacter* S1005. | + | – |
| *per1*, *per2* | **Periplasmic proteins,** $M_r$ **37.5 and 37 kd respectively,** and different IEP's. Encoded outside the T-DNA region; unknown function. | ? | + |

[a] + = present, – = absent, ? = uncertain or unknown.

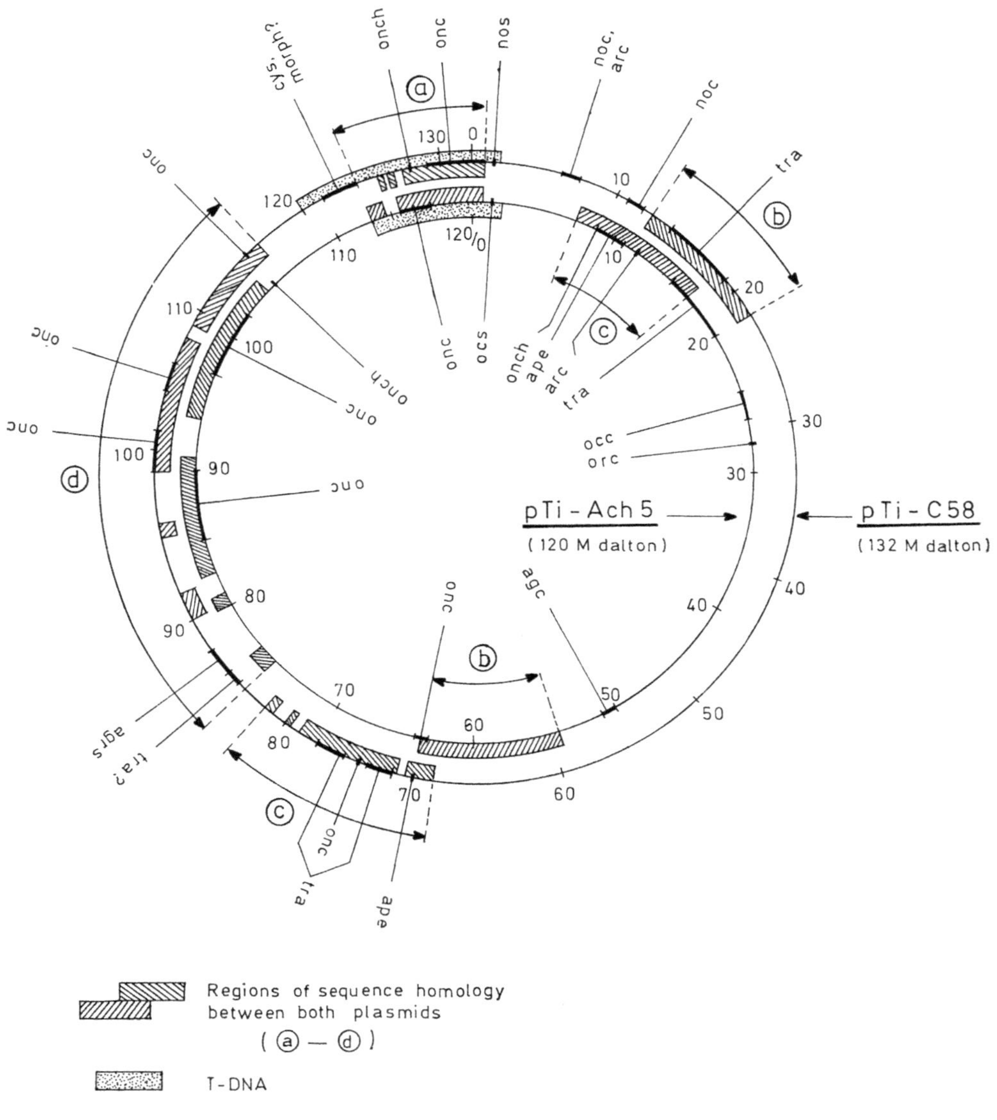


**Fig. 18.7:** Functional organization of an octopine (pTi-Ach 5) and a nopaline plasmid (pTi-C58) (after Schell et al. [4] and van Montagu, M., & J. Schell [5]).

Map coordinates correspond to DNA length in M dalton. A Sma I restriction site common to both plasmids in the T-DNA region is chosen as zero point. For description of the indicated functions see Table 18.2. There are four regions of sequence homology between both plasmids, they are labelled (a)–(d) and are localized as hatched areas between the plasmid circles. (a) belongs to the T-DNA found in CGT; (b) and (c) are obviously inverted on pTi-Ach5 in comparison to pTi-C58. The maps were established by restriction analysis, transposon insertion and deletion mutations, and electron microscopic heteroduplex analysis (cf. text).

detailed functional map of the Ti plasmids (cf. Fig. **18**.7 and Table **18**.2) as well as for localizing the homologous regions of octopine-type and nopaline-type Ti plasmids. The most important step towards the functional mapping of Ti plasmids was undoubtedly mutagenesis by transposon insertion and deletion [2, 5].

The Ti plasmid to be studied is transferred by conjugation to a pTi-free strain of *A. tumefaciens*. In the acceptor strain, containing an antibiotic resistance conferring transposon in the chromosome (transferred from *Escherichia coli* by a special plasmid that is unstable in *A. tumefaciens*), the Ti plasmid is mutagenized i.e., the transposon is spontaneously integrated into the Ti plasmid at a number of different sites. After back transfer into another genetically labelled, likewise originally pTi-free *A. tumefaciens* strain, transposon-containing Ti plasmids can be selected, since they alone exhibit the antibiotic resistance imposed by the special transposon. The site at which the transposon is integrated into the Ti plasmid can be determined by restriction analysis and electron microscopic heteroduplex analysis. In the former method the restriction fragment into which the transposon DNA has been incorporated disappears, or it changes its position upon agar gel electrophoresis. In the second case heteroduplices of the mutated Ti plasmid and a reference Ti plasmid containing two mapped insertions as position markers are studied under the electron microscope. The transposon DNA can be recognized and localized by loop formation (cf. the physical mapping of intervening sequences, Fig. **7**.3). However, the most efficient method for localizing the transposon insertion sites is Southern blotting hybridization, in which the electrophoretically separated restriction fragments of the total DNA of a *A. tumefaciens* mutant (chromosomal plus transposon--containing plasmid DNA) are hybridized one after the other with the three following DNA samples, radioactively labelled by nick translation (cf. **2**.7):

1. with pure Ti plasmid DNA: all pTi-derived DNA bands become visible;
2. with pure transponon DNA: those pTi DNA bands that carry the transposon sequence can be localized and
3. with the thus determined pTi restriction fragment: this last hybridization serves as confirmation in the identification of the point of insertion of the transposon into the Ti plasmid.

A few micrograms of DNA from the mutant to be analyzed are sufficient for such an analysis.

The integration sites of transposons are hot spots for spontaneous deletions, so that further mutants can arise from one insertion pTi mutant site and the extension of the deletions are identified in a manner similar to that described above.

After the transposon insertions or deletions on the Ti plasmid have been mapped the mutants are tested for loss of phenotypical properties for which Ti plasmid genes are responsible. In this way the majority of the known functions of a typical octopine--type Ti plasmid and a typical nopaline-type Ti plasmid could be mapped (see Fig. **18**.7). The individual functions are described in Table **18**.2. Here we shall limit our discussion to the general conclusions about CGT induction that can be drawn from the functional organization of the Ti plasmids.

Several genetic loci situated in four regions spread over the whole Ti plasmid (Fig. **18**.7a, b, c, d) are responsible for the oncogenicity of the plasmid (*onc* loci). These regions are to a large extent homologous in the octopine- and nopaline-type Ti plasmids. Only region *a* is found in CGT cells as T-DNA. It contains a highly conserved DNA sequence (ca. 5 Md DNA) that has been found in all Ti plasmids so far studied. It is apparently responsible for the tumor character, and its maintenance and perpetuation from one generation of CGT cells to the next (cf. **18**.3.1). The exact functions of the *onc* loci in the homologous regions *b*–*d* are unknown but some possibilities are listed in Table **18**.2. For successful CGT induction all *onc* loci have to be intact. This means that tumor induction is based on a sequential or concerted, but in any case differential expression of Ti plasmid genes. Although the exact course of this gene expression program is unknown, it results in the development of the competence of *A. tumefaciens* for CGT induction. It can only be realized under the normal chromosomal background of *A. tumefaciens* (cf. Table **18**.2) and requires diverse metabolic activities of the bacterial cell. For example, a number of auxotrophic mutants of

*A. tumefaciens*, although containing an intact Ti plasmid, are only virulent to the extent to which the auxotrophic factors (e.g., amino acids, nucleosides etc.) are made available by the plant's wound sap. Their oncogenicity fluctuates according to the degree of normalization of their metabolism. In the early stages of CGT induction rifampicin, an inhibitor of prokaryotic RNA polymerase, prevents the achievement of the stage of competence (up to about 14 h after wound infection), thus indicating that synthesis of (specific) RNA is an essential part of the bacterial gene expression program. That there is also synthesis of specific proteins, whose function within the CGT induction process is in some cases temporarily limited, can be safely assumed. This can be inferred, for example, from the heat sensitivity of certain phases of induction, i.e., following completion of the RNA synthesis described above (cf. Fig. **18**.4). The critical temperature is about 30 °C [44]. Certain cloned pTi DNA fragments that contain *onc* functions can cause protein synthesis in *E. coli* minicells.

It is of interest that the host plant is able to influence the gene expression program leading to competence for CGT induction. Certain loci for oncogenicity, the so-called *onch* loci (cf. Table **18**.2), which may be situated either in or outside the T-DNA region (cf. Fig. **18**.7), are only necessary for CGT induction in certain host plants.

There are many reports that phytohormones, especially auxin, play an essential or modifying role in CGT induction. Young plants, rich in hormones, are the most susceptible to CGT induction, whereas older organs often exhibit reduced sensitivity. However, the latter can be partially reversed by administration of phytohormones, e.g., of kinetin to old *Kalanchoë* leaves or of gibberellins to ageing castor bean cotyledons. It is not known whether these phytohormones normally arise in the wound cells, since *A. tumefaciens* is also able to produce auxin, cytokinins and gibberellin-like substances (cf. **18**.3.1 and **18**.3.2).

In summing up it can be said that CGT induction is brought about by transfer of the T-DNA region of the Ti plasmid. For this transfer to take place a multitude of plant and bacterial genetic potencies have to be realised in exact spatial and temporal sequence. However, details of the gene expression programs in the bacterial and plant cells and their possible coordination by exchange of signals have still to be elucidated.

## 18.3 The Transformed CGT Cell

### 18.3.1 The T-DNA in the CGT Cell

The hypothesis of gene transfer from the Ti plasmid to the plant cell was put forward as soon as the plasmid was discovered. Substantial support was provided by the discovery that the specific capacity of the individual CGT's to synthesize opines is dependent on the character of the Ti plasmid actually inducing CGT formation (octopine- or nopaline-type, cf. Table **18**.1). But it was the demonstration of the accelerated reassociation kinetics of certain restriction fragments of the Ti plasmid by addition of CGT-DNA, but not of normal cell DNA, that constituted the first real evidence for the correctness of the hypothesis and made possible the determination of the extent and number of the DNA sequences (T-DNA sequences by definition) transferred from the Ti plasmid to the CGT [22]. Using Southern blotting hybridization of the DNA restriction fragments from different CGT's with nick-translated cloned fragments of the corresponding inducing Ti plasmids (cf. **2**.6, **2**.7) it was possible to confirm the results and demonstrate their general validity. The T-DNA of a CGT is a piece of the inducing Ti plasmid, with a central, continuous, highly conserved DNA region of

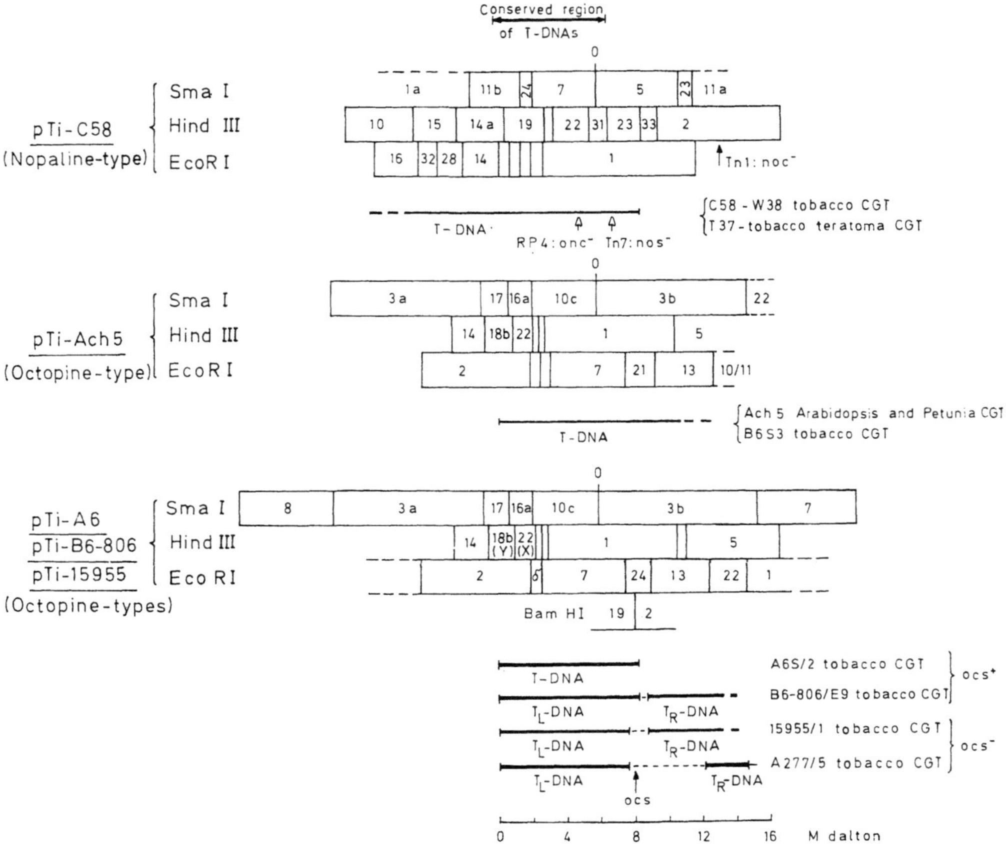


**Fig. 18.8:** Restriction maps of T-DNA's of different CGT's (after Schell et al. [4], van Montagu & Schell [5] and Tomashow et al. [1]).

The figure shows restriction maps of T-DNA regions of three groups of plasmids as indicated on the *left site.* Restriction fragments obtained with enzymes Sma I, Hind III and Eco RI (cf. Table **2.1**) are characterized by numbers deriving from their relative positions after electrophoretic separation on agarose gels. Their size can be estimated from the scale given below. The zero point of the Sma I restriction maps corresponds to the zero point of the pTi maps of Fig. **18.7.** For the T-DNA of pTi-C58 three transposon insertion points are indicated causing the *noc⁻* (Tn 1), the *onc⁻* (RP 4) and the *nos⁻* phenotypes (Tn 7). The position of the *ocs* gene derives from comparison of different octopine-type plasmids *(below)* containing deletions *(broken, thin lines)* between the left ($T_L$) and right ($T_R$) parts of the T-DNA.

about 5 Md, flanked to the left and/or right with Ti plasmid sequences of varying size ([20, 21], cf. Figs. **18**.7 and **18**.8). CGT cell cloning experiments showed that every CGT cell possesses T-DNA, so that it must be stably replicated and transferred from one cell generation to the next. The length of the T-DNA may vary from one CGT to another (ca. 14–16 Md), and even one and the same Ti plasmid can give rise to different T-DNA complements in the same host plant. Apparently the number of copies of the central, common part of all T-DNA's amounts to 1 or 2, whereas the non-conserved parts of some T-DNA's can be present in several copies (e.g., right-hand portion, $T_R$-DNA, of tobacco CGT B6-806/E9 with 15–30 copies per diploid plant cell genome [1]). We do not known whether in CGT induction several T-DNA sequences from different *A. tumefaciens* cells are transferred to one host cell or whethe ra single primary T-DNA sequence is selectively amplified. The T-DNA is not integrated in the plastidal or mitochondrial DNA's. It can be localized in the cell nucleus [32]. Either the T-DNA forms an independent replicon – possibly as a tandem repeat – or it is integrated in the chromosomal DNA. So far most results support the latter alternative [33]. For example, the terminal restriction fragments of the T-DNA regions of different pTi's (cf. Fig. **18**.8) could be hybridized with restriction fragments from CGT T-DNA's that differed in size from the T-DNA border fragments employed and thus were obviously composite fragments of the terminal portions of the T-DNA and the plant DNA. Such "composite border fragments" can also yield information about the specificity of the site of T-DNA integration in the host DNA. The occurrence of multiple left-end and right-end border fragments in certain CGT DNA's supports the idea of different sites of integration, although in other CGT DNA's only one left--end and one right-end border fragment could be found in each DNA, suggesting only one integration site [1, 4, 48].

It is still too early to draw general conclusions concerning the integration of the T-DNA in CGT cells. Furthermore, investigations of the detailed molecular architecture of the T-DNA sequences so far point to the possibility of considerable heterogeneity in the fine structure of the T-DNA of individual CGT lines involving rearrangements and/or deletions of the original pTi-derived DNA sequence (cf. for instance, deletions between $T_L$- and $T_R$-DNA in several tobacco CGT's, Fig. **18**.8).

On the basis of its known functions T-DNA can be divided into three sections. The central and invariant part of the T-DNA's is responsible for the tumor character and its persistence. Ti plasmid mutations in this region invariably result in the loss of oncogenicity, e.g., by insertions of the wide host range replicative plasmid RP 4 into the central T-DNA region of pTi-C58 (cf. Fig. **18**.8). The pTi-C58: : RP4 cointegrate is non-oncogenic, whereas after dissociation into the parental molecules pTi-C58 regains oncogenicity [4]. Mutations in the T-DNA sequences to the immediate left or right of the central invariant part do not affect oncogenicity but lead to

- changes in the phenotypic traits of the CGT's (left-hand end of T-DNA portion of certain nopaline-type Ti plasmids which might possibly code for cytokinin synthesis, cf. Table **18.2** and Fig. **18**.7), or
- loss of the capacity to synthesize opines [mutations in the right-hand end of the T-DNA portion of both nopaline- and octopine-type Ti plasmids, e.g., evoked by transposon Tn 7 insertion into pTi-C58 *($nos^-$)*; or by deletions in the T-DNA of tobacco CGT's 15955/1 and A 277/5 *($ocs^-$)* (cf. Fig. **18**.8)].

The latter results prove that the ability to synthesize a certain opine is coded by the T-DNA and that the loss of this genetic function does not lead to loss of oncogenicity. In other words, opine synthesis is not an essential requirement for CGT transformation, which is also confirmed by the natural occurrence of strains of *A. tumefaciens* that are oncogenic but negative with respect to opine synthesis (e.g., strains AT 181 and EU 6).

If such strains lack the opine-catabolizing function as well as that for opine synthesis, its Ti plasmids are termed "O-type plasmids", since they cannot be allocated to either of the two opine types (e.g., plasmids AT 1 and AT 4). Restriction analysis of these plasmids, or in some cases the continued ability to induce agropine-synthesizing CGT's (e.g., strain 542, cf. Tables **18.1** and **18.2**), nevertheless make it possible to determine the starting groups of such mutant plasmids.[2] The opines are reliable biochemical markers for CGT's, since they are not synthesized by any other

[2] Recently "O-type" plasmids with agropine-synthesizing and -catabolizing functions have been grouped in a separate family, so-called agropine family [34].

plant cell, not even by habituated cells, i.e., by cells that have attained a pseudo-tumor state by means of spontaneously arising phytohormone autotrophy (cf. **18.3.2**).

The right-hand ends of the T-DNA's from nopaline- and octopine-type pTi's exhibit no sequence homology, which is to be expected. They could carry the genes for nopaline or octopine synthase and the enzymes for lysaline or agropine synthesis respectively. Alternatively they might only code for one subunit of the enzymes mentioned, the other subunit(s) being made by the host cell, or they trigger the expression of latent genes in the plant genome, or they code for gene products activating latent plant enzymes. The progress being made in the characterization of opine synthases and the in vitro translation of mRNA derived from the right-hand end of T-DNA [23] will make it possible to decide between these different possibilities.

It has been shown that all parts of the T-DNA region are actively transcribed in CGT cells [16, 47]. Of vital importance for advances in CGT research is the exact characterization of the transcript from the central part of the T-DNA, which is responsible for tumor transformation, and of its oncogenic gene product, which recently has been synthesized by in vitro translation of the mRNA [23]. It is a protein of $M_r$ 15,000, which is common to all CGT's so far examined. Its further characterisation should play a key role in elucidating the causal relationships between T-DNA transfer and expression on the one hand, and CGT transformation on the other.

In the meantime we have to limit our discussion to models which may help explain the transforming potential of T-DNA. CGT transformation of normal cells could be caused by T-DNA via one of the following alternatives or combinations thereof:

- interruption of plant genes by T-DNA insertion, producing silencing or polarization effects,
- creation of novel promotor and/or operator sites in plant DNA,
- binding of plant regulatory macromolecules and
- direct influence of T-DNA products (RNA's or proteins or products of the enzymic activity of the latter) on the regulation of cell division of plant CGT cells.

The variability in length, copy number and integration site(s) of T-DNA in different CGT's favours the latter possibility and points to molecular diversity among the primary events of CGT transformation. The possibility of phenotypic reversion of the CGT transformation without loss of the T-DNA (cf. **18**.4) shows that (1) no part of the plant DNA is lost or irreversibly modified and (2) the tumorigenic functions of T-DNA can be repressed by the plant under certain circumstances.

In any case the wounded normal plant cell is forced to change the gene expression program of wound healing into a gene expression program of a CGT cell. Since the exact molecular mechanisms underlying these changes are still unknown, we can only give a more or less phenomonological description of the characteristics of the transformed CGT cell.

### 18.3.2 Unlimited Cell Division Caused by Hormone Autotrophy

Tumor cells are autonomous with regard to division. They are independent of the regulatory mechanisms that determine, so precisely, the mitotic activity of individual groups of cells in the normal multicellular organism. This trait is common to all tumors (cf. Chap. **19**), and applies to CGT cells, too. In the normal cells of higher plants cell growth and division are mainly regulated by two groups of phytohormones, auxins and cytokinins. Since CGT's show unrestricted proliferation in the plant and, unlike the cells of normal plants, grow in cell cultures without addition of phytohormones (cf. Fig. **18**.9), it seems very likely that they are able to produce the latter themselves in optimal concentrations. This has in fact been confirmed by hormone analyses of CGT cells, which showed that indolylacetic acid, the main auxin of CGT's, and trans-zeatin riboside, their main cytokinin, are present in higher than normal concentrations [50]. An additional class of factors affecting plant cell division (cytokinesins) have also been identified in CGT's. Since they inhibit phosphodiesterases it is believed that they may exert their action via the regulation of the cAMP level. The same conclusion holds for another inhibitor of phosphodiesterases, theophyllin, also found to stimulate cell division in plants.

Recently a cAMP-like substance has been isolated from normal and CGT cells of tobacco in concentrations that correlated well with the cell enlargement and rate of DNA replication. A similar correlation had previously been described for exogenously applied 8-bromo cyclic 3′,5′-AMP. Apparently cAMP and its analogues fulfill the function of intra- or extracellular signals influencing or controlling the process of differentiation in all kinds of organisms, although the actual regulatory mechanism are different in the various groups of organisms (cf. **4.2.3.1**, **4.2.4.1**, **5.3** and Fig. **17.5**).

In other kinds of plant tumors, too, one of the main features of the tumor is the permanent activation of the systems synthesizing the regulators necessary for cell division. This applies to virus-induced tumors, to so-called *genetic tumors*, arising from chromosome imbalances in intra- or interspecific crosses in which each parent contributes the autonomy for synthesis of one growth factor, and to *habituation tumors*, arising, with a frequency $10^2$–$10^3$ times higher than that of spontaneous mutations, in plant cell cultures by spontaneous autotrophy of growth factor formation. Their pseudo-tumor character is stably expressed and can be handed on to normal plants by grafting, sometimes resulting in spontaneous reversion. Habituated cells can be distinguished from CGT cells by their lack of opine synthesis, the presence of intact *A. tumefaciens* receptors and by the lack of T-DNA [51].

Genetic and habituation tumors prove that autonomous cell division in plant cell tumors may be due to the differential expression of the genetic information of the plant alone. In the case of CGT's, however, participation of bacterial genetic information is possible. The left-hand end of the T-DNA portion of certain nopaline-type Ti plasmids codes for the release and/or synthesis of bacterial trans-zeatin and is possibly involved in the cytokinin synthesis of the CGT cell as well. Mutations in this region cause morphogenetic alterations of the respective CGT's (cf. Table **18.2** and Fig. **18.7**). Likewise, indolylacetic acid production has been reported for *A. tumefaciens*. It is directed or regulated by pTi-encoded functions, too [35].

### 18.3.3 CGT Phenotypes

CGT's can develop different phenotypes. The actual form is determined by
1. the strain of *A. tumefaciens* involved in CGT induction and
2. the host plant, in other words, the physiological condition of the normal cell at the time of infection.

A few examples of the way in which these two factors influence the phenotype are given below. It seems that phytohormones frequently play a decisive role.

1. Nopaline-strains of *A. tumefaciens* usually induce teratoma CGT's whereas octopine-strains mostly cause CGT's exhibiting unorganized growth. The influence of bacteria-derived phytohormones on teratoma formation has been postulated. Teratoma-CGT's possess distinct morphogenetic potentials. They are capable of developing pseudo-organs (shoots, leaflets, roots), although these are anatomically abnormal. A possible explanation is that the normal cell-cell interactions and the processes involved in signal exchange between the plant cells are disturbed. Strain-specificity of the morphogenetic behaviour is further illustrated by observations that (1) octopine and nopaline CGT's of *Kalanchoë* leaves have typically rough and smooth surfaces respectively; and (2) different teratogenic strains of *A. tumefaciens* lead to different teratoma-types, e.g., root-like or shoot-like. Mixed infection of one and the same plant wound with two such strains results

in the formation of both teratoma phenotypes, the frequency of each being determined by the relative concentration of each strain in the inoculum.

2. The phenotype of a CGT is governed by the position of the tumor in the host plant. It is likely that the actual concentration of the various morphogenetic substances at the site of infection determines the phenotype. If, for example, both cut surfaces of a transversely cut tobacco shoot are infected with a particular strain of *A. tumefaciens*, a teratoma CGT forms on the upper surface, whereas the CGT developing on the lower surface exhibits unorganized growth. In principle, however, both possess the same morphogenetic potential, i.e., in cell cultures both develop the teratoma phenotype, although if grafted into the cambial region of normal plants both develop in an unorganized manner. The teratoma phenotype can be suppressed by exogenous application of auxin.

One and the same strain of *A. tumefaciens* evokes unorganized, rapidly growing CGT's in certain varieties of *Bryophyllum calycinum* during the vegetative stage, whereas in the flowering stage, characterized, inter alia, by a change in the spectrum of morphogenetic substances, teratoma CGT's develop.

A phenotypic property of a CGT is its individual growth rate in cell culture. This is chiefly dependent on the stage of competence of the wound cell at the time of infection with *A. tumefaciens* (cf. 18.2.1). Infection at the time of full competence leads to rapidly growing CGT's (type I), whereas infection at an earlier or at a later stage leads to slowly growing CGT's (type II). CGT induction immediately after the end of the initial resistant phase of wound healing leads to tumors with a minimum growth rate (type III). The different phenotypes are stably expressed in cell culture and their growth rates appear to be correlated with their degree of autotrophy for certain hormones and nutrients. In order to achieve the same growth rate as type I on a basic medium, type II requires addition of auxin, myoinositol and glutaminate, and type III needs, in addition, aspartic acid or asparagin, GMP and CMP, whereas normal cells require all of these substances plus a cytokinin and high salt concentrations. The gene expression program during wound cell division evidently includes the transitional synthesis of these growth promoting compounds, in a carefully regulated temporal sequence. At the point of tumor induction the actual gene expression state is fixed and passes unchanged to the following generations of CGT cells.

Each of these phenotypes of the CGT cells differs markedly from the cell type from which it originates. This means that transformation leads to a general change in the differentiation status. Since all reported phenotypes are stably expressed under defined conditions, the underlying specific gene expression programs must also be stable.

### 18.3.4 Other Characteristics of CGT Cells

Apart from the above-mentioned stable expression of metabolic systems directly or indirectly involved in cell growth and division, CGT cells are further characterized by changes in other aspects of metabolism [45]:

- the specific synthetic programs expressed in the original cells are switched off;
- the template activity of chromatin of CGT cells is increased;
- the redox potential is relatively low, as in all rapidly growing cells;
- the membrane permeability of the CGT cell is considerably higher than that of the normal cell;
- certain isoenzyme patterns are altered (e.g., peroxidase pattern);
- the phospholipid content of the CGT cell drops to about one-third of the normal concentration (reported for *Vinca rosea*) [36];
- new isoacceptor tRNAs appear, e.g., of $tRNA_{Phe}$ [37];
- specific tRNA methylases are induced [38].

The first five of these characteristics also appear temporarily in the differentiation program "wound cell healing", but in the CGT cell they are permanently expressed. Additionally, environmental signals may induce, in CGT cells as in every other living cell, temporarily limited, adaptive processes of differentiation.

If, for example, a nopaline teratoma CGT, grown in cell culture with nitrate as sole nitrogen source is supplied with glutaminate, a new phenotype described as "snowy white" develops (reduced growth rate, chlorophyll deficiency, formation of cortex and core cells). On return to the nitrate medium the original teratoma phenotype reappears. α-Ketoglutarate, malate and ammonium ions have the same effect as nitrate, whereas glutamate, asparaginate, aspartate and γ-aminobutyrate have the same effect as glutaminate. Changes in nitrogen metabolism may be responsible for the change of phenotype, a regulatory role apparently being played by glutamate dehydrogenase and aspartate-glutamate transaminase. The activities of both enzymes rise during the transition to the "snowy white" phenotype, presumably due to de novo synthesis, whereas the activity of glutaminate synthetase drops. The return to the teratoma phenotype is accompanied by a reversal of these processes.

The fact that the above mentioned events have no primary connection with T-DNA and that the basic tumor character of the cells involved remains unaltered support the postulate that the transformed CGT cell can be regarded as a new, individual genetic system with its own specific program of differentiation. An analysis of the phenotypic reversion of teratoma CGT's leads to the same conclusion.

## 18.4 Reversion of CGT Cells to Normal Cells

One of the most striking properties of tumor cells is the stability of the malignant state. Certain CGT cells that have been growing for 25 years in cell culture still exhibit a stable geno- and phenotype. However, under defined experimental conditions, reversion of the CGT state can take place. Likewise, the reversion of certain animal tumors has repeatedly been reported in the literature (cf. **19**.4). Such reversions are of fundamental significance in elucidating the causality of the tumor concerned. They indicate that the mechanism underlying transformation cannot be mutations, deletions or other irreversible alterations of the genetic information of the tumor cell. They rather emphasize the decisive role of differentiation states, which, although usually very stable in the case of tumors, nevertheless are potentially reversible.

### 18.4.1 Reversion of Teratoma CGT's

The basic experimental system for reversion of teratoma CGT's is shown in Fig. **18.9**. If cloned cell cultures of teratoma CGT's of tobacco are grafted into the cambial region of decapitated tobacco plants, vascular connections between CGT tissue and that of the normal plant are established in 5% of the cases, and relatively normal shoots and leaves develop. If the grafting procedure is repeated two or three times using the most "normal" of the resulting shoots, the shoots and leaves that develop finally are macro- and microscopically, as well as physiologically, normal. This process is termed phenotypic reversion.

However, such plants still produce opines and in cell culture express the original teratoma CGT phenotype, i.e., possess hormone autotrophy, thus indicating that the tumor character is still latently present. This was also demonstrated by the detection of T-DNA in all diploid cells of such plants, which, in addition, remain resistant to renewed CGT induction. All of these findings can be explained by repression of T--DNA expression (with the exception of its opine-synthesizing function) under the morphogenetic restraints exerted by the cambial cells of the stock plant. The use of phenotypically different tobacco varieties ensured that the "normalized" shoots did in fact arise by reversion of the teratoma CGT cells and not by sprouting of the stock plants.

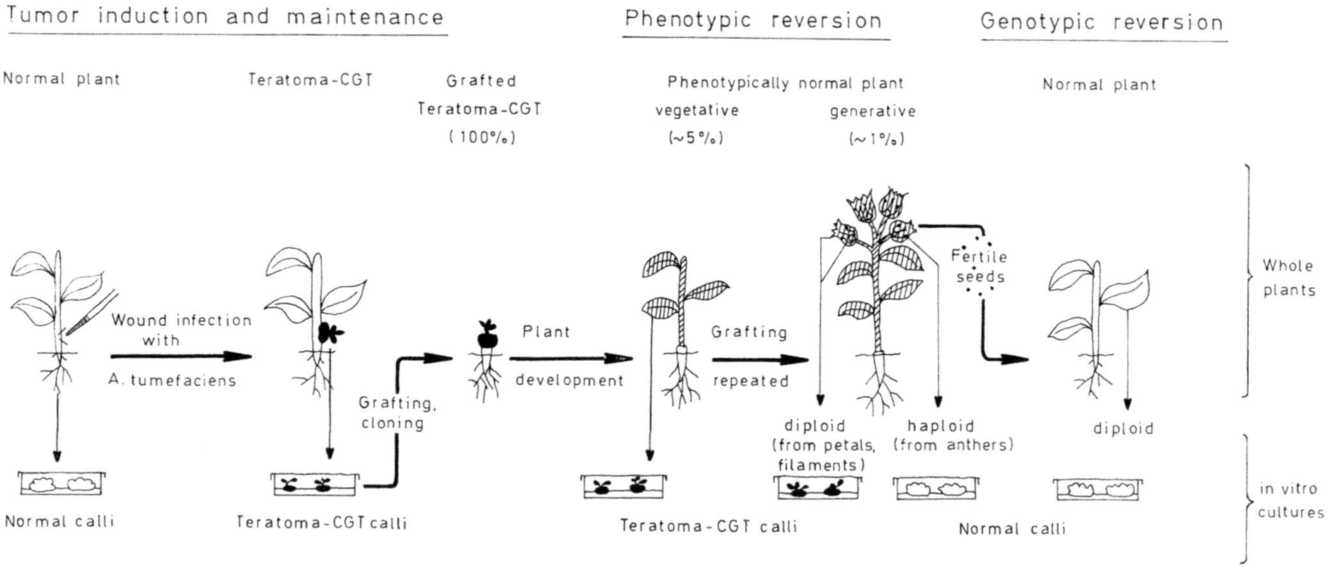


| | | | | | | | |
|---|---|---|---|---|---|---|---|
| + | − | − | − | − | + | + | Hormone dependency in cell culture |
| − | + | + | + | + | − | − | Opine synthesis |
| − | + | + | + | + | − | − | T-DNA in plant cells |
| + | − | − | − | − | + | + | Sensitivity against CGT-induction |

**Fig. 18.9:** Scheme of induction and reversion of teratoma-type CGT's (after Drummond [25]).
Experimental details for the scheme are derived from papers by Braun & Wood [24] and Turgeon et al. [26]. *White*, normal plant-tissues; *black*, CGT-tissues; *hatched*, inherently neoplastic tissues.

In about 1% of the initial grafts the normalized shoots produce flowers and viable seeds from which plants can be grown that are in no way distinguishable from normal, non-transformed plants of the same species. The tumor character is still inherent in all diploid flower organs (e.g., petals, filaments), whereas haploid organs (e.g., haploid calli obtained from anther cultures) and diploid cells of the $F_1$ generation may be fully normalized (cf. Fig. **18**.9). This process, known as genotypic reversion, is characterized by loss of the T-DNA, probably during meiosis, although the exact mechanism of elimination has still to be elucidated. Furthermore, it must be added that there are examples of T-DNA markers retained through meiosis and expressed in the $F_1$ progeny [46].

These results show that the CGT cell remains totipotent and that tumor transformation does not lead to loss or irreversible modification of the plant's genetic material. Phenotypic reversion indicates that the genetic material of the CGT cell (cell DNA plus T-DNA) can, under certain environmental conditions, undergo a fundamental reprogramming of its expression. The repression of phytohormone synthesis undoubtedly plays a key role in this respect.

### 18.4.2 Reversion of Unorganized CGT's

Reversion phenomena have also been reported for CGT's of the unorganized type [27]. In cell cultures, cloned cells of CGT's of this type from *Nicotiana tabacum* lose their autonomy with respect to cell division under the influence of light and high doses of cytokinin, and regenerate to phenotypically normal plants, some of which are even fertile. Exact details of the reversion process with respect to opine synthesis and the presence or elimination of T-DNA are not yet available. Since, however, newly initiated cell cultures of leaf cells of younger regenerated leaves only grow if phytohormones are added, and the regenerated plants are sensitive to renewed CGT induction, it is obvious that in this case genotypic reversion takes place in the early stages of plant regeneration, meiosis being apparently not a prerequisite. Although very rare, such genotypic reversions have been reported for teratoma CGT's as well. Unorganized CGT's that have been maintained for years in cell culture only regenerate, if at all, to sterile plants with abnormal phenotype, probably as a result of an abnormal chromosome complement (aneuploidy). This is not necessarily connected with tumor transformation or reversion, since even "normal cells" and their regenerates are often aneuploid after long-term cultivation.

## 18.5 Differentiation of *A. tumefaciens* under the Influence of Opines

Opines, released into the rhizosphere by CGT's under field conditions or added in vitro to bacterial cultures, lead to the induction of an apparently highly complex differentiation program in *A. tumefaciens*, mainly involving expression of certain gene groups of the Ti plasmid. The genes concerned are those necessary for opine catabolism and/or those responsible for conjugative Ti plasmid transfer (*opc*, *opp*, *arc*, *orc* and *tra* functions, cf. Table **18**.2). Both groups of genes are regulated by the same regulatory principle (*ocr* or *nor*), but details of the mode of action are still unknown.

### 18.5.1 Ti Plasmids as Catabolic Plasmids

Nopaline- and octopine-strains of *A. tumefaciens* utilize only those opines whose synthesis they induce in CGT. Nopaline- and octopine-type pTi's each code for their own opine permease *(nop* or *ocp)* and opine oxidase *(noc* or *occ)*, but they have the same arginine- and ornithine-degrading functions (*arc* and *orc*, for symbols see Table **18**.2). Mapping of these genes on the pTi's was achieved by using deletion and transposon insertion mutagenesis (see Fig. **18**.7). On both types of pTi these functions form regulated gene groups whose expression is induced by the appropriate opine. Apparently the specificity of the nopaline-degrading system resides in the regulatory region of the genes since, once induced, nopaline permease and nopaline oxidase themselves also accept octopine. This phenomenon is especially well developed in constitutive *nop*/*noc* mutants (which also occur naturally). The enzymes of the octopine system (*ocp*/*occ*) only utilize opines of the octopine family, although in the case of agropine a special system that can only be induced by agropine itself is responsible for its uptake and degradation (*agp*/*agc*, cf. footnote [2] and Tables **18**.1 and **18**.2). There may also be other "unusual" opines to which the same situation applies.

Although opine permeases are normally induced and expressed together with the opine-catabolizing functions, they are somehow independent of these, as was shown by the finding of $opp^+opc^-$ mutants.

The opine oxidases or dehydrogenases are probably membrane-bound, cytochrome c-dependent enzymes which split opines into their original components. The arginine or ornithine released in this process is broken down further, not only by the appropriate catabolic functions of the pTi's (*arc* or *orc*) but also by chromosomally coded functions. Lysine and histidine catabolism is not coded by the pTi.

The opine catabolizing functions can account for a substantial part of the coding capacity of the pTi's (ca. 7%). The close proximity of some of these genes on the Ti plasmid may possibly reflect organization in an operon (cf. Fig. **18**.7). In any case their expression is regulated by one and the same mechanism, which probably contains both negative and positive control elements, and occupies a key position with respect to the gene expression program "opine utilization". This program apparently also includes chromosomal functions, since transfer of Ti plasmids to *E. coli* does not confer on the latter the ability to utilize opines (cf. Table **18**.2).

The opine catabolizing functions of the Ti plasmid are expressed independently of the functions for opine synthesis and oncogenicity. Spontaneous and induced mutants with the phenotypes $opc^+onc^-$, $opc^+ops^-$ or $opc^-onc^+ops^+$ have been described. The strict correlation between synthetic and breakdown functions of the octopine- and nopaline-type pTi's suggests, however, that Ti plasmids consist of only two building blocks, an oncogenic part and one involved in opine metabolism (synthesis and breakdown), the latter part differing according to the opine family involved. Each of these portions might be of different evolutionary origin. Their integration into one genetic unit and their concerted regulation and expression enables *A. tumefaciens* harbouring the Ti plasmid to use CGT-derived opines for growth and reproduction, thereby gaining a selective advantage over other soil bacteria (cf. Fig. **18**.3).

### 18.5.2 Ti Plasmids as Conjugative Plasmids

Ti plasmids can be transferred by means of conjugation between *A. tumefaciens* cells. The rate of spontaneous conjugation is very low (ca. $10^{-8}$) but can be greatly increased by addition of opines. The *tra* functions of nopaline- and octopine-type pTi's can only be induced by the corresponding opine. Conjugational transfer of the Ti plasmid can be shown to take place in situ in the infected plant wound following opine release by the induced CGT [17], but also in vitro following addition of opines to the bacterial cultures. The rate of conjugation rises with increasing time after opine addition (or synthesis by CGT's), but takes place more rapidly in vitro than in situ [8]. According to the donor and acceptor strains the rate can rise to $10^{-4}$–$10^{-1}$ for octopine-

-type pTi's, although for nopaline-type pTi's these values are considerably lower. Transformation of *A. tumefaciens* with purified Ti plasmid DNA can only be achieved after special treatment of the bacterial cells [7].

On the Ti plasmid the *tra* functions responsible for conjugational transfer are apparently subject to the same regulatory mechanism as the opine-catabolizing functions. For the octopine-type pTi-B6 three groups of regulatory mutants have been found (regulation defective mutants, aberrant repressor mutants, operator mutants, cf. Table **18.2**). They can best be explained by the existence of two separate operons [(1) *uad*, with *uadA* = octopine oxidase and *uadB* = octopine permease and (2) all *tra* functions] and a common regulatory gene ("ocr"?). In nopaline-type pTi's regulation of the opine-catabolizing functions and of the *tra* functions is only linked in certain strains, e.g., in pTi-T37 but not in pTi-C58 [6].

Induction of the propagation of a catabolic plasmid in the presence of its substrate seems to be logical, since in this way all individuals of a mixed population are enabled to utilize this substrate after induction of the appropriate enzymes. Both processes are based on differential gene expression following reception of signals from the environment.

## 18.6 Conclusions

In many gymnosperms and dicotyledonous angiosperms *Agrobacterium tumefaciens* can induce autonomously growing tumors known as crown gall tumors by infection of wounded sites of the plant. The process involves the transfer of a part of the genes (T-DNA) of the tumor-inducing plasmid into the plant cell nucleus, followed by their selective expression under the new genetic background of the host's genome. For this process to take place certain conditions have to be fulfilled: there must be (1) changes in the host cell gene expression patterns during the differentiation program "wound healing", (2) interactions between specific bacterial and plant receptors, and (3) presumably exchange of signals between both organisms involved. The T-DNA, which is responsible for transformation and for the maintenance of the tumor character, is stably replicated and transcribed in the plant cell nucleus. CGT's thus resemble to some extent virus-induced tumors in animals (cf. **19**.2.2).

Directly or indirectly, T-DNA brings about the synthesis of opines (which can be regarded as markers for CGT's) and autonomous cell division of the tumor cells, i.e., the permanent activation of all those hormonal and metabolic systems of the host plant that are necessary for cell growth and division. The molecular basis of CGT transformation is not yet completely understood. Developping of experimental systems for synchronous induction of CGT's is one of the most important methodological prerequisites for solving the following problems:

- What is the connection between the gene expression program "wound healing" and CGT induction?
- How is the Ti plasmid mobilized after interaction of the receptors?
- How, when and where is the T-DNA excised from the Ti plasmid?
- What are the functions of the other *onc* loci not localized in the T-DNA region?
- How does T-DNA enter the cell nucleus?
- How does T-DNA become integrated into the plant genome?
- How is T-DNA replicated?
- What are its gene products?
- What are the molecular mechanisms leading to the physiological effects of T-DNA?
- What are the reasons for and consequences of different T-DNA complements and copy numbers in different CGT's?

T-DNA expression in CGT cells leads to stable changes in the gene expression programs of the host cell, which are reflected by autonomous division, opine synthesis and abnormal phenotypes. This stable gene expression pattern may be modulated by temporary, reversible differentiation processes of an adaptive nature and can, under certain circumstances, be switched back to that of the normal cell. Such reversions indicate that the tumor cell remains totipotent, i.e., that transformation does not involve loss or irreversible modifications of the genetic material of the plant cell.

Reversions of plant and animal tumors are fundamentally important for an understanding of the entire problem of tumor formation. They focus attention on the importance of differentiation states in transformation and tumor reversion. The malignant state is revealed to be an atypical gene expression pattern that in some cases and under certain defined conditions can be reversed.

During phenotypic reversion of teratoma-CGT's normalization takes place without loss of T-DNA. Expression of the T-DNA is simply repressed under the influence of morphogenetic substances of the plant. This finding justifies the hope that the *A. tumefaciens*-CGT system will provide a suitable host-vector system for genetic engineering in higher plants. Positive results have already been reported (insertion of transposon Tn 7 into T-DNA region and its detection in CGT cells, cf. Fig. **18**.8), although this does not necessarily mean that the system will prove to be of practical application for planned stable genetic changes in higher plants. Whatever may happen, CGT research is providing a decisive contribution to our understanding of molecular mechanisms involved in the differentiation programs of plants and bacteria.

## 18.7 Selected References


1. Tomashow, M. F., R. Nutter, A. L. Montoya, M. P. Gordon & E. W. Nester: Integration and organisation of Ti plasmid sequences in crown gall tumors. Cell *19* (1980): 729–739.
2. Rhaese, P., H. DeGreve, H. Decramer, J. Schell & M. van Montagu: Rapid mapping transposon insertion and deletion mutations in the large plasmids of *Agrobacterium tumefaciens*. Nucl. Acids Res. *7* (1979): 1837–1849.
3. Hooykaas, P. J. J., R. A. Schilperoort & A. Rörsch: *Agrobacterium* tumor inducing plasmids: Potential vectors for the genetic engineering of plants. In: Setlow, J. K., & A. Hollaender (eds.) Genetic Engineering, Principles and Methods, Vol. 1. Plenum Press, New York 1979, pp. 151–180.
4. Schell, J., M. van Montagu, M. de Beuckeleer, M. de Block, A. de Picker, M. de Wilde, G. Engler, C. Genetello, J. P. Hernalsteens, M. Holsters, J. Seurinck, B. Silva, F. van Vliet & R. Villarroel: Interactions and DNA transfer between *Agrobacterium tumefaciens*, the Ti plasmid and the plant host. Proc. Royal Soc. London B *204* (1979): 251–266.
5. Montagu, M. van & J. Schell: The plasmids of *Agrobacterium tumefaciens*. In: Timmis, K. N., & A. Pühler (eds.) Plasmids of Medical, Environmental and Commercial Importance. Elsevier/North-Holland Biomedical Press, Amsterdam 1979, pp. 71–95.
6. Schilperoort, R. A., P. M. Klapwijk, P. J. J. Hoykaas, B. P. Koekman, G. Ooms, L. A. B. M. Otten, E. M. Wurzer-Figurelli, G. J. Wullems & A. Rörsch: *A. tumefaciens* plasmids as vectors for genetic transformation of plant cells. In: Thorpe, T. A. (ed.) Proc. 4th Intern. Congr. Plant Tiss. and Cell Culture, Calgary, Alberta, Canada 1978, pp. 85–95.
7. Holsters, M., D. de Waele, A. de Picker, E. Messens, M. van Montagu & J. Schell: Transfection and transformation of *Agrobacterium tumefaciens*. Molec. Gen. Genet. *163* (1978): 181–187.
8. Kerr, A., P. Manigault & J. Tempé: Transfer of virulence in vivo and in vitro in *Agrobacterium*. Nature *265* (1977): 560–561.
9. Smith, V. A., & J. Hindley: Effect of agrocin 84 on attachment of *Agrobacterium tumefaciens* to cultured tobacco cells. Nature *276* (1978): 498–500.

10. Lippincott, J. A., & B. B. Lippincott: Cell walls of crown-gall tumors and embryonic plant tissues lack *Agrobacterium* adherence sites. Science *199* (1978): 1075–1078.
11. DeCleene, M., & J. DeLey: The host range of crown gall. Bot. Review *42* (1976): 389–411.
12. Zaenen, I., N. van Larebeke, H. Teuchy, M. van Montagu & J. Schell: Supercoiled circular DNA in crown gall inducing *Agrobacterium* strains. J. Mol. Biol. *86* (1974): 109–127.
13. Braun, A. C.: Plant tumors. Biochim. Biophys. Acta *516* (1978): 167–191.
14. Lippincott, J. A. & B. B. Lippincott: The genus *Agrobacterium* and plant tumorigenesis. Ann. Rev. Microbiol. *29* (1975): 377–405.
15. Braun, A. C.: The relevance of plant tumor systems to an understanding of basic cellular mechanisms underlying tumorigenesis. Progr. Exptl. Tumor Res. *15* (1972): 165–187.
16. Gurley, W. B., J. D. Kemp, M. J. Albert, D. W. Sutton & J. Callis: Transcription of Ti plasmid-derived sequences in three octopine-type crown gall tumor lines. Proc. Natl. Acad. Sci. USA *76* (1979): 2828–2832.
17. Kerr, A.: Transfer of virulence between isolates of *Agrobacterium*. Nature *223* (1969): 1175–1176.
18. Larebeke, N. van, C. Genetello, J. Schell, R. A. Schilperoort, A. K. Hermans, J. P. Hernalsteens & M. van Montagu: Aquisition of tumor-inducing ability by non-oncogenic agrobacteria as a result of plasmid transfer. Nature *255* (1975): 742–743.
19. Larebeke, N. van, G. Engler, M. Holsters, S. v. d. Elsaker, I. Zaenen, R. A. Schilperoort & J. Schell: Large plasmid in *Agrobacterium tumefaciens* essential for crown-gall-inducing ability. Nature *252* (1974): 169–170.
20. DePicker, A., M. van Montagu & J. Schell: Homologous DNA sequences in different Ti--plasmids are essential for oncogenicity. Nature *275* (1978): 150–153.
21. Chilton, M.-D., M. H. Drummond, D. J. Merlo & D. Sciaky: Highly conserved DNA of Ti plasmids overlaps T-DNA maintained in plant tumors. Nature *275* (1978): 147–149.
22. Chilton, M.-D., M. H. Drummond, D. J. Merlo, D. Sciaky, A. L. Montoya, M. P. Gordon & E. W. Nester: Stable incorporation of plasmid DNA into higher plant cells: The molecular basis of crown gall tumorigenesis. Cell *11* (1977): 263–271.
23. McPherson, J. C., E. W. Nester & M. P. Gordon: Proteins encoded by *Agrobacterium tumefaciens* Ti plasmid DNA (T-DNA) in crown gall tumors. Proc. Natl. Acad. Sci. USA *77* (1980): 2666–2670.
24. Braun, A. C., & H. N. Wood: Supression of the neoplastic state with the acquisition of specialized functions in cells, tissues, and organs of crown gall teratomas of tobacco. Proc. Natl. Acad. Sci. USA *73* (1976): 469–500.
25. Drummond, M.: Crown gall disease. Nature *281* (1979): 343–347.
26. Turgeon, R., H. N. Wood & A. C. Braun: Studies on the recovery of crown gall tumors. Proc. Natl. Acad. Sci. USA *73* (1976): 3562–3564.
27. Sacristán, M. D., & G. Melchers: Regeneration of plants from "habituated" and "*Agrobacterium*--transformed" single cell clones of tobacco. Mol. Gen. Genet. *152* (1977): 111–117.
28. Kemp, J. D.: In vivo synthesis of crown gall specific *Agrobacterium tumefaciens*-directed derivatives of basic amino acids. Plant Physiol. *62* (1978): 26–30.
29. Gautheret, R.: Les transformations tumorales chez les végétaux. Bull Acad. Nat. Méd. *164* (1980): 358–374.
30. Hack, E., & J. D. Kemp: Purification and characterization of the crown gall-specific enzyme, octopine synthase. Plant Physiol. *65* (1980): 949–955.
31. Coxon, D. T., A. M. C. Davies, G. R. Fenwick, R. Self, J. L. Firmin, D. Lipkin & N. F. Janes: Agropine, a new amino acid derivative from crown gall tumours. Tetrahedron Letters *21* (1980): 495–498.
32. Chilton, M.-D., R. K. Saiki, N. S. Yadav, M. P. Gordon & F. Quetier: T-DNA from *Agrobacterium* Ti plasmid is in the nuclear DNA fraction of crown gall tumour cells. Proc. Natl. Acad. Sci. USA *77* (1980): 4060–4064.
33. Yadav, N. S., K. Postle, R. K. Saiki, M. F. Tomashow & M.-D. Chilton: T-DNA of crown gall teratoma is covalently joined to host plant DNA. Nature *287* (1980): 458–461.
34. Guyon, P., M.-D. Chilton, A. Petit & J. Tempé: Agropine in "null type" crown gall tumors: Evidence for generality of the opine concept. Proc. Natl. Acad. Sci. USA *77* (1980): 2693–2697.
35. Liu, S.-T., & C. I. Kado: Indolacetic acid production: Function of *Agrobacterium tumefaciens* C-58. Biochem. Biophys. Res. Comm. *90* (1979): 171–178.

36. Cockerham, L. E., & C. V. Lundeen: A transformation-induced alteration of cellular membranes in crown gall tumor cells. Plant Physiol. *64* (1979): 543–545.
37. Cornelis, P. E., E. Claessen & J. Claessen: Reversed phase chromatography of isoaccepting tRNAs from healthy and crown gall tissue from *Nicotiana tabacum*. Nucl. Acids Res. *2* (1975): 1153–1161.
38. Dubois, E. G., & J. H. Weil: A 7-methylguanine tRNA methylase present in crown gall cells but absent in normal *Parthenocissus tricuspidata* L. cells. Plant Sci. Letters *8* (1977): 385–394.
39. Braun, A. C., & P. R. White: Bacteriological sterility of tissues derived from secondary crown-gall tumors. Phytopathology *33* (1943): 85–100.
40. Petit, A., S. Delhaye, J. Tempe & G. Morel: Recherches sur les guanidines des tissus de crown gall. Mise en évidence d'une relation biochimique spécifique entre les souches d'*Agrobacterium* et les tumeurs qu'elles induisent. Physiol. Vég. *8* (1970): 205–215.
41. Lippincott, B. B., M. H. Whatley & J. A. Lippincott: Tumor induction by *Agrobacterium tumefaciens* involves attachment of the bacterium to a site on the host plant cell wall. Plant Physiol. *59* (1977): 388–390.
42. Wood, H. N., A. H. Pomerantz, A. N. Binns, V. G. Allfrey & A. C. Braun: Inhibition of mammalian protein kinase and phosphodiesterase activities by a cyclic AMP-like compound isolated from higher plants. Proc. Natl. Acad. Sci. USA *75* (1978): 4301–4305.
43. Yang, F., A. L. Montoya, D. L. Merlo, M. H. Drummond, M.-D. Chilton, E. W. Nester & H.M. Gordon: Foreign DNA sequences in crown gall teratomas and their fate during the loss of the tumorous traits. Molec. Gen. Genet. *177* (1980): 707–714.
44. Rogler, C. E.: Plasmid-dependent temperature-sensitive phase in crown-gall tumorigenesis. Proc. Natl. Acad. Sci. USA *77* (1980): 2688–2692.
45. Braun, A. C. (ed.): Plant Tumor Research. In: Progress Exptl. Tumor Research. Vol. *15*. S. Karger, Basel 1972.
46. Wullems, G. J., L. Molendijk, G. Ooms & R. A. Schilperoort: Retention of tumor markers in F1 progeny plants from in vitro induced octopine and nopaline tumor tissues. Cell *24* (1981): 719–727.
47. Yang, F., J. C. McPherson, M. P. Gordon & E. W. Nester: Extensive transcription of foreign DNA in a crown-gall teratoma. Biochem. Biophys. Res. Commun. *92* (1980): 1273–1277.
48. Tomashow, M. F., R. Nutter, K. Postle, M.-D. Chilton, F. R. Blattner, A. Powell, M. P. Gordon & E. W. Nester: Recombination between higher plant DNA and the Ti-plasmid of *Agrobacterium tumefaciens*. Proc. Natl. Acad. Sci. USA *77* (1980): 6448–6452.
49. Montagu, M. van, M. Holsters, P. Zambrynski, J. P. Hernalsteens, A. de Picker, M. de Beukeleer, G. Engler, M. Lemmers, L. Willmitzer & J. Schell: The interaction of *Agrobacterium* Ti-plasmid DNA and plant cells. Proc. Royal Soc. London B *210* (1980): 351–365.
50. Einset, J. W.: Cytokinins in tobacco crown gall tumors. Biochem. Biophys. Res. Commun. *93* (1980): 510–515.
51. Meins, F. Jr., J. Lutz & R. Foster: Factors influencing the incidence of habituation for cytokinin of tobacco pith tissue in culture. Planta *150* (1980): 264–268.

**Acknowledgement**

The author whishes to thank Prof. Dr. J. Schell for helpful discussions and for making manuscripts available prior to publication.

# 19. Aberrations in the Regulation of Cell Division and Differentiation as a Cause of Malignant Tumors

G. Butschak

## 19.1 Introduction

The molecular basis of cell differentiation and the problems connected with it have been discussed extensively in the preceding chapters of this book. Malignant tumors originate in cells that no longer, or only to a limited extent, respond to the complex regulatory processes which are finely tuned to one another during the development of an organism. Although the intimate connections between tumorigenesis and aberration in cell differentiation had long ago been recognized by many investigators, it was not until recently that advances in the field of molecular biology and in research on differentiation processes, as well as in cancer research, opened up the possibility of a more exact approach.

At present tumor cells can only be characterized on the basis of their biological behaviour, using the property of relatively autonomous proliferation, which makes it possible for them to be transplanted into suitable host animals. The tumor that arises is termed benign if its growth is exclusively expansive and does not lead to the destruction of neighbouring tissue, and malignant if it invades, infiltrates and finally destroys the surrounding tissues. Cells detaching from the latter type of tumor are capable of forming secondary tumors (metastases) at very different sites in the organism.

Despite intensive efforts it has not yet been possible to find any morphological or biochemical characteristics common to all cancer cells and by means of which they could be distinguished from normal body cells. Nevertheless, there are many differences between the cells of any tumor and the cells of the normal tissue from which it originates. In no case, however, is it possible to say whether any one of these differences is of immediate significacne for the malignant properties.

On account of their relatively autonomous proliferation and disturbances in their differentiation most malignant cells are partially or even totally incapable of fulfilling the functions of the corresponding normal cells. In tumors arising from stem cells or other cells that have not yet become specialized, such as certain epidermal tumors, carcinomas of the intestine, or in leukemia, normal differentiation is greatly reduced or does not take place at all. In malignant cells genes that are otherwise only active during embryonal or fetal development of the organs frequently become reactivated. Gene products may even appear which are not normally formed during development of the organs from which the tumors derived (cf. **19**.3.1). The tendency to form metastases points to changes in cell adhesion properties, probably due to alterations in the cell membrane (cf. **19**.3.2). The interconnections between proliferation, adhesion and differentiation of cells are of the utmost importance for an understanding of malignant properties of tumor cells.

In most cases malignant tumors do not arise in one step but are formed during a lengthy process in which cell development gradually progresses in the direction of

---

Abbreviations: $M_r$, relative molecular weight; kd, kilodalton; SV40, ASV, oncogenic viruses (cf. Table **19**.1); MGI, macrophage-granulocyte inducer.

greater autonomy and malignancy. The existence of potential autonomy with respect to a particular regulatory system may suffice for tumor formation if supplemented by another factor, such as stimulation of cell proliferation. The processes of growth and differentiation are regulated by a large number of genes. Therefore it is very likely that the mutation, repression or derepression of a considerable variety of genes can lead to the formation of malignant cells, every tumor thus having its individual history of development. This also implies that every tumor has its own specific biochemical and morphological characteristics, which may change continuously throughout the course of its development. Perhaps it will prove possible to find biochemical parameters which, even if not common to all tumor cells, are crucial for the malignant behaviour of the cells in which they occur.

## 19.2 The Origin of Cancer Cells

### 19.2.1 Carcinogenic Effects of Physical and Chemical Factors

Transformation of normal cells into tumor cells can result from physical, chemica or viral factors. Under the heading of physical factors an important place is occupied by irradiation of many kinds. Both UV light and ionizing rays can trigger the formation of malignant tumors. It is probably the diversity of action of various kinds of rays on DNA and the subsequent repair processes that are decisive. Some of the

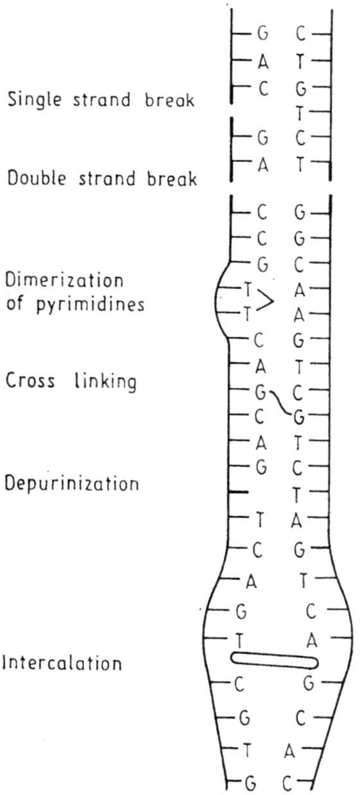


**Fig. 19.1:** Damages of the DNA double-strand caused by physical or chemical factors.

The *lower part* of the figure shows schematically the insertion and binding of an ultimate carcinogen leading to a local denaturation of the DNA.

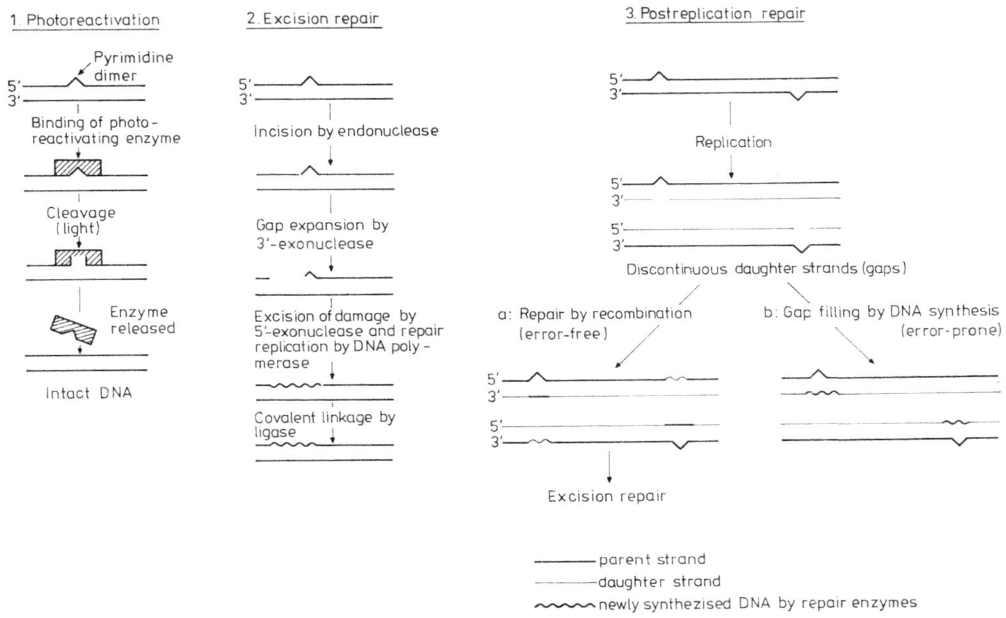


**Fig. 19.2**: Mechanisms of DNA repair (from Witkin, E. M.: Bacteriol. Rev. *40* (1976): 869–907, modified).

**1.** Photoreactivation (specific for pyrimidine dimers): The pyrimidine dimers formed after UV-irradiation are removed by a photoreactivating enzyme under the action of light (310 to 440 nm).

**2.** Excision repair: By this repair mechanism damaged sites are excised from the DNA strand and the gaps are filled by DNA synthesis using the complementary strand as template.

**3.** Postreplication repair: Damaged DNA sites can be repaired also after replication. In the course of replication gaps arise in the daughter strands opposite to the damaged DNA sites. These gaps can be filled by recombination between the parent and daughter strands *(a)*. The new gaps arising from recombination processes will be filled then by error-free DNA synthesis and the remaining pyrimidine dimers are removed by excision repair. Besides this, error-prone gap filling mechanisms can also occur *(b)*. This is the case in the so-called SOS repair, which is induced if DNA replication is arrested in consequence of relatively large damages. Other mechanisms of postreplication repair, for example, DNA double-strand exchange, are also discussed. Fore more detailed discussion of results and problems in the field of DNA repair see: Körner, L., & W. Malz: Studia biophysica *71* (1975): 115–126. Lehmann, A. R., & B. A. Bridges: Essays Biochem. *13* (1977): 71–119. Hofemeister, J.: Biol. Zentralbl. *97* (1978): 667–700. Hanawalt, P. C., P. K. Cooper, A. K. Granesan & C. A. Smith: Annu. Rev. Biochem. **48** (1979): 783–836.

damages caused to DNA by physical or chemical means are shown diagrammatically in Fig. **19**.1. DNA repair processes eliminating such lesions can be error-prone leading to mutations (Fig. **19**.2).

Malignant tumors can also be caused by foreign bodies such as plastic implants. The tumor-inducing action in this case is solely dependent on the size, form and surface properties of the implanted material. The tumors begin in the mesenchymal stem cells of vascular origin (pericytes), but the basic defect responsible for the malignant transformation of these cells is totally unknown. One possibility is that the implants mainly exert a cocarcinogenic effect (cf. **19**.2.3.3) stimulating the proliferation of certain cells, some of which are already transformed.

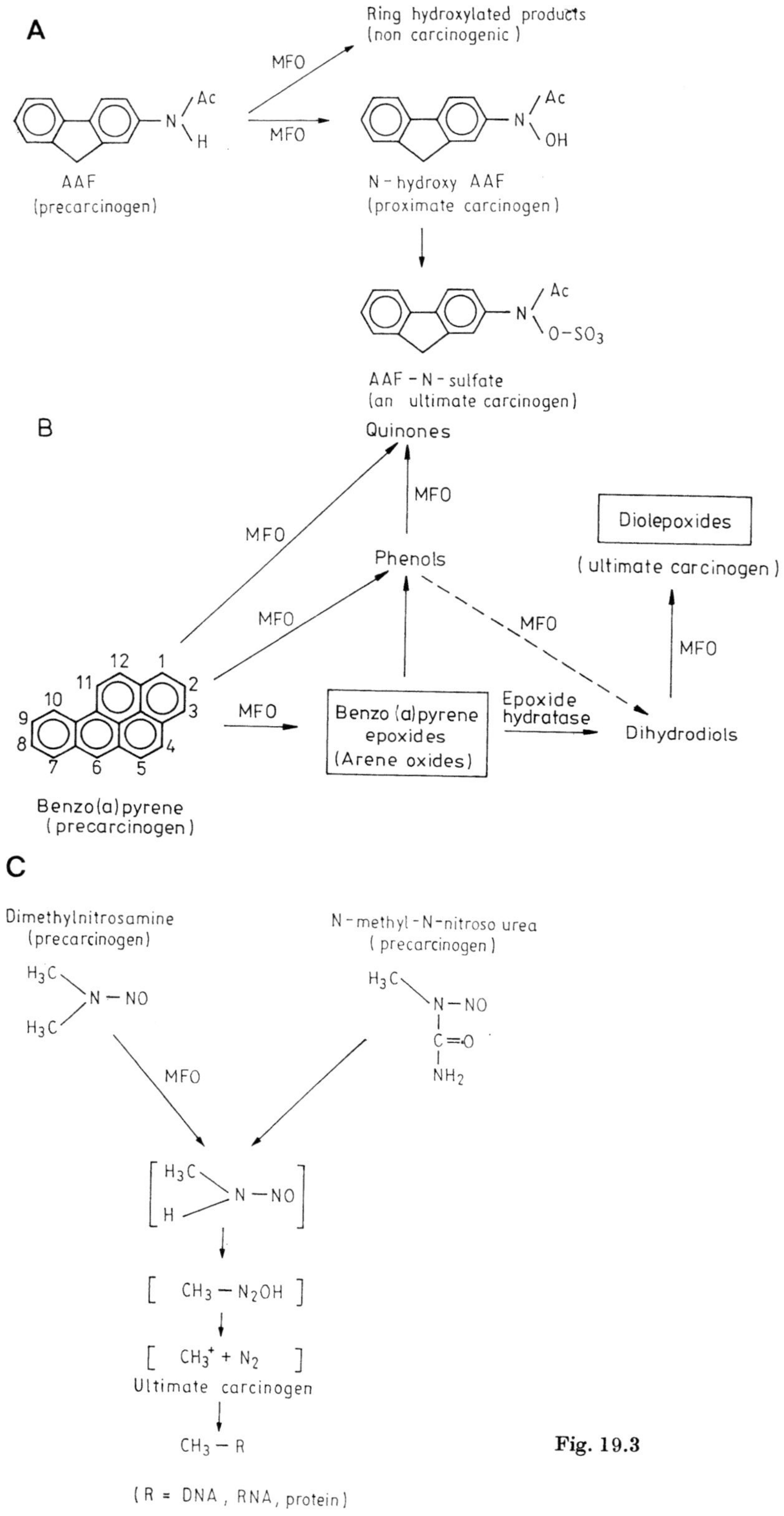
A
Ring hydroxylated products
(non carcinogenic)
MFO
MFO
Ac
N
H
AAF
(precarcinogen)
Ac
N
OH
N-hydroxy AAF
(proximate carcinogen)
Ac
N
O-SO3
AAF-N-sulfate
(an ultimate carcinogen)
B
Quinones
MFO
MFO
Diolepoxides
(ultimate carcinogen)
Phenols
MFO
MFO
MFO
12
1
11
2
10
3
9
8
4
7
6
5
MFO
Benzo (a)pyrene
epoxides
(Arene oxides)
Epoxide
hydratase
Dihydrodiols
Benzo(a)pyrene
(precarcinogen)
C
Dimethylnitrosamine
(precarcinogen)
N-methyl-N-nitroso urea
(precarcinogen)
H3C
N — NO
H3C
H3C
N — NO
C=O
NH2
MFO
H3C
N — NO
H
[ CH3 — N2OH ]
[ CH3+ + N2 ]
Ultimate carcinogen
CH3 — R
(R = DNA, RNA, protein)

**Fig. 19.3**

Epidemiological studies have yielded considerable evidence that some chemicals play an important role in the induction of tumors in man. Experiments with animals have shown a wide variety of chemical compounds to lead to tumor formation. A large number of these compounds do not act directly but have first to be metabolized within the organism itself to produce the ultimate carcinogen. In recent years considerable advances have been made in this field with the result that a number of questions, such as the species- and organ-specificity of the action of chemical carcinogens, are nearer to being answered. In addition, such studies have contributed to a better understanding of the mechanism of how chemical carcinogens act. It has been shown that, despite their great variety, all such carcinogens act in principle in the same way, i.e., they react with biological macromolecules, in most cases becoming covalently bound to proteins or nucleic acids, either directly or after chemical or metabolic transformation. Most of the ultimate carcinogens so far recognized are electrophilic and react with nucleophilic sites of the macromolecule. A few examples of enzymatic and non-enzymatic transformations of precarcinogens to ultimate carcinogens are shown in Fig. **19**.3.

Most of the enzymes involved in the metabolism of chemical carcinogens can be induced by xenobiotics, such as phenobarbital, certain polycyclic aromatic hydrocarbons or polychlorinate biphenyls. The inducibility of these enzymes varies from species to species and from individual to individual and plays an important part in the genetic disposition to development of malignant tumor under the influence of certain chemical carcinogens. Fig. **19**.4 shows diagrammatically the current concepts about regulation of the induction of aryl hydrocarbon hydroxylase. This is an enzyme system dependent in a special subspecies of cytochrome P-450 (cytochrome P-448 or $P_1$-450) and plays an important role in the metabolism of polycyclic aromatic hydrocarbons.

Many investigations have been made on the binding of carcinogenic compounds to proteins. First of all, different chemical carcinogens were shown to be preferentially bound to certain proteins localized in the cytosol. The role of these proteins in carcino-

---

◀

**Fig. 19.3**: Examples for the transformation of precarcinogens to ultimate carcinogens.

**A**: 2-Acetylaminofluorene *(AAF)* (from Conney, A. H., & W. Lewin: In IARC Scientific Publications 10, Lyon 1974: 3–22): The first step in metabolic activation of this and many other precarcinogens is a hydroxylation catalysed by cytochrome P-450 dependent monooxygenase system, also called mixed function oxidase *(MFO)*. This enzyme system is localized mainly in the membranes of the endoplasmic reticulum. From the hydroxylation products only N-hydroxy--AAF is carcinogenic (proximate carcinogen) and can be further metabolized to ultimate carcinogens, for instance, by sulfatation catalysed by a cytosolic sulfotransferase.

**B**: Benzo(a)pyrene (from Gelboin, H. V., et al.: In IARC Scientific Publications 12, Lyon 1976: 225–247): In this case the oxidation by the MFO-system leads to several epoxides (arene oxides), phenols and quinones. In this reaction a special form of cytochrome P-450 (cytochrome $P\text{-}450_1$ or P-448) is involved. Some of the arene oxides are reactive electrophils showing high mutagenic capacities in various test systems, but they have only a low carcinogenic activity. These epoxides can be converted by epoxide hydratase to dihydrodiols which can subsequently be oxidized by the MFO-system producing strongly mutagenic and carcinogenic diolepoxides. There is evidence that certain stereoisomers of them may act as ultimate carcinogens.

**C**: Dimethylnitrosamine and N-methylnitroso urea (from Magee, P. N., A. E. Pegg & P. F. Swann: In Altmann, H.-W., et al. (eds.), Handbuch der Allgemeinen Pathologie, Vol. 6, part 6 Tumours II. Springer Verlag, Berlin 1975: 328–419): The dimethylnitrosamine is N-demethylated by the MFO-system or by a special microsomal amine oxidase. The N-methylnitroso urea is hydrolysed non-enzymatically to the same intermediate. After a sequence of non-enzymatic reactions the ultimate reactive methyl cation is formed, leading to alkylations of proteins and nucleic acids.

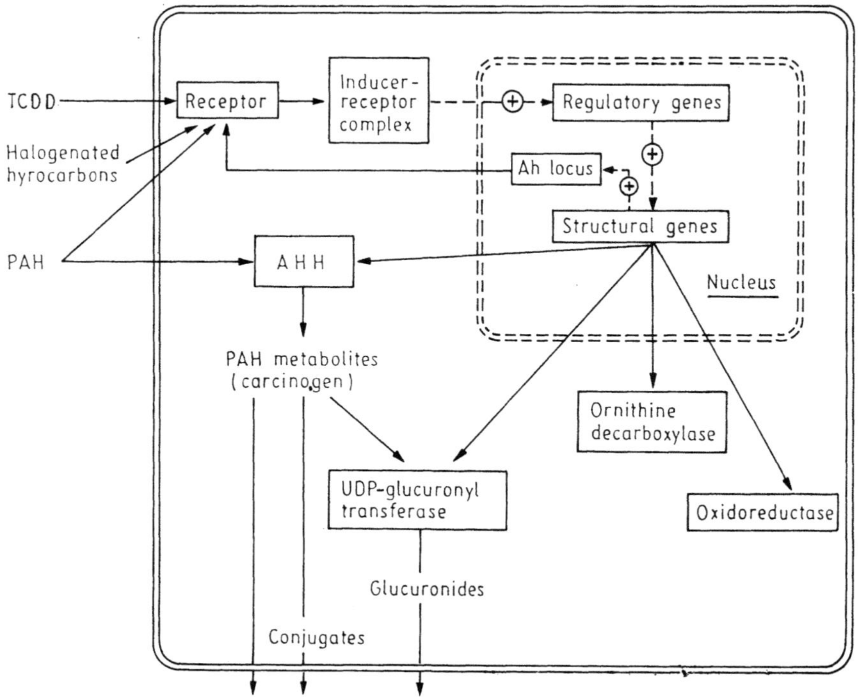


**Fig. 19.4**: Diagram of aryl hydrocarbon hydroxylase induction by sequential gene activation (from Nebert, D. W.: Pharmac. Ther. *6* (1979): 395–417, modified).

The aryl hydrocarbon hydroxylase *(AHH)* is induced by certain polycyclic aromatic hydrocarbons *(PAH)* like benzo(a)pyrene, 3-methylcholanthrene, benzo(a)anthracene and also by 2-naphthoflavone. Other inducers are halogenated hydrocarbons, e.g., polychlorinated biphenyls and 2,3,7,8-tetrachlorodibenzo-p-dioxin *(TCDD)*. The inducer is bound to the receptor, a product of the *Ah*-locus. The resulting inducer-receptor complex activates a set of regulatory genes, whose products control the activity of structural genes coding for AHH (cytochrome P-448) and some other enzymes (UDP-glucuronyl transferase, ornithine decarboxylase and a specific cytosolic oxidoreductase). Simultaneously the *Ah*-locus is activated to produce more receptor protein.

genesis is still totally unexplained. Some of them might be receptor proteins for the polycylic aromatic hydrocarbons important in the induction of the aryl hydrocarbon hydroxylase (cf. Fig. **19**.4). In addition, chemical carcinogens can be bound to histones and to non-histone proteins of the chromatin (cf. **1**.2.3) and can bring about dephosphorylation of phosphoproteins by alkylation of the phosphate residues. Alterations in regulatory proteins can lead to changes in cell differentiation and to stimulation of cell proliferation. However, no definite connection between carcinogenesis and the binding of chemical carcinogens to nuclear proteins has been observed so far.

Binding of chemical carcinogens to RNA can, via disturbances in translation, also lead to changes in regulatory proteins and thus influence gene expression. Here again, no unequivocal connections with carcinogenesis have so far been demonstrated.

The results of many investigations suggest that the binding of chemical carcinogens to DNA may be of special importance. Alkylation of DNA can result in erroneous base pairing, to single-strand breaks and to deletions. It can take place at various sites of the DNA bases. In recent investigations good correlation between alkylation in the $O^6$-position of guanine and the formation of primary tumors was found, although exceptions have been reported. Apart from alkylation itself, the persistence of alkylated products, which is dependent on the activity of repair processes, is also of importance. The repair processes exhibit a high degree of organ specificity. In the case of $O^6$-methyl guanine a positive correlation was shown between tumorigenesis and the persistence of the methylated base in the DNA.

In addition to simple alkyl groups, larger residues of ultimate carcinogens can become bound to DNA and the resultant compounds can lead to various types of mutations such as base substitution or "frame shift" mutations. Chemical carcinogens can be bound at different sites of the DNA. Binding to phosphate groups leads to single-strand breaks which, via error-prone repair mechanisms, can cause mutations. Chemical carcinogens can also be bound at nucleophilic sites of DNA bases. The carcinogenic metabolites of polycyclic aromatic hydrocarbons, for example, react mainly with the 2-amino group of guanine, whereas the ultimate carcinogens of certain aromatic amines are mainly bound to the C-8 of guanine. In both cases the substances are intercalated between two adjacent base pairs, and a short length of the double helix is denatured (insertion-denaturation model, cf. Fig. **19**.1). The consequences of this partial denaturation are as yet unknown, but they might possibly contribute to the triggering of repair processes. It is also conceivable that these sites in DNA are more readily accessible to certain regulatory proteins (e.g., non-histone proteins) and thus lead to activation of the corresponding chromatin region. This would provide an explanation for the commonly observed reactivation of genes during the development of tumor cells (cf. **19**.3.1). Binding of DNA or RNA polymerases may also be facilitated after partial denaturation of DNA (cf. **19**.2). Other effects, such as locally unwinding of the DNA double helix or changes in the tertiary structure of DNA, may have the same consequences. It can be anticipated that a more detailed study of the interaction of chemical carcinogens and DNA and its effects on structure and function of chromatin will reveal more about mechanisms of chemical carcinogenesis.

## 19.2.2 Action of Oncogenic Viruses

### 19.2.2.1 Oncogenic DNA Viruses

Malignant tumors can be caused by DNA- and RNA-viruses (cf. Table **19**.1). Oncogenic DNA viruses which have been thoroughly investigated are Simian virus 40 (SV40) and polyoma viruses. These viruses contain double-stranded circular DNA (2.5–3 Md) which corresponds to about 4,500 nucleotide pairs (cf. Fig. **10**.8). Complete reproduction of the oncogenic DNA viruses leads to lysis of the (permissive) host cell, whereas malignant transformation is confined to non-permissive cells in which virus multiplication is suppressed (cf. Fig. **19**.5). Sequencing of SV40 and polyoma virus DNA led to the conclusion that cell transformation is mediated by different genes. A prerequisite for transformation by these viruses in most cases is incorporation of the virus DNA into the cell genome. It is then replicated together with the cellular DNA so that the viral genome is present in every new cell generation.

On the surface of cells transformed by these viruses, virus-specific antigens and special antigens coded by the cell genome, including the S antigen, normally expressed only during ontogenesis, have been detected. The cell nucleus itself contains so-called T antigens, i.e., tumor-specific proteins, which, like virus-specific transplantation antigens, are coded by the virus genome. So far, two T antigens have been identified for SV40 and three for polyoma viruses, each with a different molecular weight. Cell transformation depends on certain T antigens, the middle T antigen and, in the case of the polyoma viruses, probably also the small T antigen. There are indications for certain T antigens to possess protein kinase activity. It is assumed that, by modification of various as yet unknown cellular regulatory proteins, they can interfere with cell proliferation and differentiation (cf. **10**.2.3.7.2). Furthermore, T proteins are supposed to associate directly with host cell DNA. Thus, they might trigger cell

**Table 19.1**: Survey of oncogenic viruses (from Bierwolf, D.: Arch. Geschwulstforschg. *50* [1980]: 506–514).

| Genome | Virus family | Particle size (nm) | Viruses | Isolated from | Oncogenic in natural host |
|---|---|---|---|---|---|
| DNA | *Papovaviridae* | 40–50 | Papillomaviruses | Diverse species | yes |
| | | | Polyoma virus | Mouse | yes |
| | | | SV40 | Monkey | no |
| | | | BK-, GBM-, JC-viruses | Man | ? |
| DNA | *Adenoviridae* | 70–90 | Several types | Man and other species | no |
| DNA | *Herpetoviridae* | 120–180 | Lucké virus | Frog | yes |
| | | | Marek's disease virus | Chicken | yes |
| | | | HV saimiri, HV ateles | Monkey | yes |
| | | | Baboon HV | Baboon | yes |
| | | | Epstein-Barr virus | Man | (yes) |
| DNA | *Poxviridae* | 200–300 | Molluscum cont. virus | Man | yes |
| | | | Fibroma viruses | Diverse species | yes |
| | | | Yaba monkey tumor virus | Monkey | yes |
| RNA | *Retroviridae* | | | | |
| | Type B | 100–120 | Mammary tumor virus | Mouse | yes |
| | Type C | 100 | ALV, ASV | Chicken | yes |
| | | | MuLV, MuSV | Mouse | yes |
| | | | FeLV, FeSV | Cat | yes |
| | | | GALV, SSV, BAEV | Monkey | yes (exept BAEV) |
| | Type D | | Mason Pfizer virus | Monkey | ? |
| | | | Langur virus | Monkey | ? |
| | | | Squirrel monkey virus | Monkey | ? |
| | | | Hela-, HEp-2-, PMF-virus | Certain human cell lines | ? |
| | not yet classified | 100–120 | Bovine leukemia virus | Cattle | yes |
| | | | Guinea pig leukemia virus | Guinea pig | yes |

SV: simian virus; BK, GBM, JC: human SV40-like papova viruses; ALV: avian leukemia virus; ASV: avian sarcoma virus; MuLV: murine leukemia virus; MuSV: murine sarcoma virus; FeLV: feline sarcoma virus; GALV: Gibbon ape leukemia virus; SSV: simian sarcoma virus; BAEV: Baboon endogenous virus; HV: Herpes virus.

proliferation by binding at points of replication origin and by locally unwinding the DNA double helix to facilitate interaction with DNA polymerase and initiation of replication. Finally, the association of, e.g., the middle T antigen with the cell surface is interesting when considering molecular mechanisms of cell transformation, because modifications of cell surface proteins may play an important part in this respect (cf. **19**.3.2.). Similar to some T antigens, the *src* gene product of avian sarcoma virus is also a protein kinase closely connected with the outer cell membrane (cf. **19**.2.2.2).

In recent studies characteristic host cell coded phosphoproteins ($M_r$ 54 kd) with unknown function have been found in cells transformed by different DNA and RNA viruses as well as by chemical carcinogens. Such proteins are also present in much smaller quantities in dividing normal cells. Finally, oncogenic DNA viruses can trigger mutations in their host cells, probably during incorporation into cellular DNA. It is not yet possible to say whether this contributes to their oncogenic effect or not.

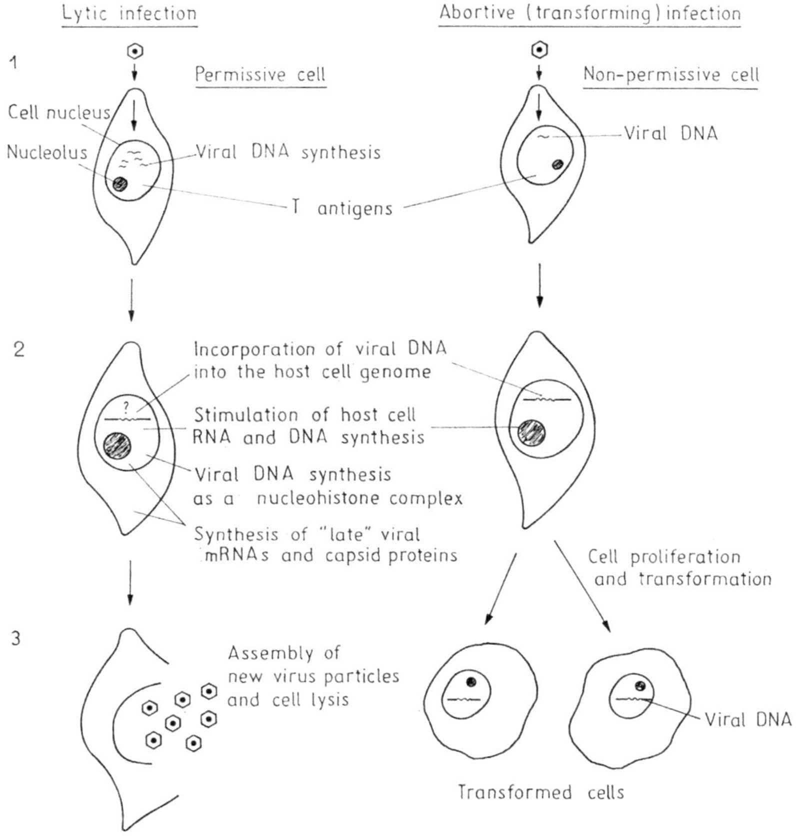


**Fig. 19.5**: Infection of cells by SV 40 or polyoma virus (for references cf. [16, 19]).

*Lytic infection (virus multiplication)*: (1) The lytic infection only occurs in permissive cells. After adsorption at the cell membrane the virus enters the cytoplasm of the host cell. The viral DNA is released into the nucleus and starts its replication. Simultaneously "early" viral mRNAs and proteins (T antigens) are synthesized. (2) If T antigen concentration has reached a threshold, a marked increase in the size of nucleoli is observed. Host cell RNA synthesis is strongly stimulated followed by activation of cellular chromatin replication. Simultaneously viral DNA is synthesized as a nucleohistone complex. There are many observations supporting the hypothesis that viral DNA at this time is associated with host cell chromatin. In parallel with this cellular and viral DNA replication "late" viral mRNAs and the capsid proteins are synthesized. (3) The viral DNA-nucleohistone complex is encapsulated and virus particles are released by cell lysis (about $10^5$ virus particles per cell).

*Abortive infection (cell transformation)*: Cell transformation by polyoma virus or SV 40 only occurs in non-permissive cells. In these cells synthesis of "late" viral mRNAs and proteins is suppressed. In most cases the viral DNA is incorporated into the host cell genome and is only replicated in the cell cycle. (1) Virus adsorption, releasing of viral DNA and synthesis of "early" viral mRNAs and proteins (T antigens); (2) activation of cellular RNA and DNA synthesis, incorporation of viral DNA into the host cell genome; (3) cell proliferation and transformation.

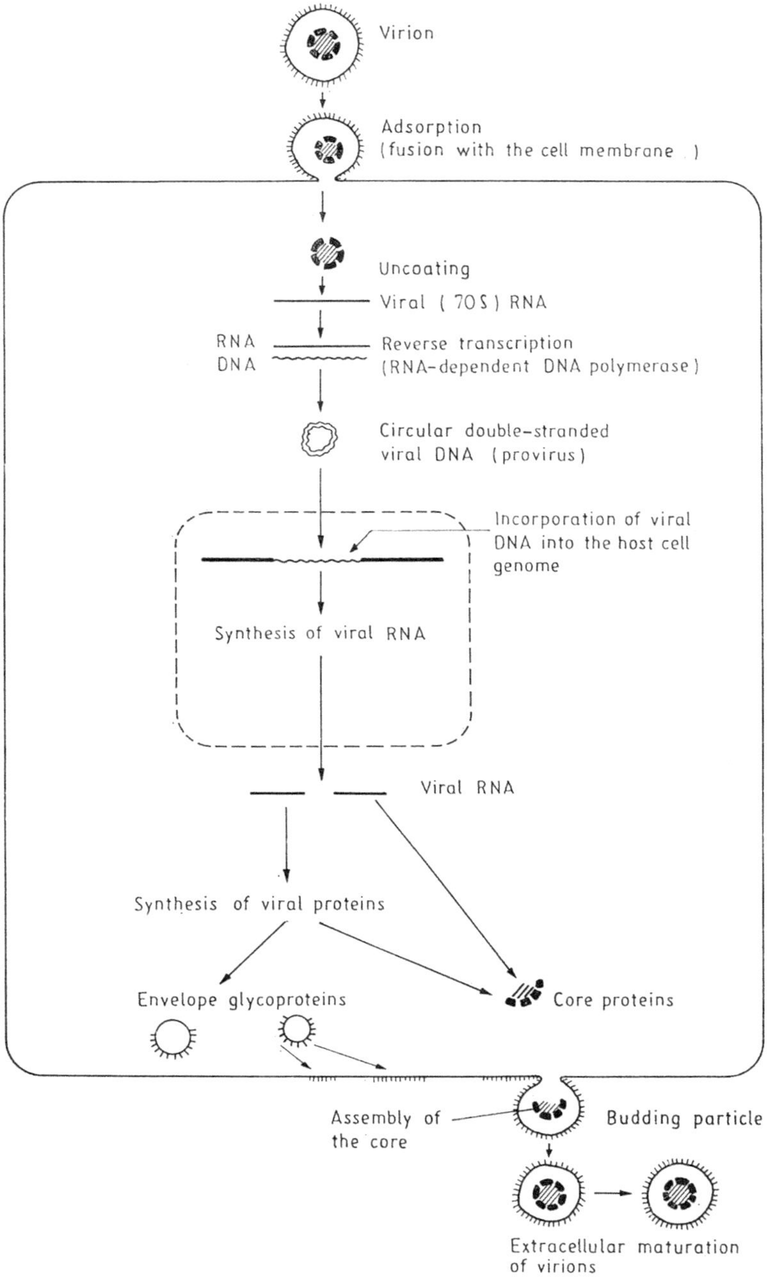


**Fig. 19.6:** Replicative cycle of oncogenic RNA viruses (for references cf. [13, 16]).

After adsorption at the cell membrane the virus particle penetrates into the cytoplasm. The viral RNA is released and the reverse transcriptase starts synthesis of a complementary DNA copy. The events between adsorption of the virus and the onset of DNA synthesis are not yet clear. The circular double stranded DNA (provirus) is incorporated into the cell genome and

### 19.2.2.2 Oncogenic RNA Viruses

Oncogenic RNA viruses are retroviruses that occur in many animal species (cf. Table **19**.1). They consist of a nucleoid (core), varying in shape according to the morphological type, surrounded by an outer envelope. Unlike the DNA viruses, they can also multiply in transformed cells, the mature virus particles being formed by budding from the cell membrane. All oncogenic RNA viruses are able to synthesize DNA complementary to the viral RNA. This process is catalysed by an enzyme known as reverse transcriptase. Here, too, the DNA copy of the viral RNA is integrated into and replicated with the host cell genome. During formation of new virus particles the viral DNA copy (provirus) serves as template for synthesis of viral RNA. The reproduction of oncogenic RNA viruses is shown diagrammatically in Fig. **19**.6. Although quite often in cells transformed by oncogenic RNA viruses no virus particles are formed, their synthesis can in many cases be induced by a variety of factors.

The virus genome consists of two identical subunits, each with a MW of about 3 Md, carrying, like mRNAs (cf. Fig. **1**.8), a typical cap—structure at the 5'- and a poly(A) sequence at the the 3'-end. The coding capacity of about 10,0000 nucleotides comprises three or four genes (Fig. **19**.7). Virus proteins arise by co-translational processing of large polyproteins which constitute the primary translation products (cf. details given for picornaviruses in Fig. **4**.32).

In cells infected by avian sarcoma viruses (ASV), transformation is induced by the product of a specific oncogene (*src* gene). In some leukaemia viruses other oncogenes have recently been found, whose location in the genome is not identical to that of the *src* gene. But DNA sequences with unknown function corresponding to the viral *src* genes and termed *sarc* genes have also been found in the host cell genome. Nucleotide sequences of *src* genes so far studied differ considerably from one virus to the other. Apparently, cell transformation can be induced by widely differing gene products. The reaction of the host cells to the *src* gene product is pleiotropic consisting not only of stimulation of cell proliferation but also of effects on various differentiation processes. The products of *src* genes themselves are phosphoproteins and, like the T antigens of oncogenic DNA viruses, possess protein kinase activity. In both cases of viral transforming proteins phosphorylation was shown to affect tyrosine residues. This unusual specificity of protein kinases may be of considerable value for indentifying their protein substrates and, thus, possible sites of primary attack for malignant transformation.

In chicken embryo fibroblasts one such protein was detected ($M_r$ 36 kd). It exists unphosphorylated in normal cells, but undergoes rapid phosphorylation of tyrosine residue(s) upon malignant transformation by Rous sarcoma virus, or after shifting a virus-infected cell culture from non--permissive to permissive conditions. In addition, six further phosphoproteins with similar properties were detected in two-dimensional gel electropherograms; but so far there is no evidence for their function within cells. An exception is a component of the cytoskeleton, vinculin ($M_r$ 130 kd), which is part of the membrane-associated focal adhesion plaques, i. e., of the sites of cell--substratum contact. Its phosphotyrosine content increases 10-fold upon transformation of cells with RSV [19b].

---

◀

serves as a template for viral RNA synthesis. Synthesis and processing of viral proteins take place in the cytoplasm (for details cf. **4**.4.5.1.2 and **4**.4.5.3). Assembly of the viral core and formation of the envelope by budding from the cell membrane are essential steps of virion synthesis that undergoes further maturation in the extracellular compartment (for structural details see Fig. **19**.7).

Host cell transformation and stimulation of their proliferation are independent of virus multiplication. Virus replication can take place in transformed cells, too.

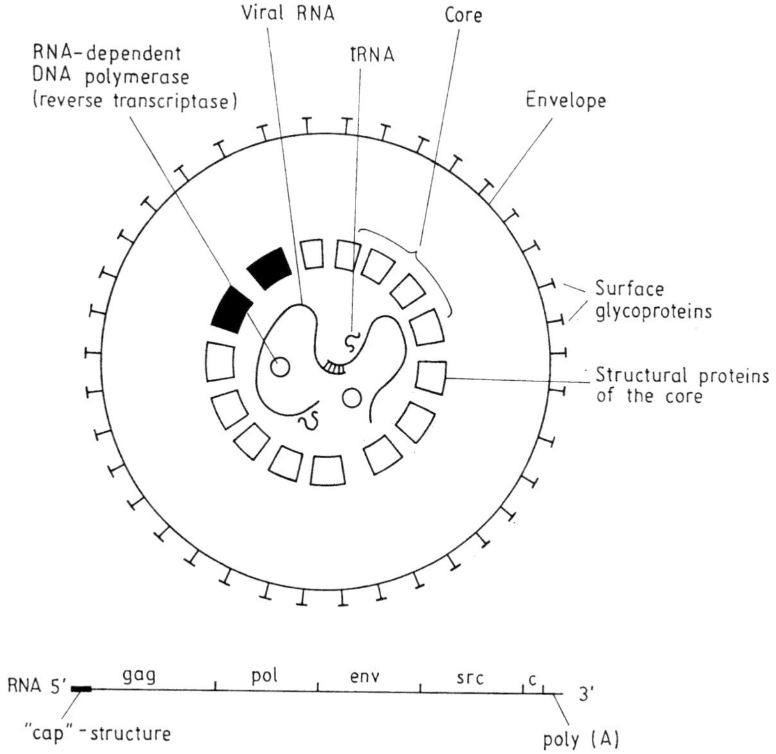


**Fig. 19.7**: Composition and genetic map of an oncogenic RNA virus (from Bauer, G.: Naturwiss. Rdsch. *31* [1978]: 445–455, modified).

The virion (upper part of figure) consists of the core and the envelope. The core contains the two identical RNA subunits and the RNA-dependent DNA polymerase (reverse transcriptase). Frequently, there are also small amounts of cellular RNA, e.g., tRNA, in the virion. On the surface of the envelope there are virus-specific glycoproteins.

The genetic map (lower part of figure) refers to avian sarcoma viruses (ASV). *Gag, pol,* and *env* are the genes coding for the core proteins, the reverse transcriptase and the glycoproteins of the envelope respectively. The *src*-gene is necessary for malignant cell transformation. A short sequence "c", identical in all avian sarcoma or leukemia viruses, is localized near the 3′-end. Furthermore, both ends contain some redundant sequences. Like mRNAs, the 5′-end has a typical cap structure and the 3′-end carries a poly(A) sequence.

There are also other results [19a] on *src* gene products acting directly on elements of the cytoskeleton and on cell membrane constituents. Immunoflurescence studies locate the *src* gene product at the inner face of the plasma membrane, especially in areas of cell-cell contact and cell-substratum contact. These sites are also known to serve the attachment of microfilaments. Modification of any protein involved in the interaction of cytoskeletal elements with the cell membrane could have decisive influence on cellular morphology and adhesion and lead to the achievement of more or less autonomy with respect to regulation of poliferation as well as to the differentiation of the cells concerned (cf. **19**.2.1 and **19**.3.2). In connection with the mechanism of cell transformation by leukemia viruses, some particularly interesting studies have shown that further differentiation of the target cells, e.g., lymphoblasts, erythroblasts, myeloblasts, is blocked by the virus in question, so that, although the cells still proliferate, they cannot develop into the specialized, non-dividing type of cell (cf. **4**.5.4.2).

A major new aspect of the retrovirus-induced malignant transformation originated from the detection of genetic loci homologous to viral oncogenes in practically all normal cells. Most of them are unique loci at constant chromosomal positions and unlinked to any endogenous viruses. These *sarc* genes or proto-oncogenes possibly belong to a multigene family and are assumed to code for tissue-specific regulators of normal growth and development. Three probable explanations of the tumor-inducing effect of retroviral oncogenes can be given [19c]: (1) By introduction of the viral oncogene the cell may be overloaded with the normal gene product (gene dosis effect) resulting in a severe disregulation of cell proliferation. (2) Only those cells can be transformed by the oncogene product that, due to the differentiation state, normally do not express the corresponding proto-oncogene. (3) Though both products of the proto-oncogene and oncogene were shown to be very similar, they might exhibit subtle but significant differences influencing their substrate specificity and/or cellular localization. It may be highly significant in this respect that malignant transformation could be induced by cellular DNA derived from tumor cells, from cells transformed by chemical carcinogens or even from normal cells. Thus, the existence and function of cell-inherent proto-oncogenes, as detected by analysing the life cycle of retroviruses, may provide a unifying concept of tumorigenesis in general [19c].

### 19.2.3 Significance of Genetic and Epigenetic Mechanisms in the Origin of Malignant Tumors

#### 19.2.3.1 Role of Mutations in Tumorigenesis

The assumption that malignant transformation has its roots in somatic mutations is one of the oldest concepts concerning the origin of cancer. The idea is largely supported by the fact that malignant properties are passed on to the daughter cells during cell division. Further support for this theory is also drawn from the mutagenic action of most of the carcinogens known to-day (radiation and many chemical carcinogens).

A convincing piece of evidence for the tumorigenic role of DNA lesions caused by irradiation with UV light was found in experiments with a species of fish, *Poecilia formosa*, that can be cloned by parthenogenetic propagation. Cells of the thyroid gland were irradiated in vitro with UV light, after which some were preincubated in light and some in the dark. Subsequent transplantation into isogenic recipient individuals showed that cancer cell induction was much higher with cells preincubated in dark than with those preincubated in light (Fig. **19**.8). Apparently, the cleavage of pyrimidine dimers in the DNA under the influence of light (photoreactivation, cf. Fig. **19**.2) results in a lower tumor incidence, i.e., DNA lesions are directly involved in tumorigenesis.

In hereditary diseases associated with defects in the DNA repair system, e.g., *Xeroderma pigmentosum*, there is an elevated incidence of tumors following treatment with UV light, for example. Chemical mutagens and carcinogens (e.g., N-acetyl derivatives of aromatic amines or some expoxides of polycyclic aromatic hydrocarbons) exert stronger cytotoxic and mutagenic effects on fibroblasts isolated from patients suffering from *Xeroderma pigmentosum*, than on normal human fibroblasts.

The characteristic properties of the transformed phenotype, which develop after treatment of cell cultures with chemical carcinogens, are often temperature-sensitive. These findings point to the formation of altered gene products and thus indicate somatic mutations.

The occurrence of hereditary tumors also gives evidence for the role of mutations in tumor formation. Familiar examples of such tumors in man are certain retinoblastomas, neuroblastomas, and the Wilms' tumor. In some cases there is a strong genetic disposition to certain tumors and it was assumed that they resulted from two mutations, the first of which was prezygotic, i.e., took place in the germ cells;

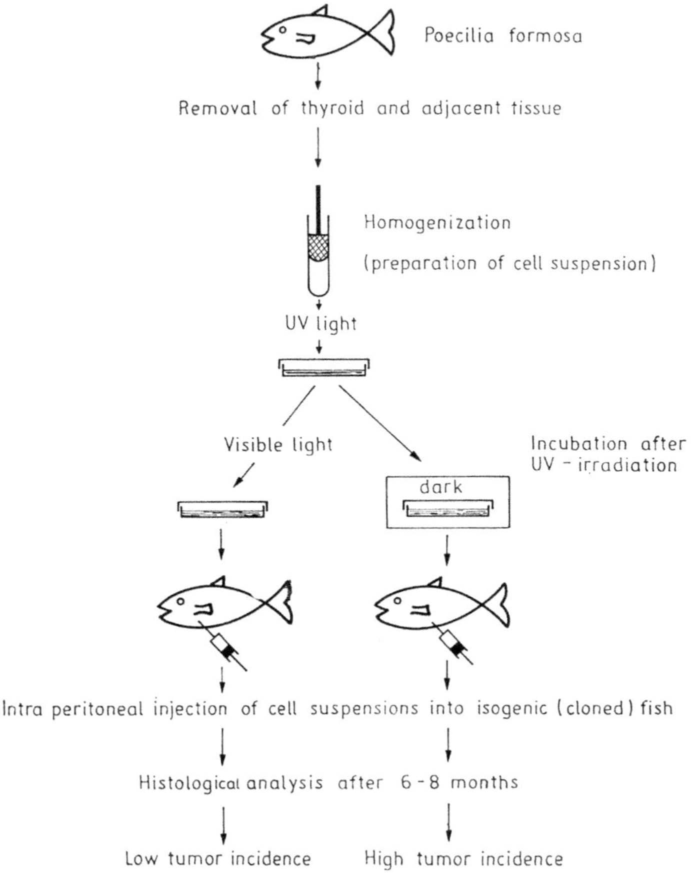


**Fig. 19.8:** Scheme of the photoreactivation experiment using the Amazon molly *(Poecilia formosa)* (for reference see Hart, R. W., R. B. Setlow & A. D. Woodhead: Proc. Natl. Acad. Sci. USA *74* [1977]: 5574–5578).

The results of this experiment, i.e., a significant decrease of the tumor incidence after preincubation of the implanted cells in light, strongly suggest that pyrimidine dimers may lead to tumors (cf. **Fig. 19.2** for photoreactivation repair). That means tumors can arise by mutations.

the second occurring postzygotically, as a somatic mutation. Other forms of genetic disposition to malignant diseases may be the consequence of very different genetic defects or differences, affecting, for example, the metabolism of chemical carcinogens (cf. Sect. **19**.2.1 and Fig. **19**.4), the alteration of DNA repair mechanisms or gene balance due to changes in chromosome number.

Extensive genetic studies on fishes of the genus *Platypoecilus* and *Xiphophorus* led to the conclusion that a "tumor gene" is governed by a complex regulatory system. Mutations of regulatory genes and interspecific crosses can result in imbalance of the regulatory processes and to expression of the tumor gene, whereby the transformation depends on the competence, i.e., a particular state of differentiation of the cells concerned. The expression of oncogenes coded by endogenous viruses might be induced in a similar way. In higher vertebrates the dependence of tumor gene expression on the state of differentiation is shown by the organ-specificity of inheritable tumors.

### 19.2.3.2 Role of Epigenetic Factors in Tumorigenesis

The idea of malignant tumors developping as a result of changes in gene expression even in the absence of damage to the genetic material has been discussed by numerous authors. The binding of chemical carcinogens to certain regulatory proteins (cf. **19**.2.1), for example, can bring about changes in gene expression and thus possibly lead to a relatively stable state of differentiation that prevents the response of cells to the regulatory mechanisms active at a given time. Once a particular state of differentiation has been attained, it is passed unchanged to the daughter cells as if it was due to a genetic alteration. The reversibility of normal differentiation processes can only be demonstrated under special experimental conditions, for example, employing nuclear transplantation (cf. **16**.2.2) or cell fusion, or following implantation into embryonic tissues. Under these conditions even the malignant status of many tumor cells is reversible. In many cases they differentiate to cells fulfilling a normal function in the body.

Good examples for this behaviour are given by certain skin tumors, neuroblastomas and teratocarcinomas. Under special conditions or by addition of certain factors, such as dibutyryl-cAMP, inhibitors of the cAMP-phosphodiesterase or sodium butyrate, cells of neuroblastoma cell cultures can develop structural and functional properties corresponding to normal nerve cells. Such differentiated tumor cells are no longer tumorigenous. Even in vivo neuroblastomas can regress by differentiation of their cells. Cells of teratocarcinomas can realize many differentiation programs when losing their malignant capacity. The same has been observed in plant tumors (cf. Chap. **18**).

Several experiments were carried out transplanting amphibian tumors into regenerating blastemas. Under these conditions tumor cells can differentiate into a variety of normally functioning cells that no longer exhibit malignant properties. Also co-cultivation of tumor pieces with embryonic tissues can induce differentiation processes in the tumor cells even if the two tissues are separated by a millipore filter. The induction of differentiation in leukemia cells is reported in Sect. **19**.4.2. Nevertheless, in experiments of this kind the possibility cannot be excluded that, via the differentiation processes the mutated regulatory genes are repressed or, in other words, the differentiated cells lose their competence to respond to certain regulatory proteins, e.g., stimulators of cell proliferation. In such cases reversion of the malignant state, i.e., suppression of the malignant properties, is only phenotypical, the genetic lesions remaining.

Investigations in which different organ systems or entire organs were formed from tumor cells support the idea that some tumors may arise solely as a result of altered gene expression. For example, the transplantation of cell nuclei from the renal adenocarcinoma of the leopard frog *(Rana pipiens)* into enucleated zygotes resulted in the development of free-swimming larvae in which all organ systems were normally differentiated. The experiments show that a malignant tumor contains all genes necessary for normal embryonic development. However, since this was a case of a virus--induced tumor it can be assumed that the virus genome was repressed. Because development of adult frogs has never been observed in such experiments it is not known whether tumors might develop after metamorphosis.

More convincing experiments were performed with teratocarcinoma of mice. If malignant cells from tumors, transplanted over many generations, were implanted into blastocysts of genetically defined host mice, the tumor cells participated in the development and functioning of all organs in the resulting mosaic animals. Some of the teratoma cells even differentiated to functional sperm cells from which offspring resulted. Experimental details are shown in Fig. **19**.9.

Most of the tumors in man and animals arise from precancerous alterations which appear simultaneously at many sites within the tissues concerned. Only a small portion of the altered

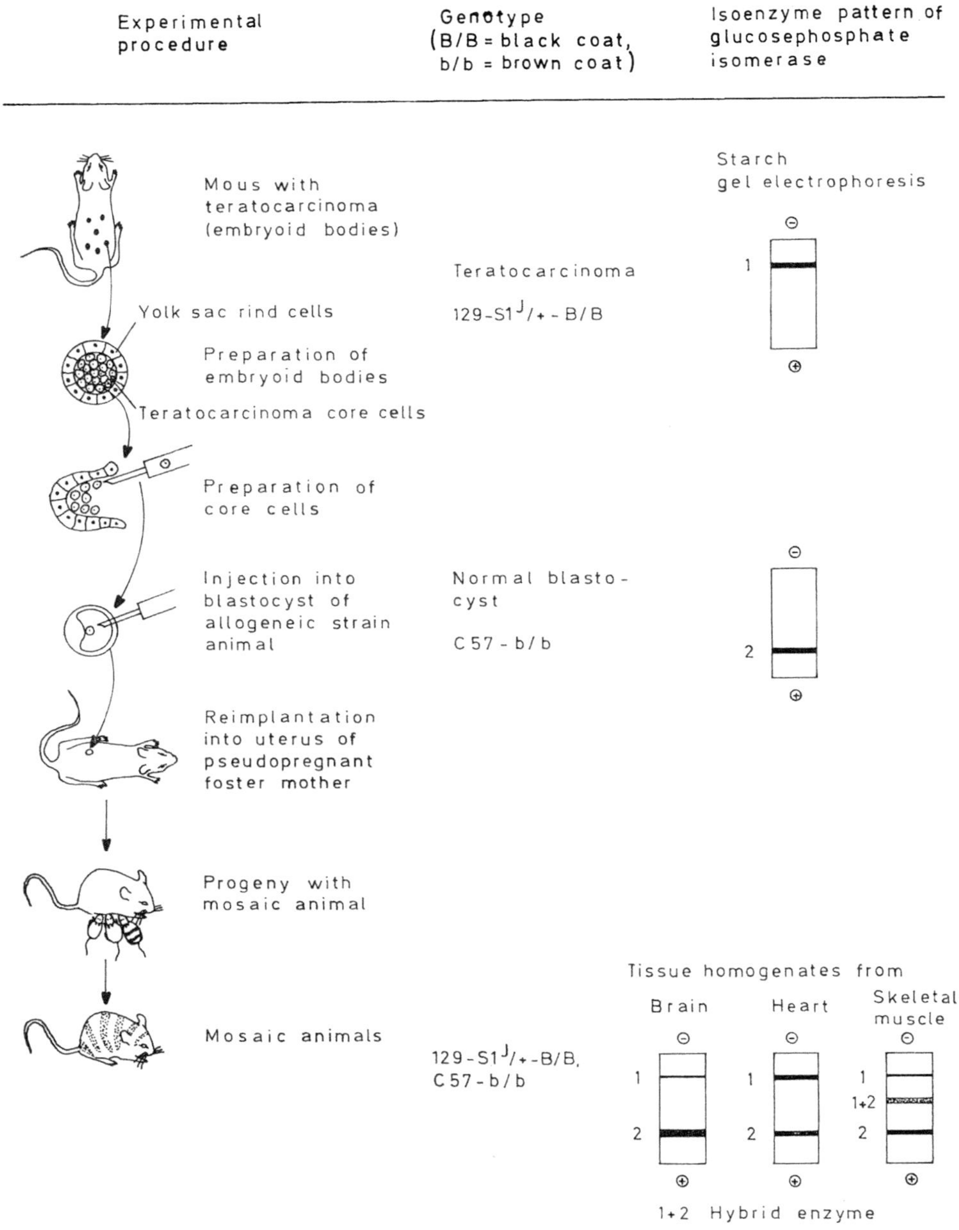


**Fig. 19.9:** Reversion of the malignant state of mouse teratocarcinoma tumor cells (from Mintz [31]).

The *left site* of the figure documents the experimental protocol. Cells originating from the teratocarcinoma cell are present in practically all tissues of the mosaic animal shown at the end. But they have lost their malignant character. Genetic markers for coat pigmentation (B/6) and for glucose-6-phosphate isomerase are used to discriminate between cells derived from the blastocyst and from the teratocarcinoma cell respectively.

cells later develop into malignant tumors. In rats, for example, hepatomas arising after administration of N-nitrosomorpholine originate from preneoplastic nodules which form in large numbers in the liver tissue, and first become recognizable on account of their conspicuous accumulation of glycogen. As carcinogenesis proceeds, the glycogen content of these cells is reduced again and, at the same time, changes in the ergastoplasm of the cells become noticeable. The studies show a sequence of disturbances in cell differentiation that may end in tumor formation. The large number of preneoplastic nodules and their characteristic stepwise development into microtumors suggest a direct effect on certain processes of differentiation rather than a somatic mutation to be the primary cause of carcinogenesis. However, tumorigenesis may depend on alterations of many different genes. If this is true, multiple precarcinogenous alterations can also be interpreted in terms of the mutation theory.

Following the treatment of cell cultures with chemical carcinogens, the frequency of transformation very often exeeds the frequency of mutation by 2–3 orders of magnitude. This high frequency of transformation is also interpreted as evidence of a direct influence on gene expression. On the other hand, in a special line of hamster cells (BHK) and in diploid human cells, the transformation frequency was very low and corresponded to the mutation frequency of genes. The high transformation frequencies may also result from activation of oncogenic viruses or by mutations of several genes.

### 19.2.3.3 Tumorigenesis as a Multistep Process Involving Genetic and Epigenetic Alterations

In the preceding sections it has been outlined that malignant tumors can be caused by alterations at the genetic level (mutation or introduction of viral DNA) as well as by epigenetic processes. Development of tumors can be triggered by even small disturbances in the regulatory mechanism. Since processes of cell division and differentiation are controlled by a complex network of interactions between numerous regulatory systems with a large number of genes involved, the activation or inactivation as well as mutation of one of these genes might lead to partial autonomy of the cell concerned. This could then be reinforced by the alteration of the function of still other genes or by their mutation. Triggering and development of malignant tumors can be caused by alterations in an variety of regulatory systems, so that each tumor follows its own individual line of development. Moreover, even within one and the same tumor it is not uncommon to find different cell clones with their own biological and biochemical properties.

Three, often overlapping phases can be distinguished in the development of malignant tumors, i.e., initiation, promotion and progression (Fig. **19**.10). That the development of malignant tumors is a complex process involving many steps has been demonstrated in a wealth of studies employing a variety of models. Thus, application of a small dose of a carcinogen, such as benzo(a)pyrene or 7,12-dimethylbenz(a)-anthracene, to the skin of mice produced irreversible changes in the epidermal cells, although no tumor formed (tumor initiation). Realization of this alteration could then be brought about by application of further substances that, by themselves, cause no tumor. Such substances are termed co-carcinogens or tumor promotors. Highly effective tumor promotors have been isolated, for example, from the seed oil of *Croton tiglium* L. One in particular, 12-O-tetradecanoylphorbol-13-acetate (TPA), has been widely employed experimentally. Such experiments led to a distinction between two phases in the development of tumors, i.e., the phases of initiation and of promotion. The resulting benign skin papillomas can be transformed into carcinomas by additional alterations (tumor progression).

Mutations or alterations in expression of genes involved in regulation of cell division can only affect proliferating cells, so that any stimulation of cell proliferation leads to a considerable increase in the incidence of malignant tumors, e.g., the generation of liver tumors by means of chemical carcinogens after partial hepatectomy. In fact,

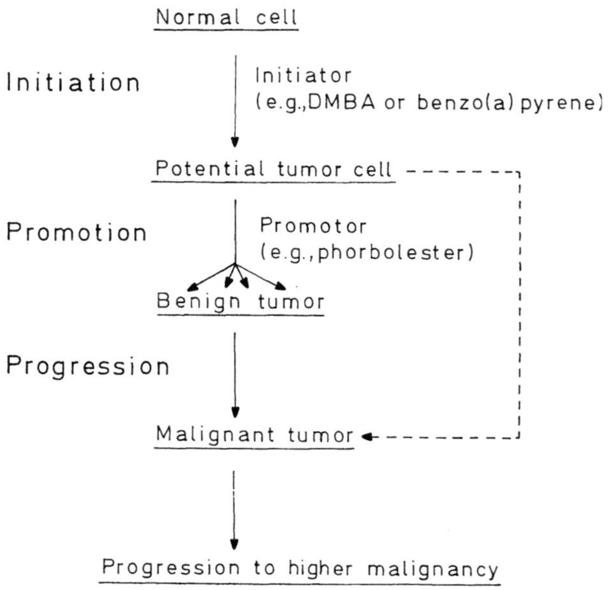


**Fig. 19.10:** Multistep model of tumor development.
The potential tumor cell is altered irreversibly but may not develop a tumor without further action of carcinogenic or co-carcinogenic factors. The mechanism of action of co-carcinogens (promotors) is not yet clear, but most of them stimulate cell proliferation, an important condition for realizing alterations in the genetic material. Changes of differentiation state can also be expressed only after cell division. (For further information see text and Hecker, E.: Naturwissenschaften *67* [1978]: 640–648.) *DMBA*, 7,12-dimethylbenz(a)anthracene.

the effect of tumor promotors may be due mainly to their stimulation of cell proliferation. As was described in Sects. **4**.5.4.2 and **19**.2.2, transformation by oncogenic viruses is also associated with or even dependent on the activation of cell proliferation. Remarkably, the initiation phase was shown to require DNA synthesis. Similar observations hold true for the initiation of tumors in plants by wound infection with *Agrobacterium tumefaciens* (cf. Chap. **18**).

## 19.3 Properties of Tumor Cells

### 19.3.1 Reprogramming of Gene Expression in Tumor Cells

The connection between the generation of malignant tumors and disturbances in the regulation of cell differentiation is most clearly emphasized by the fact that all malignant tumor cells exhibit greater or smaller alterations in gene expression. As already mentioned in the introduction, tumor cells, in the course of their development, progressively lose special properties of the corresponding normal cell type, while genes that were active during embryonic or fetal development of the tissue of origin become reactivated. In addition, in certain cases gene products are observed never occuring during the development of the organ concerned. Some of these so-called ectopic proteins are normally formed only in cells derived from the trophoblast.

In experimentally produced mammary tumors an increased alteration of gene expression in various stages of tumor development (hyperplastic alveolar nodules, primary carcinoma, transplanted carcinomas) could be demonstrated. In every one of the above-mentioned stages a number of nuclear RNA species were synthesized that distinctly differed from those of preceding stages. Further, the carcinomas produced an enzyme (tRNA-guanosine-7-methylase) that does not occur in normal mammary tissue but is nevertheless active in certain other normal tissues. At the same time numerous genes, active in normal mammary glands, are repressed. Similar results were obtained in liver carcinoma. Whereas the so-called minimum deviating hepatomas only exhibit relatively small but clearly demonstrable alterations in expression of some genes, derepression of genes active in embryonic or fetal development increases in rapidly growing hepatomas. Simultaneously, synthesis of proteins necessary for the various specific functions of the liver is blocked.

Embryonic or fetal isoenzymes and other proteins, detectable by immunological methods during ontogenesis, also appear in tumor cells. A well-known example is the α-fetoprotein, whose demonstration in serum has become important in the diagnosis and control of therapy of certain liver tumors. Another example is provided by the carcinoembryonic antigen produced by gastrointestinal tumors (CEA), although the diagnostic applicability of these two antigens is limited by the fact that under certain conditions both can be formed in larger quantities by non-neoplastic tissues, too. Particularly, genes that were active during the late fetal period can become reactivated also in normally proliferating cells, e.g., following partial hepatectomy. However, in normally dividing cells the quantity of fetal gene products produced is usually much smaller than that formed in a corresponding malignant tumor with the same rate of proliferation.

Little is known so far about the mechanisms of reprogramming of gene expression in tumor cells. Apart from the blockage and reactivation of genes, gene amplification and alterations in regulation of translation and post-translational processes probably also play a role.

### 19.3.2 Alterations of the Cell Membrane

An important role in the regulation of cell division and cell differentiation is played by the cell membrane, which is intimately connected with the fibrillar cytoskeleton (cf. **4**.1.4). Especially the behaviour of cells towards one another, their mutual influence, their adhesion as well as their attachment to extracellular substances such as collagen are all dependent on the cell membrane. In many cases the response to humoral factors and other extracellular signals requires the presence of membrane-bound receptors (cf. Figs. **4**.4, **8**.12 and **10**.3.3.2). Interference with any of the various functions of the cell membrane can result in a behaviour characteristic of malignant tumor cells, e.g., defective proliferation control or invasion of foreign tissues. Changes in the cell membrane observable in tumor cells and following transformation in vitro can be regarded partly as a consequence of reprogramming of gene expression discussed in the preceding section. Here, too, characteristics develop that are typical of the membrane of embryonic cells or of certain other cells of the adult organism that have taken a different line of ontogenetic development. In addition, changes of the membrane composition in distinct phases of the cell cycle, if accompanied by loss of their reversibility, may also contribute to malignant transformation.

In tumor cells and following transformation in vitro considerable alterations in the cell membrane are detectable. Demonstration of the presence of specific tumor-associated transplantation antigens points to changes in the composition of membrane proteins. Whereas these antigens are specific for every tumor produced by chemical carcinogens, they are identical in all tumors produced by a particular virus. Furthermore, in virus-induced tumors viral antigens can be detected on the cell surface. Embryonic surface antigens have also been demonstrated in tumor cells and following transformation in vitro.

An important indication of changes in the cell membrane of tumor cells is their ability to be agglutinated by lectins. Since the same amount of lectin is bound by normal and tumor cells the receptors themselves cannot differ. Electron microscopical studies employing ferritin-labelled lectins have shown the receptors to aggregate to patches in the cell membrane. This happens relatively quickly in tumor cells, whereas in normal cells the receptors remain more or less evenly distributed for a longer time. The greater mobility of receptors in tumor cells might possibly be connected with profound changes of the cytoskeletal system.

In addition there are many further observations supporting the significance of membrane alterations for malignant transformation:

- In some cases hormone receptors are lacking or additional receptors appear. Thus, changes in the responsiveness to extracellular signals and in the differentiation behaviour can be anticipated.
- High levels of proteinases have been detected, e.g., of the membrane-associated plasminogen activator protein in different tumors and on in vitro transformed cells. This activator is an arginine-specific serum protease normally involved in the initiation of fibrinolysis. Abnormally high activities of the enzyme are also found on the surface of rapidly dividing normal cells capable of penetrating tissues, e.g., macrophages. Although normal tissue culture cells can be stimulated to proliferate by means of proteolytic enzymes, the mechanism involved is not yet known.
- Alterations of the glycolipid and glycoprotein composition are supposed to result from enzymatic changes in the ergastoplasm, especially in the glycosyl transferases localized in the Golgi system.
- One glycoprotein, assumed to mediate adhesion of cells to the extracellular matrix, is lacking on the surface of several types of transformed cells. This is the LETS protein or fibronectin ($M_r$ 200–240 kd) exhibiting high affinity for collagen, fibrin and heparin.
- Outer membrane proteins of normal cells, e.g., of melanocytes or fibroblasts, were shown to normalize growth behaviour of in vitro transformed cells and tumor culture cells.
- Typical changes in the ion and metabolite transport systems have been reported. In practically all tumor cells the transport capacity for sugars and amino acids is increased, which under certain conditions gives them an advantage as compared with normal cells. Alterations in ion transport, especially if involving changes in intracellular calcium content, could lead to interference with the regulation of cell proliferation and differentiation.

## 19.4 Tumor Therapy by Means of Biological Regulation Factors

### 19.4.1 Therapy of Malignant Tumors by Means of Hormones

Recently, chemotherapy of malignant tumors is gaining ground besides surgical and radiological treatments. Of primary interest in this approach are the numerous and variously acting inhibitors of nucleic acid and protein synthesis, as well as compounds that block mitosis. Such compounds mainly affect proliferating cells and thus also cause damage to dividing normal cells. Since tumors often grow relatively slowly and usually contain a considerable proportion of non-dividing cells, chemotherapy aimed at killing tumor cells is of only limited applicability. By using combinations of compounds with different modes of action and different pharmacokinetic behaviour, however, it has been possible to make substantial advances in tumor chemotherapy.

An alternative to killing tumor cells is to inhibit their proliferation as well as to induce various differentiation processes via biological regulators. In this way, tumor cells may lose their malignant properties and develop into specialized cells, which in the majority of cases die sooner or later, thus leading to regression or total remission of the tumor. At the present time this therapeutic concept is being exclusively applied as hormone therapy, employing both the increase and decrease in concentration of particular hormones within the organism. In addition, antihormones, i.e., substances that either inhibit or block certain hormonal acitivities in the organism, are being employed (Table **19.2**).

**Table 19.2**: Hormonal therapy of malignant tumors (from Heise [47]).

1. Therapeutic range of endocrine gland extirpation

| Operation | Hormone deficiency | Therapeutic range |
|---|---|---|
| Ovariectomy | Estrogens, progesterone (androgens) | Mammary carcinoma |
| Adrenalectomy | Estrogens, gestagens, androgens, corticosteroids | Mammary and endometrial carcinomas |
| Orchidectomy | Testosterone | Carcinoma of the prostate |
| Hypophysectomy | Gonadotropins, somatropin, adrenocorticotrophic and thyrotrophic hormones | Tumors of hormone dependent organs controlled by the pituitary |

2. Additive hormone applications

| Hormone | Therapeutic range |
|---|---|
| Estrogens | Carcinoma of the prostate, mammary carcinoma of postmenopausal women |
| Androgens | Carcinomas of the prostate and kidney, mammary carcinoma of premenopausal women |
| Corticosteroids | Mammary and prostate carcinomas, lymphosarcoma |

The concentration of a hormone can be lowered by extirpation or irradiation of the appropriate endocrine gland, or, via feedback mechanisms, certain hormones can be employed to depress the production of others. In addition to this, some hormones have a direct influence on the growth of malignant tumors. Antihormones can influence the synthesis of certain hormones as well as directly compete for their receptor sites. The latter mechanism, for instance, is the basis of the treatment of metastasizing mammary carcinoma with antiestrogens. Other examples of therapy of malignant tumors by altering hormone concentrations are shown in Table **19.2**.

Hormone therapy is most successful in tumors whose growth is dependent upon hormones. A prerequisite for every hormone effect is the presence of specific receptors on the cell membrane or in the cytoplasm and cell nucleus. Hormonal treatment prevents cell proliferation and may trigger further differentiation. The hormone dependence of malignant tumors can be lost in the course of their development, in which cases hormone therapy becomes ineffective.

In judging whether or not a malignant tumor is hormone-dependent the demonstration of the existence of receptors has become of major clinical importance. If receptors are lacking, hormone treatment of the tumor is useless. On the other hand presence of receptors indicates only that the tumor may possibly respond to hormone therapy. The binding of the hormone to the receptor is only the first step in a chain of signal transformations involved in the hormone action (cf. Figs. **4.4** and **10.6**), but since the chain can be interrupted in other places as well, even tumors with hormone receptors may be independent of the hormone concerned. Further ways of determining the hormone responsiveness of tumors is (1) by testing the inhibition of DNA synthesis in tumor tissue cultivated in vitro (hormone biogram) and (2) by determining the extend to which synthesis and activity of certain enzymes can be influenced by hormones.

### 19.4.2 Role of Cell-Specific Inhibitors of Cell Proliferation (Chalones) in Tumor Therapy

Chalones are tissue-specific substances that play a part in the regulation of cell division by inhibiting at the $G_1$ or $G_2$ phase of cell division (cf. **15**.1.5). Proliferation of tumor cells can also be inhibited by chalones. Their application in tumor therapy would involve much less general damage to the organism as a whole than cytotoxic drugs since, apart from their action in suppressing proliferation of tumor cells, they suppress proliferation of only the normal cell type from which the tumor cells derived.

Employing chalones from granulocytes, therapeutic effects have been obtained on subcutaneously transplanted chloromas and in generalized myeloid leukemias in rat. In 9 of 40 rats complete remission of the leukemia was observed and the animals survived to normal age. The remaining rats survived longer than untreated controls. All blood cells not belonging to the granulocyte system developed normally in the chalone-treated rats. Preliminary therapeutic experiments on myeloid leukemia in man have yielded encouraging results. A definite regression was observed, and the resistance of patients to infections was good. In these experiments, too, inhibition of normal blood cells was limited to the granulocytes. Unfortunately, large-scale treatment of myeloid leukemias with granulocyte chalones is not yet feasible owing to the small quantities available. It is to be expected, however, that larger quantities of chalones will be produced in the future, providing the basis for their application in cancer therapy.

Apart from their direct application in bringing about remission of certain tumors or leukemias, chalones could also be used in combination with cytotoxic drugs. For example, cytostatics could be applied after withdrawal of chalones, at a point where tumor cell proliferation has already started, but the division of normal cells is still inhibited. This may be possible because, in most cases, normal cells were stronger inhibited by chalones than the corresponding tumor cells.

It has also been suggested that chalones could be used during therapy of certain tumors with cytotoxic drugs, to protect proliferating normal cells of other types than those the tumors derived from. Under these conditions the usual chemotherapeutics could be administered over a short period of time in relatively high doses.

### 19.4.3 The Ability of Tumors to Differentiate: a Basis for Cancer Therapy of the Future?

In many cases tumor cells exhibit some of the properties of embryonic cells, and under certain conditions can be stimulated to differentiate (cf. Sects. **19**.2.3.2 and **19**.3.1). The differentiated tumor cells mostly lose their malignant properties since they can no longer proliferate, and become subject to control by the neighbouring cells or the organism as a whole.

At present, differentiation capacity has been studied in most details in various leukemia cells. The development of normal stem cells of the myeloid lineage into granulocytes and macrophages is induced by a protein known as the macrophage--granulocyte inducer (MGI). Certain myeloid leukemia cells from man and mouse can be stimulated, both in vivo and in vitro, by MGI, to develop into apparently normal granulocytes and macrophages. Leukemia cells exhibiting this property are termed

$MGI^+D^+$, the $D^+$ standing for the ability to differentiate. Just like normal granulocytes and macrophages, these differentiated leukemia cells are no longer capable of dividing. They fulfill the function of the corresponding normal cells and exhibit no malignant properties. The response of these $MGI^+D^+$ cells to MGI is the same as that of normal stem cells and even the sequence of events in differentiation is identical with that in the developing normal cell. But unlike the normal stem cell, leukemia cells are capable of remaining alive and dividing without MGI. However, since they possess the ability to differentiate they can only become manifest leukemia cells in the absence of MGI, or if its concentration is very low. Another possible means for development of leukemias by $MGI^+D^+$ cells would be the formation of an MGI inhibitor. In $MGI^+D^+$ leukemia cells a chromosomal anomaly has been demonstrated, pointing to a genetic lesion in such cells. If the malignant behaviour of these cells is due to the chromosome aberration, the reversion to non-malignant, normally differentiated cells would be exclusively phenotypic since the chromosomal anomaly cannot be reversed by differentiation. Additional genetical alterations in leukemia cells can lead to variant clones non-responsive to MGI.

Regulation of differentiation processes induced by MGI is a very complex matter, as shown by the fact that certain lines of cells with reduced competence for MGI have also been isolated. In some of the cell lines this alteration was reversible but not in others. Specific chromosomal anomalies were demonstrated in those cell lines exhibiting a stable reduction in competence for MGI. In addition, a series of mutants could be isolated whose differentiation is inducible by MGI, although there is a blockage at one or another of the developmental stages. Other mutants, despite a blockage in differentiation, were dependent on MGI with respect to viability and ability to divide· Cell lines of this type exhibited malignant properties. The various mutants of myeloid leukemia cells are of great value in studying differentiation processes during the development of granulocytes and macrophages, but they provide also evidence of the difficulties associated with a form of therapy employing inducers of differentiation.

In $MGI^+D^+$ leukemia cells normal differentiation processes can also be induced by a number of non-specific substances and factors such as peptide hormones, steroids, lectins, polycyclic aromatic hydrocarbons, tumor promotors, antibiotics, 5-bromo-2--deoxyuridine, N-methyl-N-nitro-N-nitrosoguanidine, dimethylsulphoxide, and X--rays. However, in each of these cases it is the induction of single steps in differentiation or of longer or shorter stretches of normal development that is observed. The entire course of development up to the mature macrophage or granulocyte has so far only been triggered by MGI. Extremely interesting results have been obtained from experiments in which $MGI^+D^+$ leukaemia cells could be stimulated by certain lipopolysaccharides to produce MGI, which, in turn, induced differentiation of cells producing it.

In murine erythroleukemias generated by the Friend virus complex, normal differentiation processes can be induced by a number of agents, although erythropoietin is without effect (cf. **4.5.4.2**). Differentiation includes distinct parts of erythropoiesis but rarely proceeds to the non-nucleated stage characteristic of normal erythrocytes. In some cases, also patterns of erythroid cell gene expression was observed different from those in normal erythropoiesis. Differentiation leads to prolongation of the cell cycle. By treatment of leukemia cells in vitro with dimethylsulphoxide, hexamethylene bisacetamide, N,N-dimethylacetamide, N-methylpyrrolidinone, butyric acid or other substances their oncogenicity decreased. However, in vivo, none of the substances tested significantly increased survival, but a prolonged disease-free period was observed. Friend erythroleukemia cells are a very successful model for studying erythroid differentiation (cf. **4.5.4.2**).

Results so far available on the capacity of malignant tumor cells to differentiate with concomitant loss of malignancy suggest the future possibility of tumor therapy, i.e., application of various differentiation factors to differentiate malignant cells into phenotypically normal cells. At the same time the experiments point at problems connected with such therapy. The major difficulty will be presented by formation of tumor cells that behave indifferently towards the factors employed. To what extent the differentiation of such cells could be induced by other factors is not yet predictable. The most important prerequisite for this kind of therapy is an exact understanding of the normal processes of differentiation and their control by regulatory proteins involved. As this book endeavours to show, first steps in this direction are being taken. It may be possible in a not so far future to treat certain malignant diseases with the help of differentiation factors, e.g., MGI, in combination with irradiation or cytostatic drugs.

## 19.5 Conclusions

The development of malignant tumors can be triggered and promoted by physical, chemical or viral factors, the effect of which leads to genetic or epigenetic interference with the regulation of cell division and differentiation. This is a complex multistep process and may have its beginnings in the damage of one particular regulatory system. Further alterations of the same or other regulatory systems of cell division or differentiation can then lead to an increasing autonomy of the cells concerned. Stimulation of cell proliferation considerably accelerates this process. While irradiation chiefly acts via triggering mutations, chemical carcinogens can bring about changes in the DNA as well as in RNA and proteins. By means of oncogenic viruses new genetic material is incorporated into the cellular genome, and its translation products can affect expression of the host cell genes either directly or via the alteration of cellular regulation proteins.

All malignant tumor cells differ from their corresponding normal cells by more or less alterations in their gene expression pattern, whereby quite frequently genes, otherwise only active during ontogenesis, are reactivated. Alterations in the cell membrane play a decisive role in the malignant character of tumors, particularly for autonomous and invasive growth and for giving rise of metastases.

The use of biological regulatory factors in tumor therapy is at present confined to hormones. In investigations involving animal experiments and in preliminary therapeutic trials with human subjects remission of malignant tumors has also been achieved with chalones. Further induced differentiation of malignant cells to phenotypically normal cells suggests the possibility of employing differentiation factors in tumor therapy. However, for a practical application of these methods a detailed understanding and command of normal processes of differentiation is necessary. Investigation of these fundamental processes is not only of theoretical but also of utmost practical importance.

## 19.6 Selected References

*Books and Reviews of General Aspects*

1. Braun, A. C.: The Cancer Problem: A Critical Analysis and Modern Synthesis. Columbia Univ. Press, New York 1969.
2. Butschak, G.: Biochemische Grundlagen der Teilung und Differenzierung von Normal- und Tumorzellen. G. Fischer Verlag, Jena 1976.

3. Graffi, A.: Betrachtungen zu den Theorien der Krebsentstehung. Arch. Geschwulstforsch. *22* (1963): 13–41.
4. Reinert, J., & H. Holtzer (eds.): Cell Cycle and Cell Differentiation. Results and Problems in Cell Differentiation, Vol. 7. Springer Verlag, Berlin 1975.
5. Saunders, G. F.: Cell Differentiation and Neoplasia. Raven Press, New York 1978.
6. Sherbet, G. V.: Neoplasia and Cell Differentiation. S. Karger, Basel 1974.
7. Theile, M., & S. Scherneck: Zellgenetik. Akademie-Verlag, Berlin 1978.
8. Wunderlich, V.: Molekulare Aspekte der Karzinogenese. Arch. Geschwulstforsch. *49* (1979): 551–564.

*Physical and Chemical Carcinogens*

9. Brand, G.: Karzinogene Wirkungsweise gewebseingelagerter Partikel. Zbl. Bakt. Hyg. I., Abt. Orig. B. *166* (1978): 159–168.
10. Daune, M.: Changements de conformation du DNA modifié par des agents chimiques. Biochimie *60* (1978): 1097–1110.
11. Graffi, A., E. Magdon, Th. Matthes & St. Tanneberger (eds.): DNA Repair and Cancer Research. Akademie-Verlag, Berlin 1979.
12. Miller, E. C.: Some current perspectives on chemical carcinogenesis in human and experimental animals; presidential address. Cancer Res. *38* (1978): 1479–1496.

*Oncogenic Viruses*

13. Bishop, J. M.: Retroviruses. Annu. Rev. Biochem. *47* (1978): 35–88.
14. Burr, J. G., G. Dreyfuss, S. Peuman & J. M. Buchanan: Association of the src gene product of Rous sarcoma virus with cytoskeletal structures of chicken embryo fibroblasts. Proc. Natl. Acad. Sci. USA *77* (1980): 3484–3488.
15. Graf, T., & H. Beug: Avian leukemia viruses: interaction with their target cells in vivo and in vitro. Biochim. Biophys. Acta *516* (1978): 269–299.
16. Klein, G. (ed.): Viral Oncology. Raven Press, New York 1980.
17. Van Beveren, Ch., J. G. Goddard, A. Berns & I. M. Verma: Structure of Moloney murine leukemia viral DNA: Nucleotide sequence of the 5'long terminal repeat and adjacent cellular sequences. Proc. Natl. Acad. Sci. USA *77* (1980): 3307–3311.
18. Viral Oncogenes. Cold Spring Harbor Symp. Quant. Biol., Vol. *44*, Cold Spring Harbor Lab., New York 1980.
19. Weil, R.: Viral "tumor antigens" a novel type of mammalian regulator protein. Biochim. Biophys. Acta *516* (1978): 301–388.
19a. Hynes, R. O.: Cellular location of viral transforming proteins. Cell *21* (1980): 601–602.
19b. Hunter, T.: Proteins phosphorylated by the RSV transforming function. Cell *22* (1980): 647–648.
19c. Bishop, J. M.: Enemies within: the genesis of retrovirus oncogenes. Cell *23* (1981): 5–6.

*Mutation Theory*

20. Ames, B. N.: Identifying environmental chemicals causing mutations and cancer. Science *204* (1979): 587–593.
21. Anders, A., & F. Anders: Etiology of cancer as studied in the platyfish-swordtail system. Biochim. Biophys. Acta *516* (1978): 61–95.
22. Bouck, N., & G. di Mayorca: Somatic mutation as the basis for malignant transformation of BHK cells by chemical carcinogens. Nature *264* (1976): 722–727.
23. Giannelli, F.: Xeroderma pigmentosum and the role of DNA repair in oncogenesis. Bull. Cancer *65* (1978): 323–334.
24. Mulvihill, J. J., R. W. Miller & J. F. Fraumeni: Genetics of Human Cancer. Progr. Cancer Res. Therap., Vol. 3, Raven Press, New York 1977.
25. Robbins, J. H.: Significance of repair of human DNA: Evidence from studies of Xeroderma pigmentosum. J. Natl. Cancer Inst. *61* (1978): 645–656.
26. Strong, L. C.: Genetic and environmental interactions. Cancer *40* (1977): 1861–1866.

*Gene Expression and Differentiation in Tumor Cells*

27. Bannasch, P.: The Cytoplasm of Hepatocytes during Carcinogenesis. Recent Results in Cancer Research, Vol. 19. Springer-Verlag, Berlin 1968.

28. Ibsen, K. H., & W. H. Fishman: Developmental gene expression in cancer. Biochim. Biophys. Acta *560* (1979): 243–280.
29. Marks, P. A., & R. A. Rifkind: Erythroleukemic differentiation. Annu. Rev. Biochem. *47* (1978): 419–448.
30. McKinnell, R. G.: The pluripotential genome of the frog renal tumor cell as revealed by nuclear transplantation. Int. Rev. Cytol. Suppl. *9* (1979): 179–188.
31. Mintz, B.: Normalized gene expression in differentiation of mouse teratocarcinoma cells. In: Nover, L., & K. Mothes (eds.): Cell Differentiation in Microorganisms, Plants and Animals. G. Fischer Verlag, Jena 1977, pp. 351–360.
32. Parodi, S., & G. Brambilla: Relationship between mutation and transformation frequencies in mammalian cells treated "in vitro" with chemical carcinogens. Mutation Res. *47* (1977): 53–74.
33. Sachs, L.: Control of normal cell differentiation and the phenotypic reversion of malignancy in myeloid leukemia. Nature *274* (1978): 535–539.
34. Spandidos, P. A., & L. Siminovitch: The relationship between transformation and somatic mutation in human and Chinese hamster cells. Cell *13* (1978): 651–662.
35. Turkington, R. W.: Regulation of gene expression in normal and neoplastic mammary cells: a review. J. Natl. Cancer Inst. *48* (1972): 1231–1234.

*Alteration of the Cell Membrane*

36. Bramwell, M. E., & H. Harris: An abnormal membrane glycoprotein associated with malignancy in a wide range of different tumors. Proc. Royal. Soc. London, B. *201* (1978): 87–106.
37. Burger, M. M.: Cell surface and neoplasia. Int. Cell Biology, Rockefeller Press, New York 1977, pp. 131–137.
38. Hynes, R. O.: Cell surface proteins and malignant transformation. Biochim. Biophys. Acta *458* (1976): 73–107.
39. Lipkin, G., M. Rosenberg & M. E. Knecht: Factors affecting growth of normal and malignant cells in vitro. Biochem. Pharmacol. *25* (1976): 1333–1337.
40. Nicolson, G. L.: Trans-membrane control of the receptors on normal and tumor cells. II. Surface changes associated with transformation and malignancy. Biochim. Biophys. Acta *458* (1976): 1–72.
41. Pasternak, G.: Transplantations- und Tumorimmunologie. In: Ambrosius, H.., W. Rudolph (eds.): Grundriß der Immunologie. G. Fischer Verlag, Jena 1978, 341–378.
42. Unkeless, J. C., & S. Gordon: Secretion and regulation of plasminogen activator in macrophages. In: Reich, E., D. B. Rifkin & E. Shaw (eds.): Proteases and Biological Control. Cold Spring Harb. Lab., New York 1975, pp. 495–514.
43. Vaheri, A., & D. F. Mosher: High molecular weight, cell surface-associated glycoprotein (fibronectin) lost in malignant transformation. Biochim. Biophys. Acta *516* (1978): 1–25.
44. Wallach, D. F. H.: Membrane Anomalies of Tumor Cells. Progress in Experimental Tumor Research, Vol. 22. S. Karger, Basel 1978.
45. Waren, L., C. A. Buck & G. P. Tuszynski: Glycopeptide changes and malignant transformation: a possible role for carbohydrate in malignant behavior. Biochim. Biophys. Acta *516* (1978): 97–127.
46. Whittenberger, B., D. Rarfn, M. A. Lieberman & C. Glaser: Inhibition of growth of 3T3 cells by extract of surface membranes. Proc. Natl. Acad. Sci. USA *75* (1978): 5457–5461.

*Therapy of Malignant Tumors by Biological Factors*

47. Heise, E.: Hormone und antineoplastische Chemotherapie. In: Tanneberger, St. (ed.): Experimentelle und klinische Tumortherapie, Vol. 1. Akademie-Verlag, Berlin 1980, pp. 169–189.
48. Rytömaa, T.: The chalone concept. Int. Rev. Exp. Pathol. *16* (1976): 155–206.
49. Rytömaa, T., & H. Toivonen: Chalones, concepts and results. Mechanisms of Ageing and Development *9* (1979): 471–480.

# 20. Subject Index[1]



[1] For abbreviations frequently used in the text see appropriate places of the subject index.

Additional material from *Cell Differentiation*,
ISBN 978-3-642-68022-9, is available at http://extras.springer.com

GPSR Compliance

*The European Union's (EU) General Product Safety Regulation (GPSR) is a set of rules that requires consumer products to be safe and our obligations to ensure this.*

*If you have any concerns about our products, you can contact us on ProductSafety@springernature.com*

In case Publisher is established outside the EU, the EU authorized representative is:

Springer Nature Customer Service Center GmbH
Europaplatz 3
69115 Heidelberg, Germany

**Batch number: 10620881**

Printed by Printforce, the Netherlands